This book is in the
ADDISON-WESLEY SERIES IN PHYSICS

Cover photo (upper) of Copernicus manuscript courtesy of Charles Eames.

ISBN 0-201-02971-5
BCDEFGHIJK-HA-7987654

# INTRODUCTION TO
# CONCEPTS AND THEORIES IN PHYSICAL SCIENCE

## SECOND EDITION

**by GERALD HOLTON**
Harvard University

revised and with new material by
**STEPHEN G. BRUSH**
University of Maryland

**ADDISON-WESLEY PUBLISHING COMPANY**
Reading, Massachusetts · Menlo Park, California
London · Amsterdam · Don Mills, Ontario · Sydney

To E.C.K.
Colleague and Teacher

# INTRODUCTION
# TO THE SECOND
# EDITION

The publication in 1952 of the first edition of this book was a landmark in science education. As Professor Holton indicated in his Introduction at the time, the book was inspired in large part by the discussions of the role of science in general education during the preceding decade. At several points credit was given to his colleagues at Harvard and elsewhere for information and for their suggestions about the presentation of particular topics. Nevertheless the originality of the book was clear: It was the first modern textbook in physics (or in any science for that matter) to make full and effective use of the history and philosophy of science in presenting for both the general and the science-oriented student an account of the nature of physical science.

During the past two decades, the importance of a general-education approach to science has been widely recognized; many colleges and universities have initiated courses in "physics for the nonscience major," and many of these courses have used this book as a primary or second text to give future scientists the proper orientation to their own heritage. The need to teach physics to students with widely varying interests and abilities has called forth a number of different textbooks. Some of these texts have followed the lead of Holton's *Concepts and Theories* by attempting an historical approach with a more or less similar set of topics selected from physics, astronomy, and chemistry. Others have been frankly un-historical and have focused on completely different topics. This divergence merely points up the fact that students themselves are attracted or repelled by different ways of presenting science, and no single approach can hope to be universally successful.

In 1958, Holton in collaboration with Duane H. D. Roller published a new textbook, *Foundations of Modern Physical Science*, which has sometimes been regarded as a "second edition" of *Concepts and Theories* and which also remains in print. While it utilizes some of the same material, *Foundations* is in fact directed toward a slightly different audience, and attempts a more comprehensive (though introductory) treatment of physical science by adding new topics, making room for them by not going into some of the deeper analyses which had characterized *Concepts and Theories*. The earlier book, far from being superseded by the later one, remained in demand and has been reprinted several times (most recently in 1962), generating a remarkably loyal group of teachers and "alumni." The wide influence of both books may be judged by the large number of references to them in books as disparate as college physics texts, high school texts, and works on the history or philosophy of science.

Parenthetically, it may be of interest that an offspring of these books was a new physics course for nation-wide use, designed by a group working at Harvard under the direction of Gerald Holton, James Rutherford, and Fletcher Watson, with financial support from government agencies and private foundations. This undertaking to make a nationally used course, known as Harvard Project Physics, began in 1962, with a group of physicists, educators, high-school teachers, and historians of science, taking Professor Holton's texts as a starting point, adding new material, and using a number of innovations in teaching methods and equipment. They developed an integrated combination of text, laboratory experiments, films, transparencies, and teacher resource books which in 1970 was published in a high-school version as *The Project Physics Course* by Holt, Rinehart and Winston, with a version for college use. My own involvement in general education in science is a direct result of my participation in the work of Harvard Project Physics, and several ideas used in the present revision of *Concepts and Theories* were developed in discussions with Gerald Holton and the staff of Project Physics.

In weighing the proposal for a second edition of *Concepts and Theories*, it was necessary to take into account the present existence of the many textbooks designed for physics courses for non-science majors, books which to some extent satisfy the demand created by the success of the general education movement (and in part by the success of the first edition of this book). We concluded that *Concepts and Theories* had certain qualities that are still unique, and that a new edition would be justified if these qualities could be enhanced and kept in the forefront, in spite of the changes that would inevitably be needed to bring the book up to date.

Perhaps the most striking characteristic of this book is the emphasis it puts on the nature of discovery, reasoning, and concept-formation in science as a fascinating topic in its own right. This means that the historical and philosophical aspects of the exposition are not merely sugar-coating to enable the reader to swallow the material as easily as possible, but are presented for their own inherent interest. Indeed, this is not intended to be an "easy" book; it does not pretend that physics can be understood without any mathematics or without logical reasoning. At the same time, it is not assumed that there is any obligation to "cover" a standard amount of material. There is no apology for omitting several "fundamental" equations and even entire subfields of physical science that are conventionally included in textbooks. The selection of topics (and depth of treatment of each topic) is to a considerable extent governed by the goal of presenting a comprehensible account—a continuous story line, as it were—of how science evolves through the interactions of theories, experiments, and actual scientists.

In revising the book, I have had the active cooperation of the original author, and I have preserved to a large extent the viewpoint and much of the actual text of the first edition. There has been a major rearrangement of the material of Parts A, B, and C, so that the development of astronomical systems up to Kepler and Galileo now comes at the beginning, as Part A, providing a cosmic motivation for the study of motion in Part B. The Newtonian synthesis (Part C) then appears in its proper place as an outgrowth of the new physics applied to long-established problems in astronomy. This manner of approaching Newtonian mechanics in historical context has become sufficiently familiar to physics teachers during the past few years so that the apparent anomaly of starting a physics text with a section on astronomy is no longer cause for alarm.

Part D (On Structure and Method in Physical Science) has been left essentially unchanged from the first edition. Part E (The Laws of Conservation) has a new chapter on the dissipation of energy which, in addition to introducing some geological ideas, provides a link to the statistical interpretation of irreversibility discussed in Chapter 22 (Kinetic-Molecular Theory of Gases). To make room for this new material, some of the more technical applications of Newtonian mechanics (e.g., to rotational motion and collisions) have been omitted. Aside from some changes in the later parts of Chapters 19 and 22, Part F (Origins of the Atomic Theory in Physics and Chemistry) covers substantially the same material as in the first edition: the gas laws, Dalton's atomic theory, the periodic system, and the kinetic-molecular theory of gases.

The long final part in the first edition has been divided into two parts: Part G (Light and Electromagnetism) and Part H (The Atom and the Universe in Modern Physics). A short chapter on the wave theory of light (Chapter 23) and a new section on the interaction of currents and magnets (Section 25.2) have been added in order to provide some background for the electromagnetic theory of light. Part G closes with a chapter on the quantum theory of light, beginning the 20th-century revolution in physics; some aspects of this revolution are then explored in more detail in Part H. Much of the detailed material on the application of the Bohr model (Chapter 28) has been omitted. Instead there is a brief discussion of quantum mechanics and its philosophical interpretation (Chapter 29). The last chapter of the book, also newly written for this edition, is on Einstein's theory of relativity; it indicates how physics has returned to a concern with the cosmological problems that stimulated its early development. (I am indebted to my colleague Charles Misner for several suggestions used in this chapter.)

As in the first edition, the *Problems* play an important part. Many details of subject matter are referred to them, and some are designed to carry the student as far as his resources and imagination will permit. A few problems have no cut-and-dried solution. It is suggested that the majority of problems be either assigned or discussed in class.

*References to further reading*, collected for the most part at the ends of chapters (except where a specific quotation or fact needs to be identified), have been completely revised. There are now two categories: "Recommended for further reading"— articles or short books whose technical level is generally no higher than that of the text itself; and "Sources, interpretations, and reference works." The latter are intended for teachers and advanced students, and give some indication of the great flowering of research in the history of science during the last few years. It should also be noted that most of the important original sources are now available in English translation, often in inexpensive paperback editions. A list of general works on the history of ancient and early modern science is given at the end of Chapter 1, and another list of works dealing primarily with the development of modern physical science may be found at the end of Chapter 26. All of these references are collected in a general bibliography at the end of the book. The student should also have available one or more conventional physics textbooks for reference purposes, e.g., chosen from those by Arons, Holton and Roller, Kemble, Resnick and Halliday, Rogers, or the Project Physics Text.

At a time when many students and scientists are looking again for the greater "relevance" of the study of science, this book may furnish the needed evidence of the intellectual excitement, the uses of science in historic development, and the actual lives of real men and women who made this magnificent construction.

*College Park, Maryland*                    S.G.B.
*December 1972.*

# FROM THE
# INTRODUCTION
# TO THE
# FIRST EDITION

In the last decade or two, the function and form of introductory science courses in our colleges have been the subject of general discussion. We have already begun to see some stimulating consequences of this widespread interest in re-analysis and re-evaluation, principally the inauguration of general education courses in their several forms.

In this book I have attempted to incorporate some of the conclusions drawn from an illuminating experience, that of teaching simultaneously for a number of years a conventional type of introductory college physics course and also in general education courses in physical science. Perhaps the novelty of some features of the book will excuse a few introductory remarks concerning the guiding ideas that have shaped it.

The old postulate has remained: the principal task for an introductory course in the physical sciences must surely be a clear or, rather, a dedicated presentation of the fundamental concepts and theories. A firm knowledge of fundamentals is the basis for whatever is to be accomplished in addition. Morever, by presenting neither an encyclopedic account of all details nor a survey treatment, by concentrating instead on the meaning and power of some basic ideas—even though there can probably be no general agreement on just what topics one should include—one may hope to leave the student with a view of physical science

as an interconnected achievement; one can engage his attention in those lengthier, more rigorous, penetrating arguments and derivations without which an honest understanding of the processes of science is hardly attainable.

While therefore the focus in this book is on the subject matter of physical science, there are also other aims—above all the presentation of science as *experience*, as an integrated and exciting intellectual adventure. Alfred North Whitehead, in the preface of his collection of essays, *The Aims of Education*, protested "against dead knowledge, that is to say, against inert ideas." Indeed, the record would tend to show that in the past it has often been found difficult to avoid the cataloguing of inert ideas, particularly in science textbooks. This does not refer to the text directed at those students who concentrate or major in one of the physical sciences; they bring to their studies a commitment and a motivation that give life to the basic concepts and theories; furthermore, they can look forward to a fulfillment in later courses of all promises implicit in the introductory treatment. But this commitment, motivation, and long-range interest is much attenuated in the student who does not plan to devote his life to physical science, for whom the first and perhaps the last formal contact with the subject is a single course. Yet he too is eager to understand the achievements and methods of science. He responds warmly to sympathetic guidance.

His dilemma was well analyzed in the searching Report on *General Education in a Free Society* (Harvard University Press, 1945, pp. 220–222):

"From the viewpoint of general education the principal criticism to be leveled at much of present college instruction in science is that it consists of courses in special fields, directed toward training the future specialist and making few concessions to the general student. Most of the time in such courses is devoted to developing a technical vocabulary and technical skills, and to a systematic presentation of the accumulated fact and theory which the science has inherited from the past. Comparatively little serious attention is given to the examination of basic concepts, the nature of the scientific enterprise, the historical development of the subject, its great literature, or its interrelationships with other areas of interest and activity. What such courses frequently supply are only the bricks of the scientific structure. The student who goes on into more advanced work can build something from them. The general student is more likely to be left simply with bricks. Eventually he constructs his educational edifice elsewhere with other materials. . . .

"The body of science includes not only special knowledge and skills but conceptual interrelations, a world-view, and a view of the nature of man and knowledge, which together constitute the philosophy of science; a history which forms a continuous and important segment of all human history; and writings which include some of the most significant and impressive contributions to all literature. These aspects of the sciences are frequently almost entirely neglected in the college teaching of science . . ."

Certainly, on their own merits, the inclusion of history and of philosophy of science is justified in an introductory course for the nonspecialist. But beyond that, their use in this book has a further threefold purpose. First, they prepare the appropriate setting in which a particular idea came to have meaning and importance. Second, they provide insight into the sources, motivations, and methods of approach of the founders of science, illustrating the human triumph behind the bare abstraction. And third, they present science as one facet of the great quest for knowledge which our free society holds to be a most cherished right and a most respected driving force. At the end of such a course, the student will be in a position to know the principal laws and the evolution of key conceptual schemes, but it is then also hoped that he, as a responsible citizen, will understand the criteria of validity in scientific thought, the conditions that aid the fruitful growth of science—and perhaps even the exhilaration that binds his instructor to the scientific profession.

A few specific remarks may now be made about the form of presentation. It was not dictated by any single pedagogic method. The sequence follows an eclectic pattern, most frequently a series of modified case accounts involving the great conceptual schemes. But while the latter are analyzed, as far as practical, in terms of original source materials, a contemporary mode of exposition has been adopted instead of strictly historical and methodological ones wherever these would not aid materially in clarifying the fundamental ideas.

Being directed primarily to students not planning to become professional physicists or chemists, the book will find use in two rather different contexts: the course explicitly of the general education type and the liberal arts physics course for nonscience majors. As to the first of these, I have in mind particularly a course analogous in many ways to the one described in the Report on General Education in the following words (*op. cit.*, pp. 226–227):

". . . Rather than provide the student with a systematic presentation of the materials of one science, this course should develop particular aspects of the scientific enterprise within the whole range of the physical sciences. To give the course greater unity, it should be built about a core of physics. Materials from other sciences—chemistry, astronomy, and geology—should be introduced only to the degree that they are pertinent to the problems under discussion. The course would probably omit, for example, descriptive chemistry and descriptive astronomy. It should, however, explore basic chemical concepts: atomic theory, the periodic system, laws of chemical combination, valence, and so on. Similarly, celestial mechanics might provide the material for much of its discussion of dynamical principles.

"Such a course must discard at the outset any attempt to survey the material of the sciences which compose it. Rather, it must look to some dominant intellectual pattern to guide its selection of material. In the present case the pattern is to be found in the development of basic physical principles and concepts, and the methods and approaches by which they have been developed.

"This is not intended to be merely a course about science. It will contain much solid scientific content. The student will learn fundamental facts and laws and will solve problems theoretically and in the laboratory. He will do so, however, with a highly selected subject matter, which in every case is chosen to subserve the major aims of the course.

"The emphasis on historical development in this course is in no sense to constitute merely a humanistic garnishing of its factual material. On the contrary, it is introduced to illuminate and vitalize the content with which it is integrated. The attempt should be made in this course to teach science as part of the total intellectual and historical process, of which, in fact, it has always been an important part. The student should gain thereby an insight into the principles of science, and appreciation of the values of the scientific enterprise; and he should also learn much of the subject matter of the physical sciences."

The second group to which this book is directed is that of students in the liberal arts course in physics for whom the further curriculum does not offer another opportunity to obtain the general education background. Here the instructor may possibly wish to omit such chapters as do not primarily involve physics. Or he may supplement this selected account with whatever topics he chooses from an orthodox type of physics text, in accord with the particular needs of his class; I myself have used preliminary editions of this material in conjunction with such a text in an introductory physics course attended largely by premedical students. I am inclined to agree with the growing opinion that the influence of general education programs upon the teaching of just such groups will eventually be counted as one of its most significant contributions to the educational pattern.

As reference material, these studies should also aid in teaching the strictly technical course for physics majors, for whom the Report on General Education pleads (*op. cit.*, p. 221):

". . . It frequently happens that even the student who concentrates in a science is preoccupied with his specialty to such a degree that he fails to achieve a view of science as a whole and of the interrelationships of the special fields within it. A general education in science needs to be provided for the future scientist or technologist as well as for the general student . . ."

. . .

I wish to acknowledge with pleasure my great debt to friends and colleagues whose advice aided in the preparation of this work—principally to Professors E. C. Kemble, P. LeCorbeiller, P. Frank, L. K. Nash, I. B. Cohen; Drs. T. S. Kuhn and C. L. Clark; Mr. D. H. D. Roller; to Professor F. W. Sears of M.I.T.; and to those indispensable critics and companions, my teaching assistants, particularly Mr. S. J. Smith, who prepared the major part of the problem answers.

To Professor Kemble goes an extraordinary measure of gratitude. In nine years of association, often teaching side-by-side with him in the same physics or general education course, I have not only learned from him, I fear I may unwittingly have appropriated from him whatever is of merit in this account.

*Cambridge, Massachusetts*                    G.H.
*January 1952.*

# CONTENTS

PART A

# THE ORIGINS OF SCIENTIFIC COSMOLOGY

1
## THE ASTRONOMY OF ANCIENT GREECE

2
## COPERNICUS' HELIOCENTRIC THEORY

3
## ON THE NATURE OF SCIENTIFIC THEORY

4
## KEPLER'S LAWS

5
## GALILEO AND THE NEW ASTRONOMY

# THE ORIGINS OF SCIENTIFIC COSMOLOGY

Our purpose in this book is to explore the development and content of the major ideas that have contributed to the understanding of the physical universe. Although most of the book deals with physics, we will begin with astronomy, for several reasons.

The methods of science were perhaps first suspected when man attempted to reduce the apparently chaotic yet recurring motions of the stars, sun, moon, and planets to an orderly system. Second, from early astronomical schemes are derived many of our guiding ideas in science, not only in methodology but also in such specific concepts as time and space, force and motion. We shall see that a continuum of ideas leads from the early naive speculations about stellar motions right to twentieth-century science, paralleling the expansion of our universe from the idea of a near sky to the rotating celestial sphere, the recognition of the solar system as a huge galaxy, and finally the universe of innumerable galaxies scattered through unbounded space. Third, the world system involves many of the men responsible for the strength and content of present physical theory. Their place in science and the atmosphere of their times are all part of our story. Fourth, the work to be discussed —the planetary systems, the law of universal gravitation—had a most profound effect on Western culture. For example, without an understanding of the Newtonian revolution and its background, one cannot hope to comprehend fully the upheavals in eighteenth-century history, or the development of modern theories of economics and government, or the philosophies of Locke, Berkeley, and Hume. And finally, we may use the opportunity now to watch the rise and fall of physical theories, to analyze their structure, and to sharpen our faculties of discrimination between theories that are useful and those that are misleading. Thus a close examination of this one topic not only will reveal new and significant physical laws, but also will give us a brief chance to inspect those several other aspects of physical science, so important and yet so often inaccessible.

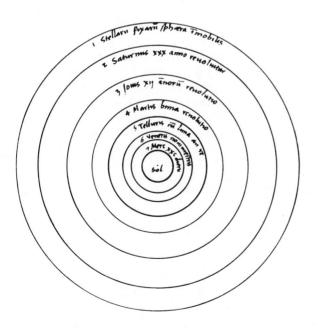

3

# Chapter 1

# THE ASTRONOMY OF ANCIENT GREECE

## 1.1 THE MOTIONS OF STARS, SUN, AND PLANETS

Although there is now ample evidence that civilizations in earlier centuries and in other parts of the world had already gathered significant pieces of astronomical and mathematical knowledge, we can trace the beginnings of science as we know it to the imaginative minds of the great Greek thinkers. In order to have a definite starting place, let us try to imagine ourselves in the position of the leading scientific community of the ancient world around 400 B.C., in Athens.

Although the use of optical instruments of precision was still 2000 years away, simple observation of the night sky had by about 400 B.C. established enough data and interpretations concerning the motions of the heavens to nourish several world theories. The "fixed stars" and the Milky Way seem to move throughout the night as if they were rigidly attached to an invisible bowl which rotates around a fixed center in the sky (now called the *North Celestial Pole*). From observations made from different points on the earth's surface, it could be inferred that this bowl is really more like a very large sphere surrounding the earth, and that the earth itself is also a sphere.* As we will see in Section 1.3, it was possible by the third century B.C. to obtain a fairly accurate estimate of the size of the earth; but it was not until the nineteenth century A.D. that scientists could determine even approximately the distance of some of the stars from the earth.

The Greeks were quite familiar with the fact that the hypothetical "celestial sphere" containing the stars appears to rotate uniformly from east to west, returning to its starting point every 24 hours. We will refer to this motion as the *diurnal* (daily) rotation. Of course we now know that it is the earth rather than the stars that rotates on its axis every 24 hours, and that the appearance of stars arranged on a big sphere is an illusion. But that is not something you should simply accept "on authority." Instead you should consider carefully which of the following observations can be explained equally well by assuming either that the stars move or that the earth moves.

---

* Here as elsewhere, the very brief summary of early astronomy presented in this book may well be supplemented by some of the books and articles listed in the Bibliography at the end of the chapter.

There is now a particular star, called the Pole Star or Polaris, which happens to be very close to the North Celestial Pole (the apparent center of rotation of the stars in the northern hemisphere). However, in 400 B.C., Polaris was several degrees away from the North Celestial Pole; there was no obvious Pole Star at that time. The Greeks knew that the North Celestial Pole does move very slowly with respect to the stars, producing the phenomenon of *precession of the equinoxes* (see below); but the available observations did not extend over a long enough period of time for them to understand the precise nature of this motion. It is now known that the North Celestial Pole itself moves in a small circle, returning to its original position after about 26,000 years.

It was also well known to the Greeks (though, it appears, not so well known to the layman today) that while the sun shares the diurnal motion of the stars it does not quite keep up with them. By observing the stars just before sunrise and just after sunset, one can discover that the sun slowly changes its position relative to the stars every day. In fact it follows a path from west to east through the stars, going through the "signs of the Zodiac," and returning to its starting point after 365 days. More precisely, it does not simply go eastward, but it also has a north-south motion: On about March 21 (vernal equinox) the sun is directly overhead at noon for places on the earth's equator, and it then moves northward every day until about June 21 (summer solstice) when it is directly overhead at noon for places $23\frac{1}{2}°$ north of the equator (Tropic of Cancer). The sun then moves southward, so that about September 23 (autumnal equinox) it is directly overhead at noon on the equator again, and about December 21 (winter solstice) it is directly overhead at noon for places $23\frac{1}{2}°$ south of the equator (Tropic of Capricorn). The sun then moves northward again and the cycle repeats itself (see Fig. 1.1).

The north-south motion of the sun is of course the major factor that determines temperatures on

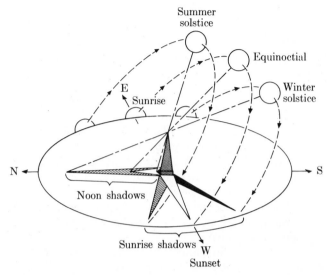

**Figure 1.1**

the surface of the earth. Between March 21 and September 23 the day will be more than 12 hours long in the northern hemisphere, and the sun will rise relatively high in the sky (depending on latitude). Between September 23 and March 21 the day will be less than 12 hours long in the northern hemisphere, and the sun will not rise very high. (These circumstances will simply be reversed for the southern hemisphere.) On March 21 and September 23, both day and night will be 12 hours long everywhere, hence the term *equinox* (Latin for *equal night*).

The correlation between the seasons and the sun's motion through the stars was a scientific finding of vital importance to the ancient agricultural civilizations. By setting up a calendar of 365 days, the ancient astronomers could predict the coming of spring and thus tell the farmer when to plant his crops. Eventually it was found that such a calendar would become more and more inaccurate unless occasional extra days were added to take account of the awkward fact that the year (as measured by the time from the sun's leaving a spot among the stars to its return to that spot) is about a quarter of a day longer than 365 days. Another difficulty was that the position of the sun in the Zodiac at the vernal equinox gradually changed over the centuries. For more than a thousand years before the birth of Christ the sun, on March 21, was among the stars in the constellation Aries, but by and by it moved into Pisces, where it still is each year on that date. In a few more centuries it will be in Aquarius.* This phenomenon is called the *precession of the equinoxes* and will later on be seen to be simply another aspect of the gradual motion of the North Celestial Pole mentioned above.

Finally, the ancients noticed that certain special stars did not stay in fixed positions on the celestial sphere, but wandered around in a complicated but regular manner. These stars became known as *planets* (from a Greek word meaning *wanderer*), and the study of their motions was one of the chief occupations of astronomers up to the seventeenth century A.D.

The ancient Greek theories of planetary motion were not narrowly technical attempts to correlate a particular set of observations in the fashion that might nowadays be called "scientific." The Greek philosophers favored a broader approach. On the basis of preliminary observations, an astronomer might formulate a scheme of planetary motion which could account for the data available at the time; but for such a scheme to have lasting value it did not so much need to be in accord with all subsequent observations, but instead should dovetail with other, more strictly philosophical or theological ideas. The followers of Pythagoras, for example, conceived that the relative sizes of the orbits of planets were proportional to the lengths of the successive strings on a harmoniously-

---

\* Hence the assertion, derived from popular astrology, that the "Age of Aquarius" is about to arrive.

tuned stringed instrument. This was thought to assure a "harmony of the spheres," more satisfying and important to the general philosophical requirements of the Pythagoreans than would be the relatively narrower demands of quantitative predictability of physical events made by the modern astronomer. Or, to cite an example involving an entirely different school in Greek science, Aristotle found in qualitative observations of the free fall of large and small bodies enough evidence to satisfactorily integrate these physical phenomena with his philosophically more wide-reaching, and therefore to him more satisfying, scheme of elements and of natural place.

## 1.2 PLATO'S PROBLEM

Nothing is easier and more erroneous than to underestimate the point of view of the early Greeks. Not only did the mechanics of the Greeks "work" within the limits of current interest, but in some schools of thought it was also intellectually a majestic and deeply meaningful construct. Above all, it presents, as it were, the childhood of science, not to be judged in terms of the maturer point of view of contemporary knowledge. It may be true that eventually it outlived its usefulness, but it gave modern men the direction along which they could develop fruitfully. The influence of Greek thought is present in any contemporary activity, in science no less than in art or law, government or education.

Astronomy is a case in point. It is related that Plato (427–347 B.C.) set a problem for his students along these lines: The stars—eternal, divine, unchanging beings—move around the earth, as we can see, in that eminently perfect path, the circle. But a few "stars" appear to wander rather disturbingly across the sky, tracing perplexingly irregular figures in their yearly paths. These are the planets. Surely they too must *really* move in uniform, ordered circles, or rather, in their case, combinations of circles. Now, how can we account for the observations on planetary motion, and so "save the appearances"? Plato's important question may be paraphrased as follows: "Determine what *uniform* and *ordered* motions must be assumed for each of the planets to account for the apparent, more irregular movements." The very formulation of this historic problem strikingly summarizes for us the two main contributions of Greek philosophers to the topic of our discussion:

a) Physical theory (e.g., the theory of planetary motions) is intelligible only in the context of a prior metaphysical assumption (e.g., the assumption that heavenly bodies must execute "perfect," circular motions).* This proved to be an unfruitful doctrine on which to base science. By the sixteenth and seventeenth centuries, after the great struggle which we will discuss in later chapters, it began to be abandoned in favor of the experimental sciences.

b) Physical theory is built on observable and measurable phenomena (e.g., the apparent motion of the planets), concerns itself with uniformities of behavior underlying the apparent irregularities, and expresses itself in the language of number and geometry. This guiding idea, which Plato did not extend beyond astronomy and which was derived in part from the Pythagoreans, was a treasured hint which reappeared when Kepler and Galileo fashioned their experimental science. But Plato held that physical laws can be found from directly intuited principles, the aim being to explain specific phenomena in the context of a philosophic system. The truth of a principle was not measured, as it is today, by its usefulness in every conceivable known or predicted physical situation. Consider this sentence from Plato's *Phaedon*: "This was the method I adopted: I first assumed some principle,

---

* The term *metaphysics* is used in a specific sense: the discipline which studies principles of knowledge or of being in terms of intuitive, self-evident concepts, concepts of direct "everyday" experience, and analogies.

which I judged to be the strongest, and then I affirmed as true whatever seemed to agree with this, whether relating to the cause or to anything else; and that which disagreed I regarded as untrue."

Plato's specific astronomical question, which he did not seriously attempt to answer himself, became the prime concern of astronomers to the time of Galileo. Let us see how they tried to construct a system of the world in accordance with the axiom of "uniform and ordered motion," that is, a system which allows the celestial objects to move only in uniform (constant) circular motions or in combinations of several such motions. We shall discover that ultimately all such systems proved unsatisfactory; but modern science was born out of this failure.

## 1.3 THE ARISTOTELIAN SYSTEM

Obviously, the first guess to investigate in rationalizing the motion of celestial objects is the most simple geocentric* system, which places the earth at the center of the celestial sphere carrying the fixed stars. The observed daily motion of those fixed stars will result from our simple model at once if we require the large celestial sphere to rotate uniformly on a north-south axis once a day. We may then attempt to explain the apparent motion of the sun, the moon, and the five visible planets about the fixed earth by letting each be carried on a transparent sphere of its own, one within the other, all seven enclosed by the sphere of fixed stars, and the earth at the center of the whole scheme (Fig. 1.2). But since those seven celestial bodies do not rise and set always at the same point of the horizon, since in fact the planets travel on most complicated paths against the fixed stars, sometimes even briefly reversing the direction of motion, we must give each of their spheres a whole set of simultaneous rotations about different axes,

---

* Earth-centered, from the Greek *gē*, earth.

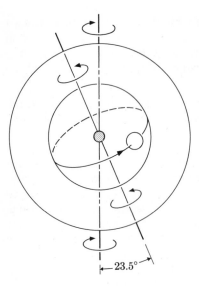

← 23.5° →

**Fig. 1.2** The annual north–south (seasonal) motion of the sun was explained by assuming it to be fixed on a sphere whose axis was tilted $23\frac{1}{2}°$ from the axis of the stars.

each rotation of the proper speed and direction, each axis of the proper inclination, to duplicate the actually observed paths.

Here the mathematical genius of the Greek was indeed put to the test! The sun and the moon were relatively simple to discuss, but some of the planets offered much difficulty. Plato's pupil Eudoxus thought that 26 simultaneous uniform motions would do for all seven celestial bodies. Others proposed to abandon the assumption that the sun and planets are fixed on celestial spheres—which means that they must always remain at the same distance from the earth—and developed the more complicated combinations of circles which we shall briefly summarize in Sections 1.6 and 1.7.

But is the task of the astronomer (or any scientist) strictly limited to constructing mathematical systems, no matter how artificial, that will agree with observations? Is his job done when he can "save the phenomena" (or "fit the data" as we would now say)—or should he also look for a physically plausible explanation of how nature works?

A clear answer to this question was given by Aristotle (384–322 B.C.), Plato's successor and the greatest philosopher of antiquity (some would say the greatest of all time). Aristotle's writings on cosmology were closely integrated with his overall philosophy; he united in one conceptual scheme elements that we now separate into separate components—scientific, poetic, theological, ethical. It was precisely because his theory of the universe concentrated on physical understanding rather than mathematical computation that it was so widely adopted, especially in the medieval period just before the birth of modern science.

What does the universe consist of? Proceeding from a then current notion, Aristotle postulated that all actual matter within our physical reach is a mixture of four elements: Earth, Water, Air, and Fire. Actually one would never see the pure elements; it was understood that a lump of earth or a stone might contain mostly the element Earth, mixed with smaller amounts of the other three elements. A vessel full of the purest water that man can obtain would be expected to contain beside the element Water also some earthy substance. Indeed, on boiling away the liquid, one does find a solid residue. (So convincing was this experiment and this interpretation that when the great chemist Lavoisier by ingenious experimental demonstration first proved in 1770 that most of the residue came from the walls of the vessel itself, he also reported that what was "thought by the old philosophers, still is thought by some chemists of the day.")

A second postulate prescribed that each of these four elements has a tendency or desire to reach its "natural place" of repose: Earth at the bottom (or center of the universe), next Water, then Air, and finally Fire at the top.

A third postulate directed that the actual motion of an object is determined by the tendency of the element most abundantly present. So the behavior of steam rising from a boiling vessel was explained as the upward motion owing to the introduction of the element Fire into the heated water; when, on cooling, the steam gave up its Fire, the then predominant element, Water, could assert itself again, and the condensed moisture could precipitate down to its natural place below.

One consequence of this view was that the motion of an object up or down to its natural place, the so-called "natural motion," was so governed by the balance of its constituent elements that its speed of motion must be proportional to the amount of the predominant element. A big stone, containing evidently more Earth than a small one, consequently was expected to descend that much faster when allowed to execute its natural motion of free fall. We will return to this point in Chapter 7, where we will see that this prediction about the speed of falling objects provided one of the handles by which Galileo could overturn the Aristotelian system.

Aristotle also postulated that these four elements are found only in the terrestrial or "sublunar" (below the moon) domain. Beyond that, there is an entirely different kind of element, the *Ether*,* from which the heavens are constructed. Whereas the four terrestrial elements are always involved in processes of change—"generation and corruption," in Aristotle's vivid phrase—the Ether is by definition pure and immutable. It has its own natural motion appropriate to its nature, a motion which has no beginning and no end, and which always keeps it in its natural place: *circular* motion. Thus the use of circles to explain the motions of heavenly bodies, as suggested by Plato, is not merely a mathematical convenience: it became a philosophical necessity in Aristotle's system.

Aristotle conceived of the system of celestial spheres in a mechanical way that is not entirely alien to the physics of the seventeenth, eighteenth,

---

* Also known as the "quintessence," from the Latin term for "fifth element." Ether, sometimes spelled Aether, comes from a Greek word which means to burn or glow.

**Fig. 1.3** A late medieval conception of the world. The sphere of the moon (*lune*) separates the terrestrial region (composed of concentric shells of the four elements Earth, Water, Air, and Fire) from the celestial region. Beyond the moon are the concentric spheres carrying Mercury, Venus, Sun, Mars, Jupiter, Saturn, and the fixed stars. (From a woodcut of 1508.)

and nineteenth centuries. Outside the largest sphere (in which the stars are embedded) is the divine "prime mover" (labeled as *primum mobile* in Fig. 1.3). The prime mover turns the starry sphere at a regular rate; this motion is transferred, by friction with some loss, to the spheres of the outer planets, and thence to the spheres of the sun and inner planets. While Aristotle accepted the 26 spheres of Eudoxus mentioned above, he prescribed

29 additional ones, partly to account for some of the most obvious discrepancies between Eudoxus' system and the observed paths of the planets, and partly as intermediate spheres needed to "neutralize" the reversed motion of some of the spheres.

And yet, there remained easily observed features of the sky which were unexplained by Aristotle's system—notably the fact that the sun, the moon, Venus, Mars, and Jupiter at times seemed brighter or nearer and at other times farther away from the earth. A set of uniform rotations of the celestial bodies on spheres *concentric* with the earth could never allow them to change their distance from the earth. Aristotle was aware of this, but discounted the importance of the simple, but eventually fatal, argument against the fundamental hypothesis of his system. He did not take the consequences which to us seem deceptively obvious, mainly because it was then, as it is now, incomparably easier to leave in doubt the testimony of a few contrary observations than to give up a whole world scheme which seemed conclusive and necessary from the point of view both of his philosophy and his mechanics. Of course, this does not mean that he proposed a theory he knew to be false. The original Aristotelian science was not simply bad modern science but an activity fundamentally different from it. Perhaps we may also see here an example of an important human trait which colors all scientific work and which even the greatest thinkers can never hope to overcome entirely—we all tend to deny the importance of facts or observations not in accord with our convictions and preconceptions, so that sometimes we ignore them altogether, even though, from another point of view, they stand before our very eyes. Moreover, even the most general and modern scientific theory does not hope or even seriously attempt to accommodate every single detail of every specific case. One is always forced to idealize the observations before attempting a match between "facts" and theory—not only because there are usually unavoidable experimental uncertainties in observation, but because

conceptual schemes are consciously designed to apply to *selected* observations rather than to the totality of raw experience. As a consequence, the history of science is studded with cases in which it turned out later that the neglected part of a phenomenon was actually its most significant aspect. But on the other hand, if there had been no tentative, half-true, or even plainly wrong theories in science, we should probably never see a correct one evolving at all. Since the task cannot humanly be done in one jump, we must be satisfied with successive approximations.

## 1.4  HOW BIG IS THE EARTH?

Alexander the Great (356–323 B.C.), who as a youth had been tutored by Aristotle, conquered most of the world known to the Greeks during his short life. In 331 B.C. he added Egypt to his empire and founded there the city of Alexandria. During the following decades the center of Greek civilization shifted from Athens to Alexandria. Although Alexander's own empire broke up after his death, the kingdom of the Ptolemies in Egypt maintained Greek civilization at a high level for the next two centuries. Alexandria itself became a cosmopolitan mixture of races and creeds, the population being mostly Egyptian with a Greek-Macedonian upper class, many Jews, Africans, Arabs, Syrians, Hindus, etc. George Sarton, who has written extensively about Greek and Alexandrian science, compares the relation between Alexandria and Athens to that between New York and London.

Two of the most famous cultural institutions in the ancient world were the Museum and the Library at Alexandria. The Museum was the center for scientific and mathematical research, conducted by Euclid, Apollonius, Eratosthenes, Hipparchus, perhaps Aristarchus, and (much later) the astronomer Ptolemy (not to be confused with the kings named Ptolemy). Some of these names will appear again later in this chapter.

Eratosthenes (273–192 B.C.) was educated in Athens but spent more than half his life in Alexandria. At a time when many scientists were already becoming overspecialized in mathematics or astronomy, Eratosthenes, like Aristotle, dared to be a generalist. As chief librarian of the Library at Alexandria, he was responsible not only for collecting, arranging, and preserving several thousand rolls of papyrus manuscripts, but also for knowing what was in them. He applied mathematics and his knowledge of the world to found the science of geography. His contemporaries gave him the deprecatory nicknames "beta" (meaning "second-rate") and "pentathlos" (the modern equivalent would be a 5-letter man in athletics). They assumed that anyone who was Jack of all trades must be master of none. Nevertheless Eratosthenes accomplished one outstanding feat: He measured the size of the earth!

A brief sketch of Eratosthenes' calculation will give some indication of the level of sophistication that Greek science had reached by the third century B.C. Eratosthenes made the following assumptions:

1) The earth is spherical.

2) The sun's rays are parallel when they strike the earth (which means that the sun must be very far away from the earth).

3) The two places on the earth's surface where he made his observations, Alexandria and Syene (now Aswan, Egypt) lie on the same north-south line.

4) Syene is exactly on the Tropic of Cancer (latitude $23\frac{1}{2}°$ N), so that the sun is directly overhead at noon on the day of the summer solstice.

None of these assumptions is exactly correct, but they are so nearly correct that Eratosthenes was able to obtain a surprisingly accurate result.

The calculation was based on two measurements: the angle which the sun's rays make with a vertical stick at noon at the summer solstice at Alexandria (an easy measurement); and the linear distance ($D$) between Alexandria ($A$) and Syene ($S$)

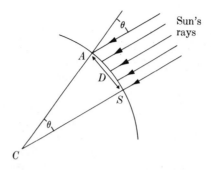

**Fig. 1.4**  How Eratosthenes estimated the size of the earth.

(a difficult one, done by timing the march from $A$ to $S$ of a well-trained group of soldiers whose rate of marching was known and fairly constant). According to the fourth assumption, the angle at $A$ must be the same angle $\theta$ subtended at the earth's center, $C$, by the arc $AS$; in other words, it is the difference in latitude between Alexandria and Syene. Eratosthenes could then set up a proportion between the angle $\theta$ and the distance $D$: the ratio of $\theta$ to a complete circle (360°) is the same as the ratio of $D$ to the circumference of the circle, $c$.

$$\frac{D}{c} = \frac{\theta}{360},$$

where $\theta$ = angle $ACS$, $c$ = circumference of earth, and $D$ = length of arc $AS$.

According to Eratosthenes, the angle $\theta$ was, by measurement, exactly the one-fiftieth part of a whole circle:

$$\theta = \tfrac{360}{50} = 7.2°$$

and the measured distance was

$$D = 5000 \text{ stadia.}$$

Hence, solving the above equation for the circumference $c$, he found

$$c = \frac{360D}{\theta} = 250,000 \text{ stadia.}$$

To compare this result with modern values, all we need to know is the conversion factor between the Greek unit of distance, the stadium, and some modern unit (mile or kilometer). Unfortunately there is some doubt about this conversion factor; the best guess is that a stadium was about one-tenth of a mile, or perhaps a little larger. Depending on what factor is used, Eratosthenes' result comes out to within 1 or 2% of the modern value of 24,860 miles for the circumference of the earth, or 3957 miles for its radius $R$. In any case, it was a remarkable demonstration that man's intellect could survey portions of the world at least an order of magnitude larger than could be subdued by physical force, even by Alexander the Great.

---

**Problem 1.1.** The data reported by Eratosthenes look as if they had been rounded off. Assuming that the value of the angle $\theta$ is really between $\frac{360}{49}$ degrees and $\frac{360}{51}$ degrees, and that the distance $D$ is really between 4900 and 5100 stadia, find the upper and lower limit for the measurement of the circumference and radius of the earth in stadia. Convert these values to miles, assuming 1 stadium = $\frac{1}{10}$ mile.

## 1.5  THE HELIOCENTRIC* THEORY

Of course, the problem of planetary motion persisted. There were two different major types of attack after Aristotle—the heliocentric theory and the modified geocentric theory. Let us now discuss the first of these. Aristarchus of Samos (third century B.C.), perhaps influenced by the work of Heraclides of Pontus (fourth century B.C.), suggested that a simple world system would result if the sun were put at the center of the universe and if the moon, the earth, and the five then known planets revolved around the sun in orbits of different sizes and speeds. We do not know many details; his work on this subject is known to us only through references in other old writings. But

---

\*    From the Greek *helios*, sun.

evidently he assumed that the earth has a daily rotation on its north-south axis, as well as a yearly revolution in the orbit around the sun, and he placed the whole system in the sphere of fixed stars, which thereby could be considered at rest with respect to the center of the universe.

This heliocentric hypothesis has one immediate advantage. It explains the bothersome observation that the planets are at times nearer to the earth and at other times farther away. But the ancient world saw three very serious flaws in Aristarchus' suggestion:

*First*, it did violence to philosophical doctrines (e.g., that the earth, by its very "immobility" and position is differentiated from the "celestial bodies," and that the natural "place" of the earth is the center of the universe). In fact, his contemporaries considered Aristarchus impious "for putting in motion the hearth of the Universe." Also, the new picture of the solar system contradicted common sense and everyday observation; the very words used in astronomy (sunrise, progression of planets, etc.) reflect the intuitive certainty that the earth must be at rest.

---

**Problem 1.2.** List the common-sense observations concerning sun, stars, and planets from which the geocentric theory sprang, apparently so convincingly. Be careful not to *assume* the geocentric or any other theory in your description.

---

*Second*, Aristarchus does not seem to have fortified his system with detailed calculations and quantitative predictions of planetary paths, by our present standards an obvious condition for assuring recognition in the physical sciences. This work seems to have been purely qualitative, although in some other accomplishments he is said to have showed considerable mathematical powers.

*Third*, the Greek thinkers offered an ingenious bit of reasoning to refute Aristarchus. If the earth is to move around the sun, its large orbit will carry it sometimes near to a given fixed star on the celestial sphere, and sometimes farther away from it. Thus the angle at which we have to look for this star will be different as seen from the various points in the earth's annual sweep (Fig. 1.5). This phenomenon, called the annual *parallax* of the fixed stars, should occur on the basis of Aristarchus' heliocentric hypothesis, but it was not observed by the Greek astronomers.

In explanation we may say either (a) that the stellar parallax is so small as to be unobservable with the naked eye—which in turn requires the fixed stars to be unimaginably distant compared with the diameter of the yearly orbit of the earth, or (b) that Aristarchus was wrong and the earth does not move around within the celestial sphere. It seems natural that the ancients, predisposed to reject the heliocentric system, and also loath to consider an infinitely extensive universe, chose the second of these alternatives. The first, however, proved eventually to be the correct one. The parallax was indeed present, although so small that even telescopic measurements did not reveal it until 1838.

**Problem 1.3.** The annual parallax of a given star may be crudely defined as half the angle between the two lines of sight drawn from the center of the earth to the star from opposite ends of a diameter across the earth's orbit. Since the Greeks could not detect the parallax, and since their accuracy of measurement was often only about $\frac{1}{2}°$, what must be the least distance from the earth's orbit to the nearest fixed star? (Express in astronomical units, or A.U., one A.U. being the sun's mean distance from us, or about $93 \times 10^6$ miles.) F. W. Bessel in 1838 first observed the annual parallax, which for the nearest bright star, $\alpha$ Centauri, is of the order of $\frac{3}{4}$ second of arc. What is its approximate distance from us? (Modern telescopes can measure parallaxes of about 0.01″, and therefore can directly determine distances of stars about 100 times as far away as $\alpha$ Centauri.)

**Fig. 1.5** Parallax of a star as seen from the earth. The diagram is not to scale; the star is much farther away, relative to the size of the earth's orbit.

The heliocentric theories of Aristarchus and others seem to have been so uninfluential in Greek thinking that we should ordinarily spend no time on them. But these speculations stimulated the crucial work of Copernicus 18 centuries later. Fruitful ideas, it is evident, are not bound by time or space, and can never be evaluated with *final* certainty.

## 1.6 MODIFIED GEOCENTRIC THEORIES

We now turn to the other and more vigorous offshoot of the early astronomical schemes. To allow the planets to have variable distances from the earth while still keeping to the old belief of an immovable earth, the system of concentric spheres was modified in ingenious ways, principally by

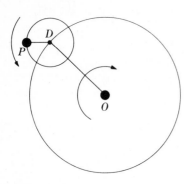

**Fig. 1.6**  An example of epicyclic motion of a planet *P*.

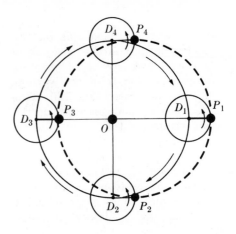

**Fig. 1.7**  Eccentric path represented by epicyclic motion.

Apollonius,* Hipparchus,† and finally the influential astronomer and geographer, Claudius Ptolemy (second century A.D.).**

a) *Eccentric motion.* If the stationary earth were not exactly at the center of rotation of a uniformly

moving celestial object, the latter would move along an *eccentric* path as seen from the earth, and vary in distance at different times. This scheme fits the apparent yearly motion of the sun fairly well, since it appears larger (and consequently nearer) at noon in our winter compared with our summer. Note that by admitting eccentric motions the astronomers really were dodging the old principle that required the planet motions to be circular around the center of the earth.

b) *Epicyclic motion.* Figure 1.6 represents an object *P* (such as the sun or a planet) having two simultaneous uniform rotary motions, one a rotation of *P* about a point *D* (in space) with radius *PD*, the other a rotation of the line *OD* about point *O* (the position of the earth). The small circle is called an epicycle, the large circle the deferent. The two motions may have entirely independent speeds, directions, and radii. Figure 1.7 indicates the special case in which the epicyclic scheme yields an eccentric path (dashed line), and Fig. 1.8 shows, by the dashed line connecting 24 successive positions of *P*, the complicated motion resulting when *P* rotates about *D* several times while *D* moves once around *O*.

---

*    Apollonius (third century B.C.) suggested the epicycle; he also worked out, without any reference to astronomy, the properties of conic sections such as the parabola and the ellipse, which became important later in the Kepler-Newton theory of the seventeenth century (see Chapters 4 and 11).

†    Hipparchus (second century B.C.) invented a method of calculation equivalent to what is now called trigonometry. His work on the theory of the motions of the sun, moon, and planets was incorporated into Ptolemy's system; since most of what is known of Hipparchus' work comes from ambiguous references by Ptolemy, it is difficult to know how much credit each should get for creating the "Ptolemaic system." Hipparchus is also credited with having discovered the precession of the equinoxes, although something was probably known about this phenomenon long before him.

**   Note the enormous time interval between Hipparchus and Ptolemy: We have no record of any major contributions to astronomy during the period which saw the rise of the Roman Empire and of Christianity.

**Problem 1.4.** What is the ratio of the speeds of revolution along the epicycle and the deferent in Fig. 1.8? What might be the path if the ratio were exactly 3 : 1?

This last type of motion does, in fact, exhibit most of the observed complications in the paths of planets as seen from the earth against the background of the fixed stars; note particularly the reversal in the direction of motion (*retrograde motion*) at positions $P_4$, $P_{11}$, and $P_{18}$. Jupiter's path, one with 11 such loops, is covered in approximately 12 years per complete revolution around the earth.

By a proper choice of radii, speeds, and directions, an eccentric path can be traced just as well by an epicyclic motion. Ptolemy used one or the other, and sometimes mixtures of both types, depending on the problem. We must also note that the device of epicyclic motion suffers from the same philosophical difficulties as the eccentric. The rotation of $P$ about $D$ is, if we wish to be strictly logical, just as much in conflict with old axioms as was the circular motion of $P$ about a center other than the earth.

**Problem 1.5.** Construct as well as you can with compass and ruler the path of $P$ (*similar* to Fig. 1.7), if $P$ revolves around $D$ twice while $D$ goes around $O$ once.

**Problem 1.6.** From observations we may form the following hypotheses. The sun moves around the earth from east to west once a day, just as the celestial sphere does, but about four minutes per day less quickly. The sun also approaches the earth during our winter (i.e., for observers in the northern hemisphere) and recedes in summer, and in addition it slowly travels north of the equator in summer and south of the equator in winter. Which of these features can be incorporated into the spherical model shown in Fig. 1.2, and which

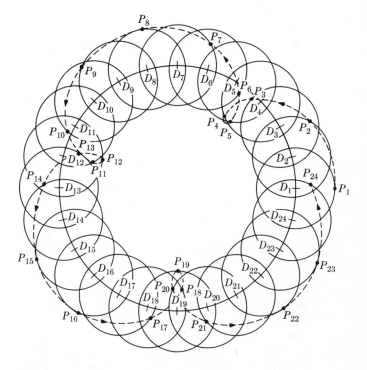

**Fig. 1.8** Epicyclic motion of $P$, showing three temporary reversals of direction of motion at $P_4$, $P_{11}$, and $P_{18}$. (Adapted from M. Cohen and I. Drabkin.)

cannot? Suggest and draw the outlines of a qualitative geocentric model for these motions of the sun about the earth, using the devices described in this section.

Now we can summarize what in the Ptolemaic system has been retained from Plato and Aristotle, and what has been amended. Use is still made of *uniform circular motions*, and of a *stationary earth*. Vanished is the scheme of spheres all concentric at the earth, and with it the need to have all rotations exactly earth-centered. This is even more strikingly evident when we discover that Ptolemy found it necessary to add still another device to the list in order to represent more faithfully some

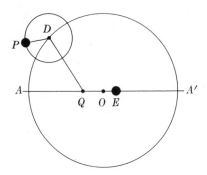

**Fig. 1.9**  Motion with respect to equant at $Q$.

features of celestial motion.  This device is the *equant*.

c) *The equant.*  The sun moves halfway around its path through the stars (i.e., 180° out of a full cycle of 360°) between the vernal equinox, March 21, and the autumnal equinox, September 23.  It then completes the other half of the cycle between September 23 and March 21.  Thus the sun appears to move a little more slowly during the summer, and a little faster during the winter; it takes almost 6 days more to cover the same *angular* distance (measured against the celestial sphere) in the summer than in the winter.  This fact could not be explained by any combination of rotating spheres or epicycles, since it was always assumed that the rate of circular rotation is constant.  Similar but less obvious difficulties arose in accounting for planetary motions.

Ptolemy's equant device is shown in Fig. 1.9.  An object $P$ is in cyclic motion around $D$, which in turn moves on a circle whose center is at $O$.  The earth may be at $O$ or, if this is a mixture of epicyclic and eccentric motion, the earth may be anywhere along the line $AA'$, say at position $E$.  So far the motion of $D$ has been specified as uniform with respect to $O$, but to represent the irregular motions of the sun and planets mentioned above, Ptolemy proposed to let $D$ revolve uniformly with respect to

$Q$, a point called the equant.  That is, the angle $DQA$ changes at a constant rate while $D$ executes its circular sweep.  Now $D$ is no longer in strictly uniform circular motion, although its motion is still *uniform* (seen from $Q$) and *circular* (seen from $O$).

To those astronomers whose only concern was to "save the appearances," that is, to establish a system that would provide accurate predictions of all celestial phenomena, the equant was a brilliant success.  But those who wanted a system that would be compatible with general philosophical principles, such as the necessity of uniform circular motion, would be unhappy about the artificial character of this device.  Which is the more important criterion for a scientific theory: accuracy or intelligibility?

Ptolemy himself thought that his system was based on a set of reasonable assumptions which any reader could accept, even if he could not follow all the details of the calculations.  In his great work, which became known through Arab translations as the *Almagest*, he stated these assumptions as follows.

1) That the heaven is spherical in form and rotates as a sphere.

2) That the earth, too, viewed as a complete whole, is spherical in that form.

3) That it is situated in the middle of the whole heaven, like a center.

4) That by reason of its size and its distance from the sphere of fixed stars the earth bears to this sphere the relation of a point.

5) That the earth does not participate in any locomotion.

Implicit in his work is also the old, now somewhat distorted doctrine of uniform circular motion as the only behavior thinkable for celestial objects.

---

**Problem 1.7.**  What is the significance of the fourth point in Ptolemy's preliminary assumptions?

## 1.7 THE SUCCESS OF THE PTOLEMAIC SYSTEM

By adjusting the respective axes, directions of motions, rates and radii of rotations, number and size of epicycles, eccentrics, and equants, by fitting his devices to the observed paths in a cut-and-try fashion, by utilizing partial results of generations of previous astronomers—so did Ptolemy manage to assemble the apparatus which proved still useful to astronomers, navigators, and astrologers more than 14 centuries later. It was an answer to Plato's original question, and on the whole a magnificent piece of work. One pleasing feature was that centers of the epicycles of the moon, Mercury, and Venus fell on the same line between earth and sun (see Fig. 1.10).

But the complexity and inaccuracy of detail in the whole scheme was considerable, and subsequent observations required amendment of some features of the model from time to time, e.g., the change of an assigned equant or the addition of another epicycle. By the time of Copernicus, this geocentric system required more than 70 simultaneous motions for the seven celestial bodies. But against this were five powerful reasons why the system was generally accepted wherever it became known.

a) It gave an accurate enough description of what could be observed with the instruments of the times.

b) It predicted the future paths of planets well enough for the purposes of the times, if only after cumbersome calculations; and when a serious discrepancy between predictions and observations occurred, it could be resolved by tampering a little with the gears of the flexible apparatus. Today, even as at that time, the geocentric system is still preferred for calculations in navigation—and in astrology!

c) It explains naturally why the fixed stars show no annual parallax.

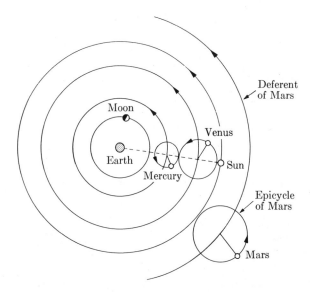

**Fig. 1.10** Partial and schematic diagram of Ptolemaic system of planetary motion.

d) In most details it coincided with Greek philosophical and physical doctrine concerning the nature of the earth and the celestial bodies. Later, when reintroduced to Europe by the Arabs, the Ptolemaic system* was given theological significance by the scholastics. Furthermore, it was also in line with contemporary physics (e.g., projectile motion), based on the same philosophical doctrine of natural motion, natural place, and so forth. (See Chapter 7.)

e) It had common-sense appeal. It is almost unavoidable to hold the opinion that we actually can "see" the sun and stars moving around us, and it is comforting to think of ourselves on a stable, immovable earth.

---

\*   In connection with our study of the Ptolemaic system there is surely no more thrilling experience than to read that famous excursion through the medieval universe, the *Paradiso* of Dante (for example, in the 1944 edition of the *Modern Library*).

But for all that, Ptolemy's theory was eventually displaced by a heliocentric one. Why did this happen? What were the most significant deficiencies in Ptolemy's picture? When is a scientific theory, from our present point of view, successful or unsuccessful? We shall want to answer these questions in detail after we have looked at an outline of the great rival scheme of planetary motion.

## RECOMMENDED FOR FURTHER READING

*Note:* The books and articles mentioned in these pages are intended to guide you to stimulating sources if you wish to carry this subject further. Those in the "Recommended for Further Reading" category are generally written at a technical level no higher than that of the text itself, while "Sources, Interpretations, and Reference Works" are intended primarily for the use of more advanced students and for instructors. The excerpts from the original writings are especially recommended, for, as the scientist-philosopher Ernst Mach wrote, "There is no grander nor more intellectually elevating spectacle than that of the utterances of the fundamental investigators in their gigantic power."

M. Clagett, *Greek Science in Antiquity*, New York: Abelard Schuman, 1955; Collier Books; paperback

B. Farrington, *Science in Antiquity*, New York: Oxford University Press, second edition, 1969; paperback

D. L. Hurd and J. J. Kipling, *The Origins and Growth of Physical Science*, Baltimore Md: Penguin Books, 1964. See especially the extracts from the writings of Aristotle and Ptolemy (Vol. 1, pages 28–32 and 64–71).

E. C. Kemble, *Physical Science: Its Structure and Development*, Cambridge, Mass., M.I.T. Press, 1966; Vol. I, Chapters 1, 2, and 3

T. S. Kuhn, *The Copernican Revolution*, Cambridge, Mass.: Harvard University Press, 1957; paperback; Chapters 1, 2, 3

R. M. Palter (editor), *Toward Modern Science*, New York: Farrar, Straus & Cudahy, 1961; paperback. See the articles on Greek and medieval science in Volume I.

G. Sarton, *Ancient Science and Modern Civilization*, Lincoln, Neb.: University of Nebraska Press, 1954; paperback. Three essays dealing with the world of Hellenism as personified by Euclid and Ptolemy, and with the decline of Greek science and culture.

## SOURCES, INTERPRETATIONS, AND REFERENCE WORKS

Aristotle, *On the Heavens*, translated by Guthrie; Cambridge, Mass.: Harvard University Press, 1939

A. Berry, *A Short History of Astronomy*, London, 1898; Dover paperback reprint; Chapters I, II, III

M. R. Cohen and I. E. Drabkin, *A Source Book in Greek Science*, New York: McGraw-Hill, 1948; reprinted by Harvard University Press. A wealth of source material and penetrating comments; see pages 90–143 for astronomy.

D. R. Dicks, *Early Greek Astronomy to Aristotle*, Ithaca, N.Y.: Cornell University Press, 1970

J. L. E. Dreyer, *A History of Astronomy*, New York: Dover, 1953, Chapters I–XI (more detailed than Berry's book listed above).

Pierre Duhem, *To Save the Phenomena, an Essay on the Idea of Physical Theory from Plato to Galileo*, translated from the French edition of 1908 by E. Doland and C. Maschler. Chicago: University of Chicago Press, 1969; Chapters 1, 2, 3

W. Jaeger, *Aristotle*, Oxford: Clarendon Press, second edition, 1948

M. K. Munitz (editor), *Theories of the Universe*, New York: Free Press (Macmillan), 1957; paperback. A valuable collection of sources and historical articles; see pages 1–138.

O. Neugebauer, *The Exact Sciences in Antiquity*, Providence: Brown University Press, second edition, 1957; Dover paperback reprint. Egyptian, Babylonian, and Greek mathematics and astronomy.

Ptolemy, *The Almagest*, translated by R. C. Taliaferro, in *Great Books of the Western World* (editor R. M. Hutchins), Chicago: Encyclopedia Britannica, 1939, 1952; Vol. 16

G. Sarton, *A History of Science*, Cambridge, Mass.: Harvard University Press, 1952, 1959. Two volumes on Greek science; unfortunately Sarton was unable to complete this project before his death.

F. Solmsen, *Aristotle's System of the Physical World*, Ithaca, N.Y.: Cornell University Press, 1960

## GENERAL BOOKS ON THE HISTORY OF SCIENCE AND THE HISTORY OF PHYSICS

(These books deal primarily with the pre-twentieth-century period. See the end of Chapter 26 for a list of books on the history of modern physical science.)

J. D. Bernal, *The Extension of Man. A History of Physics before 1900*, London: Weidenfeld and Nicolson, 1972

J. D. Bernal, *Science in History*, New York: Watts, third edition, 1965

S. G. Brush (editor), *Resources for the History of Physics*, Hanover, N.H.: University Press of New England, 1972

I. B. Cohen, *The Birth of a New Physics*, Garden City, N.Y.: Doubleday, Anchor Books, 1960; paperback. A brief, readable account covering most of the same topics as Parts A, B, and C of this book.

R. G. Collingwood, *The Idea of Nature*, New York: Oxford University Press, 1945; paperback. A stimulating survey of the metaphysical ideas of ancient and early modern scientists.

W. C. Dampier, *A History of Science*, New York: Cambridge University Press, 4th edition, 1948; paperback

C. C. Gillispie, *The Edge of Objectivity*, Princeton, N.J.: Princeton University Press, 1960; paperback. Presents a sophisticated viewpoint for readers already familiar with the major events in the history of science.

C. C. Gillispie (editor), *Dictionary of Scientific Biography*, New York: Scribner, 1970– . When completed, this work will provide the most comprehensive and authoritative treatise on the history of science available, and an invaluable source of information on the lives and accomplishments of thousands of scientists.

A. R. Hall, *The Scientific Revolution 1500–1800*, Boston: Beacon Press, 1954; paperback

O. Lodge, *Pioneers of Science*, London, 1893; Dover paperback reprint

S. Mason, *A History of the Sciences*, Collier Books (Macmillan), 1962

G. de Santillana, *The Origins of Modern Scientific Thought, from Anaximander to Proclus, 600 B.C. to 500 A.D.*, Chicago: University of Chicago Press; Mentor Books, 1961

C. J. Schneer, *The Search for Order*, New York: Harper, 1960; reprinted as *The Evolution of Physical Science* by Grove Press; paperback

L. W. Taylor, *Physics the Pioneer Science;* Dover paperback reprint of the 1941 edition

S. Toulmin and J. Goodfield, *The Fabric of the Heavens*, New York: Harper, 1961; *The Architecture of Matter*, New York: Harper, 1962; paperback. The first two volumes of a four-volume series on the history of science.

W. P. D. Wightman, *The Growth of Scientific Ideas*, New Haven: Yale University Press, 1951; paperback

A. Wolf, *A History of Science, Technology, and Philosophy in the XVIth and XVIIth Centuries*, and same for *XVIIIth Century*, New York: Macmillan, 1935 and 1939; Harper Torchbook reprint, paperback. Generally useful for its historical material (arranged by subjects) and illustrations.

# Chapter 2

# COPERNICUS' HELIOCENTRIC THEORY

## 2.1 EUROPE REBORN

Our study takes us now to Renaissance Europe of the years around 1500 A.D. Astronomical theory had not progressed significantly since Ptolemy. St. Thomas Aquinas (1225–1274) had joined the Aristotelian ideas of celestial motions with Christian theology. Thus the geocentric theory had received new meaning in terms of current philosophical doctrine; to question one was to attack the other, and so far there had appeared no one who thought it necessary or had the daring to battle seriously against such formidable allies.

But the spirit was changing; the Renaissance movements spread out from Italy to sweep the Western world, and within a few generations there arose a new ideal of man, the confident individual full of curiosity and *joie de vivre*. Concern with spiritual abstractions was balanced by a new enthusiasm for exploring the natural world, both among scientists and among artists (and especially when combined in one person who was both, such as Leonardo da Vinci). The invention of printing revolutionized the intellectual world by assuring that the works of a writer could be rapidly distributed to an ever-growing audience, rather than restricted to the few lucky enough to obtain copies

laboriously transcribed by hand. Even the works of the ancient Greek scientists and philosophers were more widely available than they had been at the time they were first written; but their authority was soon to be undermined by new ideas and discoveries.

In the topography of world history this great period of summing up old wisdom and of forming new attitudes is analogous to a watershed on which great rivers have their sources. It was the time in which lived as contemporaries or within a generation of one another most of the men whose work heralded the new age: Gutenberg and da Vinci, Columbus and Vasco da Gama, Michelangelo and Dürer, Erasmus, Vesalius and Agricola, Luther and Henry VIII. One cannot help comparing this short time interval of decisive change with a similar break in the traditions of Western thought four hundred years later—a period which, between the publication of Darwin's *The Origin of Species* (1859) and the first largescale release of atomic energy, bracketed such names as Mendel and Pasteur, Planck and Einstein, Rutherford and Bohr, and also such new forces as Marx and Lenin, Freud and Pareto, Picasso and Stravinsky, Shaw and Joyce.

## 2.2 THE COPERNICAN SYSTEM

In the year that the New World was discovered, Nicolaus Copernicus (1473–1543) was a young student in Poland. During his life he watched in his old world the gathering of great cultural changes. And it is said that on the very day he died he saw the first copy of his great book, the *Revolutions*, which gave us a whole new universe.*

The full title of Copernicus' main work, *Six Books Concerning the Revolutions of the Heavenly Spheres*, startles us at the very outset with the implication of an Aristotelian notion of concentric spheres. He was indeed concerned with the old problem of Plato, the construction of a planetary system by combination of the fewest possible uniform circular motions. Far from being a revolutionary who wanted to replace traditional theories by a radical new one, he wanted to eliminate some of the innovations which Ptolemy had introduced—in particular, the equant, described above in Section 1.6—and return to the principles established by earlier Greek astronomers. He was careful to point out that his proposal of a heliocentric system—letting the earth as well as the other planets revolve around a fixed sun— was not a new idea, but was sanctioned by some of the authorities of antiquity.

The early ideas of Copernicus can be traced in a brief sketch, called *Commentariolus*, which he wrote several years before composing the *Revolutions*.†

---

* A highly readable and stimulating presentation is *Sun, Stand Thou Still* by Angus Armitage (New York: Henry Schuman, 1947), reprinted in a paperback edition under the title *The World of Copernicus*. The account reveals to us a most remarkable astronomer, mathematician, churchman, administrator, diplomat, physician, and able student of the classics and of economics— and yet a humble and compassionate man.

† According to Edward Rosen, who has published an English translation of the *Commentariolus* (now available in a Dover paperback edition entitled *Three Copernican Treatises*), it was not printed during the author's lifetime, but a number of handwritten copies circulated among astronomers, and then disappeared for three centuries. It was first published in 1878.

**Fig. 2.1**  Nicolaus Copernicus (1473–1543) (from a woodcut of about 1590).

He begins by attacking Ptolemy's equant:

"...the planetary theories of Ptolemy and most other astronomers, although consistent with the numerical data, seemed ... to present no small difficulty. For these theories were not adequate unless certain equants were also conceived; it then appeared that a planet moved with uniform velocity neither on its deferent nor about the center of its epicycle. Hence a system of this sort seemed neither sufficiently absolute nor sufficiently pleasing to the mind.

"Having become aware of these defects, I often considered whether there could perhaps be found a more reasonable arrangement of circles, from which every apparent inequality would be derived and in which everything would move uniformly about its proper center, as the rule of absolute motion requires."

To Copernicus, any type of celestial motion other than uniform circular motion was "obviously" impossible: ". . . the intellect recoils with horror" from any other suggestion; ". . . it would be unworthy to suppose such a thing in a Creation constituted in the best possible way." These arguments are of the same type as those of his scholastic opponents, except that to them the immobility of the earth was equally "obvious."

Copernicus then argued that a more reasonable arrangement of circles could be constructed by putting the sun rather than the earth at the center. He reported later in the *Revolutions* that he found this idea by reading the classics:

"...according to Cicero, Nicetas had thought the earth moved,... according to Plutarch certain others [including Aristarchus] had held the same opinion...when from this, therefore, I had conceived its possibility, I myself also began to meditate upon the mobility of the earth. And although it seemed an absurd opinion, yet, because I knew that others before me had been granted the liberty of supposing whatever circles they chose in order to demonstrate the observations concerning the celestial bodies, I considered that I too might well be allowed to try whether sounder demonstrations of the revolutions of the heavenly orbs might be discovered by supposing some motion of the earth.... I found after much and long observation, that if the motions of the other planets were added to the motions [daily rotation, and yearly revolution about the sun] of the earth,...not only did the apparent behavior of the others follow from this, but the system so connects the orders and sizes of the planets and their orbits, and of the whole heaven, that no single feature can be altered without confusion among the other parts and in all the Universe. For this reason, therefore,...have I followed this system."

Thus, in essence, Copernicus proposed a change of point of view for the representation of celestial motions along the lines of Aristarchus' heliocentric system. By highly gifted calculations, using some observations of his own, Copernicus then proved what Aristarchus had failed to do: that the motion of celestial objects as then known could indeed be represented by a combination of a few uniform circular motions in a sun-centered system. All planets, the earth included, could be imagined as moving on concentric spheres, with relatively few and small epicycles and eccentrics still needed to account for the finer details of motion.* Moreover, the *same direction* of motion along almost all deferents and epicycles could be assumed, which had not been true for the geocentric model. But above all, the odious equant could be discarded. All motions were now truly circular and uniform with respect to their own centers. Plato's question was answered again, in an alternative way.

The most obvious advantage of the heliocentric system is that it gives a much more natural explanation of the retrograde motion of the planets.

---

* The common center of the main spheres actually had to fall a little to one side of the unmoving sun, so that this system is not *quite* heliocentric, just as Ptolemy's earth position did not coincide with the center of the sun's motion.

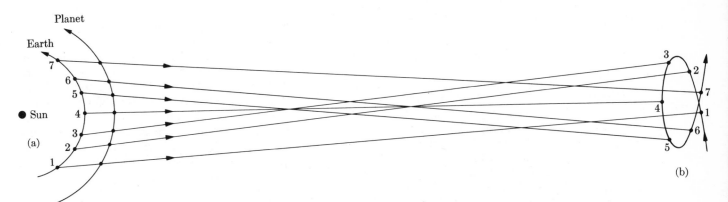

**Fig. 2.2** (a) Actual configurations of sun, earth, and planet, and changing line of sight during the planet's apparently retrograde motion. (b) Apparent path of the planet against the background stars at far right, as seen from the earth.

This explanation can be visualized by imagining first that you are riding in a fast car which is passing a slower-moving car on a highway. Forgetting about the motion of your own car, observe the changing position of the slower car *as seen against the background of trees and buildings* (these correspond to the fixed stars). Initially, when it is still distant, the other car appears to be moving forward, but as your car passes it, it appears to be moving backward for a short time, simply because of the rotation of the line of sight (see Fig. 2.2). This phenomenon is precisely analogous to the retrograde motion of a planet, for example Mars, as seen from the earth. As Copernicus explained it in the *Commentariolus*,

"This happens by reason of the motion, not of the planet, but of the earth changing its position in the great circle [its orbit around the sun]. For since the earth moves more rapidly than the planet, the line of sight directed toward the firmament regresses, and the earth more than neutralizes the motion of the planet. This regression is most notable when the earth is nearest to the planet."

But the earth has two distinct motions in Copernicus' system: it rotates around its own axis, and it revolves around the sun. Not only does this simplify the explanation of the planetary motions, it allows the outer sphere of fixed stars to be at rest. Copernicus did not foresee that this apparently trivial change could lead to enormous consequences: if the stars no longer have to rotate together, they need not all be placed at the same distance from the earth; perhaps they are scattered through space out to infinity.*

Copernicus argued that the precession of the equinoxes could also be more easily explained by attributing the motions of rotation and revolution to the earth. In the geocentric system this phenomenon could be explained by assuming a gradual change in the relative rotations of the celestial spheres carrying the sun and stars, a change which could be rationalized only by adding yet another celestial sphere to the system. Copernicus wrote:

"Hence it is the common opinion that the firmament has several motions in conformity with a law not sufficiently understood. But the motion of the earth can explain all these changes in a less surprising way."

(*Commentariolus*, Rosen translation, pages 64–65)

---

* For an analysis of the Copernican revolution from this viewpoint, see Alexandre Koyré, *From the Closed World to the Infinite Universe.*

Although Copernicus did not in fact accomplish any significant improvement over the geocentric system in his theory of the precession of the equinoxes, one historian of science has argued that it was his attempt to deal with this particular problem that led him to work out the details of the heliocentric system.* It is worth noting that the problem had theological as well as astronomical aspects, and that Copernicus was a canon of the Roman Catholic Church. At the Council of Nicaea in 325 A.D., the Church had defined Easter as the first Sunday after the first full moon on or after the vernal equinox. The vernal equinox had been originally set at March 21, but because of the inaccuracy of the Julian calendar (with the leap year occurring once every four years) the date of the vernal equinox had actually slipped back to March 11 by the thirteenth century. Furthermore, the computation of full moons was somewhat inaccurate. With the growth of commerce and communications in Europe in the late Middle Ages, the problem of calendar reform was becoming increasingly urgent, and in 1514 Pope Leo X asked Copernicus and other astronomers for their assistance. Copernicus declined the invitation to come to Rome at that time, on the grounds that he must first complete his work on the motions of the sun and moon before he could help straighten out the calendar. Later, in dedicating his book, *The Revolutions of the Heavenly Spheres*, to Pope Paul III, Copernicus suggested that his results may be of some assistance to the "ecclesiastical commonwealth" in connection with the calendar problem. The Copernican system was in fact used in the astronomical computations on which the Gregorian calendar (introduced by Pope Gregory XIII in 1582) was based. But that did not necessarily mean that the Church had accepted the physical assumptions—in particular the motion of the earth—from which the Copernican system had been developed; it was

---

* J. Ravetz, "The origins of the Copernican revolution," *Scientific American*, October 1966.

simply admitted to be a more efficient system for doing calculations.

Copernicus himself emphasized that the heliocentric system provided a unique pattern into which all the planets must fit in a definite way, in contrast to the Ptolemaic system which required a special and separate construction for each planet. For example, Copernicus found that his system provided a definite ordering of the planets in terms of their orbital *period*, which agreed with the order of orbital *sizes*:

"Saturn, first of the planets, which accomplishes its revolution in thirty years, is nearest to the first sphere [the immobile sphere of the fixed stars]. Jupiter, making its revolution in twelve years, is next. Then comes Mars, revolving once in two years. The fourth place in the series is occupied by the sphere which contains the earth and the sphere of the moon, and which performs an annual revolution. The fifth place is that of Venus, revolving in nine months. Finally, the sixth place is occupied by Mercury, revolving in eighty days."

In the Ptolemaic system, on the other hand, Mercury, Venus, and the sun all had the same period (one year) so that there was some disagreement as to how to order their spheres. Moreover, in the geocentric systems there might be two or more equally good ways of "saving the phenomena"—the observed motions of the planets might be explained by eccentrics, by epicycles, by equants, or by various different combinations of these devices. Hence it was difficult to believe in the physical reality of any one particular combination. Copernicus claimed that his own system "binds together so closely the order and magnitudes of all the planets and of their spheres or orbital circles and the heavens themselves that nothing can be shifted around in any part of them without disrupting the remaining parts and the universe as a whole."

Actually this last claim is a little exaggerated: The Copernican system in its full elaboration

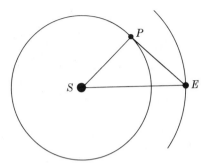

**Fig. 2.3** Method for computing relative distances between an inferior planet *P* and the sun *S*, observed from earth *E*. (The planet may be Mercury or Venus.)

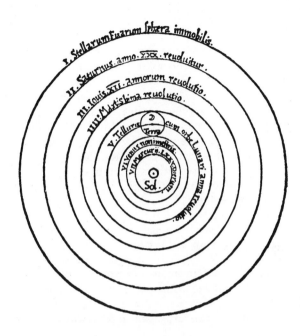

**Fig. 2.4** The heliocentric system as shown in Copernicus' book *The Revolutions of the Heavenly Spheres* (1543). The spheres around the sun (*Sol*) are identified as follows: I, immobile sphere of the fixed stars; II, Saturn, revolves in 30 years; III, Jupiter, revolves in 12 years; IV, Mars, revolves in 2 years; V, earth, revolves in 1 year, with moon's orbit; VI, Venus, revolves in 9 months; VII, Mercury, revolves in 80 days.

suffers from some of the same ambiguities as the Ptolemaic. But it does provide a better and more direct method for determining the relative sizes of the planetary orbits. The method is most simply illustrated for the *inferior* planets (those closer to the sun than is the earth, i.e., Venus and Mercury).

Let $< SEP$ be the angle of maximum *elongation* of the planet away from the sun, as viewed from the earth. Then the angle $< SPE$ must be a right angle; hence the ratio $SP/SE$ must be equal to the sine of the angle $SEP$. A direct observation of the maximum elongation of Mercury or Venus can therefore be used to compute the ratio of its distance from the sun to that of the earth from the sun. The relative distances of the outer planets from the sun can be found by a somewhat more complicated method.* The results are shown in Table 2.1.

There was still no accurate method for estimating the absolute distances, but at least one now had, for the first time, an idea of the relative distances. Copernicus did not try to improve on the value of the absolute distance of the earth as estimated by the Greek astronomer Hipparchus for the earth–sun distance, namely 1142 times the radius of the earth. (The modern value is about 92 million miles, or 23,000 times the radius of the earth.)

---

\* See for example T. S. Kuhn, *The Copernican Revolution*, pages 175–176.

**Table 2.1**    Relative distances of planets from sun

|          | Copernicus | Modern value |
|----------|-----------|--------------|
| Mercury  | 0.3763    | 0.3871       |
| Venus    | 0.7193    | 0.7233       |
| Earth    | 1.0000    | 1.0000 (by definition) |
| Mars     | 1.5198    | 1.5237       |
| Jupiter  | 5.2192    | 5.2028       |
| Saturn   | 9.1742    | 9.5389       |

**Problem 2.1.** One of the impressive features of the Copernican model of the solar system is its ability to find the relative radii of the planetary orbits. Thus, one should be able to find how close the earth comes to any of the other planets. Using the data available to Copernicus (Hipparchus' estimate of the earth–sun distance and Eratosthenes' estimate of the size of the earth), calculate the minimum possible distance between the earth and any other planet, assuming circular orbits. Compare this to the modern value.

## 2.3 BRACING THE SYSTEM

Knowing well that to many his work would seem absurd, "nay, almost contrary to ordinary human understanding," Copernicus attempted to fortify it against anticipated criticism in four ways.

a) He tried to make plausible that his assumptions coincided with dogma at least as well as Ptolemy's. He has many passages (we have seen samples) on the deficiencies of the Ptolemaic system, on how harmonious and orderly his own seems, and how pleasingly and evidently it reflects the mind of the Divine Architect. To Copernicus, as to all men of his period (and to many of our own), the observable world was but a symbol of the working of God's mind; to find symmetry and order in the apparent chaos of sense data was to him an act of reverence, a renewed proof of the Deity. He would have been horrified if he had known that his theory was ultimately responsible for the sharp clash, in Galileo's time, between science and dogma. Copernicus, we must not forget, was a highly placed and honored church dignitary. In matters of doctrine he might have regarded himself as a conservative who was objecting to current scholastic teaching only to the extent of wishing to make it more consistent with Aristotle's principles of mechanics. (Compare the forms of the systems in Figs. 1.3 and 2.4.)

b) Copernicus prepared enough quantitative material to put his book mathematically on the same footing as Ptolemy's, that is, he calculated the relative radii and speeds of the elements in his system so that tables of planetary motion could be made. (Note how this compares with Aristarchus' more qualitative work.) The two theories were thus equally correct with respect to prediction of future planet positions within the then current error of observation of at least $\frac{1}{6}$ degree of arc. However, on looking at Copernicus' treatise, we must not expect to find a mathematical treatment of the type seen in modern texts. Remember that even our simplest trick of mathematical notation, + and − signs, did not come into use until after Copernicus' death.

c) With considerable ingenuity and much success, Copernicus tried to answer several objections that were as certain to be raised against his heliocentric system as they had been, long ago, against Aristarchus'. To the argument that his earth, rotating so rapidly about its own axis, would surely burst like a flywheel driven too fast, he countered, "Why does the defender of the geocentric theory not fear the same fate for his rotating celestial sphere—so much faster because so much larger?" To the argument that birds in flight and the like should be left behind by the rapidly rotating and revolving earth,* he answered that the atmosphere is dragged along with the earth. As to the old question of the absence of parallax among the fixed stars, he could give only the same answer as Aristarchus (a good answer even though the Greeks did not accept it):

"...the dimensions of the world are so vast that though the distance from the sun to the earth appears very large compared with the size of the orbs of some planets, yet compared with the dimensions of the sphere of fixed stars, it is as nothing."

---

* The orbital speed around the sun is about 70,000 mi/hr, and the speed of west-east rotation alone of a point on the equator is more than 1000 mi/hr.

This distance to the fixed stars, he says elsewhere, is "so immense as to render imperceptible to us even their apparent annual motion . . . ."

He cites several other good arguments (and a few erroneous ones) to support his system, but these will sufficiently illustrate his point.

d) To us the most impressive success of Copernicus is the reduction in the number of elements needed in his system (to about 34) and the consequent greater ease with which his apparatus can be used by the astronomer for the solution of practical problems. Science has learned to cherish these features: economy of concepts and assumptions, simplicity of formulation, applicability to a variety of problems. Copernicus knew well the advantage he held over the rival theory, and he hoped that he would be accepted on this basis, irrespective of current prejudices.

---

**Problem 2.2.** Make a list of the assumptions underlying Copernicus' theory, analogous to this list for Ptolemy's (see Section 1.6).

**Problem 2.3.** Copernicus postulated that the atmosphere is dragged along by the rotating earth, presumably being unaware of any observable effects of rotation on the atmosphere. Find out and explain one such effect which is now considered well established.

## 2.4  THE OPPOSITION TO COPERNICUS' THEORY

But Copernicus' hope was not quickly fulfilled. It took more than 100 years for the heliocentric system to be generally accepted by astronomers; in the meantime the theory and its few champions were faced with powerful opposition, some of it the very same which antagonists had used against the heliocentric ideas of the Greek philosophers.

a) First came the argument from dogma concerning the immobility and central position of the earth. For all his efforts, Copernicus was in general unsuccessful in persuading his readers that the heliocentric system was at least as close as the geocentric one to the mind and intent of the Deity. All religious faiths in Europe, including the newly emerged Protestants, found enough Biblical quotations (e.g., Joshua 10:13) to assert that the Divine Architect had worked from a Ptolemaic blueprint.

---

**Problem 2.4.** Read Joshua 10:13, and interpret the astronomical event referred to in terms of the two rival planetary systems.

---

Martin Luther himself branded Copernicus a fool and a heretic. The Roman Catholic Church eventually put the *Revolutions* on the index of forbidden books as "false and altogether opposed to Holy Scriptures," withdrawing its approval of an earlier outline of Copernicus' work. Some Jewish communities forbade the teaching of the heliocentric theory. It was as if man's egocentric philosophy demanded the middle of the stage for his earth, the scene both of his daily life and prayer in a world created especially for his use, and of the drama of salvation with the expected eventual coming of the Savior or Messiah.

Although it was allowed that for mathematical simplicity the Copernican scheme might indeed be useful, and although even St. Thomas Aquinas had grave misgivings about the Ptolemaic system because it did not stick to strictly uniform circular motions as interpreted by Aristotle, yet it seemed "philosophically false and absurd," dangerous, and fantastic to abandon the geocentric hypothesis. And in all truth, what other reaction could one have expected? We must try to appreciate the frame of reference of the time. In general, Europe then recognized as its two supreme sources of authority the Bible (either the literal text or its interpretation by Rome) and Aristotle. Physical science as we now think of it was only gradually evolving; that was still the period in which scientists like Kepler and Paracelsus could also practice the pseudosciences of astrology and alchemy.

b) A further explanation for the resistance which the Copernican theory had to overcome is to be sought in the damage it brought even to contemporary physics. This point is well discussed by the historian Herbert Butterfield:

"...at least some of the economy of the Copernican system is rather an optical illusion of more recent centuries. We nowadays may say that it requires smaller effort to move the earth round upon its axis than to swing the whole universe in a twenty-four-hour revolution about the earth; but in the Aristotelian physics it required something colossal to shift the heavy and sluggish earth, while all the skies were made of a subtle substance that was supposed to have no weight, and they were comparatively easy to turn, since turning was concordant with their nature. Above all, if you grant Copernicus a certain advantage in respect of geometrical simplicity, the sacrifice that had to be made for the sake of this was nothing less than tremendous. You lost the whole cosmology associated with Aristotelianism—the whole intricately dovetailed system in which the nobility of the various elements and the hierarchical arrangement of these had been so beautifully interlocked. In fact, you had to throw overboard the very framework of existing science, and it was here that Copernicus clearly failed to discover a satisfactory alternative. He provided a neater geometry of the heavens, but it was one which made nonsense of the reasons and explanations that had previously been given to account for the movements in the sky."*

**Problem 2.5.** Read Milton's *Paradise Lost*, Book VIII, lines 1–202, in which Adam and Raphael

---

* H. Butterfield, *Origins of Modern Science*, page 27.

discuss the two systems from a classic point of view which may well have mirrored the general opinion of educated Englishmen until the full impact of Newton's work was established.

---

c) Third, argument arose from the oft-mentioned lack of observable parallax among fixed stars. Copernicus' only possible (and correct) reply was still unacceptable because it involved expanding the celestial sphere to practically an infinite distance away from the earth. This is to us no longer intellectually difficult, but it was a manifestly absurd notion at the time. Among many reasons for this, perhaps the intensely religious mind of the period could not be comfortable with a threatening Hell so close below the ground and a saving Heaven so infinitely far above. If we consider for a moment, it is evident that as a consequence of this attitude an actual observation of annual parallax at that time would not necessarily have settled the dispute in favor of Copernicus, for an obvious and simple Ptolemaic assumption of an additional yearly epicyclic motion superposed on the great daily rotation of the fixed stars would explain away such an observation in full accord with the geocentric system.

d) A fourth important argument against the Copernican astronomy at the time was that apart from its powerful simplicity it offered to contemporary astronomers no other overruling *scientific* advantages over their geocentric astronomy, i.e., there was then no important observation which was explainable only by one and not by the other, no experiment to pit one against the other in a clear-cut decision. Copernicus introduced no fundamentally new experimental facts into his work, nor was the accuracy of his final predictions significantly better than previous estimates.

To us it is clear, although it did not enter the argument then, that the scientific content of both theories, the power of prediction of planetary motion, was about the same at that time. As

Francis Bacon wrote in the early seventeenth century: "Now it is easy to see that both they who think the earth revolves and they who hold the *primum mobile* and the old construction, are about equally and indifferently supported by the phenomena." In our modern terminology we would say (although this is not what Bacon had in mind) that the rival systems differed mainly in the choice of the coordinate system used to describe the observed movements. As measured with respect to the earth, the sun and stars *do* move, and the earth, of course, is at rest. On the other hand, measured with respect to the sun, the earth is not at rest. Any other system of reference, for instance one placing the moon at the origin of the coordinate system and referring all motions to that new center of the universe, is quite conceivable and equally "correct," although, of course, evidently most complex.

Bacon's empirical relativism, which led him to downgrade the Copernican system, is refuted by certain observations made later on. That it is primarily the earth, and only to a much smaller extent the sun, which moves relative to the fixed stars, was shown by James Bradley's discovery of stellar aberration in 1727.* The propagation of weather conditions in the earth's atmosphere from east to west can be explained most simply by thinking that the atmosphere is not completely dragged along by a rotating earth, and thus makes it more plausible that it is indeed the earth that is rotating. The earth's rotation in space is made more plausible still by the famous pendulum experiment first performed by J. B. L. Foucault in the middle of the nineteenth century, and now repeated daily in science museums all over the world.

But at the beginning of the seventeenth century every newly discovered peculiarity of planetary motion could equally easily have been accommodated in some geocentric representation as well

---

* See A. B. Stewart, "The discovery of stellar aberration," *Scientific American,* March 1964, pages 100–108.

as in a heliocentric one. In fact, to many people it seemed that the compromise system proposed by Tycho Brahe (1546–1601) combined the best features of both: All the planets except the earth revolved around the sun, and the sun then revolved around the earth, which remained at rest.

## 2.5  HISTORIC CONSEQUENCES

Eventually the vision of Copernicus did triumph; in a moment we shall follow the work leading to the universal acceptance of the heliocentric theory, even though his specific *system* of uniform circular motions will have to be sacrificed along the way. We shall see that the real scientific significance and the reason for the ultimate glorification of Copernicus' work lie in a circumstance he would never have known or understood—in the fact that only a heliocentric formulation opens the way for an integration of the problem of planetary motion with the simple laws of "ordinary" (terrestrial) mechanics as developed during the following 150 years—a synthesis of two sciences, even of two methods, which was achieved through Newton's Theory of Universal Gravitation. Consequently, it became possible to explain, on the basis of the supposition that the earth turns and travels, such diverse phenomena as the diurnal and the annual apparent motion of the stars, the flattening of the earth, the behavior of cyclones, tradewinds, and gyroscopes, and much else that could not be bound together into so *simple* a scheme in a geocentric physics.

But apart from this historic triumph, the memory of Copernicus is hallowed for two more reasons. First, he was one of those giants of the fifteenth and sixteenth centuries who challenged the contemporary world picture and thereby gave life to that new and strange idea which later was to grow into science as we now know it. Secondly, his theory proved to be a main force in the gathering intellectual revolution which shook man out of his self-centered preoccupations. Out with the

Ptolemaic system has to go also the self-confident certainty that a symbolic uniqueness about our position among planets shows that man is the sum and summit, the chief beneficiary of creation to whom the whole is given as if it were his toy or his conquest. If this certainty still exists, at least it cannot claim that science backs it up.

An important purpose of our story of planetary motion to this point has been preparation for the following digression before returning to our main topic. We may now ask briefly: By what standards shall we judge a scientific theory?

## RECOMMENDED FOR FURTHER READING

A. Armitage, *Sun Stand Thou Still*, New York: Henry Schuman, 1947; reprinted as *The World of Copernicus*, Mentor Books paperback. The life and work of the great astronomer. Revised edition, *Copernicus, The Founder of Modern Astronomy*, New York: A. S. Barnes & Co., 1957; Perpetua edition, 1962

Marie Boas, *The Scientific Renaissance 1450–1630*, New York: Harper, 1962; paperback; Chapters I–IV

H. Butterfield, *The Origins of Modern Science*, London: Bell, 1957; Free Press paperback; Chapter 2

Nicolaus Copernicus, extract from *De Revolutionibus*, translated in Hurd and Kipling, *The Origins and Growth of Physical Science*, Vol. I, pages 99–111

E. C. Kemble, *Physical Science*, Chapter 4

T. S. Kuhn, *The Copernican Revolution*, Chapters 4 and 5

R. M. Palter (editor), *Toward Modern Science*, paperback. See the articles on Renaissance science in Volume II.

J. Ravetz, "The origins of the Copernican Revolution," *Scientific American*, October 1966, pages 88–98

## SOURCES, INTERPRETATIONS, AND REFERENCE WORKS

A. Berry, *A Short History of Astronomy*, Chapters IV and V

Nicolaus Copernicus, *Commentariolus*, translated with introduction and notes by E. Rosen in *Three Copernican Treatises*, New York: Dover, 1959

Nicolaus Copernicus, *On the Revolutions of the Heavenly Spheres*, translated by C. G. Wallis, in *Great Books of the Western World* (editor R. M. Hutchins), Chicago: Encyclopedia Britannica, 1939, 1952; Vol. 16

J. L. E. Dreyer, *A History of Astronomy*, Chapters XIII and XIV

Pierre Duhem, *To Save the Phenomena*, Chapters 4, 5, and 6

E. Grant, "Medieval and seventeenth-century conceptions of an infinite void space beyond the cosmos," *Isis*, Vol. 60, pages 39–60 (1969)

M. B. Hall (editor), *Nature and Nature's Laws*, New York: Harper, 1970; paperback; pages 1–52, extracts from sixteenth-century writings on astronomy.

N. R. Hanson, "Contra-equivalence: a defense of the originality of Copernicus," *Isis*, Vol. 55, pages 308–325 (1964)

A. Koyré, *From the Closed World to the Infinite Universe*, Baltimore: Johns Hopkins Press, 1952; Harper Torchbook paperback

M. K. Munitz (editor), *Theories of the Universe*, pages 141–189. Includes extracts from the writings of Copernicus and Bruno.

O. Neugebauer, "On the planetary theory of Copernicus," *Vistas in Astronomy*, Vol. 10, pages 89–103 (1968)

D. J. deS. Price, "Contra-Copernicus: a critical re-estimation of the mathematical planetary theory of Ptolemy, Copernicus, and Kepler," in *Critical Problems in the History of Science* (editor M. Clagett), Madison, Wisconsin: University of Wisconsin Press, 1962, pages 197–218

# Chapter 3

# ON THE NATURE OF
# SCIENTIFIC THEORY

As we now turn to the standards by which to judge scientific theories—and the rival planetary theories might serve as models—it is obvious that we cannot pretend to lay down criteria by which a working scientist should check his own progress during the construction of a theory. This cannot be done any more meaningfully than constructing a scientific method for his use. What we shall do here is something else, namely, to attempt a clarification of our thoughts on how we, as students of science and its development, may evaluate a particular theory, past or present.

## 3.1 THE PURPOSE OF THEORIES

We have argued that the task of science, as that of all thought, is to penetrate beyond the immediate and visible to the unseen, and thereby to place the visible into a new, larger context. For like a distant floating iceberg whose bulk is largely hidden under the sea, only the smallest part of reality impresses itself upon us directly. To help us grasp the whole picture is the supreme function of theory. On a simple level, a theory helps us to interpret the unknown in terms of the known. It is a conceptual scheme which we invent or postulate in order to explain to ourselves, and to others, observed phenomena and the relationships

between them, thereby bringing together into one structure the concepts, laws, principles, hypotheses, and observations from often very widely different fields. These functions may equally well be claimed for the hypothesis itself. In truth, we need not lay down a precise dividing line, but might regard theory and hypothesis as differing in degree of generality. Therefore at one extreme we might find the *limited working hypothesis* by which we guide our way through a specific experiment, placing at the other end of the spectrum the *general theory*, which guides the design and interpretation of all experiments in that field of study.

Examples of general theories suggest themselves readily, even if we decide, perhaps rather arbitrarily, to use the word "theory" only for those few historic and general schemes of thought like the theories of planetary motion, of universal gravitation, of the nuclear atom, and the like. Galileo's theory of projectile motion welded together the laws of uniformly accelerated motion, including free fall and the principle of superposition of velocities, to produce one conceptual scheme, one over-all method of attacking, predicting, and interpreting every conceivable problem involving bodies moving under the influence of a constant force. Similarly, when Charles Darwin had pondered the results of such fields as paleontology and comparative anatomy, he could find an inspired connection in widely different types of observations, explaining previously separate truths in different disciplines as part of one grand scheme of thought, his theory of evolution.

---

**Problem 3.1.** Examine some one theory discussed in other courses with which you are thoroughly familiar—perhaps Gregor Mendel's theory of heredity, the Marxian theory of society, Sigmund Freud's theory of the unconscious, or a theory of "business cycles." Differentiate clearly in each case between the *data* available to the theorizer on one hand and the *hypotheses* used by him to explain the data on the other.

**Problem 3.2.** Do the same for Ptolemy's and Copernicus' theory of planetary motion.

---

If we inquire more deeply into the main purposes of theories, we find that there are, on the whole, three functions. Let us summarize them here briefly.

1) A theory correlates many separate facts in a logical, easily grasped structure of thought. By correlating, by putting in juxtaposition and order previously unrelated observations, we can understand them; we always explain by pointing to a relationship. The Platonists explained the motion of planets by relating planets to certain "necessary" attributes of divine beings. Ptolemy and Copernicus explained it by relating the planetary paths to the mathematical combination of certain more or less simple geometrical figures. And not only will a fruitful theory explain the laws which it relates within its framework, it also will show where and why these laws may not hold precisely in practice, as, for example, Galileo's specific use of his theory of projectile motion to explain why projectiles are not expected actually to follow parabolas for very large paths.

It is a primitive fact about human beings that when they look at the moving parts of some intriguing new toy or gadget, they usually try to visualize a mechanical model of the hidden mechanism, to explain by one theoretical system the separate motions observed. This type of activity seems to be a necessity for the human mind. It is as if the intellect were restless until it can subordinate individual events to more inclusive schemes. It may be a matter of economy of thought, for a good theory allows us to grasp, remember, and deduce a large number of otherwise elusive facts. (Think of the wealth of facts summarized by the Copernican theory.)

And simple theories in physics are often based on mechanical models. We recall Lord Kelvin's famous remark, "I never satisfy myself until I can make a mechanical model of a thing. If I can make a mechanical model I can understand it."* This is, of course, not true for all scientists or all branches of science. Conceptual schemes can certainly not always be phrased in mechanical models, and there are several examples in the history of physics which prove that at times progress may be seriously delayed by a too strong belief in a mechanical model. Particularly, the last 100 years have shown that such models, like all analogies, while often helpful as guides to the imagination, can cause it to fall into traps. For example, it undoubtedly was— and is—easier to think of light as a vibration moving through a material ether than as energy propagated through a void, and yet the ether picture carried with it some eventually inescapable fallacies. Similarly, we owe modern chemistry in large part to the simple, almost naive pictorial representation which Dalton used to order his thoughts about atoms and molecules; but, again, much early trouble in his atomic theory can be traced to his too concrete and prematurely complete mental picture, as we shall see in the section on the chemical atom.

---

* *Notes of Lectures on Molecular Dynamics and the Wave Theory of Light* (Baltimore: Johns Hopkins University, 1884), p. 270.

In fact, it has sometimes been suggested that the science of the ancients, in relying frequently on analogies involving the behavior and drives of living beings, was *organismic*, whereas science from Newton on became *mechanistic*, to turn finally in our century more and more to abstract, *mathematical* sets of mental images. But neither is the mathematical type of model free from the dangers inherent in the mechanistic science. Yet our thoughts proceed, as it were, on crutches, and so depend on these schemes and pictures no matter how incomplete.

2) Whether general or limited, theories and hypotheses are expected to suggest new relations, to start the imagination along hitherto unsuspected connecting paths between old and new facts. In this sense even the theories that were eventually abandoned or found altogether faulty, such as the Ptolemaic theory of planetary motion, the phlogiston theory of chemistry, or the caloric theory of heat, played a vital and positive role in the life of science, for they tended to propose problems, focus the attention, and integrate the effort of many scientists along the same lines rather than allowing it to be completely dispersed in random directions. Even a fallacious theory, if actively and widely pursued, can soon lead to the key observations needed for a better theory, for, as Francis Bacon remarked, truth arises more easily from error than from confusion.

---

**Problem 3.3.** The following table indicates the period of revolution about the sun (the so-called *sidereal period*) of the planets in the Copernican system. Instead of Copernicus' values we shall list approximate modern results.

a) Examine these values and the order of spheres in the two systems (consult the diagrams) and then explain why Copernicus felt that there was exhibited order and harmony in his scheme, "a precise relation between the motion and sizes of the orbs."

b) If a new planet were discovered with a sidereal period of about 1830 days, what would the theory *suggest* about the limits of its orbit? If it should turn out that the actual orbit is larger than Jupiter's, is the whole theory wrong? Explain.

Sidereal period (days)

| | |
|---|---|
| Mercury | 88 |
| Venus | 225 |
| Earth | $365\frac{1}{4}$ |
| Mars | 687 |
| Jupiter | 7,330 |
| Saturn | 10,760 |

---

**Problem 3.4.** Compare the orbit of Venus in the two rival theories by referring to Figs. 1.10 and 2.4. If it were found that Venus was not self-luminous but obtained its light from the sun, what would these two theories suggest concerning the nature of the shadow on Venus that might be seen at different times through a telescope from the earth? [*Hint:* Where would the sun have to be with respect to Venus and the earth when we see Venus fully illuminated? Make sketches of planetary position.]

---

3) The foregoing paragraphs hint at a further purpose of theories: the prediction of specific new observable phenomena and the solution of *practical* problems. In Shakespeare's great tragedy, Hamlet's theory of his uncle's guilt yields him a prediction of how the suspect will react to the little play-within-the-play; conversely, the expected reaction proves Hamlet's theory to be very likely correct, and so gives him the desired solution to the problem of the old king's death. In astronomy and physics, where the problems and predictions needed are usually numerical ones, the theories are, as a rule,

quantitative. The problems to be solved may be quite abstract, e.g., why the planets sometimes seem near, at other times far, or may be very practical ones. For example, just as some of the earliest speculations on celestial motions seem to have been prompted by the need to make calendars and to predict eclipses, so one of the problems Copernicus' theory helped to solve was the exact length of the year and of the lunar month, data needed for the revision of the calendar.

**Problem 3.5.** Read Copernicus' treatise in extensive extract (see the sources listed at the end of the preceding chapter) and determine to what extent his work seems to exhibit these three purposes of theory.

## 3.2 CRITERIA FOR A GOOD THEORY IN PHYSICAL SCIENCE

When we examine why scientists as a whole, in the historical development of science, have favored or rejected a given theory, we may discern a few criteria which seemed to have implicitly or unconsciously dominated the slowly evolving process of decision. (But by no means must we suppose that a theory is necessarily ever rejected solely because it fails to answer perfectly to one or another of the questions in the listing that follows.)

Three qualifications have already been cited. (1) A fruitful theory correlates many separate facts, particularly the important prior observations, in a logical, preferably easily grasped structure of thought. (2) In the course of continued use it suggests new relations and stimulates directed research. (3) The theory permits us to deduce predictions that actually check with experience by test, and it is useful for solving long-standing puzzles and quantitative problems.

To illustrate: In Einstein's work on relativity, he first shows that those observations explained by contemporary physics are not in conflict with his brief set of assumptions, but that a large variety of facts from previously separate fields can all be deduced from these assumptions, thereby gathering them into one structure. Second, he demonstrates that a few hitherto unlooked-for effects should exist (such as the deflection of light beams passing near large masses) and he stimulates the experimental astrophysicists to search for these phenomena. And lastly, not only does he use his theory to explain some old, previously unexplainable observations—e.g., the "erratic" component of the planetary motion of Mercury—but his work has also found amazingly useful applications in almost all branches of physics, from optics to nuclear physics.

The history of science has shown us that prosperous theories frequently have some of the following additional properties:

4) When the smoke of initial battle has lifted, the successful theory often appears to have simple and few hypotheses—to the extent that the words "simple" and "few" have any meaning at all in science. The test of a theory comes through its use; hence the survival value of economy. A theory which needs a separate mechanism for each fact that it wishes to explain is nothing but an elaborate and sterile tautology.

5) Ideally, the assumptions should be plausible—to contemporary scientists, of course—even if they are not immediately subject to test; and the whole tenor of the theory should not be in conflict with current ideas. Where this could not be arranged, the theory faced an often stormy and hostile reception and had to submit to long and careful scrutiny before its general acceptance. Therefore it did not always grow as quickly and widely as one might have hoped. On the other hand, this cannot be helped, if nature disagrees with current common-sense preconceptions; in fact, precisely these preconceptions may block the road to the core of the difficult problems. As the work of Copernicus and later Galileo shows us (a strange truth which

will become even more evident in subsequent discussions), major advances in scientific theories often depend on the daring and persistence of one investigator who questions the obvious and stubbornly upholds the unbelievable.

This problem of the plausibility of new theories is difficult and fascinating. As Whitehead reminds us, all truly great ideas seemed somewhat absurd when first proposed. Perhaps we call them *great* ideas just because it takes an unusual mind to break through the pattern of contemporary thought and to discern the truth in implausible, "absurd" forms.

Why then do we not drop the tentative requirement of conformity? Why did almost all great innovators in science, from Copernicus to Einstein, initially meet with skepticism—or worse—from most of their colleagues? Is it just unreasoning conservatism on the part of the scientific fraternity? Not at all. In this little collection of cases, we are discussing only revolutionary theories, but we must not allow this to distort our perceptions: Great revolutionary ideas arise only very rarely, compared with the large number of workable or even great ideas conceived within the *traditional* setting, so that the individual scientist is naturally predisposed to favor the traditional type of advance which he knows and believes in from personal experience. He quite rightly must defend his fundamental concept of nature against large-scale destruction, particularly at the earlier stages, where the innovators cannot present very many results and confirmations of their new ideas. Indeed, sometimes the discoverers themselves are so strongly committed to the established ideas, so startled by the implications of their own work, that they predict the storm of condemnation or even simply fail to draw the final conclusions.

A beautiful and now famous instance of this sort appears in the original publication (1939) by Hahn and Strassmann, who are usually credited with the experimental discovery of nuclear fission. At that time it was still axiomatic in the thinking of all scientists that atomic nuclei are indivisible,

that bombardment with small particles (e.g., neutrons) may at best change their internal atomic structure only superficially or slightly. But these two men, after firing neutrons into pure uranium, were left with material which by chemical test was proved to contain barium (Ba) and other elements whose atoms are about half as large as uranium atoms—"evidently" as we would say now, the result of splitting uranium atoms. But in an almost agonized expression of the labor pains attending the birth of all great recognitions, Hahn and Strassmann could not quite dare to accept publicly the evidence of their own chemical experiments, and so they wrote: "As 'nuclear chemists,' in many ways closely associated with physics, we cannot yet bring ourselves to make this leap in contradiction to all previous lessons of nuclear physics. Perhaps, after all, our results have been rendered deceptive by some chain of strange accidents."*

Later, after having developed our view that science has its laws of evolution analogous to those governing the evolution of species, we shall appreciate more fully the significance of the bitter and long struggles that new theories may create within science. *The fitness of truths is most advantageously shaped and most convincingly demonstrated in vigorous contest.* The situation is, after all, not so very different in other fields. The predominant religious and social concepts of our time have not all developed quietly or been accepted spontaneously. If the physical sciences sometimes seem to have progressed so very much faster, if the recent struggles have been relatively short, we might, to a great extent, credit this to the emer-

---

* Contrast this caution and the willingness to be proved wrong by their fellow scientists, expressed in the previous quotation, with the sensational claims and impatient pomp of our modern pseudoscientific best-sellers, one of whose authors begins his treatise with the sentence that the work "is a milestone for Man comparable to his discovery of fire and superior to his invention of the wheel and arch... The hidden source of all psychosomatic ills and human aberration has been discovered, and skills have been developed for their invariable cure."

gence, from the seventeenth century onward, of a more or less tacit agreement among scientists by what standards to judge a new conceptual scheme—above all, of course, the pragmatic test of predictions.

Even so, scientists are, above all, human; a really new and startling idea may in retrospect clearly have fulfilled all these requirements without being widely accepted. Max Planck, with perhaps only a little too much bitterness about his own early struggles, says, "An important scientific innovation rarely makes its way by gradually winning over and converting its opponents: it rarely happens that Saul becomes Paul. What does happen is that its opponents gradually die out, and that the growing generation is familiarized with the ideas from the beginning: another instance of the fact that the future lies with youth."*

6) History points out another feature of successful theory: It is flexible enough to grow, and to undergo minor modifications where necessary. But if, after a full life, it eventually dies, it dies gracefully, leaving a minimum of wreckage—and a descendant.

Let us consider these interesting points. In any field a prediction is no more than a guess unless it follows from some theory. And just as the meaning of a prediction depends on a theory, so does the validity of a theory depend on the correctness of the predictions. Like the very laws it correlates, a theory must be based on a finite, though perhaps small, number of observations; yet to be useful it may have to predict correctly a very large number of future observations. Think again of the Ptolemaic theory; it was built on many separate observations of planetary positions, but it yielded enough reasonably correct predictions of positions to be still useful more than 1400 years later.

---

* See *The Philosophy of Physics,* page 97; New York: W. W. Norton, 1936. We shall meet Planck again as the founder of quantum theory (Chapter 26).

Yet we see at once that physical theories are not likely to last forever. Sooner or later there will appear observations to contradict the predictions, probably in a region for which the conditions had to be extrapolated from the original range of the theory (e.g., the region of very small or very large quantities, or of much more precise measurements than served initially as a basis for the theory), or perhaps as a result of a general widening of the scientific horizon. With ingenuity one can usually find that the old theory can be modified to apply also to the new phenomena. Like an apple tree in an orchard, we keep a theory for the sake of its fruits; when the apple crop becomes poor, we then shall try to save the tree by judicious pruning and the like. In the endangered theory, perhaps a re-examination of some assumptions or concepts will save the scheme. But otherwise the only alternatives are to retain the old theory as is, with the clear understanding of the restricted range to which it still applies, as before, and to develop a new structure to take care of the new range; or, if the flaws appeared within the proper jurisdiction of the old theory and the assumptions stand challenged beyond repair, to abandon it as soon as a new, better one can be contrived.

These three choices—expansion, restriction, or death—appear in the history of theories and of laws generally. We tend to remember and to honor most those schemes which indicated the solution in a crisis, or left some hint of a new start at their demise. In the words of Niels Bohr, "The utmost any theory can do [is] to be instrumental in suggesting and guiding new developments beyond its original scope."

But finally, lest our brief paragraphs on these criteria of good theories be mistaken for scientific dogma, we should echo the very humble opinion of Einstein on this same topic. He distinguishes between two main criteria: (a) the *external confirmation* of a theory, which informs us in experimental checks of the correctness of the theory, and (b) the *inner perfection* of a theory which judges its "logical simplicity" or "naturalness." He then qualifies these remarks as follows: "The meager precision of the assertions [(a) and (b) above] ... I shall not attempt to excuse by lack of sufficient printing space at my disposal, but confess herewith that I am not, without more ado, and perhaps not at all, capable to replace these hints by more precise definitions."

## ADDITIONAL PROBLEMS

**Problem 3.6.** Investigate the death of a theory from another field with which you are very well acquainted (such as the phlogiston theory in chemistry or the theory of spontaneous generation in biology) in terms of our six criteria. What were the predictions which failed to come true, or the phenomena which contradicted the assumptions of the theory? Could the theory have been modified to serve for a time? Did a new theory rise directly from the old?

**Problem 3.7.** P. W. Bridgman writes in *Reflections of a Physicist* (Philosophical Library, New York, 1950, page 252) that in the current flux in ideologies, moral ideas, and other social concepts the intelligent scientist sees "... an exemplification of what his physical experience has taught him; namely, ideas are to be inspected and re-examined when the domain of application is extended beyond that in which they arose. In one very important respect he recognizes that the present epoch differs from former epochs in that the enormous increase in invention, bringing peoples nearer together and increasing their command over forces more advantageous to man, effectively provides just that extension in the domain of application of social concepts which he is prepared to expect would demand fundamental revision." Write a short essay on your opinion about this quotation, and also examine what other guiding ideas you think may be transferred to the field of social studies from the study of physical theories.

**Problem 3.8.** Was the geocentric theory as set forth by Ptolemy false? Was it true?

**Problem 3.9.** Discuss the relative advantages of the Ptolemaic and Copernican theories of celestial motion, comparing both theories according to each of the six criteria developed in the preceding section.

## FURTHER READING

*Note:* For further reading, consult the bibliography at the end of Chapter 14.

# Chapter 4

# KEPLER'S LAWS

## 4.1 THE LIFE OF JOHANNES KEPLER

The theory of planetary motion now develops with gathering momentum. We have reached the years around 1600 A.D. The Renaissance and Reformation are passing. Copernicus has been read by a few astronomers who see the computational advantages of his system but will not take seriously its physical and philosophical implications. Through this silence we hear one striking voice sounding the first cries of a coming battle. The antiorthodox pantheist and evangelizing Copernican, Giordano Bruno, is traveling through Europe, announcing that the boundaries of the universe are infinitely far away and that our solar system is one of infinitely many. For his several outspoken heresies he is tried by the Inquisition and burned at the stake in 1600.

But the seeds of a new science are sprouting vigorously here and there. In England there are Francis Bacon (1561–1626) and William Gilbert (1540–1603); in Italy, Galileo Galilei (1564–1642). And at Copenhagen, Tycho Brahe (1546–1601), the first man since the Greeks to bring real improvements into observational astronomy, spends nearly a lifetime in patient recording of planetary motion with unheard-of precision. His data are often accurate to less than half a minute of arc, more

than twenty times better than those of Copernicus, even though the telescope has not yet been invented.

After Tycho's death, his German assistant Johannes Kepler (Fig. 4.1) continued the observations and above all the theoretical interpretation of the voluminous data. Whereas Tycho had developed a planetary system of his own (mentioned at the end of Section 2.4) which became widely accepted for a time, Kepler was a Copernican. The announced purpose of his work was the construction of better tables of planetary motion than were then available on the basis of the more dubious data of Copernicus' own period. But Kepler's motivation and main preoccupation was the perfection of the heliocentric theory, whose harmony and simplicity he contemplated "with incredible and ravishing delight." Throughout his long labors he was strongly influenced by the metaphysical viewpoint associated with the Pythagorean and "neo-Platonic" tradition; this tradition had revived in the Renaissance as one of the challenges to the hegemony of Aristotle.

To Kepler even more than to Copernicus, the clue to God's mind was geometric order and numerical relation, expressed in the features of the simple heliocentric scheme. Among his earliest works we find, typically, an enthusiastic attempt to link the six then known planets and their

**Fig. 4.1** Johannes Kepler (1571–1630).

relative distances from the sun (see Section 2.2) with the relationships between the five regular solids of geometry. The best result of this work was that it served to bring Kepler to the attention of Tycho and Galileo.

In attempting to fit the accurate new data on Mars' orbit to a Copernican system of simple uniform motion (*even* if equants were used), Kepler was completely unsuccessful. After four years of calculations he still found that the new data placed the orbit just eight minutes of arc outside the scheme of Copernican devices. This is many minutes less than the error which Copernicus knew his own data to contain; therefore Copernicus would not have worried about such a deficiency.

But Kepler knew that Tycho Brahe, with his unfailing eye and superb instruments, had recorded planetary positions with uncertainty much smaller than the eight minutes of arc. With an integrity that has become the characteristic attitude of scientists in the face of quantitative fact, he would not allow himself to hide this fatal discrepancy behind some convenient assumptions. To Kepler, these eight minutes meant simply that the Copernican scheme of concentric spheres and epicycles failed to explain the actual motion of Mars when the observations were made accurately enough.

## 4.2 KEPLER'S FIRST LAW

Kepler must have been stunned, for after all he was a convinced Copernican. After years of continuous labor he at length discovered how the Copernican theory could be amended to make it applicable to both the old and the new observations—by dropping the one assumption which bound it most explicitly to the doctrines derived from ancient Greece. When he examined the paths of planets on the basis of a heliocentric representation, he found that they corresponded to a simple type of figure whose properties had been known to mathematicians since the second century B.C. (It is ironical that Apollonius, who proposed the epicycle device discussed in Section 1.6, developed the theory of ellipses without any thought for their possible application in astronomy.) If, therefore, the ellipse is recognized as the natural path of celestial bodies, then we have a geometrically truly simple world scheme. "Planets move in elliptical paths, with the sun at one focus of the ellipse." This may be our statement of the *Law of Elliptical Paths*, Kepler's first of three great laws.

Kepler's first law, by amending the Copernican heliocentric theory, gives us a wonderfully simple mental picture of the solar system. Gone are all the epicycles and eccentrics; the orbits are clean and elliptical. Figure 4.2 shows a schematic representation of the present conception of the solar system—in essence Kepler's, but with the addition

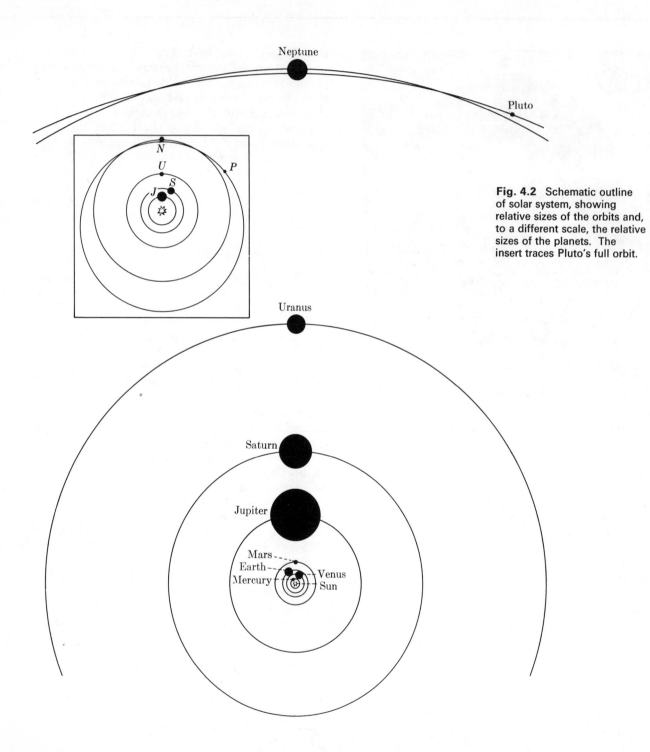

**Fig. 4.2** Schematic outline of solar system, showing relative sizes of the orbits and, to a different scale, the relative sizes of the planets. The insert traces Pluto's full orbit.

of the planets Uranus, Neptune, and Pluto, discovered much later. An attempt has been made to represent on the figure the relative sizes of the planets; the approximate relative sizes of the orbits, to a different scale, are indicated also, but since most of these ellipses are quite nearly circular, they are shown as circles where possible. All orbits are nearly in the same plane, except for the more pronounced tilt of Pluto's (which means that Neptune's and Pluto's paths do not cross at any point in space, although Fig. 4.2 may give that impression).*

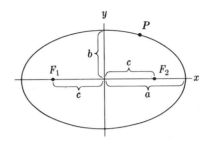

**Fig. 4.3**  Ellipse, with foci at $F_1$ and $F_2$.

**Problem 4.1.** (The important properties of the ellipse; refer to Fig. 4.3.) The equation for an ellipse in rectangular coordinates is

$$\frac{x^2}{a^2} + \frac{y^2}{b^2} = 1,$$

where $a$ and $b$ are respectively half the major and half the minor axis. On graph paper, construct a large ellipse by deciding on some constant values for $a$ and $b$ and by then solving for a few $\pm y$ values for assumed $\pm x$ values. (Note that if $a = b$, the figure is a circle.) The distance $c$ from the origin of the coordinates to one or the other focus ($F_1$ or $F_2$) is given by

$$c = \sqrt{a^2 - b^2}.$$

Enter the foci on your ellipse. The eccentricity† $e$ is a quantity indicating by how much an ellipse differs from a circle. By definition,

$$e = \sqrt{1 - \frac{b^2}{a^2}}.$$

Find $e$ for your ellipse. Test by measurements that for a point $P$ *anywhere* on the ellipse, $PF_1 +$

$PF_2$ = a constant. Lastly, note that for a given set of coordinates an ellipse in a given plane is completely defined (i.e., can be correctly drawn) if just its values for $a$ and $b$ are given; these two data suffice (Fig. 4.3).

**Problem 4.2.** Determine the approximate value of $e$ for Pluto from measurements on the projection of its orbit shown in Fig. 4.2. (Note: $e$ for Mercury is about 0.2, and for the remaining planets even less.)

Although Kepler was delighted to find that he could replace the complicated combinations of epicycles and eccentrics previously used to describe a planet's orbit by a simple ellipse, he may have asked himself the obvious question, "Is it not rather mysterious that of all possible types of paths the planets all select just the ellipse?" We could perhaps understand Plato's predisposition for uniform circular motions, but we cannot readily comprehend nature's insistence on the ellipse! It is no longer a man-made mystery, but therefore all the darker. In fact, there was no rational answer until a reticent genius in England almost 80 years later showed the law of the ellipse to be one of the many surprising consequences of a much more far-reaching law of nature. But we are not yet ready to take up his reasoning.

---

\*  For a readable account of the structure of the solar system, see F. L. Whipple's *Earth, Moon and Planets*.

†  Do not confuse with the word "eccentric" as used in Section 1.6.

If for the present we accept Kepler's first law purely as a summary of observed facts—an *empirical* law—we note that by describing the paths as elliptical, the law gives us all the possible locations of a given planet—but it does not tell us *when* it will be at any of these positions; it talks about the shape of the orbit, but not the changing speed along the orbit. This makes the law rather inadequate for an astronomer who wishes to know where he should expect to look for a planet at a given time, or even for the layman who already knows (as we noted in Section 1.6 in connection with the equant) that the sun appears to move faster through the stars in winter than in summer. Of course Kepler was well aware of all this, and in fact even before finding what we now call his "first" law he had already established another law governing changes in the speed of a planet.

## 4.3  KEPLER'S SECOND LAW

Kepler knew that he needed a mathematical relation between the speed of a planet in any position of its orbit and the speed in any other position. If such a relation could be found, the motion of any one planet would be specifiable by just a few separate figures: two data to specify the ellipse (e.g., the major and minor axes); one more to give the speed at a certain portion of the path, for instance when the planet is nearest to the sun, at the *perihelion*; and enough additional data to place this ellipse at the correct tilt with respect to those for the other planets. Thus we could hope to summarize all features of planetary motion in a compact, elegant way.

But does the hoped-for relation between speed and path actually exist? There has appeared so far no necessary reason that it should. Kepler is said to have been in ecstasy when by sheer labor and ingenuity he finally did read the needed, "second" law out of his own voluminous data. Well he might have been; his whole labor would have been of little use without this discovery.

Kepler's route to the second law was an amazing performance in which the correct result emerged as a deduction from three incorrect assumptions. First, Kepler assumed that planets are driven around their orbits by a force emanating from the sun, and that the strength of this force is inversely proportional to the planet's distance from the sun. (The reason for this, in Kepler's thinking, and using his imagery, was that the force at any given distance $r$ must be uniformly spread over the circumference of a circle in the orbital plane; at a greater distance, say $2r$, the same total force must be spread over a circle whose circumference is twice as great, hence the strength of the force at any given point on that circle can only be half as much.) He then assumed that the speed of the planet must be proportional to the force pushing it, and hence inversely proportional to the distance:

$$v \propto 1/r. \qquad (4.1)$$

The assumption that speed is proportional to the net force is of course incompatible with modern principles of physics, as we shall see in Chapters 7 and 9; it was simply one of the Aristotelian or common-sense ideas that Kepler shared with all his contemporaries.

According to Kepler's first assumption, the *time* it takes a planet to go a small distance along its path would be proportional to its distance from the sun. This is roughly correct, and happens to be exactly correct at certain special points in the orbit. Kepler then proposed to compute the time it takes the planet to cover a large segment of the path (during which the distance from the sun will be changing) by adding up the planet–sun distances for each of the small arcs composing this large segment. He assumes that the sum of these distances will be equal to the area swept out by the line drawn from sun to planet (see Fig. 4.4):

$$t \propto \text{area swept out by planet–sun line.} \quad (4.2)$$

This is a good approximation for the actual orbits Kepler was dealing with, but the mathematics

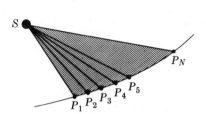

**Fig. 4.4** Kepler assumed that the sum of all the lines drawn from the sun to the planet as it moves along its path, $SP_1 + SP_2 + SP_3 + \cdots + SP_N$, could be approximated by the shaded area, segment $SP_1P_N$ (eccentricity of orbit is exaggerated).

**Fig. 4.5** Elliptical path of planets around the sun $S$ (at left focus), illustrating Kepler's Second Law (eccentricity much exaggerated).

needed to make it exact (the "calculus" of Newton and Leibniz) was not to be invented for another half century.

Now Kepler introduced a third assumption, that the orbit is a circle. This is again only an approximation which happens to be fairly accurate for almost all planetary orbits (Kepler had not yet established his "first" law, which required the orbits to be elliptical); but in fact it is not even necessary to make such an approximation.

Kepler's second law, which he stumbled on by a line of reasoning that is unconvincing to a modern reader, is already contained in Eq. (4.2) above: the area swept out by the planet–sun line is proportional to the time elapsed. Or, to rephrase it in the way that has now become standard: *During a given time interval a line drawn from the planet to the sun sweeps out an equal area anywhere along its path.* This is also referred to as the *Law of Equal Areas.* In spite of the inaccuracy of the assumptions used in its original derivation, the law itself describes exactly the motion of any planet around the sun;* it also applies to the motion

of the moon around the earth, or of a satellite around any planet.

Figure 4.5 illustrates the meaning of the second law. We are looking down on a greatly exaggerated representation of a planetary path around the sun $S$ at one focus of the ellipse. The planet may be our earth; it is found to move from position $P_1$ to $P_2$ (say during a week in our winter). During an equal time interval in the spring the journey proceeds from $P_3$ to $P_4$, and from $P_5$ to $P_6$ in summer. The implication from the drawing is that the planet's speed along the orbit is greatest when nearest the sun, less when more distant. This is qualitatively in agreement with the first assumption, $v \propto 1/r$, which Kepler used to derive the law.

The fact that the earth does move faster (or that the sun as seen from the earth appears to move faster against the background of stars) in the winter than in the summer had been well known to astronomers long before Kepler; it was an effect that could be explained by the introduction of the "equant" device in the geocentric system (see Section 1.6), and it was one reason why the Copernican system without equants was not quite adequate to represent the details of planetary motion. Kepler's second law serves the same purpose as the equant, but in a way that ultimately

---

*    This statement must be qualified slightly, for reasons which will become apparent when we discuss Newton's derivation in Chapter 9: the law is accurate provided the influence of a third body can be neglected.

turns out to be much more satisfactory. However, in Kepler's own work it is only an empirical rule which, though accurate, does not yet have any theoretical explanation.

**Problem 4.3.** As the text mentioned, the earth is actually closer to the sun in winter than in summer. Explain this apparent paradox.

## 4.4 KEPLER'S THIRD LAW

Kepler's first and second laws were published together in 1609 in his *New Astronomy*. But he was still unsatisfied with one aspect of his achievement: It had not revealed any connection between the motions of the different planets. So far, each planet seemed to have its own elliptical orbit and speeds, but there appeared to be no overall pattern for all planets. Nor was there any good reason why one should expect such a relationship. Kepler, however, was convinced that on investigating different possibilities he might hit on a simple rule linking the whole solar system. He looked for this rule even in the domain of musical theory, hoping like the Pythagoreans (Section 1.1) to find a connection between planetary orbits and musical notes; his last great work (1619) is entitled *The Harmonies of the World*.

This conviction that a simple rule exists, so strong that it seems to us like an obsession, was partly a remnant of his earlier numerological preoccupations, and partly the good instinct of genius for the right thing to work on. But it also indicates a deep undercurrent running through the whole history of science: the belief in the simplicity and uniformity of nature. It has always been a source of inspiration, helping scientists over the inevitable unforeseen obstacles, and sustaining their spirit during periods of long and fruitless labor. For Kepler it made bearable a life of heartbreaking personal misfortunes, so that he could

write triumphantly, on finally finding the great third law:

"...after I had by unceasing toil through a long period of time, using the observations of Brahe, discovered the true distances of the orbits, at last, at last, the true relation...overcame by storm the shadows of my mind, with such fullness of agreement between my seventeen years' labor on the observations of Brahe and this present study of mine that I at first believed that I was dreaming...."*

The law itself, in modern terminology, states that if $T$ be the sidereal period of any chosen planet (i.e., the time for one complete orbital revolution about the sun), and $\overline{R}$ be the mean radius of the orbit of that planet,† then

$$T^2 = K(\overline{R})^3,$$

where $K$ is a constant having the same value for *all* planets. But if $T^2/(\overline{R})^3$ is the same for all planets, we can calculate its numerical value for *one* planet (for the earth: $T_E = 1$ year, $\overline{R}_E \doteq 9.3 \times 10^7$ miles), and thereafter we shall always be able to compute for any other planet its $T$ if its $\overline{R}$ is given, and vice versa.

**Problem 4.4.** Using the third law and the data for $T_E$ and $\overline{R}_E$ for the earth, compute values of $\overline{R}$ for a few other planets, taking the corresponding values of $T$ from the quotation from Copernicus' *Revolutions* in Section 2.2.

* *Harmonies of the World,* Book V, Section 3. For an extended extract which includes this passage, see D. L. Hurd and J. J. Kipling, *The Origins and Growth of Physical Science,* Penguin Books, 1964, Vol. I, pages 122–134.

† The value of $\overline{R}$ for an elliptical path is half the length of a straight line from perihelion to aphelion (see Fig. 4.5); conveniently, most planetary paths are nearly enough circular so that $\overline{R}$ is then simply the radius of the circular orbit.

Appropriately, Kepler's third law is often called the *Harmonic Law* because it establishes a beautifully simple relationship among the planets. From this peak we may survey our progress so far. Starting from the disconnected multitude of Ptolemaic gears we have reached a heliocentric formulation that views the solar system as a simple, logically connected unit. Our mind's eye grasps the Keplerian universe at one glance, and recognizes the main motions as the expression of simple mathematical laws.

## 4.5  THE NEW CONCEPT OF PHYSICAL LAW

With Tycho's work, his own observations, and his powerful three laws, Kepler was able to construct the long-needed accurate tables of planetary motion which remained useful for a century. We honor him for all these great accomplishments in astronomy (and these are only a few of the achievements of this prodigious man), but we must also not forget to single out two features of his work which had a profound effect on the development of all physical sciences. One, which has been discussed, is the rather new attitude to *observed facts*. We noted the change in Kepler's work from his earlier insistence on a geometric model and its form as the main tool of explanation, to the discussion of the movement itself and of the numerical relations underlying it. The other is his successful attempt to formulate physical laws in mathematical form, in the language of geometry and algebra. From then on, after the period of Kepler, the *equation* develops naturally as the prototype of the laws in physical science. (We shall see some of this development when we come to the work of Galileo and Descartes in the following chapters.)

In this sense, Kepler's science was truly modern; he, more than anyone before him, bowed to the relentless and supreme arbiter of physical theory, namely, the evidence of *precise* and *quantitative* observation. But in the Keplerian system the planets no longer are considered to move in their orbits because of their divine nature or the continual effort of angels, as in scholastic teaching, nor because their spherical shapes themselves were self-evident explanation for their circular motion, as in Copernicus' thoughts; and so we are left without any physical agency to "explain" or give plausibility to the planetary motion so well described in these three laws.

Kepler himself felt a need to back up his mathematical descriptions with physical mechanism. In one of his later books he tells how his own views have changed:

"Once I firmly believed that the motive force of a planet was a soul.... Yet as I reflected that this cause of motion diminishes in proportion to distance, just as the light of the sun diminishes in proportion to distance from the sun, I came to the conclusion that this force must be substantial— 'substantial' not in the literal sense but...in the same manner as we say that light is something substantial, meaning by this an unsubstantial entity emanating from a substantial body."*

Although it was left to Newton to work out the theory of gravitational forces, and thereby to tie Kepler's three laws together with the heliocentric conception and the laws of terrestrial mechanics in a monumental synthesis, Kepler entertained a then quite promising hypothesis: The recent work of the Englishman William Gilbert (1544–1603) on magnetism had greatly intrigued him, and his rich imagination could picture magnetic forces emanating from the sun to "drive" the planets in their orbits.

Magnetism in fact does not explain Kepler's laws; Newton later felt it necessary to prove in some detail that this hypothetical agency would not account for the quantitative observations. But in a more general sense Kepler did anticipate the *type* of explanation that Newton was to es-

___

*    *Mysterium Cosmographicum,* second edition.

tablish.  As he writes to a friend in 1605:

"My aim is to show that the heavenly machine is not a kind of divine, live being, but a kind of clockwork (and he who believes that a clock has a soul, attributes the maker's glory to the work), insofar as nearly all the manifold motions are caused by a most simple, magnetic, and material force, just as all motions of the clock are caused by a simple weight.  And I also show how these physical causes are to be given numerical and geometrical expression."*

We have here an example of the enormous change in outlook in Europe that had begun more than two centuries before.  More and more, events ceased to be regarded as *symbols* and were allowed to stand for themselves.  Man ceased to be preoccupied with anthropomorphic riddles in an organismic world, and slowly became a factual observer and theorizer in a mechanistic world. Without this new attitude, there could have been no modern science, for if we are to start our science from experimental observables, we must have faith that we are dealing with the raw material of experience, not with symbols of complex mysteries. We must become enthusiastic about the observable world for its own sake, we must attain a tacit faith in the meaningfulness of nature and her direct accessibility to our understanding, before we can expect generations of scientists to devote themselves to the minute and often most tedious quantitative investigations of nature.  In this sense Kepler's work heralds the change toward the modern scientific attitude—to regard a wide variety of phenomena as explained when they can all be described by one simple, preferably mathematical, pattern of behavior.

It seems at first astonishing that Kepler should be the one to follow this road.  He had begun his career as a symbol-seeking mystic, but now we may discern a reflection of the great change in

his complex soul: he fashions his physical laws— and looks for his symbolism afterwards.  Philosophical speculation, still often rather colorful, follows factual analysis, not the reverse; and to this day many scientists have found it quite possible to reconcile their physics and their personal philosophy on the basis of this sequence.

## ADDITIONAL PROBLEMS

**Problem 4.5.** Evaluate Kepler's contribution in terms of our criteria for a good theory as far as you can.

**Problem 4.6.** Trace the changes and evolution of the Copernican heliocentric theory in the hands of Kepler, and then compare this with the development and expansion of some important theory from another field familiar to you (e.g., the theory of evolution, from its exposition in 1859 to the recent modifications of evolutionary genetics).

**Problem 4.7.**  Does it seem possible that a revision of the geocentric theory can be made, doing away with equants, eccentric motions, etc., and instead describing the planetary paths *about the earth* directly by simple geometric figures?  Is it likely that laws could be found for this simplified geocentric system which relate the paths, speeds, and periods of revolution of the different planets?

## RECOMMENDED FOR FURTHER READING

Marie Boas, *The Scientific Renaissance 1450–1630*, Chapter X

E. A. Burtt, *The Metaphysical Foundations of Modern Physical Science*, second edition, 1932; Doubleday Anchor Books, paperback reprint; Chapters I and II

---

\*    *Ibid.*, page 155.

O. Gingerich, "The computer versus Kepler," *American Scientist*, Vol. 52, pages 218–226 (1964). The numerical problems involved in calculating the orbit of Mars are attacked by modern methods, with surprising results.

Johannes Kepler, extract from *Harmonies of the World*, Book V, translated in Hurd & Kipling, *The Origins and Growth of Physical Science*, Vol. 1, pages 122–134

*Project Physics Reader*, New York: Holt, Rinehart & Winston, 1970, Vol. 2. Excerpts from Kepler, and articles by Cohen and Holton.

C. Wilson, "How did Kepler discover his first two laws?" *Scientific American*, Vol. 226, No. 3, pages 92–106 (March 1972).

## SOURCES, INTERPRETATIONS, AND REFERENCE WORKS

E. J. Aiton, "Kepler's second law of planetary motion," *Isis*, Vol. 60, pages 75–90 (1969)

R. H. Baker, *Astronomy*, New York: D. Van Nostrand, 1930 and later editions. Excellent introduction.

A. Berry, *A Short History of Astronomy*, Chapter VII

J. L. E. Dreyer, *A History of Astronomy*, Chapters XV and XVI

N. R. Hanson, "The Copernican disturbance and the Keplerian revolution," *Journal of the History of Ideas*, Vol. 22, pages 169–184 (1961)

G. Holton, "Johannes Kepler's universe: its physics and metaphysics," *American Journal of Physics*, Vol. 24, pages 340–351 (1956), reprinted in *Toward Modern Science* (editor R. M. Palter); Vol. II

Johannes Kepler, *Epitome of Copernican Astronomy*, parts IV and V; *The Harmonies of the World*, part V, translated by C. G. Wallis, in *Great Books of the Western World* (editor R. M. Hutchins), Chicago: Encyclopedia Britannica, 1939, 1952, Vol. 16

Johannes Kepler, *Somnium, Sive Astronomia Lunaris*, translated as *Kepler's Dream* by P. F. Kirkwood, Berkeley: University of California Press, 1965. Probably the first science-fiction novel.

S. Nakayama, "On the alleged independent discovery of Kepler's third law by Asada Goryu," *Japanese Studies in History of Science*, no. 7, pages 55–59 (1968)

R. Small, *An Account of the Astronomical Discoveries of Kepler*, Madison, Wisconsin: University of Wisconsin Press, 1963 (reprint of the 1804 edition)

# Chapter 5

# GALILEO AND THE
# NEW ASTRONOMY

"There are more things in heaven and earth,
　Horatio,
Than are dreamt of in your philosophy . . . ."

*Hamlet*, Act I

One of the few friends and fellow scientists with whom Kepler corresponded and exchanged news of the latest findings was Galileo Galilei (1564–1642). Although the Italian's *scientific* contribution to planetary theory is not so substantial as that of his friend across the Alps, he nevertheless was projected into history as a key figure in this subject. In a sense, Kepler and Galileo complemented each other in preparing the world for the eventual acceptance of the heliocentric theory—the one laying the scientific foundation with his astronomical work, the other fighting the dogmatic objections and, in his work on mechanics, helping to overturn the whole structure of scholastic physics with which the old cosmology was entwined.

For it was Galileo more than any other man who challenged the fruitfulness of the ancient interpretation of experience, and focused the attention of physical science on the productive concepts—time and distance, velocity and acceleration, force and matter—and not on qualities or essences, ultimate causes or harmonies, which were still the motivation of a Copernicus and at times the ecstasy of a Kepler. Galileo's insistence, so clearly expressed in his work on freely falling bodies (see Chapter 7), on fitting the concepts and conclusions to observable facts, on expressing his results in the concise language of mathematics, are now accepted as central achievements, reenforcing the same traits then emerging from parts of the work of Kepler.

But perhaps the greatest difference between the work of Galileo and that of his scholastic contemporaries was his orientation, his viewpoint, the kind of question he considered important. To most of his opponents, Galileo's specific problems were not general enough, since he excluded the orthodox philosophical problems. Then, too, his procedure for discovering truth seemed to them fantastic, his conclusions preposterous, haughty, often impious. There exists an almost perfect parallel between these objections to Galileo's point of view and the outraged derision and even violence initially heaped on the discoverers of new ways of viewing the world in art, e.g., the French painter Édouard Manet and his fellow impressionists in the late nineteenth century.

## 5.1 THE LIFE OF GALILEO

Galileo Galilei, one of our chief symbols of the great struggle out of which modern science arose, was born in Pisa, Italy, in 1564, the year of Shakespeare's birth and Michelangelo's death. His father was a poor but noble Florentine from whom he acquired an active and competent interest in music, poetry, and the classics. Galileo's mechanical ingenuity began to show itself early when, as a medical student at the University of Pisa, he invented a simple pendulum-type timing device for the accurate measurement of pulse rates. It appears that even at that time he made himself conspicuous by constantly challenging the authoritative opinions of his elders.

Lured from medicine to physical sciences by reading Euclid and Archimedes, he quickly became known for his unusual powers. Although he had dropped out of the University of Pisa without taking his degree, he soon acquired such a reputation that he was appointed professor of mathematics at the same university at the age of 26. His characteristic traits now showed themselves: an independence of spirit and an inquisitive and forthright intellect unmellowed by tact or patience. He protested against the traditional attire of the faculty by circulating a poetic lampoon, "Against wearing the toga." He attacked the scientific views of his colleagues, who were of course almost all dogmatic Aristotelians and, as disciples too often are, were rabid defenders of their often erroneous interpretations of the master's works.

It was around 1590, while Galileo was at Pisa, that he may have made a public experiment on the speeds of unequal weights dropped from the famous Campanile of Pisa, though it is more likely that the story is only a legend. Because the incident is so widely known and is popularly supposed to have been a crucial experiment in the history of physics, we may, for what it is worth, look at a source of the story, the biographical notes on Galileo by one of his latest and closest students,

Vincenzo Viviani—an account whose accuracy has often been challenged.

"As it seemed to him [Galileo] that a true knowledge of the nature of motion was required for the investigation of the natural effects, he gave himself completely to the contemplation of that [motion]: and then to the great confusion of all the philosophers, very many conclusions of Aristotle himself about the nature of motion, which had been theretofore held as most clear and indubitable, were convicted of falseness, by means of experiments and by sound demonstrations and discourses; as, among others, that the velocity of moving bodies of the same composition, unequal in weight, moving through the same medium, do not attain the proportion of their weights, as Aristotle assigned it to them, but rather that they move with equal velocity, proving this by repeated experiments performed from the summit of the Campanile of Pisa, in the presence of all other teachers and philosophers and of all students."*

It should be noted that the speeds of bodies of the *same composition* are being compared. Apparently in 1590 Galileo believed that bodies of the same *density* would fall at the same rate, but that the rate of fall might still depend on the difference in density between the object and the medium through which it falls. Galileo's writings on mechanics during this period indicate that he had not yet developed the theory presented in his definitive work on mechanics published in 1638, according to which all bodies of whatever composition must fall at the same rate in a vacuum.† So Galileo's interpretation of the famous Leaning

---

* For this translation we are indebted to Dr. Arnolfo Ferruolo of Harvard University.

† See *Galileo Galilei on Motion and on Mechanics,* translated by I. E. Drabkin and S. Drake (University of Wisconsin Press, 1960), pages 6–7; E. A. Moody, "Galileo and Avempace: The Dynamics of the Leaning Tower Experiment," *Journal of the History of Ideas,* Vol. 12, pages 163, 375 (1951).

Tower experiment, if in fact it was actually performed at this time, would not have been quite the same as the modern one.

In 1591, Galileo's father died, leaving a large family to be supported. Galileo's salary at Pisa was inadequate, and moreover his appointment was not likely to be renewed when the three-year contract expired, because he had made too many enemies by his blunt attacks on the Aristotelians. Fortunately he was able to obtain an appointment at Padua, where he spent the next 18 years in more congenial surroundings and with a larger salary. Marina Gamba bore him three children; but they separated in 1610, when Galileo returned to Florence in his native Tuscany.

At Padua, Galileo had begun his work in astronomy. The first evidence that he had accepted the Copernican system is found in two letters written in 1597, one of them to Kepler in response to the latter's book, *Mysterium Cosmographicum*, of 1596. Galileo told Kepler that he had been a Copernican for several years, and had found several physical arguments in favor of the earth's motion. (Probably these arguments were based on the periodicity of ocean tides.) But Galileo paid little attention to the details of Kepler's work on planetary orbits; he never adopted Kepler's ellipse in place of the traditional circle.

In 1609, Galileo learned that someone in Holland had discovered how to magnify the appearance of distant objects by putting together two glass lenses. On the basis of this report Galileo constructed his own "telescope" and pointed it toward the heavens, with results that we shall describe in the next section. He also demonstrated it to the leaders of the Venetian republic, showing how it could be used to detect ships approaching the city long before they were visible to the naked eye. They were sufficiently impressed to give Galileo a lifetime contract as professor, at a considerable increase in salary.

Yet Galileo was not satisfied with his position. He dedicated his book, *The Starry Messenger*, an account of his first discoveries with the telescope, to Cosimo de Medici, Grand Duke of Tuscany, and even called the newly discovered satellites of Jupiter the "Medicean stars." The reason was obvious: he hoped to flatter the duke into giving him a job, or rather, into giving him financial support for his research without time-consuming obligations. As he explained his reasons for leaving the position he had just been given at Padua in the republic of Venice:

"It is impossible to obtain wages from a republic, however splendid and generous it may be, without having duties attached. For to have anything from the public one must satisfy the public and not any one individual; and so long as I am capable of lecturing and serving, no one in the republic can exempt me from duty while I receive pay. In brief, I can hope to enjoy these benefits only from an absolute ruler."*

But when Galileo left the republic of Venice to accept the patronage of the authoritarian ruler of Florence, he was making himself much more vulnerable to the attacks of his future enemies in the Church. Venice (so we are told by historians) would never have surrendered Galileo to the Inquisition.

At any rate, in 1610 Galileo became Court Mathematician and Philosopher (a title he chose himself) to the Grand Duke of Tuscany. From then, until his death at 78 in 1642, his life, despite recurring illness, and family and money troubles, was filled with continuous excellent work, with teaching and writing, and of course with the relentless and tragic fight with his enemies that is to be described in this chapter.

## 5.2 THE TELESCOPIC EVIDENCES FOR THE COPERNICAN SYSTEM

In 1610, Galileo published a little booklet entitled *Sidereus Nuncius*, which may be translated *Starry*

---

* *Discoveries and Opinions of Galileo*, translated by S. Drake, page 65.

*Messenger* (or "message from the stars").* In it he announced the discoveries he had made with his telescope:

1) The planet Jupiter has four smaller planets revolving around it. These were later called *satellites* by Kepler and other astronomers, and Jupiter is now known to have at least 12 of them. But the existence of even one such satellite is a blow to traditional ideas, for two reasons. First, many philosophers had convinced themselves that there *must* be exactly *seven* heavenly bodies (not counting the stars); hence the discovery of any more would be a metaphysical impossibility.† Second, whereas all the other celestial objects *appeared* to revolve around the earth with a motion that could only be explained away or understood by rather sophisticated theories, the satellites of Jupiter slowly but obviously revolved around Jupiter; hence the earth cannot be the center of rotation for *all* objects in the universe.

2) By telescopic observation, "The surface of the moon is not smooth, uniform, and precisely spherical as a great number of philosophers believe it (and the other heavenly bodies) to be, but is uneven, rough, and full of cavities and prominences, being not unlike the face of the earth, relieved by chains of mountains and deep valleys" (page 31). "Some of the mountains are as high as four miles" (page 41). Galileo alludes here to the fact that the same set of doctrines which then still generally upheld the geocentric system as the only possible choice also required that the celestial objects be "perfect" —spherical and unblemished—yet he could see not only mountains on the moon but also spots on the sun.

3) The fixed stars do not look much larger when viewed through a telescope; in fact, almost all of them are still just pinpoints of light. Galileo concludes that their apparent size as seen by the naked eye must be misleadingly large. We can imagine them as being extremely far away without having to attribute an incredibly immense size to them, and thus put aside for the moment the vexing problem of the absence of parallax (see Sections 1.5 and 2.4).

4) The Milky Way, which appears as a continuous region of light to the naked eye, is resolved by the telescope into many individual stars. Indeed, hundreds if not thousands of stars can be seen with the telescope which are not visible to the naked eye. Such facts could not easily be explained by those who believed that the universe had been created solely for the benefit of man. Why would God put so many invisible things in the sky?

Many of Galileo's contemporaries refused to accept the scientific validity of his telescopic discoveries. After all, it was known that all kinds of visual tricks could be played with lenses. The only scientist who publicly supported Galileo at the time was Kepler, who wrote a pamphlet titled *Conversation with the Starry Messenger*, pointing out that the new discoveries were consistent with his own theories. Having gained the endorsement of the leading astronomer of Europe, Galileo could

---

* An English translation of this charming work is readily available in Stillman Drake's book, *Discoveries and Opinions of Galileo*; page references are to this translation.

† The Florentine astronomer Francesco Sizzi argued in 1611 that there could not be any satellites around Jupiter for the following reasons: "There are seven windows in the head, two nostrils, two ears, two eyes and a mouth; so in the heavens there are two favorable stars, two unpropitious, two luminaries, and Mercury alone undecided and indifferent. From which and many other similar phenomena of nature such as the seven metals, etc., which it were tedious to enumerate, we gather that the number of planets is necessarily seven.... Besides, the Jews and other ancient nations, as well as modern Europeans, have adopted the division of the week into seven days, and have named them from the seven planets: now if we increase the number of planets, this whole system falls to the ground.... Moreover, the satellites are invisible to the naked eye and therefore can have no influence on the earth and therefore would be useless and therefore do not exist."

not be ignored; yet even Kepler and Galileo together could not immediately convert the rest of the world.

A year after his discoveries, Galileo had to write to Kepler:

"You are the first and almost the only person who, after a cursory investigation, has given entire credit to my statements.... What do you say of the leading philosophers here to whom I have offered a thousand times of my own accord to show my studies, but who, with the lazy obstinacy of a serpent who has eaten his fill, have never consented to look at the planets, or moon, or telescope?"

In his characteristic enthusiasm, Galileo had thought that through his telescopic discoveries everyone would see, *as with his own eyes*, the absurdity of the assumptions which prevented a general acceptance of the Copernican system. But man can believe only what he is ready to believe. In their fight against the new Copernicans the scholastics were convinced that they were surely "sticking to facts," that the heliocentric theory was obviously false and in contradiction with both sense observation and common sense, not to speak of the theological heresies implied in the heliocentric view. They had made Aristotelian science their exclusive tool for understanding facts, just as today most laymen make their understanding of physical theory depend on their ability to visualize it in terms of simple mechanical models obeying Newtonian laws. But at the root of the tragic position of the Aristotelians was, in part, the fact that an acceptance of the Copernican theory as even a *possible* theory would have had to be preceded by a most far-reaching re-examination and re-evaluation of their personal beliefs. It would have required them to do the humanly almost impossible, to discard their common-sense ideas, to seek new bases for their old moral and theological doctrines, and to learn their science anew (which was of course what Galileo himself did to an amazing degree, for which his contemporaries called him fool, or worse, and for which we call him

genius). Being satisfied with their system, the Aristotelians were, of course, unaware that history would soon prove their point of view to be far less effective in man's quest to understand nature.

---

**Problem 5.1.** Explain carefully the implications of Galileo's discovery that the stars do not appear any larger when viewed through a telescope.

**Problem 5.2.** Another of Galileo's telescopic discoveries was that the planet Venus sometimes seemed to be fully illuminated by the sun and at other times not at all; it has phases, like the moon. Referring to your answers to Problem 3.4, explain the significance of this discovery. How would you account for the fact that Venus, as seen by the naked eye, shows little change in brightness throughout the year?

## 5.3 TOWARD A PHYSICAL BASIS FOR THE HELIOCENTRIC SYSTEM

During the next two decades Galileo developed further his arguments for the Copernican system, and presented them at length in his great work, *The Dialogue on the Two Chief Systems of the World*, published in 1632. Here he stressed the rational arguments which seemed to him to be conclusive, apart from observational evidence of the type presented in the *Starry Messenger*. Observations alone, says Galileo, do not decide uniquely between a heliocentric and a geocentric hypothesis, "for the same phenomena would result from either hypothesis." This would be close to our modern view, but where we would make a choice between the two mainly on the basis of convenience in use, without regarding one as true and the other as false, Galileo understandably thinks of the earth's motion as "real," real just *because* it seems more reasonable and simplifies the picture. He then adduces several arguments, largely following Copernicus:

"When we merely consider the immensity of the starry sphere in comparison with the smallness of the terrestrial ball...and then think of the rapidity of the motion which completes a whole rotation in one day and night, I cannot persuade myself how anyone can hold it to be more reasonable and credible that it is the heavenly sphere which rotates, while the earth stands still."*

As a second point, Galileo reminds his readers that in the geocentric model it is necessary to ascribe to the planets a motion opposite to that of the celestial sphere (why?), again an unreasonable, we might almost say unharmonious or unesthetic, assumption. Third, Jupiter's four moons had shown him that there, too, existed the rule that the larger the orbit of the rotating body as reckoned from the center of rotation, the longer the period of rotation (qualitatively, Kepler's third law). This Copernicus had pointed out long before for the case of the planets themselves, and it held true even in the Ptolemaic system—but with this disharmony: In the Ptolemaic system the characteristic periods of revolution around the earth increase from the short one ($27\frac{1}{3}$ days) for the moon, to the very large one (30 years) for Saturn—and then suddenly drop back to 24 hours for the celestial sphere. In the Copernican system, however, "the rapidity of the periods is very well preserved; from the slowest sphere of Saturn we come to the wholly motionless fixed stars. We also escape thereby a fourth difficulty...I mean the great unevenness in the movement of these very stars [on the Ptolemaic hypothesis], some of which would have to revolve with extraordinary rapidity in immense circles, while others moved very slowly in small circles, since some of them are at a greater, others at a less distance from the pole."

Fifth, owing to the slightly changing tilt of the earth's axis, the apparent paths of the stars on the celestial sphere change slowly over the centuries,† again an improbable or at any rate an unreasonable feature of a geocentric theory that claimed to be based on the immutable, ideal, eternal characteristics of the heavenly bodies. Next, Galileo found it impossible to conceive in what manner the stars could be fixed in the celestial sphere to allow them rapid rotation and even motion as a group (cf. last point), while also having them preserve their relative distances so perfectly. "It seems to me much easier and more convenient to make them motionless instead of moving" and to ascribe the apparent motions to that of one "small insignificant body in comparison with the whole universe," the earth itself.

Lastly Galileo points out "I could also not explain why the earth, a freely poised body, balancing itself about its center [on the geocentric scheme] and surrounded on all sides by a fluid medium [postulated at the time to provide means for communicating the force and power needed to keep up the celestial motions] should not be affected by the universal rotation."

In the text, Galileo's hypothetical opponent answers these points:

"You support your arguments throughout, it seems to me, on the greater ease and simplicity with which the said effects are produced. You mean that as a cause [of the observed motion of planets and stars] the motion of the earth alone is just as satisfactory as the motion of all the rest of the universe with the exception of the earth; you hold the actual event to be much easier in the former case than in the latter."

These are meant to be accusing or scoffing remarks, as though, in modern terminology, one would say, "You are just a relativist, willing to

---

*    From the translation by Dannemann.

†    This refers to the precession of the equinoxes mentioned in Sections 1.2 and 2.2.

throw to the winds, for the sake of convenience, all the sacred implications of our geocentric hypothesis." And, in fact, the fictional opponent then dismisses all of Galileo's arguments with this quite characteristic counter-argument:

"For the ruler of the Universe, however, whose might is infinite, it is no less easy to move the universe than the earth or a straw balm."

Analyze carefully this part of Galileo's modern-sounding and conclusive reply (and incidentally notice the sly final phrases):

"What I said...does not refer to Him who causes the motion, but to that which is moved.... If we consider the moving bodies, we must unquestionably regard the motion of the earth as a *much simpler process* than that of the universe. If, furthermore, we direct our attention to *so many other simplifications* which may be reached *only* by this heliocentric theory, the daily movements of the earth must appear much more probable than the motion of the universe without the earth; for according to Aristotle's just axiom, 'It is vain to expend many means where a few are sufficient'."

But no matter how simple it might be to assume that the earth moves, Galileo realized that common sense must rebel at the idea, for we do not observe with our eyes the phenomena that we would expect to occur if the surface of the earth were in fact rushing through space at 1000 miles per hour. Salviati, Galileo's spokesman in the *Dialogue*, recalls such an argument:

"...Aristotle says, then, that a most certain proof of the earth's being motionless is that things projected perpendicularly upward are seen to return by the same line to the same place from which they were thrown, even though the movement is extremely high. This, he argues, could not happen if the earth moved, since in the time during which the projectile is moving upward and then downward it is separated from the earth, and the place from which the projectile began its motion would go a long way toward the east, thanks to the revolving of the earth, and the falling projectile would strike the earth that distance away from the place in question."

Salviati then attacks Aristotle's reasoning in an indirect way, by reconstructing the description that Aristotle would have to give of a stone's motion when dropped from a tower, supposing the earth *did* move. In order to account for the observation that the stone does fall along the side of the tower, Aristotle would have to compound this motion out of a natural motion downward toward the center of the earth, and a circular motion around the center; Aristotle would then have to claim that such a combination of two different natural motions in the same object is impossible.

Simplicio, another character in the *Dialogue* who represents the Aristotelian position,* now argues that the impossibility of such motion is confirmed by the behavior of a stone dropped from the mast of a moving ship,

"...which falls to the foot of the mast when the ship is standing still, but falls as far from that same point when the ship is sailing as the ship is perceived to have advanced during the time of fall, this being several yards when the ship's course is rapid."

Salviati and Sagredo (playing the role of the "neutral observer") help Simplicio to reconstruct the Aristotelian argument, showing that although a rock dropped from a tower on land might conceivably share the "natural" circular motion of the tower if the earth did rotate, the stone dropped from the ship could not do so because the ship's motion is not natural. The supposed falling behind in the latter case is thus set up as the crucial proof of the earth's immobility.

---

* The suspicion that Simplicio's name is used because it means "simpleton" is only partly allayed by the fact that there was an early medieval Aristotelian commentator named Simplicius, whose writings were well known in Galileo's time.

Now we might expect that Galileo will turn to the test of experiment, and portray his opponent's credulous reliance on authority. And this seems to be the case, at first:

"*Salviati:* ...Now, have you ever made this experiment of the ship?

"*Simplicio:* I have never made it, but I certainly believe that the authorities who adduced it had carefully observed it. Besides, the cause of the difference is so exactly known that there is no room for doubt.

"*Salviati:* You yourself are sufficient evidence that those authorities may have offered it without having performed it, for you take it as certain without having done it, and commit yourself to the good faith of their dictum. Similarly it not only may be, but must be that they did the same thing too—I mean, put faith in their predecessors, right on back without ever arriving at anyone who had performed it. For anyone who does it will find that the experiment shows exactly the opposite of what is written; that is, it will show that the stone always falls in the same place on the ship, whether the ship is standing still or moving with any speed you please. Therefore, the same cause holding good on the earth as on the ship, nothing can be inferred about the earth's motion or rest from the stone falling always perpendicularly to the foot of the tower.

"*Simplicio:* If you had referred me to any other agency than experiment, I think that our dispute would not soon come to an end; for this appears to me to be a thing so remote from human reason that there is no place in it for credulity or probability."

But suddenly Simplicio realizes that Salviati has not actually done the experiment either!

"*Simplicio:* So you have not made a hundred tests, or even one? And yet you so freely declare it to be certain? I shall retain my incredulity, and my own confidence that the experiment has been made by the most important authors who make use of it, and that it shows what they say it does.

"*Salviati:* Without experiment, I am sure that the effect will happen as I tell you, because it must happen that way; and I might add that you yourself also know that it cannot happen otherwise, no matter how you may pretend not to know it—or give that impression. But I am so handy at picking people's brains that I shall make you confess this in spite of yourself."

Have we, then, come up to the great turning point that divides ancient, scholastic, verbalistic, qualitative, a-prioristic Aristotelian science from modern, experimental, inductive science—only to find that the transition from Aristotelian to (eventually) Newtonian dynamics is effected by the methods of the Socratic dialogue without actually *doing* any experiment at all? Should Galileo's *Dialogue* be banned for its heresy against what even today is so commonly thought to be the "scientific method"?

Galileo's proof that the stone would fall at the foot of the mast of a moving ship introduces what we would now call the *law of inertia* or *Newton's first law of motion*. The proof is based on what the modern physicist calls a *Gedankenexperiment* (*thought experiment*), a device often used by Albert Einstein and others in discussing fundamental problems. Rather than either performing an actual experiment or reasoning syllogistically from general principles (in the Aristotelian fashion), Galileo asks us to consider what would happen in certain specified situations, and then shows that the conclusion of his opponent cannot be right because it leads to self-contradictory statements, or else that the conclusion Galileo is after is inherent in, and a necessary consequence of, sound convictions about those situations, convictions which we formed on the basis of ideas already tested (e.g., by daily experience).

"*Salviati:* ...Suppose you have a plane surface as smooth as a mirror and made of some hard material like steel. This is not parallel to the horizon, but somewhat inclined, and upon it you have placed a ball which is perfectly spherical and of some hard and heavy material like bronze. What do you believe this ball will do when released?"

Simplicio agrees that the ball will roll down the plane spontaneously, continuously accelerating; to keep it at rest requires the use of force; and if all air resistance and other obstacles were removed, it would keep moving indefinitely "as far as the slope of the surface extended."

Already the suggestion of motion to infinity in a vacuum has started to sneak in, but without arousing Simplicio's Aristotelian suspicions.

Now Salviati asks whether the ball would also roll upward on the slope, and of course Simplicio replies that this could happen only if a force were impressed on it; and moreover, "The motion would constantly slow down and be retarded, being contrary to nature." He also agrees that (in accord with Aristotelian physics) the greater the downward slope, the greater the speed.

"*Salviati:* Now tell me what would happen to the same movable body placed upon a surface with no slope upward or downward."

In this case, says Simplicio, the body will not move at all if it is simply set down at rest on the plane; but if it is given any initial impetus in a particular direction, there is no cause for either acceleration or deceleration.

"*Salviati:* Exactly so. But if there is no cause for the ball's retardation, there ought to be still less for its coming to rest; so how far would you have the ball continue to move?"

"*Simplicio:* As far as the extension of the surface continued without rising or falling.

"*Salviati:* Then if such a space were unbounded, the motion on it would likewise be boundless? That is, perpetual?

"*Simplicio:* It seems so to me...."

So, starting out from the Aristotelian postulate that a force is always needed to sustain an unnatural motion, Simplicio has been forced to concede that a motion which is not natural can nevertheless continue indefinitely without a force!

In other words, Galileo has found a loophole in the apparently impregnable structure of Aristotelean physics—even if it seems not a very large loophole. Galileo concentrates on the single point on the boundary between natural motions downward and unnatural motions upward. Nevertheless this point will prove to be a hole big enough for Galileo to sneak through, carrying with him the stone dropped from the mast of the ship, the moving earth, and in fact the foundations of modern physics.

Continuing the discussion, Galileo seems to abandon his infinite horizontal plane, which we all know does not exist in nature anyway, and returns to local motion on the surface of the earth. Now such motion is, in this context, equivalent to motion on the infinite horizontal plane, since motion on the surface of a truly spherical earth is neither upward nor downward. Hence, he argues, the stone must continue to move along with the ship even when it is no longer physically connected with it, and will therefore fall directly at the foot of the mast, whether the ship itself is moving or not.

Galileo is thus able to complete the original argument about the stone dropped from the mast of a moving ship, and thereby refute Aristotle's major objection to the rotation of the earth; he has shown that no conclusions about the motion of the earth can be drawn directly from observations of falling objects.* Crucial to this proof is his

---

* In fact he has proved a little too much. Strictly speaking it is *not* true that a stone dropped from the top of a tower on land, for example, will land precisely at the bottom if the earth is rotating. Galileo himself recognized this when later in the *Dialogue* he discussed the fall to earth of an object placed in the moon's orbit, revolving around the earth so as to maintain itself at the same height until it falls: "And far from failing to follow the motion of the earth and necessarily falling behind it, it would even go ahead of it, seeing that in its approach toward the earth the rotational motion would have to be made in ever smaller circles, so that if the same speed were conserved in it which it had within the orbit, it ought to run ahead of the whirling of the earth, as I said." The question is taken up again in Problem 10.4 in this book.

principle of inertia, which he states again in his later book *Two New Sciences* (1638):

"Furthermore we may remark that any velocity once imparted to a moving body will be rigidly maintained as long as the external causes of acceleration or retardation are removed, a condition which is found only on horizontal planes; for in the case of planes which slope downward there is already present a cause of acceleration, while on planes sloping upward there is retardation; from this it follows that motion along a horizontal plane is perpetual; for, if the velocity is uniform, it cannot be diminished or slackened, much less destroyed."*

Since the law of inertia was to become a cornerstone of Newton's theory of planetary motion, one might ask why Galileo himself applied it only to local motion at the surface of the earth, and not to the motion of the earth and planets around the sun. Perhaps he would have done so, but before writing his *Two New Sciences* he had solemnly sworn not to speak of such matters.

---

**Problem 5.3.** The experiment described by Galileo, of dropping an object from the mast of a moving ship, was first performed in 1641 (by the French philosopher Gassendi). Within the error of observation, the object did land directly at the bottom of the mast, just as if the ship were not moving. What does this result by itself, apart from Galileo's arguments presented in the *Dialogue*, prove about the motion of the earth?

## 5.4 SCIENCE AND FREEDOM

The tragedy that descended on Galileo is described in many places, and it is impossible to do justice to the whole story without referring to the details. Briefly, he was warned in 1616 by the Inquisition

---

* *Two New Sciences*, translated by H. Crew and A. de Salvio (1914, reprinted by Dover Publications, New York, 1954) page 215.

to cease teaching the Copernican theory, for it was now held "contrary to Holy Scripture." At the same time Copernicus' book itself was placed on the *Index Expurgatorius*, and was suspended "until corrected." But Galileo could not suppress what he felt deeply to be the truth. Whereas Copernicus had still invoked Aristotelian doctrine to make his theory plausible, Galileo had reached the new point of view in which he urged acceptance of the heliocentric system on its own merits of simplicity and usefulness, and apart from such questions as those of faith and salvation. This was the great break.

In 1623, Cardinal Barberini, formerly his friend, was elevated to the papal throne, and Galileo seems to have considered it safe enough to write again on the controversial topic. In 1632, after making some required changes, Galileo obtained the Inquisitor's necessary consent to publish the work *Dialogue Concerning the Two Chief Systems of the World* (from which the previous arguments for the Copernican theory were drawn), setting forth most persuasively the Copernican view in a thinly disguised discussion of the relative merits of the Ptolemaic and Copernican systems. After publication it was realized that he may have tried to circumvent the warning of 1616. Furthermore, Galileo's forthright and tactless behavior, and the Inquisition's need to demonstrate its power over suspected heretics, conspired to mark him for punishment.

Among the many other factors in this complex story a prominent role is to be assigned to Galileo's religious attitude—which he himself believed to be devoutly faithful, but which had come under suspicion by the Inquisition. Galileo's letters of 1613 and 1615 showed he held that God's mind contains all the natural laws, and that the occasional glimpses of these laws which the human investigator may laboriously achieve are proofs and direct revelations of the Deity quite as valid and grandiose as those recorded in the Bible. "From the Divine Word, the Sacred Scripture and Nature did both alike proceed.... Nor does God less admirably discover himself to us in Nature's actions than in the Scripture's sacred dictions." These opinions—held, incidentally, by many present-day scientists—can, however, be taken for symptoms of pantheism, one of the heresies for which Galileo's contemporary, Giordano Bruno, had been burned at the stake in 1600. Nor did Galileo help his case by such phrases as his quotation of Cardinal Baronius' saying: "The Holy Spirit intended to teach us in the Bible how to go to heaven, not how the heavens go."

The old, ailing man was called to Rome and confined for a few months. From the partly still secret proceedings we gather that he was tried (in absentia), threatened with torture, induced to give his oath that he had formally renounced the Copernican theory, and finally sentenced to perpetual confinement. None of his friends in Italy dared to defend Galileo publicly. His book was placed on the Index (where it remained along with Copernicus' and one of Kepler's until 1835). In short, and this is the only point of interest to us here, he was set up as a warning for all men that the demand for spiritual and ideological obedience indisputably entails intellectual obedience also, and that there can be no free science where there is no free conscience. His famous *Abjuration*, later ordered to be read from the pulpits throughout Italy and posted as a warning, has an ominously modern sound.*

But without freedom, science cannot flourish for long. It is perhaps not simply a coincidence that after Galileo, Italy, the mother of outstanding men till then, produced for the next 200 years only a few great physicists, while elsewhere in Europe they arose in great numbers. To scientists today this famous facet in the story of planetary theories is not just an episode in passing. Not a few teachers and scientists in our time have had to face powerful enemies of open-minded inquiry and of free teaching,

and again must stand up before those men who fear the strength of unindoctrinated intellect.

Even Plato knew that an authoritarian state is threatened by intellectual nonconformists, and he recommended for them the now time-honored treatment: "re-education," prison, or death. In the Soviet Union, geneticists at one time were expected to reject well-established theories, not on grounds of conclusive new *scientific* evidence, but because of doctrinal conflicts. This same struggle explains the banishment of the discussion of relativity theory from Nazi Germany's textbooks in the 1930s, because according to racist metaphysics, Einstein's Jewish origin invalidated his work for Germans. In the United States, too, there has not always been freedom to express scientific views. One of the most notorious examples was the 1925 "Monkey Trial" in Tennessee, at which the teaching of Darwin's theory of evolution was suppressed because it conflicted with certain types of Bible interpretations. More recently, in 1954, a governmental agency publicly humiliated Robert Oppenheimer for having hesitated to embark vigorously on the development of nuclear weapons systems along lines desired by military planners.

The warfare of authoritarianism against science, like the warfare of ignorance against knowledge, has not diminished since Galileo's day. Scientists take what comfort they can from the verdict of history. The attitude that got Galileo in trouble has since become quite acceptable to the religious authorities. Although the Vatican did not announce until 1968 that it might be appropriate to reverse its 1633 condemnation of Galileo's theories,* the scientific usefulness of his work was not so long delayed. Less than 50 years after Galileo's death, Newton's great book, the *Principia*, had appeared, integrating the work of Copernicus, Kepler, and Galileo so brilliantly with the principles of mechanics that the triumph of their ideas

---

* The full condemnation and abjuration is printed as Appendix E in Sedgwick and Tyler, *A Short History of Science* (New York: Macmillan, second edition, 1939).

* See the announcement by Franz Cardinal König of Vienna, reported in the *New York Times,* July 2, 1968.

was irrevocable, and, as we shall soon see, more significant than they themselves might have hoped.

---

**Problem 5.4.** In J. B. Conant's book, *On Understanding Science* [New Haven: Yale University Press (1947)] are summarized "Certain Principles of the Tactics and Strategy of Science" and "The Interaction of Science and Society." The following main points are made: (a) New concepts evolve from experiments or observations and are fruitful of new experiments and observations. (b) Significant observations are the result of "controlled" experiments or observations; the difficulties of experimentation must not be overlooked. (c) New techniques arise as a result of experimentation (and invention) and influence further experimentation. (d) An important aspect of scientific work is the interaction between science and society.

Examine the development of planetary theory in terms of each of the four points (preferably after reading at least pages 101–109 in Conant's book).

## RECOMMENDED FOR FURTHER READING

Marie Boas, *The Scientific Renaissance 1450–1630*, Chapter XI

H. Butterfield, *The Origins of Modern Science*, Chapter IV

Galileo Galilei, *The Starry Messenger* (1610) and other short pieces, translated with introduction and notes by Stillman Drake in *Discoveries and Opinions of Galileo*, Garden City, N.Y.: Doubleday Anchor, 1957; paperback. (An extract from *The Starry Messenger* may also be found in Hurd and Kipling, *The Origins and Growth of the Physical Sciences*, Vol. I, pages 137–162, together with a selection from Galileo's *Dialogue on the Two Chief World Systems*.)

T. S. Kuhn, *The Copernican Revolution*, Chapter 6

G. de Santillana, *The Crime of Galileo*, Chicago: University of Chicago Press, 1955; Phoenix Books paperback

## SOURCES, INTERPRETATIONS, AND REFERENCE WORKS

P. Ariotti, "From the top to the foot of a mast on a moving ship," *Annals of Science*, Vol. 28, pages 191–203 (1972)

A. Berry, *A Short History of Astronomy*, Chapter VI

S. Drake, "Galileo and the law of inertia," *American Journal of Physics*, Vol. 32, pages 601–608 (1964)

Galileo Galilei, *Dialogues Concerning the Two Chief World Systems—Ptolemaic and Copernican*, translated from Italian by S. Drake. Berkeley: University of California Press, 2nd edition, 1967; paperback. Another translation (originally by Salisbury, revised by G. de Santillana) is *Dialogue on the Great World Systems*, Chicago: University of Chicago Press, 1953.

L. Geymonat, *Galileo Galilei, a Biography and Inquiry into his Philosophy of Science*, translated from Italian with notes and appendix by S. Drake, foreword by G. de Santillana, New York: McGraw-Hill, 1965; paperback

W. Hartner, "Galileo's contribution to astronomy," *Vistas in Astronomy*, Vol. 11, pages 31–43 (1968)

*Kepler's Conversation with Galileo's Sidereal Messenger*, translated by E. Rosen, New York: Johnson Reprint, 1965

E. McMullin (editor), *Galileo, Man of Science*, New York: Basic Books, 1968

# Chapter 6

# MATHEMATICS AND THE DESCRIPTION OF MOTION

## 6.1 RENÉ DESCARTES

The Aristotelian philosophy was rapidly fading away under the attacks of such men as Galileo in Italy and Francis Bacon (1561–1626) in England; but what was to replace it? A collection of experimental facts and mathematical theories, no matter how precisely they fit together, does not constitute a satisfying philosophical system for interpreting the cosmos and man's place in it. The cosmology that was ultimately to be associated with the name of Isaac Newton owes much of its general character to a French philosopher, René Descartes (1596–1649), whose detailed scientific theories were overthrown by Newton. It was Descartes who first attempted to provide a general philosophical framework for the new science of the seventeenth century, defined some of the basic problems that scientists were concerned with, and suggested methods for solving them. Yet his name does not figure prominently in the history of physics, because most of his detailed solutions were rejected by later generations. What remained, aside from a general attitude toward the physical world, was the beginnings of a powerful mathematical technique for representing geometric forms and physical processes in symbolic language which greatly facilitated logical deductions: Des-

cartes' *analytic geometry*, later to be wedded to the *differential and integral calculus* of Newton and Leibniz.

In this chapter, we will summarize only a few of the more elementary concepts of the mathematical description of motion: not in their historical sequence, but from a viewpoint that now seems best adapted to physical applications. But first let us glance briefly at the career of Descartes himself.

Descartes was educated in a French Jesuit school from the age of 8 to 16, but his health was poor, so he was allowed to stay in bed late in the morning. Perhaps this helped him to develop his lifelong habits of prolonged meditation. At 17 he went to Paris, but it is said that he "had sufficient strength of character to hold himself aloof from the distractions of the capital." He traveled for several years, serving as an officer in various armies. In 1619 he conceived his reform of philosophy, taking mathematics as a model since that discipline alone seems to produce certain results. Although some of his discoveries in optics and mathematics were first worked out during this period, it was not until 1628 that he settled in Holland and devoted himself wholeheartedly to science.

In 1633, Descartes was about to publish a book on his view of the system of the world when he

# THE STUDY OF
# MOTION

Galileus Galileus     Florentinus

Superior<br>1 6     licentia<br>2 4

Eques Octauius Leoni     Roman° pictor fecit

Historically as well as logically mech
resents the foundation of physics and the
for the study of other physical scie
concepts developed in this field will
and again in this book. Moreover, m
physics what the skeleton is to the h
at first glance it may appear stiff,
what ghastly, but after even a br
functions one experiences with
ment the discovery of an astonis
design, of a structure that is in
yet so simple as to be almost in

We shall begin with the ke
the laws governing some of th
objects—movement along a
curved trajectory. For even
was this axiom: *ignorato m*

s
p
su
do
ph
rej
asi
wor
ema
form
which

63

heard of Galileo's condemnation by the Church at Rome. Descartes himself was a good Catholic and did not wish to come in conflict with the Church, so he suspended the publication of his book even though it would not have been officially censored or suppressed in Protestant Holland. Four years later he published his first work, actually a collection of four treatises: the best known of these, the *Discourse on Method*, was intended as a preface to the other three. The second part, *Optics*, includes the first publication and first proof of "Snell's law" of refraction, together with a description of how the eye works, and instructions for making better telescope lenses. We shall discuss Descartes' *Optics* later in connection with Newton's theory of light (Section 23.1). The third part, on *Meteorology*, discusses weather, clouds, snow, etc. and in particular gives a very good theory of the rainbow. The fourth part, *Geometry*, lays the foundations for analytical geometry, one of the essential mathematical tools of modern physical science.

In 1644, Descartes published his major work, *Principles of Philosophy*. Here he made a magnificent attempt to construct a complete theory of the world out of nothing but the concepts of matter and motion. It was a failure, but it left an indelible impact on subsequent thinking about the nature of the physical world. Even after the triumph of Newtonian physics, many scientists still shared Descartes' preference for avoiding the concept of *force*—i.e., of "action at a distance"—and for postulating instead that space is filled with pieces of matter that can interact only when they touch. What appear in Newton's theory to be long-range forces acting across empty space, such as gravity, would on Descartes' view be explained by the propagation of impulses through an invisible etherial matter that is imagined to fill the intervening space.

For Descartes, all motion is relative: one piece of matter can be said to be moving only with respect to other pieces of matter in its vicinity. This makes it possible to assert that the earth is "at rest" without abandoning the Copernican system! For Descartes considers the heavens to be undergoing a continuous circular motion with respect to the sun; the heavens carry along with them all the bodies that they contain, including the stars, planets, and earth, each whirling in its own vortex. Each body is thus at rest with respect to its own local vortex, while the vortex moves around the sun. Perhaps, with this doctrine of relativity of motion, Descartes hoped to avoid conflict with his Church, whose theologians sometimes asserted that science should concern itself only with describing the "appearances" and leave to theology the task of determining "reality."

Descartes' cosmology, though permeated with references to God, nevertheless represents a major step toward the elimination of God from any active role in moving the celestial objects or other matter. For example, Descartes asserts that God always conserves the same amount of "movement" in the world; this means in effect that while each piece of matter may have been created with any arbitrary movement, the transfer of movement from one piece to another by collisions is governed by a deterministic rule. From a scientific viewpoint, this means that Descartes not only asserts the principle of inertia—each body remains in the same state of motion in a straight line, or of rest, until acted on by another body—but that he also has some inkling of the principle of conservation of momentum (see Chapter 16). From a philosophical viewpoint, it means that the Cartesian world is very much like a machine. Indeed, Descartes admits that he made use of the example of machines in constructing his model of how the world acts, and he asserts that there is no real difference between machines made by men and objects found in nature, except that in the former all the moving parts must be large enough for us to see them. Even animals may be like machines. On the other hand, mechanistic theories cannot explain *human* behavior because, according to Descartes, the world of spirit and thought is completely different and separate from the world of matter, and is not governed by the same laws. With this dualistic approach, Descartes attempted to mark off a realm of natural phenomena that could be studied by the scientist without interference from the theologian, while leaving the theologian supreme in the realm of the spirit.

In 1649, Descartes was summoned to Sweden to serve as tutor to Queen Christina. The Swedish climate was too much for him. Within a few months he caught a chill, suffered inflammation of the lungs, and died.

## 6.2  CONSTANT VELOCITY

According to Galileo and Descartes, the natural state of a physical object is rest or *motion along a straight line, at constant speed.* Let us see how the physicist describes such motion in a way that can easily be generalized to other kinds of motion.

**Table 6.1**

| Station | Distance from station A in meters | Time of travel from station A in seconds |
|---|---|---|
| A | 0 | 0 |
| B | 30 | 2 |
| C | 60 | 4 |
| D | 90 | 6 |
| E | 120 | 8 |

Suppose we are watching the motion of a car traveling along a perfectly straight and smooth road, and that we wish to find the exact relationship between the *distance* covered and the *time* required to cover it. The progress of this car may be timed at stations regularly spaced at, say, 30 meters* all

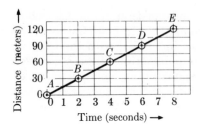

**Fig. 6.1**  Graph for uniform speed.

along the line. The car has reached full speed before it goes past the first station, and all our time measurements will be expressed in terms of the time elapsed since the first station was reached, the moment at which our experiment begins. If the motion has been uniform (constant speed), a tabular arrangement of the time schedule for the first five stations might read as in Table 6.1.

A more useful way of representing the same information is a time-vs.-distance graph (Fig. 6.1). Along the horizontal axis we plot the time elapsed in seconds, and along the vertical axis the distance covered in meters.† On this graph the five known data are entered, and since they fall on a straight line, we draw it in as shown to indicate our not-unreasonable belief that any additional data taken would also have fallen along that line. For example, we are fairly confident that at a point 45 meters from station A the car would have passed by 3 seconds after the beginning of our series of observations. This conclusion is an *interpolation*, and is most conveniently done on a graph of this type.

On the other hand, we are also fairly sure that after 9 seconds, the vehicle, if it continued to move at the same speed, reached a position 135 meters beyond station A. This is an *extrapolation*. Ob-

---

\*    From now on we shall use the metric system of units, which has been adopted in all countries for scientific work. It may be helpful to remember that one meter (1 m) = 39.37 inches, or a little over 1 yard; one centimeter ($\frac{1}{100}$ of a meter) is about $\frac{1}{4}$ inch. See Appendix II for other conversion factors.

---

†    The interval between the origin and a point measured along the horizontal axis is sometimes called the *abscissa,* and the interval along the vertical axis is called the *ordinate.*

# PART B

# THE STUDY OF MOTION

# THE STUDY OF MOTION

Galileus Galileus    Florentinus

Superioã         licentia
  16                24

Eques Octauius Leonii    Roman° pictor fecit

Historically as well as logically mechanics represents the foundation of physics and the prototype for the study of other physical sciences. The concepts developed in this field will appear again and again in this book. Moreover, mechanics is to physics what the skeleton is to the human figure—at first glance it may appear stiff, cold, and somewhat ghastly, but after even a brief study of its functions one experiences with mounting excitement the discovery of an astonishingly successful design, of a structure that is ingeniously complex, yet so simple as to be almost inevitable.

We shall begin with the key topic in mechanics, the laws governing some of the simpler motions of objects—movement along a straight line and a curved trajectory. For even in Galileo's time there was this axiom: *ignorato motu, ignoratur natura.*

# Chapter 6

# MATHEMATICS AND THE DESCRIPTION OF MOTION

## 6.1 RENÉ DESCARTES

The Aristotelian philosophy was rapidly fading away under the attacks of such men as Galileo in Italy and Francis Bacon (1561–1626) in England; but what was to replace it? A collection of experimental facts and mathematical theories, no matter how precisely they fit together, does not constitute a satisfying philosophical system for interpreting the cosmos and man's place in it. The cosmology that was ultimately to be associated with the name of Isaac Newton owes much of its general character to a French philosopher, René Descartes (1596–1649), whose detailed scientific theories were overthrown by Newton. It was Descartes who first attempted to provide a general philosophical framework for the new science of the seventeenth century, defined some of the basic problems that scientists were concerned with, and suggested methods for solving them. Yet his name does not figure prominently in the history of physics, because most of his detailed solutions were rejected by later generations. What remained, aside from a general attitude toward the physical world, was the beginnings of a powerful mathematical technique for representing geometric forms and physical processes in symbolic language which greatly facilitated logical deductions: Des-

cartes' *analytic geometry*, later to be wedded to the *differential and integral calculus* of Newton and Leibniz.

In this chapter, we will summarize only a few of the more elementary concepts of the mathematical description of motion: not in their historical sequence, but from a viewpoint that now seems best adapted to physical applications. But first let us glance briefly at the career of Descartes himself.

Descartes was educated in a French Jesuit school from the age of 8 to 16, but his health was poor, so he was allowed to stay in bed late in the morning. Perhaps this helped him to develop his lifelong habits of prolonged meditation. At 17 he went to Paris, but it is said that he "had sufficient strength of character to hold himself aloof from the distractions of the capital." He traveled for several years, serving as an officer in various armies. In 1619 he conceived his reform of philosophy, taking mathematics as a model since that discipline alone seems to produce certain results. Although some of his discoveries in optics and mathematics were first worked out during this period, it was not until 1628 that he settled in Holland and devoted himself wholeheartedly to science.

In 1633, Descartes was about to publish a book on his view of the system of the world when he heard of Galileo's condemnation by the Church at Rome. Descartes himself was a good Catholic and did not wish to come in conflict with the Church, so he suspended the publication of his book even though it would not have been officially censored or suppressed in Protestant Holland. Four years later he published his first work, actually a collection of four treatises: the best known of these, the *Discourse on Method*, was intended as a preface to the other three. The second part, *Optics*, includes the first publication and first proof of "Snell's law" of refraction, together with a description of how the eye works, and instructions for making better telescope lenses. We shall discuss Descartes' *Optics* later in connection with Newton's theory of light (Section 23.1). The third part, on *Meteorology*, discusses weather, clouds, snow, etc. and in particular gives a very good theory of the rainbow. The fourth part, *Geometry*, lays the foundations for analytical geometry, one of the essential mathematical tools of modern physical science.

In 1644, Descartes published his major work, *Principles of Philosophy*. Here he made a magnificent attempt to construct a complete theory of the world out of nothing but the concepts of matter and motion. It was a failure, but it left an indelible impact on subsequent thinking about the nature of the physical world. Even after the triumph of Newtonian physics, many scientists still shared Descartes' preference for avoiding the concept of *force*—i.e., of "action at a distance"—and for postulating instead that space is filled with pieces of matter that can interact only when they touch. What appear in Newton's theory to be long-range forces acting across empty space, such as gravity, would on Descartes' view be explained by the propagation of impulses through an invisible etherial matter that is imagined to fill the intervening space.

For Descartes, all motion is relative: one piece of matter can be said to be moving only with respect to other pieces of matter in its vicinity. This makes it possible to assert that the earth is "at rest" without abandoning the Copernican system! For

Descartes considers the heavens to be undergoing a continuous circular motion with respect to the sun; the heavens carry along with them all the bodies that they contain, including the stars, planets, and earth, each whirling in its own vortex. Each body is thus at rest with respect to its own local vortex, while the vortex moves around the sun. Perhaps, with this doctrine of relativity of motion, Descartes hoped to avoid conflict with his Church, whose theologians sometimes asserted that science should concern itself only with describing the "appearances" and leave to theology the task of determining "reality."

Descartes' cosmology, though permeated with references to God, nevertheless represents a major step toward the elimination of God from any active role in moving the celestial objects or other matter. For example, Descartes asserts that God always conserves the same amount of "movement" in the world; this means in effect that while each piece of matter may have been created with any arbitrary movement, the transfer of movement from one piece to another by collisions is governed by a deterministic rule. From a scientific viewpoint, this means that Descartes not only asserts the principle of inertia—each body remains in the same state of motion in a straight line, or of rest, until acted on by another body—but that he also has some inkling of the principle of conservation of momentum (see Chapter 16). From a philosophical viewpoint, it means that the Cartesian world is very much like a machine. Indeed, Descartes admits that he made use of the example of machines in constructing his model of how the world acts, and he asserts that there is no real difference between machines made by men and objects found in nature, except that in the former all the moving parts must be large enough for us to see them. Even animals may be like machines. On the other hand, mechanistic theories cannot explain *human* behavior because, according to Descartes, the world of spirit and thought is completely different and separate from the world of matter, and is not governed by the same laws. With this dualistic approach, Descartes attempted to mark off a realm of natural phenomena that could be studied by the scientist without interference from the theologian, while leaving the theologian supreme in the realm of the spirit.

In 1649, Descartes was summoned to Sweden to serve as tutor to Queen Christina. The Swedish climate was too much for him. Within a few months he caught a chill, suffered inflammation of the lungs, and died.

## 6.2 CONSTANT VELOCITY

According to Galileo and Descartes, the natural state of a physical object is rest or *motion along a straight line, at constant speed.* Let us see how the physicist describes such motion in a way that can easily be generalized to other kinds of motion.

**Table 6.1**

| Station | Distance from station A in meters | Time of travel from station A in seconds |
|---------|-----------------------------------|------------------------------------------|
| A | 0 | 0 |
| B | 30 | 2 |
| C | 60 | 4 |
| D | 90 | 6 |
| E | 120 | 8 |

Suppose we are watching the motion of a car traveling along a perfectly straight and smooth road, and that we wish to find the exact relationship between the *distance* covered and the *time* required to cover it. The progress of this car may be timed at stations regularly spaced at, say, 30 meters* all

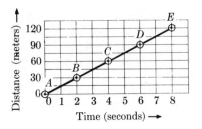

**Fig. 6.1**  Graph for uniform speed.

along the line. The car has reached full speed before it goes past the first station, and all our time measurements will be expressed in terms of the time elapsed since the first station was reached, the moment at which our experiment begins. If the motion has been uniform (constant speed), a tabular arrangement of the time schedule for the first five stations might read as in Table 6.1.

A more useful way of representing the same information is a time-vs.-distance graph (Fig. 6.1). Along the horizontal axis we plot the time elapsed in seconds, and along the vertical axis the distance covered in meters.† On this graph the five known data are entered, and since they fall on a straight line, we draw it in as shown to indicate our not-unreasonable belief that any additional data taken would also have fallen along that line. For example, we are fairly confident that at a point 45 meters from station A the car would have passed by 3 seconds after the beginning of our series of observations. This conclusion is an *interpolation*, and is most conveniently done on a graph of this type.

On the other hand, we are also fairly sure that after 9 seconds, the vehicle, if it continued to move at the same speed, reached a position 135 meters beyond station A. This is an *extrapolation*. Ob-

---

\*   From now on we shall use the metric system of units, which has been adopted in all countries for scientific work. It may be helpful to remember that one meter (1 m) = 39.37 inches, or a little over 1 yard; one centimeter ($\frac{1}{100}$ of a meter) is about $\frac{1}{4}$ inch. See Appendix II for other conversion factors.

---

†   The interval between the origin and a point measured along the horizontal axis is sometimes called the *abscissa,* and the interval along the vertical axis is called the *ordinate.*

viously, if it is not already known that the graph describes a uniform motion, one must not place too much confidence in results obtained by interpolation or extrapolation, particularly when the basic data are few or widely spaced.

From Table 6.1 it should already be evident that the car travels twice the distance in twice the time, three times the distance in three times the time, and so forth; in other words, the distance covered is directly proportional to the time elapsed. The straight line in the graph is another way of representing the same thing. Descartes' system of analytic geometry enables us to translate the straight line into an algebraic equation: Let the time elapsed since station A was passed be represented by the letter $t$, and let the distance covered during that time interval be called $s$; then for the motion here described we can write

$$s \propto t,$$

where the symbol $\propto$ means "is proportional to." But this is equivalent to the equation

$$s = kt,$$

where $k$ is a constant which depends on neither of the variables $s$ and $t$.

In analytic geometry, the equation $s = kt$ represents a straight line passing through the origin of coordinates. The constant $k$ is called the *slope*. It is also the *tangent* of the angle between the line and the horizontal axis. In this case, its value is simply the ratio of distance covered to time elapsed,

$$k = s/t.$$

This ratio of distance covered to time elapsed is, by definition, the *speed*; the fact that it is equal to a constant brings us back to (and is a consequence of) our original assumption that the car travels with constant speed.

Consider for a moment what an amazing thing has actually happened. First we watched the actual motion of a car along a straight road. Then, from the multitude of ever-changing impressions—

the blur, the noise, the turning of wheels, the whole chaos of events progressing in time and space—we have rescued two measurable quantities, $s$ and $t$, both of which take on different values every instant, and we have found that their ratio is a constant, an unchanging theme underlying the flux of otherwise meaningless, unrelated data. We have defined a concept, speed, and so have been led to discover a simple feature in an otherwise complex situation. Perhaps familiarity with the concept of *speed* prevents you from appreciating this experience of *creating order* from a *chaos* of sense impressions by abstracting from it some measurable data and by perceiving or inventing or intuiting a suitable concept to describe that portion of the total phenomenon; however, we shall return to this method, the very heart of scientific procedure, again and again.

The exact numerical value of the speed is to be found by substituting corresponding values for $s$ and $t$. For example, from the data taken at station B, $s = 30$ m and $t = 2$ sec; from this, it follows that the speed, $s/t$, is 15 m/sec, which is about 44 ft/sec or 30 miles per hour (mi/hr).* As a check on the constancy of the speed, you may wish to recompute it, using data for stations C, D, or E.

We have solved the problem concerning the motion of the car. The speed is constant, and is 15 m/sec. These two conclusions must seem obvious, but there are a few details which may be worth an extra word. To compute the speed we do not really need to rely on the raw data in Table 6.1. We can use the time-vs.-distance graph directly if we have any confidence in it at all. Consider, for example, the uniform motion indicated by the line $OQPU$ in Fig. 6.2. From the previous paragraphs we know that for such a case the speed is given by $s_1/t_1$ or by $s_2/t_2$ (since $s_1/t_1 = s_2/t_2$ by similar triangles), but it is equally well given by $(s_2 - s_1)/(t_2 - t_1)$, i.e., by the ratio of *any* small distance along the

---

*    The relationship 30 mi/hr $\doteq$ 44 ft/sec $\doteq$ 13 m/sec is a very useful one to memorize.

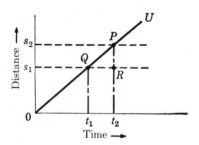

**Figure 6.2**

line of motion to the time needed to pass through that distance. This statement may seem intuitively acceptable, but we should check it rigorously.

---

**Problem 6.1.** Prove that $(s_2 - s_1)/(t_2 - t_1) = s_1/t_1$.

---

*Solution:* Note that in Fig. 6.2 the distance $QR = t_2 - t_1$, and $PR = s_2 - s_1$; also that the following three are similar triangles: $Os_1Q$, $Os_2P$, and $QRP$. Therefore corresponding distances have equal ratios, that is,

$$\frac{s_1}{t_1} = \frac{s_2}{t_2} = \frac{s_2 - s_1}{t_2 - t_1} \qquad \text{Q.E.D.}$$

Before you leave these relatively harmless findings, be sure that you can answer to your own full satisfaction the following questions:

What is the definition of constant speed (use algebraic notation)?

What experiment must I make to find out whether a certain body moves with constant speed?

How do I find the value of the speed once I know that it is constant?

How small a distance $(s_2 - s_1)$ may I choose in order to calculate the speed?

After that, try to tackle the superficially deceptive questions upon which, as we shall find, depends the understanding of a large section of our physics:

For what reason do we here accept these arguments from Euclidean geometry or algebra?

What does mathematics tell us about real physical events, such as the motion of material bodies?

Another problem is raised by a second critical look at our concept of constant speed along a straight line. Table 6.1 presented raw experimental data in tabular form and was our first step in formulating the idea of constant speed. Figure 6.1 represented these data in the form of a graph and expanded their usefulness by permitting convenient interpolation and extrapolation. The proportion $s \propto t$ abstracted from this graph a concise statement which we expressed by the equation $s/t =$ a constant (15 m/sec in this specific case).

What has happened here is that in four stages we have proceeded from a bulky table of raw data to a generalization ($s/t = 15$ m/sec). If on repeated observations of this car we always arrive at this same equation, in other words, if the car always moves at a constant speed of 15 m/sec, we would regard the equation as a *law* applicable to this restricted type of observation. As it is, our equation arose from a single experiment and is simply a general description of this one motion.

At first glance all four stages (the table, the graph, the proportion, and the equation) seem to contain the same information, with the last stage appearing vastly preferable because of its more economical demands on paper and on mental effort. However, we must not let our innate love for conciseness misguide us; the equation $s/t = 15$ m/sec does in fact include everything that the table offers, but it also tells indiscriminately a great deal more that may actually be quite unwarranted. For example, unless you had seen the tabulated data, you might believe on the basis of the equation alone that surely $s = 1.5$ mm if $t = 0.001$ sec, or that $s = 4.7 \times 10^8$ m if $t = 1$ year.

Both statements are correct only if the motion under investigation is truly uniform within the limits of such very exact measurements as 0.001 sec and for such a span of time as 1 year, but in our actual, limited experience neither of these conditions is likely to hold. The lesson to be drawn here is that equations in physics, other than a few purely axiomatic definitions, must be thought to have attached an unseen "text," a statement which describes the actual limitations and other implied assumptions under which the equation may be applied. These conditions, this text, may simply be of the following form. "The equation $s/t = 15$ m/sec describes the result of a certain experiment in which $t$ did not exceed 8 seconds, and in which only a few measurements were made, each with such and such accuracy." Without the clear understanding that equations in physics almost always carry hidden limitations, we cannot expect to understand a single physical law; we would make unwarranted extrapolations and interpolations. We would be in the catastrophic position of a navigator who has to negotiate a rocky channel without having any idea of the length, width, and draft of his ship.

## 6.3 THE CONCEPT OF AVERAGE SPEED

Prolonged motion with constant speed occurs so rarely in nature (despite its importance as an "ideal" motion in the theories of Galileo and Descartes) that, as you may have noticed, we have denied it the honor of a special letter symbol of its own. (We did not introduce the letter $v$ which you might have expected to be used for constant velocity or constant speed; this will instead be used for *instantaneous* speed, in the next section.)

We turn now to motions with *non*uniform speed (accelerated or decelerated motions) but still taking place only along a straight line. A general case of this type is shown in Fig. 6.3, where a car starts from rest, moves in a straight line gaining speed until time $t_1$, and proceeds with constant speed until time $t_2$ (as indicated by the straight line portion between $t_1$ and $t_2$), after which it slows down to rest ($t_3$), waits without moving until time $t_4$, and then again accelerates to a new,

**Fig. 6.3** Graph for nonuniform speed.

higher constant speed. To deal with such variations in speed, it is necessary to introduce the familiar term, *average speed* or average velocity, symbolized by $\bar{v}$.

**Definition.** *The average speed $\bar{v}$ of an object during a time interval t is the distance covered during this time interval, divided by t.*

Note that for any motion other than prolonged constant speed, the value of $\bar{v}$ depends on the length of the time interval chosen. For example, a glance at Fig. 6.3 suggests that since the curve is becoming steeper between $t_1$ and $t_2$ (its slope is increasing), $\bar{v}$ for the first $t_1$ seconds is smaller than $\bar{v}$ computed for the first $t_2$ seconds. On the other hand, during the interval $(t_4 - t_3)$ the average speed is zero. Although the speed from $t_1$ to $t_2$ might be 30 m/sec and between $t_5$ and $t_6$ 50 m/sec, the average speed $\bar{v}$ for the entire time to $t_6$, which is calculated by $\bar{v} = (s_6/t_6)$, is perhaps 20 m/sec. We recognize that the meaning of $\bar{v}$ is this: No matter how irregular the motion actually is during a time interval $t$, if the object had traveled instead with a constant speed of magnitude $\bar{v}$ it would have traveled the same distance during those $t$ seconds.

Instead of writing out "the interval from $t_1$ to $t_2$" we will often use the abbreviation $\Delta t$. The Greek letter $\Delta$, delta, means "a small change in (whatever quantity follows)." Similarly for distance intervals: $\Delta s$ does not mean "delta times $s$" but reads "delta $s$," i.e., a small increment in distance,

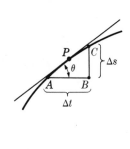

**Fig. 6.4**  Determination of instantaneous speed.

a small displacement covered during the corresponding short time interval $\Delta t$. Therefore we can write the definition of average speed as:

$$\bar{v} = \Delta s / \Delta t.$$

## 6.4  INSTANTANEOUS SPEED

All our speed measurements so far have involved more or less extended time measurements. However, we certainly also want to have a concept which tells us the speed of a moving object *at one instant*, the type of information which we read off a speedometer. For example, we may ask, "What is the speed at time $t_a$ for the motion pictured in Fig. 6.4?" Now we *do* know how to find the average speed $\bar{v}$ during a time interval from $t_1$ to $t_2$, which includes the instant $t_a$, i.e. $\bar{v} = (s_2 - s_1)/(t_2 - t_1)$. However, since the speed is not constant from $t_1$ to $t_2$, we certainly cannot identify the speed *at the instant* $t_a$ with the average speed $\bar{v}$ *during* $(t_2 - t_1)$. Therefore we must resort to what undoubtedly will look like a trick: we calculate $\bar{v}$ for a very short time interval $\Delta t$ encompassing the instant $t_a$; so short an interval that the value of $\bar{v}$ would not change materially if the interval were made even shorter. The instantaneous speed (symbolized simply by $v$) at $t_a$ is therefore calculated by successive approximations from the average speed $\bar{v}$, and may be defined by

$$v = \lim_{\Delta t \to 0} \bar{v} = \lim_{\Delta t \to 0} \left( \frac{\Delta s}{\Delta t} \right).$$

Putting this equation into words, the instantaneous speed is the value which the average speed ($\Delta s/\Delta t$) attains when the time interval ($\Delta t$) is chosen sufficiently small, approaching zero in the limiting case.

A concrete example will help here. From the data on which Fig. 6.4 is based we might construct Table 6.2 for space and time intervals including point $P$ on the curve. The intervals are nested inside $t_2 - t_1$ and $s_2 - s_1$, converging on the point $P$ as they become smaller and smaller. For each pair $\Delta s$ and $\Delta t$ we can compute the quotient, the average speed $\bar{v} = \Delta s/\Delta t$ for that interval. Although both $\Delta s$ and $\Delta t$ become indefinitely small, $\bar{v}$ approaches a definite limiting value as long as the

**Table 6.2**  Data for estimation of instantaneous speed at the point $P$ ($t = 7$ sec, $s = 20$ m)

| $t_B$ | $t_A$ | $s_B$ | $s_A$ | $\Delta t = t_B - t_A$ | $\Delta s = s_B - s_A$ | $\bar{v}$ |
|---|---|---|---|---|---|---|
| 12 | 2 | 30 | 5 | 10 | 25 | 2.5 m/sec |
| 7.5 | 6.5 | 21 | 19 | 1.0 | 2.0 | 2.0 m/sec |
| 7.05 | 6.95 | 20.10 | 19.92 | 0.10 | 0.18 | 1.8 m/sec |
| 7.005 | 6.995 | 20.010 | 19.993 | 0.010 | 0.017 | 1.7 m/sec |

curve is smooth. In the table the subscripts $A$ and $B$ are used to identify the variable lower and upper limits of the intervals.

As you perhaps remember from mathematics courses, what we have just done is the algebraic equivalent of finding the slope of the straight line tangent to the curve at the point $P$ in Fig. 6.4. As $\Delta t$ and $\Delta s$ become smaller and smaller, the section of the curve which they include is more nearly a straight line, playing the role of the hypotenuse in the right triangle $ABC$ (see insert in Fig. 6.4). But by trigonometry, $\Delta s/\Delta t = \tan\theta$, where $\theta$ is the angle made by the hypotenuse with the horizontal; so that ultimately, when $\Delta t$ is sufficiently small, $\Delta s/\Delta t$ becomes simultaneously both the numerical value of the instantaneous speed $v$ at point $P$ *and* the value of $\tan\theta$ of the line tangent at point $P$, i.e., the slope of that line.*

We gratefully accept this unexpected result because it cuts short the labor involved in finding instantaneous speeds for nonuniform motions. In the future we need not construct such tables as Table 6.2 in order to find the instantaneous speed at a given point during the motion of an object. We now simply refer to the distance-vs.-time graph, find the line that is tangent to the curve at the point in question, and calculate its slope. In fact, this last full sentence may stand as the definition of the concept of instantaneous speed in terms of actual operations needed for its determination. If ascertaining speeds by looking at a speedometer seems far simpler, realize that in the initial construction and calibration of the speedometer we must still take good care that it can automatically perform tasks equivalent to all those outlined in our definition of instantaneous speed.

It is this concept which we shall find most useful from now on, and we shall discuss typical problems

---

* Strictly speaking the equation $\Delta s/\Delta t = \tan\theta$ is incorrect since the left-hand side must have units (or "dimensions") m/sec, while the right-hand side is a pure number having no units. This is why we have to specify "the numerical value" of $\Delta s/\Delta t$, as distinct from the physical quantity.

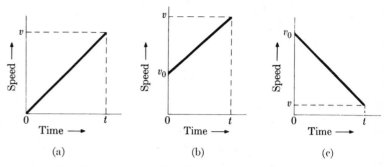

**Fig. 6.5** Three cases of uniformly changing velocity.

in which the (instantaneous) speed of a body changes from some initial value $v_0$ to some other value $v$ during a specified time interval $t$. We shall sometimes make plots of (instantaneous) speed vs. time. For the case shown in Fig. 6.1, such a plot would evidently be simply a straight horizontal line, but for Fig. 6.3 it is more complex.

---

**Problem 6.2.** Copy Fig. 6.3 on a large scale and superpose on it a plot of the speed vs. time.

---

## 6.5  ACCELERATION

Of all motions, the one most important for our purposes is motion with *uniformly changing velocity*, i.e., motion with constant acceleration or constant deceleration. To a fair approximation this is the characteristic behavior of ordinary freely falling bodies, objects sliding down a smooth inclined plane, carts coasting to a stop on level ground, and generally of all bodies which experience a constant force.

Graphically, there are three simple cases involving uniformly changing velocity. In Fig. 6.5(a) the speed at the start is zero ($v_0 = 0$) and reaches a value $v$ after $t$ sec; in Fig. 6.5(b) $v_0$ is not zero, and the speed increases uniformly; and in Fig. 6.5(c) the speed decreases from a large initial value $v_0$ to nearly zero during time $t$. Concrete experimental situations corresponding to each drawing will suggest themselves readily.

For all such cases it is easy to define the concept of acceleration, for which we shall use the letter $a$: acceleration is the ratio of the change of speed $(v - v_0)$ to the time $(t)$ during which this change occurs, or

$$a = \frac{v - v_0}{t}. \qquad (6.1)$$

As we have agreed to restrict our attention for the time being to motion with constant acceleration, we need not be concerned with such refinements as average acceleration and instantaneous acceleration, for, as in the case of constant speed, the average and instantaneous values are numerically equal.

To illustrate this new concept let us consider some very straightforward cases.

**Example 1.** A skier starts down an incline and gains speed with uniform acceleration, reaching a speed of 10 m/sec after 5 sec. What is his acceleration?

*Solution:* As in every problem, we must first *translate* each separate phrase into mathematical symbols, or else extract from each some clues as to which physical concepts, laws, and limitations are here applicable. This is not a trivial process; most students find that a kind of intuition gained from experience in working out such problems is more important than facility in solving mathematical equations. The phrase "a skier *starts* down an incline" is translated to $v_0 = 0$; "gains speed with uniform acceleration" means $a$ is constant, therefore Eq. (6.1) will hold for this case; "reaching a speed of 10 m/sec after 5 sec" can be simply restated as $v = 10$ m/sec at $t = 5$ sec. The final phrase does *not* mean "what is the cause or physical nature of his acceleration" (that will come later), but simply asks for a numerical value, $a = ?$

Now we put all our symbolic statements side by side and read:

$$v_0 = 0; \qquad v = 10 \text{ m/sec}; \qquad t = 5 \text{ sec};$$

$$a = \frac{v - v_0}{t} \text{ holds here}; \qquad a = ?$$

Evidently,

$$a = \frac{10 \text{ m/sec} - 0}{5 \text{ sec}} = 2 \text{ (m/sec)/sec}.$$

This may be read "2 meters per second, per second, and may be written more concisely 2 m/sec². Although this abbreviation can be read "2 meters per second squared," remember that the unusual concept of "square seconds" arises merely from the fact that we have defined acceleration as a change in speed (units *m/sec*) divided by a time interval (units *sec*).

Although this example must still appear exceedingly simple, you will find that the steps taken are the same here as in every single problem, no matter how sophisticated. If you train yourself from the very start to *translate each phrase separately*, you will have avoided half the initial troubles with physics problems.

**Example 2.** A car has been traveling on a level road at a speed of 70 km/hr when suddenly the motor fails and the car coasts along, gradually slowing down. During the first minute it loses 30 km/hr. What was the acceleration during that time?

*Solution:* Translating, we read $v_0 = 70$ km/hr, $v = 40$ km/hr at $t = 1$ min $= 60$ sec; $a = ?$ (Evidently it will be a *negative* acceleration, sometimes called deceleration, but that is a superfluous term.)

It is very tempting to plunge ahead and say that

$$a = \frac{40 \text{ km/hr} - 70 \text{ km/hr}}{60 \text{ sec}} = -\tfrac{1}{2} \text{ (km/hr)/sec}$$

However, this is true only if we assume that the car coasted with uniform acceleration, a guess, but not necessarily the only correct one. Therefore we must say $a = -\frac{1}{2}$ (km/hr)/sec *if* the acceleration was uniform.

**Example 3.** A car on a roller coaster with many curves changed its speed from 5 km/hr to 25 km/hr in 20 sec. What was the acceleration?

*Translation:* $v_0 = 5$ km/hr, $v = 25$ km/hr, $t = 20$ sec, $a = ?$ However, since there is little likelihood that the acceleration was constant, we must simply confess that we cannot solve this problem until further data are given from which the type of motion may be deduced.*

The units for acceleration have been variously given as (m/sec)/sec or (km/hr)/sec; others are evidently also possible; for example, (m/sec)/min. (If you happen to live in one of the few countries in the world which has not yet adopted the metric system, you may have to use the table of conversion factors in Appendix II in order to translate these units into others that are more familiar to you.) In all these cases, the units correspond to the dimension of (length/time)/time, or in a symbolic shorthand frequently encountered, $L/T^2$.

## 6.6 ORESME'S GRAPHICAL PROOF OF THE MEAN-SPEED THEOREM

In the last four sections we have temporarily set aside our historical account in order to introduce some basic concepts in modern notation. Many of the same problems of accelerated motion were discussed as long ago as the thirteenth century by mathematicians at Oxford and Paris, though many

of their arguments, expressed verbally without the aid of algebra and analytic geometry, are hard for modern readers to follow. One particular result of their work which is of great importance for physics is the *mean-speed theorem*, sometimes called the Merton theorem, after Merton College, Oxford, where it was first developed.

In modern notation, the mean-speed theorem refers to a motion that is uniformly accelerated with initial speed $v_0$ and final speed $v$, and occupies a time interval $t$. The theorem states that the distance traveled is the same as that traveled in another motion which takes place at the mean speed (that is, at a constant speed equal to the average of $v_0$ and $v$) during the same time interval $v$. In other words, if the entire motion had taken place at a constant speed $v_0$ for a time $t$, the distance would be $v_0 t$; if it had taken place at a constant speed $v$ for a time $t$ it would be $vt$; but in the case in which the speed changes uniformly from $v_0$ to $v$, the distance is the same as if the speed had been equal to a constant speed of amount $\frac{1}{2}(v_0 + v)$, namely $\frac{1}{2}(v_0 + v)t$.

This result may seem uninteresting, even trivial. It can easily be derived from the algebraic equations to be discussed in the next section. More significant, however, is a particular *method* used to derive it by Nicolas Oresme at the University of Paris, early in the fourteenth century.† Oresme realized that since a quantity like $v_0 t$ is a product of two numbers, it can be represented as the area of a rectangle whose sides are $v_0$ and $t$. This would be the rectangle $OABC$ in Fig. 6.6. Similarly, $vt$ is the area of the rectangle $OADE$. Therefore, Oresme argued, the actual distance traveled when the velocity is changing uniformly from $v_0$ to $v$

---

* It would be a mistake to regard such a statement as an admission of dishonorable defeat; on the contrary, in science, as in every other field, the possibility of solving a problem offhand is the exception rather than the rule, and it is a good enough preliminary victory to know what additional information is needed for an eventual solution.

---

† Oresme (*circa* 1320–1382), chaplain to King Charles V and later Bishop of Lisieux, has recently attracted the attention of historians of science because of his ideas on kinematics and astronomy, which anticipated to some extent those established by Galileo in the seventeenth century.

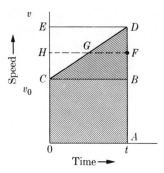

**Fig. 6.6** Oresme's graphical representation of the relation between speed, time, and distance.

must be equal to the area of the shaded part of the figure, namely the rectangle $OABC$ plus the triangle $CBD$. (If this is not obvious, after a little study of the figure it should seem at least reasonably plausible.)

Once it is accepted that the distance covered in the accelerated motion is equal to the shaded area $OABDC$, the proof of the mean-speed theorem is simple enough. If to this area we add the triangle $CGH$, and subtract the triangle $GFD$, we obtain the rectangle $OAFH$. Now if the line $HGF$ is parallel to $ED$ and $CB$ and is exactly halfway between them, then the point $G$ will be exactly halfway between $C$ and $D$, and the line segment $CH$ will be equal to $DF$. Also, $HG = GF$ and $\angle DGF = \angle HGC$. Therefore the triangles $HCG$ and $GDF$ are congruent, and have equal area, so that the area added is equal to the area subtracted. So the area of the rectangle $OAFH$, which represents the distance traveled at a constant speed halfway between $v_0$ and $v$, is equal to the area of the original figure (rectangle $OABC$ plus triangle $CBD$) representing the area covered in uniformly accelerated motion with speed increasing constantly from $v_0$ to $v$.

The mathematical reasoning presented above is about as difficult as any we shall need in this book, aside from the solution of a few simple algebraic

equations. In fact, the first step of Oresme's proof hints at the fundamental theorem of the calculus of Newton and Leibniz: The area under the speed-time curve is equal to the distance traveled. Since the velocity itself is defined as the rate of change of distance with time, we could generalize this theorem to obtain a general relation between any quantity (such as distance) which varies with changes in another quantity (such as time), and the rate of change of the first quantity with respect to the second; this general relation would involve finding the area under the curve of "rate of change" vs. the second quantity, and may be valid even when the change of the rate of change (in this case the acceleration) is not uniform. Although the calculus based on this theorem is an extremely powerful tool, used throughout all of theoretical physics since the seventeenth century, we shall not find it necessary for our basic explanations of physical concepts and theories.

## 6.7 EQUATIONS OF MOTION FOR CONSTANT ACCELERATION

The reason for our present preoccupation with uniformly accelerated motion is not only its occasional appearance in nature but also, and primarily, the *ease* with which we can formulate some simple equations of motions, i.e., equations which relate the five important variables:

>distance covered ($s$)
>time taken ($t$)
>initial velocity ($v_0$)
>final velocity ($v$)
>acceleration ($a$).

Equation (6.1) is a good example. We may rewrite it as

$$v = v_0 + at, \qquad (6.2)$$

or, in words, the velocity at the end of a time interval $t$ is the sum of the initial velocity and the change in velocity, which is the constant acceleration times the time elapsed.

There are other similar equations relating the five variables. Let us now construct three of them. We can begin with the property of speed vs. time graphs mentioned in the previous section: The area under the curve is numerically equal to the distance ($s$) covered. Thus, in Fig. 6.6, the shaded area under the curve is that of a rectangle of sides $v_0$ and $t$ plus a triangle of base $t$ and height ($v - v_0$):

$$s = v_0 t + \tfrac{1}{2}(v - v_0)t. \qquad (6.3)$$

We could interpret this as the distance $v_0 t$ covered during time $t$ if the motion had proceeded with constant speed $v_0$, plus the distance $\tfrac{1}{2}(v - v_0)t$ added by virtue of the acceleration of the motion.

If we did not originally know the final velocity $v$, but were given the initial speed $v_0$ and the acceleration $a$, we could use Eq. (6.1) to eliminate $v$ from the above equation. The quickest way to do this is to notice that the expression ($v - v_0$) would be the same as the right-hand side of Eq. (6.1) if it were divided by $t$. So let us multiply and divide the second term of Eq. (6.3) by $t$ in order to put it in this form:

$$s = v_0 t + \frac{1}{2}\left(\frac{v - v_0}{t}\right) t^2.$$

We can now use Eq. (6.1) to replace the quantity in parentheses by $a$:

$$s = v_0 t + \tfrac{1}{2}at^2. \qquad (6.4)$$

Equation (6.4) is a very useful equation. It lets us compute the distance covered in time $t$, starting from initial speed $v_0$, whenever the motion is uniformly accelerated with acceleration $a$. It is often used in the special case in which the initial speed is zero; then:

$$s = \tfrac{1}{2}at^2 \qquad \text{(when } v_0 = 0\text{)}. \qquad (6.5)$$

On the other hand, sometimes a somewhat more general form of (6.4) is needed to cover situations in which the distance variable $s$ does not begin at a zero mark when the time variable $t$ is zero. This is best illustrated by an example.

**Example 4.** An electric automobile was 10 km west of Boston, going west at a speed of 50 km/hr, when a uniform acceleration of 1 km/hr$^2$ was applied. How far from Boston was the car an hour later?

*Translation:* We must define the initial distance $s_0 = 10$ km; we are given $v_0 = 50$ km/hr, $a = 1$ km/hr$^2$, $t = 1$ hr. If we use the letter $s$ to represent total distance from Boston, then the distance covered in this time interval $t$ will be $s - s_0$; it is this quantity rather than $s$ which must be put on the left-hand side of Eq. (6.4):

$$s - s_0 = v_0 t + \tfrac{1}{2}at^2;$$

hence we have

$$s = s_0 + v_0 t + \tfrac{1}{2}at^2. \qquad (6.6)$$

Substituting our numerical values, we find

$$s = 10 \text{ km} + (50 \text{ km/hr}) \times (1 \text{ hr})$$
$$+ \tfrac{1}{2}(1 \text{ km/hr}^2) \times (1 \text{ hr})^2$$
$$= (10 + 50 + \tfrac{1}{2}) \text{ km} = 60.5 \text{ km}.$$

Equation (6.6) is the most general equation of this type which we shall need, as long as we restrict ourselves to cases with uniform acceleration.

So far we have been assuming that the motion proceeds for a given time interval $t$, and we have derived equations for the distance and speed. But we might also want to deal with problems in which the distance is given and we are required to find the time or speed. In the simplest such case, if the initial velocity and initial distance are both known to be zero, we can use Eq. (6.5) and simply solve for $t$ in terms of $s$:

$$t = \sqrt{(2s/a)}.$$

In more general cases we might have to solve a quadratic equation such as Eq. (6.6), with given values of $s$, $s_0$, and $v_0$.

A problem that will turn out to have considerable importance in physics is: Given the initial speed, the distance covered, and the accelera-tion, find the final speed. Now that you have seen the general procedure, you should try to derive an expression to find the answer for this yourself.

---

**Problem 6.3.** Prove that if $s_0 = 0$,

$$v^2 = v_0^2 + 2as. \qquad (6.7)$$

[*Hint:* According to the mean-speed theorem, $s = \frac{1}{2}(v_0 + v)t$. Combine this with Eq. (6.1).]

---

Now let us collect together the main results of this section, four equations relating the six variables $s_0$, $s$, $t$, $v_0$, $v$, and $a$, starting with the mean-speed theorem:

$$s = \tfrac{1}{2}(v_0 + v)t \qquad \text{(I)}$$

$$s = s_0 + v_0 t + \tfrac{1}{2}at^2 \qquad \text{(II)}$$

$$v = v_0 + at \qquad \text{(III)}$$

$$v^2 = v_0^2 + 2as \qquad \text{(IV)}$$

Unless otherwise stated, it can be assumed that $s_0 = 0$.

Now it is possible to step back, so to speak, and to study the overall pattern of our work so far. We are now convinced that the equations I through IV contain all the important and interesting relations between distances traveled, time needed, and so forth, for those motions which specifically are uniformly accelerated. Whenever we encounter this type of motion, be it executed by a falling stone or whatever, we shall at once know how to predict any two of the five variables when given the re-maining three. For example, if it is known that a stone dropped from the Leaning Tower of Pisa strikes the ground 60 m below with a measured speed of 36 m/sec, we may at once calculate its time of fall and its acceleration. Translating the data to $v_0 = 0$, $v = 36$ m/sec, $s = 60$ m, $t = ?$, $a = ?$, we scan our four trusted equations and note that $t$ can be found from equation I [$t = (2s)/(v_0 + v) = 3.33$ sec] and the acceleration can be

found either from IV [$a = (v^2 - v_0^2)/2s$] or, now that $t$ is known, from II or III.*

The fact is that equations I through IV become our *tools* for dealing with this type of problem, and now that we have derived and established them, we shall simply remember and apply them as occasion arises, in much the same manner as a craftsman, once having carefully selected a proper tool kit for his work, ceases to reinvent or justify his tools at every turn, but is content to reach into his bag for the proper instrument demanded by the situation at hand. Before our tool kit begins to fill up, we must take the important step of gaining complete mastery over and confidence in each new concept or law as it is introduced. We can then apply it when called upon without feeling a need to re-examine it in detail. For example, when later we are to find $v$ from $v_0$, $a$, and $s$ for a uniformly accelerating body, we can use equation IV (that is, $v^2 = v_0^2 + 2as$, therefore $v = \sqrt{v_0^2 + 2as}$) even though undoubtedly one can have only little intuitive feeling about this relation. If you have initially derived equation IV to your satisfaction, you must then believe in it.

This is, of course, not at all the same thing as the self-defeating practice of just learning equations by heart or of looking them up in the notes to apply them when called upon. Such a ruse cannot work, because one cannot apply an equation properly without knowing all its intimate strengths and weaknesses, the implied "text."

To give a final example on this belabored but important point, let us imagine that we are asked to find the time of fall of a brick from the top of the Empire State building. Given are $v_0$, $a$, and $s$. The temptation to "plug in" the readily memorized

---

\* The assumption that $a$ is indeed constant for free fall of real bodies at Pisa must be checked first if reason for doubt exists. If the motion should in fact turn out not to be uniformly accelerated, we should investigate how good an approximation our idealizations make toward the true case of free fall. This will be discussed in the next chapter.

equation II must here yield to the knowledge which springs from closer study that the four equations undoubtedly do not apply in this case, for such large distance of fall the air friction will prevent constancy of acceleration. As everyone knows by observation of falling leaves, snow, rain, etc., each body sooner or later may reach a constant terminal velocity of fall. This velocity, beyond which there is no further acceleration at all, depends on the surface area of the body, its weight, and the properties of the medium (air, etc.) through which it falls.

This completes our formal study of simple motion, the basis for our coming discussion of the laws of projectile motion.

## ADDITIONAL PROBLEMS

*Note:* As a rule, do not spend more than about half an hour at the very longest on any problem at one session. If you are unable to solve it in that time, plan to come back to it later, or else finish it by stating what you foresee is likely to be the subsequent procedure and final solution. Note also that not all problems *have* simple numerical answers, even if the statement of the problem is numerical.

**Problem 6.4.** a) Find by graphical interpolation the approximate position of the car described in Fig. 6.1 corresponding to a time of travel of 5 sec, 0.5 sec, 7.005 sec. Extrapolate to find the position corresponding to 10 sec, 50 sec, $-2$ sec (do this work on graph paper).
b) Discuss the limitations of the process of interpolation and of extrapolation.

**Problem 6.5.** On graph paper, plot a graph corresponding to Fig. 6.1, but use data which were taken at stations $A$, $B$, $C$, $D$, and $E$ when a particle with a constant velocity of 50 mi/hr ($= 22.2$ m/sec) was passing by. You must first construct a table corresponding to Table 6.1. (Keep your eyes open; the numerous calculations can be simplified.)

**Problem 6.6.** Describe the possible experimental situations which gave rise to each of the eight graphs of distance vs. time (Fig. 6.7) for motion along one line. Then try to express each of four or five of these graphs in terms of an equation relating $s$ and $t$ in algebraic fashion.

**Problem 6.7.** A leisurely vacation trip from Cambridge, Mass., to Berkeley, Calif., took 15 days, including several long stops en route. Calculate the average speed for this trip in miles per hour. [*Note:* Whenever the data given seem insufficient, you will find the necessary additional information in pertinent books in your library. If you cannot find such books, use an intelligent guess.]

**Problem 6.8.** Make some reasonable assumption about the speed which a car can pick up during the first 10 sec after starting from rest, then calculate the acceleration.

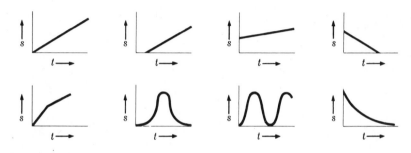

**Figure 6.7**

[*Note:* In all future problems, assume that the acceleration of freely falling bodies at the surface of the earth is 9.80 m/sec². This happens to be true, within fairly narrow limits, over the whole earth. The symbol employed for this "constant" (the acceleration due to gravity) is the letter $g$ instead of $a$.]

**Problem 6.9.** Five observation posts are spaced 20 m apart along a straight vertical line. A freely falling body is released at the topmost station and is timed as it passes each of the others. Compute a table of the data expected, and represent this information in a graph (use graph paper).

**Problem 6.10.** Make a table of the value of $s$, $v$, and $\bar{v}$ for a freely falling body if $v_0 = 0$ and $t$ is, in turn, 1 sec, 2 sec, 3 sec, 5 sec, and 10 sec. Represent this information in three graphs (on graph paper).

**Problem 6.11.** A boy violently throws a stone vertically downward from a building 50 m high, imparting to it a substantial initial speed. If the stone takes 1.8 sec to hit the pavement below, what was its initial speed?

**Problem 6.12.** Make a list of all the relevant limitations, conditions, and idealizations which were implied in your solution of Problem 6.11.

**Problem 6.13.** The nucleus of a helium atom is to travel along the inside of a straight hollow tube 2 m long, which forms a portion of a particle accelerator. If the nucleus is to enter at 1000 m/sec and to leave at the other end at 10,000 m/sec, how long is it in the tube? State your assumption.

## RECOMMENDED FOR FURTHER READING

E. T. Bell, *Men of Mathematics*, New York: Simon & Schuster, reprint of the 1937 edition; paperback, Chapters 1, 2, and 3

Marie Boas, *The Scientific Renaissance 1450–1630*, Chapter VII

E. A. Burtt, *The Metaphysical Foundations of Modern Physical Science*, Chapter IV

A. C. Crombie, "Descartes," *Scientific American*, pages 160–173 (October 1959)

*Project Physics Reader*, Vol. 1, article by Gerhart and Nussbaum

## SOURCES, INTERPRETATIONS, AND REFERENCE WORKS

A. C. Crombie, *Medieval and Early Modern Science* (revised edition of *Augustine to Galileo*), Garden City, N.Y.: Doubleday Anchor Books, 1959, paperback, Vol. II, Chapter I

René Descartes, *Discourse on Method, Optics, Geometry, and Meteorology* (1637), translated by P. J. Olscamp, Indianapolis: Bobbs-Merrill, 1965, paperback. (The famous "Discourse on Method" was originally published as an introduction to the three other works; "Geometry" contains the methods of analytic geometry.)

S. Drake, "The uniform motion equivalent to a uniformly accelerated motion from rest," *Isis*, Vol. 63, pages 28–38 (1972)

J. F. Scott, *Scientific Works of René Descartes*, London: Taylor & Francis, 1952

# Chapter 7

# GALILEO AND THE KINEMATICS OF FREE FALL

## 7.1 INTRODUCTION

The material in the preceding chapter on velocity and acceleration for simple types of motion, which is the logical and historical groundwork of physics, is to a large extent the work of Galileo Galilei. It is a small but critical part of the voluminous work of this arresting man, who is often called the father of modern science. Now we shall examine in some detail Galileo's contribution to the problem of freely falling bodies as a historical, scientific, and methodological achievement.

Galileo is important and interesting to us for several reasons, quite apart from his dramatic role in the history of astronomy described in Chapter 5. First, if we understand what he had to struggle against, what contemporary attitudes he had to discard in founding the new "Mechanical Philosophy," then perhaps we shall be clearer about the nature of our present-day structure of science and our own attitudes and preconceptions. To cite an analogy, we certainly must not fail to study the period of the American Revolution if we wish to understand our present democratic institutions which grew out of it.

Second, after the historical setting of Galileo's work has been clarified, he will speak to us as a scientist at his best, as an investigator whose virtuosity in argument and discovery illustrates clearly the profound excitement of scientific research. He will be for us an example of a physicist at work, and although in his case circumstances were extraordinary, that fact does not make him unrepresentative: If he had been born in the twentieth century, we would find it not at all hard to imagine him making major new discoveries and playing a leading part in many branches of modern science.

Third, Galileo's approach to the problem of motion of freely falling bodies will provide us, both in this chapter and later, with an opportunity for a short introspective discussion on procedure in science. We need this. Within recent years, many people have displayed misunderstanding and even hostility toward science and technology, while at the same time workers in the life sciences and social sciences have debated vigorously whether the methods of the physical sciences are applicable to their own fields. Since the physical sciences have in fact been spectacularly successful in discovering the laws of nature (whether or not one approves of the use that society has made of those discoveries), a clear understanding of how the physicist works must be central to the larger question of the nature of science. Without an awareness of the place of definition, hypothesis, experimentation and intuition in physics, without a knowledge of the vital processes by which this field functions and grows, science must seem a dry formalism, to be memorized without understanding, reproduced on examinations without enthusiasm, and perhaps applied to practical problems without due consideration of its limitations.

To begin with, let us consider what kinematics* Galileo might have learned at his university, what type of physics he eventually had to reject. It was, of course, the science of the scholastics, the heirs and often misinterpreters of that greatest of scientist-philosophers of antiquity, Aristotle. But at this point, a warning: The history of science is a sophisticated field in its own right. By the very nature of its subject matter, there is a good deal of inconclusiveness and uncertainty at important points, and we shall in this book neither attempt nor presume to evaluate such delicate questions as, for example, to what extent Galileo shared many opinions and methods with the Aristotelians, or drew on the work of other investigators—Archimedes, the whole tradition of scholars at the venerable University of Paris, Tartaglia, Benedetti, or Stevinus.† For science, like any other intellectual enterprise, cannot start from scratch without any roots in some tradition. Galileo, too, had the often unacknowledged encouragement of many enlightened bits of contemporary physics in his struggle to shape a consistent whole. But his fight, judging by his own words, does appear to have been against a group of exponents of the stricter contemporary scholastic view. In short, the statement that Galileo started modern science is so abrupt and leaves out so much that it can justly be called wrong, but no more wrong than the statement

---

*    *Kinematics* (from the Greek *kinema*, motion) is the branch of physics that deals with the description of motion, excluding the study of the specific forces responsible for the motion.

†    For details on the state of mechanics before Galileo, see some of the books mentioned in the bibliography at the end of this chapter.

that the American Republic dates from June 1788, when the Constitution went into effect.

## 7.2 ARISTOTELIAN PHYSICS

Despite the widespread story of Galileo's experiment from the Leaning Tower of Pisa, most people believe (erroneously) that heavy bodies gain speed far more quickly when falling than do lighter bodies. This was the view recorded by Aristotle (Section 1.3), much of whose works, through the mediation of St. Thomas Aquinas and his followers, became the basis of Christian thinking in Europe from the thirteenth century on. However, Aristotle's view on the motion of falling bodies was not simply an error or the result of miscarried experimentation, but a characteristic consequence of a great overall scheme; indeed it was a rather minor member of an imposing structure which Aristotle erected as a result of his search for a unity of *all* thought and experience. His system embraced at the same time elements that are now separated into component parts: scientific, poetic, theological, ethical.

According to Aristotle's system, described in Section 1.3, the natural motion of an object must depend on how much it contains of the elements Water and Earth, whose natural place is at the center of the universe. Thus a big stone would be expected to fall faster than a small one, and this expectation seems to be borne out by common experience, at least if we compare extremes (a boulder and a pinch of dust).

A modern scientist would immediately object that Aristotle's theory must fail any *quantitative* test. In his book *On the Heavens*, Aristotle states that the speed of fall is proportional to the weight of the object:

"A given weight moves a given distance in a given time; a weight which is heavier moves the same distance in less time, the time being inversely proportional to the weights. For instance, if one weight is twice another, it will take half as long over a given distance."*

Casual or careful observation (which Aristotle did indeed use well) does prepare us to note differences in speed of fall. But one would think that even with a crude method for actually measuring times of fall, anyone could discover that a stone twice as heavy as another one does not fall twice as fast; and certainly that a stone ten times as heavy as another one does not fall ten times as fast. But that is the wisdom of hindsight. It was then not natural to think in terms of such actual tests on earth.

Aristotle took into account not only the weight of the falling object but also the resistance offered by the medium through which it falls. Thus a stone falls faster in air than in water. It was perhaps natural to suggest, therefore, that the speed is inversely proportional to the resisting force. One could think of the weight of the object and the resistance of the medium as two opposing forces: There is motion only if the weight exceeds the resistance (a stone cannot fall through solid earth), and the greater the weight in proportion to the resistance, the greater the speed.†

Stated verbally, this sounds like a plausible rule, although it too may not stand up to experimental test. But as soon as we try to express it in mathematical terms, unexpected consequences emerge. If speed $v$ is directly proportional to weight $W$ and inversely proportional to resistance $R$, we can write

$$v \propto W/R \quad \text{or} \quad v = k(W/R),$$

where $k$ is an unspecified proportionality constant.

We also have the proviso that

$$v = 0 \quad \text{unless} \quad W > R$$

(no motion unless the weight exceeds the resisting force). It was not until the fourteenth century that Thomas Bradwardine, at Merton College, Oxford, realized that these two equations do not go together harmoniously. He considered the following imaginary (or thought) experiment: Start with a given weight and a very small resisting force, and gradually let the resistance $R$ increase until it finally exceeds the weight $W$. From the qualitative description of motion according to Aristotle, one would expect that the speed $v$ would gradually decrease to zero in this experiment. But the equations tell us that $v$ gradually decreases to a value slightly greater than $k$, as $R$ approaches $W$, but then $v$ suddenly drops to zero when $R$ is equal to $W$. This discontinuous change in speed is not at all consistent with the basic viewpoint of Aristotle's physics, and therefore, according to Bradwardine, the relation $v \propto W/R$ must be rejected.

It is actually possible to find a mathematical function of $W$ and $R$ which has the desired property —decreasing continuously to zero as $R$, initially less than $W$, approaches it.‡ But that was not to be a fruitful approach, and so we shall not pursue it. Perhaps it is enough to point out the useful role of mathematics in the development of physical theories: By forcing the physicist to make his ideas precise, it sometimes exposes his inconsistencies and leads to the rejection of hypotheses without the need for any experimental test. Yet in the search for valid hypotheses, mathematics cannot be a substitute for *all* experiments; a growing theory must somehow be fertilized by contact with the real world.

---

* Aristotle's *De Caelo,* translated by J. L. Stokes, Oxford University Press, Book I, Chapter 6, page 273b. Note that Aristotle does not write out equations—nor did most physicists until after the time of Galileo.

† Aristotle apparently thought of $v$ as remaining constant, once the motion begins, unless there is some change in the resistance or other forces acting on the object.

‡ Bradwardine himself suggested a logarithmic function: $v = \log W/R$ (recall that $\log 1 = 0$). For an account of Bradwardine's life and work, see the article on Bradwardine by J. E. Murdoch in the *Dictionary of Scientific Biography.* Another possibility was suggested by John Philoponus of Alexandria in the fifth century: $v = W - R$.

Another consequence of Aristotle's relation $v \propto W/R$ is that motion in a vacuum, with no resistance at all ($R = 0$), would have to take place with infinite speed, no matter what the value of the weight $W$. This absurdity was sometimes cited by the Aristotelians as a reason why a vacuum itself is inconceivable! It was unthinkable to them that the paradox might be attributed to some inadequacy in the original postulate.

By the end of the sixteenth century, Aristotle's theory of motion had been strongly criticized by a number of eminent scientists, and it was perfectly clear that it could not hope to provide quantitative agreement with the results of the simplest experiments. Yet the theory was still firmly entrenched in the learned world. According to modern tenets of scientific method, this should have been a scandal: Scientists are not supposed to adhere dogmatically to a theory that did not measure up in experimental tests. But before we rush to condemn our ancestors (for behavior that is in truth not so much different from that observed in all fields even in our own time), let us inquire into the reasons for the continuing attractiveness of the Aristotelian philosophy.

For, according to Aristotle, a theory that merely describes and predicts the facts of observation, no matter how accurately, is not good enough: it must also give them meaning in a wider sense, by showing that the facts are in accord with the general postulates of the whole philosophical system. To the Aristotelian of the late sixteenth century, the postulates concerning elements and tendencies (like the postulates about circular motions in astronomy), and all conclusions from them, were true ("philosophically true"), clear, certain, because they were also part of a larger, important, and satisfying scheme, reaching into theology and other fields. The postulates could not be given up just because the detailed motion of two weights dropped from a tower contradicted one of their many conclusions. He would give "philosophic truth" precedence over contradictory "scientific truth"—though the latter did not yet exist

as a distinct concept. His physics was the investigation of the nature of things in a context so large that the preoccupation of later science—for example, with detailed measurement of falling bodies—would seem to him like a small, artificial, incomplete subject with little significance by itself. As the historian F. S. Taylor put it in *Science Past and Present*:

"The science of Aristotle and of the scholastics had been primarily a science of purpose. It was transparently clear to the latter that the world and all that's in it was created for the service of man, and that man had been created for the service of God. That was a perfectly intelligible scheme of the world. The sun was there to give us light and to tell the time and mark out the calendar by his motions. The stars and planets were a means of distributing beneficent or maleficent influences to the things on earth to which they were sympathetically linked; plants and animals were there to give us food and pleasure, and we were there to please God by doing His will. Modern science does not produce any evidence that all this is not perfectly true, but it does not regard it as a scientific explanation; it is not what science wants to know about the world."

Conversely, in general (though there are notable exceptions), students of nature before Galileo's time would necessarily have found it unreasonable or even inconceivable to be seriously interested only in the type of questions our science asks. That such questions are indeed interesting or important, that the methods employed in our scientific investigations can give results at all, this could not have been apparent before the results themselves began to pour forth in the seventeenth century. But so overwhelming was the unsuspected, evergrowing stream that we, who are caught in the flood, can hardly think in prescientific terms. We must deplore this unresourceful imagination of ours. If each of us even for a moment could transcend his own pattern to appreciate clearly the point of view from which an Aristotelian would have proceeded, we should be better equipped for dealing with the everyday conflicts of differing temperaments and opinions.

It was not easy for men to give up the Aristotelian view of the world, in which everything had its own purpose and place, carefully planned in relation to the whole. To many devout and sensitive intellects of the early seventeenth century, it must have been extremely disturbing to learn of new discoveries of stars randomly distributed through the sky with no evident relation to the earth, which itself no longer had a privileged position, and to read of theories that, like the old atheistic doctrines of Lucretius and Epicurus, seemed to describe the world as a meaningless, fortuitous concourse of atoms moving through empty space. The poet John Donne lamented the changing times in some famous lines, written in 1611:

"And new Philosophy calls all in doubt,
The Element of fire is quite put out;
The Sun is lost, and th'earth, and no man's wit
Can well direct him where to looke for it.
And freely men confesse that this world's spent,
When in the Planets, and the Firmament
They seeke so many new; then see that this
Is crumbled out again to his Atomies.
'Tis all in peeces, all coherence gone;
All just supply, and all Relation..."*

## 7.3 GALILEO'S *TWO NEW SCIENCES*

Galileo's early writings on mechanics were in the tradition of the standard medieval critics of Arisstotle, and as we mentioned in Section 5.1, at the time when the legendary Tower of Pisa experiment might have been performed, he had not yet developed the general theory for which he is now best known. During his mature years, his chief interest

was in astronomy, and it was in connection with the theoretical effects of the earth's rotation that he proposed his principle of inertia (Section 5.3). When his *Dialogues on the Two Great World Systems* (1632) was condemned by the Roman Catholic Inquisition and he was forbidden to teach the "new" astronomy, Galileo decided to concentrate on mechanics. This work led to his book, *Discourses and Mathematical Demonstrations Concerning Two New Sciences Pertaining to Mechanics and Local Motion*, usually referred to as the *Two New Sciences*. It was written while Galileo was technically a prisoner of the Inquisition, and was published surreptitiously in Holland in 1638, for in his own Italy his books were then being suppressed.

Galileo was old, sick, and nearly blind at the time he wrote *Two New Sciences*; yet his style in it is spritely and delightful. As in his early work on astronomy, he uses the dialogue form to allow a lively conversation between three speakers: Simplicio, who represents the Aristotelian view; Salviati, who presents the new views of Galileo; and Sagredo, the uncommitted man of good will and open mind, eager to learn. Let us listen to Galileo's three speakers as they discuss the problem of free fall:

"*Salviati:* I greatly doubt that Aristotle ever tested by experiment whether it be true that two stones, one weighing ten times as much as the other, if allowed to fall, at the same instant, from a height of, say, 100 cubits,† would so differ in speed that when the heavier had reached the ground, the other would not have fallen more than 10 cubits.

"*Simplicio:* His language would seem to indicate that he had tried the experiment, because he says: *We see the heavier*; now the word *see* shows that he had made the experiment.

---

* John Donne, "The First Anniversary," reprinted in several modern editions of Donne's poems.

† 1 cubit = about 20 inches; 100 cubits happens to be nearly the height of the Tower of Pisa. The cubit is a measure derived from the length of the human forearm (Latin *cubitum* = elbow, from *cumbere* = to recline).

"*Sagredo:* But I, Simplicio, who have made the test, can assure you that a cannon ball weighing one or two hundred pounds, or even more, will not reach the ground by as much as a span* ahead of a musket ball weighing only half a pound, provided both are dropped from a height of 200 cubits."

Here, perhaps, one might have expected to find a detailed report on an experiment done by Galileo or one of his colleagues. Instead, Galileo presents us with a thought experiment—an analysis of what would happen in an imaginary experiment. Galileo also ironically uses Aristotle's own method of logical reasoning to attack Aristotle's theory of motion (just as he had done in his proof of the law of inertia, discussed in Section 5.3):

"*Salviati:* But, even without further experiment, it is possible to prove clearly, by means of a short and conclusive argument, that a heavier body does not move more rapidly than a lighter one provided both bodies are of the same material and in short such as those mentioned by Aristotle. But tell me, Simplicio, whether you admit that each falling body acquires a definite speed fixed by nature, a velocity which cannot be increased or diminished except by the use of force [*violenza*] or resistance.

"*Simplicio:* There can be no doubt but that one and the same body moving in a single medium has a fixed velocity which is determined by nature and which cannot be increased except by the addition of momentum [*impeto*] or diminished except by some resistance which retards it.

"*Salviati:* If then we take two bodies whose natural speeds are different, it is clear that on uniting the two, the more rapid one will be partly retarded by the slower, and the slower will be somewhat hastened by the swifter. Do you not agree with me in this opinion?

---

*    1 span = 9 inches (distance from tip of thumb to tip of little finger when extended).

"*Simplicio:* You are unquestionably right.

"*Salviati:* But if this is true, and if a large stone moves with a speed of, say, eight, while a smaller moves with a speed of four, then when they are united, the system will move with a speed less than eight; but the two stones when tied together make a stone larger than that which before moved with a speed of eight. Hence the heavier body moves with less speed than the lighter; an effect which is contrary to your supposition. Thus you see how, from your assumption that the heavier body moves more rapidly than the lighter one, I infer that the heavier body moves more slowly.

"*Simplicio:* I am all at sea.... This is, indeed, quite beyond my comprehension."

As Simplicio retreats in confusion, Salviati presses forward with the argument, showing that it is self-contradictory to assume that an object would fall faster if its weight were increased by a small amount. Simplicio cannot refute Salviati's logic. But, on the other hand, both Aristotle's book and his own casual observations tell him that a heavy object *does*, at least to some extent, fall faster than a light object:

"*Simplicio:* Your discussion is really admirable; yet I do not find it easy to believe that a bird-shot falls as swiftly as a cannon ball.

"*Salviati:* Why not say a grain of sand as rapidly as a grindstone? But, Simplicio, I trust you will not follow the example of many others who divert the discussion from its main intent and fasten upon some statement of mine that lacks a hairsbreadth of the truth, and under this hair hide the fault of another which is as big as a ship's cable. Aristotle says that 'an iron ball of one hundred pounds falling from a height of one hundred cubits reaches the ground before a one-pound ball has fallen a single cubit.' I say that they arrive at the same time. You find, on making the experiment, that the larger outstrips the smaller by two fingerbreadths.... Now you would not hide behind these two fingers the 99 cubits of Aristotle, nor would you mention my small error and at the same time pass over in silence his very large one."*

This is as clear a statement as we would wish of the idea that a naïve first glance at natural events is not at all a sufficient basis for a physical theory; different freely falling bodies in air indeed do *not* arrive at the very same instant at the foot of the tower, but this, on further thought, turns out to be far less significant than the fact that they do *almost* arrive at the same instant. In a vacuum the fall would be equal for both objects. Attention to this last point, a point which is actually far removed from the immediate sense impressions gained in this experiment, is suggestive and fruitful, because it regards the failure of equal arrivals in fall through air as a minor discrepancy instead of a major truth as an experimental circumstance explainable by air friction. And indeed, when the air pump was invented soon after the time of Galileo, this hypothesis was confirmed by the observation of the truly simultaneous fall of a light feather and a heavy coin inside an evacuated glass tube.

As Galileo reveals his attitude toward experiments, we are reminded of the old quip that "science has grown almost more by what it has learned to ignore than by what it has had to take into account." Here again we meet the recurring theme in science: Regard observable events with a penetrating eye to search behind the immediate confusion of appearances for an underlying simplicity and mathematical lawfulness. In this particular case, it was easier first to sweep away a contrary hypothesis (Aristotle's postulate that speed of fall depends on weight) by reasoning in a thought experiment, than to bother with extensive experiments that could never be completely convincing. To arrive at the correct result, it would

---

* This and the preceding quotations are taken from the translation by Henry Crew and Alfonso de Salvio, originally published in 1914 and now available as a paperback reprint by Dover Publications; see pages 62–65.

have been futile to rely on the observational methods available at the time; instead, everything depended on being able to "think away" the air and its effect on free fall.

This is easy enough for us who know of pumps and the properties of gases; but it was at that time conceptually almost impossible for several reasons —some evident, but some quite subtle. For example, Aristotle, who had examined the question of the existence of a vacuum, had denied its possibility as contrary not only to experience but to reason. One of his arguments was: Where there is a vacuum, there is *nothing*. If nothing remains, he continued, then also space itself and the physical laws can no longer exist in this vacuum; for how could a stone, for example, know which way is up or down except in relation to and through the influence of a medium?

The answer depended on dissociating the words *vacuum* and *nothingness*, and so preventing *space* from becoming annihilated merely because air has been withdrawn. But though Galileo found it possible to do this in thought, Descartes and his followers continued to reject the possible existence of a vacuum, and we may believe that other scientists eventually accepted it mainly on the basis of subsequent *experimental demonstrations* that bodies could fall, and that heat, light, magnetism, etc., could be propagated through vessels from which air had been withdrawn.

## 7.4 GALILEO'S STUDY OF ACCELERATED MOTION

Others had known before Galileo that the Aristotelians were wrong about free fall, but it is to his credit that he proceeded to discover the details of the correct description of this motion and to make it part of a more general system of mechanics. The task he set himself in the *Two New Sciences* was to invent concepts, methods of calculation and measurement, and so forth, to arrive at a description of the motion of objects in rigorous mathematical form.

In the following excerpts we must not lose sight of the main plan. First, Galileo discusses the mathematics of a *possible*, simple type of motion, defined as uniformly accelerated. Then he proposes the hypothesis that the motion of free fall occurring in nature is actually uniformly accelerated. However, it is not possible to test this hypothesis directly by experiment with the instruments available in Galileo's time. He therefore argues that the hypothesis will be confirmed if it is successful in describing another kind of motion closely related to free fall: the rolling of a ball down an inclined plane. (In modern terms, the same force of gravity will act on the ball whether it is dropped vertically downward or allowed to roll down an inclined plane, but in the second case the actual motion is slowed down so that it can be more easily measured.) Finally, he describes an experimental test of quantitative conclusions derived from his theory of accelerated motion, as applied to the particular experimental situation.

Galileo introduces the discussion with these remarks:

"My purpose is to set forth a very new science dealing with a very ancient subject. There is, in nature, perhaps nothing older than motion, concerning which the books written by philosophers are neither few nor small; nevertheless I have discovered some properties of it that are worth knowing and that have not hitherto been either observed or demonstrated. Some superficial observations have been made, as, for instance, that the free motion of a heavy falling body is continuously accelerated; but to just what extent this acceleration occurs has not yet been announced;...

"It has been observed that missiles and projectiles describe a curved path of some sort; however, no one has pointed out the fact that this path is a parabola. But this and other facts, not few in number or less worth knowing, I have succeeded in proving; *and what I consider more important*, there have been opened up to this vast and most excellent science, of which my work is merely the beginning, ways and means by which other minds more acute than mine will explore its remote corners."

There follows first a very thorough and searching discussion of uniform (unaccelerated) motion, along the lines indicated in our previous chapter. Then begins the section on naturally accelerated motion:

"We shall consider now naturally accelerated motion, such as that of falling bodies.

"First of all it seems desirable to find and explain a definition that best fits natural phenomena. Although certainly it is quite permissible to invent an arbitrary type of motion and study its course (as, indeed, some have proposed to themselves helices and conchoids as the paths described in certain motions—it being granted that nature does not make use of such—and have very commendably deduced the properties of these from their definitions), we however have decided to consider the phenomena of freely falling heavy bodies with an acceleration such as actually occurs in nature, and to make our definition of accelerated motion exhibit the essential features of this type of natural accelerated motion. And this, at last, after repeated efforts we trust we have succeeded in doing. In this belief we are confirmed mainly by the consideration that experimental results are seen to agree with and exactly correspond with those properties which have been described by us.

"Finally, in the investigation of naturally accelerated motion we have been led, as if by the hand, by careful observation of the usage and precept of nature herself in all her other actions, in which she is accustomed to employ the simplest and easiest means ...

"... When, therefore, I observe a stone initially at rest falling from an elevated position and continually acquiring new increments of velocity, why should I not believe that such increases take place in a manner which is exceedingly simple and fairly easily apprehended by everybody? If now we examine the matter carefully, we find no addition or increment more simple than that which repeats itself always in the same manner. This we readily understand when we consider the intimate relationship between time and motion; for just as uniform motion is defined by and thought of in terms of equal time intervals and equal distances (thus we call a motion uniform when equal distances are traversed during equal time intervals), so also we may, in similar manner, by thinking in terms of equal time intervals, conceive additions of velocity as taking place without complication..., hence the definition of the motion which we are about to discuss may be stated as follows:

"*A body is said to be uniformly accelerated when, starting from rest, it acquires equal increments of velocity during equal time intervals.*"

In the preceding passage Galileo has been speaking for himself, as it were, in the role of an "Academician" who has written a treatise on mechanics which the characters in the dialogue are discussing. He has introduced what is in fact the modern definition of uniformly accelerated motion (and of acceleration itself): Just as velocity was defined as the change in distance divided by change in time,

$$v = \frac{\Delta s}{\Delta t},$$

so will acceleration be defined as change in velocity divided by change in time,

$$a = \frac{\Delta v}{\Delta t},$$

and a uniformly accelerated motion is simply one for which $a$ is constant for any time interval. But from his own earlier struggles with the concept of acceleration, Galileo knows that it is far from obvious that this is the only possible definition. He therefore inserts at this point a critical discussion which gives us considerable insight into the problem of constructing a reasonable mathematical description of nature:

"*Sagredo:* Although I can offer no rational objection to this or indeed to any other definition, devised by any author whosoever, since all definitions are arbitrary, I may nevertheless without offense be allowed to doubt whether such a definition as the foregoing, established in an abstract manner, corresponds to and describes that kind of accelerated motion which we meet in nature in the case of freely falling bodies..."

The discussion now turns toward the only correct answer to this excellent question, namely, that Galileo's "arbitrary" definition of acceleration happens to be most useful to describe the experimental facts of real, observable motion. But there is first a significant little excursion when Sagredo proposes:

"From these considerations perhaps we can obtain an answer to a question that has been argued by philosophers, namely, what is the *cause* of the acceleration of the natural motion of heavy bodies..."

Salviati sternly turns away this persistent preoccupation of the previous two millennia with a simple and modern type of statement which, roughly speaking, directs us to ask first about the "how" instead of the "why" of motion, not to waste energy on a theory of motion while the *descriptive law* of motion is yet unknown:

"*Salviati:* The present does not seem to be the proper time to investigate the cause of the acceleration of natural motion concerning which various opinions have been expressed by various philosophers, some explaining it by attraction to the center, others to repulsion between the very small parts of the body, while still others attribute it to a certain stress in the surrounding medium which closes in behind the falling body and drives it from one of its positions to another. Now, all these fantasies, and others too, ought to be examined; but it is not really worth while. At present it is the purpose of our Author [Galileo] merely to investigate and to demonstrate some of the properties of accelerated motion (whatever the cause of this acceleration may be)."

Galileo then scrutinizes his decision that for uniformly accelerated motion he should regard the increase in velocity proportional to time elapsed rather than distance covered, a proposition which equally well fulfills his requirement of simplicity of relationships.    He admits having at one time entertained the latter alternative, but shows by a set of plausible experiments which he asks us to imagine (in a set of *thought experiments*, one of the very fruitful devices in the scientist's kit of methods) that such a hypothetical motion does not correspond to actual free fall.* In modern terms, a motion for which $\Delta v \propto \Delta s$ may arbitrarily be *called* uniformly accelerated, but in this case the word acceleration cannot mean $a = \Delta v/\Delta t$, and cannot be the same constant which we have postulated to be the constant during actual free fall.

Having disposed of this alternative, Galileo might now have produced a set of experimental data to show that in free fall, say for a stone dropped from a high tower, the measured quantity $\Delta v/\Delta t$ does indeed remain constant for various time intervals $\Delta t$ during the descent. This would have settled everything.  But consider the experimental troubles involved in such a direct attack!  Today we might use high-speed motion picture equipment to record the successive positions of the falling stone, from which a graph of distance vs. time could be constructed.  From this graph one could calculate the necessary data for an instantaneous velocity vs. time graph, and we could then check the hypothesis that $\Delta v/\Delta t$ is constant throughout the motion.  But in Galileo's time, not even a good clock for timing rapid motions was available.

Experiments on fast free fall being impossible, he now characteristically turned to a more easily testable consequence of his hypothesis, "the truth of which will be established when we find that the inferences from it correspond to and agree exactly with experiment."    First he convinced himself analytically that a ball rolling down a smooth incline obeys the same rules which apply to motion in free fall, that it is in short a "diluted" or sloweddown case of that motion, whatever its law. If the ball is found to move with constant acceleration, so must also a freely falling body (though the numerical value of the acceleration will of course be different). His attention therefore turned finally to a simple experiment.  (Since this experiment is often done, with modern apparatus, in elementary physics courses, you may have an opportunity to judge for yourself whether Galileo could have attained the accuracy he claimed.)

"We took a piece of wooden scantling,† about 12 cubits long, half a cubit wide, and three fingerbreadths thick.  In its top edge we cut a straight channel a little more than one finger in breadth; this groove was made smooth by lining it with parchment, polished as smooth as possible, to facilitate the rolling in it of a smooth and very round ball made of the hardest bronze.  Having placed the scantling in a sloping position by raising one end some one or two cubits above the other, we let the ball roll down the channel, noting, in a manner presently to be described, the time required for the descent.  We repeated this experiment more than once in order to be sure of the time of descent and found that the deviation between two observations never exceeded one-tenth of a pulse beat. Having performed this operation until assured of its reliability, we now let the ball roll down only one-quarter the length of the channel; and having measured the time of its descent, we found it to be precisely one-half of the former. Next we tried other distances, comparing the time for the whole

---

*    His argument at this point, though correct in its conclusion, is erroneous in some details.  A thorough discussion may be found in an article by Stillman Drake, in the June 1970 issue of the *British Journal for the History of Science.*

†    A beam of small cross section, used by carpenters.

length with that for the half, or for two-thirds, or for three-fourths, or indeed for any fraction. In such experiments, repeated a full hundred times, we always found that the distances traversed were to each other as the squares of the times, and this was true for any inclination of the ... channel along which we rolled the ball . . . .

"For the measurement of time, we employed a large vessel of water placed in an elevated position; to the bottom of this vessel was soldered a pipe of small diameter giving a thin jet of water, which we collected in a small cup during the time of each descent, whether for the whole length of the channel or for a part of its length; the water thus collected was weighed on a very accurate balance; the differences and ratios of these weights gave us the differences and ratios of the time intervals, and this with such accuracy that, although the operation was repeated many, many times, there was no appreciable discrepancy in the results."

Do not fail to notice the further ingenious twist in the argument. Even in these inclined-plane experiments it was not possible to test directly whether $\Delta v/\Delta t$ is constant, because this would have demanded direct measurements of instantaneous velocities, whereas Galileo could measure directly only distances and time intervals. It is for just this reason that Galileo devoted a considerable portion of the *Two New Sciences* to deriving various relations between distance, time, speed, and acceleration, equivalent to the equations we have already presented in Section 6.7. In particular, he was able to use a theorem stating that *in uniformly accelerated motion starting from rest, the distance covered is proportional to the square of the time of descent* (Eq. 6.5). This was precisely the result Galileo claimed to have found in his inclined-plane experiment, and therefore he concluded that the motion is indeed uniformly accelerated.

It is easy to criticize Galileo's account of his experiment, on the grounds that the results may not be applicable to free fall. For example, if the plane (or scantling) is tilted more steeply toward the vertical, the ball starts to slip as well as roll, and it is not evident that one can extrapolate the values of the acceleration found for various angles of inclination to the free-fall value for vertical planes. For that matter, Galileo does not even report a numerical value for acceleration, either for the rolling ball or for free fall.*

Although the complete experimental verification of Galileo's view of the way bodies move would have to come later, the important thing was that it had been clearly stated, bolstered by plausible reasoning and demonstrations, and put into forms that could be quantitatively tested. Let us now recapitulate these results.

1) Any body set into motion on a horizontal frictionless plane will continue to move indefinitely at the same speed (law of inertia).

2) In free fall through a vacuum, all objects—of whatever weight, size, or constitution—will fall a given distance in the same time.

3) The motion of an object in free fall, or while rolling down an inclined plane, is uniformly accelerated, i.e., it gains equal increments of speed in equal times.

These laws are not enough by themselves to constitute a complete science of motion, but they are certainly a good start. As Sagredo said to Salviati in concluding the discussion of the "Third Day" of the *Two New Sciences*:

". . . The theorems set forth in this brief discussion, if they come into the hands of other investigators, will continually lead to wonderful new knowledge. It is conceivable that in such a manner a worthy treatment may be gradually extended to all the realms of nature . . . .

---

* A value roughly equivalent to the modern value of $g = 9.8$ m/sec² for free fall was first obtained by Christian Huygens.

**Figure 7.1**

acceleration in free fall $g$ as the vertical height of the plane $h$ to the length of the plane $l$, or, in short, that $a = g \sin \theta$ (Fig. 7.1).

**Problem 7.4.** Summarize what specific methods of attack and of solution you may have discerned in this discussion of Galileo's work on freely falling bodies.

"During this long and laborious day, I have enjoyed these simple theorems more than their proofs, many of which, for their complete comprehension, would require more than an hour each; this study, if you will be good enough to leave the book in my hands, is one which I mean to take up at my leisure after we have read the remaining portion which deals with the motion of projectiles; and this if agreeable to you we shall take up tomorrow.

"*Salviati:* I shall not fail to be with you."

## PROBLEMS

**Problem 7.1.** Read through Galileo's "Third Day" in the *Two New Sciences* in order to find out what his arguments were for thinking that the laws of motion on an inclined plane are the same as for free fall. (For a translation use that of H. Crew and A. de Salvio.)

**Problem 7.2.** List the steps outlined in this chapter by which Galileo progressed from his first definition of what he means by uniformly accelerated motion to his final confirmation that his definition coincides with the actually observable motion of free fall. What limitations and idealizations entered into the argument?

**Problem 7.3.** Read through Galileo's "Third Day" to Theorem IV. Then restate as concisely as possible the argument that the acceleration $a$ of an object sliding on a smooth inclined plane is to the

## RECOMMENDED FOR FURTHER READING

C. B. Boyer, "Aristotle's Physics," *Scientific American*, pages 48–51 (May 1950)

E. A. Burtt, *The Metaphysical Foundations of Modern Physical Science*, Chapter III

Galileo Galilei, extract from *Two New Sciences*, translated in Hurd and Kipling, *The Origins and Growth of Physical Science*, Vol. 1, pages 163–177

L. Geymonat, *Galileo Galilei, a Biography and Inquiry into his Philosophy of Science*, translated from the Italian with notes and appendix by S. Drake, foreword by G. de Santillana; New York: McGraw-Hill, 1965; paperback

E. C. Kemble, *Physical Science*, Chapter 6

M. Wertheimer, *Productive Thinking*, New York: Harper, 1945. The thought processes which may have guided Galileo in his development of the laws of motion are discussed from the point of view of Gestalt psychology.

A. N. Whitehead, *Science and the Modern World*, New York: Macmillan Co., 1925; Mentor paperback reprint, Chapter I

## SOURCES, INTERPRETATIONS, AND REFERENCE WORKS

Aristotle's *Physics*, translated by H. G. Apostle. Bloomington, Indiana: Indiana University Press, 1969; paperback

M. Clagett, *The Science of Mechanics in the Middle Ages,* Madison, Wisconsin: University of Wisconsin Press, 1959

A. C. Crombie, *Medieval and Early Modern Science*, Garden City, N.Y.: Doubleday Anchor Books, paperback, Vol. II, pages 121–166

E. J. Dijksterhuis, *The Mechanization of the World Picture*, translated by C. Dikshoorn, New York: Oxford University Press, 1961; paperback

S. Drake and I. E. Drabkin, *Mechanics in Sixteenth Century Italy: Selections from Tartaglia, Benedetti, Guido Ubaldo and Galileo*, translated and annotated, Madison, Wisconsin: University of Wisconsin Press, 1969

S. Drake, "Uniform acceleration, space, and time," *British Journal for the History of Science*, Vol. 5, pages 21–43 (1970)

S. Drake, *Galileo Studies, Personality, Tradition, and Revolution*, Ann Arbor, Michigan: University of Michigan Press, 1970

Galileo Galilei, *On Motion* and *On Mechanics*, translated by I. E. Drabkin and S. Drake, Madison, Wisconsin: University of Wisconsin Press, 1960

Galileo Galilei, *Dialogues Concerning Two New Sciences* (First and Third Days), translated by H. Crew and A. de Salvio, New York: Dover Publications, 1952

E. Grant, "Bradwardine and Galileo: equality of velocities in the void," *Archive for History of Exact Sciences*, Vol. 2, pages 344–364 (1962–66)

W. C. Humphreys, "Galileo, falling bodies, and inclined planes. An attempt at reconstructing Galileo's discovery of the law of squares," *British Journal for the History of Science*, Vol. 3, pages 225–244 (1967)

A. Koslow, *The Changeless Order*, New York: Braziller, 1967 (extracts from the writings of Plato, Aristotle, Descartes, and Galileo), pages 1–63

A. Koyré, *Documentary History of the Problem of Fall from Kepler to Newton*, Philadelphia: American Philosophical Society, 1955

A. Koyré, *Metaphysics and Measurement*, Cambridge: Harvard University Press, 1968

Ernst Mach, *The Science of Mechanics*, translated by T. J. McCormack from German, LaSalle, Illinois: Open Court Publishing Co., 6th edition 1960; paperback (the first German edition was published in 1883) pages 151–181

W. Wallace, "The enigma of Domingo de Soto: Uniformiter difformus and falling bodies in late medieval physics," *Isis*, Vol. 59, pages 384–401 (1968)

M. W. Wartofsky, "All fall down: the development of the concept of motion from Aristotle to Galileo," pages 419–473 in his *Conceptual Foundations of Scientific Thought*, New York: Macmillan, 1968

# Chapter 8

# PROJECTILE MOTION

Turning now to the more general motion of projectiles, we leave the relatively simple case of movements along a *straight line* only and expand our methods to deal with motion *in a plane*. Our whole understanding of this field, a typical and historically important one, will hinge on a far-reaching discovery: The observed motion of a projectile may be thought of as the result of two *separate* motions, combined and followed *simultaneously* by the projectile, the one component of motion being an unchanging, unaccelerated horizontal translation, the other component being a vertical, accelerating motion obeying the laws of free fall. Furthermore, these two components do not impede or interfere with each other; on the contrary, the resultant at any moment is the simple effect of a superposition of the two individual components.

Again it was Galileo who perfected this subject, in the section "The Fourth Day" immediately following the excerpts from the *Two New Sciences* which were presented in our last chapter. Let us refashion his original arguments to yield somewhat more general conclusions, and to illustrate by one painstaking and searching inquiry how an understanding of a complex problem may grow. In later chapters it will usually be necessary that you

supply some of the intermediate steps of a problem, therefore this effort now will be a good investment. Thus, if Chapter 7 represented a historically inclined chapter, this one may be called a factually directed one.

## 8.1 PROJECTILE WITH INITIAL HORIZONTAL MOTION

We may start with two simple but perhaps surprising experiments.

a) If we watch a plane in steady horizontal flight drop a small, heavy parcel, we realize that (except for air friction) the parcel *remains directly below the plane* while, of course, dropping closer and closer to the ground. This is represented in Fig. 8.1. If we watched this event from a balloon, very high up and directly above this region, we should, of course, see only the horizontal part (component) of the motion; and if from there we now could see the parcel at all, we should think it traveled directly along with the plane, instead of falling away from it. The clear implication is that the horizontal component of the motion of the parcel remains what it was at the moment of release from the plane (unchanged, as suggested by the law of inertia),

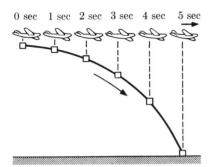

0 sec  1 sec  2 sec  3 sec  4 sec  5 sec

**Fig. 8.1**  Successive positions of a parcel dropped from a plane.

**Fig. 8.2**  The horizontal and vertical components of motion are independent.

even though there is superposed on it the other, ever-increasing component of vertical velocity.

b) In the second experiment we shall place two similar small spheres at the edge of a table (cf. Fig. 8.2). At the very moment that one is given a strong horizontal push, so that it flies off rapidly and makes a wide, curved trajectory,* the other is touched only very gently and therefore drops straight down to the floor. Which will land first?

The experimental fact is that they both land together, no matter what the initial horizontal speed of the pushed ball; furthermore, the balls remain at equal levels throughout their fall, though of course they separate in the horizontal dimension, one having a large (and by the previous experiment, presumably undiminishing) horizontal velocity and the other having none. The conclusion here is that the vertical component of the motion is quite independent of any additional horizontal movement.

The joint result of both experiments is that motion in a plane can be resolved into two com-

ponents, the horizontal and vertical, and in the case of projectiles (parcel, spheres, etc.) moving near the earth, these two components of motion are independent. This allows us to answer some inquiries into the motion of our projectiles, the simplest being "How far away from the starting point does the parcel or the second sphere land?" During the time taken for the fall or flight, say $t$ seconds, the horizontal, unchanging component of motion will transport the projectile a distance equal to the initial velocity imparted in the forward direction at the moment of release multiplied by the time $t$; simultaneously the growing vertical component of motion contributes a downward displacement by an amount corresponding to the distance of free fall during that time $t$ under the influence of the gravitational acceleration, $g$. If we agree to give the subscript $x$ to all horizontal components, and $y$ to all vertical components, as indicated in Fig. 8.3, then we may rewrite the previous sentence more concisely:

$$s_x = v_{0x}t \quad \text{or} \quad s_x = v_x t, \tag{8.1}$$

because $v_{0x} = v_x$, the horizontal motion being by assumption unaccelerated; and second,

$$s_y = \tfrac{1}{2}gt^2, \tag{8.2}$$

---

* By trajectory we mean here the path followed by a projectile under the influence of gravity.

**Fig. 8.3** Trajectory and total displacement of a projectile.

**Fig. 8.4** Graph of the equation $y = $ (positive constant) $\cdot x^2$.

because $v_{0y}$ was here assumed to be zero, the initial motion at the time of release having been altogether in the horizontal direction.

If $t$ is the actual total time of flight, $s_x$ so calculated is the so-called "range" of the projectile and $s_y$ is the distance of drop, but our question above inquired about neither $s_x$ nor $s_y$. If we still wish to find the total displacement from the starting point (a quantity we might call $s$), a glance at Fig. 8.3 will at once solve our problem. By the Pythagorean theorem, for displacement in a plane,

$$s = \sqrt{s_x^2 + s_y^2}. \qquad (8.3)$$

We may now substitute, and find that

$$s = \sqrt{(v_{0x}t)^2 + (\tfrac{1}{2}gt^2)^2}.$$

In order to produce a numerical answer, we must therefore be given the initial velocity and time of flight, quite apart from the value of $g$, which in all our work will be assumed to be numerically equal to 9.8 m/sec$^2$ or 32.2 ft/sec$^2$.

Furthermore, the angle which direction $s$ makes with the horizontal, $\beta$, is clearly given* by $\tan \beta = s_y/s_x$. It is at this angle that a "sight" will

---
* See Appendix V for a summary of trigonometric relations.

have to be set to ascertain the proper instant of release of this projectile from the plane. It would be a more difficult problem, and one which we shall not try to solve, to discover not the total *displacement* but the length of the actual path along the curved trajectory. It happens to be a fairly unimportant question to us, the range $s_x$ and the distance $s_y$ being the more sought-after quantities; yet we may deduce an interesting result about the shape of the path. From Eqs. (8.1) and (8.2),

$$s_x = v_{0x}t, \qquad \therefore\ t = \frac{s_x}{v_{0x}} \qquad \text{and} \qquad t^2 = \frac{s_x^2}{v_{0x}^2}.$$

But

$$s_y = \tfrac{1}{2}gt^2 = \tfrac{1}{2}g \left( \frac{s_x^2}{v_{0x}^2} \right),$$

which may be written

$$s_y = \left( \frac{g}{2v_{0x}^2} \right) s_x^2. \qquad (8.4)$$

Since the quantities in parentheses are constant throughout a particular projectile motion, it appears that $s_y$ is proportional to $s_x^2$; hence $s_y = ks_x^2$ where $k = (g/2v_{0x}^2)$. But this equation, according to the analytic geometry of Descartes, corresponds simply to a parabola, as shown in Fig. 8.4.

**Problem 8.1.** Plot on graph paper the equation $y = (-\frac{1}{3})x^2$. (a) Give $x$ only $+$ values; (b) give $x$ both $+$ and $-$ values. Recall the useful convention that to distances above the origin or to the right of the origin, positive values are assigned; for distances below or to the left of the origin, negative values.

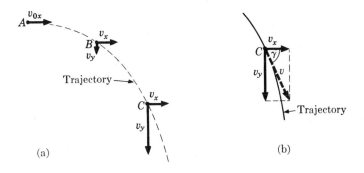

**Fig. 8.5**  The actual velocity $v$ can be compounded from its components $v_x$ and $v_y$.

After Galileo had deduced the parabolic nature of the trajectory (by an original and similar argument), projectile motion immediately became much simpler to understand; for the geometrical properties of the parabola had long been established by the mathematicians in their disinterested pursuit of abstract geometrical worlds. We find here a first clue to three important facts of life in science: (1) If we can express phenomena quantitatively and cast the relation between observables into equation form, then we can grasp the phenomenon at one glance, manipulate it by the laws of mathematics, and so open the way to the discovery of new truths concerning this set of phenomena. For example, having found that our trajectories are parabolic, we could confidently calculate, if required, the length of the actual path along the curve by means of some formula proposed to us by a mathematician, one who may never even have seen an actual projectile motion, but who has studied parabolas thoroughly. (2) Consequently there is always an imperative need for a well-developed system of pure mathematics from which the physicist may draw. (3) We can see why the physical scientist always tries to cast his problem into such a form that established approaches, procedures, or tricks from another branch of science or from mathematics will aid in the solution. As an example of the last point, just as Galileo applied the mathematics of parabolas to actual projectile motions, so does the modern acoustical engineer solve his problems by means of the mathematical schemes developed quite independently by electrical engineers for their own very complex field. Whatever the methods of science may be, they have shown

themselves to be transferable from one specialty to the other in a remarkable and fruitful way.

Yet another question to be asked about our projectile motion is this: What is the actual velocity of the moving object at some time $t$ after release? We may be prompted by our initial two experiments and by the discussion leading to Eq. (8.3) to represent the two observed components of the actual velocity at several portions along the path, as has been done in Fig. 8.5(a). At each of the three selected points the horizontal velocity component $v_{0x} = v_x = $ constant, but the vertical component grows linearly from $v_{0y} = 0$ to a larger and larger value, in accordance with the law for falling bodies, $v_y = v_{0y} + gt$, which here becomes simply $v_y = gt$. Our experiments showed us that $v_x$ and $v_y$ will not disturb each other, but this does not tell us yet what is the actual velocity $v$ at any chosen point on the trajectory. Now it was Galileo's very important discovery that the total effect of both velocity components, i.e., the actual velocity $v$, could be compounded from the two components $v_x$ and $v_y$ by the very same simple scheme which enabled us to find $s$ when $s_x$ and $s_y$ were known. In Fig. 8.5(b), the velocity at point $C$ is constructed by completing the parallelogram (in this case simply a rectangle, since $v_x$ and $v_y$ are at 90°) formed by the arrows corresponding to $v_x$ and $v_y$; the total, the "real" velocity, is then indicated by the diagonal $v$, which at every point is tangent to the momentary path of the projectile, and which has the value

$$v = \sqrt{v_x^2 + v_y^2}. \tag{8.5}$$

Consequently, the momentary direction of motion at point $C$ can be specified in terms of the angle $\gamma$ between $v$ and the horizontal direction, and is given by

$$\tan \gamma = \frac{v_y}{v_x}. \tag{8.6}$$

This, to repeat, is a postulate that is *checked by experiment*. We may cast it into formal language: The instantaneous velocity of a projectile, and its direction of motion at any instant, may be obtained by *superposing the two independent components $v_x$ and $v_y$ in a simple parallelogram construction*, where $v$, the resultant, is given in both direction and magnitude by the diagonal, as shown in Fig. 8.5(b).

This *principle of the superposition of velocity components* is so simple as to seem now self-evident, but that is deceptive. The *only* justification for postulating the crucial equations (8.5) and (8.6) is that they are essentially experimentally confirmed discoveries. In one form or another the superposition principles are such simplifying devices that they may seem to have been too much to hope for; yet natural phenomena graciously do exhibit such features at times, and it is practically unthinkable that physics could have embarked on its great journey of vigorous discovery from Galileo on without his having intuited many such "simple" fundamentals.

**Example 1.** A stone is dropped from a plane flying level with the ground at a speed of 200 m/sec. If the stone needs 5 sec to reach the ground, how high did the plane fly, what is the range of this projectile, what is its total displacement, and what is its final speed just before striking? As we now translate

phrase by phrase, we write: $v_{0y} = 0$, $v_{0x} = 200$ m/sec, $t = 5$ sec, $s_y = ?$ $s_x = ?$ $s = ?$ $v = ?$ We must assume that $g = 9.8$ m/sec$^2$ (often, 10 m/sec$^2$ will be accurate enough), and that air friction on the projectile is so small that its horizontal motion is not affected and its vertical motion remains uniformly accelerated. Both these assumptions are, incidentally, experimentally found to hold quite well for the first few seconds of motion, but increasingly less well thereafter.

For a solution we find $s_y = \frac{1}{2}gt^2 = 125$ m, $s_x = v_{0x}t = 1000$ m, $s = \sqrt{125^2 + 1000^2}$ m, and so forth.

## 8.2  INTRODUCTION TO VECTORS

The foregoing discussion has exhausted every worthwhile question we may put at present about projectile motion, *provided the motion started with a straight horizontal velocity* and no vertical component of velocity at all. Now we shall go on to the general case, i.e., to projectile motion with all conceivable initial conditions of velocity, not neglecting projectiles shot from a gun pointing at any angle above or below the horizontal line. Our interest here is not so much a practical one—those aspects of science which men abuse to direct weapons against one another are surely the most dismal. Instead, we shall regard this exposition as a typical formulation of a large and successful *general scheme to handle all possible specific instances* of a problem. Furthermore, in this lengthy effort we shall of necessity have to develop or invent some new physical concepts of permanent usefulness.

Imagine now the progress of a projectile which leaves the muzzle of a gun with an initial velocity $v_0$ and at an angle $\theta$ with the horizontal [Fig. 8.6(a)], rises to some height $h$, and then drops into a valley below the horizon. The horizontal and vertical components of the initial velocity, $v_{0x}$ and $v_{0y}$, can be computed by *resolution* from $v_0$ and the angle $\theta$, by means of the same trigonometric

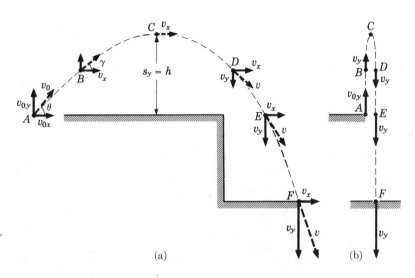

(a)

(b)                    **Fig. 8.6**  Generalized projectile motion.

approach which led us to the converse proportion, the *composition* of $v$ from $v_x$ and $v_y$.

Horizontal component of $v_0 = v_{0x} = v_0 \cos \theta$,

Vertical component of $v_0 = v_{0y} = v_0 \sin \theta$,  (8.7)

As before, $v_{0x}$ remains also the value of the horizontal component $v_x$ at all later points, including the six positions shown. ($F$ refers to the state of affairs just *before* the projectile hits.) On the other hand, $v_{0y}$ at point $A$ is subsequently diminished as the projectile rises against the opposing, downward acceleration of gravity, until, at point $C$, $v_y = 0$. From then on, $v_y$ no longer points upward, but reverses and becomes larger and larger in the previously examined manner. *So far as this vertical component alone is concerned*, its history would have been exactly the same if $v_{0x}$ had been zero, i.e., if we had been thinking about a stone thrown vertically upward at the edge of a cliff with that same initial value of $v_{0y}$ [Fig. 8.6(b)].

The question now arises how to calculate the total actual velocity $v$ of the projectile in Fig. 8.6(a) at some point $B$. We still have faith that at $B$,

$$v_y = \uparrow \; + \; \downarrow gt$$

(at $B$)

$$v_y = \uparrow \; + \; \downarrow gt$$

(at $E$)

**Figure 8.7**

as elsewhere, $v = \sqrt{v_x^2 + v_y^2}$ (and in the absence of faith an experiment would prove the point). $v_x$ at $B$ is still, in magnitude, equal to $v_{0x}$, and therefore equal to $v_0 \cos \theta$. But when we now turn to the vertical component, we find that its initial magnitude has been decreasing between points $A$ and $B$ through constant deceleration due to gravitational pull in a direction opposite to the motion. We may regard $v_y$ at $B$ as a constant speed component $v_{0y}$ diminished by a contrary, growing component of magnitude $gt$, where $t$ is the time needed to reach point $B$ (see Fig. 8.7, "at $B$"). At point $C$, evidently $v_y = 0$, since there the term $gt$ has become exactly

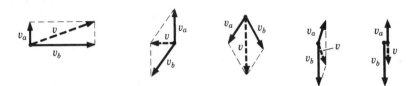

**Fig. 8.8**  Graphical method for adding two velocity vectors.

as large as $v_{0y}$. Beyond that point we may become somewhat uncertain how to calculate $v_y$ from $v_{0y}$, $t$, and $g$, and here we must have recourse to a geometric scheme.

Consider Fig. 8.6(b). Each arrow representing $v_y$ at any one point may itself be thought of as the result of simple addition or subtraction of the length of two other arrows, representing respectively the constant velocity $v_{0y}$ (upward), and the cumulative effect of gravity, $gt$, downward (Fig. 8.7). But this is just a special case of a general scheme of additions of quantities such as displacement (for which we used it in Section 8.1) and velocities, the scheme due to Galileo which we previously employed to compose $v$ from $v_x$ and $v_y$: no matter whether two velocities simultaneously at work on one body are active at right angles to each other (as before), or along the same line (as now while we consider only the *vertical* component), or make any other angle with each other, we can find the resulting total velocity by the so-called parallelogram method of construction. In this construction, the arrows representing the two velocity components are drawn to some scale (e.g., 1 cm = 10 m/sec), placed on paper in correct relation end to end, the parallelogram is completed, and the diagonal is drawn which, pointing in the direction away from the meeting point of the component arrows, gives to the same previous scale the magnitude and direction of the resultant total velocity.

Figure 8.8 shows several such hypothetical cases, $v$ being the resultant of two components $v_a$ and $v_b$. The first of these examples is familiar to us. The last is a special case of the one preceding; the

two components lie in opposite directions along one line and the drawing of a clear parallelogram becomes practically impossible, although the result is obviously that shown. Similarly, if the angle between $v_a$ and $v_b$ in the third sketch were to shrink to zero, the resultant, $v$, would grow to the plain numerical sum of the components. Quantities which permit such a graphic method of addition are called *vectors*; besides displacement and velocity we shall find many other important vector quantities, notably force. In each case the method for vector addition (that is, for finding the resultant of two separate components) is this parallelogram method, based on a simple superposition law.*

### 8.3 THE GENERAL CASE OF PROJECTILE MOTION

We shall soon have ample opportunity to make use of vector addition in other contexts, but for the moment it is enough to consider the scheme in Fig. 8.7 as a restatement of the special case of vector addition shown in the last drawing of Fig. 8.8. Now a simple convention suggests itself which circumvents the unavoidably inaccurate and somewhat messy scheme of drawing these arrows to scale to find the resultant in each case. If we agree to the convention to give *positive* values to quantities represented by an upright arrow, and *negative* values to those oppositely directed, then we can employ the simple equation

$$v_y = v_{0y} + gt \qquad (8.8)$$

---

\*    For further discussion of vectors, see Appendix IX.

in its original form to compute $v_y$ at any time $t$ after the projectile motion starts. A glance at Fig. 8.7 will show qualitatively that this equation, together with that convention, assures the correct calculation of the numerical values for $v$.

**Example 2.** A projectile is shot from a gun tilted at 60° above the horizontal; the muzzle velocity $v_0$ is 1000 m/sec. (a) What is $v_y$ after 20 sec of flight, (b) what is the total velocity $v$ of the projectile after 20 sec of flight, (c) at what angle with the horizontal does it then travel?

*Solution:* (a) $v_y = v_{0y} + gt$, where $v_{0y} = v_0 \sin \theta =$ 870 m/sec (positive value); $gt$ is numerically about 10 (m/sec$^2$) × 20 sec = 200 m/sec, but representing always a *downward directed* quantity, it is by our convention to be used with a negative sign, that is, as $(-200)$ m/sec. Therefore $v_y = 870$ m/sec + $(-200)$ m/sec = 670 m/sec. [The fact that our answer has a positive sign itself means that the velocity is still directed upward; this means we might be now at some position on the trajectory like point $B$ in Fig. 8.6(a).] (b) To compute $v$ by $\sqrt{v_x^2 + v_y^2}$, and then the angle from $\tan \gamma = v_y/v_x$, we now find $v_x$ from $v_x = v_{0x} = v_0 \cos \theta$. (Continue as exercise.)

**Example 3.** How long will the above-mentioned projectile take to reach the highest point? The translation of this question depends on the realization that $v_y = 0$ at the top point of the trajectory, i.e., at point $C$ in Fig. 8.5(a). Then $gt = -v_{0y}$ by Eq. (3.8), or $(-10 \times t)$ m/sec = $-870$ m/sec, and consequently $t = -870/(-10)$ sec $\doteq$ 87 sec. (We enter at once our suspicion that for such conditions the physical realities of air friction undoubtedly make our calculations quite academic.)

**Example 4.** What is $v_y$, the vertical component of velocity of this projectile, after 174 sec?

*Solution:* From Eq. (8.8) we find $v_y = 870$ m/sec + $(-10 \times 174)$ m/sec $\doteq$ 870 m/sec − 1740 m/sec = $-870$ m/sec (negative = downward). In short, after a time as long again as it took to reach the top of the trajectory, the velocity component downward is

numerically equal to but oppositely directed to the initial vertical component. This is pictured at point $E$, Fig. 8.6.

The last two examples make it clear that it might be wise to assign the minus sign permanently to the value for $g$, to write $g = -9.8$ m/sec$^2$ once and for all, and so banish the temptation to change the plus sign in Eq. (8.8) at any time. We shall do this from now on.

But you may have wondered why we did not adopt these conventions before, in Section 8.1, or, since we failed to do so at that time, whether conclusions obtained there are not now erroneous. The answer is that our convention of + and − signs is arbitrary and could be turned around completely, as long as we somehow distinguish in our numerical work between "up" and "down" components. In Section 8.1, where $v_{0y}$ was kept jealously at zero value, only components of one variety, namely "downward," appeared in any problem; hence there was no need to introduce a differentiating sign convention then.

With this treatment of one type of vector components as a background, we need perhaps only point out that the same sign convention, if applied to *displacement* components, will allow us to solve in a similar manner all problems involving the height of a trajectory, the range, etc. After all, displacements also are vectors in the sense that they can be represented by arrows and can be added (composed) or resolved according to the same simple parallelogram scheme which we used for velocities. For example, our use of the Pythagorean theorem in Eq. (8.3) shows this vector aspect of displacements clearly. Another example might be the graphical derivation of $s_y$ at, say, point $B$ in Fig. 8.6, as shown in Fig. 8.9.

$$s_y \uparrow\; = \; \uparrow (v_{0y}t) \; + \; \downarrow (\tfrac{1}{2} gt^2)$$

**Figure 8.9**

Therefore $s_y$ will also attain positive values if represented by an arrow pointing upward, i.e., when the displacement is to a point *above the level of release* of the projectile; and $s_y$ will assume negative values for displacements measured in the other direction. Then we can use in their original form the pertinent equations involving displacement, such as

$$s_y = v_{0y}t + \tfrac{1}{2}gt^2, \tag{8.9}$$

provided again that we substitute in all our calculations for $g$ the value $-9.80$ m/sec$^2$.

**Example 5.** The projectile mentioned in the previous examples falls into a valley 300 m below the level of the gun. How long was it in the air, and what was the range? Translating, we note that the old data are $v_0 = 1000$ m/sec at $60°$, therefore $v_{0y} = 870$ m/sec, $v_{0x} \doteq 500$ m/sec; $g \doteq 10$ m/sec$^2$. To this we add $s_y = -300$ m (a negative sign to indicate that the vertical displacement is to a point *below* the level of release), $t = ?$ $s_x = ?$

Equation (8.9) can be written: $\tfrac{1}{2}gt^2 + v_{0y}t - s_y = 0$, a quadratic equation easily solved for $t$:

$$t = \frac{-(v_{0y}) \pm \sqrt{(v_{0y})^2 - 4(\tfrac{1}{2}g)(-s_y)}}{2(\tfrac{1}{2}g)}.$$

So far, we have not called in our sign convention. Now we may substitute our values with their proper signs to obtain a numerical answer. (Do this.*) Having calculated $t$, we can solve for $s_x$, since it is still true that $s_x = v_{0x}t$.

*To summarize,* we now feel justified in using the following equations of motion in projectile problems, where all symbols have their discussed meaning and the new convention is used to substitute numerical values with correct signs.

---

* At such times it is well to recall that of two possible solutions of such quadratics, one *may* be physically irrelevant or even absurd (for example, if $t$ comes out zero or negative or if the quantity under the square root is negative). What does this reveal about the role of mathematics in physical science?

$$s_y = \left(\frac{v_{0y} + v_y}{2}\right) t \tag{I}$$

$$s_y = v_{0y}t + \tfrac{1}{2}gt^2 \tag{II}$$

$$v_y = v_{0y} + gt \tag{III}$$

$$v_y^2 = v_{0y}^2 + 2gs_y \tag{IV}$$

$$s_x = v_{0x}t = v_x t \tag{V}$$

$$v_{0x} = v_x = v_0 \cos \theta \tag{VI}$$

$$v_{0y} = v_0 \sin \theta \tag{VII}$$

$$v = \sqrt{v_x^2 + v_y^2}, \ \tan \gamma = \frac{v_y}{v_x} \tag{VIII}$$

$$s = \sqrt{s_x^2 + s_y^2}, \ \tan \beta = \frac{s_y}{s_x} \tag{IX}$$

*Conventions regarding sign:*

Vector components in the $+y$ direction are given $+$ values.

Vector components in the $-y$ direction are given $-$ values.

Consequently, $s_y$ has $+$ values if the displacement is above the level of release; $s_y$ has $-$ values if the displacement is below the level of release; and $v_{0y}$ and $v_y$ have $+$ values while the corresponding motion is upward, whereas $v_{0y}$ and $v_y$ have $-$ values while the corresponding motion is downward.

The value of $g$ is taken to be about $-9.80$ m/sec$^2$.

This set is not quite as formidable as it might seem at first glance. We recall that Eq. (I) follows from the definition of average velocity for uniformly accelerated motion, that is, $\bar{v} = (v_{0y} + v_y)/2 = s/t$. Equation (II) is an experimental law (from free-fall experiments), and therefore, so to speak, on a different plane. Equations (III) and (IV) can be derived from I and II. Equation (V) is a second experimental law (i.e., the horizontal component of motion $v_x$ is unaccelerated). The remaining equations (VI) to (IX) simply refer to the mathematical properties of vectors: Eq. (VI) includes

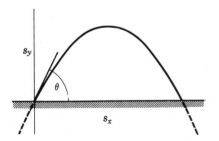

**Fig. 8.10** This curve represents Eq. (8.10).

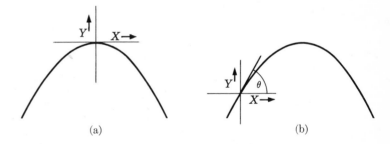

**Figure 8.11**

the rule defining how to find the $x$ component of vector $v_0$, while Eq. (VII) provides for the $y$ component. Equations (VIII) and (IX) give the rule for vector addition (magnitude and direction of resultant, first for the total actual velocity, second for the total actual displacement). Finally, the sign conventions, which quickly become second nature in use, reflect only a commonsensical way of differentiating between addition and subtraction of quantities.

When a practicing physicist has to solve projectile problems, he does not really need to have before him all these defining equations and the conventions. He understands these tacitly; what he will regard as important and essential about ideal projectile motion are Eqs. (II) and (IV), modified as follows:

$$s_y = (v_0 \sin \theta)t + \tfrac{1}{2}gt^2; \qquad s_x = (v_0 \cos \theta)t.$$

He will ordinarily combine these equations by eliminating $t$ between them (i.e., in the left equation, replacing $t$ by $s_x/v_0 \cos \theta$), thus obtaining

$$s_y = (\tan \theta)s_x + \tfrac{1}{2}g \left( \frac{s_x^2}{v_0^2 \cos^2 \theta} \right). \qquad (8.10)$$

This last equation he will consider to be *the* equation of the trajectory *in vacuo*, and he can attack most problems with this formula (see Fig. 8.10). For us it must suffice to realize that the whole

truth about projectile motion, through mathematics, can be so economically expressed and so conveniently handled. Also, we should realize that Eq. (8.10) is of the following form:

$$s_y = (\text{a constant}) \times s_x + (\text{another constant}) \times s_x^2.$$

This once more implies that the *trajectory is parabolic* [no longer simply the right half of the parabola in Fig. 8.4, which applied where $\theta = 0$, whence $\tan \theta = 0$ and $s_y = (\text{constant}) \times s_x^2$ only].

---

**Problem 8.2.** The two curves of Fig. 8.11 differ only in the placement of the origin of the coordinates, so that in (a) $\theta = 0$, in (b) $\theta = 60°$. Copy the two parabolas on a sheet of graph paper and show by computation for several points that the equation is for (a) $Y = (\text{a constant}) \cdot X^2$, for (b) $Y = (\text{a constant}) \cdot X + (\text{another constant}) \cdot X^2$. Once more we see that a knowledge of parabolas is all we should really need to solve practical projectile problems. Nevertheless, at our level it will prove easier for us if we continue to solve our problems by means of Eqs. (I)–(IX) rather than, say, by Eq. (8.10) plus the theory of parabolas.

## 8.4 APPLICATIONS OF THE LAW OF PROJECTILE MOTION

Let us review what we have done. First, starting from some experimental observations, we have

**Figure 8.12**

found the separate equations for the horizontal and the vertical components of a general projectile motion. In particular, the horizontal component is found to be unaccelerated, and the vertical one obeys the laws of motion of uniformly accelerated objects. Second, the total actual motion (displacement $s$ and speed $v$) is obtained in direction and magnitude by a vector addition of the components, following an empirically justified procedure of adding by the parallelogram method. Now comes the reward. We see what a wide variety of projectile problems we can solve with this set of equations (which we may regard as the algebraic expression of *the general law of projectile motion*).

**Example 6.** A gun pointed at an angle $\theta$ above the horizon shoots a projectile with muzzle velocity $v_0$. What is the time $t$ needed for the projectile to return to level ground? That is, given $v_0$, $\theta$, and, of course, $g$, find $t$ if $s_y = 0$ (Fig. 8.12).

*Solution:* Since

$$s_y = v_{0y}t + \tfrac{1}{2}gt^2,  \qquad\text{(II)}$$

$$0 = (v_0 \sin \theta)t + \tfrac{1}{2}gt^2,$$

and

$$t = \frac{-2v_0 \sin \theta}{g}.$$

Since $g$ takes on a negative value, $t$ will come out positive, as it surely must.

**Example 7.** In the previous example, what is the maximum height reached? That is, given $v_0$, $\theta$, $g$, find $s_y \,(\equiv h)$ if $v_y = 0$.

*Solution:* By Eq. (IV), $0 = (v_0 \sin \theta)^2 + 2gs_y$, and

$$s_y \,(\equiv h) = \frac{-(v_0 \sin \theta)^2}{2g}.$$

Again $s_y$ is positive, since $g$ has a negative value.

**Example 8.** In the first example, what is the range, i.e., if $v_0$, $\theta$, and $g$ are given, and $s_y = 0$, what is $s_x$?

*Solution:* $s_x = v_{0x}t = (v_0 \cos \theta)t$. However, since $t$ is not explicitly known, it must first be calculated, and this can be done as shown in Example 1. (This may be called a two-step problem.) By direct substitution, it follows that

$$s_x = \frac{-v_0^2 2 \cos \theta \sin \theta}{g} = \frac{-v_0^2 \sin (2\theta)}{g}.$$

It should be emphasized again that these results would hold strictly only for projectiles moving in a vacuum above a flat earth. The actual effects of air resistance alone may decrease the effective range by 50%. Evidently, for practical applications, we should have to investigate how such effects modify these results through theoretical and empirical corrections.

## 8.5  GALILEO'S CONCLUSIONS

Galileo himself carried his work to the point at which he computed fine tables of ranges and heights of trajectories over level ground for different angles $\theta$. These calculations proved that for a given initial velocity the range was a maximum if $\theta = 45°$, as we can ascertain by inspection of the general equation $s_x = [-v_0^2 (\sin 2\theta)]/g$, for $\sin (2\theta)$ takes on its maximum value, 1, if $\theta = 45°$. Galileo also used this result to show that $s_x$ for a given type of projectile is equally large for any two values of $\theta$ differing from 45° by an equal amount in each direction (e.g., 52° and 38°). Convince yourself of this unexpected fact by substitution, if no simpler way suggests itself. Galileo very penetratingly remarks at this point:

"The force of rigorous demonstrations such as occur only by use of mathematics fills me with wonder and delight. From accounts given by gunners, I was already aware of the fact that in the use of cannon and mortars, the maximum range, that is, the one in which the shot goes farthest, is obtained when the elevation is 45°...; but to understand *why* this happens far outweighs the mere information obtained by the testimony of others or even by repeated experiment.... The knowledge of a single effect apprehended through its cause opens the mind to understand and ascertain other facts without need of recourse to experiment, precisely as in the present case, where, having won by demonstration the certainty that the maximum range occurs when the elevation [θ] is 45°, the Author [Galileo] demonstrates what has perhaps never been observed in practice, namely, that for elevations which exceed or fall short of 45° by equal amounts, the ranges are equal ...."

Note the pleasure which he finds in discovering a mathematical law, and that he regards it as the way "to understand *why*" projectiles move as they are observed to do. Note also especially the phrase "without need of recourse to experiment"; Galileo, traditionally regarded as the patron saint of experimental science, clearly warns us that the continual experimental verification of a prediction from a law is unnecessary once the law is sufficiently well established. After the initial mandatory doubts are satisfied, one must sufficiently believe in the law in order to obtain any benefit from it at all.

A second important conclusion from the work on projectiles may seem almost trivial at first glance. In support of the Copernican thesis that the earth moves about its axis and around the sun, Galileo offered an answer to critics who argued that if the earth moved, a stone dropped from a

tower would be left behind while falling through the air, and consequently would not land directly at the foot of the tower as observed, but much beyond. Galileo assumes that during the *short* time of fall, the earth, and the top and the foot of the tower may be thought to move forward equally far with uniform velocity. If, then, the whole tower moves with the same speed $v_{0x}$, the dropped stone must fall along the tower, because it will retain this "initial" horizontal component of speed, as truly as a parcel dropped from a moving plane lands below the plane or, in Galileo's analogy, as an object falling from the mast of a moving ship lands at the foot of the mast. From this and equivalent observations concerning the other laws of mechanics has been developed a most valuable generalization, usually called the *Galilean relativity principle*: Any mechanical experiment, such as on the fall of bodies, done in a stationary "laboratory" (e.g., on a stationary ship) will come out precisely the same way when repeated in a second "laboratory" (say on a moving ship) as long as the second place of experimentation moves with *constant velocity* as measured from the first.

To express this more elegantly, let us use the words "coordinate system" instead of "laboratory," since all that counts in the description of experiments is the system of reference used for measurements. Some corner of the rectangular laboratory table might be the origin for all our measurements of distance, and the $x$, $y$, and $z$ directions may correspond to the directions of the three edges. Then we may restate the Galilean relativity principle: "All laws of mechanics observed in one coordinate system are equally valid in any other coordinate system moving with a constant velocity relative to the first." If you drop a suitcase in a compartment in a train, be it at rest or speeding at 40 km/hr or at 80 km/hr across level terrain, the case will always hit the floor below its point of release, and will fall downwards with the usual acceleration as measured in the car. Consequently, from the motion of a projectile in a car you cannot decide whether the car itself is moving. If the principle is true in its general form, it follows that no *mechanical* experiment made on our earth can decide such intriguing questions as whether the solar system as a whole proceeds with some steady forward motion through the space about us.

## 8.6 SUMMARY

Evidently, this chapter has carried us far from the starting point of our discussion. Recall that we began with the simple case of projectile motion in a plane, $\theta$ being $0°$ at first. Then we deduced the superposition principles for velocities, learned how to add vector components by the simple parallelogram method, and extended our treatment to the general case, obtaining the *law of projectile motion* without any restrictions on the initial direction of motion. Finally we heard that even more general principles were induced from these laws. We also noted in passing the power and meaning of the mathematical formulation of physical events, and the multifold predictions flowing from such mathematical laws—whether theoretical, as the determination of a maximum range for $\theta = 45°$, or practical, as the construction of gunnery tables. And from this example we can already perceive the moral problem that faces the modern scientist: the obvious possibility that his discoveries may be applied by others in warfare or for other purposes, whether or not he likes it.

With these examples of the development and power of physical law in mind, we shall later return to a discussion of "scientific methods." But throughout the rest of this study we shall encounter influences of Galileo's work from time to time, as one might find an ancestral trait running through generations of one family.

For the moment, an urgent task remains. In this section, Part B of the text, we have studied motions of various sorts, but without regard to the forces that cause them. To kinematics, we must

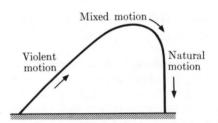

**Fig. 8.13** An early sixteenth-century conception of projectile motion using the Aristotelian terminology. ("Violent" motion was considered to be that produced by a propelling force acting on the object.)

now add the concept of force to achieve a *dynamics*. These are as two sides of a coin; without both, we cannot purchase the full understanding of any mechanical phenomenon.

## ADDITIONAL PROBLEMS

**Problem 8.3.** What is meant by saying that the paths of projectiles *in vacuo* are parabolic? When we find by experiment that the paths *are* parabolic, what help is that in solving projectile problems? *Why* should trajectories follow parabolic paths instead of perhaps that shown in Fig. 8.13?

**Problem 8.4.** A stone is thrown virtually vertically upward with an initial speed $v_{0y} = 10$ m/sec (Fig. 8.14). Find (at the end of the first, third, and seventh second), (a) the displacement $s_y$ of the stone, (b) the instantaneous velocity. Use your data to plot a graph of $v$ vs. $t$ (on graph paper), and find from it (*graphically*), (c) when the stone will have a speed of 20 m/sec (numerically), (d) what the acceleration is at the instant that the stone has reached its topmost position and just begins to turn downward.

**Figure 8.14**

**Problem 8.5.** How long will the stone in Problem 8.4 take to reach again the level of release? Obtain this result first from your graph for Problem 8.4, then check it by independent direct calculation.

**Problem 8.6.** A gun fires a shell at a muzzle velocity of 1000 m/sec and at an angle of 30° above the horizon. How long will it take to return to earth, how high will the projectile rise, and what is its range on level ground? (Neglect air friction.)

**Problem 8.7.** A baseball is thrown by a pitcher horizontally with a speed of 40 m/sec from a vertical height of 2 m above home plate. The batter is 20 m away from the pitcher. Will this pitch be called a strike if the batter does not swing?

**Problem 8.8.** The famous gun known as Big Bertha in World War I had a maximum range on flat territory of about 100 km. What would have been the muzzle velocity of the projectile if it had exhibited such a range *in vacuo*? (The actual muzzle velocity under ordinary circumstances would be somewhat larger.)

**Problem 8.9.** A baseball is thrown with $v_0 = 15$ m/sec at an elevation ($\theta$) of 60°. Show that after 2 sec it has risen 6.6 m, has traveled 15 m horizontally, and that its direction of motion is now tilted downward, making an angle of about −40.9° (that is, 40.9° below the horizontal).

**Problem 8.10.** Modern cosmic ray research sometimes employs rockets to sample automatically the radiation at high altitudes. For the sake of simple calculation, consider that a self-propelled rocket constantly increases its speed on a straight vertical path so that at an altitude of 25 km, when the fuel is exhausted, it has attained a speed of 1500 m/sec. At this point the stabilizers automatically turn the rocket at an angle of 60°, after which the missile continues as an ordinary projectile. What time interval (in seconds) is available for measurements in the relatively air-free region above 25 km? What height does the rocket reach? What is the total time from the firing of the rocket to the instant it hits the ground? What assumptions have you introduced?

**Problem 8.11.** A hunter aims his gun barrel directly at a monkey in a distant palm tree. Where will the bullet go? If the animal, startled by the flash, drops out of the branches at the very instant of firing, will it then be hit by the bullet? Explain. Would this last answer hold if the acceleration of gravity were not 9.8 m/sec² but only $\frac{1}{6}$ as great, as on the moon?

**Problem 8.12.** For each of the following cases sketch two graphs, one of vertical displacement vs. time elapsed, the other of vertical velocity vs. time elapsed. Carefully use the convention concerning positive and negative values. (a) A parachutist falls from a plane, after a while opens the 'chute, and floats to the ground. (b) A marble drops from your hand and bounces three times before coming to rest on the floor. (c) A shell is fired at 60° elevation and falls into a deep valley.

**Problem 8.13.** Read through the Fourth Day (Motion of Projectiles) in Galileo's *Two New Sciences,* and on this basis briefly discuss the following: (a) Galileo's examination of the role of air resistance in projectile motion; (b) Galileo's use of experiments to further his argument; (c) Galileo's interest in practical applications of his work.

## RECOMMENDED FOR FURTHER READING

A. R. Hall, *From Galileo to Newton 1630–1720*, New York: Harper, 1963; Chapters I–V. Covers several aspects of the development of seventeenth-century science not mentioned in this book.

P. Kirkpatrick, "Bad physics in athletic measurements," *American Journal of Physics*, 1944, reprinted in *Project Physics Reader*, Vol. 1

P. Wiener and A. Noland (editors), *Roots of Scientific Thought*, New York: Basic Books, 1957. The articles by Koyré, Nicolson, Johnson, Strong, Boas, and Thorndike discuss various topics in seventeenth-century science.

## SOURCES, INTERPRETATIONS, AND REFERENCE WORKS

René Dugas, *Mechanics in the 17th Century*, New York: Central Book Co., 1958; Chapters I–IV, and the first part of Chapter XV

Galileo Galilei, *Dialogues Concerning Two New Sciences* (Fourth Day)

Ernst Mach, *The Science of Mechanics*, pages 181–191

# PART C

# NEWTON'S LAWS AND HIS SYSTEM OF THE WORLD

# NEWTON'S LAWS AND HIS SYSTEM OF THE WORLD

"This subject of the formation of the three laws of motion and of the law of gravitation deserves critical attention. The whole development of thought occupied exactly two generations. It commenced with Galileo and ended with Newton's *Principia*; and Newton was born in the year that Galileo died. Also the lives of Descartes and Huygens fall within the period occupied by these great terminal figures. The issue of the combined labors of these four men has some right to be considered as the greatest single intellectual success which mankind has achieved."

A. N. Whitehead, *Science and the Modern World*

As we look into history, it seems that sometimes progress in a field of learning depended on one man's incisive formulation of the right problem at the right time. This is so with the section of mechanics called *dynamics*, the science that deals with the effects of forces on moving bodies. The man was Isaac Newton, and the formulation was that of the concepts *force* and *mass*, expounded in three interconnected statements that have come to be called *Newton's three laws of motion*. These, together with an introduction to rotational motion, complement and extend the study of pure kinematics in Part B. We can then not only understand a great many simple mechanical phenomena, but also, with the help of Newton's law of universal gravitation, we can solve some of the outstanding problems in astronomy introduced in Part A. Once we have successfully dealt with such problems, Newtonian dynamics becomes a powerful tool for advancing to the next level of problems in physical science.

# Chapter 9

# NEWTON'S LAWS OF MOTION

## 9.1 SCIENCE IN THE SEVENTEENTH CENTURY

Between the time of Galileo's death and the publication of Isaac Newton's *Principia** lie barely 44 years, yet an amazing change in the intellectual climate of science has taken place in that brief interval. On the one side, the "New Philosophy" of experimental science is becoming a respected and respectable tool in the hands of vigorous and inventive investigators; and on the other, this new attitude is responsible for a gathering storm of inventions, discoveries, and theories. Even a very abbreviated list of these, covering less than half of the seventeenth century and only the physical sciences, will show the justification for the term "the century of genius": the work on vacuums and pneumatics by Torricelli, Pascal, von Guericke, Boyle, and Mariotte; Descartes' great studies on analytical geometry and on optics; Huygens' work in astronomy and on centripetal force, his perfection of the pendulum clock, and his book on light; the establishment of the laws of collisions by John Wallis, Christopher Wren, and Christian Huygens; Newton's work on optics, including the interpreta-

tion of the solar spectrum, and his invention of calculus almost simultaneously with Leibniz; the opening of the famous Greenwich observatory; and Hooke's work, including that on elasticity.

Men of science are no longer isolated. Although we must not suppose that suddenly all scientists are of a modern bent of mind, the adherents of the New Philosophy are many and growing in numbers. They have formed several scientific societies (in Italy, England, and France), of which the most famous is the Royal Society of London (as of 1662). They meet and communicate with one another freely now, they cooperate and debate, write copiously, and sometimes quarrel healthily. As a group, they solicit support for their work, combat attacks by antagonists, and have a widely read scientific journal. Science is clearly becoming well defined, strong, and international.

What is behind this sudden blossoming? Even a partial answer would lead us far away from science itself, to an examination of the whole picture of quickening cultural, political, and economic changes in the sixteenth and seventeenth centuries.† One aspect is that both craftsmen and men

---

* Full title: *Philosophiae Naturalis Principia Mathematica;* published in London, 1687.

† Some of these problems are skillfully analyzed in G. N. Clark, *Science and Social Welfare in the Age of Newton*, Oxford University Press (second edition 1949, reprinted 1969).

of leisure and money begin to turn to science, the one group for improvement of methods and products, the other for a new and exciting hobby. But availability of money and time, the need for science, and the presence of interest and organizations do not alone explain or sustain such a thriving enterprise. Even more important ingredients are able men, well-formulated problems, and good mathematical and experimental tools.

Able scientists—these were indeed at hand. Some, like Newton, found freedom to work even in universities still dominated by medieval scholasticism. Others, like Wren and Hooke, found time to pursue scientific research in the midst of busy careers in other fields such as architecture, if they did not, like Robert Boyle, possess the initial advantages of inherited wealth.

Well-formulated problems—these had recently become clear in the writings of Galileo and others. It was Galileo, above all, who had directed attention to the fruitful language of science, who had presented a new way of viewing the world of fact and experiment. While Francis Bacon's emphasis on observation and induction was acclaimed by many, especially in the Royal Society, Galileo had also demonstrated the fruitfulness of bold hypotheses combined with mathematical deduction. His scorn of sterile introspection and blind subservience to ancient dogma echoed through all fields of science.* Plato's old question, "By the assumption of what uniform and ordered motions can the apparent motions of the planets be accounted for?" had lost its original meaning in the new science. The new preoccupation is illustrated by what may be called the two most critical problems in seventeenth-century science: "What forces act on the planets to explain their actually observed paths?" and "What accounts for the observed facts of terrestrial motions now that Aristotelian doctrines have failed us?"

Good mathematical and experimental tools were also being created. With mathematics finding expression in physics, the two cross-fertilized and gave rich harvests, as might be expected when the same men (Descartes, Newton, and Leibniz) were making major discoveries in both fields. Analytic geometry and the calculus are parts of the rich legacy of the seventeenth century which will undoubtedly still be useful in science even if many of the scientific theories of that century are ultimately superseded. The telescope, the microscope, and the vacuum pump opened up vast new realms of phenomena for scientists to study, and the need for more accurate measurement of these phenomena stimulated the invention of other ingenious devices, thus initiating the collaboration between scientist and instrument maker which has become typical of modern science.

But in trying to solve the riddle of the spectacular rise of science after Galileo, perhaps we must simply realize that science has an explosive potential for growth once the necessary conditions are given.† Around the time of Galileo the necessary conditions for growth were established: At last there were many men with similar attitudes working in the same fields; they could communicate more freely with one another and with the accumulated achievements of the past (partly through the relatively young art of printing); they had become impatient with qualitative reasoning and began to be more and more intrigued with the quantitative approach. To use a modern analogy, the critical mass was reached, and the chain reaction could proceed.

---

* For a delightful example, read the first thirty or forty pages in the Everyman's Library edition of Robert Boyle's *Sceptical Chymist* (originally published 1661).

† Historians in recent years have been debating the role of religious factors, in particular the rise of Puritanism in England, in creating a favorable social climate for science. Some articles summarizing this controversy are cited at the end of the chapter. Another interpretation, from the viewpoint of the new discipline of *sociology of science*, is well presented in Joseph Ben-David's recent book, *The Scientist's Role in Society* (Prentice-Hall, 1971).

This, from the point of view of science, was the new age in which Newton lived. But a brief word of caution before we follow his work. The history of ideas, in science as in any other field, is not just an account of visits with the most important names. The work of each genius is made possible, is stabilized, sometimes is even provoked, but is always connected with the whole structure of science only through the labors of lesser-known men, just as a brick is surrounded by fortifying mortar. A house is not just a heap of bricks—and science cannot be made by giants only. As Lord Rutherford, a founder of modern nuclear physics, said: "It is not in the nature of things for any one man to make a sudden violent discovery; science goes step by step, and every man depends on the work of his predecessors.... Scientists are not dependent on the ideas of a single man, but on the combined wisdom of thousands of men." Therefore, properly, we should trace in each man's contribution the heritage of the past, the influence of his contemporaries, and the meaning for his successors. It would be rewarding, but here we can touch on only the barest outlines.

## 9.2  A SHORT SKETCH OF NEWTON'S LIFE

Isaac Newton was born on Christmas day, 1642, in the small village of Woolsthorpe in Lincolnshire. He was a quiet farm boy who, like young Galileo, loved to build and tinker with mechanical gadgets and who seemed to have a secret liking for mathematics. Through the fortunate intervention of an uncle he was allowed to go to Trinity College, Cambridge University, in 1661 (where he appears to have initially enrolled in the study of mathematics as applied to astrology!). He proved an eager and excellent student. By 1666, at 24, he had quietly made spectacular discoveries in mathematics (binomial theorem, differential calculus), optics (theory of colors), and mechanics. Referring to this period, Newton once wrote:

"And the same year I began to think of gravity extending to the orb of the Moon, and...from

Kepler's Rule [third law]...I deduced that the forces which keep the Planets in their orbs must [be] reciprocally as the squares of their distances from the centers about which they revolve: and thereby compared the force requisite to keep the Moon in her orb with the force of gravity at the surface of the earth, and found them to answer pretty nearly. All this was in the two plague years of 1665 and 1666, for in those days I was in the prime of my age for invention, and minded Mathematicks and Philosophy more than at any time since."

From his descriptions we may conclude that during those plague years, having left Cambridge for the time to study in isolation at his home in Woolsthorpe, he had developed a clear idea of the first two laws of motion and of the formula for centripetal acceleration, although he did not announce the latter until many years after Huygens' equivalent statement.*

After his return to Cambridge, he did such creditable work that he followed his teacher as professor of mathematics. He lectured, and contributed papers to the Royal Society, at first particularly on optics. His *Theory of Light and Colors*, when finally published, involved him in so long and bitter a controversy with rivals that the shy and introspective man even resolved not to publish anything else. As Bertrand Russell pointed out, "If he had encountered the sort of opposition with which Galileo had to contend, it is probable that he would never have published a line." Newton now concentrated most on an extension of his early efforts in celestial mechanics, the study of planetary

---

* This too must have been the time of the famous fall of the apple. One of the better authorities for this story is a biography of Newton written by his friend Stukely in 1752, where we read that on one occasion Stukely was having tea with Newton in a garden under some apple trees, when Newton told him he realized "he was just in the same situation, as when formerly, the notion of gravitation came to his mind. It was occasioned by the fall of an apple, as he sat in a contemplative mood."

motions as a problem of physics. In 1684 the devoted
Halley came to ask his advice in a dispute with
Wren and Robert Hooke as to the force which
would have to act on a body executing motion along
an ellipse in accord with Kepler's laws; Newton
had some time ago found the rigorous solution to
this problem ("and much other matter"). Halley
persuaded his reluctant friend to publish the work
which touched on one of the most debated and
intriguing questions of the time. In less than two
years of incredible labors the *Principia* was ready
for the printer; its publication in 1687 established
Newton almost at once as one of the greatest
thinkers in history.

A few years afterward, Newton, who had always
been in delicate health, appears to have had what
we would now call a nervous breakdown. On re-
covering, and for the next 35 years until his death
in 1727, he made no major new discoveries, but
rounded out earlier studies on heat and optics, and
turned more and more to writing on theology.
During those years he received honors in abundance. In 1699 he was appointed Master of the
Mint, partly because of his great interest in, and
competent knowledge of, matters concerning the
chemistry of metals, and he seems to have helped
in re-establishing the debased currency of the
country. In 1689 and 1701 he represented his
university in Parliament. He was knighted in 1705.
From 1703 to his death he was president of the
Royal Society. He was buried in Westminster
Abbey.

## 9.3 NEWTON'S *PRINCIPIA*

In the original preface to Newton's work—probably
the greatest single book in the history of science—
we find a clear outline:

"Since the ancients (as we are told by Pappus)
esteemed the science of mechanics of greatest importance in the investigation of natural things, and
the moderns, rejecting substantial forms and occult
qualities, have endeavored to subject the phenomena of nature to the laws of mathematics, I have in

this treatise cultivated mathematics as far as it
related to philosophy [we would say 'physical
science'] ... for the whole burden of philosophy
seems to me to consist in this—from the phenomena of motions to investigate [induce] the forces of
nature, and then from these forces to demonstrate
[deduce] the other phenomena, and to this end the
general propositions of the first and second Books
are directed. In the third Book I give an example
of this in the explication of the System of the
World; for by the propositions mathematically
demonstrated in the former Books, in the third I
derive from the celestial phenomena the forces of
gravity with which bodies tend to the sun and the
several planets. Then from the forces, by other
propositions which are also mathematical, I deduce
the motions of the planets, the comets, the moon,
and the sea [tides] ...."

The work begins with a set of definitions: mass,
momentum, inertia, force, centripetal force. Then
follows a section on *absolute* and *relative* space,
time, and motion. Since critiques of these definitions have been of some importance in the development of modern physics, some of them are worth
noting here.

*Mass.* Newton was very successful in using the
concept of mass, but less so in clarifying its meaning. He states that mass or "quantity of matter"
is "the product of density and bulk" (bulk =
volume). But what is density? Later on in the
*Principia,* he defines density as the ratio of "inertia"
to bulk; yet at the beginning, he had defined inertia
as proportional to mass. So the definition appears
to be circular, as Ernst Mach pointed out toward
the end of the nineteenth century.* There have
been various attempts to justify or replace Newton's
definition of mass, but they are unnecessary, since
the logical structure of the *Principia* does not really
depend on this definition. In fact, it is now recognized by most scientists that *any* new theory is

* Ernst Mach, *The Science of Mechanics* (first published
in German in 1883, translation available in a paperback
edition by the Open Court Publishing Co.).

likely to postulate a certain number of concepts whose meaning must first of all be grasped intuitively (though later they may be defined operationally). What is more important is that Newton clearly established the modern distinction between *mass* and *weight*—the former being an inherent property of a body, whereas the latter depends on the acceleration due to gravity at a particular location (see next section).

*Time*. "Absolute, true, and mathematical time, of itself, and from its own nature, flows equably without relation to anything external, and by another name is called duration: relative, apparent, and common time, is some sensible and external (whether accurate or unequable) measure of duration by the means of motion, which is commonly used instead of true time; such as an hour, a day, a month, a year."

No one dreamed of disputing this definition until Mach complained that absolute time is "an idle metaphysical conception,"* and few took Mach's criticism seriously at first. Albert Einstein reported that he was deeply influenced by reading Mach's book, and his theory of relativity, published in 1905, is in agreement with Mach at least on the point of rejecting absolutes such as time. Still we shall accept Newton's definition provisionally until Chapter 30, where, following Einstein, we shall argue that in a strict sense there cannot be any such thing as absolute time in physical theory.

*Space*. "Absolute space, in its own nature, without relation to anything external, remains always similar and immovable. Relative space is some movable dimension or measure of the absolute spaces; which our senses determine by its position to bodies . . . because the parts of space cannot be seen, or distinguished from one another by our senses, therefore in their stead we use sensible† measures of them . . . instead of absolute places and motions, we use relative ones; and that without any inconvenience in common affairs; but in philosophical disquisitions, we ought to abstract from our senses, and consider things themselves, distinct from what are only sensible measures of them. For it may be that there is no body really at rest, to which the places and motions of others may be referred."

From the last sentence it might seem that Newton was not sure if there is such a thing as absolute space. That would be a mistake. He thought he had proved by experiment that rotation is a kind of absolute motion.‡ The well-known Foucault pendulum experiment, first done in the middle of the nineteenth century, seems like another demonstration that the earth moves with respect to something. At least we should recognize that Newton's belief in absolute space was plausible; at any rate it was not damaging to the structure of his system, since he hardly anywhere made explicit use of it.

Immediately thereafter, still in the introductory section of the *Principia*, Newton stated his famous three laws of motion (which we shall discuss in the following section) and the principles of composition of vectors (e.g., of forces and of velocities). Book I, titled "The motion of bodies," applies these laws to problems of interest to theoretical astronomy: the determination of the orbit described

---

* *Ibid.*, page 273.

† The word "sensible" in these quotations means "perceptible by the senses."

‡ A bucket, half filled with water, is suspended by a long rope which is twisted so much that, when released, the bucket starts spinning rapidly. A few seconds later the water begins to follow the motion of the bucket, and also rises up on the sides, thus showing that it is rotating in space, though at rest relative to the bucket with which it is in contact. Mach later countered that this is no proof that the water is rising owing to forces brought about while rotating in absolute space. He speculated, for example, that the same observation would be made if the bucket of water were kept still and the "fixed stars" were set in rotation with respect to the bucket.

by one body around another, when they are assumed to interact according to various force laws; and mathematical theorems concerning the summation of gravitational forces exerted by different parts of the same body on another body inside or outside the first one. Another kind of application is relevant to the particle model of light: Newton computes the action of surfaces on small particles reflected or refracted by them. Book II, titled "The motion of bodies (in resisting mediums)," seems to have as its main purpose the proof that Descartes' vortex model cannot adequately account for the observed motions of the planets; but along with this there are a number of theorems and conjectures about the properties of fluids.

Book III, "The system of the world (in mathematical treatment)," makes use of the general results derived in Book I to explain the motions of the planets and other gravitational phenomena such as the tides. It begins with a remarkable passage on "Rules of reasoning in philosophy." The four rules, reflecting the profound faith in the uniformity of all nature, are intended to guide scientists in making hypotheses, and in that function they are still up to date. The first has been called a *principle of parsimony*, the second and third, *principles of unity*. The fourth is a faith without which we could not use the processes of logic. Because of their signal importance in the history of science we quote these rules almost in full:

### RULE I

"*We are to admit no more causes of natural things than such as are both true and sufficient to explain their appearances.*

"To this purpose the philosophers say that Nature does nothing in vain, and more is in vain when less will serve; for Nature is pleased with simplicity, and affects not the pomp of superfluous causes.

### RULE II

"*Therefore to the same natural effects we must, as far as possible, assign the same causes.*

"As to respiration in a man and in a beast; the descent of stones in *Europe* and in *America*; the light of our culinary fire and of the sun; the reflection of light in the earth and in the planets.

### RULE III

"*The qualities of bodies which admit neither intensification nor remission of degrees,* * *and which are found to belong to all bodies within the reach of our experiments, are to be esteemed the universal qualities of all bodies whatsoever.*

"For since the qualities of bodies are only known to us by experiments, we are to hold for universal all such as universally agree with experiments; and such as are not liable to diminution can never be quite taken away. We are certainly not to relinquish the evidence of experiments for the sake of dreams and vain fictions of our own devising; nor are we to recede from the analogy of Nature, which is wont to be simple, and always consonant to itself. We no other way know the extension of bodies than by our senses, nor do these reach it in all bodies; but because we perceive extension in all that are sensible, therefore we ascribe it universally to all others also . . . . The bodies which we handle we find impenetrable, and thence conclude impenetrability to be an universal property of all bodies whatsoever . . .

### RULE IV

"*In experimental philosophy we are to look upon propositions inferred by general induction from phenomena as accurately or very nearly true, notwithstanding any contrary hypotheses that may be imagined, till such time as other phenomena occur, by which they may either be made more accurate, or liable to exceptions.*

"This rule we must follow, that the argument of induction may not be evaded by hypotheses."

---

\* That is, qualitative properties such as impenetrability, as distinguished from quantitative properties such as temperature.

At the end of Book III is a famous "General Scholium" which has had as much influence on subsequent discussions of scientific method as the "Rules of reasoning," because of one phrase which occurs in it. Having concluded that a satisfactory system of the world can be based on the postulate that there exists a universal force of gravitation between all pieces of matter in the universe, Newton confesses himself unable to discover any deeper explanation of the *cause* of this force which stands to *proof*—despite much effort he seems to have expended on finding such a cause. Unwilling to put forward an artificial hypothesis unsupported by sufficient evidence, he simply states: "*Hypotheses non fingo*"—"I do not feign* hypotheses."

Following Book III in most editions of the *Principia* is an essay on "The system of the world" which provides a nonmathematical summary of the principal results of the third book. Now, as in the past, few readers ever get all the way through Books I, II, and III, so we recommend that this concluding section be read immediately after the Introduction in order to get an overview of the main results of the work from its author's viewpoint.

## 9.4  NEWTON'S FIRST LAW OF MOTION

Since we have already discussed the historical background of this law (Sections 5.3 and 6.1) we can proceed directly to a fairly didactic treatment from the modern viewpoint.

We may phrase the first law of motion, or law of inertia, as follows: *every material body persists in its state of rest or of uniform, unaccelerated motion in a straight line, if and only if it is not acted upon by a net (i.e., unbalanced external) force.*

The essence is this: If you see a moving body deviating from a straight line, or accelerating in

---

*    The Latin word *fingo* is sometimes translated "frame," but this obscures the connotation which Newton seems to have intended: To *feign* a hypothesis is to suggest it without really believing in it. We shall discuss the significance of Newton's assertion in Section 11–10.

any way, then you must assume that a net force (of whatever kind) is acting on the body; here is the criterion for recognizing qualitatively the *presence* of an unbalanced force. But note well that this law does not help you to discover either how large the force is or its origin. There is implied only the definition of force as the "cause" of change of velocity, a definition we already alluded to in our discussion of Galileo's work (see Section 7.3). We recall that the Aristotelian scholastics had a rather different view in this matter; they held that force was also the cause of uniform (unaccelerated) motion.

As an example, consider the two opposing opinions that would exist when watching a cart pulled by a horse across a plane in steady motion. The Newtonian commentator would hold that the horse's efforts serve solely to equal and cancel the force of friction on the wheels, so that the *net* force on the cart, i.e., the pull of the animal on it minus the force of friction between wheels and ground, is zero, the cart being consequently in a state of equilibrium, a state which by definition includes uniform unaccelerated motion as well as rest. On the other hand, the Aristotelian scholar would explain that since the natural state of the cart is rest, a force has to be supplied to the cart by the horse to keep it in uniform motion—even in the absence of friction under the cart, if such a circumstance could be approached. We have here not so much a dispute about experimentally observable facts as a difference in viewpoint, complicated by the use of the same word, *force*, in two such disparate meanings.

As a matter of fact, most people who have not studied a little physics are Aristotelians, not Newtonians. The Aristotelian view, here as in some other contexts, is the one that is closer to contemporary common-sense opinion. Since friction is in actuality never absent and is often a very real hindrance in moving an object, it is natural that one develops the intuitive idea that a force is necessary to "keep things moving," and, as a next step, to define force as "the cause of continued

motion." The Newtonian turns from this anthropomorphic position to consider the net forces
on the moving body, and even dares to cast his
definition of force in terms of the *vector sum* (i.e.,
*net* force) without having first identified the individual components separately. Only the great
practical success which followed it could have
justified a step that logically is so precarious.

Other points are also raised by the first law of
motion. It has been made plain that a net force
must be supplied to change a body's state from rest
to motion or from motion to rest, from one speed to
another or from one direction of motion to another
even at the same speed. In somewhat unrigorous
language still, we may say that material bodies are
endowed with a sluggishness, a laziness toward
change, an *inertia*. There is very little profit in
trying to account for this experimental fact in
terms of some physical model or picture—let us
simply accept that physical bodies are characterized
by inertia just as they are by such other fundamental attributes as volume, chemical constitution,
and so forth.

If the change of direction of motion implies the
action of a net force, we cannot think—as the
ancients did—that no force need act on the planets
in their orbits to keep their movement circular; on
the contrary, we must consider that they are subjected to forces which draw them continually from
a straight-line motion.

**Fig. 9.1**  Cutaway view of a spring balance.

block is placed on it and given a small push. What
will be the motion of the block? How will this be
affected by a straight-line motion of the whole
laboratory during the experiment? What if the
block moves in an arc sideways? How could you
decide whether or not it is your laboratory that is
moving in a curve?

**Problem 9.3.** An object weighing $X$ units hangs
on a spring balance (Fig. 9.1) in an elevator.
Decide in each of the following cases whether the
balance will read more than $X$ or less, and why:
the elevator accelerates upward; downward; moves
without acceleration down; decelerates on the way
down.

**Problem 9.1.** Explain in terms of Newton's first
law of motion the common experience of lurching
forward when a moving train suddenly decelerates
and stops. Explain what happens to the passengers
of a car that makes a sharp and quick turn. A
penny is put on a record, and the motor is started;
when the record has reached a certain speed of
rotation, the coin flies off. Why?

**Problem 9.2.** Assume that a horizontal, perfectly
frictionless table is set up in your laboratory. A

## 9.5  NEWTON'S SECOND LAW OF MOTION

So far, we have only a qualitative notion of the
concept of force and, incidentally, of the property
of bodies called *inertia*. But to develop a science of
dynamics we must be able to determine forces and
inertias quantitatively, instead of merely calling
a body's inertia large because, in a qualitative way,
a large force is needed to change its state of motion.

In Newton's own formulation of the second law, he states that the force acting on a body is equal to the change of its quantity of motion, where "quantity of motion" (later called "momentum") is defined as the product of mass and velocity. This is a simple generalization arising naturally from observations on collisions, in which a sudden blow produces a finite change of motion in a short period of time. But for continuously acting forces, such as gravity, it was far more convenient to define force differently, i.e., to use the *rate of change* of motion, which thus brings in Galileo's concept of acceleration; it is this version of the second law, formalized by the Swiss mathematician Leonhard Euler in 1750, that was eventually adopted in physics, and that we shall be using.*

It is customary to begin by stating the second law of motion in the form: *The net (external unbalanced) force acting on a material body is directly proportional to, and in the same direction as, its acceleration.*

Here the point is this: Once a net force is qualitatively detected by noticing changes of velocity (first law), let us define it precisely by the exact rate of change of velocity. If we adopt the symbols $F_{net}$ and $a$, we can put the sentence above in the following form:

$$F_{net} \propto a \text{ (for a given body)},$$

or

$$\frac{F_{net}}{a} = \text{a constant for a given body.} \quad (9.1)$$

The constant so defined is a measure of the body's inertia, for clearly a large ratio of $F_{net}$ to $a$

means that a large force is needed to produce a desired acceleration, which is just what we expect to find true for large, bulky objects, to which we intuitively assign a larger inertia than to small, light objects. If we now symbolize the constant in Eq. (9.1), the measure of inertia, by a letter $m$ of its own, and give it the alternative name *mass*, we can write the second law of motion:

$$\frac{F_{net}}{a} = m; \quad \text{or} \quad F_{net} = m \times a. \quad (9.2)$$

Equation (9.2) enables us in principle to assign a numerical value to a net force if we measure the acceleration it produces on a body of known mass, or conversely we could in principle obtain a numerical value for the mass in terms of the acceleration and the net force. But of course you notice the vicious circle: Unhappily we find that inertia and force, the two concepts we are trying to quantify rigorously, are mutually interconnected. To find the one we seemingly would have to know the other in advance.

There is one straightforward solution for this apparent dilemma. Before we try to measure any masses or forces, we first choose some one unique object, perhaps a certain small piece of polished rock or metal, as the universal *standard of mass*, and regard it arbitrarily as having an inertia of unity, or *unit mass*; for example, name it the *standard mass* of *one kilogram* (abbreviated 1 kg). Then, by Eq. (9.2), any net force can be evaluated by means of the observable acceleration it produces on our standard object.

This important point will be clearer if we resort to an imaginary, idealized experiment ("thought experiment"): Place the standard 1-kg object on a smooth horizontal plane where friction is so negligible that we can speak of it as a frictionless plane.† Now attach to the standard an uncalibrated

---

* It might appear that there would be a conflict between these two versions, yet Newton was able to use both of them in appropriate situations. He glossed over the difference in the definitions implied in the two equations $F = \Delta mv$ and $F = \Delta mv/\Delta t$. He did this either by taking $\Delta t = 1$ or by implicitly absorbing the factor $\Delta t$ into the definition of $F$ in the first case. Although Newton's second version is equivalent to the modern equation $F = ma$ discussed below, that equation itself never appeared in the *Principia*.

---

† For demonstration purposes, a strikingly close approximation to this ideal is reached when a smooth piece of "dry ice" is placed on a flat, horizontal sheet of plate glass. The thin layer of vapor between the two keeps friction to a minimum.

spring balance (of negligible mass) and pull on it horizontally, as shown in Fig. 9.2. Here the applied force is indeed the net force. If this net force is constant, the acceleration of the object is also constant, and therefore ascertainable without fundamental difficulties; let us say that it comes to 4 m/sec² for a given pull in this concrete example. But that result tells us at once the magnitude of the pull also, for by Eq. (9.2), $F_{net} = m \times a$; thus

$$F_{net} = 1 \text{ kg} \times 4 \text{ m/sec}^2 = 4 \text{ kg-m/sec}^2.$$

The pointer reading corresponding to that pull may now be marked off* on the spring balance as 4 kg-m/sec²; or, since the unit kg-m/sec² has traditionally been given the shorter name *newton*, the marking should be 4 newtons.†

A single point will, of course, not serve to fix the calibration of a spring balance. Ideally, therefore, we repeat the whole procedure several times with different applied forces, by observing in each case the corresponding acceleration of the 1-g object. Eventually we shall thus develop a fully calibrated balance; with its aid we may not only measure other forces directly, but—more important—may reverse the argument to determine quickly the unknown mass $m_x$ of any other object. For example, the now calibrated balance is applied to $m_x$ and drawn out to read 15 newtons while accelerating $m_x$ on the frictionless plane. If the acceleration, measured simultaneously, is 6.2 m/sec², then, by Eq. (9.2),

$$m_x = \frac{F_{net}}{a} = \frac{15 \text{ newtons}}{6.2 \text{ m/sec}^2} = \frac{15 \text{ kg-m/sec}^2}{6.2 \text{ m/sec}^2} = 2.4 \text{ kg}.$$

*In summary:* Newton's second law, in conjunction with the essentially arbitrary choice of one

---

* That the force was constant can, strictly speaking, be judged only from the constancy of the observed acceleration. However, happily, there is a separate law of nature governing the behavior of elastic bodies like springs in spring balances, by which the displacement of the spring, and therefore the position of the pointer, is steady and unique for constant forces.

† Another unit sometimes encountered is the *dyne*, defined as 1 g-cm/sec²; note that 1 newton = 10⁵ dynes.

**Fig. 9.2**  Thought experiment to define the unit of force.

standard of mass, conveniently fixes the unit of force, permits the calibration of balances, and gives us an operational determination of the mass of all other bodies.

---

**Problem 9.4.**  Recount in detail what steps you must take (in idealized experimentation) to determine the unknown mass $m_x$ (in kilograms) of a certain object if you are given nothing but a frictionless horizontal plane, a 1-kg standard, an uncalibrated spring balance, a meter stick, and a stop watch. [*Note:* Your answer represents a first approach toward an operational definition of mass, and will prove a great aid in comprehension of dynamical problems.]

**Problem 9.5.**  What is the mass $m_x$ if the observed acceleration of the object on the frictionless horizontal plane is 0.85 m/sec² while the balance reads off a net accelerating force of 22 newtons? What acceleration will a net force of 1.0 newton produce on a 3.8-kg mass?

**Problem 9.6.**  On a certain horizontal plane the friction under a moving body is not zero but 7.4 newtons, while the balance pulls horizontally with 13.6 newtons. If the observed acceleration of the body is 2.2 cm/sec², what is its mass?

**Problem 9.7.**  The mass of a student is about 75 kg. As he sits in his car and starts from rest, his acceleration may be about 2 m/sec². What force must the car seat exert to accelerate him, and in which direction?

**Fig. 9.3** Standard kilogram, together with a standard meter. (Courtesy of National Bureau of Standards)

are obvious experimental difficulties owing to the effects of evaporation, the additional inertia of the necessary containers, relatively poor accuracy for measuring volumes, and so on. Therefore, it became accepted custom to use as the standard of mass a certain piece of precious metal, a cylinder of platinum alloy kept under careful guard at the *Bureau Internationale des Poids et Mesures* at Sèvres, a suburb of Paris (alongside the standard of length, a metal bar regarded as representing the distance of 1 meter*). For use elsewhere, accurate replicas of this international standard of mass have been deposited at the various bureaus of standards throughout the world; and from these, in turn, auxiliary replicas are made for distribution to manufacturers, laboratories, etc.

Again, it was a matter of convenience and accuracy to make this metal block have an inertia 1000 times that of a 1-g mass; and thus the standard object (see Fig. 9.3) is a cylinder $1\frac{1}{2}$ in. high of mass 1000 g, called one kilogram (abbreviated 1 kg). When we think of 1 g we can think of the $\frac{1}{1000}$ part of that large kilogram standard. For all our discussions and thought experiments, we shall speak and think of our standard as being indeed a metal block having the mass of 1 kg. This is about 2.2 times as large as the pound, the unit of mass generally used in the U.S.A. for commercial purposes.†

## 9.6 STANDARD OF MASS

It will readily be appreciated that the Standard of Mass that represents 1 kilogram, though essentially arbitrary, has been chosen with care. For scientific work, 1 gram was originally defined as the mass of a quantity of 1 cubic centimeter (1 cm³) of distilled water at 4°C. This decision, dating from the late eighteenth century, is rather inconvenient in practice. Although standardizing on the basis of a certain amount of water has the important advantage that it permits cheap and easy reproduction of the standard anywhere on earth, yet there

## 9.7 WEIGHT

Objects can be acted on by all kinds of forces; by a push from the hand; or by the pull on a string attached to the object; or by a magnetic attraction if the object is made of iron or other susceptible

---

\* As explained in Appendix VIII, this bar is no longer the *primary* standard for the meter.

† More strictly, 1 kg = 2.204622 lb avoirdupois, or 1 lb = 0.4535924 kg. But the easily remembered relation 1 lb = 0.454 kg is sufficient for us. For further discussion and tables of conversion factors, see the appendixes.

materials; or by the action of electric charges; or by the gravitational attraction that the earth exerts on bodies. But no matter what the origin or cause of the force, its effect is always given by the same equation, $F_{net} = ma$. Newton's second law is so powerful precisely because it is so general, and because we can apply it even though at this stage we may be completely at a loss to understand exactly why and how a particular force (like magnetism or gravity) should act on a body.

If the net force is, in fact, of magnetic origin, we might write $F_{mag} = ma$; if electric, $F_{el} = ma$; and so forth. Along the same line, we shall use the symbol $F_{grav}$ when the particular force involved is the gravitational pull of the earth. Because this case is so frequently considered, a special name for $F_{grav}$, namely *weight*, and a special symbol, *W*, is generally used.

Of all the forces a body may experience, the gravitational force is the most remarkable. Let us discuss it in detail. First of all, we are struck by the fact that gravitational forces arise between two bodies (such as a stone in my hand and the earth below) without any need to prepare them for it—simply by placing them near each other. Before two pieces of steel attract each other by magnetism, we must specially treat at least one of them to magnetize it; and before electric forces can act between two spheres, at least one has to be sprayed with charges. But gravitational pulls are inherently present, although ordinarily the mutual attraction is too small to become noticeable between objects lying on a table. At any rate, we shall concentrate at first on the more spectacular case of the earth's pull on bodies at or near its surface. There, $F_{grav}$ acts whether we wish it or no, whereas we can apply or remove at will mechanical, magnetic, and electric influences.

Our second point is this: It is easy to change the mechanical and electric forces on a material particle—by pulling harder on the string to which it is tied, or by depositing more charges on it, etc.—but it is not at all like that with its weight. No matter what we may do to it, a body's weight $F_{grav}$ at a given locality is essentially unaffected. To be sure, the farther the two bodies are separated, the less the gravitational attraction between them, as borne out by experiment, so that the weight of an object disappears completely as we remove it to a spot extremely far from all other objects, for example to interstellar space.* But if *at a certain spot* on this globe the weight of a block of metal is 1 newton, nothing will change this result short of destroying the identity of the object by cutting something off or by adding to it. Even if we interpose screens of all kinds between block and earth, the gravitational pull of the earth remains unchanged, as has been proved long ago by careful experimentation.

Presupposed throughout the previous paragraphs was some accurate method for measuring $F_{grav}$. We might simply drop the object, allowing $F_{grav}$ to pull on it and to accelerate the body in free fall, and then find the magnitude of $F_{grav}$ = mass (*m*) × acceleration owing to gravity (*g*). Knowing *m* by our previous type of experiment (Section 9.5), and observing experimentally the acceleration *g* in free fall, we can at once calculate $F_{grav}$.

Happily there is another method, easier and more direct. We need only our previously calibrated spring balance; from it we hang the body for which $F_{grav}$ is to be determined, and then wait until equilibrium is established. Now we do not allow $F_{grav}$ to be the only force on the body, but instead we balance it out. When the pointer comes to rest—say on the 5-newton reading—then we know (by Newton's first law) that the upward pull of the spring, $F_{bal}$, just counterbalances the downward pull of $F_{grav}$ on the object. The net force is zero. While oppositely directed, these two forces on the same object are numerically equal, and therefore $F_{grav}$, the weight in question, must also

---

*    The same result, though for an entirely different reason, is obtained by placing the object—in thought—at the center of our planet, where the gravitational pulls from all sides will fairly well cancel.

**Fig. 9.4**  Weighing with a spring balance.

be 5 newtons (Fig. 9.4). In passing we note the curious fact that in "weighing" we did not "read off" the weight, but rather the magnitude of an oppositely directed balancing force. This might be called a *static* determination of weight, whereas the previous one was *dynamic*. As a matter of experimental fact, both procedures, though so different, give equal values at the same locality.

Yet another remarkable attribute of the gravitational force is revealed by Galileo's famous observation that (again at a given locality) all objects fall with the same gravitational acceleration $g$. This can mean only that the gravitational attraction $F_{grav}$ is simply proportional to the mass of the attracted object, that its composition, shape, etc. do not influence $F_{grav}$ at all. On reflection, this is an astonishing discovery. The magnitude of other forces (e.g., electric and magnetic ones) is not at all simply proportional to the mass of the affected object. On the contrary; two spheres of equal mass but of different material generally behave entirely differently when in the presence of the same magnet or electric charge. That there should exist a universal and strictly linear proportionality between weight and mass is logically as unexpected as if we had found that the weights of objects are proportional exactly to the square root of their volumes, or, to use an analogy, if it were discovered that on some far-off planet a man's wealth is pro-

portional to the size of his person. Certainly, wealth and size do not thereby become identical concepts, but such a simple empirical relationship between them should allow us quickly to judge a man's bank account by surveying his bulk, and vice versa. Similarly, although $F_{grav}$, the weight, and $m$, the mass, of an object refer to entirely different concepts, the simple relationship between the corresponding values soon puts us in the habit of judging the relative weight from an estimate of the mass, and vice versa.

In summary, the dynamic method of measuring weights by $F_{grav} = m \cdot g$ involves a prior determination of mass $m$ and also a measurement of $g$. Now, while $g$ is constant for all types of objects at a given locality and may for most purposes be taken as 9.80 m/sec$^2$ or 32.2 ft/sec$^2$ on the surface of the earth, the *exact* value is measurably different at different localities. As Table 9.1 shows, the variation is from about 9.7804 m/sec$^2$ at the equator to 9.8322 m/sec$^2$ at the North Pole—no less than $\frac{1}{2}\%$. A 1-kg mass will weigh about 9.81 newtons in Maine and 9.79 newtons in Florida. Even at the same latitude, $g$ will decrease with increasing height from sea level; and although the effect is generally quite feeble, so accurate are the methods of measurement that the minute change in $g$ from one floor of a laboratory to the next is actually discernible!

**Table 9.1.**  Gravitational acceleration

| Location | Value of $g$ (m/sec$^2$) |
|---|---|
| Cambridge, Mass. | 9.804 |
| Chicago | 9.803 |
| Denver | 9.796 |
| Key West | 9.790 |
| San Francisco | 9.800 |
| Washington, D.C. | 9.801 |
| Paris | 9.809 |
| Latitude 0°, sea level | 9.78039 |
| Latitude 90°, sea level | 9.83217 |
| Sun's surface | 274.40 |
| Moon's surface | 1.67 |

**Problem 9.8.** You are given a frictionless horizontal plane, a 1-kg standard mass, an *uncalibrated* spring balance, a meter stick, stop watches, and assistants. Without further assumptions or aids, and calling only upon Newton's laws of motion, recount in complete detail what steps you should take in idealized experimentation to determine the unknown weight $F_{grav}$ (in newtons) of an object in two essentially different ways.

**Problem 9.9.** A standard mass is taken from Paris, France, to our National Bureau of Standards in Washington, D.C. How much is the mass in Washington if it was 1 kg in Paris? What is its weight in Paris and in Washington? What is the percentage change in its weight? (Consult Table 9.1 for necessary data.)

**Problem 9.10.** At the equator (latitude 0°), the gravitational acceleration $g$ is 9.78039 m/sec² at sea level and 9.77630 m/sec² at a height 1000 m above sea level. If these results were to be confirmed by comparing the weight of a 500-g object at these two levels on the same spring balance, what must be the magnitude of the smallest division on the scale that one can accurately estimate?

## 9.8 THE EQUAL-ARM BALANCE

Before we leave the crucial—and initially perhaps troublesome—concept of mass, we must mention a third way of measuring the mass of objects, in everyday practice by far the most favored and accurate method.

By way of review, recall first that we need the essentially arbitrary standard. Once this has been generally agreed on, we can calibrate a spring balance by using it to give the standard measurable accelerations on a smooth horizontal plane. Then the calibrated balance can serve to determine other, unknown masses, either (a) by a new ob-

**Fig. 9.5** Weighing with an equal-arm balance.

servation of pull and resulting acceleration on the horizontal plane, or (b) by measuring on the balance $F_{grav}$ for the object in question and dividing this value by the measured gravitational acceleration $g$ at that locality.

To methods (a) and (b) we now add method (c), known of course as "weighing" on an equal-arm balance—at first glance seemingly simple and straightforward, but in fact conceptually most deceptive. We place the unknown mass $m_x$ on one pan (Fig. 9.5), and add a sufficient quantity of marked auxiliary-standard masses on the other side to keep the beam horizontal. When this point is reached, the gravitational pull on the unknown mass, namely $m_x g$, is counterbalanced by the weight of the standards, $m_s g$. (Note that we assume here the sufficiently verified experimental result concerning the equality of gravitational accelerations for all masses at one locality.) But if $m_x g = m_s g$, then $m_x = m_s$. Counting up the value of all standard masses in one pan tells us directly the value of the mass on the other side. The whole procedure is much to be preferred over methods (a) and (b), for the range, the accuracy of calibration, and the consistency of a spring balance are all quite limited, whereas in method (c) the equal-arm balance can handle accurately a very large range of masses, and the accuracy of individual measurements (which can be as good as $10^{-7}$ g with special care) even without refinements is easily better than $\frac{1}{1000}$ of a gram. And the results are repeatable and consistent, there being only these three sound and verifiable assumptions implied: that the arms of the balance are straight and equally long, that the value of $g$ is the same under each pan, and that the initial marking of the auxiliary standard masses is permanently correct.

## 9.9  INERTIAL AND GRAVITATIONAL MASS

We probably take for granted that a result obtained by method (a), described in the previous section, is always equal to the result, for the same object, involving methods (b) or (c). And yet, a moment's thought should convince us that this identity is completely unforeseeable—either an incredible coincidence or a symptom of a profound new law of nature. To begin with, consider that the mass was obtained in case (a) in a dynamic measurement: The object actually is allowed to "give in" to the force, is actually accelerated by the pull of the horizontally moving spring balance, and the mass so determined is truly a measure of the inertia of the object. Gravitational forces do not enter at all into the picture; in fact, the experimental measurement could be carried out just as well or better in gravitation-free space. On the other hand, methods (b) and (c), the static procedures of weighing on a vertically held spring balance or on an equal-arm balance, could not be invoked at all in the absence of gravity. Here the whole measurement depends on one property of material bodies: that they are attracted to other objects such as the earth; and inertia plays no role whatever. We see that case (a) on one side and (b) and (c) on the other measure two entirely different attributes of matter, to which we may assign the terms *inertial mass* and *gravitational mass*, respectively.

For practical purposes we shall make little distinction between the two types of mass. But in order to remind ourselves how essential and rewarding it may be to keep a clear notion of the operational significance of scientific concepts, and that, historically, considerable consequences may follow from a reconsideration of long-established facts in a new light, there are these words from Albert Einstein's and Leopold Infeld's book *The Evolution of Physics*:*

"Is this identity of the two kinds of mass purely accidental, or does it have a deeper significance? The answer, from the point of view of classical physics, is: the identity of the two masses is accidental and no deeper significance should be attached to it. The answer of modern physics is just the opposite: the identity of the two masses is fundamental and forms a new and essential clew leading to a more profound understanding. This was, in fact, one of the most important clews from which the so-called general theory of relativity was developed.

"A mystery story seems inferior if it explains strange events as accidents. It is certainly more satisfying to have the story follow a rational pattern. In exactly the same way a theory which offers an explanation for the identity of gravitational and inertial mass is superior to the one which interprets their identity as accidental, provided, of course, that the two theories are equally consistent with observed facts."

---

**Problem 9.11.** It is conceivable that, in another universe, some bodies have only gravitational mass and others only inertial mass. Describe how some everyday objects would behave if made only of one or only of the other type of material.

**Problem 9.12.** Assume that some gigantic equal-arm balance is firmly set up halfway between the earth and the moon, so that one pan hangs directly above the surface of the earth, the other directly above the *moon's surface*. If an earth-man with a

---

*    A. Einstein and L. Infeld, *The Evolution of Physics*, page 36. New York: Simon and Schuster (1938).

mass of about 80 kg (about 170 lb) steps on one pan, what must be the mass (in kilograms) of a moon-man on the other side to bring the balance to equilibrium?

## 9.10 EXAMPLES AND APPLICATIONS OF NEWTON'S SECOND LAW OF MOTION

Newton's second law, on which were built the previous paragraphs, is so central, so indispensable in physical science, that it will be well worth our looking at a few examples of its application to simple mechanical problems.

**Example 1.** An object of mass $m$ kg hangs from a calibrated spring balance in an elevator (Fig. 9.6). The whole assembly moves upward with a known acceleration $a$ m/sec². What is the reading on the balance?

*Solution:* As long as the elevator is stationary, the upward pull of the balance, and hence its reading $F_1$, will be equal in magnitude to the weight $mg$ of the object. The same is true if the elevator is moving *up* or *down* with constant speed, which are also conditions of equilibrium and of cancellation of all forces acting on $m$. However, in order to accelerate upward with $a$ m/sec², $m$ must be acted on by a net force of $ma$ newtons in that direction. In symbol form, $F_{net} = ma$; but

$$F_{net} = F_1 - mg,$$

so

$$F_1 = ma + mg = m(a + g).$$

The reading $F_1$ will be larger than before. Conversely, on decelerating by $a$ m/sec² while on the way up or accelerating on the way down, the spring balance will relax a little so that there remains on $m$ some net force $(mg - F_1)$ in the downward direction and, of course, of magnitude $ma$. Hence the reading $F_1 = mg - ma$. Finally, if the elevator were to fall freely ($a = g$), the net force on all

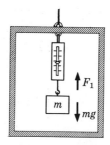

**Fig. 9.6**  Weight of an object suspended in an elevator.

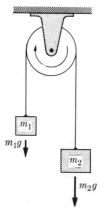

**Fig. 9.7**  Atwood's machine.

objects within must be simply their weight. That is, the last equation becomes $F_1 = mg - mg$, i.e., the spring balance reads zero. The object has only one force acting on it, its own weight $F_{grav}$.

Once more we are impressed with the need to speak clearly: A spring balance evidently does not always simply register the weight of an object hanging from it. This would be true only if the whole system were in equilibrium; in other cases, the balance shows a larger or smaller reading depending on the accelerating forces it must supply to the object.

**Example 2.** A string, thrown over a very light and frictionless pulley, has attached at its ends the two known masses $m_1$ and $m_2$, as shown in Fig. 9.7. Find the magnitude of the acceleration of the masses.

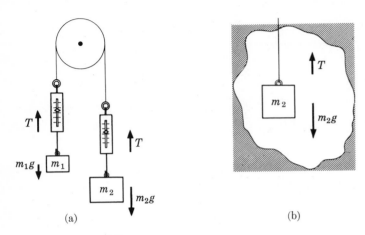

**Figure 9.8**

*Solution:* In this arrangement, called Atwood's Machine after the eighteenth-century British physicist who originated it, the net external force on the system of bodies is $m_2g - m_1g$ (again assuming $m_2$ to be larger than $m_1$). The total mass being accelerated is $m_1 + m_2$, and, consequently,

$$a = \frac{(m_2 - m_1)g}{(m_1 + m_2)}, \qquad (9.3)$$

which we may solve with the known values of $m_1$, $m_2$, and $g$ for that locality. Conversely, if $m_1$, $m_2$, and $a$ are experimentally determined with accuracy, and if corrections are made for the effects of string, pulley, and air resistance, a good value for $g$ is obtainable.

**Example 3.** In the Atwood's Machine problem above, Example 2, we may wish to inquire what the tension is in the string while the masses accelerate, a problem not unlike that of Example 1. First of all we must define clearly the meaning of "tension in a rope": for our purposes it will simply be the force reading if a spring balance were cut into the rope [Fig. 9.8(a)]. For a rope of negligible mass (the usual specification in our problem), this force is the same at each end.

To obtain the tension $T$ in terms of the other data or observables, consider Fig. 9.8(b) closely. If we fasten our attention on one or the other of the two masses only, say $m_2$, to the exclusion of the rest of the whole system, then it is evident that $m_2$ is acted on by a net force $(m_2g - T)$; therefore $(m_2g - T) = m_2a$, for $m_2$ is the only mass acted on by that net force. Rewriting, we obtain

$$T = m_2g - m_2a = m_2(g - a). \qquad (9.4)$$

On the other hand, exclusive concentration on $m_1$ gives $F_{net} = T - m_1g$, or

$$T = m_1a + m_1g = m_1(g + a). \qquad (9.5)$$

In answering the question as to the tension in the rope, either of these two results is usable. Incidentally, juxtaposing the two expressions for $T$, we obtain $m_2(g - a) = m_1(g + a)$, which yields again

$$a = \frac{(m_2 - m_1)}{(m_1 + m_2)} g.$$

This we had already derived on the basis of more direct arguments, which gives us confidence in the method of "isolation" adopted here. Conversely, the last equation may be used to determine the magnitude of the mass on one side of the pulley (say $m_1$) exclusively by observations made on the other side (i.e., the magnitude of $m_2$ and $a$). For example, $m_1 = m_2(g - a)/(g + a)$.

In a casual way this last example has hinted at a most powerful method of argument in physical science, namely, *the isolation of a specific and small part of the total situation.* As Fig. 9.8(b) symbolized, $T$ (and from it, $m_1$) could be found by observations made solely on one small portion of the assembly, the rest being perhaps hidden in the inaccessible darkness of a black box. We can thereby tell something about the "inside" of the box though we must remain "outside."

In a sense, nature always confronts the scientist with a "black-box problem": He is given some general laws of nature and a few observable manifestations of a complex scheme or hidden mechanism, and his task is to know as much as possible about the scheme or mechanism despite the necessarily scant and incomplete direct information. The problem may be as humble as this one—finding $m_1$ by observation made on $m_2$—or it may be as general as deducing the temperature, motion, size, mass, and composition of a distant star by observations made on a thin beam of its weak light.

Figure 9.9

**Problem 9.13.** A lamp hangs vertically from a cord in a descending elevator. The elevator decelerates by 300 cm/sec$^2$ before coming to a stop. If during this event the tension in the cord is 20 newtons, what is the mass of the lamp?

**Problem 9.14.** A 160-lb man stands on a platform (spring) balance in a moving elevator. Under what conditions will the balance read $\frac{1}{4}$ more than when at rest?

**Problem 9.15.** On a frictionless horizontal table rests a 15-kg mass $m_2$ connected as in Fig. 9.9 to hanging masses $m_1 = 10$ kg and $m_3 = 25$ kg. Find the acceleration of the system, and the tension in the rope above $m_1$.

**Problem 9.16.** A man tries to obtain a value for $g$ with an Atwood's Machine but, without his knowing it, he and his whole laboratory are freely falling through space. What will be his result for $g$, and why?

**Problem 9.17.** If tomorrow the whole earth were to move with large acceleration in a direction toward the North Pole Star, how would this manifest itself mechanically in different parts of the world?

## 9.11 NEWTON'S THIRD LAW OF MOTION

Newton's first law defined the force concept qualitatively, and the second law quantified the concept while at the same time providing a meaning for the idea of mass. To these, Newton added another highly original and important law of motion, the third law, which completes the general characterization of the concept of force by explaining, in essence, that each existing force has its mirror-image twin. In Newton's words,

"To every action there is always opposed an equal reaction: or the mutual actions of two bodies upon each other are always equal, and directed to contrary parts.

"Whatever draws or presses another is as much drawn or pressed by that other. If you press a stone with your finger, the finger is also pressed by the stone. If a horse draws a stone tied to a rope, the horse (if I may so say) will be equally drawn back towards the stone: for the distended rope, by the same endeavour to relax or unbend itself, will draw the horse as much toward the stone, as it does the stone toward the horse, and will obstruct the progress of the one as much as it advances that of the other...."

The point is rather startling: a single particle, all by itself, can neither exert nor experience any force at all. Forces spring up only as the result of interaction of two entities, and then the one pushes or pulls the other just as much as it is being pushed or pulled in return. The earth is attracted upward to a falling apple as much as the falling apple is attracted downward toward the earth. We may choose to call the one force the action and the other the reaction, but the order of naming is really arbitrary. It is not that one of the two forces *causes* the other. They both cause each other simultaneously, as two men, caught in a crowd, may collide and, at the same moment, complain it was the other one who had pushed. In the same manner, it is as meaningful to say it was the borrowing of some money that caused the debt as it is to reply that the lending of it was the cause of the credit. Action and reaction do stand to each other in the same relation as debit to credit: The one is impossible without the other, they are equally large but in an opposite sense, the causal connection is introduced only artificially, and, most important, they happen respectively to two different bodies.

To emphasize all these points, we might rephrase the third law of motion as follows: *Whenever two bodies A and B interact so that body A experiences a force (whether by contact, by gravity, by magnetic interaction, or whatever), then body B experiences simultaneously an equally large and oppositely directed force.*

## 9.12 EXAMPLES AND APPLICATIONS OF NEWTON'S THIRD LAW

It is easiest to understand the third law in terms of worked-out examples. The first two involve bodies in equilibrium, the last two accelerating systems.

**Example 1.** The simplest case concerns a box (body $A$) standing on the earth (body $B$). Let us enter the forces that act on each. Probably the one that comes to mind first is the weight of the box, $F_{grav}$. We name it here $F_{1A}$ and enter it as a vertically downward arrow, "anchored" to the box $A$ at its center of gravity [see Fig. 9.10(a)]. The reaction that must exist simultaneously with this pull of the earth on the box is the pull of the box on

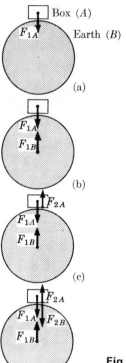

**Fig. 9.10** Forces on box and earth in contact.

the earth, equally large (by the third law) and entered as a vertical, upward arrow, $F_{1B}$, at the center of the earth in Fig. 9.10(b). This completely fulfills the third law. However, as the second law informs us, if this were the complete scheme of force, the box should fall down while the earth accelerates up. This is indeed what can and does happen while the box drops to the earth, settles in the sand, or compresses the stones beneath it. In short, the two bodies do move toward each other until there are built up enough mutual elastic forces to balance the previous set. Specifically, the earth provides an upward push on the box at the surface of contact, as shown in Fig. 9.10(c) by $F_{2A}$, an upward arrow "attached" to the box, while, by

**Figure 9.11**

the law now under discussion, there exists also an equal, oppositely directed force on the ground, indicated by $F_{2B}$ in part (d) of the figure.

There are now two forces on each body. Equilibrium is achieved by the equality in magnitude of $F_{1A}$ and $F_{2A}$ on $A$, and by $F_{1B}$ and $F_{2B}$ on $B$. But beware! $F_{1A}$ and $F_{2A}$ are not to be interpreted as action and reaction, nor are $F_{1B}$ and $F_{2B}$. The reaction to $F_{1A}$ is $F_{1B}$; to $F_{2A}$, $F_{2B}$. Furthermore, $F_1$ and $F_2$ are by nature entirely different sets of forces—the one gravitational, the other elastic. In short, $F_{1A}$ and $F_{1B}$ are equally large by Newton's third law, but $F_{1A}$ is as large as $F_{2A}$ (and $F_{1B}$ as large as $F_{2B}$) by the condition of equilibrium, derived from the *second* law.

**Example 2.** The sketch in Fig. 9.11 involves a system of four elements—a horizontal stretch of earth $E$ on which a recalcitrant beast of burden $B$ is being pulled by its master $M$ by means of a rope $R$. Follow these four force couples: $F_{1E}$ is the push experienced by the earth, communicated to it by the man's heels (essentially by static friction). The reaction to $F_{1E}$ is the equally large force $F_{1M}$ exerted on the man by the earth. The man pulls on the rope to the left with a force $F_{2R}$, and the reaction to this is the force $F_{2M}$ with which the rope pulls on the man to the right. A third set is $F_{3B}$ and $F_{3R}$, acting respectively on the donkey and on the rope. Finally, the interaction between earth and animal is $F_{4E}$ and $F_{4B}$. In equilibrium, the separate forces on each of the four objects balance; but if equilibrium does not exist, if the man succeeds

in increasing the donkey's speed to the left, then $F_{3B} - F_{4B} = m_{\text{beast}} \times a$, and similarly for the other members of the system. And whether there is equilibrium or not, any "action" force is equal and opposite to its "reaction."

The whole point may be phrased this way: By the third law, the forces $F_{1E}$ and $F_{1M}$ are equal; similarly, $F_{2M}$ and $F_{2R}$ are equal. But the third law says nothing whatever about the relationship of $F_{1M}$ to $F_{2M}$, two forces arranged to act on the same body by virtue of the man's decision to pull on a rope, not by any necessity or law of physics. If there happens to be equilibrium, then $F_{1M}$ will be as large as $F_{2M}$ by Newton's second law (or, if you prefer, by definition of "equilibrium").

Incidentally, to the master of the beast the difference between zero velocity and a larger, though constant, velocity is of course all that matters; and so he would naturally consider Newtonian physics contrary to common sense if he heard that we lump both these so differently prized states, rest and uniform motion, under one name—equilibrium—and treat them as equivalent. The donkey driver's attitude is certainly understandable, but surely we must avoid it if, as happens again and again, science forces us to break with commonsense preconceptions. The reward is always greater insight into and power over a whole class of natural phenomena.

**Example 3.** Consider an automobile accelerating on a straight road. The force which propels it is essentially the (static) friction between the tires

**Fig. 9.12**  The reaction-car experiment.

and road at their surfaces of momentary contact, and this force is, by the third law, equal to the contrary force which the earth experiences. Consequently an observer on a fixed outpost in the universe, on a star perhaps, should ideally be in a position to observe the earth moving backward under the car while the latter moves forward over the earth. But evidently these equal forces, if they were the only ones acting, would produce very unequal accelerations. The inertia of the earth being effectively infinitely large compared with that of the car, the former will experience only an infinitely small acceleration. Nevertheless, there is implied an important conceptual clarification: Strictly speaking, the $a$ in $F_{net} = ma$, the acceleration produced by forces on the earth, should be measured not with respect to the earth itself, but rather with respect to some fixed point in space—a "fixed star" would serve. In practice, we do allow ourselves to use the earth as a frame of reference for the measurement of accelerations because the earth's large inertia ensures that it will respond only negligibly to the forces of reaction communicated to it by mutual gravitation, friction, elastic interaction, and so on.

**Example 4.** Consider two blocks at rest on a horizontal, frictionless track (or fitted as cars with light, friction-free wheels). They are connected to each other by a tightly compressed spring (Fig. 9.12). Evidently, the forces $F_1$ and $F_2$ on the two blocks owing to their mutual pressure are equal though oppositely directed, and only the balancing forces exerted by a string stretched from one block

to the other prevents them from shooting off in opposite directions. Now the string is cut or burned off, and the freed spring propels the cars to the right and the left. At any instant, the opposing forces on the two cars are still equally large; we express this by an equation,

$$F_1 = -F_2.$$

As the spring quickly relaxes more and more, the forces at its two ends diminish in magnitude until finally the mutual forces on the cars are zero as they separate completely and the spring drops out entirely. Taking an average force $\overline{F}$ for the whole performance, we have

$$\overline{F}_1 = -\overline{F}_2.$$

However, in the absence of friction, $\overline{F} = m\overline{a}$ (where $\overline{a}$ stands for average acceleration), or

$$m_1\overline{a}_1 = -m_2\overline{a}_2. \tag{9.6}$$

The average acceleration is given by $\overline{a} = (v - v_0)/t$. For both cars, $v_0 = 0$ (start from rest), and $t$, the time of interaction, is of course the same on each. Thus we rewrite Eq. (9.6):

$$m_1v_1 = -m_2v_2. \tag{9.7}$$

Considering simply the *numerical magnitudes* of the velocities, i.e., neglecting the minus sign (which only serves to remind us that the velocities are toward opposite directions), we note that

$$\frac{m_1}{m_2} = \frac{v_2}{v_1}, \quad \text{or} \quad m_1 = m_2\frac{v_2}{v_1}. \tag{9.8}$$

In words, these equations reveal that the velocities achieved by two objects by mutually interacting forces are inversely proportional to their masses. Equation (9.8) has special further importance. If one of these objects, say $m_2$, is chosen to be *a standard of mass*, then the inertia of any other object $m_1$ can be directly compared by means of the easily found ratio of measured velocities $v_2$ and $v_1$. Specifically, if in some special case $m_2 =$

1 kg (by definition), and $v_2 = 20$ cm/sec and $v_1 = 5$ cm/sec by observation, then $m_1 = 4$ kg.

This so-called *reaction-car experiment* immediately offers us a fourth and last method, which we shall call (d) in analogy with the three methods discussed in Section 9.8, for the determination of the mass of an object, namely, by using a standard as mass $m_2$ and finding the unknown mass from Eq. (9.8). As in method (a), we find here its inertial mass relative to a previously chosen standard; but this new case has the profound advantage that there enters nowhere a calibration of spring balances or even a *measurement* of forces. As long as we know, by ample experimental verification of Newton's third law, that the mutual forces between these two cars are equally large and oppositely directed, there is no further need to investigate their actual magnitudes.

---

**Problem 9.18.** A $3\frac{1}{3}$-ton truck starts from rest and reaches a speed of 40 km per hour in half a minute. What will be the change of speed of the earth during that same time? (Mass of earth = $6.6 \times 10^{21}$ tons or about $6 \times 10^{24}$ kg.) Is the effect on the earth different if the truck takes three times as long to reach the same final speed?

**Problem 9.19.** The thought experiment involving the reaction cars for the determination of relative masses was first proposed by Ernst Mach, the nineteenth-century Austrian physicist, in one of the early self-conscious examinations of the foundations of mechanics. (His works later stimulated a good deal of modern thought in physics and the philosophy of science.) Describe a detailed step-by-step procedure for using his method to find the inertial mass of an object (in grams). Then describe how you would go on to prove experimentally that the gravitational mass of the object (in kilograms) has an identical value, and finally, show how you would find the accurate weight of the object (in newtons) at that locality.

## RECOMMENDED FOR FURTHER READING

E. N. da C. Andrade, *Sir Isaac Newton, his Life and Work*, Garden City, N.Y.: Doubleday Anchor, 1958; paperback. A short, readable biography.

A. R. Hall, *From Galileo to Newton 1630–1720*, New York: Harper, 1963; Chapters VI, VII, and IX

E. C. Kemble, *Physical Science*, Chapter 8

Isaac Newton, extracts from the *Principia*, in Hurd and Kipling, *The Origins and Growth of Physical Science*, Vol. 1, pages 178–208

## SOURCES, INTERPRETATIONS, AND REFERENCE WORKS

I. B. Cohen, *Introduction to Newton's Principia*, Cambridge, Mass.: Harvard University Press, 1971

R. G. Colodny (editor), *Beyond the Edge of Certainty*, Englewood Cliffs, N.J.: Prentice-Hall, 1965. See the articles by B. Ellis and N. R. Hanson.

R. Dugas, *Mechanics in the 17th Century*, New York: Central Book Co., 1958; Chapters XI and XII

R. P. Feynman *et al.*, "Newton's laws of dynamics," in *The Feynman Lectures on Physics*, Reading, Mass.: Addison-Wesley, 1963, Vol. 1, Chapter 9; reprinted in *Project Physics Reader*, Vol. 1

R. L. Greaves, "Puritanism and science: the anatomy of a controversy," *Journal of the History of Ideas*, Vol. 30, pages 345–368 (1969)

A. R. Hall, "Merton revisited, or science and society in the seventeenth century," *History of Science*, Vol. 2, pages 1–16 (1963).

Marie Boas Hall, *Nature and Nature's Laws*, pages 97–130, 216–229, 271–347.

T. L. Hankins, "The reception of Newton's second law of motion in the eighteenth century," *Archives Internationales d'Histoire des Sciences*, Vol. 20, pages 43–65 (1967)

J. Herivel, *The Background to Newton's Principia*, New York: Oxford University Press, 1965. Includes several unpublished manuscripts.

R. W. Home, "The third law in Newton's mechanics," *British Journal for the History of Science*, Vol. 4, pages 39–51 (1968)

M. Jammer, *Concepts of Force*, Cambridge, Mass.: Harvard University Press, 1957; Harper Torchbook paperback reprint

M. Jammer, *Concepts of Mass*, Cambridge, Mass.: Harvard University Press, 1961; Harper Torchbook paperback reprint; Chapters 1 to 8

Ernst Mach, *The Science of Mechanics*, pages 226–342

F. Manuel, *A Portrait of Isaac Newton*, Cambridge, Mass.: Harvard University Press, 1968

L. T. More, *Isaac Newton, a Biography*, New York: Scribner, 1934; Dover paperback reprint

Isaac Newton, *Mathematical Principles of Natural Philosophy*, third edition, London, 1726; A. Motte's translation revised by F. Cajori, Berkeley: University of California Press, 1934, paperback

M. R. Perl, "Newton's justification of the laws of motion," *Journal of the History of Ideas*, Vol. 27, pages 585–592 (1966)

C. Truesdell, "Reactions of late Baroque mechanics to success, conjecture, error, and failure in Newton's Principia," pages 138–183 in his *Essays in the History of Mechanics*, New York: Springer-Verlag, 1968

R. S. Westfall, *Force in Newton's Physics*, New York: American Elsevier, 1971

# Chapter 10

# ROTATIONAL MOTION

At this point, we have at our command enough knowledge and tools to deal with a large variety of problems concerning motions and forces. The basic structure has been set up for an understanding of the type of concepts, questions, and methods of answer in the repertoire of the physical scientist. But there is still one dangerous gap to be filled, one main supporting pillar to be moved into place before the next level can be constructed.

In the previous chapters we first acquainted ourselves with the description of uniformly accelerated motions along a straight line, and in particular with that historically so important case of free fall. Next came general projectile motion, an example of motion in a plane considered as the superposition of two simple motions. Then we turned to a consideration of the *forces* needed to accelerate bodies along a straight line. But there exists in nature another type of behavior, not amenable to discussion in the terms that we have used so far, and that is *rotational motion*, the motion of an object in a plane and around a center, acted on by a force that continually changes its direction of action. This topic subsumes the movement of planets, flywheels, and electrons alike.

We shall follow the same pattern as before: concentrating on a simple case of this type, namely, circular motion. We shall first discuss the kine-

matics of rotation without regard to the forces involved, and finally study the dynamics of rotation and its close ally, vibration.

## 10.1 KINEMATICS OF UNIFORM CIRCULAR MOTION

Consider a point whirling with constant speed in a circular path about a center $O$; the point may be a spot on a record turntable, or a place on our rotating globe, or, to a good approximation, the planet Venus in its path around the sun.

Before we can investigate this motion, we must be able to describe it. How shall we do so with economy and precision? Some simple new concepts are needed:

a) The *frequency* of rotation is the number of revolutions made per second (letter symbol $n$), expressed in 1/sec (or $\sec^{-1}$), since the phrase "number of revolutions" does not belong among such fundamental physical quantities as mass, length, and time. A record that revolves at 78 revolutions per minute (rpm) therefore has a frequency of rotation of 78/60 or 1.3 $\sec^{-1}$.

b) Next, we define the concept *period of rotation* (symbol $T$) as the number of seconds needed per

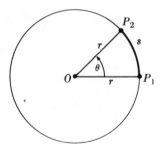

**Fig. 10.1** Definition of an angle in radian measure: $\theta = s/r$.

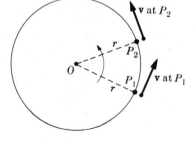

**Figure 10.2**

complete revolution, exactly the reciprocal of $n$, and expressed in units of seconds.

$$T = \frac{1}{n}. \qquad (10.1)$$

The record would consequently have a period of rotation of 0.77 sec.

c) An angular measure is required. The angle $\theta$ swept through by a point going from $P_1$ to $P_2$ (Fig. 10.1) can, of course, be measured in degrees, but it is more convenient at this point to express $\theta$ by the defining equation

$$\theta = \frac{s}{r}. \qquad (10.2)$$

This ratio of arc to radius is a dimensionless quantity; however, the name *radians* is nevertheless attached to this measure of angle, partly to distinguish it from degrees of arc.

To fix the rate of exchange between these two types of angular measure, consider the special case of $\theta = 360°$. For that angle, $s = 2\pi r$, namely, the circumference of the circle, and therefore $\theta = 2\pi r/r = 2\pi$ radians. If $360° = 2\pi$ radians, $1° = 0.0175$ radian and 1 radian $= 57.3°$.

d) We now inquire into the velocity of a particle executing uniform circular motion. The word

"uniform" means, of course, that the rate of spin (the speed $s/t$) does not change. Nevertheless, for future reference, it is well to remember that the velocity vector associated with the rotating point does change in *direction* from one instant to the next, although its magnitude, represented by the length of the arrows in Fig. 10.2, is here constant.

Let us now concentrate entirely on the magnitude of the velocity, the speed, given again by the ratio of distance covered to time taken; if we know the period or frequency of the motion and the distance $r$ from the spot to the center of the circle, $v$ is directly found (usually in cm/sec) by realizing that $s = 2\pi r$ if $t = T$, that is,

$$v = \frac{2\pi r}{T} = 2\pi n r. \qquad (10.3)$$

e) The quantity $v$ defined in this last equation refers to the magnitude of the tangential or *linear* velocity, that is, to the velocity of the point along the direction of its path. Analogous to this time rate of change of distance stands the powerful concept *angular velocity* [symbolized by the Greek letter $\omega$ (omega)], which is the time rate of change of angle. By definition, for the type of motion we are here considering,

$$\omega = \frac{\theta}{t}, \qquad (10.4)$$

a quantity to which we may assign units of $sec^{-1}$ or, more colloquially, radians/sec. If we happen to know $n$ or $T$, we can find the magnitude of the angular velocity from the fact that $\theta = 2\pi$ if $t = T$, or

$$\omega = \frac{2\pi}{T} = 2\pi n. \qquad (10.5)$$

The formal relation between $\omega$ and $v$ is evident if we compare Eqs. (10.3) and (10.5):

$$v = \omega r. \qquad (10.6)$$

But direct inspection of the defining equation (10.4) for $\omega$ reveals a significant difference between $v$ and $\omega$. Every point on a rigid rotating body (e.g., a disk), no matter where it is placed on that body, has the *same* $\omega$ at the same instant, whereas those different points take on very different values for $v$, depending on their respective distances $r$ from the axis of rotation. The concept of angular velocity $\omega$ therefore introduces into problems of rotational motion the possibility of constancies not attainable directly in terms of $v$ alone. There is no better reason for coining a new concept in any field of science.

---

**Problem 10.1.** How great an angle (in radians and in degrees) is covered by a spot on a 78-rpm record in 5 min of constant rotation? What are the linear speeds of two points, respectively 3 and 12 cm from the center of rotation? What are the corresponding angular speeds?

**Problem 10.2.** How long is the groove that travels under the needle while playing a 12-in. LP (33 rpm) record? (You must make and explicitly state certain reasonable assumptions in this problem.)

**Problem 10.3.** Find the frequency of revolution, the linear velocity, and the angular velocity for a spot on the equator (earth's radius = 6380 km; period of rotation about its own axis = 23 hr 56 min per rotation).

**Problem 10.4.** Suppose a high tower stands at a point on the earth's equator. Using the data given in Problem 10.3, calculate the linear velocity of an object at the top of the tower. Is an object at the foot of the tower going with exactly the same linear velocity? If the object at the top were dropped, would you expect it to land exactly at the base of the tower (neglecting air resistance, and assuming an absolutely vertical tower)? If not, in which direction would you expect a deviation? (You are not expected to calculate the actual amount of deviation, but you should be able to explain qualitatively the reason for a deviation, and evaluate Galileo's argument given in a footnote near the end of Section 5.3.)

**Problem 10.5.** Find the approximate linear speed of our planet in its yearly path around the sun (assuming it to be a circular orbit with radius $1.5 \times 10^{11}$ m).

## 10.2 CENTRIPETAL ACCELERATION

It has been noted that motion with constant speed around a circle implies that the velocity vector is continually changing in direction though not in magnitude. According to Newton's laws of motion, a force must be acting on a body whose velocity vector is changing in any way, since if there were no net force it would continue to move at constant velocity in a straight line. And if there is a force, there must be an acceleration. So when in circular motion with constant speed, a body is in the seemingly paradoxical situation of being accelerated but never going any faster (or slower)!

To clarify this situation we have to emphasize the distinction that the physicist makes between velocity and speed. As we saw in Section 8.2, velocity is a *vector* quantity, having both magnitude and direction; speed is a *scalar* quantity, having magnitude only.* Any change in the velocity is

---

\*    See Appendix IX.

Figure 10.3

**Fig. 10.4** A centripetal force acts on the stone, while a centrifugal force acts on the string and hand.

an acceleration, whether or not there is a change in the speed.

In the case of circular motion, the change in direction of the velocity vector has been shown in Fig. 10.2; we now need to analyze this change in a little more detail (Fig. 10.3). The vector labeled "$v$ at $P_2$" is the resultant of two other vectors which must be added together: the vector "$v$ at $P_1$" and the vector "$\Delta v$" which represents the change in velocity that occurs during the time interval $\Delta t$ as the body moves along the circle from $P_1$ to $P_2$.

As can be seen from the diagram, the vector $\Delta v$ is directed *toward the center of the circle*. (A mathematical proof of this fact could be given, but it seems unnecessary.) The acceleration is defined as the change in velocity divided by the time interval, in the limit as the time interval becomes very small:

$$a = \lim_{\Delta t = 0} \left( \frac{\Delta v}{\Delta t} \right).$$

Acceleration is also a vector, directed toward the center of the circle; hence it is called the *centripetal acceleration* ("centripetal" means "seeking the center"). The corresponding force that must be

acting on the body to produce the acceleration is a vector in the same direction, $F = ma$ according to Newton's second law; it is called the *centripetal force* ($m$ is the mass of the moving body).

Unfortunately the ordinary language that we use to describe motion causes confusion at this point. We are accustomed to hear talk about "centrifugal" force, a force that is said to act on a whirling body in a direction *away* from the center. We shall not use this term, because there is no such force, despite the illusion. If you tie a weight on a string and twirl it around your head (Fig. 10.4), you think you can feel such a force, but that is a force acting on you at the center, not the force acting on the whirling body; it is in fact the third-law reaction. Or, if the string breaks, you see the object flying away from the circular path it was following before. But no force pulls it in flight (not counting gravity, downward) and it does not move off along a radius either.

Isaac Newton was one of the first to recognize that all these phenomena are due to the natural tendency—*inertia*—of any body to keep on moving in the same direction if it is not constrained to do otherwise. If the string breaks when the weight is at the point $P_1$ (Fig. 10.3), it will "fly off on a tangent" (not along the radius)—that is, it will continue to move in the direction indicated by the arrow of the velocity vector at $P_1$. (In this case, at least, the popular expression is accurate, since the velocity vector is indeed along a line tangent to the circle at $P_1$.) While the object is still attached to the string and moving in a circle, you have to provide a force on it toward yourself—the centripetal force—to prevent it from flying off. And since you are exerting a force on the string, the string must also, by Newton's third law, exert an equal and opposite force on your hand. The outward force that *you* feel is the reaction to the force you apply.

What is the magnitude of the centripetal acceleration? In other words, how does it depend on the speed of rotation and the size of the circular orbit? The answer to this question was crucial to Newton's theory of the solar system; as we shall see in the next chapter, it was needed as a step to the law of gravitation. Newton worked out the theory of centripetal acceleration during his burst of creative activity in the Plague years 1665–66, although his publication of it, in 1687, was 14 years later than the published derivation of his great contemporary, the Dutch physicist Christian Huygens. The result for the magnitude of the centripetal acceleration is very simple:

$$a = \frac{v^2}{r}.$$  (10.7)

Thus the centripetal acceleration increases with the square of the linear speed of the moving body in its orbit, but decreases inversely as the radius of the circle.

Before turning to the derivation of this result, let us summarize its physical meaning by quoting Newton's very clear discussion in the *Principia* of 1687:

"Definition: A centripetal force is that by which bodies are drawn or impelled, or in any way tend, towards a point as to a centre. Of this sort is gravity, by which bodies tend to the centre of the earth; magnetism, by which iron tends to the lodestone; and that force, whatever it is, by which the planets are perpetually drawn aside from the rectilinear motions, which otherwise they would pursue, and made to revolve in curvilinear orbits. A stone, whirled about in a sling, endeavours to recede from the hand that turns it; and by that endeavour, distends the sling, and that with so much the greater force, as it is revolved with the greater velocity, and as soon as ever it is let go, flies away. That force which opposes itself to this endeavour, and by which the sling perpetually draws back the stone towards the hand, and retains it in its orbit, because it is directed to the hand as the centre of the orbit, I call the centripetal force. And the same thing is to be understood of all bodies, revolved in any orbits. They all endeavour to recede from the centres of their orbits; and were it not for the opposition of a ... force which restrains them to, and detains them in their orbits, which I therefore call centripetal, would fly off in right lines, with an uniform motion . . . . Nor could the moon, without some such force, be retained in its orbit. If this force was too small, it would not sufficiently turn the moon out of a rectilinear course: if it was too great, it would turn it too much, and draw down the moon from its orbit towards the earth. It is necessary, that the force be of a just quantity, and it belongs to the mathematicians to find the force, that it may serve exactly to retain a body in a given orbit, with a given velocity . . . ."

## 10.3 DERIVATION OF THE FORMULA FOR CENTRIPETAL FORCE

In Fig. 10.3, a point moves uniformly through an angle $\theta$ from $P_1$ to $P_2$ along the arc $s$. The speeds at $P_1$ and $P_2$ are equal, but the direction of the velocity vectors changes through angle $\theta$ between $P_1$ and $P_2$. The change of velocity, $\Delta v$, is obtained in the usual manner in Fig. 10.3(b). Now note that the triangle there, and the one contained by $P_1OP_2$

in Fig. 10.3(a), are similar isosceles triangles. Therefore $\Delta v/v = x/r$ and $\Delta v = vx/r$. On dividing both sides by $\Delta t$, the time interval needed for this motion, we obtain $\Delta v/\Delta t = vx/r\,\Delta t$. The left side represents the average acceleration $\bar{a}$ during $\Delta t$, and if, as in Chapter 6, we restrict $\Delta t$ to smaller and smaller values, in the limit as $\Delta t$ approaches 0, the average acceleration becomes equal to the instantaneous acceleration $a$:

$$a = \lim_{\Delta t \to 0}\left(\frac{\Delta v}{\Delta t}\right) = \lim_{\Delta t \to 0}\left(\frac{vx}{r\,\Delta t}\right).$$

At the same time, however, as $\Delta t$ (and with it, $\theta$) decreases, the line $x$ becomes more and more nearly equal to the arc $s$, so that in the limit, $x = s$, and we write

$$a = \lim_{\Delta t \to 0}\left(\frac{v}{r}\frac{s}{\Delta t}\right).$$

We note finally that as $\Delta t \to 0$, $s/\Delta t \to v$, the instantaneous velocity of the moving point, so that

$$a = \frac{v^2}{r}.$$

Earlier we found $v = \omega r$; thus $a$ is also expressed by

$$a = \omega^2 r.$$

As can be seen from Fig. 10.3, the direction of $\Delta v$, and hence the direction of $a$, is perpendicular to the chord marked $x$. In the limit as $P_1$ and $P_2$ coalesce, it will therefore be perpendicular to the instantaneous velocity vector, which points along the tangent to the curve. Since the acceleration is directed toward the center, the term "centripetal" can be used, and we shall henceforth attach a subscript $c$. Similarly, the centripetal force will be denoted by $F_c$; it is, by the second law that always applies, equal to $ma_c$.

The physical causes and means for providing a rotating object with the necessary centripetal force vary greatly. A rotating flywheel is held together by the strength of the material itself, which can supply the stress, at least up to a point. A car rounding a curve on a level road is kept from flying off tangentially in a straight line by the force of friction, which supplies the centripetal force by acting (sideways) on the tires. The moon, as Newton first explained, is kept captive in its orbit about the earth by the gravitational pull continually experienced by it. The electron circulating around an atomic nucleus does so by virtue of electric attraction to the center. But in each case, insofar as the motion is circular and uniform, $F_c = ma_c$, or, to clothe this expression in all its alternative forms,

$$F_c = ma_c = \frac{mv^2}{r} = m\omega^2 r = m\frac{4\pi^2}{T^2}r = m4\pi^2 n^2 r.$$

$$(10.8)$$

**Example.** Find the centripetal acceleration of an object on the equator.

*Solution:* This object makes a circular path, covering an angle of $2\pi$ radians per day ($= 8.6 \times 10^4$ sec); $\omega = (2\pi/8.6 \times 10^4)$ rad/sec $= 7.3 \times 10^{-5}$ rad/sec. The radius of the earth is about $6.4 \times 10^6$ m. Therefore $a_c = \omega^2 r \doteq 0.034$ m/sec$^2$.

**Problem 10.6.** Find the centripetal acceleration owing to the earth's rotation at a place with latitude 30°. Find the centripetal acceleration for the whole earth, owing to its yearly swing around the sun. (Approximate distance $= 1.5 \times 10^{11}$ m; the assumption of a circular orbit is here necessary, though not entirely correct.)

**Problem 10.7.** Derive, step by step, from first principles, the relation $F_c = m(4\pi^2/T^2)r$, and explain at each point what law of physics, concept, approximation, etc., has been introduced.

**Problem 10.8.** The rim of a flywheel 3 m in height is a metal band of 1000 kg mass. The spokes can support a tension of up to $10^6$ newtons. At what speed of rotation will the wheel burst?  Make a drawing to show how the pieces will fly away.

**Problem 10.9.**  Mention several other examples of rotational motion, and explain in each case how centripetal force is supplied.

## SOURCES, INTERPRETATIONS, AND REFERENCE WORKS

J. Herivel, "Newton's discovery of the law of centrifugal force," *Isis*, Vol. 51, pages 546–553 (1960). See also his book, *The Background to Newton's Principia*, New York: Oxford University Press, 1965, for further discussion and comments.

# Chapter 11

# NEWTON'S LAW OF UNIVERSAL GRAVITATION

In the second half of the seventeenth century, as science was beginning to take a recognizably modern form, one long-standing problem was ripe for solution: the structure and dynamics of the solar system. Most of the preceding chapters in this book have been concerned with assembling the pieces of the puzzle—not because the major goal of physics is to understand the solar system, but because the successful solution of that problem provided at the same time a powerful method for attacking the whole range of problems in terrestrial physics, and even for inquiring into how the universe as a whole is constructed.

In this chapter we will see how Newton, in his *Principia*, was able to apply his theory of forces and motion to the astronomical system developed by Copernicus, Kepler, and Galileo. In this application the formula for centripetal acceleration derived in Chapter 10 will play an essential part. After showing how Newton's law of universal gravitation provides a convincing explanation for some of the phenomena known in the seventeenth century, we shall briefly survey its role in later developments in astronomy. Finally, we shall attempt to assess Newton's place in the overall development of science in the light of his own theories, as well as their impact on the modern view of the universe.

## 11.1 DERIVATION OF THE LAW OF UNIVERSAL GRAVITATION

The three main sections of the *Principia* contain an overwhelming wealth of mathematical and physical discoveries; included are the proofs leading to the law of universal gravitation, proofs so rigidly patterned* that it seems better to present here a derivation of this historic law in another plausible sequence. The arguments are sometimes rather subtle, and thus they afford a splendid illustration of the interplay between established laws, new hypotheses, experimental observations, and theoretical deductions in physical theory. The purpose of the following pages is to gain an understanding of the process, rather than to invite the memorizing of the individual steps.

a) Planets and satellites are not in equilibrium. An unbalanced (net) force acts on them. If they were in equilibrium, if no net force were acting on them, they would be moving in straight lines instead of elliptical paths, in accord with Newton's first law of motion.

---

\* It is not without interest to note that the *form* of the arguments in Newton is along the traditional deductive pattern of Euclid, as, for that matter, was the work of St. Thomas Aquinas or Spinoza.

b) Whatever the nature or the magnitude of the net force acting on a planet or on a satellite, its direction must be toward the center of motion at every instant. Newton derived this conclusion directly from Kepler's second law (*Principia*, Book I, Propositions I and II), and we may paraphrase his own argument as follows:

A body moves along a straight line at constant velocity; at equal time intervals $\Delta t$ it traverses equal spaces, i.e., in Fig. 11.1(a), $\overline{PQ} = \overline{QR} = \overline{RS}$, etc. With respect to any fixed point $O$, the line from $O$ to the moving body sweeps out equal areas during equal time intervals, for the triangles $PQO$, $QRO$, $RSO$, etc., are all of equal area, having equal bases and equal heights. Now imagine that this body experiences a brief and sudden blow at $Q$, a force directed exactly along $QO$. Of course, the direction of motion is modified. A velocity component has been added which, during time $\Delta t$, by itself would move the body from $Q$ to $Q'$ [Fig. 11.1(b)], but which, in conjunction with the original velocity toward $R$, adds up to a total displacement from $Q$ to $R'$. But the area swept out during time $\Delta t$ is *not* affected by this reorientation! The area of $OQR$, which would have been covered in time $\Delta t$ if no blow had been applied, is the same as the area of $OQR'$ that actually was covered. (*Proof:* $RR'$ is parallel to $QQ'$, thus $\Delta OQR$ and $\Delta OQR'$ have the same base and equal heights.) Therefore, area $OQR'$ is also equal to area $OPQ$.

If no further force acts, the motion during equal time intervals $\Delta t$ then proceeds from $Q$ to $R'$, from $R'$ to $S'$, etc. But a second blow at $R'$, again directed exactly to $O$, modifies the motion once more [Fig. 11.1(c)]. By the very same argument as before, we know that area $OR'S'' = OR'S'$. In general, we conclude that *centrally directed* forces applied at equal time intervals do not affect the areas swept out per unit time. There is no reason why those time intervals should not be chosen as we please, so that in the limit as $\Delta t$ approaches zero, the centrally directed force becomes a continuously acting centripetal force and the broken

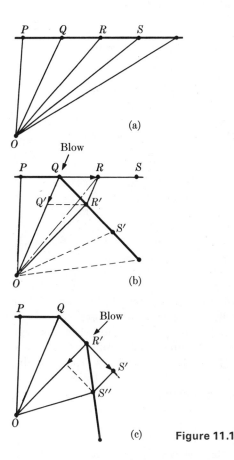

(a)

(b)

(c)     Figure 11.1

curve melts into a smooth one. Finally, we turn the argument around and, with Newton, say that since planets by Kepler's empirical second law do indeed sweep out equal areas per unit time, the force acting on each must be a continuous, centrally directed force. In the case of the ellipse, this center of force is one of the foci; for the circle, the center of the figure.

**Problem 11.1.** Strictly speaking, we should also prove that blows *not* all directed to the same center

$O$, but instead to any other points, will fail to yield an equal area law. Attempt such a proof.

---

c) Now that we accept that the force must be centrally directed, a *centripetal* force, there appears the following crucial problem: "If a body revolves in an ellipse [including the special case of a circle], it is required to find the law of the centripetal force tending to the focus of the ellipse." Newton's mathematics proved for the first time that for paths along conic sections such as hyperbolas, parabolas, ellipses, and circles, the centripetal force at any instant must be proportional to the inverse square of the distance of the body to the focus. In short, any body obeying Kepler's first law of elliptical paths must be acted on by a force following the law $F = C/R^2$, where $C$ is a constant for that body and $R$ is measured from the center of the body to the center of forces, the focus of the ellipse.

We cannot at this point follow the general proof, but we shall show that if for a planet in a *circular* path the centripetal force *is* granted to be equal to $F = C/R^2$, there follows by derivation, without further assumptions, that the celestial body should also obey the law

$$T^2 = KR^3.$$

Conversely, since we can observe the latter to be actually true—it is Kepler's third law—we judge that the hypothesis $F = C/R^2$ is well founded for the case of planetary motion.

The derivation proceeds as follows. The centripetal force $F_c$ (or now simply $F$), on the planet which, as assumed, is given by $C/R^2$, is also by Newton's second law equal to $m_p a_c$, where $m_p$ is the mass of the planet and $a_c$ is the centripetal acceleration. For circular paths around the sun, as are actually followed by almost all planets with but little deviation,

$$a_c = \frac{v^2}{R},$$

$v$ being the speed of the planet in its orbit. But

$$a_c = \frac{v^2}{R} = \frac{4\pi^2 R^2}{T^2 R} = \frac{4\pi^2 R}{T^2},$$

where $T$ is the period of orbital revolution of the planet. It follows [see Eq. (10.8)] that

$$F = m_p a_c = m_p \frac{4\pi^2 R}{T^2}. \tag{11.1}$$

Combining the last result with our assumed value for $F$, we have

$$\frac{C}{R^2} = m_p \frac{4\pi^2 R}{T^2}, \quad \text{or} \quad T^2 = \left[ m_p \frac{4\pi^2}{C} \right] R^3. \tag{11.2}$$

Because $m_p$ and $C$ are constant, at least for a given planetary orbit, the bracketed term in Eq. (11.2) is constant for a given planet, no matter what the size of its orbit. $T^2$ here is proportional to $R^3$. This is reminiscent of the form of Kepler's third law, but unless we can prove (as we have not yet done above) that the bracketed term in Eq. (11.2) is really the same constant for *all* planets, the quantity $T^2/R^3$ from Eq. (11.2) may have a different value for different planets.

But we must recognize for this reason also that this was no proof of the inverse square law for the centripetal force, either. Kepler's law requires that

$$\left[ \frac{m_p 4\pi^2}{C} \right] = K,$$

where $K$ is, for our solar system, the same constant for all planets in all their orbits around the sun. Not until we discover what $C$ contains can we know whether the bracketed term actually yields the same constant for all planets. We shall strengthen and complete the above proof as soon as we can, for we have decided not to by-pass the difficult points. Let us note in passing that use has been made of Newton's second law of motion and of the equation for centripetal acceleration.

Historically, Newton's demonstration that elliptical planetary paths imply an inverse square law of the force came at a time when the idea of such a law was generally "in the air." In fact, Halley had come to Newton in 1684 just in order to ask whether he could supply the proof which others were looking for in vain!

d) The *origin* of the centripetal force needed to keep the planets in their orbits has not been touched on so far. We recall that Kepler speculated that some driving magnetic force reached out from the sun to move the planets. He was wrong, but at least he was the first to regard the sun as the controlling mechanical agency behind planetary motion. Another picture had been given by the great French philosopher Descartes (1596–1650), who proposed that all space was filled with a subtle invisible fluid of contiguous material corpuscles; the planets of the solar system were supposed to be caught in a huge vortex-like motion of this fluid about the sun. This idea was attractive to the minds of the day, and consequently was widely accepted for a time, but Newton proved that this mechanism could not account for the quantitative observations on planetary motion as summarized, for example, in Kepler's laws. The problem remained.

At this point Newton proposed a dramatic solution: the centripetal force on the planets is nothing but a gravitational attraction of the sun, and the centripetal force on satellites revolving around planets is also completely given by the gravitational pull by the planets. (Less than a century before it would have been impious or "foolish" to suggest that terrestrial laws and forces regulated the whole universe, but now, after Kepler and Galileo had unified the physics of heaven and earth, it had become a natural thing to suspect.) If the earth attracts the moon with the same type of force with which it attracts a falling apple or a projectile, and if the sun attracts earth, moon, and all other celestial bodies with the same type of force, then there is no need for any addi-

tional cosmic force or prime mover, then gravity becomes a universal, unifying principle which, while in fundamental contradiction to the axioms of the scholastics, would have gladdened the heart of Kepler.

But we must follow the proof. First paralleling young Newton's thoughts, let us see whether the centripetal force $F$ needed to keep the moon in its (nearly enough) circular orbit about the earth *can* be identified with terrestrial gravity. The *direction* of $F$ is by definition toward the center of the earth, and this checks with the direction of the force of gravity. But what about the *magnitude* of $F$? We apply the equation for centripetal force to this case, and find

$$F = m_m \frac{4\pi^2 R}{T^2}, \qquad (11.3)$$

where $m_m$ is the mass of the moon, $R$ is now its distance from the center of rotation about the earth, and $T$ its period of revolution.

Does this value for $F$ actually coincide with the gravitational pull which the earth exerts on our satellite, as Newton proposed? That depends on the nature of the gravitational force. If gravity is propagated undiminished through all space, the weight of the moon will be simply $m_m \cdot g$, the same as a stone of the moon's mass when placed on giant scales somewhere on the surface of the earth. But not only does it seem unlikely that the gravitational acceleration stays the same no matter how far away from the earth we go, we also recall that in part (c) above we found evidence that the centripetal force (whatever its final nature) must fall off as the square of the increasing distance. If gravity is to account completely for the centripetal force, it too will have to follow an inverse square law. Let us therefore assume that the weight of an object falls off according to such a law, and consider now whether the gravitational pull of the earth on the moon just equals the centripetal force in Eq. (11.3).

This is our argument: An object with the same mass as the moon, $m_m$, has the weight $m_m g$ when weighed at the surface of the earth, i.e., at a distance $r$ (the radius of the earth) from the earth's center. That same object, when removed to a great distance $R$ from the earth's center, will assume a smaller weight, which we shall call $W_R$, and which must fulfill the following proportion if the inverse square law is to be obeyed:

$$\frac{m_m g}{W_R} = \frac{1/r^2}{1/R^2}$$

or

$$W_R = m_m g \, \frac{r^2}{R^2}. \tag{11.4}$$

If the centripetal force $F$ acting on the mass $m_m$ rotating about the earth at distance $R$ with a period $T$ is really equivalent to the gravitational force $W_R$ at that distance, the terms on the right side of Eqs. (11.3) and (11.4) should also be equivalent:

$$\frac{m_m 4\pi^2 R}{T^2} = m_m g \, \frac{r^2}{R^2}$$

or

$$T^2 = \frac{4\pi^2}{g r^2} \, R^3. \tag{11.5}$$

Conversely, if we substitute observed values of $T$, $g$, $r$, and $R$ in Eq. (11.5) and find that the equation yields equal numbers on each side, then we are justified in regarding our hypotheses as valid; then the gravitational force *does* fall off as the square of the increasing distance, and *does* fully account for the needed centripetal force. [Incidentally, we note with satisfaction that Kepler's third law is implied in our result, Eq. (11.5).]

---

**Problem 11.2.** Substitute the needed values in consistent units and check to what extent Eq.

(11.5) holds. The period $T$ of the moon is 27 days, 7 hr, 43 min; $g = 9.80$ m/sec$^2$; $r = 6380$ km; $R = 380{,}000$ km.

---

This was the calculation which Newton, with contemporary data, found to "answer pretty nearly," probably within a few percent. The assumption of a strictly circular path and somewhat inaccurate values of $r$ and $g$ made it clear from the start that no *perfect* agreement could have been expected.

It has been a source of much speculation why Newton did not tell anyone of this remarkable result when he first conceived it, or for nearly 20 years thereafter. Apart from his reticence and his fear of litigations with jealous men, he seems to have been unable at the time to account clearly for one implied assumption in the argument, namely, that the earth's gravitational force acts as if originating at the very center of the globe, and that consequently the measurement of distances must be made, not to the globe's surface, but to its center. He understood this requirement later when writing the *Principia*, where he proved in general that two homogeneous spheres attract each other as though their masses were concentrated at the center. Another reason for the delay may have been that Newton did not know or properly appreciate the significance of Kepler's law of areas until about 1684; before that time his manuscripts show that he was trying to base his planetary theory on the equant device, applied to elliptical orbits.

As a summary for part (d) we may use Newton's own statement "that the moon gravitates toward the earth, and by the force of gravity is continually drawn off from a rectilinear motion, and retained in its orbit." Apart from the assumptions and approximations alluded to, we have made use of only the following new arguments: that the gravitational force falls off as the square of the distance, and that by Rules I and II of Section 9.3 we can identify terrestrial gravity with the centripetal force on the moon.

## 11.2 GRAVITATING PLANETS AND KEPLER'S THIRD LAW

The previous paragraphs have not involved the forces between the sun and the planets, but we should again suspect that these ideas may be extended to the whole solar system. In Newton's words:

"The force which retains the celestial bodies in their orbits has been hitherto called centripetal force; but it being now made plain that it can be no other than a gravitational force, we shall hereafter call it gravity. For the cause of that centripetal force which retains the moon in its orbit will extend itself to all planets, by Rules I, II, and IV."

But such rules are just guiding, not prescriptive ones. They suggest, but do not prove; this labor remains to be done.

We remember that the centripetal force on planets in circular orbits around the sun was given by Eq. (11.1) as

$$F = m_p \frac{4\pi^2 R}{T^2}.$$

Does this actually also correspond in each case to the gravitational attraction of the sun on the particular planet? If we knew the value of $g$ on the surface of the sun we could fashion the arguments as in the previous section; but of course we do not know it at this point, and so we have recourse to another argument. Perhaps we can decide on a theoretical formula for the gravitational force of any one body on another. With the confidence gained by our success in the previous discussion of the earth and moon, we now boldly suggest that the gravitational force $F_{\text{grav}}$ between *any* two spherically symmetric bodies is proportional to the inverse square of the distance between the two centers, everything else being kept constant:

$$F_{\text{grav}} \propto \frac{1}{R^2}.$$

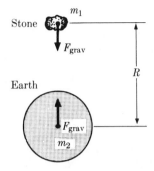

**Fig. 11.2** The force of attraction of the earth on a stone is equal in magnitude to the attraction of the stone on the earth.

Next, let us consider two specific solid bodies, quite isolated from the rest of the universe, say a stone ($m_1$) and the earth ($m_2$) at a distance $R$ between their centers. The pull of gravity, the weight of $m_1$ at the given distance $R$, is $F_{\text{grav}}$. But by Newton's law of action and reaction, the pull of the earth ($m_2$) on $m_1$ is just as large as the pull of $m_1$ on $m_2$; the weight of a stone $F_{\text{grav}}$ as measured by its attraction toward the earth *is equal* to the "weight of the earth" as measured by its attraction to the stone, strange though this may sound at first, and *either one may be called $F_{\text{grav}}$* (Fig. 11.2).

However, we know by experiment that at a given locality the weight of the stone grows directly with the mass of the stone, or $F_{\text{grav}} \propto m_1$. On the other hand, if the mass of the planet changed, the weight of a given stone would change also (in fact, you would weigh much less on our little moon than on earth). In short, if experiments prove that $F_{\text{grav}} \propto m_1$ at a constant distance $R$, then we must also accept $F_{\text{grav}} \propto m_2$, for otherwise we should have to make the assumption that the mutual gravitational pull $F_{\text{grav}}$ depends on something besides the magnitude of the masses and the distances between them.

Combining these three proportionalities,

$$F_{\text{grav}} \propto \frac{1}{R^2}, \quad F_{\text{grav}} \propto m_1, \quad F_{\text{grav}} \propto m_2,$$

we have

$$F_{\text{grav}} \propto \frac{m_1 m_2}{R^2} \quad \text{or} \quad F_{\text{grav}} = G\frac{m_1 m_2}{R^2}, \quad (11.6)$$

where $R$ is the center distance between the two bodies (let us assume them to be homogeneous spheres) and $G$ is a constant of proportionality.

If we have confidence that Eq. (11.6) gives, in fact, the correct attraction which two masses exert on each other, and if one were the sun ($m_s$), the other any planet ($m_p$), then the sun's pull would be

$$F_{\text{grav}} = G\frac{m_p m_s}{R_{ps}^2}.$$

If the moon and the earth were taken separately, their mutual force would be

$$F_{\text{grav}} = G\frac{m_m m_e}{R_{me}^2}.$$

We seem to have in Eq. (11.6) a *law of universal gravitation*.

But now we begin to be assailed by doubts. If we are right, this newly found law must certainly be compatible with all three of Kepler's laws. There is no trouble about the first two—elliptical orbits and the equal area relations must indeed result from this type of gravitational force, since it is centrally directed and is proportional to the inverse square of the distances in accordance with just those requirements, laid down in parts (b) and (c). But what of Kepler's third law? We remember that our previous argument, centering on Eq. (11.2), was a little shaky. We had not really proved, with the force assumed there, that $T^2/R^3$ had to be a truly universal constant, with the same value for *all* planets in our universe. Now that we know more about the force acting in the solar

system, we can try more rigorously to obtain such a result; for if we do *not* obtain it, gravity cannot be the sole force among the celestial bodies.

Let us test our law of universal gravitation against the nearly circular orbital motion of our own planet. The centripetal force that must exist on our mass $m_p$ at distance $R_{ps}$ from the sun is given by

$$\frac{m_p 4\pi^2 R_{ps}}{T^2}.$$

The gravitational force supposedly available is

$$\frac{G\, m_p m_s}{R_{ps}^2}.$$

If the two coincide, if the gravitational force "supplies" and so "explains" the centripetal force, then

$$m_p \frac{4\pi^2 R_{ps}}{T^2} = G\frac{m_p m_s}{R_{ps}^2} \quad \text{or} \quad T^2 = \left[\frac{4\pi^2}{Gm_s}\right] R_{ps}^3.$$
$$(11.7)$$

Now we really can see whether $T^2/R_{ps}^3$ is a constant for all planets, as required by Kepler's third law. The brackets in Eq. (11.7), which is to be identified with that constant, contain only the constant of proportionality $G$, the mass of the sun, and a numerical factor; none of these is variable from planet to planet. Therefore $T^2/R_{ps}^3$ is truly constant, Kepler's law is fulfilled, and the assumption of universal gravitation is vindicated.

## 11.3 THE CAVENDISH EXPERIMENT: THE CONSTANT OF GRAVITATION

But there remains yet another doubt. Is it proper to have assumed just now so glibly that $G$ does in fact have the same value for all planets? (If it does not, the bracketed term above is, after all, not constant for all bodies.) Here we turn to experiment. We can first measure $G$ only for various

materials on earth; using the equation

$$F_{\text{grav}} = G\,\frac{m_1 m_2}{R^2}\,,$$

we get

$$G = F_{\text{grav}}\,\frac{R^2}{m_1 m_2}\,.$$

We may propose to measure $F_{\text{grav}}$ with spring balances, neglecting all other disturbing attractions except that between two known masses $m_1$ and $m_2$ at measured distance $R$ from each other (see Fig. 11.3). It turns out that for any reasonably small masses in our laboratories, $F_{\text{grav}}$ is so exceedingly small that special techniques and precautions have to be adopted to measure any attraction between two such masses; for example, two 1-kg masses separated by 10 cm pull on each other with a force smaller than $10^{-8}$ newton! The most serious technical problems of measurement were solved by Henry Cavendish more than 100 years after the publication of the *Principia*, and the best present value for $G$ is about $6.67 \times 10^{-11}$ newton-m²/kg² for all substances. (Cavendish's measurements were made with a delicate torsion balance.)

**Fig. 11.3** $F_{\text{grav}} = G(m_1 m_2/R^2)$.

---

**Problem 11.3.** If a remeasurement of $G$ were attempted with the arrangement shown in Fig. 11.3, and if the error inherent in our measuring devices is about 0.0002 newtons, what is the smallest usable radius for each of two similar gold spheres which almost touch? (Density of gold = 19.3 g/cm³. This is a fairly representative example of what one has to go through before designing apparatus for experimentation.)

**Problem 11.4.** If by chance our standard of mass were not 1 kg, but some other unit $1.225 \times 10^5$ times larger (call it 1 tm), and if all other standards (of force, distance, etc.) stayed the same, what would be the value of $G$ in that other system of units? Does this result seem to teach any particular lesson?

Cavendish's results, and all the more accurate ones since, have shown that $G$ has indeed the same value, no matter what the composition of $m_1$ and $m_2$, even for meteorite material. Being disposed to apply terrestrial laws to celestial bodies, we extend our findings, and maintain that in the absence of evidence to the contrary all materials in the world, including sun, planets, and satellites, obey the same law of gravitation.

Although Newton knew in principle how to go about measuring $G$ accurately, he lacked the precise instruments; nevertheless he devised an ingenious proof for the *constancy* of $G$, which is the main point at issue here. Consider a mass $m_1$ on the surface of our earth (of mass $m_e$), i.e., at a distance $r$ from the center of our globe. Its weight, which we called $F_{\text{grav}}$, is of course also given by $m_1 g$; therefore, by our new law,

$$m_1 g = G\,\frac{m_1 m_e}{r^2} \quad \text{or} \quad G = \frac{r^2}{m_e}\,g. \quad (11.8)$$

At a given locality, $r^2/m_e$ is, of course, a constant, regardless of its numerical value. If at that locality all substances show precisely the same value for $g$ (the gravitational acceleration), then we have established that $G$, too, is constant, regardless of chemical composition, texture, shape, etc. This is just what Newton showed experimentally. His measurements of $g$ were made not by just dropping small and large bodies, from which even Galileo had concluded that $g$ cannot vary significantly, but by the much more accurate method of timing pendulums of equal lengths but of different materials. It was known by then that for a given length the period $T$ of a simple pendulum is proportional only to $1/\sqrt{g}$. After exhaustive experiments, all pointing to the constancy of $G$, he could write: "This is the quality of all bodies within the reach of our experiments; and therefore (by Rule III) to be affirmed of all bodies whatsoever."

Thus $G$ attains the status of the *universal* constant of gravitation (one of the very few truly universal constants in nature), and the proposed law of universal gravitation in Eq. (11.6) can be applied on a cosmic scale. And quite incidentally, Eq. (11.8) clears up a question on which Aristotelians and many present-day students harbor the strongest doubts; namely, the reason why at a given locality the acceleration of gravity is constant for all bodies. The answer is that $g$ depends only on $G$, $m_e$, and $r^2$, and all these parameters are constant at one locality.

## 11.4 THE MASSES OF THE EARTH, SUN, AND PLANETS

Once a value for $G$ is at hand, we may determine the mass of the earth [from Eq. (11.8)] and of the sun [from Eq. (11.7)]. Moreover, Eq. (11.7) should apply in the same form for a satellite moving around a planet of mass $m_p$, with a period $T_{sat}$ and a radius (or a half-major axis of the elliptical orbit) $R_{p\text{-}sat}$:

$$T_{sat}^2 = \frac{4\pi^2}{Gm_p} R_{p\text{-}sat}^3. \tag{11.9}$$

This yields the mass of any planet whose satellites we can observe.

---

**Problem 11.5.** Determine the mass of the earth and of the sun as indicated. (Answer: $m_{earth} = 5.98 \times 10^{24}$ kg, $m_{sun} = 333{,}000 \times m_{earth}$.)

**Problem 11.6.** The innermost of Saturn's nine satellites, Mimas, has a fairly circular orbit of 187,000 km radius and a period of revolution of about 23 hr. Find the mass of Saturn.

---

For the masses of the satellites themselves (including our own moon), and of the planets which boast of no moons (Mercury, Venus), no such simple calculations exist. There we may use the relatively minor and elaborate interactions between the celestial bodies, which we have conveniently neglected until now. Some of these are called *perturbations*, i.e., slight digressions from the regular path of a body because of the pull of other celestial objects. Another detail which may be used to derive the relative mass of a satellite is the fact that Kepler's laws hold, strictly speaking, only if the center of mass of the system is considered to be at the focus of the ellipse. Furthermore, it is not the earth which moves in accordance with the first law in an elliptical orbit around the sun, but the center of mass of earth and moon together (a point about 1000 miles below the surface of the earth and on a line between the centers of those two spheres).

Except for such mass determinations (or for accurate astronomical and nautical tables), we usually forget about the complications. But one of the results obtained is that the moon's mass is about $\frac{1}{81}$ of the earth's, and that the sun's mass, determined in this indirect fashion, has the same value as followed from Eq. (11.7).

It follows also that, owing to perturbations, clearly Kepler's laws cannot be expected to hold *exactly*. (In fact, astronomers before Newton had begun to worry more and more about these discrepancies, particularly for Saturn and for the moon.) Once more we find a set of laws breaking down when the precision of observations is increased beyond the original range. But this breakdown is not serious; it does not invalidate the overall outlines of the theory of planetary motion, particularly now that it is possible in principle to account in detail for each discrepancy by means of the law of gravitation.*

As a summary for the preceding paragraphs, we may state that the law of gravitation, which we formulated originally by arguments for terrestrial bodies, was found to hold also among the planets, because by its assumption Newton could predict relationships (such as Kepler's third law) well established by observation. Furthermore, the law opened up ways of calculating the mass of the sun, the planets, and the satellites. Table 11.1 gives some approximate relative masses.

**Table 11.1.** Mass relative to earth
$(1.00 = 5.98 \times 10^{24}$ Kg)

| Sun | 333,000 | Jupiter | 318 |
|---|---|---|---|
| Moon | 1/81.3 | Saturn | 95.3 |
| Mercury | 0.045 | Uranus | 14.6 |
| Venus | 0.82 | Neptune | 17.3 |
| Mars | 0.108 | Pluto | 0.1 (?) |

**Problem 11.7.** With the data given in Table 11.1, compare the gravitational pulls of sun and earth on the moon. Why does the earth not lose the moon to the sun? Can the path of the moon ever be convex as seen from the sun? [For hints you may wish to read F. L. Whipple, *Earth, Moon and Planets*, Cambridge, Mass.: Harvard University Press, 1963.]

**Problem 11.8.** Equation (11.7) can be written as

$$K = \frac{4\pi^2}{Gm_s},$$

where $K$ is the constant in Kepler's third law as

---

* We are reminded of the 8 min of arc which Copernicus did not know of, and are tempted to marvel again how significant it is that these secondary effects were too small to appear earlier and confuse the original investigators. But in this case there seems to be an extra twist; if the perturbations were any larger, we might well not have a solar system to contemplate, because it would hardly be stable enough to last these billions of years without catastrophic collisions.

obtained by observation of $T$ and $R$. You have taken this equation to find $m_s$, using also the measured value of the universal constant $G$. The value of $m_s$ so obtained actually checks against independent calculations of $m_s$ using perturbation effects. Exactly in what way is our whole argument strengthened by this gratifying confirmation?

**Problem 11.9.** (a) Using the values of the masses of the earth, moon, and sun given in Table 11.1, the distances of sun and moon from earth (Problems 10.6 and 11.2), and the fact that the angle from our eye to either edge of both sun and moon is about $\frac{1}{2}°$, find the density of the earth, moon, and sun. (The density of most rocks is about $3 \times 10^3$ kg/m³.) (b) What do these results suggest about the internal structure of the sun, earth, and moon? (c) From Eq. (11.8), find the local acceleration of gravity experienced by someone (i) on the moon, (ii) near the sun.

**Problem 11.10.** An artificial satellite or floating observation platform is to be shot up to revolve around the earth like a little moon. What must be its period and speed of revolution around the center of the earth if it is to stay at a height of 1000 miles from the surface? Outline qualitatively the requirements for the initial velocity if this satellite is to be shot from a big gun on the earth; are these requirements reasonable?

---

We have extended our law of universal gravitation in *one* direction—to all planets and satellites; it seems logical to extend it in the other direction —to all *parts* of every body. Newton writes in Book III of the *Principia*:

"Proposition VII. Theorem VII: That there is a power of gravity tending to all bodies proportional to the several quantities of matter [i.e., the product of the masses] they contain. That all the planets mutually gravitate one towards another, we have proved before; ... [we now introduce the idea that] the force of gravity towards any whole planet arises from, and is compounded of, the forces of gravity toward all its parts .... If it be objected, that, according to this law, *all* bodies with us must mutually gravitate one toward another, whereas no such gravitation anywhere appears, I answer, that ... the gravitation toward them must be far less than to fall under the observation of our senses."

With the aid of his own calculus Newton proceeds to show that assuming the same universal law of gravitation for each smallest particle of a body (say of a sphere), we obtain in sum a resultant force of gravity for the whole body of such characteristics as are actually observable.

There offers itself now an overall view of the enormous range and sweep of the simple formula for the gravitational attraction of two bodies. *Postulating* the law just for each of the parts of a body, it gives us the attraction of the whole for some outside object. For the special case of the spherical celestial bodies, it provides a force necessary and sufficient for the description of all observed motions, for the derivation of all three laws of Kepler, and for the small, long-observed deviations therefrom. This was the vindication of Copernicus, Kepler, and Galileo—in the context of the greater, more far-reaching theory of universal gravitation. The whole structure is often referred to as the *Newtonian Synthesis*.

**Problem 11.11.** List the fundamental hypotheses of the theory of universal gravitation and the experimental consequences. Which hypothesis is not *directly* confirmable by experiment (although its consequences are)?

## 11.5 SOME INFLUENCES ON NEWTON'S WORK

What were the main intellectual tools, concepts, and attitudes with which Newton worked? If we wish to claim any insight into his time and his

work we must at least summarize these facts now; most of them have already appeared scattered here and there through the discussion.

a) Newton was not a "hard-boiled" scientist of the twentieth-century type; rather he shared many characteristics of the late-Renaissance scholar. He was not free from traces of what we would regard as pseudoscientific beliefs; apart from some early interest in astrology, he seems to have spent much time in his "elaboratory," cooking potions that to us would smell more of alchemy than of chemistry —though the aim there, as in all his activities, seems to have been the search for underlying general principles rather than quick practical gains.* By our present standards we might also say that his belief in absolutes and his anthropomorphic conception of a Creator carried very far into some of his scientific writings. But here we touch on the very secret of motivation for his scientific work; though we cannot enter this field here, such a study is fascinating in its own right.†

In the category of more strictly scientific influences, we should first of all consider the clear Galilean influence on Newton's formulation of the physical concepts of mass and force. The decisive attitude throughout his work is that celestial phenomena are explainable by quantitative terrestrial laws, that these laws have legitimate general meaning and are not just mathematical conveniences covering unattainable "true" laws. Second, recent historical studies have demonstrated Newton's intellectual debt to Descartes, a debt that Newton did not publicly acknowledge because after long struggles he finally found a solution radically different from the one proposed

---

* In his essay "Newton, the Man" in *Essays in Biography* (London: Rupert Hart-Davis, 1951 edition), the great economist J. M. Keynes presented an intriguing picture, based on Newton's own papers, of Newton as a Faust, a searcher for the key to all knowledge in science, theology, and magic.

† See Frank E. Manuel, *A Portrait of Isaac Newton* (Cambridge, Mass.: Harvard University Press, 1968).

by the Frenchman to the same problem, the dynamics of the solar system.

b) To his fundamental faith in the proximity and accessibility of natural law, we must add a word about his methodology. His debt to the pioneers of the new experimental science is clear (for example, he made ingenious pieces of equipment and performed good experiments when the theory needed verification), but he also successfully combined this principally inductive approach with the deductive method then most prominently displayed in Descartes. With his mathematical powers enriching the experimental attitude, he set a clear, straight course for the methods of physical science.

c) Not just his attitude toward concepts, but many of the concepts themselves, came to him from Galileo and his followers, above all, force, acceleration, the addition of vector quantities, and the first law of motion. Newton also drew, of course, from Kepler; and through their books and reports to the Royal Society he kept in touch with the work of such contemporaries as Huygens and Hooke.

d) Apart from his own experiments, he took his data from a great variety of sources; for example, Tycho Brahe was one of several astronomers, old and new, whose observations of the moon's motion he consulted. When he could not carry out his own measurements he knew whom to ask, and we find him corresponding widely with men like Flamsteed and Halley, both Royal Astronomers. There are evidences that he searched the literature very carefully when he was in need of exact data, for example, on the radius of the earth and the distance to the moon.

e) Lastly, we must not fail to consider how fruitfully and exhaustively his own specific contributions were used repeatedly throughout his work. The laws of motion and the mathematical inventions appear again and again, at every turn in the development. But he was modest about this part of his achievement, and once said (echoing what had become a well-known saying even in the seventeenth century) that if he had seen further than others *"it is by standing upon the shoulders of Giants."** 

## 11.6 SOME CONSEQUENCES OF THE LAW OF UNIVERSAL GRAVITATION

What amazed Newton's contemporaries and increases our own admiration for him was not only the range and genius of his work on mechanics, not only the originality and elegance of his proofs, but also the detail with which he developed each idea to its fullest fruition. It took almost a century for science to fully comprehend, verify, and round out his work, and at the end of a second century an important scientist and philosopher still had to confess: "Since his time no essentially new principle [in mechanics] has been stated. All that has been accomplished in mechanics since his day has been a deductive, formal, and mathematical development of mechanics on the basis of Newton's laws." (That statement is of course an exaggeration, which ignores the important and original work done by Leonhard Euler and other eighteenth-century mathematicians in subjects such as fluid mechanics and elasticity theory.)

a) An example of Newton's thorough mastery was his treatment of the moon's perturbations. The influence on its path of the gravitational forces from almost all other celestial bodies was considered, and the repeated use of the gravitational law yielded an astonishingly close approximation to every small complexity of motion. (A full list of influential factors in tabular form would now fill hundreds of pages.) For one minor variation of the path, however, Newton's published theoretical result was only half the observed value. He did not camouflage this defect, but clearly stated that

---

*    Newton actually used this phrase in connection with his theory of light, in a letter to Robert Hooke in 1676. Its earlier history has been traced by Robert Merton, in his book *On the Shoulders of Giants* (New York: Free Press, 1965).

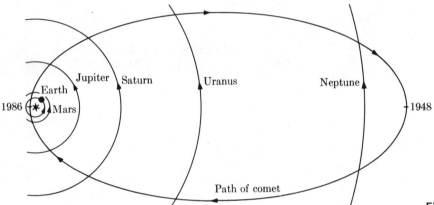

**Fig. 11.4**  Path of Halley's comet.

his calculations gave the wrong figure.  Consequently a loud, long battle arose among some scientists, several of whom called for complete abandonment of the whole theory of gravitation on account of this discrepancy.  Eventually the mathematician Clairaut noted a small error in the long calculations, clearing up the difficulty.  Later still, some unpublished notes of Newton were found, showing that he, too, 50 years before Clairaut, had discovered and corrected his own mistake!

b) Comets, whose dreaded appearance had been interpreted as sure signs of disaster throughout antiquity and the middle ages, were now shown to be nothing but passing clouds of material obeying the laws of gravitation, made visible by reflected light while near the sun.  The great comet of 1682, whose path Edmond Halley watched carefully, indicated to him a period of recurrence of *approximately* every 75 years if the comet, although an eccentric member of the solar family, nevertheless were to obey the usual laws of mechanics, including Kepler's laws.  Its returns in 1756 and twice since, after covering a wide ellipse carrying it far beyond the last planet, were heralded as significant symbols of the triumph of Newtonian science.  (See Fig. 11.4.)

c) Newton reasoned that the shape of planets and satellites might be explained as the result of the mutual gravity of the separate parts, which would pull a large quantity of initially liquid material (or a dust cloud) into a compact sphere.  Further, although a motionless body might indeed form a perfect sphere, a planet rotating about an axis ought to assume the shape of an oblate spheroid, that is, it ought to bulge at the equator and be flatter at the poles.  In fact, from the relative magnitude of the equatorial bulge, say of Jupiter as seen through a good telescope (see Fig. 11.5), one may calculate the period of rotation about the axis (the length of a day on Jupiter).

For the earth, Newton predicted that the same effect would be found, and later, accurate measurements actually revealed the diameter of the earth from pole to pole to be less by 27 miles than across the equatorial plane.  Consequently the gravitational force on an object at the pole is somewhat greater than at the equator, owing to the shorter

Fig. 11.5  The planet Jupiter.

distance from the center of gravity of the earth to the pole. This effect is relatively small, but another cause joins to diminish the acceleration of a freely falling body at the equator. At that region the surface velocity of a point on earth is about 1600 km/hr, due to the rotation of the globe about its axis. Since the centripetal acceleration $a_c$ is given by $v^2/r$, $a_c$ for any object at the equator is about 0.03 m/sec$^2$, as we calculated at the end of Chapter 10. This much acceleration is required just to keep the body from flying off along a tangent; consequently the acceleration of free fall (and therefore the apparent weight of the body) is proportionately less than it would be if our globe were not rotating. Measured values of free-fall acceleration, all at sea level, range from about 9.83 m/sec$^2$ near the poles to 9.78 m/sec$^2$ near the equator (cf. Chapter 9) with 9.80398 m/sec$^2$ the standard value for Cambridge, Massachusetts. These observed values are, of course, the true acceleration due to gravity *minus* the centripetal acceleration at that latitude. But we shall be content with the convention that the letter $g$ may stand for those *observed* values and yet be usually called "acceleration due to gravity."

**Problem 11.12.** Because of the bulging of the earth near the equator, the source of the Mississippi River, although high above sea level, is nearer to the center of the earth than is its mouth. How can the river flow "uphill"? [*Hint:* The word "uphill" needs contemplation.]

**Problem 11.13.** According to speculative theories, our moon might be material which was pulled out from the earth by a passing celestial body or which was shaken off by the rotating earth. How fast would the earth have had to rotate at that time to make the latter picture plausible?

**Problem 11.14.** What approximate percentage error does the quoted value for $g$ in Cambridge imply? How might such an accurate experiment be performed?

At present, methods of measurement have been refined to such an extent that small local fluctuations in $g$ can be detected and used to give clues of

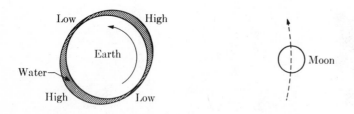

**Fig. 11.6** Tides caused by the moon. The earth is to be imagined as rotating in this (much exaggerated) envelope of water. The high tide stays relatively fixed with respect to the moon.

possible mineral deposits underground. But once one goes below the surface into a mine, $g$ is found to change, becoming zero as the center of the earth is approached. The law of gravitation in its simple form holds only if the two bodies whose mutual attraction is to be measured are separated and not entwined or intergrown. Newton predicted this excellently, and a moment's thought will persuade us that at the center of the earth, at any rate, we could have no weight because the forces of the earth's attraction from all sides would cancel.

d) The phenomenon of the tides, so important to navigators, tradesmen, and explorers through the ages, had remained a mystery despite the efforts of such men as Galileo, but at least the main features, the diurnal high tide and the semimonthly spring tide (maximum high tide), were explained by Newton through application of the law of universal gravitation. He recognized that the moon (and to a lesser extent the other celestial bodies) would pull on the nearest part of the ocean, and so tend to "heap up" the waters. We do not want to take time for the many details involved, except to look at the conclusions. The bulge of water, which actually is raised at the same time on *both* sides of the globe, does not stay directly under the moon but is displaced to a position a little ahead by the rotation of the earth. Thus a high tide at a point in

the ocean is encountered some time after seeing the moon pass overhead, and again about 12 hours later (Fig. 11.6). The sun has a similar though smaller effect,* and since sun, moon, and earth are in line twice a month (at new and full moon), the two tidal forces coincide to produce semimonthly extremes of tide level. Between spring tides, sun and moon may partially cancel each other's effects (neap tides).

The exact features of this complex phenomenon depend to a great extent on the topography of shore line and ocean floor. Long series of actual observations help to predict individual features of the tides at each locality.† In passing, we should notice that the "solid" earth has similarly explained tides, which must be allowed for in such exact experiments as refined astronomical observations, and that it is just such a tidal effect that offers one of the methods for the approximate calculation of the mass of the moon.

## 11.7 THE DISCOVERY OF NEW PLANETS USING NEWTON'S THEORY OF GRAVITY

e) One of the most intriguing aspects of Newton's work still remains to be told: how, more than 100 years after his death, Newton's law of universal gravitation helped mankind to discover new planets. Lodge refers to it as follows:

"The explanation by Newton of the observed facts of the motions of the moon, the way he accounted for precession and nutation and for the

---

* The direct pull of the sun on the earth is about 175 times that of the moon, but tides are generated by the *difference* of pull on the waters on both sides of the globe; and that is larger for the moon. (Why?)

† For the history of the tidal theory, with references to modern works, see H. L. Burstyn, "Galileo's attempt to prove that the earth moves," *Isis*, Vol. 53, pages 161–185 (1962). A more recent contribution to the historical discussion is: W. R. J. Shea, "Galileo's claim to fame: The proof that the earth moves from the evidence of the tides," *British Journal for the History of Science*, Vol. 5, pages 111–127 (1970).

tides, the way in which Laplace [whose mathematical work extended Newton's calculations] explained every detail of the planetary motions —these achievements may seem to the professional astronomer equally, if not more, striking and wonderful; ...   But to predict in the solitude of the study, with no weapons other than pen, ink, and paper, an unknown and enormously distant world, to calculate its orbit when as yet it had never been seen, and to be able to say to a practical astronomer, 'Point your telescope in such a direction at such a time, and you will see a new planet hitherto unknown to man'—this must always appeal to the imagination with dramatic intensity . . . ."*

One night in 1781 William Herschel of Bath, England, an extraordinarily energetic mixture of professional musician and gifted amateur astronomer, was searching the sky with his homemade, 10-ft telescope.  For years he had patiently looked at and re-examined every corner of the heavens, finding new stars, nebulae, and comets, and he was becoming well known among astronomers.  On that particular night he again found a celestial object, hitherto uncatalogued and of such "uncommon appearance" that he suspected it to be a new comet.  Through the Royal Society the news spread. As observations continued night after night, it became evident that this was not a comet but a newly discovered planet, more than 100 times larger than the earth and nearly twice as far from the sun as Saturn, until then the outermost member of the solar family.  So *Uranus* was discovered—an unsuspected and sensational widening of the ancient horizon, a planet just barely visible to the naked eye, but previously mistaken for a star when occasionally observed.

By that time it was known how to compute the elliptic orbit of a planet from a few widely separated observations of its varying positions.  Also, the expected small deviations from the true ellipse owing to the perturbing force of the other planets

were accurately predictable on the basis of Newton's law of gravitation.  Uranus' 84-year orbit was so mapped out for it by calculation, and all went well for many years.  But by 1830 it became more and more evident that Uranus was misbehaving, and that the assumptions on which its schedule had been worked out were in need of revision.

Some thought that Newton's theory might not apply accurately, after all, over such immense distances, but they had nothing better to offer (and as has been well pointed out, theories are in general not overthrown by isolated contradictory facts but only by other, better theories).  Others suggested that a hitherto unsuspected comet or a more distant planet might introduce additional perturbations into Uranus' path, but they, too, were merely guessing and made no concrete quantitative predictions.

But the idea of an undiscovered planet beyond Uranus intrigued John C. Adams, a young undergraduate at Cambridge University, and he undertook the immensely difficult mathematical task of locating the positions of this suspected perturbing body solely from the observed motions of Uranus, using throughout the law of gravitation in unmodified form.  He had his mathematical result of the problem two years after graduation, and for the necessary confirmation wrote to the Royal Observatory at Greenwich, asking that their powerful telescope search for the hypothetical new planet at a predicted location beyond Uranus.  Since Adams was an unknown young mathematician, however, he was not taken seriously enough to interrupt the current work there with what would have been a tedious search.

A few months later, another young man, U. J. J. Leverrier in France, published the result of similar, independent calculations, giving very nearly the same position for the suspected planet as Adams had.  While finally some observations were slowly being undertaken in England to verify Adams' theoretical conclusions, Leverrier sent his own prediction to the head of the observatory at Berlin, who, happily having at hand a new star

---

* O. Lodge, *Pioneers of Science*, page 317.

map to aid in the search, on the very evening of the letter's arrival, himself searched for and recognized the planet at very nearly its predicted position. Thus *Neptune* was added to the solar system in 1846; and here was indeed a triumph of the law of universal gravitation!

Neptune, in turn, was closely watched. The orbital radius is about 30 times the earth's, and its period is therefore given by Kepler's law (correctly) as 164.8 years. But in time the perturbations of this planet and those of Uranus again became larger than was accountable in terms of known forces, and the hypothesis of still another planet was naturally made. An arduous 25-year-long search yielded the discovery of *Pluto* in 1930, announced on the double anniversary of Herschel's discovery of Uranus and of the birthday of Percival Lowell, whose calculations had led to the search and who had founded the observatory in Arizona at which the discovery was made, by C. W. Tombaugh.

Another astronomer, W. H. Pickering, had made independent calculations and predictions of Pluto's position as far back as 1909, and had initiated a telescopic search for the planet at Mount Wilson Observatory in California. Nothing was found then, but after the discovery at the Lowell Observatory in 1930, the old Mount Wilson photographs were re-examined and showed that Pluto could have been found in 1919 if its image had not fallen directly on a small flaw in the photographic emulsion!

This story dramatizes the frequently forgotten possibility that for every overpublicized discovery which is made by "accident," without elaborate preparations, there may well be an equally important discovery which, also by accident, *failed* to be made despite careful research.

## 11.8 "BODE'S LAW": AN APPARENT REGULARITY IN THE POSITIONS OF THE PLANETS

f) This chapter would be incomplete without a short discussion of the influence of a strange and simple rule on the discovery of Neptune. This rule is known as *Bode's law* after Johann Elbert Bode (1747–1826), director of the astronomical observatory at Berlin, who published it in 1772. (The rule is also sometimes called the *Titius–Bode rule*, since it had been developed earlier by Johann Daniel Titius, a professor at Wittenberg, Germany.)

At the start of his long search for regularity in the solar system, young Kepler had discovered a crude relationship between the sizes of the various planetary orbits, but it was a rather fruitless jumble of numerological statements. The best that can be said of it was that its mathematical ingenuity first brought him to the attention of Brahe and Galileo. Later, the three Keplerian laws, unified by Newton's work, showed that there existed a simple universal relation between each planet's speed and orbit, but it left unanswered the question why a given planet did not move in some other possible orbit with a correspondingly different speed. Was there, for example, any specific reason why the earth could not pursue a course nearer to Mars', although necessarily with a relatively slower period of revolution? Nothing in Kepler's or Newton's work showed that the chosen orbit was in any way unique or necessary.

Since the whole effort of science is bent on finding simple regularities, it is no wonder that men continued to look for clues. Bode's law, obtained no doubt, like Kepler's laws, by incessant mathematical play with the data to be explained, established just such a clue, which, in the absence of physical reasons, must certainly look like a strange coincidence.

If we number the successively more distant planets from the sun with $n = 1$ (Mercury), 2 (Venus), 3 (Earth), . . . , then the radius of the orbit of any planet (in astronomical units) is given by the *empirical* formula

$$R = 0.3 \times 2^{(n-2)} + 0.4 \text{ (in AU)}.$$

But this formula works only if two exceptions are made. One is that a zero must be used instead of the factors $0.3 \times 2^{(n-2)}$ for the calculation of Mercury's orbital radius. Furthermore, although

Mars can be assigned $n = 4$, Jupiter, the next planet, has to be assigned $n = 6$; Saturn, the next and outermost planet then known, has to have $n = 7$. The number $n = 5$ could not be given to any existing planet. This at once suggested to Bode that the disproportionately large space which actually extends between Mars and Jupiter harbored a yet-unnoticed planet for which $n = 5$ and which, *according to this rule*, could be expected to have an orbital radius of 2.8 AU. (Derive this predicted radius from the last formula.)

Here was a chance to confirm whether the rule was anything more than coincidence and fantasy! Bode was convinced that a search would reveal the suspected planet, and he wrote, in a manner which reminds us again of Copernicus' and Kepler's motivations, "From Mars outward there follows a space ... in which, up to now, no planet has been seen. Can we believe that the Creator of the world has left this space empty? Certainly Not!"

At first no one discovered anything in that seemingly God-forsaken gap, and interest in the law must have lagged (one might say it did not fulfill the criterion of stimulation of further discoveries). But then, nine years after Bode's announcement, came Herschel's discovery of distant Uranus. With $n = 8$ as a logical number to assign, the law predicted an orbital radius of 19.6 AU, only about 2% larger than the actual value derived by subsequent observation.

This agreement directed attention to the other predictions inherent in Bode's law, and the search for the "missing" planet was renewed with vigor. In time, the first result came, but, as so often in science, only as the by-product of research not directly concerned with this problem. In 1801, the Sicilian astronomer Piazzi, while compiling a new star catalog, chanced on a new celestial object which was moving as rapidly as might a comet or a planet. In the exciting months that followed, astronomers and mathematicians cooperated to derive the orbit of the foundling, now christened *Ceres*. It was disappointingly puny—less than 500 miles in diameter—but the orbital radius came to 2.77 AU, only about 1% less than Bode's law pre-

dicted for the missing planet in that region between Mars and Jupiter!

Even while Ceres was being hailed as the long-sought sister planet, another, smaller one was discovered with about the same orbit, and then others still, all smaller than Ceres. Today we know of almost 2000, some with very eccentric paths, but all evidently springing from one family with a common history—perhaps the fragments which were forming into one planet when its evolution was somehow interrupted, perhaps the shattered remnants of one larger planet in the predicted orbit. It is possible to speculate that there were two (or more) planets trying to establish orbits near each other, but that they eventually collided "because" Bode's law would permit only one orbit to be there. This line of reasoning would put Bode's law into a new light; we might be tempted to search later for other clues that the law expresses simply the series of dynamically possible, stable orbits for single planets.

The discovery of these *asteroids* or *planetoids* bore out Bode's law so strikingly that it was natural to apply it when the suspicion of a planet beyond Uranus first arose. The first available value for $n$ being 9, the corresponding orbital radius would be

**Table 11.2.** Bode's law*

| $n$ | $R$ (theoretical) | Planet | $R$ (observed) |
|---|---|---|---|
| 1 | 0.4 | Mercury | 0.39 |
| 2 | 0.7 | Venus | 0.73 |
| 3 | 1.0 | Earth | 1.00 |
| 4 | 1.6 | Mars | 1.53 |
| 5 | 2.8 | (Asteroids) | (2.3 to 3.3) |
| 6 | 5.2 | Jupiter | 5.22 |
| 7 | 10.0 | Saturn | 9.6 |
| 8 | 19.6 | Uranus | 19.3 |
| | | Neptune | 30.2 |
| 9 | 38.8 | Pluto | 39.5 |

* The theoretical values are computed from the formula $R = 0.3 \times 2^{n-2} + 0.4$ (in AU), except that $R = 0$ for $n = 1$.

$$(0.3 \times 2^{(9-2)} + 0.4) \text{ AU},$$

that is, 38.8 AU. Adams and Leverrier based their calculations on the predictions of Bode's law to find the probable location, mass, and orbit of their hypothetical planet. When Neptune was afterward seen and its course plotted, the actual orbit was found to be about 20% smaller than Bode's law had predicted—the first failure of the law, but by luck one not so serious as to invalidate the calculations that led to Neptune's discovery. Later, when Pluto was found, its actual orbital radius of 39.5 AU much more seriously contradicted the value of 77.2 AU which would follow from Bode's law for $n = 10$, although it is very close to the place predicted for Neptune, with $n = 9$.

The irregularities of Neptune and Pluto with respect to Bode's law exemplify in a very simple way a general problem that is frequently encountered in testing scientific theories. If a theory is successful in explaining all known observations, and in predicting some results that were not known at the time the theory was first proposed, must we reject it when it fails to account for a further observation? Or is it legitimate to introduce, at least provisionally, an *ad hoc* hypothesis to explain the exception? For example, we might conjecture that the original solar system contained 9 planets, with orbits corresponding to the integers $n = 1$ through 9 in the formula, including Pluto as number 9 rather than Neptune. Then, just as we have suggested that planet number 5 broke up, leaving the belt of asteroids to occupy the "empty" orbit, we might propose that Neptune was captured into the solar system after its original formation, or perhaps that it was originally in orbit number 10 but suffered a near-collision with a comet that pushed it into its present orbit between numbers 8 and 9.

Some writers on scientific method would object that a theory loses its scientific value if it can be too easily modified to explain away all apparent exceptions: If such a procedure were permitted, how would we ever be able to dispose of false

theories? On the other hand, if we were too strict in requiring that a theory be abandoned as soon as it is contradicted by new data, we might lose all hope of finding rational explanations of complex phenomena, especially those which depend primarily on one or two physical factors, but weakly or occasionally on others. If a hypothesis, or even a numerical rule, is helpful in the search for new knowledge, why reject it simply because it is not yet a completely comprehensive and accurate theory? Astronomers would be extremely surprised if a modern theory of the origin of the solar system led *exactly* to a simple formula such as Bode's law; yet this or a similar algebraic formula might well be an adequate first approximation, just as the circle is a fairly good approximation for the shape of the orbit of most planets. Until we have a more fundamental theory which explains why Bode's law is approximately correct for the planets up to and including Uranus, we cannot know why it breaks down beyond Uranus. The exceptional case of Neptune, which might lead us to examine carefully whether that planet is different in some way from all the others, may yet prove to be the most interesting episode in the entire history of Bode's law; for without such a law, not only would it have been much more difficult to discover Neptune, but there would have been little reason to suspect that there is anything exceptional about it!

This may be a good place to review briefly the different types of laws which we have encountered here. At the first level of sophistication there is the type of law which is usually called *empirical*: It seems to summarize simply some fairly directly observed regularity, without attempting to provide a theoretical explanation for it. As it happens, the empirical laws we have discussed up to now all pertain to planetary orbits (Kepler's laws and Bode's law); later on we shall encounter Boyle's law of gas pressure, Gay-Lussac's law of combining volumes for chemical reactions, Mendeleev's periodic law of the elements, and Balmer's formula in spectroscopy. Since in most cases we now have a satisfactory theoretical understanding of the regularities expressed by these laws, we no longer call them merely empirical laws. But as we have already seen in our discussion of Kepler's work, even a regularity appearing rather obvious to us now may have been an insight which presupposed a difficult reorientation within science and which stimulated important further developments. At any rate, we must allow these empirical rules a vital place in science.

Next in order, we may perhaps place those laws which represent an induction of one regulating principle from a variety of apparently very different phenomena. For example, Galileo's law of projectile motion, and the fundamental equation $F = ma$ of Newtonian physics, were perceptions not obvious from direct observation. In fact, in these cases the formulation of the law was preceded by the formulation of new concepts: in Galileo's case, acceleration and the independence of horizontal and vertical velocity components; in Newton's case, mass and to some extent force. Because these laws usually carry with them the definition of fundamentally important concepts involved therein, we might call them *definitional* laws. Examples of this type might also include the laws of conservation of energy and of momentum, which we shall discuss shortly.

Third, the remaining group is formed by those laws which represent a general conclusion *derived from some postulate or theory*, whether new or old. For example, the lens laws which make possible the design of optical instruments can be derived from a theory of the propagation of light. The law of pendulum motion can be deduced from the conceptual scheme of mechanics built around Newton's three laws of motion. And the law of universal gravitation certainly was not an empirical rule, nor did it serve to exhibit a new fundamental concept (all the factors in $F = Gm_1m_2/R^2$ were previously defined concepts, $G$ being essentially a proportionality constant). Instead, as we have seen, it was a generalization derived from the conceptual scheme featuring Kepler's rules, Newton's own laws of motion, and the hypothesis of long-range attraction among the planets and the sun. This type of law, which we may perhaps name *derivative*, often seems the most satisfying of the three because, since it has been derived from some underlying theory as well as based on empirical facts, we are tempted to feel that it is also somehow more fully "explained." Indeed, scientists continually seek to reduce purely empirical laws to derivative ones, thereby giving the empirical law a measure of additional "meaning,"* while at the same time extending and fortifying the theory itself.

There is a kind of symbiotic relationship here between law and theory. A theory becomes more and more respected and powerful the more phenomena can be derived from it, and the law describing these phenomena becomes more meaningful and useful if it can be made part of a theory. Thus Newton's theory of universal gravitation gained greatly in stature because it enabled one to derive the laws that govern the moon's motion, known by empirical rules since the days of the Babylonian observers. These rules in turn, now that they were "understandable" in terms of the theory, could be reformulated more accurately; in fact, the theory even enriched them, e.g., by calling attention to hitherto unnoticed peculiarities of the moon's motion. Another important example of this kind will be presented in the section on the nuclear atom: that theory has been so enormously important in modern science because it swallowed up whole textbooks full of isolated empirical and definitional laws from all branches of physics and chemistry, and reissued them as derivative laws springing from the unifying concept of the atom.

---

* This feeling appears somewhat fallacious if subjected to rigorous analysis. After all, the theory itself is acceptable only if its derived laws can be checked by experiments, and may in fact be incorrect *even if* the derived laws turn out to be correct. We shall see such an example when we come to deal with the theory of light. Nevertheless, great theories are relatively stable for long periods, and laws derived from them are as secure as we have any right to expect.

From this discussion we may perhaps draw some general conclusions valid beyond the physical sciences. Empirical rules may be valid enough in a limited field, but they do not contain enough information to warn the naive user when some particular new case is really outside the field of applicability of the empirical rule. Not until the damage is done is it clear that the rule did not apply.

Most of our own everyday actions tend to be governed by this type of uncomprehended rule and plausible generalization. This is doubly unfortunate as these rules, unlike those in science, are usually not even based on accurate clear-headed observation. While our life thereby may become simpler, it also is made more brutal and senseless, and we open ourselves up to an invasion of pseudo-science, superstition, and prejudice, even in matters subject to test and measurement.

To return to Bode's empirical rule, recent work has given hope that its features will be explained by (derived from) Newton's law of celestial mechanics. As hinted before, the mutual perturbations of planets over billions of years would leave only a few orbits as stable possibilities; and some cosmologists and astrophysicists at present seem to be near the inevitable proof that the celestial laws predict such stability for the orbits that are actually observed. But this is a most difficult mathematical problem which from time to time has challenged some of the greatest scientists during the past 100 years.

---

**Problem 11.15.** Would the credibility of Bode's law be strengthened or weakened if a series of small planets were discovered between Venus and Mercury, with orbits whose radii are 0.55 AU, 0.475 AU, 0.4375 AU, ... ?

**Problem 11.16.** Can we conclude that an empirical law represents absolute truth if it agrees with all known facts (within limits of observational error)

at a certain time? How would your answer to this question change if the number of independent known facts were 3? 10? 1000?

**Problem 11.17.** If an empirical law works satisfactorily for five cases, but is definitely wrong for the sixth, are we justified in proposing an *ad hoc* hypothesis to account for the single discrepancy? How would your answer change if the number of successful cases was 50 instead of five?

**Problem 11.18.** Assume that in a planetary system other than our own, the distance from the central star to each of the four inner planets is given by the following table:

| Sequence number of planet ($n$): | 1 | 2 | 3 | 4 |
|---|---|---|---|---|
| Radius of orbit, in $10^6$ miles ($R$): | 3 | 5 | 9 | 17 |

Obtain from these data a "law," a formula that will relate $R$ to $n$ for each planet. At what distance do you expect the fifth planet to be?

## 11.9 GRAVITY AND THE GALAXIES

Even in Newton's work we find further consequences of the law of universal gravitation. For example, he gave an accurate explanation for the long-standing mystery of the precession of the equinoxes, i.e., a very slow rotation of the earth's axis much like the wobbling motion of a rapidly spinning top. But now we turn to a more ambitious question. Do Newton's laws, so serviceable and fruitful within the solar system, continue to apply beyond, among the "fixed" stars?

To Copernicus or Galileo this question would have been meaningless, for it was not until Newton's time that relative motions were noticed in the celestial sphere (the discoverer was Halley). In fact, our whole solar system was found to move with respect to those distant stars. William

Herschel, in 1803, also discovered that some star neighbors rotated about each other (double stars), and his son showed that their motions are compatible with the assumption of central forces between them, such as exist in the solar system.

This new picture of our living, moving universe, which the prophetic Giordano Bruno had sensed long ago, places the solar system in the uncrowded company of many billion other suns and their possible attendants (of which less than 6000 suns are visible without telescope). The whole cloud of stars forms our *galaxy*, a vaguely lentil-shaped region about 100,000 light years* across and about 1500 light years thick. At a distance of about 30,000 light years from the galactic center (not even at the center of all solar systems!) our planetary system is rather lost in this structure.

Such a picture evokes the thought that the mutual gravitational attraction of the stars should make the galaxy slowly gather up and coalesce into one solid mass at the center. Since this does not happen, motion of the stars being not all toward the center, there are then two explanations. Either the law of gravitation does not extend itself to the whole universe or, in analogy with the solar system itself, the whole galaxy is spinning around its axis, "distracting" the threatening gravitational attractions by making them do the chores of centripetal forces.

On examination, the second choice actually seems to hold.† Our own solar system revolves

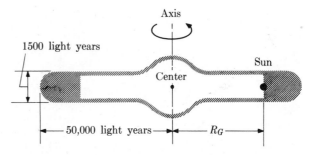

**Fig. 11.7** Position of our sun in the galaxy.

about the distant center of the galaxy at a speed of about 250 km/sec. According to our present view, the edge of the whole galaxy, whose framework we see as the Milky Way,‡ completes one revolution in perhaps a quarter-billion years.

Now let us make a bold little computation. If Newton's laws hold, we should be able to compute the approximate mass of the whole galaxy from the rate of (circular) revolution of our own sun about the galactic center.

If we neglect that part of our galaxy lying in the shaded portion in Fig. 11.7, we may say that the centripetal force on our sun is the gravitational pull exerted on it by the mass of the galaxy (acting approximately as if that were all located at the center). Thus we expect that the following equation holds:

$$m_s \frac{4\pi^2 R_G}{T^2} = G \frac{m_s m_G}{R_G^2} ,$$

where $T$ stands for the period of revolution of the sun (about $2 \times 10^8$ years), $m_G$ for the mass of the galaxy, $m_s$ for the mass of the sun. The only unknown in this equation is $m_G$, and on solving for it, $m_G = 2 \times 10^{11}$ times the mass of the sun (check this).

---

* One light year is the *distance* traveled by light in one year in vacuum, that is, $9.46 \times 10^{12}$ km or $5.88 \times 10^{12}$ mi.

† For a modern view of the structure of our galaxy and the universe beyond, see G. Abell, *Exploration of the Universe*, New York: Holt, Rinehart, and Winston, second edition, 1969; N. Calder, *Violent Universe*, New York: Viking, 1970; *Frontiers in Astronomy*, selections from *Scientific American* with introductions by O. Gingerich, San Francisco: Freeman, 1970.

---

‡ Galaxy is derived from the Greek word *gala*, milk.

**Fig. 11.8** Spiral nebula, a galaxy about a million light years from our own. (Photographed with the 200-in. telescope on Mt. Palomar.)

But when we add up the mass of all the stars which our galaxy is likely to hold, even on the most generous estimate, we can account for less than half of those 200 billion units. Either our assumptions are wrong, or we can find some galactic mass other than that contained in the stars of our galaxy. The assumptions were, of course, rather crude, but short of abandoning the law of gravitation, no tinkering with those admittedly qualitative details will explain the calculated overabundance of mass. So we turn to the alternative—and certainly we *can* find evidences that our galaxy contains matter not

congregated into suns and solar systems, namely, nebulae, interstellar dust, and most rarefied gases.

There are many such clues to the presence of this tenuous matter that would rescue the Newtonian laws, mainly its effect on starlight. The density is inconceivably low, less than for the best vacuums in the laboratory, perhaps only a few atoms per cubic centimeter. But this is spread over equally inconceivable distance, and an estimation of the total distributed material supplements the known star masses by about the missing amount. Thus our confidence in the law of gravitation rises to galactic proportions.

Beyond our world, starting at a distance of about thirty times our own galactic diameter, we find other galaxies scattered through space, as far as our biggest telescopes can reach. These island universes do not seem to differ greatly in size from ours and often appear like a spiral or pinwheel (see Fig. 11.8), probably the shape of our own galaxy also.*

Even as these distant galaxies seem to recede from us and from one another at the speed of 100 to 1000 km/sec, they rotate at the same time at rates close to our own, hence may be assumed to contain about the same mass and to obey the same laws of mechanics. And with this extension of the law of gravitation to the expanding Universe, "to all the distance of nature's infinitude," we find ourselves at the limit of our imagination, and at the height of admiration for so universal a principle. There have been few products of human genius to match its ambitious promise and its exciting fulfillment.

---

* It is clear that if Kepler's third law holds for a whole galaxy, the outlying material will revolve more slowly than the masses near the center, hence the possibility of the spiral shape. On the other hand, if the whole system were to rotate like a solid disk, each part having the same period of revolution and angular velocity, then the centripetal force holding the system together could not obey an inverse square law but would have to be linearly proportional to the distance from the center.

## 11.10 "I FEIGN NO HYPOTHESES"

The *theory* of gravitational forces, whose main hypothesis is the attraction of all particles of matter for one another, yields the derived *law* of universal gravitation, which in turn explains, as we have seen, Kepler's empirical laws and a wealth of other phenomena. Since one purpose of any theory is this type of explanation and summary, Newton's theory strikes us as eminently satisfactory. But there remained one feature which gravely bothered Newton, his contemporaries, and, indeed, students to our day. How could one account for gravity itself? What is it that causes the attraction of one body for another? Is there not some intervening medium (such as the frequently postulated *ether*) which somehow transmits the pull in a mechanical fashion?

The very statement of these questions reflects how firmly the mind is committed to physical models and to mechanical explanations, and how unsatisfied a mathematical, abstract argument leaves our emotions initially. Rather than accept *action at a distance*, i.e., the idea that bodies can exert forces on one another without the agency of a medium between them, most scientists preferred to think of all space as filled with some kind of omnipresent fluid, which, apart from being otherwise so tenuous by itself as to be undetectable by any experiment, still had to be strong and versatile enough to communicate all gravitational forces (and incidentally perhaps also serve to propagate light and electric and magnetic effects).

The search for some such medium was long and the arguments loud. At the end of Book III of the *Principia*, Newton put his often-misinterpreted remarks:

"But hitherto I have not been able to discover the cause of those properties of gravity from phenomena [observation and experimentation], and I feign no hypotheses.... To us it is enough that gravity does really exist, and act according to the laws which we have explained, and abundantly serves to account for all the motions of the celestial bodies and of our sea."

This is a famous statement, clearly an echo of Galileo's admonition in very similar circumstances (see Section 7.4). With the words, "I feign no hypotheses," Newton exempts himself from the obligation to account for the observed consequences of the gravitational theory by additional hypotheses (for example, of a mechanical ether) beyond those needed to derive the laws and observations.* The mathematical law of gravity explains a wide range of observations; that is enough justification for accepting it. It is easily conceivable that one might eventually explain the law itself in terms of something still more fundamental, but that, Newton suggests, he is at the time unprepared to do. Nor does he feel that his theory suffers by this inability, for the purpose of scientific theory is not to find final causes and ultimate explanations, but to explain observables by observables and by mathematical argument; *that* he has done.

On the other hand, his refusal, in the absence of experimental clues, to propose a mechanism by which the effect of gravity may pass from one body to another does not mean he will permit the opposite error, namely, dismissal of the question by the invention of some principle of gravity innate in matter, such as might well have satisfied a medieval scholar. This is brought out in a letter to the theologian Richard Bentley, who was preparing a popular lecture on Newton's theories and asked for clarification of their implications. Newton wrote:

"It is inconceivable, that inanimate brute matter should, without the mediation of something else, which is not material, operate upon, and affect other matter without mutual contact; as it must do,

---

* For a discussion of the significance of the word "feign" (rather than "frame" as the Latin *fingo* is sometimes translated), see A. Koyré, *Newtonian Studies*, Cambridge, Mass.: Harvard University Press, 1965, page 35.

if gravitation, in the sense of Epicurus [an ancient Greek atomist philosopher] be essential and inherent in it. And this is one reason, why I desired you would not ascribe innate gravity to me. That gravity should be innate, inherent, and essential to matter, so that one body may act upon another at a distance through a *vacuum*, without the mediation of anything else, by and through which their action and force may be conveyed from one to another, is to me so great an absurdity, that I believe no man who has in philosophical matters a competent faculty of thinking, can ever fall into it. Gravity must be caused by an agent acting constantly according to certain laws; but whether this agent be material or immaterial, I have left to the consideration of my readers."*

The question of how one body can be acted on by another across empty space has an obstinate appeal for our picture-seeking mind. Part of the great following of Descartes' vortex theory, the scheme generally accepted during most of Newton's life, was that it provided a plausible picture of a universe completely filled with a whirlpool of material corpuscles whose action on one another and on the planets was simple physical contact.† There we have an example of a plausible, easily grasped conceptual scheme which had the fault, fully appreciated by Newton but not by the majority of his contemporaries, of failing to agree quantitatively with observations, specifically, of failing to allow the derivation of Kepler's laws.

But by contrast with the most successful feature of Descartes' picture, its inherent appeal to the common sense of the muscles, Newton's lack

of an intuitive "reason" for planetary motion stood out that much more glaringly. Newton had to spend much time in his later life to uphold his less intuitively appealing system against the Cartesians, and at the same time to refute insistent attempts within his own camp to introduce metaphysical implications into the law of universal gravitation. He declared again and again that he was neither able nor willing "to adjudge natural causes" of gravitation. This was all the more remarkable (and a large step toward the modern conception of what is required of physical theory) since Newton—who, we must remember, later regarded himself as much theologian as scientist—might have introduced an explicit theistic assumption at that point. As has been hinted before, the existence of a Creator is a persistent implicit hypothesis in his work, and at times it does come out into the open, rarely in the *Principia* but more often in his other writings. Thus he could say in the *Opticks* in explanation of the term "absolute space," "God moves the bodies within his boundless uniform sensorium."

Privately, Newton, like his contemporaries, did in fact believe that some material agency would eventually be found to explain the apparent action at a distance involved in his law of gravitation. There are several strong hints in his writings that he entertained the idea of an all-pervading ether, e.g., *Opticks*, Query 18, where he asks whether heat is not "conveyed through the *vacuum* by the vibration of a much subtler medium than air, which, after the air was drawn out [as by evacuating a vessel with a pump], remained in the vacuum"; this medium he imagined as "expanded through the heavens." At the time, this was a most reasonable speculation, proceeding as it did from some simple observations on the transmission of (radiant) heat to a thermometer suspended in an evacuated vessel, and anticipating the *wave theory of heat* popular in the early nineteenth century (see Section 17.7). It is the more interesting that this tempting but vague picture was not permitted in the *Principia*.

---

*    Letter of February 25, 1693; reprinted in *Newton's Philosophy of Nature* (edited by H. S. Thayer), New York: Hafner, 1953, page 54.

†    See H. Butterfield's *Origins of Modern Science* for a brief but stimulating account of the long struggle between Newton's theory and Descartes'.

## 11.11 NEWTON'S PLACE IN MODERN SCIENCE

So impressive were the victories of Newtonian mechanics in the early eighteenth century that, through the customary extrapolation from science to philosophy, there spread a mechanistic world view which asserted that man's confident intellect could eventually reduce *all* phenomena and problems to the level of mechanical interpretation. In economics, philosophy, religion, the "science of man"—everywhere the success of the work of Newton and the Newtonians was one of the strongest influences to encourage the rising "Age of Reason."*

One of the consequences of the mechanistic attitude, lingering on to the present day, was a widespread belief that with the knowledge of Newton's laws (and those of electrodynamics later) one could predict the whole future of the whole universe and each of its parts if only one were given the several positions, velocities, and accelerations of all particles at any one instant. It was a veiled way of saying that everything worth knowing was understandable in terms of physics, and that all of physics was essentially known. Today we honor Newtonian mechanics for less all-inclusive but more valid reasons. The factual content of the *Principia* historically formed the basis of most of our physics and technology, and the success of Newton's approach to his problems revealed the most fruitful method to guide the work of the subsequent two centuries in the physical sciences.

We also recognize now that despite its breathtaking scope of applicability, Newton's mechanics holds in its stated form only within a definable region of our science. For example, although the forces *within* each galaxy may be Newtonian, one can speculate that forces of repulsion instead of attraction operate between each galaxy and its neighbors (one interpretation of the greater speed with which more distant systems recede). And at the other end of the scale, among atoms and subatomic particles, an entirely non-Newtonian set of concepts has to be presented to account for the behavior of those small-scale worlds, as we shall see in Part H.

Even within the solar system, there are a few small residual discrepancies between predictions and facts. The most famous is the erroneous value of the position of closest approach to the sun by the planet Mercury; calculations and observations differ by some 43 sec of arc per 100 years. A similar small failure of Newton's laws of motion and of gravitation exists for the path of a recurring comet that comes close to the sun and, to a progressively lesser degree, for the rest of the members of the solar system.

But these difficulties cannot be traced to a small inaccuracy of the law of gravitation (e.g., to a slight uncertainty of the exponent in the expression $Gm_1m_2/R^2$). On the contrary, as in the case of the failure of the Copernican system to account accurately for the detail and "fine structure" of planetary motion, we are here again confronted with the necessity of scrutinizing our assumptions. Out of this study has come the realization that no single revision can account for all these deviations from classical predictions, so that at present Newtonian science is joined at one end with relativity theory which, among many other achievements, covers phenomena involving bodies moving at high speeds and/or the close approach of large masses. At the other end Newtonian science borders on quantum mechanics, which gives us a physics of atoms and molecules. As for the middle ground, Newtonian mechanics still describes the world of ordinary experience and of "classical" physics as accurately and satisfyingly as it did always.

---

* A most stimulating discussion of this important aspect of science is found in J. H. Randall's *The Making of the Modern Mind*, Chapters X–XV, or in more detail in his later book, *The Career of Philosophy*, Vol. I, Book 4.

## ADDITIONAL PROBLEMS

**Problem 11.19.** Would Newton's laws of motion, as usually stated, hold in a geocentric system like Ptolemy's? [*Hint:* The term "motion along a straight line" in Newton's first law refers to a line in space such as might be drawn from the sun to a relatively "fixed" star.] If they do not, explain why not, and whether it is plausible that equivalent laws might be formulated to describe motions in a geocentric system.

**Problem 11.20.** Turn to Section 3.2 and use those paragraphs to rate the Newtonian theory of universal gravitation point by point. (Do not assume that such a conscious rating is ever seriously undertaken in scientific work.)

**Problem 11.21.** Newton proved in the *Principia* that the gravitational force on an object at a point inside a uniform sphere, at a distance $R$ from the center, would be the same as that exerted by the part of the sphere situated closer to the center (i.e., by a sphere of radius $R$, ignoring the outer spherical shell). Using this fact, show that the gravitational force inside a sphere is directly proportional to $R$.

## RECOMMENDED FOR FURTHER READING

E. A. Burtt, *The Metaphysical Foundations of Modern Physical Science*, Chapters V, VI, and VII

H. Butterfield, *The Origins of Modern Science*, Chapters 5–10

R. Feynman, *The Character of Physical Law*, London: British Broadcasting Corporation, 1965; paperback reprint by MIT Press; Chapters 1 and 2

A. R. Hall, *From Galileo to Newton 1630–1720*, New York: Harper, 1963; Chapters X and XI

E. C. Kemble, *Physical Science*, Chapters 5, 9, and 11

A. Koyré, "The significance of the Newtonian syn-

thesis" and other essays in his *Newtonian Studies*, Cambridge, Mass.: Harvard University Press, 1965; Phoenix Books paperback

T. S. Kuhn, *The Copernican Revolution*, Chapter 7

Isaac Newton, *Mathematical Principles of Natural Philosophy*, beginning of Book Three and General Scholium at the end of Book Three; extracts in H. S. Thayer, *Newton's Philosophy of Nature*, pages 3–5 and 41–46

J. H. Randall, *The Making of the Modern Mind*, Boston: Houghton Mifflin, 1940; Chapters X–XV

D. Stimson, *Scientists and Amateurs*, New York: Henry Schuman, 1948. An engaging account of the Royal Society and its illustrious members.

## SOURCES, INTERPRETATIONS, AND REFERENCE WORKS

A. Berry, *A Short History of Astronomy*, Chapters VIII–XIII

Marie Boas, "The establishment of the mechanical philosophy," *Osiris*, Vol. 10, pages 412–541 (1952). Concerned primarily with the writings and influence of Descartes and Robert Boyle. See also Marie Boas Hall, *Robert Boyle on Natural Philosophy: An Essay with Selections from his Writings*, Bloomington: Indiana University Press, 1965.

G. Buchdahl, *The Image of Newton and Locke in the Age of Reason*, New York and London: Sheed and Ward, 1961

I. B. Cohen, *An Introduction to Newton's Principia*, Cambridge, Mass.: Harvard University Press, 1971

Mary Hesse, *Forces and Fields*, London: Nelson, 1961; New York: Philosophical Library, 1962

D. L. Hurd and J. J. Kipling, *The Origins and Growth of Physical Science*, Vol. 1, pages 209–226; selections from writings of Halley and Laplace

M. Jammer, *Concepts of Space*, Cambridge, Mass.: Harvard University Press, 1954; Harper Torchbook paperback reprint; Chapters 1–5

T. B. Jones, *The Figure of the Earth*, Lawrence, Kansas: Coronado Press, 1967. The story of early-eighteenth-century attempts to test one of the predictions of Newton's theory of gravity.

R. McCormmach, "John Michell and Henry Cavendish: weighing the stars," *British Journal for History of Science*, Vol. 4, pages 126–155 (1968)

J. E. McGuire and P. M. Rattansi, "Newton and the 'pipes of Pan,'" *Notes and Records of the Royal Society of London*, Vol. 21, pages 108–143 (1966)

Isaac Newton, *Mathematical Principles of Natural Philosophy* (especially the last section, "The System of the World"), pages 549–626 in Cajori's edition, which gives a less technical survey of the applications to astronomy.

*Newton's Philosophy of Nature, Selections from his Writings*, edited by H. S. Thayer, New York: Hafner, 1953; paperback

*Newton's Unpublished Scientific Papers*, edited by A. R. and M. B. Hall, London: Cambridge University Press, 1962

*Newton's Papers and Letters on Natural Philosophy and Related Documents*, edited by I. B. Cohen *et al.*, Cambridge, Mass.: Harvard University Press, 1958

Isaac Newton: *Opticks*, London, 4th edition 1730, reprinted by Dover, New York, 1952. The famous Queries at the end of Book Three include many of Newton's speculations on the nature of matter, force, and light.

H. Odom, "The estrangement of celestial mechanics and religion," *Journal of the History of Ideas*, Vol. 27, pages 533–548 (1966)

J. H. Randall, Jr., *The Career of Philosophy*, New York: Columbia University Press, 1962; paperback. Volume 1, Book 4, discusses Newton's influence on eighteenth-century thought.

F. G. Watson, *Between the Planets*, Cambridge, Mass.: Harvard University Press, revised edition 1956

F. L. Whipple, *Earth, Moon and Planets*, Cambridge, Mass.: Harvard University Press, revised edition 1963

D. T. Whiteside, "Before the Principia: the maturing of Newton's thoughts on dynamical astronomy, 1664–1684," *Journal of the History of Astronomy*, Vol. 1, pages 5–19 (1970)

C. A. Wilson, "From Kepler's laws, so-called, to universal gravitation. Empirical factors," *Archive for History of Exact Sciences*, Vol. 6, pages 89–170 (1970)

PART D

# ON STRUCTURE AND METHOD IN PHYSICAL SCIENCE

# ON STRUCTURE AND METHOD IN PHYSICAL SCIENCE

"As in Mathematicks, so in Natural Philosophy, the Investigation of difficult Things by the Method of Analysis, ought ever to precede the Method of Composition. This Analysis consists in making Experiments and Observations, and in drawing general Conclusions from them by Induction, and admitting of no Objections against the Conclusions, but such as are taken from Experiments, or other certain Truths. For Hypotheses are not to be regarded in experimental Philosophy. And although the arguing from Experiments and Observations by Induction be no Demonstration of general Conclusions; yet it is the best way of arguing which the Nature of Things admits of, and may be looked upon as so much the stronger, by how much the Induction is more general. And if no Exception occur from Phaenomena, the Conclusion may be pronounced generally. But if at any time afterwards any Exception shall occur from Experiments, it may then begin to be pronounced with such Exceptions as occur. By this way of Analysis we may proceed from Compounds to Ingredients, and from Motions to the Forces producing them; and in general, from Effects to their Causes, and from particular Causes to more general ones, till the Argument end in the most general. This is the Method of Analysis: And the Synthesis consists in assuming the Causes discover'd, and establish'd as Principles, and by them explaining the Phaenomena proceeding from them, and proving the Explanations . . .".

Isaac Newton, *Opticks*

## ON STRUCTURE AND METHOD IN PHYSICAL SCIENCE

As long as there has been science, there has also been commentary on the tasks and procedures of science. From Aristotle to Galileo, from Roger Bacon to the present, almost all major scientists and philosophers have contributed their varying opinions. No room is left, it would seem, for remarks on a subject so lengthily and so competently examined. But even if this were true, we might permit ourselves to discuss, in an informal manner, representative opinions held by some of the men who have prominently influenced scientific thought, and so attempt to reach a plausible point of view from which to analyze the progress and procedures of science—not for the sake of a technical exposition of the problems involved, but for increased understanding and enjoyment of the scientific enterprise.

The three chapters in Part D carry on and complete a cycle begun with Chapter 3, "The nature of scientific theory." It is suggested that you reread Chapters 3, 12, 13 and 14 together when you reach the end of the book.

# Chapter 12

# ON THE NATURE OF CONCEPTS

## 12.1 INTRODUCTION: THE SEARCH FOR CONSTANCIES IN CHANGE

When you ask "What is science?" you are in effect asking "What do scientists now do at their desks and in their laboratories, and what part of their past work is still useful to men in a given field?" Because all too many physical scientists have, over the last four centuries, found it more and more difficult to communicate to nonscientists what they do, let us first visualize the scientist's tasks in terms of an analogy involving his more immediately approachable colleague, the cultural anthropologist. In that case the laboratory, says Margaret Mead's *Samoa*, is an island of primitive people. The anthropologist immerses himself in the life of the community to observe it, to study its pattern—but, of course, the eye cannot see, the mind cannot grasp, meaningful patterns within the chaos of movement and sound until one has crystalized out of one's experience the necessary concepts with which to discover, think about, and describe relationships in the community life. One set of concepts belongs to the prior common-sense background of our investigator, e.g., *family, ruler, in-group, I.Q.,* etc.; this corresponds in the physical sciences to the time-honored notions of *time, volume, speed,* and so on. Another and to us more interesting set of terms comes out of the phenom-

enon under study. The anthropologist learns the language and customs of this people and discovers and invents new constructs important to the organization of his observations and his understanding of the community, e.g., the terms "taupo" and "uso," which denote in Samoa respectively the ceremonial princess of the house and a sibling of the same sex. In the same manner the physical scientist, sometimes with great difficulty, must extract from his experience new guiding ideas like *electric field* or *valence.*

Significantly, our anthropologist may find that the group he is studying lacks the manifestations of many concepts without which our own society seemingly could not operate, e.g., some elements of the moral code of Western civilization, the knowledge of one's own exact age, or even the simple recognition and naming of most colors. Again science, too, had to learn—often the hard way—that there are no useful equivalents in inanimate nature for such common-sense terms as human longing or absolute simultaneity.

The formulation or "discovery" of new concepts helps the anthropologist gradually to reinterpret the originally meaningless, aimless, or "primitive" life of the village, and a complex, rather rigid pattern emerges. His own early common-sense concepts themselves may undergo profound changes

173

—the word *law*, for example, is surrounded by an entirely different atmosphere in different cultures, just as the idea of *force* in physics (and, in fact, every technical concept that grew out of the vocabulary of common sense) is now at variance with the original meaning of the same word as still used outside the sciences.

In the end, our anthropologist may succeed in decoding the original problem, bringing back an account of that people's political and family organization, personal esthetic values, religious beliefs and practices, economic methods, and so on. Perhaps he will also be able to reconstruct the history of that people, or even to illuminate the behavior of another group, his own, in relation to that culture.   At any rate, it is evident in this example that his job was not finished when he had collected the initial direct observables; on the contrary, that was the barest beginning, the stimulus for the really important part of his work. We may keep in mind, with all its limitations, the analogous picture of physical scientists as explorers in a universe of events and phenomena, attempting to find its pattern and meaning.

## 12.2  SCIENCE AND NONSCIENCE

Science, one may now say, is the ever unfinished quest to discover facts, establish the relationships between things, and decipher the laws by which the world runs.  But let us go beyond this.  While not trying to propose a one-sentence definition of the whole complex concept "science," we may perhaps agree at the outset that *the main business of science is to trace in the chaos and flux of phenomena a consistent structure with order and meaning*, and in this way to interpret and to transcend direct experience.  "The object of all sciences," in Einstein's words, "is to coordinate our experiences and to bring them into a logical system."  And another major twentieth-century physicist, Niels Bohr, agrees when he says "The task of science is both to extend the range of our experience and to reduce it to order."

Probably you will find these statements too all-inclusive; the same aim might well be claimed by art or by philosophy.  Thus the poet T. S. Eliot said, "It is the function of all art to give us some perception of an order in life by imposing an order upon it," and the philosopher A. N. Whitehead defined speculative philosophy as "the endeavor to frame a coherent, logical, necessary system of general ideas in terms of which every element of our experience can be interpreted."  Indeed, in science, as in art and philosophy, our most persistent intellectual efforts are directed toward the discovery of pattern, order, system, structure, whether it be as primitive as the discernment of recurring seasons or as sweeping as a cosmological synthesis.  In this sense, science is but one facet of the great intellectual adventure, the attempt to understand the world in each of its aspects.  The search for constancies in the flux of experience is so fundamental and so universal a preoccupation of intelligent life itself that, in common with many of the Greek philosophers, we may regard mind as the principle that produces order.  We note without astonishment among the great early men of science, an artist and a churchman, Leonardo da Vinci and Copernicus; indeed, the very origin of the word *science* (Latin *scire*, to know, to learn) betrays the extent of its appeal and the depths of its roots.

Of course, one must not deny the fundamental and distinct differences separating the sciences from the nonsciences.  There are obvious points which set one apart from the other, for example, the motivations of the investigators, to be discussed in more detail in Chapter 13.  To predict nature and thereby not to be at her mercy, to understand nature and so to enjoy her, these are among the main motivations for the examination of nature through science, whereas the poetic understanding of nature is primarily the self-realization, the proclamation and ennoblement of man's own spirit. But these are in a sense complementary, not contradictory, motivations, illuminating the two sides of man. Both coexist to some degree within each individual. How artificial would be a distinction between the triumph of the scientist and that of the artist within Johannes Kepler when he writes in the *Harmony of the World* (1619), at the occasion of publishing the third law of planetary motion:

" .... What I prophesied 22 years ago, as soon as I found the heavenly orbits were of the same number as the five [regular] solids, what I fully believed long before I had seen Ptolemy's Har-monies, what I promised my friends in the name of this book, which I christened before I was 16 years old, I urged as an end to be sought, that for which I joined Tycho Brahe, for which I settled at Prague, for which I have spent most of my life at astro-nomical calculations—at last I have brought to light, and seen to be true beyond my fondest hopes. It is not 18 months since I saw the first ray of light, 3 months since the unclouded sun-glorious sight— burst upon me. Let nothing confine me: I will indulge my sacred ecstasy. I will triumph over mankind by the honest confession that I have stolen vases of the Egyptians to raise a tabernacle for my God far away from the lands of Egypt. If you forgive me, I rejoice; if you are angry, I cannot help it. This book is written; the die is cast. Let it be read now or by posterity, I care not which ...".

A second point of difference between science and nonscience is much more clear-cut than the first; it is the kind of rules and concepts the scientist uses, the type of argument which will cause him to say, "Yes, I understand and I agree." This will occupy our attention to some degree, as will a third point of difference: the observation that in the course of time, despite great innovation and revolutions, there accumulates in science a set of internationally acceptable, basic, and fairly endur-ing conceptual schemes, whereas this can hardly be said for many other human endeavors.

At once the question arises as to how man, so short-lived and so fallible, can penetrate such complex truths, discover lasting patterns, and obtain general agreement. For the scientist's work, like that of any other explorer, must involve the whole man, demanding as it does reflection,

observation, experimentation, imagination, and a measure of intuition. Being human, he fails far more often than he succeeds, and even his suc-cesses and failures may in the light of further progress reverse positions. He often cannot fully explain his reasons for dedicating himself to his work and the steps by which he progresses; and if he is vocal on such matters, he very likely is contradicted by the testimony of some of his colleagues. And yet, the *result* of this uncertain human activity, namely, the growing body of science itself, is undeniably a successful, vigorous enterprise, bountiful in discoveries, in distinct contrast to the frailties and confusions of its human creators. This paradox we shall explore, for it will be the key to an understanding of the activities of the scientist and of the successful features of his work.

Immediately one answer suggests itself: There may be some trick, some *method* which scientists follow to solve their problems so well. After disposing of this opinion, we shall look quite closely at an alternative answer, namely, that there is perhaps some special merit in the idiosyncratic vocabulary, in the particular mental tools them-selves, with which scientists analyze the world around them.

## 12.3 THE LACK OF A SINGLE METHOD

If by scientific method we mean the sequence and rule by which scientists now and in the past have actually done their work, then two truths soon become obvious. First, as for every task, there are here not one but many methods and uncountable variants and, second, even those different methods are for the most part read into the story after it has been completed, and so exist only in a rather artificial and debatable way. The everpresent longing to discover some *one* master procedure underlying all scientific work is understandable, for such a discovery might enormously benefit all fields of scholarship; but like the search for the philosopher's stone, this hope had to be given up.

At about the beginning of the seventeenth century, the time of Francis Bacon and Descartes, it was still quite reasonable to hope that both these all-powerful keys to wealth, health, and wisdom existed, but the verdict of the three centuries since then is plainly negative. Even two centuries ago the chemist Joseph Priestley warned "...how little mystery there really is in the business of experimental philosophy, and with how little *sagacity*, or even *design*, discoveries (which some persons are pleased to consider as great and wonderful things) have been made...".

Priestly's words, even though an extreme generalization which does not do justice to his own carefully prepared researches, are a proper antidote to the other extreme of opinion which would present science to us as a special, infallible scheme relentlessly progressing from success to success with the precision of a smoothly moving machine. For as soon as one begins to look into the history of scientific discoveries, one is overwhelmed by evidences that there is no single, regular procedure. By temperament and by characteristics of performance, scientists have always differed from one another as widely as composers. Some proceeded from one step to the next with the certainty and restraint of a Bach, others moved among ideas with the abandon of a Schumann. Among the great scientists there have been adventurers and recluses, self-taught mechanics and aristocrats, saints and villains, mystics and businessmen, pious men and rebels. In *The Study of the History of Science*, George Sarton said about them:

"...Their manners and customs, their temperamental reactions, differ exceedingly and introduce infinite caprice and fantasy into the development of science. The logician may frown but the humanist chuckles.

"Happily such differences are more favorable to the progress of science than unfavorable. Even as all kinds of men are needed to build up a pleasant or an unpleasant community, even so we need all kinds of scientists to develop science in every possible direction. Some are very sharp and narrow-minded, others broadminded and superficial. Many scientists, like Hannibal, know how to conquer, but not how to use their victories. Others are colonizers rather than explorers. Others are pedagogues. Others want to measure everything more accurately than it was measured before. This may lead them to the making of fundamental discoveries, or they may fail, and be looked upon as insufferable pedants. This list might be lengthened endlessly."

The process of discovery itself has been as varied. While it is perfectly true that individual research projects are almost always unspectacular, each investigation being fairly routine, logically quite consistent and sound within its prescribed scope, occasionally some most important theories have come from drawing wrong conclusions from erroneous hypotheses, or from the misinterpretation of a bad experiment. Sometimes a single experiment has yielded unexpected riches, sometimes the most elaborately planned experiment has missed the essential effect by a small margin, and sometimes an apparently sound experiment was simply disbelieved by a scientist who, holding on to his "disproved" theory, emerged triumphant when the experiment was found to be faulty after all. Even the historic conceptual schemes in science themselves first captured the mind in *seemingly* the most casual or unpredictable way.

To cite two famous examples, Darwin said he got "a theory by which to work" while he happened to be reading Malthus' treatise on population "for amusement." One of Kekulé's fundamental contributions to modern organic chemistry came to him suddenly during an idle daydream. Similar things have happened to scientists in every field.

On the other hand, some men had all the "significant facts" for an important finding in their hands and drew conclusions which were trivial or proved to be mistaken; sometimes a scientist proposed a correct theory although some facts before his very eyes were seemingly in violent contradiction. Of all these we shall have further examples. Even the work of the mighty Galileo, viewed in retrospect, occasionally seems to jump ahead from error to error until the right answer is reached, as if with the instinctive certainty of a somnambulist.

In Chapter 14 we shall hear the opinions held by the men of science themselves concerning the methods of work in their field, but even this superficial recital of variety is enough to establish a feeling that the success of science lies not so much in some single method of work, but perhaps rather in some peculiar adjustment or mechanism mediating between two factors—the individual's achievement, and the body of science in which all contributions are brought together after supplementing and modifying, cooperating and competing with one another. A very crude analogy may help here; if we compare the structure of science to that of an anthill, the successful construction of that astonishingly complex, well-arranged, and successful building must be explained not by the diverse, almost erratic behavior of the individual insects, but in larger measure by the marvelous coordination of innumerable solitary efforts. The mechanism of coordination among these social insects is still largely a mystery, but in the analogous case of science, something positive might be said. We approach it by turning to the alternative possibility posed at the end of the previous section, to the astonishing and many-sided power that resides in the concepts scientists use.

## 12.4 PHYSICAL CONCEPTS: MEASUREMENT AND DEFINITION

All intelligent human endeavor stands with one foot on observation and the other on contemplation. But scientists have gradually come to limit themselves to certain *types* of observations and thought processes. One distinctly striking "limitation" is the tacit desire among scientists *to assure that they are in fact discussing the same concepts in a given argument*. This desire for clarity is, of course, general, but here we find a particularly successful scheme of meeting it, even at the cost of some sacrifices.

To take a very simplified example: If the scientific problem were to find the dimensions or volume of a body, its electric charge or chemical composition, the specialists called in to solve this problem would all appear to understand clearly what is wanted because they all would go independently through quite similar manual and mathematical operations to arrive at their answers, even though these answers may not completely coincide. (If the word *similar* seems disturbing in the previous sentence, we may go so far as to claim that these men will, on mutual consultation, very probably all agree on *exactly* the same operations to follow.) As a consequence, their findings for, say, the length of one edge of a block may read 5.01 cm, 5 cm, 4.99 cm, and 5.1 cm. The impressive thing here is not that they seem to disagree on the exact length or on the accuracy, but that they do agree on the type of answers to give. They do not, for example, say "as long as a sparrow in summer," "five times the width of my finger," "as long as it is wide," and "pretty short." Evidently the words "length of one edge of a block" mean the same things to all four experimenters, and we may be fairly sure that on consultation sooner or later all four would convince themselves of one single result, possibly 5.03 ± 0.04 cm. It should not matter which meter stick they happen to pick up, for all manufacturers of measuring equipment are under legal obligation to check their products against carefully maintained universal standards. On the whole, almost every measurement made in the laboratory is in a sense such a comparison between the observables and some well-known standard accepted by the whole fraternity. Much of the success and rapid growth of science depends on these simple truths, for it is clear that in such circumstances the energies of investigators are not regularly drained off by fruitless arguments about definitions and rules of procedure; on the contrary, the labors of many men over centuries can combine in one advancing stream.

There remain, of course, large areas of possible disagreement among scientists, but such arguments can be settled, often by having recourse to some one series of measurements which both disputants (rightly or wrongly) acknowledge at the time to be decisive one way or the other. One of the impressive features of modern physical science is the rapidity with which most major differences of opinion in the field usually disappear.* The secret of this successful harmony and continuity in physical science (which we idealized only a little in the previous paragraph) lies to a large degree in the *nature of concepts and their definitions*. For example, the concept "length of an object" as used in science is ultimately defined by the very *operations* involved in making the measurement. The problem "What is the length of a block?" is for all practical purposes identical with the question "What is the difference between those two numbers printed on a specific meter stick which stand directly below two corresponding scratches, entered there to signify local coincidence with adjacent corners of the block?"

This last sentence contained an abbreviated example of what we shall call an *operational definition*, that of length, and although it seems ridiculously involved, and may indeed never be used explicitly, our four experts engaged to measure the length can regard it as the "true meaning" of the length of a block, available for examination if any dispute should arise. Ideally, each of the concepts used in the physical sciences can be made clear in terms of some such operational definition, and that perhaps is the most important of the mechanisms whereby mutual understanding among scientists is made possible. For it is clearly more difficult to misinterpret action than words.

As has been implied throughout this discussion, not all concepts in science can be defined by *manual*

operations. Instantaneous velocity, for example, was defined by the slope of a straight line, tangent to the distance-vs.-time graph; this is a mental or mathematical operation. But each such definition becomes a fairly unambiguous set of directives to all scientists who have worked in the field, and this explains why their vocabulary is not regularly the subject of dispute.

You will find it revealing—more than that, almost indispensable to a true understanding of physical concepts—to make for yourself operational definitions of all important terms, such as we have made in previous chapters for acceleration, force, mass, and so on. As the influential American physicist and philosopher, P. W. Bridgman, said, "The true meaning of a term is to be found by observing what a man does with it, not what he says about it."†

At this point you are likely to have several disturbing questions. If you were tempted to object that the operational definition given here is far from the common-sense meaning of length, you would be quite right and would have hit on a very important truth: Everyday notions seem clear and scientific terms mysterious. However, a little thought shows the opposite to be the case. The words of daily life are usually so flexible and undefined, so open to emotional color and misunderstanding, that our first task here is to get used to the specific vocabulary of the sciences and to the apparently picayune insistence on its *rigorous* use—that enormously successful habit which scientists borrowed from scholastic logicians. Then again, you are bound to have a vaguely uncomfortable feeling that such a definition only shows us how to measure according to some man-

---

* A comparison with the state of affairs in other disciplines may be misleading, but this is a good place to interject that the writings on the history and the philosophy of science themselves, in this book as in others, are traditionally, and quite properly, ready subject matter for disagreement and debate, not only for a few years but perhaps forever.

† In his *Logic of Modern Physics*, New York: Macmillan (1927). This short and influential book outlines a modern approach to science; in particular, the operational definition of concepts is discussed fully. The first 100 pages are excellent reading for students at this stage of study.

made convention, and does not tell us what length really is; once more you are perfectly correct, and once more we must try to make our peace with the limitations of modern science. Here again we are touching on a problem that is at the very core of science, namely, what reality means to a man in a laboratory. Surely we shall have to explore this very thoroughly. For the moment let us take the answer of that great French mathematician and philosopher of science, Henri Poincaré, who at the beginning of this century illustrated the operational attitude toward physical concepts in this manner:

"When we say force is the cause of motion we talk metaphysics, and this definition, if we were content with it, would be absolutely sterile. For a definition to be of any use, it must teach us to *measure* force; moreover, that suffices; it is not at all necessary that it teach what force is *in itself* nor whether it is the cause or the effect of motion."

Lastly, you may be puzzled that a simple measurable like length should not be given with perfect exactness, that we should have said "the edge of this block is 5.03 ± 0.04 cm long." The term ± 0.04 cm is called the *probable error* in the measurement; that is to say, the chances are fifty-fifty that the next measurement of the same edge will read between 4.99 cm and 5.07 cm. Evidently all measurements, apart from simple counting, contain some error or uncertainty, no matter how carefully the job is done.

But although this uncertainty may be slight, how can we dare build an exact science on concepts defined by necessarily uncertain measurements? The paradox is resolved by two recognitions. First, this word "error" does not have in science the connotations *wrong, mistaken,* or *sinful* which exist in everyday speech. In fact, a measured value can be regarded as "exact" only if we know also just what range of values to expect on repeating the measurement. Second—and we must repeat this frequently—science is not after absolute certainties, but after relationships among observables. We

realize that the observables can be defined and measured only with some uncertainty; and we shall not demand more than that of the relationships between them, thus leaving the search for absolute truths to other fields of thought, whether we like that or not.

## 12.5 PHYSICALLY MEANINGLESS CONCEPTS AND STATEMENTS

The consequence of even such a trivial-sounding operational definition as that of length may be quite startling. In the special theory of relativity, rigorous definitions of this type directed the course of thought to unexpected results, one being that the measured length of an object depends on how fast the object moves with respect to the observer, a finding that accounted for some most perplexing previous observations. As a result of the impact of Einstein's work, scientists have become aware of the operational nature of the useful concepts; it was then discovered that some tacitly accepted ideas were leading to serious contradictions in parts of physical theory because, by their very formulation, they could not be connected with any possible activities or operations in the laboratory. For example, the concepts of time and space were defined previously not by specific and exclusive reference to manipulations with meter sticks, light signals, pendulum clocks, and the like, but in some absolute, intuitive sense. The classic reference here is Newton's statement in the first pages of the *Principia*:

"Absolute, true, and mathematical time, of itself, and from its own nature, flows equably and without regard to anything external, and by another name is called duration."

Note that phrase "without regard to anything external," that is, without necessary relation to the rotation of a second hand on a real watch. By this very definition of "true" time, we could not hope to measure it. Or again,

"Absolute space, in its own nature, without regard to anything external, remains always similar and immovable."

From the point of view of modern practical scientists, such statements, inherently without operational meaning, are sometimes called "meaningless," perhaps a rather abrupt term, but accurate in this limited sense. You may well ask whether Newton's science was not invalidated by such meaningless basic postulates, and the answer is no, simply because Newton did not, in fact, depend on the explicit use of these concepts in his own scientific work. (His reasons for stating them at all were complex and can be traced to his philosophical motivation.)

Galileo, although much nearer to that science of the ancients which he helped to displace, saw quite clearly that science should be based on concepts that have meaning in terms of possible observations. Recall again his preface to the definition of acceleration:

"We however have decided to consider the phenomena of freely falling bodies with an acceleration such as actually occurs in nature, and to make our definition of accelerated motion exhibit the essential features of this type of natural accelerated motion."

If we define as "meaningless" any concept not definable by operations, it follows that not only concepts but whole statements and even intelligent-sounding, deeply disturbing questions may turn out to be meaningless *from the standpoint of the physical scientist*. Here are a few examples: "What are the physical laws which would hold in a universe completely devoid of any material bodies?" "Which is *really* at rest, the sun or the earth?" "Will this table cease to exist while it is not being observed?" "What is length in itself, apart from measurements?" "Are there natural laws that man can never hope to discover?" and even "Is light made of corpuscles only or waves only?"

Many more such questions are listed by Bridgman, who speculatively ventures to extrapolate this *operational view* beyond science by saying,

"I believe that many of the questions asked about social and philosophical subjects will be found meaningless when examined from the point of view of operations. It would doubtless conduce greatly to clarity of thought if the operational mode of thinking were adopted in all fields of inquiry . . .".

It is necessary to enter here a warning against carrying this mode of thinking too far. Rigorous application of operationism can be and in fact has been used to attack speculations that may well develop and turn out to be fruitful after all. For example, at the end of the nineteenth century the physicist Ernst Mach argued that the *atom* is a physically meaningless concept, because there was at that time no way to observe or measure its discrete, individual properties. Nevertheless, atomic theories being developed at that time (see Chapters 19–22) were to be of great value to science even though they could not yet be based on strictly operational definitions.

## 12.6 PRIMARY AND SECONDARY QUANTITIES

If it be true that science as an established activity has its roots in operationally meaningful concepts, then there must be a large range of experience that is being neglected by science. In a sense one may compare a scientist with a man who looks at the night sky by means of a very powerful telescope and who is thus enabled to examine a small region with extreme penetration while, however, sacrificing the chance to scan a whole world at one glance. In addition, we are apt to think that with his telescope our scientist seems to observe only the number of stars, their relative brightness, and the like, instead of watching the skies for the sheer experience drawn from the beauty of the grand display that so impressed the casual observer.*

The comparison is apt. Galileo drew the distinction between those experiences and concepts which might safely serve as the foundation stones of science and those which, having a measure of more *subjective* meaning, should from the point of view of science be regarded as sources of illusion and debate. He called them respectively *primary* and *secondary quantities*. The primary quantities were direct observables that could be mathematically symbolized and measured, such as position and motion, the very elements that at the time had simple operational significance, to use the modern phrase. The quantities regarded as secondary were those not then accessible to instrumental measurement, those which were largely qualitative. By and large, the distinction, which impressed and was accepted by such followers of Galileo as Newton himself, still holds in the physical sciences. We might say that Galileo's decision reduced the extent of *eligible* experience to a small fraction of the scientist's total experience, but to precisely that fraction which he could quantify and therefore share fairly unambiguously with his fellows. The ideas of science look so stylized and unreal just because we demand of them that they help us describe those features of experience which the common-sense view of reality cares least about— measurement, mathematical manipulation, numerical prediction, and clear communicability—while failing to describe exactly those most prominent uses of everyday expressions, namely, our feelings, reactions, and other personal involvements.

Recent developments in branches of psychology and the social sciences are reminiscent of that early search for quantification in physics and chemistry. Throughout Europe, men of the early

seventeenth century were saying that the qualitative ideas inherited from the ancients would have to give way to a new, quantitative way of describing nature. A case in point is Aphorism XCVIII of Francis Bacon's *Novum Organum* of 1620, where we find this complaint concerning contemporary thinking:

"Now for grounds of experience—since to experience we must come—we have as yet had either none or very weak ones; no search has been made to collect a store of particular observations sufficient either in number, or in kind, or in certainty to inform the understanding, or in any way adequate.... Nothing duly investigated, nothing verified, nothing counted, weighed, or measured, is to be found in natural history; and what in observation is loose and vague, is in information deceptive and treacherous."

## 12.7 MATHEMATICAL LAW AND ABSTRACTION

The insistence on quantitative concepts must, of course, appear incomprehensible until we recognize that the work of the physical scientist is based on a faith as ancient as it is astonishing, namely, that nature works according to mathematical laws and that the observations are *explained* when we find the mathematical law relating the observations. For example, little more is left to be said about simple freely falling bodies *in vacuo* once we have found that they move according to the law $s = v_0 t + \frac{1}{2}gt^2$.

Galileo expressed it this way:

"Philosophy [we should call it science now] is written in that great book which ever lies before our eyes—I mean the universe—but we cannot understand it if we do not learn the language and grasp the symbols in which it is written. This book is written in the mathematical language, and the symbols are triangles, circles, and other geometrical figures [we should now add all other

---

* The analogy is suggestive in another way: The prospect of inevitable, successive refinements of his telescope reminds the researcher that his findings are always tentative, whereas the unaided (nonscientific) observer can easily persuade himself that his observations are final and can yield final truths.

mathematical devices] without whose help it is impossible to comprehend a single word of it, without which one wanders in vain through a dark labyrinth."*

To Galileo, as to his contemporaries (e.g., Kepler and, to some extent, even Harvey) and to modern physical scientists, mathematical methods provide the technique *par excellence* for ordering and comprehending nature. To logic, which for the scholastics was the main tool of investigation, is now primarily delegated the task of establishing the consistency of the scheme of mathematical demonstration, hypotheses, and observations. Although Galileo and Kepler still felt compelled to announce this faith loudly and often, by now, through the unforeseeable fruitfulness of this view, it has become a fundamental, unexamined attitude among physical scientists.

By saying that laws are mathematical relations between observables, we mean, of course, that the general descriptions of physics are usually stated in the form "variable $X$ is related to variables $Y$, $Z$, etc. by such and such a mathematical function." (Consider again, for example, the equation for free fall or Newton's law of gravitation.) Laws may state that $(X/Y) + Z$ (or some other combination of the variables) always has a certain numerical value, or tends to attain a maximum or a minimum value under given conditions. Yet another example is the superposition principle, which asserts, you recall, that the two variables $v_x$ and $v_y$ in projectile motion are *not* interdependent. Above all, there is that type of law most eagerly sought which says in effect "this function of these variables under given conditions is always *constant*." The law of freely falling bodies as conceived by Galileo was of this kind, for it could be phrased: $s/t^2$ is constant if the falling body starts from rest. Kepler's third law is an obvious example. There are numerous others: the laws of conservation of mass, of momentum, of energy, of "caloric," etc. To us it is clear why laws of constancy are so highly prized; they com-

bine the most successful features of science with its most persistent preoccupation—the mathematical formulation of concepts ($s$, $t$, etc.) that aid in the discovery of unchanging patterns in the chaos of experience.

The mathematical formulation of physical law has an incidental and initially disconcerting effect. Unless we are warned to think of a law as the relationship between variables, we are apt to regard it as a strict relationship between *cause and effect*. But in our law of the type $X = YZ$ we can just as well write $Y = X/Z$, or $Z = X/Y$, and there is no way of telling whether $X$ or $Y$ or $Z$ is a cause or an effect. As a consequence, it is on the whole more fruitful to think of an *interaction* rather than a simple causation, and to ask "to what factors is $X$ related" instead of "what causes $X$." For example, Boyle's law states that for a given mass of gas, at a constant temperature, the absolute pressure $P$ and volume $V$ of the gas are related by $PV$ = constant, that is, $P$ is proportional to $1/V$. Here the statement that pressure and volume are inversely proportional accomplishes more in a shorter space and with less ambiguity than the equivalent description of pressure changes *causing* contractions and expansions or of volume changes *causing* compressions or rarefactions.

A great aid to speedy understanding and manipulation of concepts is the simple fact that mathematically formulated ideas can be expressed symbolically in equations like the above. First consider the mere *convenience* of writing an equation like $s = v_0 t + \frac{1}{2}gt^2$, instead of expressing this in a sentence. Note how extraneous word colors disappear; how easy it becomes to communicate arguments and results clearly to your fellow men; and how such equations invite drawing further conclusions about the relationship between observables. There is a good parallel here in the field of chemistry, which was immeasurably advanced in the early nineteenth century by the simple expedient of expressing reactions in formulas.

However, still other consequences follow from the decision of scientists to restrict and direct their attention toward the measurables and the math-

---

* *Discoveries and Opinions of Galileo*, translated by S. Drake, pages 237–238.

ematical relations between them. If the important thing about a rolling ball on an inclined plane is no longer its composition, its history in the workshop, its color, or its usefulness for the sport of bowling, if all that really matters to the particular group of investigators is the relationship $s \propto t^2$, then this ball ceases to be an individual entity—it equally might have been a smooth loaf of bread or a round bottle of wine—and in the long run the real ball seems totally forgotten, its place being taken by "a point mass," an idealization abstracted from the experiment. In this sense the laws of science and the controlled experiment do not directly deal with "real bodies" but with abstractions moving in a hypothetical pure space with properties of its own, in a world which we can manipulate at will, now thinking away all air resistance, now regarding the inclined plane as perfectly smooth and straight, now changing only one aspect, perhaps the angle of inclination, and leaving all other factors untouched. As seen from the outside, this world, rendered in an esoteric language of its own and filled with seemingly grotesque simplifications and exaggerations, is quite analogous to that of the modern painter, poet, or composer.

All the features of science conspire to transform real weights in beautiful Pisa into undistinguished particles moving in idealized space. It is true that much would be lost by this transposition if that were the end of the scientific process, although even this much, the first half, yields the unique and absorbing satisfaction of being able to reduce, order, and so to understand raw sense experience. But there is more: At no point is the contact with reality completely broken; the same rules by which the transition was made are also applicable on the return trip from the world of abstraction to the world of reality. This is guaranteed by the operational nature of the concepts employed in that abstract world. For example, while you were deriving the equation $v^2 = v_0^2 + 2as$ from the other equations of motion, you may not have felt in touch with the real world of moving bodies, yet at the end the result, that equation, could again be referred to real motions. In fact—and this is most

important—*our mathematical world in which the calculations could proceed is justified and taken seriously by physical science only insofar as it does yield new knowledge about the real world around us.* If it were to fail in this function, we should have to rearrange the rules within the abstract world until they finally gave a useful harvest. As it happens, the abstraction from a real ball to a point did give us laws of motion applicable not only to this ball but to all other balls, to all simple round and sliding objects, and the like.

This is indeed making a good profit in return for sacrificing temporarily the identity of the real spherical object. The philosopher Rudolph Carnap suggested a neat analogy at this point. The symbols and equations of the physicist bear the same relation to the actual world of phenomena as the written notes of a melody do to the audible tones of the song itself. The written notes, of course, do not make sounds by themselves, and yet the sounds are not altogether lost in them. At any time one can retranslate the marks on paper into the audible melody, provided one knows the convention which relates notes to tones. Correspondingly, we might now visualize science as an arch resting on two pillars, observation and experience, with conceptualization and abstraction supported in between. The security of the arch depends on the firmness of the pillars.

A whole view of science can be developed along these lines. We find an early hint in Francis Bacon's words,

"But my course and method, as I have often clearly stated and would wish to state again, is this— not to extract works from works or experiments from experiments, as an empiric, but from works and experiments to extract causes and axioms, and again from those causes and axioms new works and experiments, as a legitimate interpreter of nature."

A modern view along such lines is to be seen in James B. Conant's definition of science: "Science is an interconnected series of concepts and conceptual schemes that have developed as the result of experimentation and observation and are fruitful of further experimentation and observations."*

## 12.8 EXPLANATION

We must now turn to answer a voice which has been questioning a fundamental point—the statement that a physical scientist finds *explanation* by discovering mathematical laws between observables. Someone might well be saying, "I can see that this convention works in the sense that it permits one to form the laws of, say, projectile motion by means of which one can accurately direct missiles. However, I do not agree that anything is *explained* thereby. True, when Galileo pointed out that the paths of projectiles coincided with a type of geometric curve called a parabola, he did intuit an elegant and simple description of a large number of separate cases, and he may in fact thereby have discovered a simple device for predicting all kinds of trajectories. But I want to know the ultimate reason *why* projectiles do go in parabolic curves, and I shall remain dissatisfied with any of your explanations until you give me a picture or some intuitive necessity for the graceful arching of projectiles." Now it is certainly simple enough to explain why trajectories are parabolic, but only in terms of the laws of gravitational force acting on massive objects, these laws being themselves mathematical relationships. Our interrogator, on the other hand, does not want more of that. He might well find satisfaction with the explanation current before the late sixteenth century, or perhaps with some other idea, such as a supposed existence of invisible curved tubes or tracks in ether which guide projectiles.

If pushed far enough, our interrogator will have to make a startling concession: His only tools for understanding physical phenomena, his imagination's vocabulary, are pictures, allusions, and analogies involving the primitive mechanical

*   J. B. Conant, *Science and Common Sense*, New Haven: Yale University Press, 1951, page 25.

events of everyday life. We might say he thinks predominantly with the concrete common sense of his muscles. For example, the gravitational or electrical attraction of one body for another across a vacuum is incomprehensible to him unless we allow him to imagine between those two bodies some invisible medium that somehow transmits a real mechanical pull or push from one to the other.

This feature of the human mind, this thirst for concreteness, characterizes not only the frequent preoccupation with mechanical models within science itself, but also the most primitive type of everyday explanation. We find it symbolized in the thought of the ancient people of India that the earth was supported in space on the backs of gigantic elephants, or in the belief of the Egyptians that Osiris weighed human souls in a hand-held balance. It would indeed be surprising if the imagery of religion, poetry, and even of early science had not found its original raw material predominantly in our ordinary, direct, everyday experiences. It is also true that some of the greatest scientific contributions were made by men whose physical intuitions were just so directed. But as modern physical science has turned to problems more and more removed from the realm of common experience, it also has become necessary to enlarge the kit of tools with which to grasp and comprehend phenomena. Our discussion of the nuclear atom will later serve as a particularly striking example of the breakdown of those more naive types of understanding that insist on intuitive, visualizable explanations. The best early statements of this theme were Galileo's insistence that motion was to be explained in terms of a particular mathematical law abstracted from experience, and Newton's refusal to bolster the law of universal gravitation by a mechanical ether hypothesis, saying: "To us it is enough that gravity does really exist, and act according to the laws which we have explained, and abundantly serves to account for all the motions of the celestial bodies and of our sea."

To take another specific example, consider the phenomenon of radioactive decay of some one isotope of radium. The experimental fact is that a given pure sample containing initially the number $N_0$ atoms will by and by change into chemically different substances, each radium atom giving off alpha radiation and changing first into an atom of the element radon. After a time $t$ seconds, only $N_t$ atoms from the original quality $N_0$ are still unchanged (radium), as shown best in a graph (Fig. 12.1). Again, it is an experimental fact, one that was originally difficult to get at and much prized, that this graph is, for all pure radioactive substances, an exponential curve, i.e., that the equation corresponding to this behavior is of the form

$$N_t = N_0 e^{-0.693t/T}. \qquad (12.1)$$

Here $N_t$, $N_0$, and $t$ have been defined above, $e = 2.718$, and $T$ is a constant whose value depends only on the substance involved and which tells of the rapidity of the decay.

Now it is clear that as the equation given describes and summarizes the facts neatly we may refer to it as the *law of radioactive decay*. Although this empirical mathematical law is indispensable and enlightening, it does not in the least explain *why* atoms of radioactive elements explode and change into other elements. That is another and as yet largely unsolved problem. However, the solution will inevitably again be through some mathematical law, such as a formula describing the fluctuation of forces binding together the structure of the nucleus. So ultimately the observations described by Eq. (12.1) will still be explained through mathematical formulations.

Perhaps it is not too frivolous to hold that "to explain" means to reduce to the familiar, to establish a relationship between what is to be explained and the (correctly or incorrectly) unquestioned preconceptions. The preconceptions of the scholastics were grandiose and sweeping (natural place, final causes, celestial vs. terrestrial phenomena, etc.), reflecting the impressive but perhaps overly ambitious universality of their questions and explanations. The modern scientist has trained himself to hold somewhat more limited and specialized preconceptions, e.g., that nature works according to simple models or mathematical

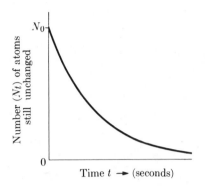

Fig. 12.1 Decay of radioactive element.

schemes. To him the law that the mathematical sum of all energies in a system can be regarded as forever constant is the satisfying explanation for a host of phenomena, whereas the same law would itself be regarded as an unexplained mystery by nonscientists. It requires training and repeated personal success in solving physical problems to be really satisfied with a mathematical answer, just as it must surely involve a strenuous program of orientation to be fully content with the rewards and the rules of behavior in any profession or, for that matter, in our own civilization.

Having examined the nature of concepts in physical science, we are ready to consider how these concepts are integrated in the advance of science.

---

**Problem 12.1** When Galileo and other supporters of the heliocentric system in the seventeenth century asserted that it is really the earth that moves rather than the sun and stars, did they violate the rule that physical concepts must have operational definitions?

---

(The bibliography for this chapter will be found at the end of Chapter 14.)

# Chapter 13

# ON THE DUALITY AND
# GROWTH OF SCIENCE

## 13.1 THE FREE LICENSE OF CREATIVITY

Our discussion of the nature of physical concepts has shown that a main reason for formulating concepts is to use them in connection with mathematically stated laws. It is tempting to go one step further and to demand that the practicing scientist deal only with ideas that correspond to strict measurables, that he formulate only concepts which can be reduced to the least ambiguous of all data: numbers and measurements. The history of science would indeed furnish examples to show the great advances that followed from the formation of strictly quantitative concepts—whether the measurements were concerned with the time of descent of rolling balls, or the tabulation of population statistics, or the weights of compounds in chemical reactions, or—in the case of Pavlov's experiments—with the rate of flow of saliva in experimental animals. The nineteenth-century scientist Lord Kelvin commended this attitude in the famous statement:

"I often say that when you can measure what you are speaking about and express it in numbers you know something about it; but when you cannot measure it, when you cannot express it in numbers, your knowledge is of a meagre and un-

satisfactory kind: it may be the beginning of knowledge, but you have scarcely, in your thoughts, advanced to the stage of *Science*, whatever the matter may be."*

Useful though this trend is within its limits, there is an entirely different aspect to scientific concepts; indeed it is probable that science would stop if every scientist were to avoid anything other than strictly quantitative concepts. We shall find that a position like Lord Kelvin's (which is similar to that held at present by some thinkers in the social studies) does justice to neither the complexity nor the fertility of the human mind, and certainly not to the needs of contemporary physical science itself—to neither scientists nor science. Quite apart from the practical impossibility of demanding of one's mind that at all times it identify such concepts as *electron* only with the measurable aspects of that construct, there are specifically two main objections: First, this position misunderstands how scientists as individuals do their work, and second, it misunderstands how science as a system grows out of the contributions of individuals.

---

* "Electrical Units of Measurement" (1883), *Popular Lectures and Address*, Vol. 1, second edition (London: Macmillan, 1891), pages 80–81.

Let us now examine the first of these two important points more closely. It would have been wrong to conclude from the previous pages that practicing scientists, while they are engaged in creative work, do—or even can—continually check the strict meaningfulness of their concepts and problems. As a rule an investigator is and should be unbothered by such methodological questions until some obstinate trouble develops. The operational view, in ordinary use, is almost instinctive; it usually becomes conscious and explicit only when something has gone wrong.

We are reminded of the case of a tightrope walker who undoubtedly is indifferent to an orderly inquiry concerning those physical laws by which he performs and survives; very likely he would fall off the rope if he worried much about them. We all have seen, too, the deplorable results when certain contemporary artists become overly conscious of their media and methods. *While a scientist struggles with a problem, there can be little conscious limitation on his free and at times audacious constructions.* Depending on his field, his problem, his training and temperament, he may allow himself to be guided by a logical sequence based on more or less provisional hypotheses, or equally likely by a "feeling for things," by likely analogy, by some promising guess, or he may follow a judicious cut-and-try procedure. The well-planned experiment is, of course, by far the most frequent one in modern science, and generally has the best chance of success; but some men or women in science have often not even mapped out a tentative plan of attack on the problems, but have instead let their enthusiasms, their hunches, and their sheer joy of discovery suggest the line of work. Sometimes, therefore, the discovery of a new effect or tool or technique is followed by a period of trying out one or the other application in a manner which superficially seems almost playful.

Even the philosophical orientation of scientists is far less rigidly prescribed than might be supposed. A rough classification of choices open to them is about as follows: Perhaps the majority of men would defend the position that the experiences which come to them through their senses are directly occasioned by and correspond exactly to an external world, a "real" world which exists regardless of our interpretations. This view is called *realism*, and it is in opposition to two major philosophical systems (among a host of others). First, *idealism*, which maintains that our ideas and impressions are the only realities, and that we use those to construct convenient concepts such as chairs, houses, and meter sticks. Second, *positivism*, which does not speak of such realities at all and regards nothing as ascertainable or meaningful beyond sense data (e.g., pointer readings) or those concepts reducible to sense data. Now it might be supposed that because the established part of physical science ideally should be directly verifiable in terms of measurements, all scientists must be positivists. The true state of affairs, however, is that the great majority of scientists, although in agreement with positivism's practical consequences for their own work, do not participate actively in the philosophical fight one way or the other. Einstein explains very well that a practicing scientist, from the point of view of philosophy, may appear to an outsider

"...as a type of unscrupulous opportunist: he appears as a *realist*, insofar as he seeks to describe the world independent of the act of perception; as *idealist* insofar as he looks upon the concepts and theories as the free inventions of the human spirit (not logically derivable from that which is empirically given); as *positivist* insofar as he considers his concepts and theories justified only to the extent to which they furnish a logical representation of relations among sense experiences. He may even appear as *Platonist* or Pythagorean insofar as he considers the viewpoint of logical simplicity as an indispensable and effective tool of his research."*

---

* From *Albert Einstein: Philosopher-Scientist* (edited by P. A. Schilpp), page 684, Evanston, Illinois: Library of Living Philosophers, 1949.

## 13.2 "PRIVATE" SCIENCE AND "PUBLIC" SCIENCE

The important problems of creative science do not present themselves with obvious routes to solutions, and so they differ in many fundamental ways from science problems which a student may encounter in homework or laboratory exercises. Above all, it is rarely known at the outset whether the problem can be solved at all with the initial conceptions or tools, or where the new complications might lurk; all of science has a unity and, unlike the usual student's exercise, the work of even a specialist in modern science may involve a large variety of different specific fields, sometimes in unexpected ways. An inquiry into the physical consistency of liquids, for example, may have to call upon such diverse topics as thermodynamics, physical chemistry, crystollography, acoustics, electronics, and even some practical metallurgy. Also, the methods and results of one field in physical science constantly suggest analogous procedures in another. Consequently, the working scientist is ever alert for the slightest hints either of new difficulties or of their resolutions. He proceeds through his problem like an explorer through a jungle, sensitive to every clue with every faculty of his being. Indeed, some of the greatest theoretical scientists have stated that during those early stages of their work they do not even *think* in terms of conventional communicable symbols and words.

Only when this "private" stage is over and the individual contribution is formalized in order to be absorbed into "public" science (often called "textbook science"), only then does it begin to be really important that each step, each concept be made clear and meaningful. The difference between these two aspects of science, what we shall have occasion to call science-in-the-making or $S_1$, and science-as-an-institution or $S_2$, has thus been characterized by the American nuclear physicist, H. D. Smyth, "We have a paradox in the method of science. The research man may often think and work like an artist, but he has to talk like a bookkeeper, in terms of facts, figures, and logical sequence of thought." For this reason we must not take very seriously the chronology or method outlined in the original publications of scientific results, including Galileo's own. It is part of the game to make the results in retrospect appear neatly derived from clear fundamentals until, in John Milton's phrase, "so easy it seemed/Once found, which yet unfound most would have thought/Impossible!" Much tortuous and often wasteful effort may be implied in a few elegant paragraphs, just as a sculptor removes his tools and clumsy scaffolds before unveiling his work.

Even in public science one will find widespread use of certain concepts that have not acquired a precise operational meaning. This situation sometimes gives rise to debates among philosophers and scientists concerned with methodology. From the seventeenth through the nineteenth centuries, there were three such controversial concepts in physical science: *atom*, *ether*, and *force*. As we mentioned at the end of Section 12.5, some of the strongest attacks on the atomic theory came toward the end of the nineteenth century, when the theory was on the verge of some of its most spectacular successes. Although Ernst Mach and his positivist colleagues may have had good reason to be skeptical about the real existence of atoms, it seems in retrospect that their criteria for even the tentative use of such theoretical concepts were unnecessarily strict. Insisting on excluding atoms from the formulation of physical theories, they deprived themselves of a good mental tool—and the rest of science of their possible additional contributions—by a doctrinal adherence to their position. Even the ether concept, though we now judge it to be superfluous, was not entirely useless in the history of science, for it helped Maxwell, for example, to formulate a successful theory of electromagnetic fields and waves. Similarly the concept of force, though often criticized as vague or superfluous, has played an important role in the development of physics, in some cases just because of its ambiguity.

Most scientists are tacitly agreed that the creative activity must be unfettered. It has been said very well "To set limits to speculation is treason to the future."

## 13.3 THE NATURAL SELECTION OF PHYSICAL CONCEPTS

It begins to appear that there are no simple rules to lead us to the invention of new concepts or to the discovery of new ideas, none by which to foretell whether our invention or discovery, once found, is going to prove useful and durable. But science does exist, and is a vigorous and successful enterprise. The lesson to be drawn from history is that science as a structure grows by a struggle for survival among ideas; there is a marvelous mechanism at work in science which in time purifies the meaning even of initially confused concepts, and eventually absorbs into what we may label $S_2$ (public science, science-as-an-institution) anything important that may be developed, no matter by what means or methods, in $S_1$ (private science, science-in-the-making).

We have seen that there are two general characteristics which are shared by those concepts that have contributed most to the growth of science. *First*, each of the guiding concepts of physical science—acceleration, force, energy, valence, etc.—has a core which is clear and unambiguous or, at any rate, which through continual application to experimental situations has attained an operational meaning that is tacitly understood and communicable. *Second*, by the same token, the great majority of physical concepts are quantitative, that is to say, they can be associated with numbers and measurements and therefore with manual or mathematical operations. Thus we say that a force is 15 newtons, a mass is 10 kg, a temperature 100°, a valence +2, the dimensions of space 3, an atomic weight 238.07, the pH 5. This observation holds even for those concepts which superficially do not seem to take on such numerical meaning. For

example, we cannot say that a body has 3, 4, or 5 units of *equilibrium*. On the other hand, equilibrium *can* be defined by saying that a body is in equilibrium when its acceleration is zero. The physical concept of "melting" can be defined as the change of state of a substance from a solid to a liquid, but this in turn has physical meaning only because the change from one state to the other can be recognized in terms of a large and abrupt change in the numerical value of measurable viscosity or hardness of the material. Consider the example of *electron*. A physicist might say it is the smallest quantity of electric charge known to exist independently, but when he uses that concept he really deals with its numerical aspects, namely, that it has $1.6 \times 10^{-19}$ units of charge, $9.1 \times 10^{-31}$ kg mass, and so forth. The name "electron" is primarily a summary term for this whole complex of measurables.

These two characteristics of concepts are joined by a third, without which science would degenerate into a meaningless conglomerate of data. Evidently the chaos of experience allows us to formulate infinitely many ideas, all of them meaningful in the above sense. For example, we might arbitrarily wish to create some new concept $X$ from observations of free fall, defining $X$, perhaps, by $X = (v - v_0)/s$. At first glance, $X$ appears as worthy of a distinguishing name of its own and of wide acceptance, as for example the quantity $(v - v_0)/t$; but pragmatically, $X$ does not happen to appear prominently in simple laws of nature. It is not, for example, a *constant* quantity during free fall. Therefore $X$, though a possible, meaningful concept, actually falls by the wayside as useless and unwanted. (This $X$ happens to be the very quantity which Galileo first and erroneously suspected of being the constant factor during free fall motion.)

*What makes certain concepts important, therefore, is their recurrence in a great many descriptions and laws, often in areas very far removed from the context of their initial formulation.* The electron, first discovered in the current within discharge tubes rather like those now used for fluorescent fixtures, later reappeared prominently in the explanation for electric currents in wires and liquids, for photoelectricity, for thermionic phenomena (as in today's radio tubes), for radioactivity, for the emission of light from hot bodies, and much besides. *That* is the only reason and the full meaning behind the statement that scientists "believe in the reality" of electrons; the concept is needed so often and in so many ways. "The only justification for our concepts," said Einstein, "is that they serve to represent the complex of our experiences."

At the inception of an idea it is, of course, hardly possible to tell whether it will fulfill this criterion, whether it will stand the test of wide applicability and so survive in the body of science. In this sense we may say Galileo's work would have been meaningless without the successive growth of physical science. Galileo sensed, but he could not have known, that his work on mechanics would turn out to be as important as it actually did, and that his special methods and special discoveries would (with Kepler's and Newton's work) find a general significance by transcending the immediate. There is a challenge in this for each individual scientist: the science of tomorrow may well need or even depend on his own work.

The most spectacular case of such generally useful ideas is, as we saw, that of the so-called "fundamental" concepts of physics (e.g., length, time, mass). There are only a small handful of them, yet they are the building blocks from which all other concepts are constructed or derived (e.g., velocity, being the ratio of a length and a time interval). One can never cease to marvel that nature should be so reducible—here must be either the basis or the result of the scientist's guiding preconception concerning the simplicity of nature of which we have spoken. We shall meet an abundance of other important though "derived" concepts (e.g., electric potential) which run like strong ropes through the maze of phenomena to give us support and direction under the most varied circumstances. Initially defined in some perhaps rather limited problem, they have become all-important over the years. In this evolution we find not infrequently that a concept changes its meaning as the context and the field of application widen; examples are such ideas as *force*, *energy*, and *element*. This possibility of change is again suggestive of the recurring discovery that the concepts of science are not absolute, that they have meaning only in terms of what scientists can do with them.

We can now attempt a preliminary summing-up of the main points so far. We see the physical scientist as a seeker of harmonies and constancies in the jungle of experience. He aims at knowledge and prediction, particularly through discovery of mathematical laws. The individual contribution, however, is only half the story, for science has two aspects: One (which we have called $S_1$) is the speculative, creative element, the continual flow of contributions by many individuals, each working on his own task by his own usually unexamined methods, motivated in his own way, and generally uninterested in attending to the long-range philosophical problems of science; and the other aspect, $S_2$, is science as the evolving compromise, science as a growing network synthesized from these individual contributions by accepting or adopting those ideas—or even those parts of ideas—which do indeed prove meaningful and useful to generation after generation of scientists. The cold tables of physical and chemical constants, the bare equations in textbooks, are the hard core of $S_2$, the residue distilled from individual triumphs of insight in $S_1$, checked and cross-checked by the multiple testimony of general experience.

This duality of science is one of its main sources of strength. In order to illuminate the duality in terms of concrete examples, we shall now briefly study two problems, that of motivation and that of factual interpretation. This will, incidentally, also prepare the ground for the bigger tasks that follow—examining the details of the mechanism whereby $S_2$ grows out of $S_1$, and explaining the vital role played by the existence of contradictory elements within science. Once an adequate distinction has been made between the several levels of meaning in the term "science," one of the central sources of confusion in methodological discussions has been removed.

## 13.4 MOTIVATION

When Whitehead said, "Science can find no individual enjoyment in Nature; science can find no aim in Nature; science can find no creativity in Nature,"* he touched on the ambivalent feeling of the present age toward science—the simultaneous fear and reverence for the reputedly emotionless yet deeply exciting enterprise. Perhaps we can see now that the opinion embodied in the quotation refers to only one of the two aspects of science, the stable, not the transient one, $S_2$ and not $S_1$. Modern science as a structure may say nothing about purposes and aims, but the scientist himself may be enchanted to some degree by nonrational preoccupations.

A few direct quotations will illustrate this point, for in other years, when scientists were freer with their human secrets, a nonrational, mystical, or religious conviction was often freely acknowledged. Galileo, a pious man, looked upon the laws of nature as proof of the Deity equal to that of the Scriptures (see Section 5.4), and ever since the Pythagoreans this faith, reflected in the recent phrase "the world is divine because it is a harmony," has been a motivating theme. Almost without exception, scientists as a group have at all times, including our own, manifested the same religious convictions or lack of convictions as did other contemporary groups of educated men; if anything, their position has more often moved them to be vocal on such questions than some other, equally well-informed groups. Until about

100 years ago, the typical scientist would quite openly assert that the physical world could not be understood without fundamental theistic assumptions. In this, too, the men of the seventeenth century were the formulators of the fundamental concepts, and though the details of the argument have changed considerably since, its form has never been put more beautifully or stated more honestly than in Newton's description of his scientific activity as a prelude to religious knowledge (specifically, that of the "first cause," the Deity):

" . . . the main Business of natural Philosophy is to argue from Phaenomena without feigning Hypotheses, and to deduce Causes from Effects, till we come to the very first Cause, which certainly is not mechanical; and not only to unfold the Mechanism of the World, but chiefly to resolve these and such like Questions: What is there in places almost empty of Matter, and whence is it that the Sun and Planets gravitate towards one another, without dense Matter between them? Whence is it that Nature doth nothing in vain; and whence arises all that Order and Beauty which we see in the World? To what end are Comets, and whence is it that Planets move all one and the same way in Orbs concentrick, while Comets move all manner of ways in Orbs very excentrick; and what hinders the fix'd Stars from falling upon one another? How came the Bodies of Animals to be contrived with so much Art, and for what ends were their several Parts? Was the Eye contrived without Skill in Opticks, and the Ear without Knowledge of Sounds? How do the Motions of the Body follow from the Will, and whence is the Instinct in Animals? Is not the Sensory of Animals that place to which the sensitive Substance is present, and into which the sensible Species of Things are carried through the Nerves and Brain, that there they may be perceived by their immediate presence to that Substance? And these things being rightly dispatch'd, does it not appear from Phaenomena that there is a

---

* A. N. Whitehead, *Modes of Thought*, New York: Macmillan, 1938, Lecture VIII.

Being incorporeal, living, intelligent, omnipresent, who in infinite Space, as it were in his Sensory, sees the things themselves intimately, and thoroughly perceives them, and comprehends them wholly by their immediate presence to himself; Of which things the Images only carried through the Organs of Sense into our little Sensoriums, are there seen and beheld by that which in us perceives and thinks. And though every true Step made in this Philosophy brings us not immediately to the Knowledge of the first Cause, yet it brings us nearer to it, and on that account is to be highly valued." (*Opticks*)

In a letter to a friend in 1692, Newton further explains his motivations for writing the *Principia*:

"When I wrote my treatise [*Principia*] about our [solar] system, I had an eye on such principles as might work with considering men for the belief of a Deity; and nothing can rejoice me more than to find it useful for that purpose."*

Consider another persistent "nonrational" trend in creative scientific work, the preoccupation with integral numbers, of which we have seen an example in Bode's law and shall see others. At the very outset of Galileo's historic work on the law of free fall, in the *Two New Sciences*, we find him drawing prominent attention to the fact that "...so far as I know, no one has yet pointed out that the distances traversed, during equal intervals of time, by a body falling from rest, stand to one another in the same ratio as the odd numbers beginning with unity." Throughout the development of physics and chemistry and to the present day we encounter this satisfaction with such simple numerical relations. Today's nuclear physicists have found that some of the characteristics of nuclei can be explained in a theory involv-

---

* Letter to Richard Bentley, reprinted in *Newton's Philosophy of Nature*, page 46.

ing a few recurring numbers; these are generally called the "magic numbers," perhaps a little less jokingly than is often realized. Examples can be multiplied indefinitely, but perhaps the following extract is sufficiently representative: The Swiss physicist, Wolfgang Pauli, on accepting the Nobel prize for physics in 1945, was speaking of his great teacher Sommerfeld and his attempts to form an explanation for the particular colors found in the spectra of glowing gases, when he said,

"Sommerfeld however preferred...a direct intepretation, as independent of models as possible, of the laws of spectra in terms of integral numbers, following as Kepler once did in his investigation of the planetary system, an inner feeling for harmony.... The series of whole numbers 2, 8, 18, 32 ...giving the length of the periods on the natural systems of chemical elements, was zealously discussed in Munich, including the remark of the Swedish physicist Rydberg that these numbers are of the simple form $2n^2$, if $n$ takes on an integral value. Sommerfeld tried especially to connect the number 8 with the number of corners of a cube."

When thus presented, the motivation of many scientists, just like the process of discovery itself, appears perhaps surprisingly "unscientific." But precisely because it has irrational elements, the drive toward discovery is powerful even under the most adverse conditions. How shall we explain that much great work has been done under the handicaps of extreme poverty and severe ill-health, sometimes even under the threat of imminent death? Or that to this day most scientists would probably reject the ever-present lure of increased standards of living in *uncreative* positions, and instead follow their chosen work without restrictions, although with relatively few material rewards? We must recognize this symbiotic relationship: The progress of science has often depended on the almost unreasonable tenacity of its devotees and, on the other hand, the scientific activity yields a unique exhilaration and deep fulfillment. In support of this point the words of Poincaré are, as always, persuasive:

"The scientist does not study nature because it is useful; he studies it because he delights in it, and he delights in it because it is beautiful. If nature were not beautiful, it would not be worth knowing, and if nature were not worth knowing, life would not be worth living. Of course, I do not here speak of that beauty which strikes the senses, the beauty of qualities and of appearances; not that I undervalue such beauty, far from it, but it has nothing to do with science; I mean that profounder beauty which comes from the harmonious order of the parts and which a pure intelligence can grasp. This it is which gives body, a structure so to speak, to the iridescent appearances which flatter our senses, and without this support the beauty of these fugitive dreams would be only imperfect, because it would be vague and always fleeting. On the contrary, intellectual beauty is sufficient unto itself, and it is for its sake, more perhaps than for the future good of humanity, that the scientist devotes himself to long and difficult labors." (*Science and Method*)

## 13.5 OBJECTIVITY

Exactly at this point we must answer a question which has been rising insistently during the last paragraphs: What of the much-vaunted objectivity of scientists? We have been told of this demand so often that we have learned to call "scientific" any inquiry that claims to be systematic and unbiased. For example, Bertrand Russell said, "The kernel of the scientific outlook is the refusal to regard our own desires, tastes, and interests as affording a key to the understanding of the world." But how can this large degree of personal involvement, just demonstrated, fail to endanger the search for objective truth? Or, since it patently does not so endanger science, how can we account for the discrepancy between the nature of motivation and of results?

The fact is that we have, by a second route, come back to the original dilemma, the apparent contradiction in function between $S_1$ and $S_2$. Now we can begin the resolution, first by pointing to an extreme example. The mathematical physicist, Sir Arthur Eddington, had such strong convictions that he once wrote a "Defense of Mysticism." Whether or not his persuasion was "meaningless" from the point of view of public science $S_2$, it may have been the mainspring, the deep cause, of his devoted search for truth. What is now important is that in his voluminous and distinguished scientific writings you will find no overt expression of personal mysticism, nothing that might not equally well have been written by a gifted scientist with exactly opposite metaphysical orientation, or even one belonging to the "hard-boiled" school and quite indifferent to such questions. In modern science, personal persuasions do not intrude explicitly into the published work—not because they do not exist, *but because they are expendable.*

We may go a step further. In a free society, the metaphysical tenets of individual scientists, though often quite strong, are generally so varied, so vague, and often technically so inept that in a sense they are canceled by one another, made ineffectual by the lack of a basis for general acceptance and agreement of such tenets. Perhaps our science would be weakened if this mechanism did not operate. Where there exists one explicit, widely accepted set of dogmas—as in some of the old scholastic universities or in modern totalitarian states—only there can extraneous metaphysics in a scientific publication survive the scrutiny of the scientists in that field and that location, with possibly detrimental consequences.

As contrasted with his largely unconscious motivations for which he has to account to no one, the intellectual disciplines imposed on a scientist, like the form of publication of research papers itself, are now quite rigorously defined; he knows that his inclinations may well not coincide with those of his colleagues and witnesses, and that the only points he and they might agree on are sound demonstrations and repeatable experiments. The personal satisfactions in $S_1$ must remain private and therefore, from the point of view of $S_2$, incidental. We can readily see how this severe convention, in such clashing contrast with the human side of the experimenter, is a kind of sociological device ensuring the rapid evolution of science. For exactly when his enthusiasm for some result runs highest, the chances for mistakes arise. Therefore, the investigator will have to force himself to the most searching re-examination before describing his work to that mercilessly impartial jury, his colleagues. Louis Pasteur gives us a good insight into this attitude:

"When you believe you have found an important scientific fact and are feverishly curious to publish it, constrain yourself for days, weeks, years sometimes; fight yourself, try and ruin your own experiments, and only proclaim your discovery after having exhausted all contrary hypotheses. But when after so many efforts you have at last arrived at certainty, your joy is one of the greatest that can be felt by the human soul."

This attitude is a fairly modern development. Previously there were other solutions. For example, in attacking the qualitative and subjective science of his day, Francis Bacon held that in order to build on a sure foundation the scientist must suppress as long as possible the human element, his prejudices, premature hypotheses, his guesses and wishes. To this end Bacon outlined a procedure which prescribed the structure of scientific work so rigidly that the human element would indeed be kept at bay. In the historical context his solution was certainly not unreasonable; it might have worked, and it does seem to solve the problem of how to remain objective. But as it turned out, scientists found another, less rigorous, less safe, but far more fruitful procedure for overcoming

human error, one that does so without stifling the enthusiasm and motivation of scientists, their main source of both error *and* truth. This discipline is *superposed* on the free creativity of the individual; his inquiry is not controlled in its progress as if it were confined in a train moving on firmly laid out rails, but rather as though it were a rider on a horse whose obedience is assured by an occasional but firm tug on the reins.

## 13.6 FACT AND INTERPRETATION

Why, you may ask, should there be any need for an imposed discipline? Is there not discipline in the very *facts* which the scientist observes? Are they not unambiguous in themselves? This raises another most interesting problem, that of the relationship between fact and interpretation.

Our very life, as well as our science, continually depends on the correct observation and classification of facts, and yet there is nothing more deceptive than facts. It is almost impossible to describe a fact without having some interpretation, some hypothesis, enter unseen—as careful cross-examination of witnesses in a courtroom often shows quite dramatically. If we follow the matter far enough, we might have to agree that the only real or basic facts are those which impress our crudest senses—the fact that we see a black or a white splotch, feel hot or cold, hear a painful or a soft sound. However, those are exactly the perceptions which Galileo called secondary quantities, and which by themselves could not conceivably have given rise to our science. Indeed, even simple animals must be able to transcend such raw observations, and thought surely does not deal with this type of isolated experience. It has been called the first law of thought that *perception* must pass into thought and knowledge via *conceptualization*. Starting from the vague impression "now-here-then-there," man had to invent—perhaps over a period of a thousand generations—such ideas as space, object, position, distance, time interval, motion, and velocity, in largely unanalyzed form, before he could even think about appearances.

"Facts" cannot be discerned by themselves, without intellectual tools for handling the sense impressions. Scientists do have preconceptions or "themes" and do use unexamined associations, although these may be justified by a long history of successful results. This admission seems to contradict the popular idea that the first step in science is to throw overboard all prejudgments. However, without preconceptions one cannot conceive new thoughts; therefore, what is required in science, as in other phases of life, is to try to become aware of one's preconceptions in an emergency, to discard those that have become meaningless, to use the others with caution, and with willingness, if found to be wrong by actual phenomena, "to pronounce that wise, ingenious, and modest sentence 'I know it not'," as Galileo said.

Without our almost axiomatic key concepts we should be largely deprived of intelligence and communication. Imagine a geologist describing his observation of a mountain range in terms of "pure fact." He could not even begin to speak of a mountain range, for strictly speaking all he sees is a change in the level of terrain. Even that implies several derived concepts, so instead of saying, "The mountain is cut by glacial valleys," he would have to describe a pattern of light and darkness, and how his eyes had to focus differently on the various features of the field of view.

Thus our thoughts, when they would deal with observation and fact, really must manipulate concepts, constructs, ideas, with all their hidden dangers—dangers of erroneous classification, of unwarranted extrapolation or analogies, of too-ambitious generalizations or too-timid specializations. No wonder physical science treasures those few concepts (e.g., length, time, mass, charge) which, through long experience, have proved to be reliable enough to build upon them the whole edifice.

The least ambiguous facts being those which involve only accounts of pointer readings and the like, physics should, as has been said, ideally start only with such descriptions. Even here is room for quarrel, for we do have to decide in advance of the experiment which are the most important significant pointer readings to be collected and described. In this sense, facts do not exist outside of the observer. Allow a layman to observe the sky through a large telescope, and he will see nothing of interest and understand nothing he sees. But give a trained man only three good looks at a comet through a mediocre instrument, and he will call upon the theories in the field to tell you to a day how soon a comet will return, how fast it is traveling at any moment, what material it is made of, and much else. In short, the pattern we perceive when we note "a fact" is organized and interpreted by a whole system of attitudes and thoughts, memories, beliefs, and learned constructs. It is thought that gives us eyes.

Not only do brute facts alone not lead to science, a program of enthusiastic compilation of facts *per se* has more than once delayed the progress of science and, in the opinion of some observers, may be doing so even now in some contemporary research on elementary particles and on certain medical problems. As the scientist-educator J. B. Conant has pointed out, "Science advances not by the accumulation of new facts...but by the continuous development of new and fruitful concepts." The picture of the experimental scientist in his well-stocked laboratory subjecting matter to intensive and undirected observation is certainly absurd, though widespread. Particularly when instruments are to be used in making observations, even the first "preliminary experiments" are preceded by much nonexperimental activity, by study, thought, and calculation. Without this, the expensively obtained facts are meaningless. We conclude once more that science cannot be made by individual scientists. To understand even the simplest observations each investigator must rely on the distilled wisdom of science as an institution. When Newton modestly pictured himself as standing on the shoulders of the giants that had done the earlier work of science, he pronounced a truth applicable to every individual scientist.

## 13.7 HOW SCIENCE GROWS

We may now coordinate several observations on the growth of science to gain one overall view. The key recognition is the distinction between $S_1$ and $S_2$—if you will, the difference between the man Galileo Galilei and the statement $s \propto t^2$. This dual view of science has an analogy in regarding the law either as *corpus juris* or as the ensemble of individual trials and juridical opinions, and in considering a tribe either as a people (a classified abstraction with recognizable stable organization, customs, etc.) or as an interacting group of individually almost unpredictable persons. Better still, compare the difference between $S_2$ and $S_1$ with the double interpretation one may give to animal species, either as calmly catalogued and described in zoology texts at the present state of evolution, or as exemplified by the diverse living, struggling individuals which it contains.

If we think about the last analogy more closely —and it is no more than a very helpful analogy— the mechanism of evolution for the species and for science appears to be similar in four ways: first of all, growth in both cases presupposes a mechanism of *continuity*—a species or a science can persist only if there is some stable means for handing on the structure from generation to generation in an unambiguous way. In biology the principle of continuity is found in the processes of heredity based on the highly specific nature of the genes— and in science it is identifiable with the specific operational and quantitative nature of the important concepts. Without this measure of unambiguous continuity, scientists could not communicate their work to one another and to their pupils.

Second, superposed on continuity is a mechanism of *mutation*, leaving open the constant opportunity for individual variations. In the case of biological species, of course, the process of mutation is made possible by various chemical and physical influences on the genes and on chromosome partition and recombination; in science, mutations are assured by the essential democracy of the institution and the boundless fertility of the free human mind.*

A third mechanism is *multiplicity of effort*. To assure continuity and growth despite the low rate of occurrence of really good modifications, and in the absence of some obvious single master plan by which to proceed, science and the species alike must rely on a large number of individual attempts, from which may ultimately come those few types that are indeed useful. The uncountable bones of bygone members of the species, like the uncountable pages of the individual scientific researches of past years, are mute testimonies to the wastefulness of the process of growth, both of the zoological type and of scientific knowledge.

Finally, there is a *selection* mechanism whereby certain of the seemingly innumerable contributions and unpredictable mutations are incorporated into the continuous stream of science, a struggle among ideas not greatly different from that fight for existence in nature which allows the species to adapt itself to the changing environment. The survival of a variant under the most diverse and adverse conditions is mirrored in science by the survival of those concepts and discoveries that find usefulness in the greatest variety of further applica-

---

*    But perhaps we must reserve a special name for those rare mutations in science, those precious occasions when, as in the work of Copernicus, Galileo, Newton, Einstein, and Bohr, a man transcends all human limitations and climbs to unsuspected heights in a "heroic effort," to use Sarton's phrase. Without this measure of heroism or prophecy or supreme dedication, the normal evolutionary progress in science might be almost as slow as that of natural evolution.

tion, of those conceptual schemes that withstand the constant check against experience.

A skeptic might add that the secret of success in science is that, contrasted with art and philosophy, science has essentially an easy and clear-cut task. There are many ways to dig a well for water, for example, some no doubt very inexpert, but that problem is clear, and sooner or later water will be found almost everywhere. The job itself, once formulated, is inherently a job that can be done. If science has limited itself to discovering facts and relationships, then patient minds will search for them, and in time facts will indeed be found and related. There may be much truth in this, but man is neither all-wise nor long-lived; hence what limited scientific knowledge he could gain by himself would die with him if it were not for its communicability to others, and would forever be limited by the frailties of individual minds if it were not for some selective growth. *Continuity or communicability, individual variability, multiplicity of free effort, and selective growth—these are the four related principles on which is based the success of science.*

As an immediate consequence we see now another relationship between $S_1$ and $S_2$: Unlike the anthill built from the tiny individual contributions of insects, science-as-an-institution does not exist *by itself* and outside the contributors, not in books or in the abstract, but in the minds of those scholars that understand and work in the field. Each physicist, each chemist, each geologist, each astronomer fulfills two functions (and it is the easy confusion between these two which has produced the common erroneous picture of him). On the one hand he is an unfettered creator in $S_1$, and on the other hand his mind, together with that of each of his colleagues, is the agency of transmission of $S_2$ from one generation to the next. Consequently the content of $S_2$, in passing through the ages from one group of rational beings to the following, continually undergoes a twofold change: it is ever more sharpened toward communicability and unambiguity; and it is incessantly expanded by tests

for continuing meaning and usefulness against the widening experience and deeper experiments.

In retrospect we can now see the signs of this mechanism of evolution in the nature of physical concepts themselves. Recall those three characteristics of scientific concepts: The first is their operational meaning, the second the preferably quantitative nature, and the third the reappearance in diverse fields of application. Each of these can be regarded as a mechanism for assuring the continuation and increase of the scientific enterprise. They aid in the unambiguous communication of problems and results, make possible unambiguous agreement (or disagreement) among different workers on facts and their interpretations, and knit together the efforts of many independent scientists, even when widely separated in subject interest, time, and space.

## 13.8 CONSEQUENCES OF THE MODEL

Several generally accepted features of scientific inquiry seem to be derivable from this view that science as an institution is continually filtered and modified by its flow through the minds of original creators, and that it exhibits the stated laws of evolution. Here is a summary of a few such points.

a) The convenience and freedom of communication is vital to the very existence of science *at each moment*. The jealous secrecy with which the alchemists hid their results doomed their efforts to stagnation and helped to delay the rise of chemistry. In our time, too, the imposition of secrecy in basic research would constrict the arteries of science. The right to pursue any promising lead, to publish and to exchange freely scientific information, must be vigorously defended. From this point of view it is also apparent what an important role is played by free, disinterested research in academic institutions, where a man may follow the unpredictable path of knowledge without serious constraint and interference. For in science, as in society, truth can be found only on the free market place of ideas. Only the filter of years, not the regulations of men, can extract the real worth from each contribution.

b) The realization that science depends on communication between its practitioners partly explains the rise of organizations among scientists. From the early seventeenth century there has been a rapidly accelerating multiplication of the channels of communication—societies, institutes, journals, etc.    National and not infrequently international meetings are held regularly each year by the men of each speciality.  Organizations have been formed for the single purpose of summarizing (abstracting), indexing, and distributing accounts of the growing avalanche of scientific developments.    The scientists band together in committees and hold international congresses to decide on uniform nomenclature, definitions, and standards of measurement.  They sponsor publication of monumental volumes of physical and chemical tables of constants.  Above all, they are ever eager to share with one another their latest findings and difficulties.

c) Multiplicity of effort is not just a phrase applicable to the sum of all past work, it is a staggering reality in science today.  Because there exists the very real danger that an account of this kind, in presenting necessarily only some of the more spectacular and historic advances of science, may well distort the sense of proportion, look at least briefly in your library at a journal summarizing current work, for example the monthly, *Physics Today*.  The point to note, above all, is the variety in topics, interests, contributors, and apparent importance of many of the papers.  No one would venture to predict which of these contributions will be best remembered and used thirty years hence.    The great revolutions, the spectacular breaks with the past, are rare indeed.  It may well be that one scientist or the other has that month planted the seeds from which a now entirely unsuspected harvest will be gathered; but the progress of science does not depend on the pursuit of the whole effort, as long as it is vigorous, free, and open.

d) Our view of science as an organism obeying laws of evolution helps to explain the truism that once "a situation is ready" for a discovery or for the formulation of a conceptual scheme, it frequently is made in several places by independent workers at the same time.  When we speak about some general advance and say that the time was ripe for it, we mean usually that at least three conditions obtained: There were *reasons* for looking into the given problem, that is, it now made sense to ask such a question, and it may in fact have been generally discussed; second, there were *means* for looking, i.e., the needed experimental tools were obtainable; and third, there were mental schemes for *understanding*, i.e., the needed concepts and wider conceptual systems were readily assembled.  When all this happens at one time, several men may unknowingly compete; it is always possible that several scientists can expose themselves to the same heritage, can encounter and interpret the problem in the same way and, borne by the same wave, may reach the same shore.

e) We have had examples, and shall have others, of the very opposite phenomenon, of the rejection or long neglect that is experienced not infrequently by precisely the most startling innovators, even though their contribution later may come to be widely accepted.  But if science is not simply the sum of individual contributions, if, as we saw, $S_2$ exists not outside but within the minds of practicing scientists, then it follows that a contribution, to become part of $S_2$, must find its way to acceptance by and into the minds of many men.  That, however, must take time, particularly if the suggestion is very radically different from the climate of opinion in the field at the moment.  In Section 3.2 we saw that this is indeed in the long run desirable for the healthy evolution of science.  And so it can happen (though rarely) that the contribution of a man regarded as something of a crackpot by his contemporary colleagues reappears, even if much modified, as an important idea later on when the horizons of the field have changed in the course of evolution.  Occasionally, science has needed the neglected recluse in the garret as much as the rare and widely acclaimed heroes and the army of competent foot soldiers.

f) By the same token that *freedom* in science increases the chances for both multiplicity of effort and individual variation, so also does a free science, ever productive of its own tools for further advancement, tend to become more and more a self-perpetuating or even an explosive chain reaction. As knowledge begets knowledge, it increases as though in a geometric series.

It is a source of constant wonder that once science (as we use the term) got its first real impetus in the early seventeenth century, it swept over Europe within a century like a long pent-up storm. The issue is, as we have seen, very complex, but one may speculate that two details contributed materially: European science had, from the fourteenth century on, expanded and progressed to the point where the principles of multiplicity of effort and of individual variations were finally in effective operation; in addition to this, the work of Galileo and of his contemporaries now provided attitudes and concepts which, being quantitative and operational, allowed the principle of continuity to operate more fully than before, thereby modifying the science of the time toward the self-sustaining character which it attained in the early eighteenth century.

g) Since the mind of man is not divided into separate compartments, and since science exists as a complex of conceptual schemes propagated through the minds of men, we see that the one must interact with the other, that science must influence the whole culture pattern of a society and in turn be shaped by it. It is only natural that, for example, Galileo's argument on the motion of heavenly bodies, like that of his adversaries also, should be colored by and again expressed in contemporary theology, or that the impact of Newton's work on the thinking of eighteenth-century Europe should have extended far beyond merely scientific questions.

When we study the long-range effects of science' we see ample evidence that every person now alive' whether or not he has studied science, is intellectually a child of Copernicus and Galileo, Isaac Newton and Faraday, Einstein and Niels Bohr.* Our imagination and intellectual tools were indeed largely shaped by the advances in the knowledge of physics they and their contemporaries made long ago. It is easy to show that, when the Copernican and Newtonian world view triumphed in the West, the recognition that a uniform law holds sway over all matter everywhere helped to overcome hierarchical thinking and to prepare the mind for self-reliant democracy. We know also that in the nineteenth century the successes of statistics and of the concepts of energy prepared the ground for the modernization of the Newtonian world view.

In addition to the kind of long-range influence that Newton's work had on the imagination of writers and philosophers from the eighteenth century on, there are also the more material long-range effects associated with the advances made by James Watt, Faraday, and Fermi. From an understanding of how the steam engine works flowed a century-long transformation of society which is now studied as the *Industrial Revolution*. From Faraday's "toys" came electric motors and generators and, in time, the electric-powered elevators, trains and subways that facilitated the upward and sideways growth of cities. Similarly, the experiments of Fermi's group on neutron-induced artificial radioactivity prepared for the study of nuclear fission, and this in turn led to the design of new energy sources that may well be the chief means for meeting our frantically growing energy needs.

It is of course even more difficult to foresee the long-range effects of science on social change than it is to see the immediate practical influences. To avoid possible negative effects and to capitalize on

---

* Most of the rest of this section is adapted from an article by Gerald Holton, "The relevance of physics," *Physics Today*, November 1970, by permission of the American Institute of Physics.

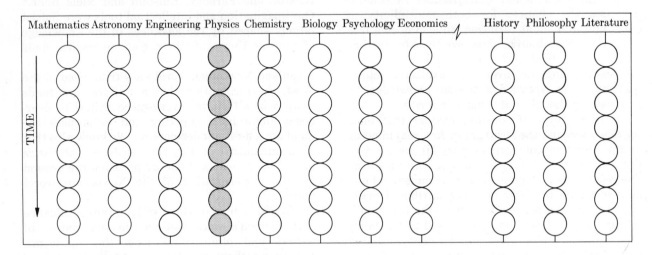

**Fig. 13.1** Traditionally one often thinks of physics as if it were separate and independent of all other subjects, and as if it were made up of separate sections, like beads on a string.

positive ones, there is only one policy available to us: the exertion of uncompromising watchfulness as both citizens and scientists. We must call attention to existing abuses of scientific knowledge or skills and keep up-to-date on scientific advance so as to be ready to keep science from being derailed and abused in the future.

To return to the meaning of science not as merely a technical study but as a part of the general humanistic development of mankind, we can illustrate this sense of humanistic interconnectedness with two simple diagrams. The physics course, as traditionally given in many classrooms, is like a string of beads. One subject follows another, from Galileo's kinematics to the most recent advances in nuclear physics. This sequence is the usual one, which more or less parallels the historical development of the science, whether this parallel is made explicit or not. But few if any

connections are shown with achievements of human beings who are not physicists, with sciences other than physics, or with studies and activities other than science. And all too often the materials studied in the other courses (for example in chemistry, in biology, in literature) also hang like so many separate strings of beads (see Fig. 13.1).

There are some advantages in such a string-of-beads presentation. It is, for example, convenient to teach and learn. But ignoring connections that do exist among all these fields does not do justice to the actual state of affairs. A research project in experimental physics, for example, sooner or later draws material not only from almost every part of a physics curriculum but also from mathematics, metallurgy, chemical thermodynamics, electronic engineering, computer technology and many other sciences, as well as group psychology, accounting and skill in writing about the work. Moreover,

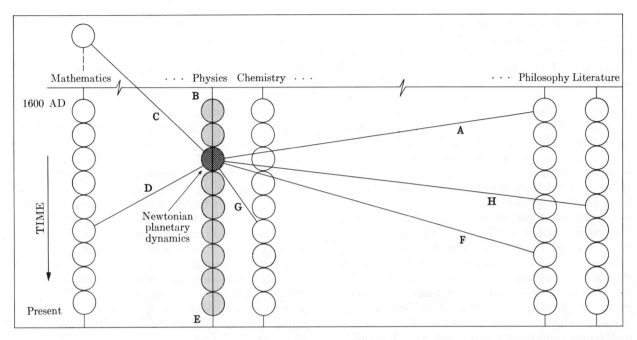

**Fig. 13.2** An example of the interconnectedness of all fields. Newtonian dynamics is linked to past and present achievements, not only in physics but in other fields as well.

nobody who has engaged in actual scientific work can fail to see the influence that scientific advances can have in terms of social and practical consequences. "Pure" physics is an invention that exists only in the most old-fashioned classrooms. If you pick up a real problem in physics (or any other science), there extends from it connections to a number of expected and unexpected problems that at first glance seem to "belong" to other professions.

We believe this is one argument in favor of including occasionally in a physics course topics not usually referred to. Think back, for example, to the study of Newtonian mechanics as applied to planetary motion, a subject that is usually one of the "beads" on the physics chain. Newton had studied theology and philosophy, and those ideas emerge in the *Principia* in his sections about the nature of time and space (see Fig. 13.2, link A to philosophy). Within physics itself, Newton brought

to a culmination the work of Kepler and Galileo (link B). Much of the established mathematics in Newton's work came from the Greeks (link C). New mathematics, particularly the basic ideas of calculus, were invented by Newton to aid his own progress, thereby advancing the progress of mathematics (link D).

Within physics, all who follow Newton will use his laws and approach (link E). His effects on the Deist theologians (link F), on John Dalton's atomic models in chemistry (link G), and on the artistic sensibilities of the eighteenth century in which Newton swayed the muses (link H) are quite easily documented.

The same kind of web extends around each of the chief topics of physics. Think of the link from philosophy to the work of Hans Christian Oersted, André Ampère, and Faraday in electricity (through their interest in *Naturphilosophie*). Think of the link reaching from nuclear physics back along the chain to the classical physics of three centuries earlier (as in the determination of the neutron mass) and of the sideways links to biology, to engineering and indeed to politics, owing to the various applications and byproducts of nuclear reactors, for example.

Such links exist between main achievements in many fields. If we drew all the links between fields on the intellectual map, we would see instead of separate strings of beads a tapestry, a fabric of ideas. Science is in dynamic interaction with the total intellectual activity of an age. In a deep sense, science *is* part of the study of history and of philosophy, and it may underlie the work of the artist, just as it penetrates into the explanation a mother gives to her child to help him understand the way things move.

If we therefore tried to think away the achievements of physics, the course of modern history would be almost incomprehensible. We could not understand, and in fact would not have had much of the work of John Locke, Voltaire and Alexander Pope, men who, among many others, were frankly inspired by the work of physicist contemporaries. Conversely philosophy, mathematics, and other fields would be far emptier studies without their fulfillment and extension through the work of philosopher-scientists such as Ernst Mach, Einstein, and Bohr. Eliminating physics would, of course, also make nonsense of the history of industrial development following Watt's steam engine, Alessandro Volta's electric battery, and Faraday's motors and generators. A neighboring science such as chemistry could not have developed without the models of gases and the theories of atomic structure that were largely the work of physicists. In short, if you pull the thread marked "physics" from the tapestry—as some antiscientific persons wish to do—the whole fabric would unravel as an old sweater does; the same would be true if any of the other threads were pulled. In this view, therefore, the relevance of any field of knowledge, including science, is that it is an integral part of the total growth of thought.

h) Let us consider one last point of universal concern, although others will suggest themselves readily. For the very same reason that one cannot foresee the course of science, one can rarely evaluate at the time of discovery whether and where it will find uses, or how important its place in pure science will be, or whether it will be twisted and abused for evil purposes by those who guide the fate of society. As one common example, consider that radioactivity was discovered rather casually by Henri Becquerel when a photographic plate became exposed by the rays from a nearby piece of uranium salt (he had put them together in a drawer in preparation for an experiment that actually was based on a plausible but erroneous notion). Consider further that Becquerel did not properly understand the meaning of his observations; that through the lonely labors of the Curies the new radioactive metals were isolated and identified—an unsuspected widening of the horizon of physics and chemistry; that the new knowledge of radioactivity then stimulated a generation of workers in all sciences, and so directly and indirectly revolutionized the conceptual models of the atom, of radiation, and of chemical reactions, with very practical consequences in such widely removed studies as the treatment for certain diseases. Then recall that today radioactivity is an integral part or indispensable tool of innumerable studies quite outside the field of discovery, in biology, geology, astrophysics, metallurgy, even archeology, and many others. Surely no one will dare prescribe limits to the process of discovery!

Finally consider the great contemporary dilemma: "atomic" (nuclear) energy may put the philosopher's stone into every man's pocket and, through the industrial *applications* of the eventual discovery of cheap and clean energy resources, may help lead us to the happy age for which we are hardly prepared—or it may help, through the applications by those who make and wish to use nuclear weapons, end the short history of our species. All governments in the "civilized" world are simultaneously developing plans for the construction of engines for industry as well as for the large-scale destruction ("dehumanization," as one official put it recently) of areas considered to be hostile.

But as our analysis has shown us, one must be careful not to confuse science with the industrial, technological, military and other applications that are made of it. Science and technology are related but separable. It is perfectly possible, and may often be wise, to limit or forbid some technological applications, as in arms-control treaties, or in the decision not to build a supersonic transport plane that would merely drain the economy and threaten the environment. The progress of mankind and even of technology itself may well depend on making such self-denying decisions, and not to produce or do everything that is possible to produce or do.

The case is quite different with the progress in the understanding of basic science itself. That progress, as even the example of Becquerel and Curie showed, often goes on in a quite unplanned and unexpected way. So it would be a grave threat to all science if fundamental research in any part of it were forbidden or discontinued. And though knowledge without deeper human wisdom may fail us, *less* knowledge hardly assures either wisdom or survival; moreover, one might at least doubt whether ignorance is now either possible or worth living for. For salvation and disaster are not reached by separate roads so that we may simply choose one and avoid the other; they both lie at the end of the same path, the choice being within ourselves. If we sought to solve the dilemma by fiat, by some official control of science, we should all have to submit to a censorship which, unable to foresee the result of curiosity and the future of discoveries, essentially would have to rule out all

intelligent speculation. Surely, this system would guarantee the loss, along with science, of our basic freedom, while also forcing us to abandon all hope for an enlightened future. Both these results to our society would be quite comparable with the vengeance of savage conquerors.

If we look for an alternative solution, we might well try to emulate the lesson found in the growth of science itself: *the marvelous ability of men to arrive at truths by the free and vigorous exchange of intelligence.*

There is another reason why it is wrong to seek relevance for science only in the immediate benefits to technology. Not only have technological advances, all too often, brought with them major social problems as unforeseen byproducts, but also these problems cannot be cured or even properly understood through existing scientific, technological or political means alone. Such cures, rather, depend to a large extent on new, basic advances, not least on advances in science itself. At the heart of social problems created by technological advance is the *absence* of some specific basic scientific knowledge. This realization gives a whole new mandate to basic scientific research.

Examples come readily to mind. It is quite customary, for instance, to say that the population explosion is in part caused by the advance of medical science (better sanitation, inoculation, antibiotics). But one can equally well claim that the population explosion is bound to overwhelm us precisely because we do not yet have at hand sufficient knowledge in pure science. That is to say, the complex problem of overpopulation is due in a large degree to our current (and long acknowledged) ignorance of the basic process of conception —its biophysics, biochemistry, and physiology. No wonder that attempts at controlling population are so halting. What is astonishing, rather, is that the first medical-school laboratory in the U.S. specifically designed to study the whole range of scientific problems in the process of reproduction is only now being built.

Similarly, it is sometimes said that progress in physics is responsible for the threatening arms race. But it is more accurate to say that nuclear arms-control treaties are difficult to achieve largely because insufficient knowledge of geophysics makes inspection of suspected, illegal weapons tests with seismographs difficult. A better understanding of geophysics, it turns out, is needed before some nations will consider it safe to enter arms-control treaties that outlaw underground weapons tests.

The problem of bringing food to hungry people in arid lands near the sea, as in Peru, India, or the United Arab Republic, is to a large extent one of logistics and management. But it is also a problem of basic science: Before we can design more economical desalination plants, we need a more fundamental understanding of liquid structure (one of the old but much neglected problems in physics and chemistry) and of flow through membranes. Pollution is, of course, a result of greed, stupidity, apathy and lack of law enforcement. But to clean smog-ridden areas more effectively will require greater basic knowledge than we now have of meteorology and of the physics and chemistry of combustion. And, in the meantime, to this day the most effective and insufficiently used device for getting rid of pollution by solid particles is the electrostatic precipitator, which essentially works on scientific principles known since 1600.

These observations serve to oppose two widely current but erroneous notions: First, that basic science is an unnecessary luxury and should be supported only if it is directed to immediate practical applicability, and second, that one way of stopping the abuses that are byproducts of technical innovation is to stop science (whereas in fact curing existing abuses often depends on scientific advances yet to be made).

# Chapter 14

# ON THE DISCOVERY
# OF LAWS

We have examined at some length the nature of concepts in physical science, and the relationship between scientists and their work. In this chapter we may now proceed to inquire more thoroughly whether anything positive may be said about the discovery of specific laws* of science. It is fitting and revealing to consult first the opinions of active scientists, and then to suggest a model to represent the process of discovery.

---

* The appellation "law" is usually reserved in physical science for those statements of generality and trustworthiness whose descriptions involve mathematical relationships. Other words are frequently used for those statements that are mainly qualitative (e.g., the statement that the horizontal and vertical components in projectile motion add up without disturbing each other) or when they are more in the nature of a model or a hidden mechanism to aid our comprehension of phenomena (e.g., the idea that chemical reactions are explainable and predictable by considering matter as made of particles). In those cases various terms may be encountered: "principle" (e.g., superposition principle), "hypothesis" in a larger sense, or "theory" in a limited sense (e.g., Avogadro's hypothesis, or the atomic theory), "postulate," "rule," and so on. We shall not find it important to make much of possible differences between these words, and shall usually refer to all of them as *laws* for the time being.

## 14.1 OPINIONS ON SCIENTIFIC PROCEDURE

We remember the warning not to expect uniformity of procedure. While there has never crystallized a representative feeling among scientists in general, there are nevertheless several types of opinions about the main elements of scientific procedure which have been held prominently, and five facets have reappeared frequently enough to be of interest here. None of these by itself tells the whole story, none is meant to be in rivalry with the others. Rather we shall consider them as five complementary aspects of the same general opinion, the proportions being variables depending on the particular problem and the particular scientist.

 a) The first of these views, symbolized in another quotation from Priestley,† emphasizes that "More is owing to what we call *chance*, that is, philosophically speaking, to the observation of events *arising from unknown* [unsuspected] *causes*, than to any proper *design* or preconceived *theory* in the business. This does not appear in the works of

---

† *Experiments and Observations on Different Kinds of Air* (1776).

205

those who write *synthetically* upon these subjects...".

Superficially one indeed cannot help but be impressed by the role of chance in scientific work. Every schoolboy is told how a falling apple directed Newton's mind toward the law of gravitation, or again how the law of the simple pendulum occurred to Galileo as he was watching a swinging lamp in the cathedral at Pisa. The initial observations leading to the epochal discoveries of the battery, many of the elements, x-rays, radioactivity, and contemporary chemotherapy, were all made more or less by chance. Speaking of related sciences, W. I. B. Beveridge writes in his excellent book, *The Art of Scientific Investigation*:

"Probably the majority of discoveries in biology and medicine have been come upon unexpectedly, or at least had an element of chance in them, especially the most important and revolutionary ones."

This dovetails with what has been said before concerning the irrational elements in the process of discovery. However, one must look at the whole picture. The huge majority of scientific work is unspectacular in achievement as well as in genesis; if chance enters at all, it is greatly outweighed by sheer hard work. Whatever else may be true of the methods of science, they are by no means short cuts; they operate within the matrix of a life spent in the laboratory, at the desk, or in field expeditions.

As to the role of chance in those several famous cases, there is the unassailable dictum, "chance favors the prepared mind"; witness the obvious fact that great discoveries in science are in general not made by people previously ignorant of these fields. It is only the master of his subject who can turn to his advantage the irrational and the unsuspected. Chance observations have usually become important if they triggered off trains of thought that from all evidences might well have been arrived at more systematically by those minds. This trigger action of chance is by no

means to be despised, but to interpret the stimulus correctly one must have had prior training and thought on the general subject of the discovery. In short, one must have acquired a sensitivity for recognizing a favorable though unexpected turn of events.

b) Although not trying to write off Priestley's statement as erroneous, we certainly do expect other elements to enter more prominently into the process of discovery. A second such main ingredient has been described by Bridgman, who stated straightforwardly,

"The scientific method as far as it is a method, is nothing more than doing one's damnedest with one's mind, no holds barred.... This means in particular that no special privileges are accorded to authority or to tradition, that personal prejudices and predilections are carefully guarded against, that one makes continued check to assure oneself that one is not making mistakes, and that any line of inquiry will be followed that appears at all promising. All of these rules are applicable to any situation in which one has to obtain the right answer, and all of them are only manifestations of intelligence."*

Recalling what has been said before about the lack of *a* method, about the variability among scientists and their methods, one can hardly fail to agree with Bridgman's statement. It finds in the habit of using one's intelligence to the utmost a common ground for the diversity of individual approaches.

c) In much the same vein, others (among them Max Planck, the discoverer of the quantum concept) regard their work simply as an extension and refinement of common sense. Initially this view must seem incomplete, if only for the reason that scientists in dealing with the common-sense

---

* P. W. Bridgman, *Reflections of a Physicist*, New York: Philosophical Library (1950).

problems of day-to-day living on the whole can hardly claim to show much more aptitude than do their nonscientific neighbors. We must admit that many relations exist between science and common sense, for example, that the primitive notions and techniques spring from prescientific experience, that science had its origin to some extent in the common sense of the early craftsmen, and that even today much research proceeds on a kind of technical common-sense or cut-and-try empiricism, particularly in the applied sciences and in the neighboring continent of technology. But when we are dealing with men working in the modern physical sciences and with such conceptual schemes as nuclear theory, their common sense is "common" only insofar as their special training and endowment allow them to find the way through their problems as naturally as the craftsman doing his own work.

d) The foregoing is supplemented by a further opinion which sees the essence of scientific procedure in some variation of Poincaré's terse characterization (duly qualified), "The scientific method consists in observing and experimenting." This widely held view has a range of interpretations and misinterpretations. In the extreme it may refer to a technique close to that advocated by Bacon—the pursuit, as we saw, of an orderly experimental search for facts while carefully keeping one's mind from the influence of prior notions and from forming early hypotheses. The available facts were to be interpreted or explained only at the very end by a process of "true" induction, that is, induction unspoiled by premature guesses and hypotheses. Now it is true that sometimes a research project in physical science reaches a stage where for months or even years theorizing on a large scale is held in check, pending the outcome of a single-minded search for specific experimental data; and it is further true that during the taking of data or the progress of a calculation scientists usually try to reduce the chances of mistakes by not becoming committed too early to a specific outcome, loving

the game itself, so to speak, a great deal more than the final score. However, this is, after all, only the intelligent thing to do, and it also coincides with what we have called the discipline imposed on scientists; sooner or later someone is sure to check and repeat your experiments, observations, and calculations, and will certainly discover faults and self-deceptions.

Whether in fact one can train one's mind to the Baconian level of control and asceticism, to completely uncommitted observation and experimentation, and whether the procedure is even fruitful as outlined, we have had reason to doubt before. Certainly the history of most scientific discoveries overwhelmingly denies these propositions. For example, Galileo's experiment with the inclined plane was surely not the result of an uncomplicated desire or plan to see what would happen to a rolling ball. On the contrary, as like as not Galileo knew rather well what he would observe. Some appreciation or foreknowledge of the type of solution to expect is almost inseparable even from the recognition of the *existence* of a problem. Galileo confessed his conviction that nature acted in the simplest possible way, hence he expected some relationship of the simplest possible kind between distances, times, and speeds. Of interest here is that some criteria are indeed needed to focus the attention on certain aspects of the total experience to the exclusion of the rest; the measurements of $s$, $t$, etc. in a controlled experiment become valuable for an eventual solution only if these factors and no others really do allow us to determine the features of free fall. Of course, not too much is expected of one's prescience at this stage; many elements may suggest themselves which will turn out to be unessential. For example, Galileo appears to have suspected initially that the material of the falling body determines its law of fall, but he finally decided against this factor. On the other hand, the major clue may appear only at the end of a long investigation. On the whole, the problem here is similar to that of a craftsman who must have at least some conception of the final product before he decides on the medium and tools for rendering the work. The experimentalist, particularly, must design his apparatus and procedure with a fair idea of the type and magnitude of the effect to be expected, including the allowable errors and the masking secondary effects. In short, the scientist cannot just observe and experiment. He cannot just *search*; he has to search *for something*, and he must have an idea of what is to be expected—otherwise he may walk by the solution without recognizing it.

Now what has remained of the opinion, "The scientific method consists in observing and experimenting"? Much; above all, this: Whether we put our theorizing at the beginning or at the end, whether we induce or deduce the important laws, or use some interplay of both processes, sooner or later we must submit our work to the verdict of experience. Even in that supreme example of "pure" theoretical work, the theory of general relativity, at the end of the theoretical development Einstein turns to the observational and experimental aspects of science by discussing several possible experimental tests of the theory. Though he was usually rather skeptical of the claims and priority of experimentation, he says bluntly of one such proposed test, "If the displacement of spectral lines toward the red ... does not exist, then the general theory of relativity will be untenable" (*Relativity*, 1920). Here we have again the persistent and striking difference between science and non-science, this appeal to quantitative observation and experiment as last authorities.

e) Finally, an element in scientific procedure held in highest esteem by those responsible for the spectacular achievements in contemporary theoretical physics is typified by Einstein's statement, "There is no logical way to the discovery of the elemental laws. There is only the way of intuition, which is helped by a feeling for the order lying behind the appearances." We hear from these men that they immerse themselves completely in the problem at hand, and either speculatively try to postulate or induce a principle by which the problematical situation may be derived (explained) or, in fact, allow their thought processes to be manipulated without imposing a conscious direction. Whereas others seem to go from step to step with logical certainty, these men make their large jumps accurately, as though borne by a guiding necessity. Sometimes they reveal that such solutions occur to them "in a sudden flash of insight" after long, even feverish study.

This method of solution by inspired postulation is rather close to the procedure advocated by Plato (e.g., in the *Phaedo*) as well as by certain scholastics, although they did not share our convention of letting the entire validity of such induced postulates depend ultimately on the verdict of quantitative experience. One can, and many do, defend the view that this was the main approach of Galileo also. In that case we may imagine that he solved the problem of free fall by postulating the principle that all bodies fall such that $\Delta v/\Delta t$ is constant, and then deducing from this the experimentally confirmable behavior of falling and rolling bodies. In the same way it used to be a fashion, particularly among French scientists of the last century, to organize their work on a model adapted from the later Greek mathematicians—beginning with clearly stated postulates or hypotheses, then demonstrating all deducible concepts rigorously and mathematically, and finally calling in experimental confirmations as though these were secondary and more in the nature of illustrations. In that type of account, and it has by no means disappeared, the important real experiment is sometimes replaced by the *thought* experiment, an argument which says in essence, "Let us imagine this or that plausible situation. Surely you must agree that things will happen thus and thus, exactly in accord with my deductions." Therefore Galileo would say he does not actually need to perform an experiment to see whether a stone dropped from the mast of a smoothly moving ship will strike at the foot of the mast no matter what the speed of the ship; he knows this to follow from the principle of composition of motions, previously verified for the case of simple projectile motion.

But again, however true and important the role of postulation, the unexplained appearance of a key principle in the exceptional mind does not by itself tell the whole story, particularly in the experimental physical sciences. For the moment let us keep in the foreground of our minds all these distinguished opinions on the requirements for successful inquiry: *an intuitive feeling for nature, particularly for its quantitative aspects, a habit of using one's intelligence to the utmost, a deepened common sense, a reliance first and last on observation and experiment, and the sensitivity for recognizing a favorable though unexpected turn of events.* Reviewing these opinions, we conclude that we may expect from the scientists themselves only the most general hints about their methods (what would a composer say if we asked by what method he created his concerti?), for in science, as in every field of endeavor, the most expert work is done by men who have passed beyond the need and interest to rationalize each step. In the sense that each problem has its own difficulties and each scientist his own approach, we may say that the scientific inquiry has its art as much as its methods.

Perhaps the best single terms for the five qualifications cited above are "scientific orientation," or "scientific outlook," or "scientific attitude"; we may even venture to say that while scientific work in general is not characterized by any one single method, to a greater or less extent *scientists do share this complex of characteristic attitudes* toward their life occupation. This is shown by the ease with which men trained in one specialized field can usually adapt themselves to another field, to new problems and a new set of collaborators; and it helps to explain why we find so frequently that a great scientist has made fundamental contributions to several widely separated branches of study.

## 14.2 A SEQUENCE OF ELEMENTS IN FORMULATIONS OF LAWS

Having allowed the several evidences to dispel the last hope for an easy formula of discovery, we must take heed not to fall into the extreme

**Fig. 14.1** Mnemonic diagram of relationships between recurring features in scientific procedure.

skepticism of Priestley. From our now suitably modest point of view, we may still analyze how scientific laws grow out of observations, concepts, and hypotheses. To be quite silent on this subject is to disregard or dismiss the distinguished and varied testimony of influential writers on the philosophy of modern science.

What we shall do is analyze a *hypothetical* case of the formulation of a law, without claiming that formulations *in general* do actually follow this specific pattern; for in truth the development of an established law cannot be traced accurately even in retrospect. Therefore the scheme to be discussed and its graphic representation in Fig. 14.1 are nothing more than a mnemonic device which may help us to think intelligently about the relationship between some recurring features in scientific procedure, particularly in experimental science. As an example we might keep in mind a concrete case, perhaps an interpretation of Galileo's discovery of the laws of motion of freely falling bodies and projectiles.

a) At the top of the list of comments we must surely put the importance of *the investigator's knowledge of contemporary science*, including the permissible tricks of the trade and the more or less tacit assumptions such as are embodied in every intellectual enterprise. Thus Galileo had available to him for his work on projectile motion the earlier and quite advanced inquiries of Tartaglia and Benedetti. Sometimes such prior knowledge proves erroneous and the assumptions misleading, but at the outset the existing conceptual schemes alone can give direction to the inquiry. They are the initial common sense from which the scientist must draw or perhaps eventually must break away.

At a later stage the investigator will undoubtedly have to go back to his books or instruments to supplement his knowledge in unfamiliar fields which turn out to touch on the developing situation. Or he may even have to supplement existing disciplines by the addition of specifically invented tools of measurements and aids of mathematics. Galileo's inventiveness, of which we have heard much, extended also to the field of mathematics. We found that other great scientists had to face the same problems—Copernicus had to invent much of his solid geometry, Newton his calculus.

b) A second point of departure, again principally in the experimental sciences, is a *first-hand acquaintance with nature through intelligent observation even before the problem has been formulated.* Not only do the great "chance" observations arise in this manner, but what is probably much more important, observations provide the raw material with which to construct the great questions, and they provide the investigator with the necessary measure of nonrational insight and an almost intuitive feeling for the behavior of nature in the field of his specialty, what the significant variables are and how they interact. To return to the work of Einstein, because it is perhaps as far removed from personal participation and experimentation as is possible in

physics, in his autobiographical notes we find this significant hint on the importance of first-hand experience with "raw" natural phenomena: "[At the Polytechnic Institute in Zurich] I really could have gotten a sound mathematical education. However, I worked most of the time in the physical laboratory, fascinated by the direct contact with experience."

c) When the setting has been prepared, *the "problematic situation" may present itself to the mind in an "occasion of reflection,"* (to use John Dewey's happy phrases), arising perhaps from some surprising chance finding, but more probably as the result either of some dissatisfaction with the degree of consistency, generality, or accuracy of the current explanation, or of some "hunch" where the rewarding new field of study lies. There is no clear biographical account to tell us why and how Galileo first decided on his study of motion, but we may gather from his remarks and from what we know of his life and work that, being dissatisfied with contemporary qualitative modes of description, he wished to find quantitative relations in motion. Other men, notably the phenomenal nineteenth-century experimentalist Michael Faraday, seemed to have been motivated more by imaginative guesses and intuitive feel for nature. To add one example to the evidences on the variability of motivation and scientific temperament, we are told by Niels Bohr that his historic theory of atomic structure originated in his recognition of a similarity in the *form* of two equations from different fields of physics.

d) Intimately connected with this recognition of the existence of the problem is the *foreknowledge of the type of solution to be expected*, as we have seen. After the initial period of wonder, the trained but curious mind can hardly prevent its first rather vague flights of speculation and anticipation. It will suggest quite naturally some further observations and attention to specific details, and so point in a general way to a suspected answer to the problem, even though the real problem itself may on deeper thought turn out to lie quite far from the initial stimulus. For example, Wilhelm Röntgen in 1895 was led to the discovery of x-rays by noting that a sample of mineral gave off a pale light while electric current was made to pass through a discharge tube 6 ft away. He quickly traced this effect to the action of previously unsuspected radiation coming from the tube. But he immediately abandoned the problem of why minerals fluoresce in this new radiation (for about 20 years this problem could not even begin to be understood), and instead he turned to the far more interesting and fruitful questions, e.g., whether x-rays are a form of invisible light.

e) This early stage of research may not only suggest new and interesting problems and some tentative hypotheses concerning the eventual solution of the problems, but it may also give rise to *new concepts with which to analyze the problematic situation.* To take our previous case, Röntgen found that objects are transparent to the passage of x-rays but that an increase in the thickness or the density of bodies decreased their transparency for x-rays materially. It occurred to him to compare whether the transparency of various metals is equal when the product of thickness and density is constant. It so happens that this proved not to be the case (the complexity of the problem is great), but evidently if this simple constancy had been found, very quickly a new name for the concept "thickness times density of metals" would have been assigned, equipped with a symbol of its own, and no doubt figuring prominently in the eventual solution of the problem of transparency of metals to x-rays. In the same vein, Galileo could analyze the problem of free fall in terms of the concept "acceleration" as formulated for that very purpose.

Evidently the great concepts, like the great problems on which physics is built, do not spring ready-made from the minds of individual scientists —far from it. They develop during the inquiry or, in the case of such concepts as force, energy, etc.,

they actually may have behind them a long history of evolution, involving many men and many false starts. Even when we give Galileo credit for defining the crucial concept "acceleration" we should not forget that in a sense he only completed a study that went back centuries before his time.

f) Proceeding from the initial preliminary experimentation and the first guiding ideas, there may begin to grow in the researcher's mind at first a rather modest *working hypothesis*—be it Röntgen's preliminary decision to regard x-rays as a form of ultraviolet light, or Galileo's thoughts that the discrepancies in the free fall of bodies might be regarded as the minor effect of air friction.

The process of inducing an hypothesis or a postulate from the limited number of known facts has been compared with the unexplainable experience one may have while playing a game of chess, when one suddenly "sees through" the whole arrangement on the board and perceives in one stroke a pattern, a long chain of moves toward a successful ending. While the mechanism by which patterns are perceived and fruitful thoughts arise is at present still being debated vigorously among psychologists of various persuasions, we can at any rate point to the purpose of hypotheses in the scheme of scientific inquiry: namely, to stimulate some intelligent, directed activity, be it theoretical or experimental. To direct action, the hypothesis does not have to be perfect—we have long ceased to be surprised about the large number of correct theories first discovered while following a hypothesis that by itself proved to be wrong. But for that very reason the initial hypothesis should not be allowed to take hold of one's mind too firmly.

The very process of induction is a hazardous art, for no logical necessity exists that the predicted consequences should actually follow. To put it more bluntly, the reasoning is usually that if the product of induction, a hypothesis—or for that matter a law—is true, then facts $a$, $b$, $c$, . . . should follow. When, on testing, these facts are actually observed, the hypothesis is *perhaps* true—but there may be numerous other and even better hypotheses which account for the same set of facts. In his book *The Scientific Outlook*, Bertrand Russell gave a rather picturesque but stimulating example to show other dangers inherent in logical processes, particularly the limitations on experimental confirmations. He asks us to imagine that the statements to be tested are "Bread is a stone" and "Stones are nourishing." An elementary syllogism informs us that on the basis of such postulates it should follow that bread is nourishing; since the conclusion can be confirmed by the simple experiment of eating bread, we might be tempted to regard the premises as correct also! The result of induction is never unique, and deduction also may easily give faulty results if only, as here, by having assumed erroneous postulates.

Likewise, science wisely never pretends to explain phenomena with absolute finality. Experiments and theories may seem to succeed or fail for an equally wide variety of wrong reasons. If you seemingly confirm a hypothesis by experiment, your measurements or the inferences from the experimental data may be faulty, or a series of canceling errors may deceive you, or the experiments may be right but the hypothesis wrong except for that singular prediction. Examples for all these combinations abound in the history of science. On the other hand, when an experiment does *not* bear out a hypothesis, the effect you looked for in vain may indeed be there but too small for the instruments at your command, or you may be deceived by masking disturbances from an unrecognized source, or again there may be faults in the method or interpretation of the test.

For all this there is no preventive. Nature's full complexity and subtlety may be far greater than our imagination can grasp or our hypotheses can accommodate. There is no substitute for the uncertain processes of thought or for the precarious business of confirming hypotheses by test, but there are several tricks which have shown themselves to be very helpful in the long history of thought: to try the simplest hypotheses first, to regard no

answer or result as true beyond all possible doubt, to rely on several independent checks before accepting an experimental result of importance, and constantly to re-evaluate and reformulate old knowledge in the light of new. We glimpse here again the fundamental truth that science depends on the continuous activities of generations.

Ernst Mach said, "Let us early get used to the fact that science is unfinished, variable." If we are reconciled to the fact that it is better to have a poor hypothesis than none, that our thoughts may originally be grossly oversimplified, that they may need later and drastic modification, that sometimes several hypotheses may coexist—then we shall not be tempted to regard as foolish all work of the past or as "certain" all work of the present. Furthermore, if we regard a hypothesis as a stimulus for action, we shall not demand of it that it be immediately and directly confirmable. In the long run, of course, hypotheses or postulates do have to lead to some testable conclusions. Even though Galileo himself had no tools to check directly the relations involved in the motion of rapidly falling bodies, his assertion that their acceleration is constant was not inherently unoperational, and has, of course, since been tested directly. (Suggest such a check. Feel free to use high-speed cameras, etc. Just how *direct* is this test?) Even the large-scale explanatory hypotheses may not lend themselves directly to experimentation, e.g., the hypothesis that the agency by which the sun pulls on all planets is the same gravitational force that pulls a stone to the earth; but at any rate we shall insist on their verification by examining the degree of agreement between observed fact and the corresponding prediction deduced from our hypothesis—in our previous example, probably some astronomical observations.

g) As our work progresses, the initial working hypotheses will permit us to design and perform *more and more specific, directed, and controlled experiments.* Slowly our insight will progress, as if on two crutches—now the hypotheses helping to interpret one experiment and suggesting the next, now the experimental result modifying the hypotheses, in continual interaction. In this sense, the law of free fall (as expressed in the equation) is the end product, the one statement that best describes all the important features concerning free fall, whether executed by large bodies or small, whether applied to stone or lead, falling through large distances or small.

## 14.3 THE LIMITATIONS OF PHYSICAL LAW

In an important sense the word "law" is very misleading. Being built on concepts, hypotheses, and experiments, laws are no more accurate or trustworthy than the wording of the definitions and the accuracy and extent of the supporting experiments. Although the laws of nature are usually called inexorable and inescapable—partly because the word suggests the erroneous analogy with divine law—the human formulation of scientific laws certainly is neither eternally true nor unchangeable. The most certain truth about the laws of science is that sooner or later the advance of knowledge and experience will prove them to be too limited or too inaccurate. This lies in the tragic role of the law itself. It is induced from a necessarily limited number of experiments or situations, and yet, to be useful, it has to pretend —cautiously—to apply to all situations of the same kind. One example of this is the set of extensive tables which Galileo compiled for the range of projectiles. The direct experimental basis for this part of his work must have been quite small, but once he had discovered the parabolic trajectory, the essence of the law of projectile motion, he could derive and believe in consequences which had never been looked for experimentally. It follows that we must be alert for a new situation in which this ancient law will prove to break down—not only because it is an idealization never completely exemplified under ordinary conditions where air friction cannot be neglected, but also because there is, for example, no logical necessity that projectiles

at extremely high speeds (as in the case of atomic particles) must obey the same laws as relatively slow cannon balls.

Here we remember the warning that all generalizations, all statements of physical law, carry a hidden "text"—including an explanation of the symbolism, a convention of how to manipulate the equations, definitions of the terms, statements of the accuracy and the range of experience supporting the law, and so forth. Almost invariably a limit is encountered sooner or later beyond which the law is no longer valid in its stated form; therefore the warning implied in the "text" should prevent continual misapplication of the law. An analogy may be found in the awareness of a party of mountain climbers concerning the limits beyond which they can put no demands on their ropes, their gear, and their own bodies. In both cases the *precise* limits can usually not be clearly drawn until some failure brings them to attention. But whereas the climbers naturally dread such a failure, scientists are not helpless if occasionally a new phenomenon does not obey these established rules. For beyond the limits of one law may lie a hitherto undiscovered part of science—new and exciting knowledge. We shall see several great theories, such as Planck's and Bohr's, that had their beginning in the insufficiency of established laws, in some "unexplainable" observations.

Galileo's laws of projectile motion contain such a "text." We find in his work a thorough examination of the major limitations involved; for example, he points out that the trajectory will not be parabolic for extremely large ranges because then the acceleration of gravity, being directed toward the center of our round earth, would no longer (as assumed in our equations) be parallel for different parts of the paths. Of course this failure applies, strictly speaking, even to small trajectories, but in such cases it is so minute as to make it permissible to neglect it for our purposes, just as we have already neglected the effect of air resistance, rotation of the earth, variation of the value of $g$ over the whole path, etc. Of these limitations within which our laws operate we certainly need not be ashamed; for if we were forced to deal with actual motions in which these secondary factors were not negligible, we are confident that we would surely be able to discover the complex laws guiding those smaller effects, and add them to our calculations.

But what if a law *is* found to break down at some point? Usually one need not become too concerned, for as a rule it is possible to rejuvenate and extend the law, perhaps by redefining some of the concepts used. Yet, one must pay the price for this flexibility in lawmaking in another way. The history of physics shows exciting cases where at one time from apparently the same set of experiments two different laws or theories were devised (e.g., the inescapable classics, namely, the nineteenth-century wave theory vs. the Newtonian corpuscular theory of light). Either one of these is useful as long as no further experiments can be devised for which the two theories would predict sufficiently different answers to permit a clear choice between the rivals. However, once such a decisive experiment has been found—no easy matter—and carried out, one or the other (or possibly both) of the conceptual schemes must be conceded to be incapable of dealing with the new observation.

Yet no general scientific theory does suddenly become regarded as completely "wrong," since it still is as useful as before in explaining and predicting the class of all the experiments prior to the decisive one that had incriminated the old theory or law. For instance, Newton's laws of mechanics as such are not "wrong" because they alone cannot deal with the motion of subatomic particles; Newton's laws are still as valid as they ever were in the huge range of experiences with larger bodies. Here a famous analogy has been suggested: If we liken the facts to be explained to fish in a pond, then the law or theory is the net with which we make the catch. It may turn out that our particular net is not fine enough to haul in *all* fish, large and small—but it may still be quite satisfactory for supplying our daily needs. We may go even further and maintain that to be useful at all, our conceptual schemes, like our nets, *must* contain some holes; if it were otherwise, we should not be able to distinguish between the significant and the trivial, the fish and the rest of the pond.

## 14.4  THE THEMATIC ELEMENT IN SCIENCE*

There remains a third element in science, at which we have only hinted briefly. So far we have stressed heavily the role of "fact" and of calculation. And it is true that regardless of what scientific statements they believe to be "meaningless," all philosophies of science agree that two types of proposition are *not* meaningless and are important, namely statements concerning empirical matters of "fact" (which ultimately boil down to meter readings), and statements concerning the calculus of logic and mathematics (which ultimately boil down to tautologies). Let us call them respectively *empirical* and *analytical* statements, and think of them as if they were arrayed respectively on orthogonal $x$- and $y$-axes; thereby we can represent these two "dimensions" of usual scientific discourse by a frank analogy, and generate terminology which will be useful as long as we do not forget that all analogy has its limits.

Now we may use the $xy$ plane to analyze the concepts of science (such as force), and the propositions of science, e.g., a hypothesis (such as "x-rays are made of high-energy photons") or a general scientific law (such as the law of universal gravitation). The *concepts* are analogous to points in the $xy$ plane, having $x$- and $y$-coordinates. The *propositions* are analogous to line elements in the same plane, having projected components along $x$- and $y$-axes.

To illustrate, consider a concept such as force. It has empirical, $x$-dimension meaning because

---

* This section is adapted from part of an article by Gerald Holton, published in *Science and Culture* (edited by G. Holton), Boston, 1965, by permission of the American Academy of Arts and Sciences.

forces can be qualitatively discovered and, indeed, quantitatively measured, by, say, the observable deflection of solid bodies. And it has analytical, $y$-dimension meaning because forces obey the mathematics of vector calculus (namely, the parallelogram law of composition of forces), rather than, for example, the mathematics of scalar quantities.

Now consider a proposition (a hypothesis, or a law): The law of universal gravitation has an empirical dimension or $x$-component—for example, the observation in the Cavendish experiment in which massive objects are "seen" to "attract" and in which this mutual effect is measured. And the law of universal gravitation has an analytical or $y$-component, the vector-algebra rules for the manipulation of forces in Euclidean space.

Whether they are arbitrary or not, the $xy$ axes have, since the seventeenth and eighteenth centuries, more and more defined the total allowable content of science, and even of sound scholarship generally. The philosopher David Hume, in a famous passage, expressed eloquently that only what can be resolved along $x$- and $y$-axes is worthy of discussion:

"If we take in our hands any volume; of divinity, or school metaphysics, for instance: Let us ask, Does it contain any abstract reasoning concerning quantity or number? No. Does it contain any experimental reasoning concerning matter of fact or criteria? No. Commit it then to the flames. For it can contain nothing but sophistry and illusion."

If we now leave the $xy$ plane, we are going off into an undeniably dangerous direction. For it must be confessed at once that the tough-minded thinkers who try to live entirely in the $xy$ plane are more often than not quite justified in their doubts about the claims of the more tender-minded people (to use a characterization made by the psychologist William James). The region below or above this plane, if it exists at all, might well be a muddy or maudlin realm, even if the names of those who have sometimes gone in this direction are distinguished. As the historian of science, Dijksterhuis, has said:

"Intuitive apprehensions of the inner workings of nature, though fascinating indeed, tend to be unfruitful. Whether they actually contain a germ of truth can only be found out by empirical verification; imagination, which constitutes an indispensable element of science, can never even so be viewed without suspicion."*

And yet, the need for going beyond the $xy$ plane in understanding science and, indeed, in doing science, has been consistently voiced since long before Copernicus, who said that the ultimate restriction on the choice of scientific hypotheses is not only that they must agree with observation but also "that they must be consistent with certain preconceptions called 'axioms of physics,' such that every celestial motion is circular, every celestial motion is uniform, and so forth."† And if we look carefully, we can find even among the most hard-headed modern philosophers and scientists a tendency to admit the necessity and existence of this additional dimension in scientific work. Thus Bertrand Russell spoke of cases in which "the premises turn out to be a set of presuppositions neither empirical nor logically necessary."**

One could cite and analyze similar opinions by a number of other scientists and philosophers. In general, however, there has been no systematic development of the point until recently. What is needed may be called thematic analyses of science (by analogy with thematic analyses that have for so long been used to great advantage in scholarship

---

* E. J. Dijksterhuis, *The Mechanization of the World Picture*, Oxford: Clarendon Press, 1961, page 304.

† Quoted from E. Rosen, *Three Copernican Treatises*, New York: Dover Publications, 1959, page 29.

** *Human Knowledge*, London: Allen and Unwin, 1948, Part 6, Chapter 2.

outside science). In addition to the empirical ($x$) dimension and the analytic ($y$) dimension, we can define a third, or $z$-axis. This third dimension is the dimension of fundamental presuppositions, notions, terms, methodological judgments and decisions—in short, of *themata**—which are themselves neither directly evolved from, nor resolvable into, objective observation on the one hand or logical, mathematical, and other formal analytic ratiocination on the other hand. With the addition of the thematic dimension, we generalize the plane in which concepts and statements were previously analyzed. It is now a three-dimensional "space"— using the terms always in full awareness of the limits of analogy—which may be called *proposition space*. A concept (such as force) or a proposition (such as the law of universal gravitation) is to be considered, respectively, as a point or as a configuration (line) in this threefold space. Its resolution and projection is in principle possible on each of the three axes.

To illustrate: The $x$- and $y$-components of the physical concept "force" (its projections in the $xy$ plane) have been mentioned. We now look at the thematic component, and see that throughout history there has existed in science a "principle of potency." It is not difficult to trace this from Aristotle's ἐνέργεια, through the neo-Platonic *anima motrix*, and the active *vis* that still is to be found in Newton's *Principia*, to the mid-nineteenth century when "Kraft" is still used in the sense of energy (Mayer, Helmholtz). In view of the obstinate preoccupation of the human mind with the theme of the potent, active principle, before and quite apart from any science of dynamics (and also with its opposite, the passive, persisting principle on which it acts), it is difficult to imagine any science in which there would not exist a conception of force (and of its opposite, inertia).

It would also be difficult to understand certain conflicts. Scholastic physics defined "force" by a projection in the phenomenic dimension that concentrated on the observation of continuing terrestrial motions against a constantly acting obstacle; Galilean-Newtonian physics defined "force" quite differently, namely, by a projection in the phenomenic dimension that concentrated on a thought experiment such as that of an object being accelerated on a friction-free horizontal plane. The projections above the analytic dimension differed also in the two forms of physics (i.e., attention to magnitudes versus vector properties of forces). On these two axes, the concepts of force are entirely different. Nevertheless, the natural philosophers in the two camps in the early seventeenth-century thought they were speaking about the same thing, and the reason was that they shared the need or desire to incorporate into their physics the same thematic conception of *anima*, or *vis*, or *Kraft*—in short, of force.

A second example of thematic analysis might be the way one would consider not a concept but a general scientific proposition. Consider the clear thematic element in the powerful laws of conservation of physics, for example the law of conservation of momentum, as formulated for the first time in useful form by Descartes. In Descartes' physics, as Dijksterhuis wrote:

"All changes taking place in nature consist in motions of . . . three kinds of particles. The primary cause of these motions resides in God's *concursus ordinarius*, the continuous act of conservation. He so directs the motion that the total *quantitas motus* (momentum), i.e., the sum of all the products of mass and velocity, remain constant."†

In Descartes' view this relation ($\sum mv$ = constant) constitutes the supreme natural law. This law, Descartes believes, springs from the invariability of God, in virtue of which, now that He has wished the world to be in motion, the variation must be as invariable as possible.

---

*    From the Greek *thema* (theme), plural *themata*.

†    *Mechanization of the World Picture*, page 410.

Since the seventeenth century, we have learned to change the *analytic* content of the conservation law—again, from a scalar to a more complex version—and we have extended the empirical applicability of this law from impact between palpable bodies to other events. But we have always tried to cling to this and to other conservation laws, even at a time when the observations seem to make it very difficult to do so. The French mathematician Henri Poincaré clearly saw this role of themata in his book *Science and Hypothesis*:

"The principle of the conservation of energy simply signifies that there is a *something* which remains constant. Whatever fresh notions of the world may be given us by future experiments, we are certain beforehand that there is something which remains constant, and which may be called *energy*"*

(to which we now add: even when we used to call it only mass). The *thema* of conservation has remained a guide, even when the language has had to change. We now do not say that the law springs from the "invariability of God," but with that curious mixture of arrogance and humility which scientists have learned to put in place of theological terminology, we say instead that the law of conservation is the physical expression of the elements of constancy by which Nature makes herself understood by us.

The strong hold that certain themes have on the mind helps to explain the stubborn faith with which some scientists cling to one explanation or another when the "facts" cannot decide which of the two is right. The passionate motivation comes out well when two scientists respond to each other's antithetical constructs. For example, Werner Heisenberg, whose physics was built on the theme of the discrete, atomistic, quantized, *discontinuous*, wrote about Erwin Schrödinger's

physics: "The more I ponder about [his] theory, the more disgusting it appears to me." And Schrödinger, in turn, whose physics was based on the idea of *continuity*, wrote that Heisenberg's approach left him "discouraged if not repelled."

A scientist's thematic preference usually also explains his commitment to some point of view that may in fact run exactly counter to all accepted doctrine and to the clear evidence of the senses. Of this no one has spoken more eloquently and memorably than Galileo when he commented on the fact that to accept the idea of a moving earth one must overcome the strong impression that one can "see" that the sun is really moving:

"Nor can I sufficiently admire the eminence of those men's intelligence [Galileo's *Salviati* says in the Third Day of the *Dialogue Concerning the Two Principal Systems*], who have received and held it [the Copernican system] to be true, and with the sprightliness of their judgments have done such violence to their own senses that they have been able to prefer that which their reason dictated to them to that which sensible experience represented most manifestly to the contrary . . . . I cannot find any bounds for my admiration how reason was able, in Aristarchus and Copernicus, to commit such rape upon their senses as, in spite of them, to make itself master of their belief."

Among the themata which permeate Galileo's work and which helped reason to "commit such rape upon their senses," we can readily discern the then widely current thema of the once-given real world which God supervises from the center of His temple; the thema of mathematical nature; and the thema that the behavior of things is the consequence of their geometrical shapes (for which reason Copernicus said the earth rotates "because" it is spherical, and Gilbert, following the lead, is said to have gone so far as to prove experimentally, at least to his own satisfaction, that a carefully mounted magnetized sphere keeps up a constant rotation).

---

* *Science and Hypothesis*, English translation reprinted by Dover, 1952, page 166. Emphasis in original.

While developing the position that themata have as legitimate and necessary a place in the pursuit and understanding of science as have observational experience and logical construction, we should make clear that we need not decide now also on the *source* of themata. Our first aim is simply to see their role in science, and to describe some of them, as a folklorist might when he catalogues the traditions and practices of a people. It is not necessary to go further and to make an association of themata with any of the following conceptions: Platonic, Keplerian, or Jungian archetypes or images; myths (in the nonderogatory sense, so rarely used in the English language); synthetic *a priori* knowledge; intuitive apprehension or Galileo's "reason"; a realistic or absolutistic or, for that matter, any other philosophy of science. To show whether any such associations do or do not exist is a task not yet fulfilled. We also do not want to imply that the occurrence of themata is characteristic only of science in the last centuries. On the contrary, we see the thematic component at work from the very beginning, in the sources of cosmogonic ideas later found in Hesiod's *Theogony* and in *Genesis*. Indeed, nowhere can one see the persistence of great questions and the obstinacy of certain pre-selected patterns for defining and solving problems better than in cosmologic speculations. The ancient Greek cosmologic assumptions presented a three-step scheme: At the beginning, in F. M. Cornford's words, there was

"... a primal Unit, a state of indistinction or fusion in which factors that will later become distinct are merged together. (2) Out of this Unity there emerge, by separation, parts of opposite things .... This separating out finally leads to the disposition of the great elemental masses constituting the world-order, and the formation of the heavenly bodies. (3) The Opposites interact or reunite, in meteoric phenomena, or in the production of individual living things ...."*

* F. M. Cornford, *Principium Sapientiae*, London: Cambridge University Press, 1952, Chapter XI.

Now the significant thing to notice is that when we move these conceptions from the animistic to the physical level, this formula of cosmogony recurs point for point, in our day, in the evolutionist camps of modern cosmology. That recent theory of the way the world started proposes a progression of the universe from a mixture of radiation and neutrons at time $t = 0$; through the subsequent stages of differentiation by expansion and neutron decay; and finally to the building up of heavier elements by thermonuclear fusion processes, preparing the ground for the later formation of molecules. And even the ancient main *opposition* to the evolutionary cosmology itself, namely, the tradition of Parmenides, has its equivalent today in the "steady-state" theory of cosmology.

So the questions persist (for example, concerning the possibility of some "fundamental stuff," of evolution, of structure, of spatial and temporal infinities). And the choices among alternative problem solutions also persist. These thematic continuities indicate the obverse side of the iconoclastic role of science; for science, since its dawn, has also had its more general themata-creating and themata-using function. James Clerk Maxwell expressed this well a century ago in an address on the subject of molecular mechanics:

"The mind of man has perplexed itself with many hard questions. Is space infinite, and in what sense? Is the material world infinite in extent, and are all places within that extent equally full of matter? Do atoms exist, or is matter infinitely divisible?

"The discussion of questions of this kind has been going on ever since man began to reason, and to each of us, as soon as we obtain the use of our faculties, the same old questions arise as fresh as ever. They form as essential a part of science of the nineteenth century of our era, as of that of the fifth century before it."†

† *The Scientific Papers of James Clerk Maxwell*, reprinted by Dover, 1965, Vol. 2, page 361.

We may add that thematic questions do not get solved and disposed of. Nineteenth-century atomism triumphs over the ether vortices of Kelvin—but then field theories rise which deal with matter particles again as singularities, now in a twentieth-century-type continuum. The modern version of the cosmological theory based on the thema of a life cycle (Beginning, Evolution, and End) may seem to triumph on experimental grounds over the rival theory based on a thema of Continuous Existence, and throw it out the window—but we can be sure that this thema will come in again through the back door. For contrary to the physical theories in which they find embodiment in $xy$ terms, themata are not proved or disproved. Rather, they rise and fall and rise again with the tides of contemporaneous usefulness or intellectual fashion. And occasionally a great theme disappears from view, or a new theme develops and struggles to establish itself—at least for a time.

Maxwell's is an unusual concession, but it is not difficult to understand why scientists speak only rarely in such terms. One must not lose sight of the obvious fact that science itself has grown strong because its practitioners have seen how to project their discourse into the $xy$ plane. This is the plane of public science ($S_2$) of fairly clear conscious formulations. Here a measure of public agreement is in principle easy to obtain, so that scientists can fruitfully cooperate or disagree with one another, can build on the work of their predecessors, and can teach more or less unambiguously the current content and problems of the field. All fields which claim or pretend to be scientific try similarly to project their concepts, statements, and problems into the $xy$ plane, to emphasize the phenomenic and analytic-heuristic aspects.

But it is clear that while there can be automatic factories run by means of programmed computers and the feedback from sensing elements, there can be no automatic laboratory. The essence of the automaton is its success in the $xy$ plane at the expense of the $z$-direction (hence automata do not make qualitatively new findings). And the essence of the scientific genius is often exactly the opposite—sensitivity in the $z$-direction even at the expense of success in the $xy$ plane. For while the $z$-dimension is never absent even in the most exact of the sciences as pursued by actual persons, it is a direction in which most of us must move *without* explicit or conscious formulation and without training; it is the direction in which the subject matter and the media for communication are entirely different from those invented specifically for discussion of matters in the $xy$ plane with which the scientist after long training can feel at home.

Therefore it is difficult to find people who are bilingual in this sense. We are not surprised that for most contemporary scientists any discussion which tries to move self-consciously away from the $xy$ plane is out of bounds. However, it is significant that even in our time the men of genius—such as Einstein, Bohr, Pauli, Born, Schrödinger, Heisenberg—have felt it to be necessary and important to try just that. For the others, for the major body of scientists, the plane of discourse has been progressively tilted or projected from $xyz$ space into the $xy$ plane. (Perhaps prompted by this example, the same thing is happening more and more in other fields of scholarship.) The themata actually used in science are now largely left implicit rather than explicit. But they are no less important. To understand fully the role a hypothesis or a law has in the development of science we need to see it also as an exemplification of persistent motifs, for example the thema of "constancy" or of "conservation"; of quantification; of atomistic discreteness; of inherently probabilistic behavior; or—to return to Newton—of the interpenetration of the worlds of theology and of physics.

We have spoken mostly of the physical sciences. We might, with equal or greater advantage, have dealt with the newer sciences, which do not have a highly developed corpus either of phenomena or of logical calculi and rational structures. In those cases, the $z$-elements are not only still relatively more prominent but also are discussed with much greater freedom—possibly because at its early stage a field of study still bears the overwhelming imprint of one or a few men of genius. It is they who are particularly "themataprone," and who have the necessary courage to make decisions on thematic grounds.

It should be clear that the most significant fact, implied in the example given, is that most and perhaps all of these themata are not restricted merely to uses in scientific context, but seem to come from the less-specialized ground of our general imaginative capacity. This view leads us at once beyond the usual antithetical juxtaposition between science and the humanities. For the laments on the separation between science and the other components of our culture depend on the oversimplification that science has only *x*- and *y*-dimensions, whereas scholarly or artistic work involves basic decisions of a different kind, with predominantly esthetic, qualitative, mythic elements. In our view this dichotomy is much attenuated, if not eliminated, if we see that in science, too, the *xy* plane is not enough, and never has been.

It is surely unnecessary to warn that despite the appearance and reappearance of the same thematic elements in science and outside, we shall not make the mistake of thinking that science and non-science are at bottom somehow the same activity. There are differences which we should treasure. As Whitehead once said about the necessity to tolerate—no, to *welcome*—national differences: "Men require of their neighbors something sufficiently akin to be understood, something sufficiently *different* to provoke attention, and something great enough to command admiration." It is in the same sense that we should be prepared to understand both the separateness that gives identity to the study of each field, as well as the kinship that does exist between them.

## 14.5 THE CONTENT OF SCIENCE; SUMMARY

In this chapter we have tried to trace out the hypothetical development of a physical law. In previous chapters we dealt with the nature of theories and concepts. Let us in summary survey the three main elements which make up the established part of physical science. On this, at least, there is fairly good agreement among contemporary philosophers of science, although detail and phrasing may differ from one to the next.

First of all there are the *concepts* or *constructs*, like velocity, mass, chemical element, etc.—the main ideas which the particular sciences use as vocabulary. They can be defined by operational definitions, the rules of correspondence between the concepts on the one hand and what can be observed (preferably measured) on the other. For example, we defined instantaneous velocity and acceleration by specific measurements and calculations, operations which involve distances, time intervals, simple algebra. But these concepts also have a thematic content, owing to the prior commitment of the scientist to a view about science which is not subject to experimental verification and falsification. Second, there are the *relationships between the concepts*. These relations may be simple factual observations (e.g., ice under pressure melts below 0°C—a humble statement involving, however, four concepts, of which at least one, temperature, is far more difficult than appears at first glance). Or they may be more general summaries of facts called laws, principles, and so forth (e.g., Newton's second law of motion, relating the net force on a body, its mass, and its acceleration by $F_{net} = ma$). Or they may even be larger systems relating laws to one another (theories, which we examined in Chapter 3). Last—although we tend to take it for granted—we must not forget at least to mention that part of science which contains the *grammar* for expressing, verbally or mathematically, definitions of concepts and relationships between concepts, i.e., the logic of language itself (rules on how to use words like *and*, *or*), and the logic of mathematics (how to use + or −, how to add vectors, etc.).

These three parts of science are obviously so completely interrelated that any one taken by itself is quite meaningless. A physical law means nothing without the definition of the concepts and the rules of mathematics or language. A concept is useless if it does not appear in relation to other concepts, or if we fail to support it with clear definitions. This has helped us to an important insight about the way science grows: An advance in any one part of science is tentative until the rest of the system absorbs the advance. For example, a newly discovered law may disturb the hitherto accepted relationships between concepts or even the usefulness of some old concept. There will be a period of rearrangement until the new discovery is fully incorporated into the ever-growing structure.

The development of a physical or chemical law is perhaps analogous to the progress of a body of water down a gentle slope where there are eddies and back currents from the advancing front, and where also here and there a sudden surge reaches out to embrace and temporarily hold some territory far ahead of the main advance. The process of gradual approximation leading toward a law may stretch over generations. How fortunate for the progress of science that individual scientists do not permit themselves to become discouraged by this prospect of lengthy struggles—that they find their reward in a wholehearted devotion to the day-to-day progress of their work.

## PROBLEMS

**Problem 14.1.** What main points of difference are there between a *law of physics* (e.g., the law of projectile motion as stated in the form of our several equations) on one hand and, on the other, a *law of magic* (e.g., the belief of the Hopi Indian that the ceremonial handling of rattlesnakes increases the chances of subsequent rainfall), a *law of phrenology*, a *civil law* (law against speeding on the highway), and a *moral* or *divine law* (the Ten Commandments)?

**Problem 14.2.** In a recent magazine article there appeared the following statement. Comment on the correctness or falsity of the *statement* and of its *implications* in a short essay.

"But the first thing to realize even about physics is its extraordinary indirectness. Physics appears to begin with very straightforward questions, but there are catches in it right from the start. For example: every high school student is told that according to Aristotle the heavier of two weights would fall faster, but that Galileo, by dropping two different weights from the leaning tower of Pisa, 'proved' that Aristotle was wrong, for the weights were found to fall in exactly the same time. And yet Aristotle was right. The heavier body does fall faster, because of air resistance, which slows up the lighter body more than the heavier. Only a very little faster, it is true, and the difference may not be detectable; but since scientists claim that they use words with such great precision, it is fair to hold them to it. If you press a physicist on this point, he will readily admit that what he means is that the two bodies would fall equally fast *in a vacuum*. If you press him further, he will very reluctantly admit that nobody has ever produced a complete vacuum. And so it turns out that not the physicist but Aristotle is talking about the actual world in which we live."

**Problem 14.3.** What exactly are the *raw observations* concerning the sun, stars, planets, satellites, etc. from which the theory of universal gravitation was derived? Try carefully to state observable facts and not to *assume* any theory in your description.

**Problem 14.4.** Examine a law outside physics with which you are *well* acquainted (e.g., a law of economics, linguistics, biology, etc.) and decide whether or not one may trace its logical development by the same type of diagram that we have used to analyze our hypothetical physical law.

**Problem 14.5.** Criticize these definitions: (a) The mass of an object is the measure of its quantity. (b) The range of a projectile is the horizontal component of its displacement. (c) The densage of an object is its volume divided by its age. (d) The factor $\pi$ is approximately equal to $\frac{22}{7}$.

**Problem 14.6.** What is meant in science by the statements: (a) Atoms exist. (b) The ether does not exist. (c) Avogadro's (or any other) hypothesis is correct. (d) The findings do not represent absolute truths.

**Problem 14.7.** Why do *you* accept as a fact that the earth is round? That the moon is spherical? That there was once a scientist named Newton? Of what facts are you certain beyond all possible doubt? Why?

**Problem 14.8.** Try to account for the fact that astronomy was one of the earliest and best-developed sciences.

**Problem 14.9.** Assume that for his experiments Galileo had been using our most modern equipment for measuring the time and distance of freely falling bodies. Let us assume that his data had been accurate to $10^{-6}$ sec or $10^{-4}$ cm, respectively. Now write out in a few sentences (a) what might have been the result of compiling a large set of experimental data, (b) what he actually would have had to do to discover the law of freely falling bodies by means of such data.

**Problem 14.10.** Write a short essay on one of these topics: (a) The scientist as a creative investigator. (b) Science as an evolving organism. (c) Important factors in the process of scientific discovery. (d) The discovery of truths by free and vigorous exchange of intelligence. (e) The complementary aspects of experience: scientific experience and poetic experience.

**Problem 14.11.** Write a short essay on Bridgman's extrapolation of the operational view to fields other than physical science (Section 12.4).

**Problem 14.12.** Read again the two "deceptive" questions on page 68 and in the footnote on page 100. Then answer them in the light of Section 12.7.

**Problem 14.13.** Examine critically the analogy cited for the evolutionary processes (Section 13.7) and show where it can be extended (e.g., dangers of overspecialization) and where it breaks down.

## RECOMMENDED FOR FURTHER READING

P. W. Bridgman, *The Logic of Modern Physics*, New York: Macmillan, 1927, reprinted 1960, paperback; Chapters I, II, and the first part of III

J. B. Conant, *Science and Common Sense*, New Haven: Yale University Press, 1951, reprinted 1961, paperback; particularly Chapters 1–3

Pierre Duhem, *The Aim and Structure of Physical Theory*, translated from the second French edition of 1914 by P. P. Wiener, Princeton, N.J.: Princeton University Press, 1954; Atheneum Books paperback

P. Frank, *Modern Science and its Philosophy*, Cambridge, Mass.: Harvard University Press, 1949; reprinted by Collier Books, 1961; paperback

E. C. Kemble, *Physical Science*, Chapter 10

Ernst Mach, "On the economical nature of physical inquiry," pages 186–213 in his *Popular Scientific Lectures*, translated by T. J. McCormack, from German (a lecture delivered in Vienna in 1882); LaSalle, Ill.: Open Court Publishing Co., fifth edition, 1943

F. S. C. Northrop, *Logic of the Sciences and the Humanities*, New York: Macmillan, 1947; Meridian Books, paperback; particularly Chapters I and II

Henri Poincaré, *Science and Method*, translated from French by F. Maitland, Dover paperback reprint; particularly Book I, Chapters 1 and 3

K. Popper, *Conjectures and Refutations*, New York: Basic Books, 1962; particularly Chapters 1–3

*Project Physics Reader*, Vol. 1, articles by Feynman, Hoyle, Bridgman, and Polya

G. Sarton, *The Study of the History of Science*, Cambridge, Mass.: Harvard University Press, 1936; Dover paperback reprint, to page 52

S. Toulmin, "The evolutionary development of natural science," *American Scientist*, Vol. 55, pages 456–471 (1967)

## SOURCES, INTERPRETATIONS, AND REFERENCE WORKS

W. I. B. Beveridge, *The Art of Scientific Investigation*, New York: W. W. Norton, 1950; paperback

P. M. Blake, C. J. Ducasse, and E. H. Madden, *Theories of Scientific Method*, Seattle, Wash.: University of Washington Press, 1960; paperback

B. Brody and N. Capaldi (editors), *Science: Men, Methods, Goals*, New York: Benjamin, 1968; paperback

N. R. Campbell, *Physics, the Elements*, London: Cambridge University Press, 1920; reprinted as *Foundations of Science* by Dover Publications, paperback

R. Carnap, *Philosophical Foundations of Physics*, New York: Basic Books, 1966

A. C. Crombie, *Medieval and Early Modern Science*, Garden City, N.Y.: Doubleday Anchor Books, 1959, Vol. II, pages 1–35, 285–333

H. Feigl and W. Sellars, *Readings in Philosophical Analysis*, New York: Appleton-Century-Crofts, 1949; particularly pages 498–514

N. R. Hanson, *Patterns of Discovery*, New York: Cambridge University Press, 1958; paperback

G. Holton, "The thematic imagination in science," pages 88–108 in *Science and Culture* (edited by G. Holton), Boston: Beacon Press, 1965

G. Holton, *The Thematic Origins of Scientific Thought: Kepler to Einstein*, Cambridge, Mass.: Harvard University Press, 1973

D. Johnson, "Mysterious craters of the Carolina coast," *American Scientist*, January 1944. An exercise in "scientific methods."

T. S. Kuhn, *The Structure of Scientific Revolutions*, Chicago: University of Chicago Press, second edition, 1970

L. Laudans, "Theories of scientific method from Plato to Mach," *History of Science*, Vol. 7, pages 1–63 (1969)

K. R. Popper, *The Logic of Scientific Discovery*, translated from the German edition of 1934, London: Hutchinson, 1959; New York: Science Editions paperback, 1961

B. Russell, *Scientific Outlook*, New York: W. W. Norton, 1931; Chapters II and III

H. Shapley *et al.*, *Readings in the Physical Sciences*, New York: Appleton-Century-Crofts, 1948; pages 401–423

S. Toulmin, *Human Understanding*, Vol. 1, Princeton, N.J.: Princeton University Press, 1972

J. Ziman, *Public Knowledge, an Essay Concerning the Social Dimension of Science*, New York: Cambridge University Press, 1968, paperback

# PART E

# THE LAWS OF CONSERVATION

# THE LAWS OF CONSERVATION

If we now look back on the structure which we have raised so far, we may perhaps realize how greatly success in the pursuit of scientific knowledge is predicated on our ability to perform three closely interrelated tasks: to isolate the phenomenon from distracting or trivial other effects, to describe unambiguously what is happening, and to discern some specific permanence in the flux of events under observation. These are the beginnings of every scientific activity. And exactly these three functions are summarized most effectively in the various laws of conservation. These laws, perhaps the most powerful—certainly the most prized—tools of analysis in physical science, say in essence that no matter what happens in detail to a set of interacting bodies in some prescribed setting, the sum of this or that measurable quantity (the mass, or the momentum, or the energy, or the charge) is *constant* throughout the period of observation.

There is a spectacular beauty about these laws. They can break through the clamor and confusion of appearances and point to an underlying constancy so convincingly that the event changes at one stroke from chaos to ordered necessity. Moreover, so simple, general, and powerful are they that by the very extensiveness of their application the conservation laws unify the various physical sciences within themselves and with one another. Indeed, they have come to be called *principles* rather than *laws*, a terminology which betrays that they are no longer merely summaries of experimental facts, but instead have become, as it were, the starting points of scientific understanding itself.

In Part E we shall discuss specifically the laws of conservation of mass, of momentum, and of energy.* This part also serves as a logical and historical bridge between the study of universal gravitation and subsequent chapters on atomic theory.

---

\* The law of conservation of charge is dealt with in Chapter 24.

# Chapter 15

# THE LAW OF CONSERVATION OF MASS

## 15.1 PRELUDE TO THE CONSERVATION LAW

In his great work *De Rerum Natura* (*On the Nature of Things*), the Roman poet Lucretius, contemporary of Julius Caesar and Cicero, recorded and embellished the nature-philosophy of Greece's Leucippus, Democritus (both *c.* 450 B.C.) and Epicurus (*c.* 300 B.C.), and he reiterated what may be considered as one of the earliest hints of a profound general principle of science: "Things cannot be born from nothing, cannot when begotten be brought back to nothing." Everything now existing must have continual existence in past, present, and future, although form, appearance, and the like may indeed change.

Yet there is a very considerable distance from the panegyric of Lucretius to the modern law of conservation of mass—that almost axiomatic basis of much in our physical sciences—which teaches that despite changes of position, shape, phase, chemical composition, and so forth, *the total mass in a given enclosed region remains constant*. Indeed, it is quite futile to search Greek thought for direct parentage of the *specific* principles of modern physical science (except perhaps for certain astronomical ones). Lucretius' ultimate purpose, for example, was not to present a scientific text but to reduce the contemporary burden of superstition and fear through Epicurean philosophy and the rational explanation of natural phenomena. Although he documents his thesis throughout with penetrating observations from nature, he is not concerned primarily with a physical problem, but rather with a philosophical one, namely the basis for a particular type of atheism. This is made clear in one of the first pages of the *De Rerum Natura*:*

"Our starting point shall be this principle:
*Nothing at all is ever born from nothing
By the gods' will*. Ah, but men's minds are
   frightened
Because they see, on earth and in the heaven,
Many events whose causes are to them
Impossible to fix; so, they suppose,
The gods' will is the reason. As for us,
Once we have seen that *Nothing comes from
   nothing*,
We shall perceive with greater clarity
What we are looking for, whence each thing
   comes,
How things are caused, and no 'gods' will'
   about it."

---

* From the translation by Rolfe Humphries, *The Way Things Are* (Bloomington: Indiana University Press, 1968), page 24.

## 15.2 STEPS TOWARD A FORMULATION

Before the law of conservation of mass (to be occasionally abbreviated to read LCM) could reveal itself and appear in terms meaningful to modern science, three separate developments had to take place, each in itself the work of the greatest physical scientist of his generation. First had to appear the concept of the ideally isolated system. The dominant tradition of the Greeks and scholastics was the vision of a unified, completely integrated universe. In such a scheme, the behavior of a single body is determined by its relation to the rest of the cosmos, by the necessary role it must play in the whole drama. It was not meaningful, therefore, to think of events in isolation, for example, to interpret the behavior of a single object in terms of a physical law applicable to that particular region and uninfluenced by events occurring simultaneously in the universe surrounding this region.

With Alfred North Whitehead, we may regard the concept of the isolated system in part as one consequence of the work of Galileo on the motion of bodies. Specifically, Galileo's prescience of the law of inertia or, as we call it now, Newton's first law of motion, speaks of the continuing, uniform, and unaccelerated motion of any object on a straight horizontal plane in the absence of external forces; and so it invites us to map out in thought a region containing only the body in equilibrium, a region at whose boundaries all causal connections with phenomena on the outside are broken.

Assuming, then, that there was now available the concept of an *isolated* or *closed system* by which to define the region of attention, there was next needed a criterion for measuring the quantity of matter before a conservation law could be formulated. And this was provided by Newton in the opening paragraph of the *Principia* where, speaking of the term "quantity of matter," he states, "It is the quantity that I mean hereafter under the name of body or *mass*." To know quantitatively the amount of matter in a system we need to know only its inertia (or else its weight, "for it is proportional to the weight, as I have found by experiments..."). With this definition in mind it was unavoidable to become aware that there is something enduring and constant about any given object—not necessarily its color or apparent size, its position or motion, or its volume or shape, or even its integrity, but rather its mass everywhere, and therefore also its weight at a given locality.

A third contribution was clearly necessary, namely, the proof that the quantity of matter in a given system or, practically speaking, the weight of some material in a closed bottle, does not change during chemical transformations. The historian of science Charles Singer* believed that Newton's explicit proof of the constancy of weight at the same place "gave a special impetus to the rationalization of chemistry" by revealing a simple and effective touchstone for checking on quantitative changes.

To us it would seem rather simple to proceed to a direct proof of the long-suspected LCM: Enclose some air and a piece of wood in a tight bottle, then focus the rays of the sun on the wood inside so that it may burn. After its transformation, the ashes and remaining gases together will be found to weigh neither more nor less than the initial components. Yet there lies a period of just about 100 years between Newton's *Principia* and the work that provided this type of final and explicit proof, the memoirs on calcination and the text *Traité Élémentaire de Chimie* (Elements of Chemistry) by Antoine Laurent Lavoisier (1743–1794), often called the father of modern chemistry (Fig. 15.1).

## 15.3 LAVOISIER'S EXPERIMENTAL PROOF

The obstacles that had to be overcome in the interval between Newton and Lavoisier can now

---

* Charles Singer, *A Short History of Science.* Oxford: Clarendon Press (1941).

**Fig. 15.1**  Antoine Laurent Lavoisier (1743–1794).

be stated so as to make them sound almost trivial. Lavoisier was one of the first to show conclusively that the most familiar of all chemical transformations, combustion of matter, is generally oxidation, i.e., the combination of the material with that part of the ambient air to which he gave the name *oxygen*; that therefore the gas taken from the atmosphere has to be reckoned into the overall calculation. Before Lavoisier's time, neither the nature of gases nor that of the combustion process itself was clear enough; there was therefore little reason or facility for working with carefully measured quantities of gases, or even for carrying out the reactions in closed vessels, isolated from

the rest of the chemical universe. Instead of a general law of conservation of all participating matter during chemical processes, there impressed itself on the scientific mind first the clear, obstinate, and confusing facts that on combustion in the open some materials, like wood, would lose weight (because, as we would now say, more gases are given off than are taken on by oxidation), whereas others, like phosphorus, would conspicuously gain weight (because, in modern terms, more oxygen is fixed than vapors given up). During the reverse type of reaction, now called *reduction*, in which oxygen may be given up, a similar variety of changes would occur.

Since scientific theories quite rightly and naturally tend to deal initially with concepts derived from the most conspicuous aspects of the phenomena, there had grown up in the eighteenth century a conceptual scheme that attempted to deal with a large number of diverse observations on combustion, including the change in the physical structure and chemical nature of the burning object, the presence of heat and of flames, the changes in the quality of the surrounding air, even the diverse changes of weight. For these purposes was invented the *phlogiston*, a substance or "principle" whose migration into or out of the transforming bodies was to account for all such observations.† The phlogiston theory died within a decade or so after Lavoisier's attack on it, not only because the concept of phlogiston had tried to explain so large a range of phenomena that it failed to achieve a quantitative focus and became self-contradictory, but also because Lavoisier proved the concept to be unnecessary, and by the incontrovertible evidence of the balance could show that shifting one's attention to the total quantity of matter undergoing a chemical reaction (including

---

†    For a complete account, see J. B. Conant's *Science and Common Sense*, New Haven: Yale University Press (1951), Chapter 7; also *The Overthrow of the Phlogiston Theory*, Case 2 in the *Harvard Case Histories in Experimental Science* (edited by J. B. Conant), Cambridge: Harvard University Press (1950).

the gases and vapors) would be rewarded by attaining a rigorous law of conservation of matter.

A few translated excerpts from Lavoisier's text of 1789 will demonstrate something of his method in this respect, and will also hint at the relative novelty of this quantitative argument:

"The elegant experiment of Mr. Ingenhouz, upon the combustion of iron, is well known. [Then follows a description of this experiment, the burning of iron wire in a closed vessel of oxygen-enriched air.] As Mr. Ingenhouz has neither examined the change [in weight] produced on iron, nor upon the air by this operation, I have repeated the experiment under different circumstances, in an apparatus adapted to answer my particular views, as follows.... [The modification is now given which permits accurate determination of weights.] If the experiment has succeeded well, from 100 grains [5.3 grams] of iron will be obtained 135 or 136 grains of ethiops [oxide of iron], which is an augmentation [of mass or of weight] by 35 per cent. If all the attention has been paid to this experiment which it deserves, the air will be found diminished in weight exactly to what the iron has gained. Having therefore burnt 100 grains of iron, which has required an additional weight of 35 grains, the diminution of air will be found exactly 70 cubical inches; and it will be found, in the sequel, that the weight of vital air [oxygen] is pretty nearly half a grain for each cubical inch; so that, in effect, the augmentation of weight in the one exactly coincides with the loss of it in the other."

Dozens of such experiments are described by Lavoisier, involving all types of reactions then known, and always repeating this lesson: In reactions within a closed system, the gain experienced by any one part of the system exactly counterbalances the loss by the rest of the system, or *the total quantity of matter within the system remains constant.*

Lavoisier's conceptual scheme of combustion and reduction was superior to the phlogiston theory in the form, the certainty, and the elegance of numerical predictions, and was therefore far more favorable to scientific progress, despite the fact that now a great many observations (e.g., the presence of flames, or the change in appearance of the material) were left out of the scheme of immediate explanations. A rather similar turn of events characterized several conceptual revolutions in science, as for example the rise and fall of the caloric theory of heat, which we shall discuss at some length in Chapter 17.

Needless to say, once the problems of the composition of air and of combustion were solved, the process of oxidation assumed its place as only one special, though sometimes unusually spectacular, case of chemical reactions which all exhibit the conservation law as a general and basic proposition. As Lavoisier wrote in 1789:

"We must lay it down as an incontestable axiom, that in all the operations of art and nature, nothing is created; an equal quantity of matter exists both before and after the experiment, ... and nothing takes place beyond changes and modifications in the combination of these elements. Upon this principle, the whole art of performing chemical experiments depends: We must [to cite one application of this principle] always suppose an exact equality between the elements of the body examined and those of the product of its analysis."

## 15.4 IS MASS REALLY CONSERVED?

Despite Lavoisier's emphatic statement of the conservation law, there was still room for doubt. A modern experimental chemist, examining Lavoisier's reports of his experiments with some appreciation of the degree of accuracy that Lavoisier could have attained with his apparatus, might be somewhat skeptical about the claim that "the augmentation of weight in the one *exactly* coincides the loss of it in the other." Nevertheless, the law was plausible, and therefore most nineteenth-century chemists were willing to go along with Lavoisier and accept it as an axiom until there was some definite reason to suspect that mass is not conserved.   As long as the law of conservation of mass was thought to be consistent with other accepted views about the nature of matter, and as long as no obvious violation had been observed, there was no incentive to undertake further experimental tests.

In 1872, the German chemist Lothar Meyer (1830–1895) suggested that the rearrangement of atoms during chemical reactions might be accompanied by the absorption or emission of particles of "ether," the substance that was believed to transmit light waves.   Such particles might leave or enter the system even though it was sealed to prevent matter from coming in or going out.   At that time it was still an open question whether such ether particles actually existed.   If they did exist, then the mass of the system could change by a very small amount, depending on the mass of these particles and how many entered or left.

Even though most chemists thought that the experiments of Lavoisier and others had established the validity of the mass-conservation law for chemical reactions, there was now interest to demonstrate it experimentally, at the highest level of accuracy.   Doubts such as those expressed by Lothar Meyer could not be simply dismissed without due consideration, since Meyer had already established his reputation as a leading chemist. Hans Landolt, another chemist, decided that further experimental tests were needed.   He conducted extensive research during the two decades after 1890, making very accurate measurements of the masses of systems in which chemical reactions were taking place.   In 1909 he stated his conclusion:

"The final result of the investigation is that no change in total weight can be found in any chemical reaction . . . .   The experimental test of the law of conservation of mass may be considered complete. If there exist any deviations from it, they would have to be less than a thousandth of a milligram."

Landolt's result is typical of the experimental data of the physical sciences: One can never claim to have proved that a difference or other measurement is *exactly* zero (or exactly any other number). One can only say that experiment shows the number to be zero within a specified margin of error.   (It may of course be convenient to assume for theoretical purposes that the exact value has been established.)

If there were no other evidence available to us than experiments on chemically reacting systems, we would have to conclude, even today, that the law of conservation of mass is correct.   A modern chemist, repeating Landolt's experiment with the best available equipment, would still reach the same conclusion, even though his margin of error would be smaller.

However, we now know that in other kinds of experiments the mass of the matter in the system *does* change.   The most important cases are those involving reactions of nuclei or elementary particles, such as radioactive decay, fission, or fusion (see Chapter 30).   In some reactions, such as the annihilation of electrons and positrons, or protons and antiprotons, the mass of the *material particles* (the *rest mass*) disappears completely. On the other hand, the mass of an object may increase considerably when it is accelerated to speeds close to the speed of light.

In all these cases, the apparent change of mass is balanced by a corresponding change of energy, in accordance with Einstein's theory of relativity (Chapter 30).   This could be interpreted by saying that energy itself has mass.   Einstein's theory predicts that there will be a very small change of rest mass even in chemical reactions.   This change is due to the fact that heat energy is absorbed or produced in the reaction, and that such energy also has mass.   However, no chemical reactions involve a sufficiently large change of mass to permit this change to be measured.   In the reactions studied by Landolt, the change of rest mass predicted by relativity theory is much less than a thousandth of a milligram. Landolt's conclusion, that the law of conservation of mass is valid for resting matter, is wrong, even though his experimental data were as good as anyone has ever been able to obtain.

Here indeed is striking testimony to the power of a theory to help overthrow a law just when it seemed to be firmly established by experiment. In Chapter 30 we shall tell in more detail how this happened, and how eventually the law of conservation of mass was restored in a new form.

## PROBLEMS

**Problem 15.1.** Read Chapter 7 in Conant's *Science and Common Sense*, then discuss three chemical reactions as they would each be explained by phlogiston chemistry and by modern chemistry (e.g., the burning of coal, the calcination of a metal, the reduction of a metallic ore).

**Problem 15.2.** Read pages 399 to 408 in *Readings in the Physical Sciences*, by H. Shapley, H. Wright, and S. Rapport (New York: Appleton-Century-Crofts, 1948), then discuss Lavoisier's experimental procedure, specifically the extent to which he used the law of conservation of mass.

**Problem 15.3.** Read the Author's Preface and Chapters III, V, and VII of Lavoisier's *Elements of Chemistry*, then discuss three reactions in which the LCM in chemical transformations is clearly demonstrated.

## RECOMMENDED FOR FURTHER READING

R. Feynman, *The Character of Physical Law*, London: British Broadcasting Corporation, 1965; paperback reprint by MIT Press; Chapter 3

## SOURCES, INTERPRETATIONS, AND REFERENCE WORKS

Ida Freund, *The Study of Chemical Composition, an Account of its Method and Historical Development*, Dover paperback reprint of the 1904 edition; Chapters II and III

P. Pomper, "Lomonosov and the discovery of the law of conservation of matter in chemical transformations," *Ambix*, Vol. 10, pages 119–127 (1962)

J. Smedslund, "The acquisition of conservation of substance and weight in children: I. Introduction," *Scandinavian Journal of Psychology*, Vol. 2, pages 11–20, 85–87 (1961), reprinted in *Readings in Child Behavior and Development* (C. B. Stendler, editor), New York: Harcourt, Brace and World, second edition, 1964

# Chapter 16

# THE LAW OF
# CONSERVATION OF
# MOMENTUM

## 16.1 INTRODUCTION

After everything has been done to describe the
motion of bodies accurately, to analyze the path
of projectiles, the orbits of planets, and the forces
needed to account for such motions, there still
remains, to the most basic common sense, a deep-
lying and mystifying puzzle. While anyone can see
intuitively that a body should move when it is
somehow pushed or pulled by an agent in direct
contact, it is far less clear to our primitive senses
why a body should continue in motion in the
absence of net forces or of any material agency in
immediate contact with the body. The source of
this organismic orientation is probably not far to
seek; after all, we each have learned from our
first days that in order to make bodies move toward
us we have to take a hand in it; and to keep up the
motion of anything on earth, we must be prepared
to supply some effort.

To most of the Greeks and to most of the Aris-
totelian-Thomistic scholars of the Middle Ages,
the proposition that the uniform straightline
motion of the bodies requires neither force nor
agent would have been patently absurd. And even
the nonuniform progress of an arrow or a bullet
was explainable only by the continual commotion
in the surrounding air set off at the initial moment
of release. The maxim was: whatever moves must

be moved by something else. Not until the days
of Galileo and Newton did uniform motion without
persistent and continual physical causation become
generally conceivable. And significantly enough,
even Newton could not free his physical universe
from the need for constant association with a
cause for the phenomena. Of course, to him planets
no longer moved by the incessant activity of God
and the angels, but at any rate motion, in His
cosmic philosophy, was thought of as proceeding
in the sensorium of the Deity, by virtue, so to speak,
of his continual though passive attention to and
knowledge of the happenings.

Actually, the attempt to liberate the idea of
motion and to discuss it as an independent phenomenon goes back to the fourteenth-century Francis-
can friar William Ockham (d. 1347). The University
of Paris, where he taught, became a center of
thinking on mechanics through the late Middle
Ages, and although it presented a minority opinion
at the time, it later influenced very strongly such
men as Leonardo da Vinci and Galileo himself.
Ockham's contribution has been succinctly described by Sir Edmund Whittaker in *From Euclid
to Eddington* (page 66):

"In the example of the arrow, he proposed to
explain the phenomena without depending on
propulsion by the air, by assuming that throughout

the flight the arrow carried a certain nonmaterial cargo, so to speak (corresponding to the modern notion of momentum), the possession of which ensured the continuance of its progress. This may be regarded as the first appearance in history of any sound dynamical principle: and, taken in conjunction with his philosophical innovations, justifies us in regarding Ockham as the chief founder of modern science. His idea was carried further by his disciple Jean Buridan, who became Rector of the University of Paris in 1327, and who asserted that the cargo in question was proportional to the product of the weight of the projectile and some function of its velocity (he did not distinguish between the concepts which we call momentum and kinetic energy). His doctrines were adopted, without improvement and indeed with some deterioration, by less able successors; however, eventually they reached Galileo, Descartes, and the other physicists of the seventeenth century, who succeeded in defining momentum precisely, . . .''

Superficially it must seem that William Ockham's own suggestion is not progress but a step back toward animism. He and his pupils after him suggest that during the act of setting-in-motion, something (whether we call it momentum or, as they did, impetus, matters little here) is imparted to the body; and if this quantity then grows or diminishes within, it makes the motion itself increase or decrease. But evidently this view, while it almost endows motion with the concreteness of matter, does help to concentrate attention on the moving object itself; and when it was eventually agreed upon how to measure this "cargo," namely, by the product of the mass $m$ and the instantaneous velocity $v$ of the object, then the pictorial element became unimportant. One could begin to think of motion in the abstract, and particularly of motion through a vacuum. One could forget about the question "What keeps a body in motion if not some continuous external push or pull," and instead ask the more fruitful question "If we postulate that a moving body has at every instant a momentum

$mv$, how does this momentum change under various interesting conditions?" Thus the original questions on motion were to be answered finally in terms of the law of inertia, by which an isolated body persists in a state of constant velocity and consequently of constant momentum, or in terms of Newton's second law, which allows one to calculate the effect of external forces on the motion and thus on the momentum. So the scientific mind in time solved an essentially barren problem by restating it from an entirely different point of view; and historically, by and large, this has always been one of the most powerful methods toward progress in science.*

## 16.2  DEFINITION OF MOMENTUM

The historical development leading from the early conception of momentum in the fourteenth century to the point which interests us now, namely, to the law of conservation of momentum, lasted almost 400 years and involved a galaxy of brilliant investigations. It will not be profitable for us to examine the complex details; instead, we shall mention only the final steps in the establishment of the law.

In Descartes' system of philosophy, widely acclaimed by scientists in the middle of the seventeenth century, the entire physical universe was likened to a clockwork mechanism. Once the machine had been constructed and set running by God, it was supposed to run indefinitely without any need for repairs or winding up (to think otherwise would be to imply that God was not a perfect clockmaker!). In order to ensure that the world machine would not run down, Descartes argued that there must be a principle of conservation of motion:

"It is wholly rational to assume that God, since in the creation of matter he imparted different

---

* In this connection read H. Butterfield, *Origins of Modern Science*, Chapter I.

motions to its parts, and preserves all matter in the same way and conditions in which he created it, so he similarly *preserves* in it *the same quantity of motion.*" (*Principles of Philosophy*, 1644.)

Although anyone might tend to agree with this statement as a general philosophical principle, we cannot judge its scientific validity until we have a precise definition of the phrase "quantity of motion." Descartes recognized that this quantity must depend on something more than the *speed* of an object: A cannonball moving at 100 km/hr has more motion than a tennis ball moving at the same speed, if we judge by how much motion each can transfer to another object which it strikes. Descartes proposed to define quantity of motion as the product: (mass) × (speed). The total quantity of motion of all the parts in the world, or in any isolated system, must then be constant from moment to moment, even though motion is constantly being transferred from one body to another by collisions within the system. In mathematical terms, the sum

$$m_1 v_1 + m_2 v_2 + m_3 v_3 + \cdots = \sum_i m_i v_i$$

remains fixed even though $v_1, v_2 \ldots$ change.*

Descartes' law of conservation of motion seemed at first to provide a correct description of collisions, but closer examination revealed two serious flaws. *First*, his proposed law, even if it were correct, is not sufficient to determine the outcome of a collision. Descartes therefore had to add various auxiliary rules, some of which were obviously wrong. (For example: If a heavy object is at rest and a light object strikes it, the light one must rebound without transferring any motion to the heavy one.) *Second*, the speed $v$ is a *scalar* quantity,

---

* The symbol Σ (Greek, capital sigma) is used to indicate a summation of the following quantities, in this case $m_i v_i$, where the index $i$ runs from 1 up to $N$, the total number of terms in the sum. This abbreviation is useful in writing equations in which the number of summands is large or indefinite.

whereas experiments indicated that the outcome of a collision depends also on the *directions* of motion of the colliding objects. (An extreme example would be the head-on collision of two blobs of clay, which stick together and remain at rest after the collision; clearly $\sum mv$ is not conserved in this case, since it is zero after collision but was a positive quantity before collision.

For those scientists who accepted Descartes' contention that the new science must be constructed mechanistically on the basis of the motions and collisions of small pieces of matter, it became imperative to find a correct version of his law of conservation of motion. When the Royal Society of London was organized in 1662, several of its members, especially Robert Hooke and Christopher Wren, were already interested in this problem, and had done some preliminary experiments. Christian Huygens, in Holland, was thought to have found a correct formulation of the laws of collision, but had not published it. The Royal Society began to focus on the problem in 1666, with Hooke demonstrating collision experiments at the weekly meetings, Wren being urged to prepare an account of his earlier work, the mathematician John Wallis being asked to look into the theoretical aspects of the problem, and the secretary of the Society, Henry Oldenburg, being instructed to write to Huygens asking for a report of his discovery. By the beginning of the year 1669 this cooperative effort had borne fruit, and the Royal Society was able to publish in its *Philosophical Transactions* the result obtained independently by Huygens, Wallis, and Wren: The total *momentum* of a system is conserved; the momentum of any object is defined as the product *mass × vector velocity* for that object. In modern notation, the definition is

$$\text{momentum } \boldsymbol{p} \equiv m\boldsymbol{v}. \qquad (16.1)$$

Being the product of a scalar quantity (mass) and a vector quantity (velocity), $\boldsymbol{p}$ is also a vector quantity, representable by an arrow pointing in the same direction as the one that signifies the corresponding velocity vector.

The usefulness of the concept momentum, considered a vector compared to the scalar quantity "quantity of motion" (as in Descartes' definition), is most clearly revealed by considering the head-on collision of two blobs of clay which stick together and remain at rest after the collision. Suppose the blobs have equal mass $m$ and move with equal speeds $v$ in opposite directions before the collision. Their vector velocities will then be $v$ and $-v$, respectively, and the total momentum of the system before the collision will be

$$mv - mv = 0.$$

If both velocities are zero after the collision, then the total momentum at that time is also zero, and the law of conservation of momentum is obeyed. Descartes' supposed law of conservation of quantity of motion, on the other hand, is shown not to work, since $\sum m_i v_i$ is finite (equal to $2mv$) before the collision, but zero afterward.

This type of collision is called *completely inelastic*; the colliding objects do not bounce. At the other extreme would be the *completely elastic* collision, in which the two objects bounce back with the same speeds as before the collision, but in opposite directions. All the intermediate cases of partial elasticity may also occur in nature; but in each case the total momentum of the system is found to be conserved.*

Unhappily for the concept of the world machine, the new law of conservation of momentum does not guarantee that the parts of the universe will keep on moving forever. Unless all collisions of pieces of matter are completely elastic, the total quantity of motion (in Descartes' sense) will continually decrease, indicating the possibility that eventually the machine will grind to a halt, leaving all its

---

* The mathematical statement of the law is: $\Sigma\,\mathbf{p}_i$ remains fixed from one moment to the next, even though the individual $\mathbf{p}_i$ may change as a result of collisions between particles in the system.

parts at rest. As we shall soon see, an additional, quite different conservation law was found to hold also, and it suggested that this is not the necessary fate. The first hint of the nature of that additional law was offered by Huygens in his rules for computing the effects of collisions. Huygens proposed that in the special case of perfectly elastic collisions, another quantity must be conserved in addition to momentum: the product (*mass*) × (*square of velocity*). In the seventeenth and eighteenth centuries this quantity was called *vis viva*, a Latin phrase which means "living force":

$$vis\ viva \equiv mv^2. \qquad (16.2)$$

The law of conservation of *vis viva*, it turned out, served to remedy—at least in the special case of elastic collisions—the first defect in Descartes' theory mentioned above, by providing a means for determining the outcome of a collision if the initial masses and velocities are known. But we must postpone this subject to the next chapter and return to our discussion of the law of conservation of momentum, a phrase to be abbreviated LCMm.

## 16.3 MOMENTUM AND NEWTON'S LAWS OF MOTION

As we noted in Section 9.5, Newton originally stated his second law in the form: The change of momentum of a body is proportional to the motive (net) force impressed, and occurs in the direction of the straight line along which that force is impressed. In symbols,

$$\mathbf{F}_{net} \propto \Delta\,\mathbf{p} \qquad (\text{where } \mathbf{p} = m\mathbf{v})$$

Although Newton did not explicitly include a time interval in stating the law—he was often dealing with impulsive forces that produce an almost instantaneous change of motion—we now find it preferable to write the law in the form

$$\mathbf{F}_{net} = \frac{\Delta\mathbf{p}}{\Delta t}, \qquad (16.3)$$

or, in words, "the net force on a body is equal to the change in momentum of the body divided by the time interval during which the net force is applied." If the force is not constant during $\Delta t$, this equation will apply only to the *average* force acting during that time interval. If we must deal with instantaneous force we must also take the value of $\Delta p/\Delta t$ in the limit $\Delta t \to 0$; in that case the right-hand side becomes the rate of change of momentum with time.

It is not difficult to see that Eq. (16.3) is precisely equivalent to the more familiar form of Newton's second law given in Section 9.5. Replacing $p$ by $mv$, we have (since $m$ is assumed constant)

$$\Delta p = \Delta(mv) = mv_2 - mv_1 = m(v_2 - v_1) = m \, \Delta v;$$

hence

$$\frac{\Delta p}{\Delta t} = m \, \frac{\Delta v}{\Delta t} \, .$$

Recalling that acceleration was defined as

$$a = \frac{\Delta v}{\Delta t}$$

in Sections 6.5 and 7.4, we see that Eq. (16.3) reduces to

$$F_{\text{net}} = ma. \tag{16.4}$$

Newton's second law, in the form of Eq. (16.3), together with Newton's third law, can now be used to *derive* the law of conservation of momentum.

Suppose two bodies with masses $m_A$ and $m_B$ exert forces on each other, during a collision or in any other circumstances. We call $F_{AB}$ the force exerted on body $A$ by body $B$, and $F_{BA}$ the force exerted on body $B$ by body $A$. No other unbalanced force acts on either body, so the objects $A$ and $B$ can be considered an isolated system.

By Newton's third law (Section 9.10), the forces $F_{AB}$ and $F_{BA}$ are equal in magnitude and opposite in direction:

$$F_{AB} = -F_{BA}. \tag{16.5}$$

For any time interval $\Delta t$ we can write the equivalent of Eq. (16.3) for each body (multiplying through on both sides by $\Delta t$):

$$F_{AB} \, \Delta t = \Delta(m_A v_A) \tag{16.6a}$$

$$F_{BA} \, \Delta t = \Delta(m_B v_B). \tag{16.6b}$$

But by Eq. (16.5), the left-hand side of Eq. (16.6a) and the left-hand side of Eq. (16.6b) are equal in magnitude and opposite in direction. Therefore the same must be true of the right-hand side; that is,

$$\Delta(m_A v_A) = -\Delta(m_B v_B). \tag{16.7}$$

Let $v_A$ and $v_B$ stand for the velocities of the two bodies at the beginning of the time interval, and let $v'_A$ and $v'_B$ stand for their velocities at some later instant. Then (assuming that the masses $m_A$ and $m_B$ do not change) Eq. (16.7) becomes

$$m_A v'_A - m_A v_A = -(m_B v'_B - m_B v_B). \tag{16.8}$$

Collecting the primed symbols on the right side, we find

$$m_A v_A + m_B v_B = m_A v'_A + m_B v'_B. \tag{16.9}$$

This is just the law of conservation of momentum, since the left-hand side of Eq. (16.9) is the total momentum of $A$ and $B$ at the beginning of the time interval, and the right-hand side is the total momentum at the end of the interval.

Since the law of conservation of momentum can be derived from Newton's laws, it would appear that it is not really an independent physical law. On the other hand, LCMm cannot be used by itself to derive Newton's second and third laws, since it does not by itself provide us with any concept of force. Would it not be preferable, therefore, to stick to Newton's three laws in an introductory general account such as this one, and leave the use of LCMm to more advanced courses dealing with special problems in mechanics?

No! We must not let the impressive structure of Newtonian physics delude us into thinking that all physical situations should or even can be analyzed in terms of forces and accelerations. Even in the most elementary collision problems, such as those to be discussed in the following section, it would be extremely difficult to measure the instantaneous forces accurately enough to permit a direct application of Newton's laws, yet the law of conservation of momentum gives us solutions simply. Later, when we discuss the interactions of subatomic particles, we shall see that instantaneous positions, velocities, and accelerations are meaningless, so that the moment-by-moment test of Newton's laws is not feasible; yet the LCMm is still applicable. Moreover, the momentum concept will be found to be still useful even when other concepts such as mass lose their ordinary meaning, as in the case of light rays.

## 16.4 EXAMPLES INVOLVING COLLISIONS

Our first example illustrates how a law of conservation of quantity of motion was suspected even before Descartes' explicit statement of it in 1644.

**Example 1.** The top part of Fig. 16.1, after Marcus Marci's *De Proportione Motus* (1639), may be taken to present the following (idealized) sequence of observable events in one diagram: A ball lies quietly on a table. A second ball, of equal mass, is shot horizontally toward the first, collides with it, and remains lying on the table while the other moves off with the same speed with which it was hit.

We may explain these observations (so well known to everyone acquainted with the games of croquet or billiards) directly from Newton's laws. For during the brief time interval $\Delta t$ of contact between the balls, the mutual forces, by the third law, are equally large though oppositely directed, therefore the decelerating (stopping) force on sphere $A$ is equal to the accelerating force on $B$; and because the two masses involved are equal, the actual deceleration of one is as large as the acceleration of the other. $B$ gains as much speed as $A$ loses; if $A$ loses it all, $B$ gains it all: $v'_B = v_A$.

But now we shall repeat this argument in terms of the LCMm:  Given are $m_A = m_B$; $v_A$; $v_B = 0$; $v'_A = 0$. Required: $v'_B = ?$

*Solution:* These two spheres form a closed system, for only the forces of interaction between them affect the phenomena.

The law states here that $\sum p$ before collision $= \sum p$ after collision (where $\sum$ stands for vectorial summation of the individual momentum vectors).

$$m_A v_A + m_B v_B = m_A v'_A + m_B v'_B. \qquad (16.10)$$

(All vectors lie along the same line and in the same direction. Therefore vector addition becomes here just algebraic addition.)

On substitution from the data, we get

$$m_A v_A + m_B \times 0 = m_A \times 0 + m_B v'_B,$$

$$v'_B = \frac{m_A}{m_B} v_A.$$

But $m_A = m_B$; therefore

$$v'_B = v_A. \qquad \text{Q.E.D.}$$

**Fig. 16.1** A cannon ball collides with a stationary sphere of equal mass. (From Marcus Marci's *De Proportione Motus*)

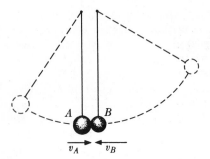

**Fig. 16.2**   Collision of pendulums of equal mass.

---

**Problem 16.1.** In the more general case, the masses may not be equal, but then $v'_A$ will not be zero. Nevertheless, the same type of calculation will apply. Show that then

$$v'_B = \frac{m_A}{m_B}(v_A - v'_A) + v_B.$$

---

**Example 2.** The previous example represented a perfectly elastic collision (e.g., between perfectly rigid bodies). Now note what happens in a perfectly inelastic collision, i.e., when the colliding bodies do not separate afterward but are wedged into each other and travel off together. For example, a car (mass $m_A$) goes at speed $v_A$ and tries to overtake a truck ($m_B$) going in the same direction at speed $v_B$. There is a misjudgment, and a collision. The car is embedded in the rear of the truck. What is their common speed just after collision?

*Solution:* The data are $m_A$; $m_B$; $v_A$; $v_B$; and $v'_A = v'_B = ?$ Here Eq. (16.10) becomes

$$m_A v_A + m_B v_B = (m_A + m_B)v'_B,$$

$$v'_B = \frac{m_A v_A + m_B v_B}{m_A + m_B}.$$

**Example 3.** When the interacting bodies, instead of moving in the same sense, approach each other

with oppositely directed momenta or, in fact, at any angle other than 0°, then the vector nature of momenta must be more carefully considered than we have done in the last two examples. Let two spheres, moving oppositely, hit; perhaps they are two pendulums released from different heights, meeting at the lowest point of their swing (Fig. 16.2). The collision is perfectly inelastic, they interlock; $v'_A = v'_B = ?$ Just as we are used to finding a net force by subtracting opposing tendencies, so it is clear also that here the sum of the momenta just prior to collision is obtained by subtracting the numerical value of $m_B v_B$ from that of $m_A v_A$. We still write the vector equation as before:

$$m_A v_A + m_B v_B = (m_A + m_B)v'_A,$$

$$v'_A = \frac{m_A v_A + m_B v_B}{m_A + m_B}. \quad (16.11)$$

But when we now substitute values for $v_A$ and $v_B$, we must take care, as before, to assign positive values for velocities or momenta toward, say, the right direction, negative ones to the opposite direction. If $m_A = 3$ g, $m_B = 2$ g, $v_A = 8$ cm/sec to the right, $v_B = 15$ cm/sec to the left, then from Eq. (16.11),

$$v'_A = \frac{(3 \text{ g})(8 \text{ cm/sec}) + (2 \text{ g})(-15 \text{ cm/sec})}{(3 + 2)\text{ g}}$$

$$= -1.2 \text{ cm/sec}$$

and the negative sign of the result informs us automatically that after collision both spheres move off together to the *left*. (This was, incidentally, just the type of experimentation by which scientists at the Royal Society established the validity of the LCMm in 1666–68.)

### 16.5   EXAMPLES INVOLVING EXPLOSIONS

In Chapter 9 we discussed the reaction-car experiment, useful for the determination of relative inertial masses. This, too, was a hidden example of the LCMm: It was assumed that the only unbalanced forces acting on the two cars was the

**Fig. 16.3**  The reaction-car experiment.

mechanical push by way of the spring. This symbolizes well what happens in an explosion (Fig. 16.3). Initially all parts of the system are at rest, the momenta are zero; then mutually opposite forces separate the members of the system. We write, therefore: from

$$(p_A + p_B)_{\text{before event}} = (p_A + p_B)_{\text{after event}}$$

follows

$$0 = m_A v'_A + m_B v'_B,$$

$$m_A v'_A = -m_B v'_B.$$

Thus we have derived the same relationship used to determine the mass $m_B$ in terms of a standard mass $m_A$ and observed velocities $v'_A$ and $v'_B$.

---

**Problem 16.2.** (a) An old-fashioned, wheel-mounted 500-kg gun, initially at rest, fires a 5-kg projectile horizontally at 1000 m/sec. If the wheels are freely turning, what will be the speed of recoil of the gun? (b) If the gun is, however, firmly anchored to the earth, what will be the recoil speed of gun and earth together ($m_{\text{earth}} = 6 \times 10^{27}$ g)? (c) If the barrel of the gun is 2 m long and the projectile is accelerated uniformly inside from rest to muzzle speed, what was the average force on the projectile inside the barrel? (d) What was the average force exerted at the same time on the gun by the expanding gases of combustion in the barrel?

## 16.6  FURTHER EXAMPLES

A great variety of other examples may be adduced to illustrate further the extent and power of the LCMm.

**Example 4.** Simply consider a man starting out on a walk. Before he begins, he and the earth are together at rest (as far as this problem is concerned). As soon as he moves forward, the earth below him must move backward, since $p_{\text{man}} + p_{\text{earth}} = 0$, and so $(mv)_{\text{man}} = -(mv)_{\text{earth}}$. To be sure, the velocity change for the earth is, as shown by calculation, extremely small, because $m_{\text{earth}}$ is so much larger than $m_{\text{man}}$. On the other hand, if the surface on which the man walks is a plank on ice, the relative velocities will be more nearly equal. The point, however, is the same in both cases: The force which "couples" the earth or the plank into the system of moving bodies in such a problem is *friction*.

As a force of interaction, friction differs in no way from other forces. In those previous examples that involved pendulums and reaction cars, the forces of interaction were elastic forces; in those involving recoiling guns and projectiles, they were forces set up by gases under pressure. The presence of friction does not reduce the applicability of the LCMm; on the contrary, it extends it by widening any system of interacting bodies to include all those bodies that experience mutual forces of friction. If, for instance, friction under the wheels of the reaction cars had *not* been negligible, we should have written, instead of the first equation in Section 16.5,

$$0 \text{ at start} = [(m_A v'_A + m_B v'_B)$$

$$+ m_{\text{earth}} v_{\text{earth}}] \text{ after separation}$$

Although this may complicate the calculation, it is still a perfectly proper application of the LCMm.

---

**Problem 16.3.** Assume that there exists a flat interstellar platform of mass 1000 kg, at rest as seen from some nearby star, and inhabited by a 70-kg man. Now we see (from the star) that he walks in 3 sec a distance of 5 m, uniformly along a straight line. How far will the observed man think he walked (by his measurement along the platform)? Is this result dependent on how fast he walked? Explain. Describe the motion of the platform while he begins to walk, proceeds and stops.

---

**Example 5.** Lastly, we may mention the role of another force of interaction in the operation of our law: mutual attraction, whether gravitational, electric, or magnetic.

If we apply the conservation law to a system consisting of an apple and the earth which attracts it, their $\sum p$ before the fall is zero. While the apple falls, it attracts the earth upward with a force as large as that by which it is being pulled down, and we can say

$$p_{\text{apple}} + p_{\text{earth}} = 0.$$

Apple and earth will meet at some intermediate point which divides the total distance of initial separation in the ratio of the two masses. (Prove as a problem.) In fact, it is this very type of problem in which we may find one of the earliest appreciations of the LCMm. In 1609, Kepler wrote in the *Astronomia nova*:

"If two stones were removed to any part of the world, near each other but outside the field of force of a third related body, then the two stones, like two magnetic bodies, would come together at some intermediate place, each approaching the other through a distance proportional to the mass of the other."

## 16.7 DOES LIGHT HAVE MOMENTUM?

Another historic application of the law of conservation of momentum, though from twentieth-century physics, involves the interaction between a beam of negatively charged electrons and positively charged atomic nuclei. When a high-speed electron passes a nearby nucleus, the mutual forces of electrical attraction can deflect the former strongly from its path in exactly the same manner

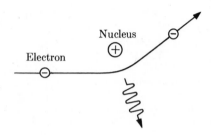

**Fig. 16.4** Radiation of energy from an electron passing near a nucleus.

as the sun affects a passing comet (Fig. 16.4). It would be expected that the change of momentum of the electron, $\Delta p_{\text{el}}$, is compensated exactly by a corresponding and opposite $\Delta p_{\text{nucleus}}$. On checking experimentally, we find that the latter change is too small to account for $\Delta p_{\text{el}}$. Is this not evidence that here at last the treasured law breaks down?

We should be loath to accept this position without looking first for an explanation elsewhere. We remember well that the irregularities in the motion of the planet Uranus also offered such an alternative: either to consider a failure of the old law (of universal gravitation), or else to search for an explanation of the puzzling behavior by a bold hypothesis beyond the immediately observed facts; and it was the latter course which triumphantly proved to be correct and led to the discovery of Neptune.

Here also: Does nothing else happen to carry away some momentum? During the deflection of the electron a flash of light is released in the general direction indicated, although of a frequency so far beyond the visible that instruments other than the eye are needed to detect it. If we could say that light has momentum, and assign it the missing quantity, all would be well and the law saved. But surely that is not so easily permissible! How can one conceive of momentum, *mass* times *velocity*, where there is nothing but disembodied wave motion, radiant energy?

While some of the detailed arguments cannot be given here and others must be delayed to Part H of this text, the main facts are these:

i)  One can no longer afford to reject any concept (such as the idea that light waves have momentum) just because it is not directly obvious to our mechanics-oriented thoughts.  It is a dangerous limitation to insist on physical pictures.  Nature has too often in the past forced us to accept what, for a time, could not possibly be imagined.  After all, there was a time when it was nearly universally considered ridiculous and obviously absurd to propose that the earth might be considered to move or that the moon has mountains.  Thus although one might, instead of the phrase "light has momentum," prefer to say "momentum is somehow *associated* with light waves," one must indeed say so if there is inescapable evidence that this is true.

ii) Apart from the experiment with high-speed electrons as just described, there was other evidence that light exerts pressure and can produce changes of momentum, most spectacularly the observation that the material in the tail of comets always points away from the sun at every point along the comet's path.  The suggestion that it was the sun's light which pushed the comet's tail was first made by none other than—again—Kepler.  Modern and most delicate measurements around 1900 showed clearly that light waves bring with them momentum, although the quantities are ordinarily small; a good beam can do no more than keep up the rotation of a little paddlewheel mounted in a vacuum with as little friction as possible.  Moreover, and most importantly, the amount of momentum in a given light beam can be predicted from simple assumptions in the theory of light, and does indeed coincide with these measured values to better than 1%, despite the serious difficulties in experimentation.  (More recently, it has been concluded that the greater part of the force acting to push a comet's tail away from the sun is due not to light pressure but to the *solar wind*, consisting of particles such as protons.)

**Fig. 16.5**  Compton effect. The x-ray proceeds with reduced energy and longer wavelength after scattering from a recoiling electron.

iii) The concept that a flash of light waves has associated with it a momentum became of importance subsequently in several other independent contexts; for example, in 1923 the American physicist A. H. Compton discovered that a beam of x-rays (also a form of very high-frequency light) can collide with an electron, and bounce off with slightly changed characteristics in another direction while the electron "recoils" (Fig. 16.5) as though it were one billiard ball hit by another. Compton found that the observable momentum of the recoil electron was numerically just equal to the change of momentum of the x-ray, as calculated by the same relation needed to fulfill the conservation law in Fig. 16.4.*  Once more the conservation law was borne out, and at the same time the nature of light itself became better understood.

### RECOMMENDED FOR FURTHER READING

R. Feynman, *The Character of Physical Law*, London: British Broadcasting Corporation, 1965; paperback reprint by MIT Press; Chapter 4

---

*  For this important work, which in turn had far-reaching consequences throughout modern physics, Compton was awarded a Nobel Prize (1927).

## SOURCES, INTERPRETATIONS, AND REFERENCE WORKS

R. J. Blackwell, "Descartes' laws of motion," *Isis*, Vol. 57, pages 220–234 (1966)

René Descartes, extract from *Principia Philosophiae* (1644) on the laws of motion, translated in M. B. Hall, *Nature and Nature's Laws*, pages 264–270

R. Dugas, *Mechanics in the Seventeenth Century*, New York: Central Book Company, 1958; Chapters VII and X

A. R. Hall, "Mechanics and the Royal Society, 1668–70," *British Journal for the History of Science*, Vol. 3, pages 24–38 (1966)

Ernst Mach, *The Science of Mechanics*, translated by T. J. McCormack from German, La Salle, Illinois: Open Court Publishing Co., sixth edition, 1960; paperback, pages 376–420

# Chapter 17

# THE LAW OF
CONSERVATION
OF ENERGY

## 17.1 CHRISTIAN HUYGENS AND THE *VIS VIVA* CONCEPT

The so-called fundamental quantities of mechanics are usually taken to be length, time, and mass. So far, we have used combinations of these three to set up derived concepts: velocity, acceleration, force, and momentum. And each of these has found a prominent role in some fundamental law of nature. Now we are about to introduce the last of the great mechanical concepts, *energy*, together with the law over which it reigns, the mighty principle of conservation of energy.

The full historical development of the concept and the law was lengthy, and more involved than usual; it took more than 150 years from the first attempts at quantitative formulation to the point at which even the basic terminology itself was fully worked out. Yet the roots lie in the very same problem that gave rise to the concept of momentum and the law of conservation of momentum, namely, the problem of how the motion of bodies changes when they collide with one another. And significantly, in introducing a first major contribution, we encounter the same man who helped formulate the law of conservation of momentum in elastic impact: Christian Huygens (1629–1695), the Dutch physicist, of whom it has often been said that he was in many respects the peer of Galileo and Newton.

**Fig. 17.1** Christian Huygens (1629–1695).

245

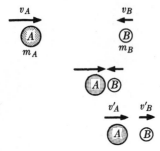

**Fig. 17.2**  Collision between two moving spheres.

Huygens was not only an outstanding mathematician (Newton referred to him as one of the three "greatest geometers of our time"), he also made an improved telescope, discovered a satellite of Saturn, and, building on Galileo's work, invented the first practical pendulum clock. Among his great accomplishments in physics are the first theorems of centripetal force, the conservation laws in elastic impact; the theory of oscillating systems; and a treatise which laid the foundations for the wave theory of light.

As we remarked at the end of Section 16.2, Huygens proposed as part of his solution of the collision problem, in 1669, the rule that a quantity equal to the product of the mass and the square of the speed of each moving body, summed over all bodies in a system, remains the same after a perfectly elastic collision as before the collision. The quantity $mv^2$ was later given the name *vis viva* and was used as the basis for mechanical theories by the German philosopher and scientist Gottfried Wilhelm Leibniz, and in other works by Huygens published around 1700 (especially in his posthumous treatise, *De motu corporum ex percussione*, 1703).

We may discuss this postulate in terms of a concrete but general example, as in Fig. 17.2. Two unequal but perfectly rigid spheres approach each other with unequal speed, collide, and sep-

arate.  To be sure, from our previous discussion we know well that the law of conservation of momentum demands that

$$m_A v_A + m_B v_B = m_A v_A' + m_B v_B', \qquad (17.1)$$

but this is not the issue any longer.  What Huygens is proposing here is that there is fulfilled at the same time *another* equation, namely,

$$m_A v_A^2 + m_B v_B^2 = m_A (v_A')^2 + m_B (v_B')^2. \qquad (17.2)$$

This is really a very astonishing proposal!  Let there be no mistake about it, Eq. (17.1) does not by itself lead to Eq. (17.2).  In the latter case the squares of the speeds are involved, not the velocity vectors themselves.  Also, Eq. (17.2) is a scalar equation; that is, the directions of motion do not enter at all.  Consequently, it holds true in this form even if body $A$ were to overtake body $B$, or if it be overtaken by $B$, or if $A$ and $B$ meet at any angle whatever.  A couple of examples will illustrate the usefulness of the new law.

**Example 1.**  Recall Example 1 in Chapter 16.  A ball $A$ of mass $m_A$ was shot at velocity $v_A$ toward a ball $B$ of equal mass ($m_B = m_A$) resting on a table ($v_B = 0$).  We inquired: $v_B' = ?$  With the LCMm alone, this could not be solved unless we added to the four data a fifth, namely, $v_A' = 0$, the reason being that the LCMm alone gives only one equation [Eq. (16.10)] with *six* quantities:

$$m_A v_A + m_B v_B = m_A v_A' + m_B v_B'. \qquad (16.10)$$

But if we also call on Eq. (17.2),

$$m_A v_A^2 + m_B v_B^2 = m_A (v_A')^2 + m_B (v_B')^2, \qquad (17.2)$$

we have two equations in six quantities; therefore we need only four data to solve for any two unknowns.  For example, using what is given above concerning the two masses and the initial velocities, we may calculate directly the values of $(v_A')^2$ and $(v_B')^2$:

From Eq. (16.10):  $v_A + 0 = v_A' + v_B'$

From Eq. (17.2):  $v_A^2 + 0 = (v_A')^2 + (v_B')^2$  $\qquad (17.3)$

Squaring the first and subtracting yields

$$0 = 2v'_A v'_B.$$

Therefore either $v'_A$ or $v'_B$ is zero, and the other one of the two final velocities, by Eq. (17.3), is equal to $v_A$. But if ball $B$ is solid, $v'_B$ cannot be zero while ball $A$ passes forward with $v'_A = v_A$. Consequently, the other alternative is the correct one, namely, $v'_A = 0$, $v'_B = v_A$. Now this is a much more satisfactory solution, because we can calculate *both* final velocities exclusively from data that refer only to the initial conditions before impact.

This applies also in general for any problem involving direct collision of two perfectly elastic objects. With the aid of the LCMm and Eq. (17.2), we can find the final velocities of both bodies from the initial conditions without further assumptions or data.

---

**Problem 17.1.** In the posthumous edition of Huygens' works (1703) we find this "Axiom": "If two equal bodies [bodies of equal mass] moving in opposite directions with equal velocities meet each other directly [in perfectly elastic collision] each of them rebounds with the same velocity with which it came." Show that this axiom is in accord (a) with the LCMm, (b) with the principle of conservation of *vis viva*.

---

**Example 2.** In a later chapter we shall need to use the answer to the following question: A very light object strikes a very heavy object, the latter being initially at rest. What are the velocities of both objects after the collision, given the initial velocity of the light object? (The collision is still assumed to be perfectly elastic.)

For the sake of concreteness, let us imagine that the light object is a rubber ball of mass $m_A$ and initial velocity $v_A$, and the heavy object is a large truck of mass $m_B$; the ball is thrown horizontally against the flat (vertical) back door of the truck, which is initially at rest ($v_B = 0$). If we

**Figure 17.3**

ignore the force of gravity, there will be no reason for the ball to deviate in any direction from the line of its original path; it can only bounce back with some velocity $v'_A$ which we have to calculate. Hence this is just a one-dimensional problem.

You can probably guess the correct answer, but it is instructive to see how it emerges from the two conservation laws. To start with, we have to define what is meant by "very heavy" and "very light." Clearly we must have in mind a limiting case $m_B/m_A \to \infty$, but if we make that substitution immediately the equations contain indeterminate quantities and cannot be solved. So let us be patient and begin by putting the ratio $m_B/m_A = n$, solve for a finite $n$, and then take the limit $n \to \infty$ at the end.

The same procedure as that used above now gives, as a result of substituting into Eqs. (16.10) and (17.2), and dividing through by $m_A$:

$$v_A + 0 = v'_A + nv'_B, \qquad (17.4a)$$

$$v_A^2 + 0 = (v'_A)^2 + n(v'_B)^2 \qquad (17.4b)$$

Squaring the first and subtracting yields

$$0 = 2v'_A v'_B + (n - 1)(v'_B)^2.$$

As in the previous case, one solution of this equation is $v'_B = 0$, which must be rejected for the same reason as before, although, as we shall see shortly, the correct solution actually converges to that value in the limit $n \to \infty$. But for finite $n$ we must choose the other solution, obtained by dividing through by $v'_B$,

$$2v'_A + (n - 1)v'_B = 0$$

or

$$v'_A = -\tfrac{1}{2}(n - 1)v'_B.$$

We can then eliminate $v'_B$ by substituting its value given in terms of $v_A$ and $v'_A$ in Eq. (17.4a), namely, $v'_B = (v_A - v'_A)/n$. Solving for $v'_A$, we find:

$$v'_A = -\left(\frac{n - 1}{n + 1}\right) v_A. \tag{17.5}$$

At last we can take the limit $n \to \infty$ and obtain the desired answer,

$$v'_A = -v_A.$$

The ball simply bounces back with the same speed in the opposite direction! To complete the solution, we note that (by solving for $v'_B$ from the above equations) the truck is still at rest after the collision.

---

**Problem 17.2.** Using the result of Example 2, show that: (a) If the truck were initially moving away from the ball, the ball would bounce back with a smaller speed. (b) If the truck were initially moving toward the ball, the ball would bounce back with a greater speed. [*Hint:* Redefine the velocities as relative to the truck's initial velocity $v_B$, for example, let $\tilde{v}_A = v_A - v_B$. You are now in a coordinate system in which the truck is at rest. After solving the problem in that coordinate system, translate the results into the original coordinate system and interpret them.)

---

The usefulness of the principle of conservation of *vis viva*, as demonstrated, is considerable, but *unlike* the LCMm *it is strictly limited to impacts between perfectly elastic bodies.* In truth, the matter is even more serious, for it is in general not possible to judge before the collision takes place whether it will be sufficiently close to a perfectly elastic one to admit using the principle; and so we often decide whether the principle applied only after a tentative calculation has shown whether the theoretical results are reasonable. This *a*

*posteriori* procedure is rather like diagnosing whether a sick man has malaria by seeing whether a malaria cure will put him on his feet.

Evidently what is needed is an extension of the principle of conservation of *vis viva* in some manner so that the extended principle will apply to all types of interactions, just as the LCMm applies to all interactions in a closed system. This was brilliantly accomplished in the nineteenth century and, leaving the main details of the historical development for later, we turn first to an account of the results.

## 17.2 PRELIMINARY QUESTIONS: THE PILE DRIVER

Consider a body in motion, perhaps a block sliding on a horizontal plane. If you tried to stop it quickly by interposing your hand in its path, you would feel that the block exerts a force on you, perhaps a considerable one. Is this some new force we have not yet discussed? Do bodies exert force on obstacles *by virtue of being in motion*, as one might say springs exhibit forces by virtue of being compressed, and magnets by virtue of having inherent magnetism?

First of all, note that this common-sensical and picturesque way of speech, which mirrors the phraseology of the seventeenth and early eighteenth century, explains little and can lead to serious semantic confusions. In the last analysis, as Newton was aware, it only serves to explain magnetism by magnetism, elasticity by elasticity, and so forth. Therefore, instead of speaking of an inherent force, we should rather seek the relationship between the force under examination and its observable effects. If we look in this light at the force which moving objects exert on an obstacle like our hand, we see at once that this is not some new "living force," as was sometimes thought, but rather that our hand experiences the force of reaction to the decelerating or braking force which, by Newton's second law, it must apply to the moving object to change its state of motion. This force of deceleration, and also the reaction to it, are of course both numerically given by the product of the mass $m$ of the object and the deceleration $a$. And that is strictly all there is to the original problem.

But it is very revealing to embellish this theme. For example, it has always been of interest to measure with some accuracy the speed of a moving object. Can this not be done by measuring the force needed to stop it? Galileo put the problem this way: Consider a falling body; after it has gathered some speed, let it fall upon a vertical stick or stake lightly impaled in some yielding sand (an arrangement that corresponds to the so-called pile driver). Then:

"It is clear that if one lifts this body a cubit or two and allows it to fall upon the same material, it will, with this impulse, exert a new and greater pressure than that caused by its mere weight; and this effect is brought about by the [weight of the] falling body together with the velocity acquired during the fall, an effect which will be greater and greater according to the height of the fall, that is, according as the velocity of the falling body becomes greater. From the quality and intensity of the blow we are thus enabled accurately to estimate the speed of a falling body. But tell me, gentlemen, is it not true that if a block be allowed to fall upon a stake from a height of four cubits and drives it into the earth, say, four fingerbreadths, that coming from a height of two cubits it will drive the stake a much less distance, and from the height of one cubit a still less distance; . . ." (*Two New Sciences*, Third Day)

Indeed, are we not thus enabled accurately to estimate the speed of the falling body? Not so simply. By stopping the body, we change its momentum by $mv$ (numerically). From Newton's laws,

$$mv = F_1 t, \quad \text{and} \quad v = \frac{F_1 t}{m},$$

**Fig. 17.4**  Pile driver.

where $F_1$ is the (average) net force which decelerates the body (and numerically also gives the force of reaction that in turn pushes on the stake), $t$ is the time during which the stake is in motion, and $m$ is the mass of the falling body. Unless we know the factor $t$ (hardly observable with ease) the force $F_1$ on the stake gives little information as to $v$. If the sand is tightly packed, $t$ will be very small, $F_1$ very large; if the soil is loose, $t$ is greater, and $F_1$ is smaller for the same value of $v$ and $m$.

But now, what can we learn about $v$ from the distance $s$ through which the stake moves while decelerating the pile driver from $v_1$ to $v_2 = 0$ (Fig. 17.4)? Let us assume that the retarding force $F_1$ on the falling mass was constant during the collision. Then, since

$$F_1 = ma,$$

$$F_1 \cdot s = mas.$$

But, from Eq. (1.5),

$$as = \frac{v_2^2 - v_1^2}{2},$$

therefore

$$F_1 \cdot s = \tfrac{1}{2}mv_2^2 - \tfrac{1}{2}mv_1^2.$$

Since $v_2 = 0$,

$$F_1 \cdot s = \tfrac{1}{2}mv_1^2 \text{ (numerically)}.$$

It appears now that for a given mass $m$, and for an obstacle that always offers the same retarding force (a condition nearly enough fulfilled by always using the same stake in the same soil), the distance through which the stake is moved is proportional to the *vis viva*, that is, to the *square* of the speed of the moving body! Using Galileo's example, and assuming that 1 cubit $\doteq$ 0.50 m and 1 finger-breadth $\doteq$ 0.02 m, we should rephrase it thus: If a block be allowed to fall on a stake from a height of 2 m (with a final velocity $v$ of $\sqrt{2gh} \doteq$ 6.2 m/sec), and drives the stake into the earth a distance $s$ of, say, 0.08 m, the same mass coming from a height of 1 m (with a final velocity $v'$ of 4.4 m/sec) will drive the same stake a smaller distance $s'$, namely (in the ideal case),

$$\frac{s'}{s} = \left(\frac{v'}{v}\right)^2, \quad \text{or} \quad s' = 0.08 \times \left(\frac{4.4}{6.2}\right)^2 \text{ m} = 0.04 \text{ m}.$$

It is only by such a calculation, and under such assumptions, that Galileo's pile driver permits a comparison of relative speeds.

## 17.3 THE CONCEPT OF WORK

The previous example, to which we shall return, has been chosen for several reasons, not the least being that it introduces, in the historical context, two concepts: *work* and *energy*. The definition of the first of these follows.

When a force $F_1$ acts on a body during a displacement $s$, then we say that *the product $F_1 \times s \times \cos \theta$ is the work done by $F_1$ on the body*, where $\theta$ is the angle between the directions of the two vectors $F_1$ and $s$. Thus in the previous discussion, the work done by the force $F_1$, the action of the pile driver on the stake, was $F_1 \times s$, $\cos \theta$ being unity in that case (why?). At the same time, the work done by the stake on the falling mass was $F_1 \times s \times (-1)$ or $-F_1 \times s$ (because the displacement $s$ and the force $F_1$ were oppositely directed, and $\cos 180° = -1$). Or again, when a person pulls on a block in the manner indicated in Fig. 17.5,

**Fig. 17.5** The work done by force $F_1$ on the block is $F_1 s \cos \theta$.

exerting a pull $F_1$ of about 45 newtons at an angle of 30° over a distance of about 0.3 m, the work done by the person on the block is 45 newtons $\times$ 0.3 m $\times$ 0.87 = 11.7 newton-m. On the other hand, if a man carries a load on his shoulders over a distance $s$ on a horizontal plane, and supplies to it a balancing force $F_1$ (upward), the work he does is zero; for here $F_1$ and $s$ may both be large, but $\cos 90° = 0$ (Fig. 17.6). Again, the centripetal force $F_c$ on an object in circular motion does no work either, for it is directed at 90° to the path at every instant.

Note carefully that, as in the example illustrated in Fig. 17.6, one may be "working hard" in the colloquial sense without doing any work in the physical sense. Similarly, if one pushes hard on a brick wall which remains stationary, $F_1$ is large, $\cos \theta = 1$, but $s = 0$, and although it might tire us out, we have not thereby done *work* as defined in physical science. Evidently, we must not let our subjective notions of activity, achievement, or fatigue become confused with this very different use of the word "work."

Before we go into the usefulness of the concept, note that work is a *scalar* quantity, although it is a product of two vector quantities. The units of work are those of force times distance, i.e. (in the MKS system) newton-meter, for which the name *joule* is used, so named in honor of the British scientist whose contributions will figure prominently in the following pages.*

---

\* Another unit of work sometimes used is the *erg*, defined as 1 dyne-cm (see footnote on page 119 for definition of *dyne*). Note that 1 joule = $10^7$ ergs.

**Figure 17.6**

**Problem 17.3.** A boy pulls a sled with a force of 0.001 newton via a horizontal string over a distance of 1 km on level ground. (a) What work does the boy do on the sled, in joules? in ergs? (b) What is the work done on the sled by the gravitational force which the earth exerts on it?

**Problem 17.4.** In the example involving the pile driver (Fig. 17.4) assume that the distance $s$ was 0.08 m and the average force needed to stop the falling body is 10 newtons. (a) Find the work done by the pile driver on the stake. (b) Find the work done by the stake on the driver. (c) If the driver fell from a height of 4 m and had a mass of 20 kg, what work was done on it during the descent before collision by the gravitational force?

## 17.4 VARIOUS FORMS OF ENERGY

Work can be done on a body in many ways, and with many results. The stake did work on the falling pile driver, thereby bringing it to rest; conversely, the driver did work on the stake and so drove it into the ground against the forces of friction. Work is done by the hand in lifting the driver to a high position before the descent, thereby setting up the condition needed for gravity to do work on the falling object and so to let it gather speed. When a spring is compressed or extended between two hands, work is done on it. When a

flywheel is started or slowed down, forces act over certain distances, and again work is done. Evidently, we should investigate what happens here systematically and quantitatively, still by analyzing simple mechanical problems first, before we are ready to handle the full power of these concepts.

a) *Work done to overcome frictional resistance.* By far the largest expenditure of work on this earth would seem to be work done "against friction." We pull on a block horizontally with an applied force $F_{ap}$ to move it through distance $s$ on a level table. (This is just the same as the case in Fig. 17.5, except we call $F_1$ now $F_{ap}$, and $\theta = 0°$.) The work done by $F_{ap}$ on the block is $F_{ap} \cdot s$. If the force $F_{ap}$ is just enough to counteract the force of friction $f$ and no net force is left over for acceleration of the block, then the work done by $F_{ap}$, namely $F_{ap} \cdot s$, is numerically just equal to $f \cdot s$. If we now remove $F_{ap}$, the block will just sit on the table. The only reward for the work we have done on the block is the transfer of the block from one spot to another, and a little heating up where the surfaces of contact rubbed against each other. If we call this "wasted" or "dissipated" work, it is so only by comparison with other circumstances where it will turn out that some or all of the work done can, on removal of $F_{ap}$, be recovered directly.

b) *Work done to overcome the inertia of bodies.* We concentrate next on a case where friction is absent, e.g., on a block on a friction-free horizontal plane, in equilibrium at some speed $v_1$ (which might be zero). Now we apply horizontally a force $F_{ap}$ through distance $s$ to accelerate the block to speed $v_2$. The work done thereby, $F_{ap} \cdot s$, is now

$$F_{ap} \times s = mas = m \left( \frac{v_2^2 - v_1^2}{2} \right)$$

or

$$F_{ap} \times s = \tfrac{1}{2} m v_2^2 - \tfrac{1}{2} m v_1^2. \qquad (17.6)$$

It appears that the work done on it serves to change the *vis viva* of the object or, more properly, a quantity that is just half the *vis viva*, and which about a century ago received its current name, *kinetic energy* (to be abbreviated to KE). Putting Eq. (17.6) more generally, we can say that here

$$F_{\text{ap}} \cdot s = \text{change in kinetic energy} \equiv \Delta\text{KE}.$$

For example, a cyclotron is to accelerate a nuclear particle, e.g., a deuteron (the nucleus of a heavy isotope of hydrogen, made of one neutron and a positive particle, a proton, of about the same mass; total mass of deuteron $= 3.3 \times 10^{-27}$ kg). The desired final speed is $\frac{1}{10}$ of the velocity of light, namely about $3 \times 10^7$ m/sec. If it starts substantially from rest, what work must be done on the deuteron?

*Solution:* $\Delta KE = \frac{1}{2}mv_2^2 - \frac{1}{2}mv_1^2$; here $v_1 \doteq 0$; $\Delta KE = \frac{1}{2} \times 3.3 \times 10^{-27}$ kg $\times (3 \times 10^7$ m/sec$)^2 = 1.5 \times 10^{-12}$ kg-m$^2$/sec$^2 = 1.5 \times 10^{-12}$ joule.* By Eq. (17.6), this corresponds also to the work needed (in the absence of other effects such as friction) and this amount of work is actually supplied by letting an electric force act upon the accelerating particle. Even though $1.5 \times 10^{-7}$ joule may seem little work, the difficulties of supplying it are as varied as the effects obtainable with such high-speed particles!

Thus work can be "converted" into energy, and work done exclusively to increase kinetic energy is in a sense directly "recoverable," for when a high-speed object is then allowed to fall upon some target, it can in turn do work on it while giving up its kinetic energy. One may say that energy is the capacity for doing work. The pile driver is a case in point. While the driver of mass $m$ falls through a distance $h$, the constant force of gravity, $mg$, does work on it ($mg \times h$), thereby increasing its kinetic energy (from Eq. (17.6), $mgh = \frac{1}{2}mv_2^2$ if there are no frictional losses on the way down and if $v_1$, the starting velocity, is zero). The work done by gravity is, to use graphic terms,

"stored up" in the moving object, and becomes available on impact with the stake to drive it into the sand.

Although such pictures are a little too concrete, these are important concepts: work on an object can be converted into kinetic energy (or, later, other forms of energy); conversely the energy of a body may be given up in doing work on something else; and often work is done against friction and may be irrecoverably lost from the mechanical system, but then it does show up as heat.

c) *Work done to change the level of an object.* Consider this third method of doing work on and imparting energy to a body: Simply lift up a heavy object from the floor to a distance $h$ above it. The constant applied force needed here is just $mg$, enough to balance the weight, provided we can neglect air friction and also proceed so slowly that there is sensibly no increase of kinetic energy. If not to friction or to KE, where does the applied work go? It is not lost, for if we let go, the pull of gravity will accelerate the body while it falls through distance $h$ and so will build up its kinetic energy at the end of the fall to exactly the same number of joules which we had expended in lifting it. Here we may have recourse to an imaginative fiction and say that the work done in lifting the body was "stored" as *potential energy* (PE) in the general region around the body, in the gravitational *field*, and furthermore, that while the object is then allowed to drop, the potential energy is drawn or given back from that region into the body and converted into kinetic energy. So on raising the level of a mass $m$ by a distance $h$, the work we do goes into potential energy ($mgh$) and, on falling, the "work done by the gravitational force" represents only a withdrawal from a bank account of energy which had been accumulated on the way up, until on returning to the original lower level all of the energy has become kinetic.

One should avoid trying to make too material a picture of this sequence or of the seat of potential energy, and regard it rather as a useful expository

---

* 1 kg-m$^2$/sec$^2$ = 1 joule, because 1 joule $\equiv$ 1 newton-m = 1 (kg-m/sec$^2$) $\cdot$ m.

device at this stage, an aid to comprehension without independent truth value. There is really only one big difficulty in the new concept, namely the *level of reference.* If we see a book of mass $m$ on a desk in the second-floor study and are asked point-blank, What is its potential energy?, it would be of no use to reply *mgh,* for there is no obvious way to decide whence to measure $h$: from the floor? or from the street below? or from the level of the shop counter where it was originally bought? The truth is that potential energy can always be reckoned with respect to an arbitrary zero or reference level, usually, for convenience, the lowest level to which the body descends in the course of a problem or example. And the reason this can be done is, as you will notice, that we shall always deal with *differences* or *changes* in the potential energy ($\Delta$PE) between two levels. So no harm is done by calling the lowest level simply $h = 0$; on the contrary, calculations are much more simplified than if we were to reckon all potential energies from some fixed point, say from the center of the earth, or from a point high above the earth.

---

**Problem 17.5.** Reconsider the case of the ball bouncing off the back of the truck (Example 2 and Problem 17.2 in Section 17.1) from the viewpoint of work. Show that in the three successive cases, (i) when the truck is stationary no work is done by either the ball or the truck; (ii) when the truck is initially moving away from the ball, the ball does work on the truck during the collision; (iii) when the truck is initially moving toward the ball, the truck does work on the ball during the collision.

## 17.5 THE CONSERVATION LAW, FIRST FORM; AND APPLICATIONS

a) *The inclined plane.* To summarize the three ways here discussed in which a body may dispose of work done on it by an applied force—work against friction, against inertia, against gravitational forces—consider this very typical problem in

**Fig. 17.7** Inclined plane.

mechanics. A block of mass $m$ approaches the bottom of an inclined plane with a speed $v_1$ (moving from left to right) and ascends it for a running distance $s$ while being pulled by an applied force $F_{ap}$, as shown in Fig. 17.7. We are also given* the frictional force $f$, and the question is: What part of the work done by $F_{ap}$ goes toward increasing the kinetic energy of the block? From this one might find the final speed $v_2$ at the top of the incline.

*Solution:* The work done by $F_{ap}$ is $F_{ap} \cdot s \cdot \cos \theta$ (but here $\theta$, the angle between the direction of $s$ and of $F_{ap}$, is zero and $\cos \theta = 1$). This amount of work is split into three not necessarily equal parts: (i) work done against friction $f$, namely, $f \cdot s$; (ii) work done to change the kinetic energy, $(\frac{1}{2}mv_2^2 - \frac{1}{2}mv_1^2)$; and (iii) work done to change the potential energy, $mgh$. Note that it is the vertical level difference $h$ which counts in calculating the potential energy,

---

\* In a typical textbook problem one would be given the value of the coefficient of friction $\mu$ and the angle of incline $\gamma$; the frictional force is the push with which the block acts perpendicularly on the surface on which it sits (here equal to the component of the gravitational force in a direction perpendicular to the motion), multiplied by $\mu$; that is, $f = \mu mg \cos \gamma$.

not the length of path $s$.  In symbols,

$$F_{\text{ap}} \cdot s \cdot (\cos \theta) = \Delta \text{KE} + \Delta \text{PE} + fs$$

or, here

$$F_{\text{ap}} \cdot s = (\tfrac{1}{2}mv_2^2 - \tfrac{1}{2}mv_1^2) + (mgh) + fs.$$

Evidently, with $F_{\text{ap}}$, $s$, $m$, $f$ at hand, $\Delta \text{KE}$ is quickly found.  Note that $F_{\text{ap}}$ refers to the pull by way of a rope or some other externally applied force, but leaves out of account entirely the force of friction and the force of gravity on the block, for those two are accommodated on the right side of the equation.

Now this was a quite plausible argument, and an experimental check would show that Eq. (17.7) is indeed correct.  But is it self-evident, or even proved?  Neither!  The fact that there are three *obvious* ways in which a body accepts and transforms work done on it does not mean that the sum of the three terms is necessarily exactly the amount of work done.  There could be other drains on the available energy than we have thought of, or perhaps the moving body may receive energy from some source other than the applied force.  That this does not happen in simple mechanical phenomena, that Eq. (17.7) is correct, must initially be an experimental discovery, a discovery which amounts to a *law of conservation of energy*; for Eq. (17.7) says that every joule of work supplied to a body can be accounted for by corresponding changes in $\Delta \text{KE}$, $\Delta \text{PE}$, and work turned into heat by frictional losses.  More generally, imagine a closed system, say a box full of elastic balls, blocks, and inclined planes. Add to the system some energy, perhaps by dropping into the box a marble having a kinetic energy of $x$ joules. After a while we look again into the box; it will be found that the bodies by mutual collision have so rearranged their speeds and positions that the energy we added is exactly equal to the sum of $\Delta \text{PE}$, $\Delta \text{KE}$, and frictional losses since the moment of interference.

*Dissipative system* is the term applied to a system in which some of the mechanical energy is indeed dissipated irretrievably, as by friction. In contrast, *conservative systems* show no such

**Figure 17.8**

losses; to them, the law of conservation of energy would apply in simpler form, for if the conservative system were totally isolated, unable to receive energy from or give it up to the rest of the world, then $\Delta \text{KE} + \Delta \text{PE} = 0$, a specialized case of the conservation law that was first expressed with most telling power and elegance in Lagrange's *Analytical Mechanics* of 1788, but was used long before in one guise or another.

**Problem 17.6.**  Re-examine Galileo's argument in Section 17.2 again in the light of the law of conservation of energy.

**Problem 17.7.**  Two crucial, almost axiomatic propositions on which many of Galileo's arguments in mechanics were based were the following: (a) "The speeds acquired by one and the same body moving down planes of different inclinations are equal when the heights of these planes are equal . . . provided, of course, that there are no chance or outside resistances, and that the planes are hard and smooth." (b) " . . . a body which descends along any inclined [frictionless] plane and continues its motion along a plane inclined upwards will, on account of the momentum acquired, ascend to an equal height above the horizontal, so that if the descent is along $AB$ the body will be carried up the plane $BC$ as far as the horizontal line $ACD$; and this is true whether the inclinations of the planes

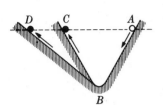

**Fig. 17.9** A ball on friction-free planes ascends to its initial level.

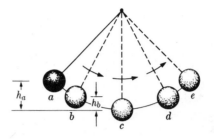

**Fig. 17.10** The pendulum.

are the same, or are different, as in the case of the planes $AB$ and $BD$."

Use the law of conservation of energy to prove each of these two propositions separately (Figs. 17.8 and 17.9).

---

b) *Pendulums*. Yet another region of application of the law of conservation of energy (to be now abbreviated LCE) relates to the varied and illuminating experiments with pendulums, favorites since the earliest days of modern mechanics. Consider an ideal pendulum, as in Fig. 17.10, released at its topmost position $a$, and swinging to the lowest position, $c$, before rising again on the other side. While it covers the left half of the arc, from $a$ to $c$, the bob undergoes a vertical displacement of $h_a$ cm, and changes its speed from $v_a = 0$ to $v_c = $ maximum. Now we apply the LCE: There is indeed a force $F_{ap}$ on the bob (it is the tension in the string) but this acts at $90°$ to the path of the bob at every instant, and therefore can do no work. If we also neglect friction, the LCE commands here

$$\Delta KE + \Delta PE = 0.$$

For any two points 1 and 2 along the path, this last equation predicts that

$$(\tfrac{1}{2}mv_2^2 - \tfrac{1}{2}mv_1^2) + (mgh_2 - mgh_1) = 0.$$

Specifically, comparing positions $a$ and $c$, and noting that $v_a = 0$, $h_c = 0$,

$$(\tfrac{1}{2}mv_c^2 - 0) + (0 - mgh_a) = 0,$$

therefore

$$\tfrac{1}{2}mv_c^2 = mgh_a \quad \text{or} \quad v_c = \sqrt{2gh_a}.$$

Here it turns out that the speed of the pendulum bob at the lowest point is exactly the same as it would be if the bob were simply to *drop* through the same level difference $h_a$. Furthermore, we recognize that the pendulum has to lift itself through the same height on each side of the swing in order to convert all its KE at position $c$ to PE at $b$ or $a$. Once started, the pendulum must incessantly go through its symmetrical motions if there is no way for its energy to escape, continually exchanging its initial energy back and forth between PE and KE.

---

**Problem 17.8.** The bob of a simple pendulum (mass = 100 g) is let go from a vertical height $h_a$ of 5 cm. (a) What will be its speed at the bottom of the swing? (b) What is its kinetic energy at that point? (c) What is its PE and KE at a point halfway up toward the top position on the other half of the arc?

Figure 17.11                                                                Figure 17.12

**Problem 17.9.** One of Galileo's famous arguments involved an arrangement now called Galileo's pendulum. First let a simple pendulum swing freely back and forth between points $B$ and $C$ (Fig. 17.11). Now put a nail or a peg in position $D$, so that the pendulum, released again from $B$, cannot reach $C$; the string is bent at $D$ and the ball or bob rises to $E$ before returning. The significant thing is that $E$ lies on the horizontal line $BC$, even though the pendulum on the right half of its trip now swings through the smaller arc $FE$ rather than $FC$. As Galileo has Salviati say in the *Two New Sciences*: "Now, gentlemen, you will observe with pleasure that the ball swings to the point $E$ on the horizontal, and you would see the same thing happen if the obstacle were placed at some lower point, say at $G$, about which the ball would describe arc $FH$, the rise of the ball always terminating exactly on the line $BC$." Explain these observations by means of the LCE.

**Problem 17.10.** In Huygens' *Horologium Oscillatorium* (1673), we find this proposition: "If a simple pendulum swings with its greatest lateral oscillation, that is, if it descends through the whole quadrant of a circle [i.e., starts, as in Fig. 17.12, from a horizontal position], when it comes to the lowest point in the circumference, it stretches the string with three times as great a force as it would

if it were simply suspended by it." Prove this with the aid of the LCE.

**Problem 17.11.** The measurement of the velocity of a fast projectile has always been difficult. One early solution was found in the so-called "ballistic pendulum" of Benjamin Robins (1742). The bullet of known mass $m_b$ and unknown velocity $v_b$ is fired horizontally toward, and embeds itself in, a bob (mass $m_p$) of a freely hanging simple pendulum. The maximum height $h$ to which the pendulum rises after the inelastic collision with the bullet is noted. (a) With the aid of the LCMm, derive the equation for the initial speed of the pendulum after impact. (b) Derive the expression for $v_b$ before impact in terms of $m_b$, $m_p$, $g$, and $h$, using the LCE for analyzing the motion of the pendulum.

c) *Collisions.* It is surely now evident that the law of conservation of *vis viva*, the original stimulus for the more general LCE, is only a special case of the latter. For if a set of spheres has a perfectly elastic collision, then *by the very definition* of the phrase "perfectly elastic" we mean that none of the kinetic energy of the colliding bodies is lost to frictional processes, i.e., that the system of bodies is a conservative one. For such collisions, either in a horizontal plane or, more generally, of such brief duration that the change in level is insignificant during the time interval, the LCE, which commands that

$$\Delta KE + \Delta PE + \text{work lost to friction} = 0,$$

becomes simply

$$\Delta KE = 0.$$

For several colliding bodies, this means

$$[\tfrac{1}{2}m_A(v'_A)^2 - \tfrac{1}{2}m_A v_A^2] + [\tfrac{1}{2}m_B(v'_B)^2 - \tfrac{1}{2}m_B v_B^2]$$
$$+ \cdots = 0.$$

Comparison with Eq. (17.2) shows that this is exactly the same as the law of conservation of *vis viva* (the factor $\tfrac{1}{2}$ may, of course, here be canceled throughout). Furthermore, we now have in principle a way to calculate what happens in all collisions, including those which are not elastic and in which *vis viva* (or kinetic energy alone) is not conserved. While it is, in general, not easy to judge how much energy is lost to friction, the case is not at all hopeless; one type of solution will suggest itself in connection with the following problem.

**Problem 17.12.** A certain ball is generally observed to bounce back from the floor to $\tfrac{1}{3}$ of the height of release after free fall. If it has a mass of 5 g and is thrown vertically downward from a height of 100 m with an initial speed of 50 m/sec, where will the ball be after 2.5 sec? What will be its KE and PE at that moment?

**Problem 17.13.** A widely used demonstration involves a series of equally heavy white balls arranged in one line on a smooth table (Fig. 17.13). Now a black ball (like the others, a perfectly elastic one) is shot against them at speed $v$. If its mass is equal to that of each of the others, the last white ball flies off with the same speed $v$. But if its mass were twice as large, we would never see

Figure 17.13

*one* white ball fly off at a greater speed, but all white balls would move at different speeds. Show that this satisfies both the LCMm and the LCE. If necessary, see Klein, *American Journal of Physics*, Vol. 28, page 102 (1960).

## 17.6 EXTENSIONS OF THE CONSERVATION LAW

For the sake of simplicity, we have so far avoided several sets of important problems to which the LCE in its present form would not apply. This does not at all mean that the LCE has only a limited validity! Rather, it means that we must extend the meaning of PE and KE a little to cover some other ways in which energy may appear or be stored. Two cases frequently encountered in elementary mechanical problems are potential energy of elastic bodies, and kinetic energy of rotation. We summarize here, for future reference, the relevant formulas for these two cases, though detailed discussion will not be necessary.

a) *Potential energy of elastic bodies.* Work or KE is transformed into PE when it serves to lift a massive object through some distance. But besides gravitational PE there are other forms of PE— for example, *elastic* PE. If a force acts on a spring (as in Fig. 17.14) to compress it for a distance $x$,

Figure 17.14

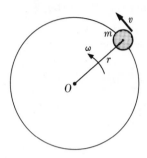

Figure 17.15

the total work done on the spring is the average force $F_{av}$ during the compression times the displacement $x$. This quantity $F_{av}x$ can be called potential energy stored in the spring, for as soon as the external applied force is removed, the spring will snap back and make available (to a machine, a clock, or in some other way) all the energy which has been given to it in the initial compression.

The amount of energy stored in the spring can be calculated by using *Hooke's law* (named for the seventeenth-century scientist Robert Hooke): Within reasonable limits, the force needed to compress a spring is proportional to the amount of compression. Since the applied force $F_e$ for a spring therefore changes linearly from $F_e = 0$ at the start (when $x = 0$) to $F_e =$ (a constant $k$) $\cdot x$ at the end, the *average* force is $F_{av} = \frac{1}{2}kx$, and the work done is $(\frac{1}{2}kx) \cdot x = \frac{1}{2}kx^2$. (The mathematical reasoning used here is similar to that employed in proving the Merton mean-speed theorem in Section 6.6.) Thus the potential energy of a spring compressed a distance $x$ is

$$\text{PE} = \tfrac{1}{2}kx^2, \qquad (17.8)$$

where $k$ is called the force constant of the spring.

The concept of elastic potential energy is also needed to understand how energy can be conserved in perfectly elastic collisions. In particular, for a head-on collision of identical bodies of mass $m$ and velocities $v_1$ and $-v_1$, the total kinetic energy of both bodies before the collision is $2 \cdot (\frac{1}{2}mv_1^2)$, and if the collision is perfectly elastic, the total kinetic

energy is the same after the collision. But at the instant of collision, when both bodies have momentarily stopped and are about to rebound, the kinetic energy is zero. If energy is to be conserved always, it must at that instant be temporarily stored as elastic potential energy. This would imply that the two bodies are compressed at the instant of collision, and will then expand, pushing each other off in opposite directions—as may indeed be observed by taking high-speed photographs of such events.

**Problem 17.14.** If an *atom* were to be defined as a particle which cannot be divided into smaller parts or change its dimensions, what would happen if two atoms approached head-on? What are the consequences of your answer for the cosmology of Descartes and the idea of the "world machine"?

**Problem 17.15.** Suppose that a cylindrical hole could be drilled straight from the North Pole to the South Pole of the earth, and assume that the earth is a perfectly uniform sphere. Describe the motion of a bullet dropped into the hole. (See Problem 11.21.)

b) *Kinetic energy of rotation.* A particle of mass $m$ rotating about a center $O$ at angular speed $\omega$ (Fig. 17.15), has at each instant kinetic energy $\frac{1}{2}mv^2$, where $v$ is its instantaneous linear speed;

according to Eq. (10.6), the kinetic energy may also be written

$$KE_{rot} = \tfrac{1}{2}mr^2\omega^2. \qquad (17.9)$$

If we define a new kind of inertia, the *rotational inertia I* of the particle, by the equation

$$I = mr^2, \qquad (17.10)$$

we can rewrite Eq. (17.9) in a form analogous to the formula for ordinary kinetic energy,

$$KE_{rot} = \tfrac{1}{2}I\omega^2. \qquad (17.11)$$

(Here $I$ plays the same role as the inertial mass, and $\omega$ the same role as the speed $v$ in the formula $KE = \tfrac{1}{2}mv^2$.)*

Equation (17.11) can be made to apply not only to a rotating particle, but to any rotating object whatever, if we realize that the rotational inertia $I$ will depend in each case on the shape and mass of the object (e.g., for a *solid disk* of radius $R$ and mass $m$ it can be calculated to be $I = \tfrac{1}{2}mR^2$). So in general the total kinetic energy of any object can be written as the translational kinetic energy of the center of mass of the object, plus the kinetic energy of rotation around its center:

$$KE_{total} = KE_{trans} + KE_{rot}. \qquad (17.12)$$

**Example 1.** A disk of mass $m$ and radius $R$ begins to roll freely from the top of an incline to the bottom, i.e., to a level of $h$ m below. What is its final speed, if rolling friction may be neglected?

*Solution:* We set up the LCE in expanded form: $F_{ap} = 0$, $v_1$ and $\omega_1$ at top $= 0$, $f = 0$. Here,

$$\Delta PE + \Delta KE_{trans} + \Delta KE_{rot} = 0,$$

$$(0 - mgh) + (\tfrac{1}{2}mv^2 - 0) + (\tfrac{1}{2}I\omega^2 - 0) = 0.$$

But $\omega$, the angular speed at the bottom, is given by $\omega = v/R$, and $I_{disk} = \tfrac{1}{2}mR^2$. Substituting and

**Figure 17.16**

transposing,

$$mgh = \tfrac{1}{2}mv^2 + \tfrac{1}{2}(\tfrac{1}{2}mR^2)\frac{v^2}{R^2},$$

$$gh = \tfrac{1}{2}v^2 + \tfrac{1}{4}v^2, \qquad v = \sqrt{\tfrac{4}{3}gh}.$$

We note, perhaps with initial astonishment, that the final translational (forward) speed $v$ of any disk is independent of its mass and radius. All disks attain the same final speed for the same descent, and $v$ for any rolling disk is less (by $\sqrt{0.07gh}$) than that for an object sliding down without friction.†

---

**Problem 17.16.** A hoop and a disk of equal radius $r$ and equal mass $m$ are released side by side along an incline. Soon the disk has outdistanced the hoop (Fig. 17.16). (a) Calculate and compare the respective final velocities. (b) Where is the hoop at the instant that the disk has reached the bottom?

**Problem 17.17.** Give a physical explanation why a sliding body reaches the bottom of an incline before rolling bodies do.

**Problem 17.18.** Discuss why the following statement of Galileo is wrong, and how it should be corrected: "All resistance and opposition having been removed, my reason tells me at once that a heavy and perfectly round ball descending along the lines, $CA$ [and] $CB$ [a vertical free fall] would reach the terminal points $A$ [and] $B$ with equal momenta." (Fig. 17.17.)

---

\*     To pursue the analogy further we could define the *angular momentum* as $I\omega$ and find a conservation law for this quantity; but this will have to be left to more advanced texts.

---

†     $\sqrt{2gh} - \sqrt{\tfrac{4}{3}gh} = \sqrt{0.07gh}.$

**Figure 17.17**

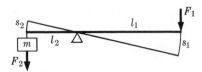

**Fig. 17.18**   The lever.

**Problem 17.19.** By experimenting with *rolling* spheres on smooth planes, Galileo could indeed confirm that the acceleration is constant (see Section 7.4), but not having the concept of rotational inertia, he could not from such data calculate the value of acceleration due to gravity ($g$) on freely falling bodies. Explain why.

Apart from the elastic potential energy and kinetic energy of rotation, there are a number of other physical processes which can add terms to the energy equation. For example, if some of the bodies are electrically charged, then there is a change in electrical potential energy as they approach or recede from one another, and this has to be included in $\Delta PE_{total}$. Or if chemical energies are set free in a closed system, they must be accounted for. Heat, too, as will be seen at once, is to be counted as a form of energy, so that heat supplied to a system, or generated within by the transformation of other forms of energy to heat, must appear in our equations. So also must sound, and light, and other radiations which carry with them energy.

Although, the complete story would seem to produce a very bulky equation for the LCE, something like this:

External work done on, or other energy supplied to, a system of bodies = $\sum$ [$\Delta$(PE of all types) + $\Delta$(KE of all types) + frictional losses + $\Delta$(chemical energies)...].

Although the complete LCE looks formidable indeed, in actual practice one encounters almost always only such cases as we have seen already, namely, cases in which only two or three of the many types of energy changes are of importance. But every time we make use of it, we are struck by the generality of meaning of the LCE. First of all, energy cannot be created or destroyed without a corresponding opposite change happening somewhere else in the universe, (e.g., the kinetic energy we impart to a cart or a baseball comes ultimately from a diminution of chemical energy within our muscles). Second, energy in any one form can be transformed into energy of some other form (a magnet pulling an iron nail toward it shows how magnetic potential energy is changed into mechanical kinetic energy).

*Application to machines.* The LCE can be applied in very simple form to machines, which are all essentially devices for transferring (more or less completely) the energy represented by work done upon one part of the system to some other part of the system, and making it available there is some convenient way. As one example, consider the lever in Fig. 17.18. The system here consists of a straight bar supported on a pivot or *fulcrum* that divides its length in the proportion $l_1 : l_2$. A load $F_2$ ($= mg$) on the left is kept balanced by an applied force $F_1$ on the right. To lift the load, one need increase $F_1$ only infinitely little, provided the friction at the fulcrum is negligible. While $F_1$

depresses one end through a small arc $s_1$, the other rises through arc $s_2$. If the angle covered is very small, $s_1$ and $s_2$ are sufficiently close to being straight lines and perpendicular to the horizontal, so that the work done on the system by $F_1$ is $F_1s_1$. This work serves to change the KE and PE of load and lever. But if, now, the speed of operation is kept very small, $\Delta$KE $= 0$, and if also the lever is relatively light, the only significant job that $F_1s_1$ can do is to increase the PE of the load ($\Delta$PE $= F_2 \times s_2 = mgs_2$). In the ideal case,

$$F_1s_1 = F_2s_2, \quad \text{or} \quad F_1 = F_2 \frac{s_2}{s_1}.$$

By similar triangles from Fig. 17.18, note that $s_2 : s_1 = l_2 : l_1$. On substituting,

$$F_1 = F_2 \frac{l_2}{l_1} \quad \text{or} \quad F_1 : F_2 = l_2 : l_1. \quad (17.13)$$

In words: The force needed to balance or slowly lift a load by means of a simple lever is to the load as the inverse ratio of the respective lever arms $l_1$ and $l_2$. This is the famous law of the lever, known empirically throughout antiquity, and derivable also from other axioms, as is best illustrated by the famous work of Archimedes of Syracuse (287–212 B.C.). The problem was also discussed by Aristotle, Leonardo da Vinci, Galileo, and many others.

---

**Problem 17.20.** Calculate first how much the earth might weigh, then calculate the dimensions and sketch the position of a lever that would have enabled Archimedes to carry out his thought of lifting the earth from its place in our solar system if he were given a long enough lever (and a fulcrum).

---

An important principle from another branch of mechanics is hidden behind Eq. (17.13). If we write it as

$$F_1l_1 = F_2l_2,$$

**Figure 17.19**

then the meaning is: For equilibrium of the lever, the torque on the right-hand member (tending to produce a clockwise motion) must be balanced by the torque on the left side (which alone would give a counterclockwise motion). More generally, one can say that a body achieves rotational equilibrium, i.e., has no angular acceleration, only if all clockwise torques balance all counterclockwise torques. This is the principle of equilibrium as regards rotational motion; it supplements the principle of equilibrium for translational motion, namely, Newton's first law, which commands that the forces themselves must balance if a body wishes to avoid linear acceleration. The latter is called the *first principle of statics*; the former is the *second principle of statics*.

---

**Problem 17.21.** In Fig. 17.19, a weightless bar is acted on by two equal forces. Show that although the forces cancel and therefore the first principle of statics (principle of translational equilibrium) is obeyed, the second principle of statics is not; that therefore the object will spin with increasing angular velocity. Then suggest where to apply additional forces, and of what magnitude, to fulfill *both* conditions of equilibrium.

---

Like the lever, the pulley is a simple machine well known in ancient times. A more useful and complex version, the simple block and tackle (Fig. 17.20), was considered by Aristotle and analyzed by Archimedes. As our hand pulls a distance $s_1$ on the free end of the string, the lower pulley and the load attached to it are raised through a

**Fig. 17.20**  Lifting a load with a block and tackle.

**Fig. 17.21**  Schematic representation of a machine.

distance $s_2$. Making the same assumptions as in the previous case, the work supplied = $\Delta$PE of load, or

$$F_1 s_1 = F_2 s_2$$

and

$$F_1 = F_2 \frac{s_2}{s_1}.$$

By measurement, we should find that in this particular pulley system, $s_2$ is always $\frac{1}{2}s_1$ (can you see why?), but in more elaborate arrangements the ratio $s_2/s_1$ may be much smaller. In every case, however, the last equation allows us to predict on the basis of an experimental determination of $s_2/s_1$ what force is ideally needed to lift a given load, even though we may be entirely mystified by the details of construction and operation of pulley systems!

We may carry this one step further and generalize on all simple mechanical arrangements in the following way. Such machines are equivalent to a black box (Fig. 17.21) to which we supply work through one opening and get work done on some load by a connection to a second opening. We may

know or understand nothing about the gears inside. Yet the LCE demands that the following be always true:

Energy supplied to machine (INPUT)
   = energy supplied by machine to load (OUTPUT)
      + energy kept by machine (its own $\Delta$KE + $\Delta$PE + friction losses).

Evidently, a good machine generally is one that does not keep much energy from the load, particularly by losses to friction. If we define the *efficiency* of a machine (symbol $\eta$, Greek letter *eta*) as the ratio of *output* to *input*, the ideal machine has $\eta = 1.00$ (or 100%), although in actual practice we may find $\eta = 0.1$ (10%) or less.

---

**Problem 17.22.** An *ideal*, weightless and frictionless block and tackle as shown in Fig. 17.20 would require $F_1$ to be 10 newtons if the load is 20 newtons, but in an actual case $F_1$ may be 13 newtons. (a) How much work must be done to raise the same load by one meter? (b) What is the efficiency of this machine in practice? (c) What is the reason for using the machine when it would evidently require so much less energy to lift the load directly?

---

## 17.7 HISTORICAL BACKGROUND OF THE GENERALIZED LAW OF CONSERVATION OF ENERGY; THE NATURE OF HEAT

There remains a final extension of the concept of energy: to problems involving *heat*. Historically it was just this extension, achieved in the middle of the last century, which in fact clarified and brought together the several separate energy concepts, and so made possible the first general law of conservation of energy. It is therefore worth undertaking a short digression on the development of ideas about the nature of heat, before discussing the quantitative aspects of transformations between heat and other forms of energy.

We recall that Aristotle regarded Fire as one of the four elements, and that by this assumption much of the observed behavior of heated bodies could indeed be accounted for in a qualitative manner. The ancient Greek atomists explained differences in the temperature of bodies—their so-called intensity or degree of heat—by a closely related conceptual scheme, picturing heat also as a special though not directly perceptible substance, atomic in structure like all the others, quick to diffuse through bodies, and presumably possessing some weight.

As a first step this picture is common-sensical, intuitively clear, even beautiful. Although we may quarrel about details, perhaps about the exact weight of the hypothetical heat substance, the concept on the whole is useful in explaining most of the casual observations, including the temperature equilibrium which two equally hot bodies will eventually attain while in contact with each other. By and by the picture became sharper: that of a tenuous fluid, free to pass in and out of the smallest "pores" of other materials, present in amounts depending on the temperature, yet otherwise imponderable and perhaps after all rather like Descartes' subtle, omnipresent fluid or the ethers then invoked by some to explain gravitation, the propagation of light and of radiant heat, the transmission of magnetic and electric forces, and the like. But those philosopher-scientists of the

seventeenth century, including Bacon, Galileo, and Boyle, who—when they accepted a fluid theory at all—did so most hesitatingly, felt inclined to distrust the fluid theory of heat altogether and instead tended to the general though yet undefined view that heat was directly explainable as a vibration or similar small-scale motion of the atomic particles of heated bodies. The arguments were vague—for example, that heat evidently could be *generated* by motion, as by friction between two sticks or between an axle and the hub of a poorly lubricated wheel.

Throughout the seventeenth century and far into the eighteenth, neither the fluid view nor the motion view of heat became firmly established and here the reason was twofold. First, certain key concepts in the study of heat were still missing (e.g., specific heat, latent heat, or even the consistent differentiation between heat and temperature), and second, reliable methods of thermometry and calorimetry were still being developed.* In the 1780s the great pioneer in the study of heat, Joseph Black of Glasgow and later Edinburgh (1728–1799), still referred uncertainly to heat as "this substance or modification of matter."

But one significant truth had begun to dawn in the first half of the eighteenth century. In experiments on mixtures, where bodies of varying temperature are brought together, *heat is neither created nor destroyed*; that is, no matter how diverse the redistribution of heat among the different bodies in mixtures or in close contact, the total amount of heat remains constant. In modern terminology we should represent this *law of conservation of heat* in the following manner for

---

* For the whole story, well told, see D. Roller, "The early developments of the concepts of temperature and heat...," Case 3 in J. B. Conant (editor), *Harvard Case Histories in Experimental Science* (Cambridge, Mass.: Harvard University Press, 1950). A more comprehensive account, dealing with developments in the early nineteenth century, is in the book by Robert Fox, *The Caloric Theory of Gases*, New York: Oxford University Press, 1971.

the specific and simple case involving two differently hot bodies $A$ and $B$, say a hot block of metal (*A*) plunged into a quantity of cool water (*B*): the loss of heat by $A$ is equal to the gain of heat by $B$, assuming, of course, that the system containing bodies $A$ and $B$ is heat-insulated from all outside influences or losses. If $A$ and $B$ change only their temperatures, but neither their chemical consistencies nor their physical states, the previous sentence can be symbolized by

$$\Delta Q_A + \Delta Q_B = 0,$$

where $\Delta Q$ is the heat change for any sample, and may be calculated from the defining equation

$\Delta Q$ (cal) = mass of sample (g) × specific heat

[(cal/g)/°C] × [final temp. − initial temp. (in °C)].

Perhaps this, the basic rule of *calorimetry* ever since, helped most at the time to sway the pendulum of opinion toward the view that heat is a material substance, for such a law of conservation of heat very naturally brought to mind the tacitly accepted concept of the *conservation of matter*. In fact the fluid theory of heat now gave a simple and plausible picture of what happens when differently heated materials are brought together: The excess of "heat fluid" would flow from the hotter to the colder body until an equilibrium had been achieved. And truly, it was almost inconceivable that such a simple observation could be accounted for by the view of heat as a mode of motion of particles.

From the 1760s on, the developments proceeded at an accelerating pace, first mainly through the work of Black and his colleagues. The hypothetical *igneous* or *heat fluid* (named *caloric* by Lavoisier in 1787) began to take on greater plausibility when two further properties were assigned to it: that the particles of the caloric, unlike *ordinary* matter, repel one another but yet are attracted to the corpuscles of ordinary matter. Therefore, if heat is applied to a material object, the caloric may be pictured as diffusing rapidly throughout the body

and clinging in a shell or atmosphere around every one of the corpuscles. If these corpuscles (whose caloric shells repel one another) are free to move apart, as indeed they are in a gas, they will tend to disperse, and more strongly so the greater their crowding or the heat applied, in accordance with the gas laws. If, however, a heated object is in solid or liquid form, the mutual attraction among the corpuscles themselves (considered to be a gravitational one) so predominates that the caloric atmospheres can provide mutual repulsion sufficient only for the well-known slight expansion on heating. Or at any rate, the attraction predominates until enough caloric has been supplied for eventual melting or vaporization.

The same model made it plausible that on sudden changes of the physical or chemical states the particles of matter will exhibit a sharply changing capacity of accommodating caloric atmospheres, that is, they would release some caloric, thus changing it from a "latent" state to a "sensible" one, where caloric would communicate itself also to neighboring bodies and heat them up. Obviously this would explain the heat generated in some chemical reactions or solutions. Similarly it was not at all unexpected to find that "sensible heat" develops during mechanical deformations: For example, the sometimes very marked rise in temperature of a gas on rapid compression, or of an iron bar by skillful and rapid hammering, was regarded simply as evidence of a squeezing-out of caloric from among the atoms as these are forced closer to one another.

Other experimental facts, including those accessible through the newly developed concepts of specific and latent heat, were subjected to similar and quite successful analyses. The conceptual schemes gradually filled out on all sides, and the gathering weight of evidence persuaded almost all Europe from the 1780s on of the usefulness of the caloric theory.

Yet this theory hid two assumptions, one of which proved fatal to it. The first, which could not by itself discredit the whole structure when brought into the open, was the old problem of whether the caloric material also possesses weight. Naturally, this question was now repeatedly and often ingeniously examined by direct experimentation. But the findings were negative, where they were not indeed contradictory, owing to the inherent difficulties of weighing accurately at varying temperatures. The second assumption was that the caloric fluid is conserved in all processes involving heat.

The first serious challenges to these assumptions were provided by the experiments of Count Rumford (1753–1814), that amazing American who began life as Benjamin Thompson of North Woburn, Massachusetts, and by 1798, a refugee from his native country, had been knighted by King George III and was in the employ of the Elector of Bavaria as Chamberlain, Minister of War, and Minister of Police, still finding time away from his many duties and schemes to perform some capable and historical experiments.* After exhaustive and to us convincing experiments† with some of the best balances in Europe, he stated in 1799: "I think we may safely conclude that *all attempts to discover any effect of heat upon the apparent weight of bodies will be fruitless.*"

Granting this conclusion, we discover that we have been brought back to the same ground that more than a century before had been fought over by the vacuists vs. the plenists, Newton vs. the Cartesians, in a sense even by Galileo vs. the Scholastics, namely, the question whether science can accommodate an imponderable fluid with material properties, endowed perhaps even with atomic structure, but inaccessible to direct experimentation or even contemplation through its lack of that most general characteristic of

**Fig. 17.22**  Benjamin Thompson, Count Rumford (1753–1814).

materiality, weight.  Those seventeenth-century philosophers who had rejected imponderables—and like them, Rumford—had done so in part because they simply could not imagine a weightless fluid, a mechanical entity lacking one of the most unforgettable properties of ordinary matter. But calorists, on the other hand, could equally well say with Black: "I myself cannot form a conception of this eternal tremor [motion of corpuscles] which has any tendency to explain even the more simple effects of heat."  Up to this point, Rumford may seem to have little cause for rejecting the imponderable calorics except a lack of imagination; and this is the more striking because the scientific world at the time generally did accept a whole set of imponderables and ethers.

Let us for a moment examine our own first impulses in this matter.  We are probably tempted to reject the concept of imponderable fluids out-

---

*   See the fascinating biography by Sanborn Brown, *Count Rumford: Physicist Extraordinary*, Garden City: Doubleday, 1962.

†   Rumford's experiments, and the discussion of them, make most enjoyable and instructive reading in Roller's account, *op. cit.*

right: and if urged to explain, we might say that a part of the very definition of the physical entity must be that we can discover it directly. But this would be a most naive attitude; we must discard it now before we go on to phenomena involving "imponderables" like electrons and photons. We must be prepared to accept not just what sounds plausible, or just what can be imagined by analogy, but we must be prepared to welcome into science any concept that may be needed in our schemes if only it fulfills these requirements: It should be definable, without internal contradiction and ambiguities, in terms of observable and measurable effects whether these be direct or not; and it should prove useful in a wide variety of descriptions and laws, long known as well as newly deduced. The caloric fluid fulfilled all these criteria at the time.

But this was not the reasoning which led the calorists to accept the imponderable, and Rumford was not just naive in rejecting it. They on the one side were not being greatly bothered by Rumford's demonstrations, and on the other side his objection to the caloric for lacking measurable weight was only part of his much more thorough experimental, thirty-year attack on that concept. In time, Rumford succeeded in discovering the real Achilles heel—a failure of that most treasured function and main justification of the concept of the caloric, namely, *its usefulness in the general law of conservation of heat.*

The famous experiments leading up to this far-reaching discovery are described by him in 1798 as follows:*

"[I] am persuaded that a habit of keeping the eyes opened to everything that is going on in the ordinary course of the business of life has oftener led—as if it were by accident or in the playful excursions of the imagination, put into action by contemplating the most common occurrences—to useful doubts and sensible schemes for investiga-

tion and improvement than all the more intensive meditations of philosophers in the hours especially set apart for study.

"It was by accident that I was led to make the experiment of which I am about to give account . . . . Being engaged lately in superintending the boring of cannons in the workshops of the military arsenal at Munich, I was struck by the considerable degree of heat [temperature] that a brass gun acquires in a short time in being bored, and with the still higher temperature (much higher than that of boiling water, as I found by experiment) of the metallic chips separated from it by the borer."

A commonplace type of observation, but heightened both by the uncommonly high temperature developed and by the exceptional gift of observation of Rumford, was therefore the effective beginning of the idea that a heat *fluid* can be generated by friction in enormous quantities. As we saw, that of itself is not altogether unexpected; on the basis of the caloric theory the mechanical treatment of metals in boring might decrease the capacity for holding caloric in the metal, thereby yielding free, "sensible" [perceptible by the senses] heat. However, experiments with the metallic chips, and comparison with the bulk metal of the cannons themselves, showed Rumford that their capacity for heat (specific heat) had not changed. What is much more significant, further experiments revealed that the source of heat generated "in these experiments appear to be inexhaustible. It is hardly necessary to add that anything which any *insulated* body, or system of bodies, can continue to furnish *without limitation*, cannot possibly be *a material substance*, and it appears to me extremely difficult, if not quite impossible, to form any distinct idea of anything capable of being excited or communicated in the manner in which heat was excited and communicated in the experiments, except it be MOTION." And throughout his voluminous writing Rumford continues to affirm those "very old doctrines which rest on the supposition that heat is nothing but a vibratory

---

*    Reprinted in S. C. Brown, *Benjamin Thompson–Count Rumford*, New York: Pergamon, 1967, page 52f.

motion taking place among the particles of the body."

A similar conclusion was drawn the following year by the 21-year-old Humphry Davy, later a distinguished chemist, in original experiments in which he allowed two pieces of ice to rub against each other until they had both melted. The large amount of heat needed for melting the ice must have been produced by friction only, particularly in view of the large increase of specific heat on changing ice into water. Davy's conclusion that "it has just been experimentally demonstrated that caloric, or matter of heat, does not exist," although very influential on the later work of Mayer and Joule, is rather an overstatement; he might have been satisfied with the claim that large quantities of caloric appeared to have been created by friction, contrary to the expectations of the calorists.*

In brief, these and a great variety of other experiments, executed mostly by the indefatigable Rumford, had struck key blows against the caloric theory by 1807. But that was not enough to convert the calorists and topple their science, for the simple reason that there was no good constructive conceptual scheme to offer instead. Even though Rumford was able to show that his model of heat as a mode of motion in the substance would account qualitatively quite well for a variety of observations such as normal thermal expansion, heat conduction, change of state, and diffusion in liquids, his theory could not explain the undisputed facts summarized by the law of conservation of heat *in mixtures* and in other such simple cases in which, as we now would put it, by careful insulation the amount of heat in the system is kept constant, and is neither generated nor converted. The caloric theory was still most practical and plausible there, and in a great region of other physical and chemical phenomena.

---

* There is an ironic twist here. Re-examination of Davy's data and methods has shown that the ice must have melted owing far more to heat leaking into his apparatus than to friction: see Roller, *op. cit.*, pages 86–87.

One of the most convincing arguments against Rumford's theory was provided by the phenomenon of *radiant* heat, on which Rumford himself had done pioneering work. The very fact that heat could travel across empty space (such as, for example, the region between the sun and the earth was thought to be) seemed to prove that heat could not simply be a mode of motion of matter, but rather a distinct substance. During the first three decades of the nineteenth century, numerous experiments by William Herschel in England, Macedonio Melloni in Italy, James Forbes in Scotland, and others indicated that radiant heat exhibits all the properties of light—reflection, refraction, interference, polarization, etc.—and is therefore probably qualitatively the same phenomenon as light. At that time the modern distinction between radiant heat and other forms of heat had not been established; it was therefore believed that any conclusion about the nature of radiant heat would be valid for the nature of heat in general.

Since many scientists in the first half of the nineteenth century accepted the principle of the qualitative identity of heat and light, we find the rather curious situation that opinions about the nature of heat depended to some extent on opinions about the nature of light. As we shall see in Chapter 23, before about 1820 the Newtonian particle theory of light was generally favored—i.e., light was thought to be a *substance*. It was therefore reasonable to suppose that heat is also a substance, probably composed of particles like light. But as a result of the work of Thomas Young in England and Augustin Fresnel in France, scientific opinion after about 1825 swung around to the view that light is not a substance but a wave motion in the ether. Hence it was now reasonable to adopt, alongside a *wave theory of light*, a *wave theory of heat*—a theory in which heat was regarded as vibrational motion of the ether. This did not yet mean a return to Rumford's theory, but it did help to phase out the caloric theory of heat, though in some writings of the 1830s and 1840s one gets the impression that caloric has been reincarnated under the name "ether"! But more significant in the long run is the fact that, thanks in part to the wave theory of light, heat was now being conceived as a form of energy.

But several other contributions at that time pointed also to a relationship between heat and other forms of energy. The conversion of kinetic energy to heat, as by the quick hammering of a nail on an anvil, was, of course, common knowledge; so was the liberation of heat in chemical reactions. About 20 years before Rumford, Lavoisier and Laplace had introduced a note from physiology by their experiments to prove that a living guinea pig developed animal heat at about the same rate as does the chemical process of burning its food intake. In 1819, the French chemist and physicist, P. L. Dulong, showed that when gas is compressed quickly the heat developed in the gas is proportional to the mechanical work done on it. Research was begun on the heat produced by electric currents.

And all the while, two other influences develop. The energy concept becomes more clearly defined; the word *energy* itself (in its modern meaning) dates from 1807. And as particularly illustrated in the work of the British physicist Michael Faraday on the interaction of electric current, magnetism, and light, there spreads increasingly a belief that the several types of natural activities are all connected and are ultimately all part of one unified story. In reviewing this period prior to the 1840s one is struck by a widespread groping toward the general law of conservation of energy from many different sides, often far beyond the context, the available evidences, or the clarity of the concepts themselves.

Heat not only changes temperature and physical state, but it can also move objects (by making a gas expand, for example) and so can be made to produce physical work. This was, of course, a commonplace after the spectacular development of steam engines in the eighteenth century, and their remarkable effect in promoting the industrial development. But the relation between heat and work was by no means clear before the 1850s.

Up to that time, it was believed that the flow of heat itself determined the motion of the parts of a heat engine in somewhat the same manner as the flow of water over a water wheel makes the mill turn, and that heat itself was no more consumed in the process than is water in the other case.

What was clearly called for (or so it seems to us now, with the advantage of hindsight) was to go beyond the qualitative idea that heat is a form of energy convertible into other forms, and to make specific determinations of the numerical equivalents or conversion factors of different forms of energy. Speculations about energy conversion, and experimental data from which conversion factors could be computed, were both available to scientists in the 1830s and 1840s; it is therefore not too surprising that the act of putting them together in a coherent way, which we now define as the "discovery" of the generalized law of conservation of energy, was accomplished almost simultaneously, and independently, by several people.* We shall discuss two of the ways in which the discovery was made, thereby also illustrating the diversity of methods that can be successful in science.

## 17.8 MAYER'S DISCOVERY OF ENERGY CONSERVATION

Several basic ideas about the interchangeability of heat and work were pulled together in 1842 in an essay that for the first time suggested what we have assumed as basic in Sections 17.5 through 17.7, that is, a general equivalence and conservation of all forms of energy, the equivalence of heat and work being only a special case. That essay, full of imaginative, almost metaphysical, but largely unsupported generalizations to such a degree that its publication was at first refused by

---

\* For a comprehensive analysis, see T. S. Kuhn, "Energy conservation as an example of simultaneous discovery," in *Critical Problems in the History of Science* (edited by M. Clagett), Madison: University of Wisconsin Press, 1959.

one of the great journals of science, was the work of Julius Robert Mayer (1814–1878), then a young doctor in the German town of Heilbronn, his native city (Fig. 17.23). He later said that his thoughts first turned to this subject when, as a ship's physician in the tropics, he discovered venous blood to be of a brighter red color than he had expected from experience in cooler climes*; and that subsequently he began to speculate on the relation between animal heat and chemical reaction, and from there to the interchangeability of all energies in nature.

In his 1842 paper "Remarks on the Energies† of Inorganic Nature," Mayer announces:

"Energies are causes: accordingly, we may in relation to them make full application of the principle—*causa aequat effectum*. If the cause *c* has the effect *e*, then *e* = *c*;.... In a chain of causes and effects, a term or a part of a term can never, as plainly appears from the nature of an equation, become equal to nothing. This first property of all causes we call their indestructibility.... If after the production of [effect] *e*, [cause] *c* still remained in whole or in part, there must be still further effects [*f*, *g*, ...] corresponding to the remaining cause. Accordingly, since *c* becomes *e*, and *e* becomes *f*, etc., we must regard these various magnitudes as different forms under which one and the same entity makes its appearance. This capability of assuming various forms is the second essential property of all causes. Taking both properties together, we may say, causes are quantitatively *indestructible* and qual-

**Fig. 17.23**  Julius Robert Mayer (1814–1878).

itatively *convertible* entities.... Energies are therefore indestructible, convertible entities."

This, then, is the first voicing of conservation and equivalence of all forms of energy. But this way of arguing seems quite false to a modern scientist. "Causes" do not have the properties derived by this private logic, nor can one equate energy with "cause" unless there is prior *experimental* proof of the indestructibility and convertibility of energy. But Mayer believed that this proof was already available from previous experiments, provided they were interpreted from his viewpoint.

Further on, Mayer makes some applications of his doctrine to chemistry and mechanics, and he continues:

"In numberless cases we see motion cease without having caused another motion or the lifting of a weight; but an energy once in existence cannot be annihilated, it can only change its form;

---

* As an illustration of the controversial nature of some of the simplest details in the history of science, we may note that it has been asserted, on the one hand, that this observation was "well known" before Mayer; and, on the other hand, that the observation must have been erroneous because venous blood is *not* bright red in the tropics! See E. Farber, *Isis*, Vol. 45, pp. 3–9 (1954); W. B. Ober, *New York State Journal of Medicine*, Vol. 68, pp. 2447–54 (1968).

† In the German original, the old word *Kraft* (force) is used throughout in this sense.

and the question therefore arises, What other forms is energy, which we have become acquainted with as potential energy and kinetic energy [modern terminology], capable of assuming? Experience alone can lead us to a conclusion."

There follows the argument along expected lines that if heat is generated by the motion of two surfaces against each other with friction, heat therefore must be equivalent to motion. Of course the idea that heat *is* a form of motion had frequently found expression throughout the history of science; Rumford and Davy had recently advocated it, and it was to be generally accepted within a decade. Curiously, Mayer held back from this conclusion; his philosophy did not lead him to make any one form of energy, such as motion, more fundamental than the others. He simply concludes: "If potential energy and kinetic energy are equivalent to heat, heat must also naturally be equivalent to kinetic energy." Consequently, "We will close our disquisition, the propositions of which have resulted as necessary consequences from the principle *causa aequat effectum* and which are in accordance with all the phenomena of nature, with a practical deduction.... How great is the quantity of heat which corresponds to a given quantity of kinetic or potential energy?"

This is indeed the crucial point. Mayer's presentation so far has been almost completely qualitative, and indeed rather reminiscent of scholastic science in many ways. This is partly understandable from the fact that Mayer himself had then no facilities for experimentation and found the physicists of his time quite uncooperative. But here, at the end of his paper, deceptively like an afterthought, is finally promised a derivation of the equivalence between heat and work, and by an approach opposite to that of Rumford. Here is the chance for empirical verification.

Mayer's solution is only sketched out, but one may reconstruct the steps. In the circumstances, he had to fashion his illustrative computation from some known data whose significance had by and large not been noted. To explain the nature of these data, we must first explain the units for measuring heat. We follow an old convention: The unit of heat, called the *calorie*,* is defined as the amount of heat which raises the temperature of a 1-g mass of water through just one degree on the Celsius† temperature scale.** After some experimentation one may then assign to every substance a *specific heat* (symbol $c$), a value that is numerically equal to the heat energy needed to raise the temperature of 1 g of that substance by 1°C. For water, $c$ is 1 (cal/g)/°C, by definition; for copper, it is by experimental test about 0.09 (cal/g)/°C; for lead, 0.03 (cal/g)/°C; and for most other solids and liquids it has some intermediate value.

In the case of gases—and here we may include air, a mixture of gases—there is an interesting and important difficulty: The specific heat depends to a marked degree on the method of measurement. If, for example, we take a sample of air in a tight enclosure at a constant volume, we must supply 0.17 cal per g per °C temperature rise and, incidentally, we shall note that the pressure in the gas rises. Now we repeat this experiment with the

---

* The name harks back to the old caloric theory of heat, but of course the unit for measuring heat is not dependent on a theory of what constitutes heat.

† Named in honor of Anders Celsius (1701–1744), a Swedish astronomer who first proposed (in 1742) a temperature scale on which the boiling point of water was defined as 0° and the freezing point as 100°. This was soon reversed, and the scale on which 0° is the *freezing* point of water and 100° the *boiling* point was known until recently as the *centigrade* (100-step) scale. However, since the Tenth General Conference on Weights and Measures in 1954 decided to redefine the temperature scale for scientific purposes by taking absolute zero and the *triple point* of water as the fixed points (see below, Section 19.3), the name centigrade is no longer appropriate and this scale should be called the *Celsius* scale instead.

** This definition is sufficiently precise for our purposes, but it may be noted that in accurate work the actual initial and final temperatures must be specified; e.g., the calorie is the amount of heat needed to raise the temperature of 1 g of water from 15°C to 16°C (which is not quite the same as that needed to raise it from, say, 95°C to 96°C).

same sample in a cylinder having one freely movable side or piston (Fig. 17.24). The pressure inside and outside the cylinder will always be the same, for the inflow of heat to the gas can now be attended by an expansion, and it will be found that one needs 0.24 (cal/g)/°C, that is, 40% more than before. We must specify two specific heats for gases, one for constant volume, another for constant pressure. These are usually denoted by the symbols $c_v$ and $c_p$, respectively.

The difference between the two specific heats indicates that the expansion of a gas is associated with an exchange of heat with the surroundings, and this fact was generally known to scientists early in the nineteenth century. What was not so generally known, however, is that if a gas is allowed to expand freely into a vacuum (instead of having to push away a piston as in Fig. 17.24), then there is no heat exchange with the surroundings, i.e., no cooling. This was shown by the French chemist Gay-Lussac (known for his law of combining volumes, see Section 20.8) in an experiment done in 1807. Gay-Lussac showed that a sample of gas will expand freely into the previously empty part of a set of connected vessels *without* measurably changing its overall temperature, whereas expansion against some surrounding medium or piston is possible without temperature drop only if heat is supplied. The experiment was repeated by J. P. Joule in 1845, using more accurate thermometers, and a few years later by Joule and William Thomson, working together. In some cases a very slight cooling effect was observed, but the temperature change was always very much less than that experienced by a gas on expansion against a piston.

Mayer used the free expansion experiment of Gay-Lussac to support his argument that heat actually disappears when it does mechanical work. According to the caloric theory, when a gas expands its temperature drops because the caloric is spread out over a larger volume, but the total amount of caloric was supposed to remain the same. However, Mayer pointed out that the temperature drops

**Fig. 17.24** Heating a sample of gas in a cylinder with freely sliding piston.

*only* when the gas does mechanical work by pushing a piston as it expands; if no work is done, the temperature remains the same. Therefore the drop in temperature which does occur when a gas does work cannot be explained by saying that the same amount of caloric is spread over a larger volume; instead, one must assume that heat actually disappears when work is done.

In order to compute the actual amount of heat that is consumed in doing a certain amount of work, Mayer used the fact, noted above, and known for some time, that 1 g of air requires 0.17 cal to gain 1°C at constant volume, but 0.24 cal if the piston is free to move and the chamber expands. Mayer assumed that in the second case also, 0.17 cal is needed to increase the temperature, but that the additional 0.07 cal is needed to move the piston out against the surrounding air (normally at the constant pressure of 1 atm or $1.013 \times 10^5$ newtons per square meter of piston area). In retrospect, one may see that this idea is inherent in Gay-Lussac's observation that no difference of specific heats exists if the gas can expand into a vacuum, against no resistance. Here, the work done by the gas on the piston (Fig. 17.25) is the perpendicular force ($F$) times the distance it moves ($s$). But since the pressure ($P$) in the gas is defined as force ($F$) ÷ area ($A$) on which $F$ acts, the work done on expansion becomes ($PA$) × $s$. Furthermore, $A \times s =$

**Figure 17.25**

change in volume during the expansion of this 1 g of air, and that was a known quantity, namely, 2.83 cm³. Thus 0.07 cal of heat appears as 1.013 × 10⁵ newtons/m² × 2.83 cm³ ~ 0.286 joules of work, or *1 cal is equivalent to about 4 joules*. (Values that are approximately equal to modern ones have been used here; the measurements known to Mayer gave him 1 cal → 3.6 joules.)

For a decade or more, little attention was paid to Mayer's work, despite the fact that similar ideas and experiments were being discussed by other scientists. Mayer's philosophical style of reasoning was unconvincing to those who had become accustomed to defining science in terms of New-tonian mathematical calculation and quantitative experimentation. Moreover, this was not a case of a generally recognized conceptual crisis which Mayer's new ideas would have resolved; instead, he was presenting a new viewpoint which could serve to unite previously separate branches of physical science, and the fruitfulness of this view-point had not yet become evident. Mayer did develop his principle of the general indestruct-ibility and strict convertibility of energy further in several subsequent essays, presenting good il-lustrations of proof by calculation, and extending the principle "with remarkable boldness, sagacity, and completeness" to chemistry, astronomy, and living processes.

But although the work in retrospect was clearly one of genius and "on the whole, a production of extraordinary merit," Mayer continued to be so utterly rejected that his mind, overtaxed by the strain of his work and his lonely fight, gave way

for a time. He recovered and witnessed growing recognition and honor in the closing years of his life.

This change of personal fortunes was by no means a sign that science had become sentimental about Mayer's condition; rather, it was the con-sequence of a general acceptance of the LCE on the basis of an entirely different series of events, above all, the remarkable and persistent experi-mental work of a British amateur of science, James Prescott Joule (1818–1889). The distinction between the two men has never been put better than by John Tyndall* in his still very readable classic *Heat Considered as a Mode of Motion* (1863 and later editions):

"True to the speculative instinct of his country, Mayer drew large and weighty conclusions from slender premises, while the Englishman aimed, above all things, at the firm establishment of facts. And he did establish them."

## 17.9 JOULE'S EXPERIMENTS ON ENERGY CONSERVATION

Joule was born near Manchester, as the son of a well-to-do brewer to whose business he succeeded. But his dedication from the start was to science. At 17, he was a pupil of the great chemist John Dalton (see Chapter 20), and at 22, he had begun a series of investigations that was to occupy the greater part of his life: the proof that when mechanical work is turned to heat under any circumstance, the ratio of work done to heat evolved has a constant and measurable value. In part, his motivation seems to have been essentially the same as that of every other proponent of a conservation law; in his words, to him it was "manifestly absurd to suppose that the powers with which God has en-dowed matter can be destroyed."

---

* From whom stem also the quotations in the previous two paragraphs.

The most powerful drive in Joule was undoubtedly the sheer love for experimental investigations. Independent of Mayer's work (with
which he said he was "only imperfectly acquainted"
as late as 1850), he announced in 1843 the first
result of his labor, the comparison of mechanical
work needed to operate an electric generator and
the heat produced by the current so generated:
$J = 4.51$ joules/cal. This was an average of 13
experiments with fairly large uncertainties, for in
all such experiments it is extremely difficult to
obtain precise and meaningful data, because of
unavoidable leakage of heat and the like. Later
the same year, Joule measured the relation between
heat of friction for water flowing through thin
pipes and the work needed to produce the flow
(4.14 joules/cal), and he concluded:

"We shall be obliged to admit that Count
Rumford was right in attributing the heat evolved
by boring cannon to friction.... I shall lose no
time repeating and extending these experiments,
being satisfied that the grand agents of nature are,
by the Creator's fiat, *indestructible*; and that
whenever mechanical force is expended, an exact
equivalent of heat is *always* obtained."

But Joule certainly did not rest now. Despite
the coolness with which his work was generally
received, he continued with almost obstinate zeal.
In 1844 he measured the ratio of work needed to
compress a gas to the heat so created (4.27 joules/
cal), in principle rather close to the method which
Mayer had considered in 1842. A little later, Joule
made the first of a long series of more and more
accurate experiments to find the mechanical
equivalent of heat by friction, e.g., by stirring
water with paddlewheels in a heat-insulated
container. This is initially a delicate experiment
because the temperature rise is fairly small, and his
result (4.78 joules/cal) is markedly different from
the others. By 1847 the technical difficulties were
in hand; stirring water and also sperm oil gave very
closely the same result ($\sim 4.20$ joules/cal).

When he wished to report this particular
experiment at a scientific meeting in 1847, a crisis
arose. Joule was requested by the chairman to
confine himself to a "short verbal description,"
partly because his previous communications, as
Joule himself recalled later, "did not excite much
general attention." "This I endeavoured to do,
and, discussion not being invited, the communication would have passed without comment if a young
man had not risen in the Section and by his intelligent observations created a lively interest in
the new theory. The young man was William
Thomson" (afterward raised to the peerage as
Lord Kelvin), then 23 years old and destined to
become one of the foremost British scientists,
collaborator and staunch supporter of Joule's
views.

The year 1849 brought further results on
friction: in water, in mercury (where the temperature change is larger by virtue of its lower
specific heat), and for two iron disks rubbing
against each other. Joule compares his methods
and results with those of Rumford, estimating a
value of about 5.6 joules/cal from the data given
for one of Rumford's experiments. Considering the
uncertainties in the available data, one must
conclude that the two equivalence values are
quite close.

Then Joule summarized all his previous work
and assigned to the mechanical value of heat the
value $J = 4.15$ joules/cal. Today, after a century
of increasingly refined precision methods, *the value
for the equivalent between work and heat is 4.184
joules/cal*, and $J$, of course, stands, with greatest
justice, for *Joule's Equivalent*.

By 1850, after 10 years of devoted activity,
Joule's work began generally to be taken more
seriously, the gratifying result of a well-planned
experimental assault. During the next 18 years,
he carried through three further determinations of
$J$. Without a doubt, it was the sum of his consistent
experimental work that made acceptable the idea
of an enlarged, general law of conservation of all
forms of energy in place of the previous specific

and limited application of the law to conservative mechanical systems alone. But, as usually happens, Joule's triumph depended in large measure on developments beyond his own participation. It will be seen in the next chapters that at about the same time the general thinking on the nature of heat and of matter itself was undergoing profound changes toward a view more favorable to the conservation of heat as a form of energy, namely, the kinetic energy of motion of the molecules. This trend removed the cause of a great deal of the original resistance to Rumford, Mayer, and Joule.

And second, the idea of a general LCE found powerful and eloquent champions, particularly at first the young German physiologist and physicist, Hermann von Helmholtz.* He did what Mayer, then still in oblivion, had not quite done and Joule had never attempted, namely, to show by mathematical demonstration the precise extent of the conservation law in various fields of science —mechanics, heat, electricity, magnetism, physical chemistry, and astronomy—to derive by its aid, in terms that the professional scientist would at once recognize, explanations for old puzzles and new—confirmable mathematical relations. Indeed, as the LCE gradually won more adherents and wider application, it was discovered that Mayer's vision had been prophetic, that the LCE provided striking general connections among the various sciences. It was perhaps the greatest step toward unity in science since Galileo and Newton, and a powerful guide in the exploration of new fields. It has been well said, by a twentieth-century historian of science (Dampier), that "Because of its practical use, and for its own intrinsic interest, the principle of the conservation of energy may be regarded as one of the great achievements of the human mind."

A footnote to this historical account may be in order to supply one rather typical detail to this sketch of the growth of a conceptual scheme in science. As Mayer's and Joule's work finally won supporters, there developed a sometimes rather bitter fight as to which of the two had been the first proponent of the general doctrine of conservation of energy in its various forms. A number of other claimants to the title were put forward, men who had expressed in some way the general idea of interconvertibility of forms of energy or had computed in particular the mechanical equivalent of heat: Ludvig Colding in Denmark; Karl Friedrich Mohr, Karl Holtzmann, and Justus Liebig in Germany; Sadi Carnot and Marc Séguin in France; Gustav Adolphe Hirn in Belgium; and W. R. Grove and Michael Faraday in England.† When a discovery is obviously "in the air," as this multiplicity of discoverers suggests, it is meaningless to argue about priorities; yet such arguments are inevitable, for the hope of getting eternal credit for a discovery is one of the motivations that drives a scientist to work overtime on a job that (until quite recently) was paid poorly or not at all. Thus the Jouleites disparaged Mayer's contribution of 1842 that gave him technical priority over Joule, whose first but quite independent paper was dated 1843; and Joule himself wrote in 1864:

"Neither in Séguin's writing (of 1839) nor in Mayer's paper of 1842 were there such proofs of the hypothesis advanced as were sufficient to cause it to be admitted into science without further inquiry.... Mayer appears to have hastened to publish his views for the express purpose of securing priority. He did not wait until he had the opportunity of supporting them by facts. My course, on the contrary, was to publish only such theories as I had established by experiments calculated to commend them to the scientific public, being well

---

* Helmholtz was then 26 years old. It is probably no coincidence that most of the protagonists were quite young, under 30, when they became interested in the LCE.

---

† The merits of these various claims have been evaluated in the article by Kuhn cited at the end of Section 17.7. The attitudes of scientists toward priority disputes have been discussed at length by R. K. Merton, *Proceedings of the American Philosophical Society*, Vol. 105, pages 470–486 (1961); *American Scientist*, Vol. 57, pages 1–23 (1969), and other articles.

convinced of the truth of Sir J. Herschel's remark that *'hasty generalization is the bane of science'*."

On the other side, and often not much more fairly, those partial to Mayer's claim to priority would say with Helmholtz:

"The progress of natural science depends on this: that new [theoretical] ideas are continually induced from the available facts; and that afterwards the consequence of these ideas, insofar as they refer to new facts, are experimentally checked against actuality. There can be no doubt about the necessity of this second endeavor; and often this second part will cost a great deal of labor and ingenuity, and will be thought of as a high accomplishment of him who carries it through well. But the fame of discovery still belongs to him who found the new idea; the experimental investigation is afterwards a much more mechanical type of achievement. Also, one cannot unconditionally demand that the inventor of the idea be under obligation to carry out the second part of the endeavor. With that demand one would dismiss the greatest part of the work of all mathematical physicists."

Here we see an old fight, or rather two fights: one for priority (and each great idea, after it is accepted, tends to give rise to such a dispute) and second, the fight concerning the relative merits of the experimentalist and the theoretician. Both illustrate some of the human element behind the printed equation.

## 17.10 GENERAL ILLUSTRATIONS OF THE LCE

As has been demonstrated, the LCE may take on any one of many different forms, depending on the intended application. But broadly speaking, we have met three main groups of formulations:

i) *For an isolated system*, the sum of all forms of energy remains constant (the net changes add up to zero) although the internal energy may take on different forms. Examples are collisions between balls or molecules, and disintegration in natural radioactivity.

ii) *For a system to which energy is given*, including work done or heat supplied to it, *but which does not in turn give energy to the world outside the system*, the energy supplied to the system is equal to the total change in internal energy of all forms. Examples are a block being pulled by a string, gas in a cylinder being compressed by a piston, induced transmutation, and fission by neutron bombardment.

iii) *For a system to which energy is supplied and which also does work on or loses energy to the outside world*, the energy supplied to a system is equal to the total change of internal energy in the system *plus* the energy given up by the system. Examples are machines and Joule's paddlewheel experiment with heat leakage.

But of course the LCE is a *single* law, and the forms it takes depend only on how large we choose the "system" under consideration to be. Thus when a man hoists a load by means of a pulley, the form (iii) applies if the pulley itself is the whole system, and the form (i) if the earth together with man, pulley, and load is considered one system. Furthermore, the first two (and one other possible form: which? examples?) are but special examples of the third form, with one or the other factor set equal to zero.

In scientific writings one encounters the LCE phrased in many different ways. One equivalent name, *first law of thermodynamics*, is the LCE in such form as applies in particular to cases in which exchanges between heat and mechanical energy enter. The first law of thermodynamics can be stated in the form:

$$\Delta E = H - W, \qquad (17.17)$$

where $E$ is the energy of a body, $\Delta E$ is the change of energy involved in going from one state (i.e., temperature and pressure) to another, $H$ is the heat added to the body, and $W$ is the work done by it on its surroundings. In addition, it is assumed in thermodynamics that the energy of a body in a particular state does not depend on how it got to that state. Thus for example in Mayer's calculation of the mechanical equivalent of heat (Section 17.8), it was assumed that the net change in energy of 1 g of air is the same whether it is simply heated by 1°C with the pressure kept constant, or whether it is first heated at constant volume (while the pressure rises) and then allowed to expand by pushing out a piston (thereby doing mechanical work and dropping the pressure) so that it ends up in the same final state. In the first case $\Delta E = H = 0.17$ cal; in the second case $\Delta E = 0.24$ cal $- W$. Since in the second case also the net energy change must be $\Delta E = 0.17$, we can infer that $W$ is equivalent to 0.07 cal.

The fact that a substance can be heated by compression or cooled by expansion—i.e., its temperature changes without any actual exchange of heat with the surroundings—is a further illustration of the first law of thermodynamics. A given $\Delta E$ (indicated by a definite change of temperature, or perhaps by a change of state such as melting or evaporation) can be accomplished by several different combinations of $H$ and $W$. (Note that if the body loses heat instead of gaining it, $H$ in Eq. (17.14) is negative. Similarly, if work is done *on* the body—e.g. by compressing it—$W$ is negative. This does not mean that heat or work are themselves negative quantities, but simply that they can either decrease or increase the energy of a body, depending on which process is being considered.)

Another, equivalent formulation of the LCE is of particular interest: It is the statement that a *perpetuum mobile* (perpetual mover) is impossible; for by *perpetuum mobile* is generally meant a machine that has an efficiency of *more* than 1.00 (100%), one which yields more output than input. One may even identify the rise of the LCE in mechanics with the continued unsuccessful attempts to make such a machine, for it should be remembered that this project was very popular

indeed for hundreds of years. The search, which engaged some of the greatest intellects, was in itself no more implausible than the earlier and equally fruitless search for the "Philosopher's Stone" (which was supposed to be able to turn all common metals into gold or silver); it may have reflected the preoccupation of an increasingly more machine-minded civilization. However, it probably points up equally that most pathetic of all human traits, the wish to get something for nothing.

The inability to make a perpetual motion machine became itself a fact of physical reality, one example of a series of so-called *postulates of impotency* which have profoundly influenced physical science. To this class belong the impossibility of transmuting matter chemically, which helped to shape the idea of chemical elements; the inability to discover the relative motion of the ether in optical experiments, which was turned into a fundamental postulate of relativity theory; the impossibility of obtaining net mechanical energy from a medium by cooling it below the temperature of its surroundings,* which is the effective content of the second law of thermodynamics; and several others.

The following examples will provide additional familiarity with the LCE in its most general form, and additional evidence of its wide-ranging power.

*The human body as an engine.* Once it was thought that animals, including man, must eat only in order to grow and to rebuild used-up tissue. The realization that the food intake itself also provides heat and energy for the functioning of the body was attained only slowly. At the danger of simplifying a complex topic far too much, we may compare the body to a chemical engine: The

---

* An engine that functioned thus, if it were not impossible for other reasons, would not be in conflict with the LCE. It is referred to as a *perpetuum mobile of the second kind,* and was once as much sought after as the *perpetuum mobile of the first kind* described above.

oxygen we breathe burns the sugars, fats, and some of the proteins from our digested food intake, converting their complex molecules mainly into carbon dioxide, water, and other waste products in an exothermic reaction which liberates heat and, in the contraction of muscles, provides mechanical work.

The balance of energy intake in the form of *combustible* food vs. work and heat output has been measured with precision, partly in an effort to check whether life processes also obey the conservation laws in the strict forms that prescribe the activities of inanimate nature. Since the 1890s this problem has been clearly settled in the affirmative. In a representative case, the daily food intake would yield about $3 \times 10^6$ cal if burned directly, and the heat given up together with the work done and the energy in the waste products come to the same figure within a small experimental uncertainty. But the body, considered as an engine, ordinarily has an efficiency of less than 20%; so if one has to do large amounts of physical work in a short time interval, the attendant heat output is correspondingly larger, and since the rate of dissipation of heat is not very flexible, the body heats up drastically.

*Photosynthesis.* If the energy of the animal body were derived from a natural fuel like oil, the food supply on this earth would have been exhausted long ago. The present form of animal life depends on the fact that the body's fuels are continually synthesized by green plants and are readily available, one might say, in the form of either vegetables or vegetarians. The green plant is uniquely able to draw carbon dioxide from the surrounding medium and water, both waste products of animal life, and to recombine the aggregates of carbon, hydrogen, and oxygen, plus those of some mineral salts, into more complex "organic" compounds, each having a large amount of built-in chemical "potential" energy that can be released later during decomposition inside the animal body. For good measure, the plant also releases oxygen into the atmosphere while it builds our food.

For each gram of carbon content, these "fuels" can deliver roughly $10^4$ cal worth of work and heat, and on this depends the life of all animals except some lower organisms that live by fermentation. But is all this not a crass contradiction of the law of conservation of energy? Do the plant and animal worlds together not constitute a *perpetuum mobile*? The answer is no, because the energy in the manufactured organic molecules is originally taken by the plant from sunlight. The reactions depend on the availability of light energy (and on the presence of chlorophyll and other catalysts); technically this is termed photosynthesis. In this sense, our food is largely sunlight.

If not eaten, the energy-rich plant material in the open will soon decay into its original compounds; but if it is submerged or otherwise suitably kept from ordinary oxidation, it may turn into coal or other heating fuels by slow chemical decompositions that favor the retention of carbon-rich materials.

*The origin of the solar system.* One of the famous old puzzles to which the law of conservation of energy in its early days seemed to give an immediate answer was how the sun could continually emit heat and light at such a staggering rate (about $3.8 \times 10^{26}$ joules or $9 \times 10^{25}$ cal every second). If this were achieved by burning ordinary fuels or some such chemical reaction, it was difficult to see why the material for it had not been exhausted long ago. Mayer, Helmholtz, and others proposed promising alternatives that the sun's radiated energy was derived from gravitational potential energy by a slow shrinking of the sun or by the capture of comets and meteors. But then it was calculated that the sun's diameter would have to shrink one part out of 10,000 every 2000 years to provide radiation in this manner at the present level. That is too fast to account for the long history of life on earth. Nor can enough meteoric matter be found to give the observed effect. The now accepted source of the sun's energy, namely a nuclear fusion process (see page 525), was of course not imaginable until much more recently.

Yet, the original proposal is not altogether without merit, although it failed for the initial application. The same general mechanism of conversion of gravitational potential energy to heat is invoked today in theories concerning the creation of the solar system.

The facts to be explained include the following At least half of all the stars in our galaxy are binaries (double stars) or stars of higher multiples. These usually revolve quickly about one another. Of the single stars, the very hot ones usually have even more enormous rates of spin, but the "cooler" ones, like our sun, always revolve quite slowly. In our own planetary system, in which the orbits of all planets are in the same direction and substantially in the same plane, the total angular momentum* of the planets about the sun is very many times that of the sun about its own axis. Now the present-day theories, with different accent on detail to be explained, begin with the same general proposition: The solar system started by the gathering up of a thinly spread cloud of dust, such as the interstellar or "cosmic" dust clouds that are observable with our telescopes. It may be that light pressure from surrounding stars pushed the dust particles into one another's shadow until the density was sufficient to allow the mutual gravitational forces to predominate.

Such a mass would coalesce in about $10^9$ years. As the cloud shrinks, any initial rotational motion, if ever so slow, will by the law of conservation of angular momentum increase progressively, and as the gravitational forces pull the particles together, work is done and heat can be generated. According to one line of thought, in the final stages of this process the central portion of the cloud gathers up into one bright, spinning central star (or, in other cases, two or more stars) leaving the outlying and still revolving portions of the cloud to condense into the planets and their satellites.

Although the details of this scheme also explain

---

*    See footnote on page 259.

the direction and plane of planetary orbits, the spinning of planets and their moons, and the general distribution of angular momentum, there seem to be theoretical objections in details. To avoid these, the alternative theory, also not yet without its adverse critics, interpolates one sequence: It prefers the fairly complete collapse of the entire initial cloud into one very rapidly spinning star. Eventually the star spins so fast that it breaks into two (or more) sections, or it becomes so hot that nuclear reactions may be set off which explode the star somewhat in the manner of an "atomic bomb" (*super-nova*). Such an explosion might also separate the original star into two (or more) sections. Whatever the mechanism of separation, the double or multiple stars still spin fairly rapidly at a close distance, but dissipate their energy in, and give their momentum to, the surrounding dust and debris. In time the rotation of the double stars slows down, therefore they coalesce once more by mutual gravitational attraction, now rotating slowly as one single "cooler" sun, while the rest of the surrounding medium gathers up into planets and satellites. Evidently this scheme orders the different types of stars into an evolutionary sequence; and incidentally allows the speculation that each exploding star may become the originator of a planetary system. On this basis it has been estimated that the known universe contains perhaps as many as 100 billion planetary systems.

From the present scientific point of view it is natural to think that our earth and all it contains were fashioned by all-pervasive laws that are in continuous operation with similar consequences elsewhere. Here, as a footnote to the powerful and intriguing chapter that the law of conservation of energy has written into science, we cannot forego a brief word of comment. Very probably one theory or another will in time prove that the origin and form of the solar system, including the regularities pointed out by Bode's Law, are explainable by straightforward physical processes. If the past is a guide, this will be interpreted by some as another serious blow of science against religion, for it has generally been held that science would not be able to provide such an explanation, that a theistic assumption was consequently necessary, and that, indeed, the lack of a scientific alternative was demonstrable proof of the existence of the Deity. Thus, speaking of the planets and comets, Newton himself said in the "General Scholium," Book III of the *Principia*:

"But though these bodies may, indeed, continue in their orbits by the mere laws of gravity, yet they could by no means have first derived the regular positions of the orbits themselves from those laws. The six primary planets are revolved about the sun [roughly] in circles concentric with the sun, and with motions directed towards the same parts, and almost in the same plane. Ten moons are revolved about the earth, Jupiter, and Saturn, in circles concentric with them, with the same direction of motion, and nearly in the planes of the orbits of these planets; but it is not to be conceived that mere mechanical causes could give birth to so many regular motions, since the comets range over all parts of the heavens in very eccentric orbits.... This most beautiful system of the sun, planets, and comets, could only proceed from the counsel and dominion of an intelligent and powerful Being...."

The same thinking existed in connection with the problem of the descent of man before it was explained by evolution, with the existence of chemical elements before nuclear physics recently began to supply us with promising theories of their genesis, and with many other examples of this kind. But surely such events teach only this lesson: It is not only folly but blasphemy to let one's belief in the Deity be based on a premature estimate of what science can *not* do. The opposite course, namely the deification of the discoveries of science, is in truth quite as precarious, for science changes, and may come to regard its old knowledge as less relevant. The secure foundation for belief, as all great religious founders have taught, is not the capability of man's mind, or the limitations of his imagination, neither the impressive power of the laws of science, nor the supposed inadequacy of these laws—but faith.

## ADDITIONAL PROBLEMS

**Problem 17.23.** Before Joule obtained a more accurate value of the mechanical equivalent of heat ($J$) he wrote (1845): "Any of your readers who are so fortunate as to reside amid the romantic scenery of Wales or Scotland could, I doubt not, confirm my experiments by trying the temperature of the water at the top and bottom of a cascade. If my views be correct, a fall of 817 feet will of course generate one degree [1°F] ... and the temperature of the river Niagara will be raised about one-fifth of a degree [F] by its fall of 160 feet." (This general prediction he later checked himself, during his honeymoon trip to a large Swiss waterfall.) Now compute from this prediction what value of $J$ he was using at the time. To convert temperature scales, recall that 0°C = 32°F, and 100°C = 212°F.

**Problem 17.24.** When a 1-kg ball of lead (specific heat = 0.03 (cal/g)/°C) falls freely through 10 m and is stopped suddenly by collision with the floor, what is the greatest temperature rise you would expect? Cite several causes which would contribute to making the observed temperature rise smaller.

**Problem 17.25.** (a) Calculate the total energy of the earth at one point of its orbit (PE with respect to the sun, plus $KE_{rev}$ in its orbit and $KE_{rot}$ about its axis). (b) Knowing the size and ellipticity of the earth's orbit ($e = 0.017$), find the difference in our actual distance from the sun at perihelion and aphelion, and compute from that plus part (a) the maximum difference in orbital speed.

**Problem 17.26.** In his widely discussed book *Worlds in Collision* (Macmillan, 1950), Immanuel Velikovsky maintains that at about 1500 B.C. the planet Jupiter ejected material which, in the form of a comet, passed close to the earth; that the subsequent shower of meteorites falling on the earth stopped its rotation, at least for a short time; that the comet itself eventually turned into the planet Venus; and that later, in 747 B.C. and again in 687 B.C., "Mars caused a repetition of the earlier catastrophes on a smaller scale." Given the mass and radius of the earth, as well as its angular speed of rotation ($m_e = 5.98 \times 10^{27}$ g, $r \doteq 6370$ km, $\omega = 2\pi$ rad/day), find the angular momentum of the earth which must be carried off before it will stop rotating and which must be supplied again to start up the rotation afterward. In the most favorable case the meteorites would have fallen on the equator at grazing incidence, traveling with a speed of perhaps 50,000 mi/hr relative to the surface of the earth. What must have been the approximate total mass of the meteorites? What is the total kinetic energy of rotation which was annihilated by the stopping of the earth and meteorites? What happened to that energy, and how could the earth's rotation start again without another collision? How could gravitational, therefore central, forces between passing spherical bodies radically and suddenly change their state of rotation? What would happen to the oceans and the atmosphere if the earth were to stop rotating by Velikovsky's mechanism?

**Problem 17.27.** Explain by means of the LCE the following observations: (a) A sample of naturally radioactive material maintains its temperature above that of its surroundings. (b) In a mass of air that is rising to greater height, the moisture content may freeze. (c) While a battery generates electric energy, it undergoes chemical decomposition at the plates or electrodes.

**Problem 17.28.** Reread the quotations from Joule and Helmholtz at the end of Section 17.9, and discuss, on the basis of the available evidence, the fairness of each.

## RECOMMENDED FOR FURTHER READING

S. C. Brown, *Count Rumford: Physicist Extraordinary*, Garden City, N.Y.: Doubleday Anchor Books, 1962; paperback

Hermann von Helmholtz, "On the interaction of natural forces" and "On the conservation of force," translated from lectures given in 1854 and 1862, in his *Popular Scientific Lectures*; Dover paperback reprint of the 1881 edition

James Prescott Joule, "On matter, living force, and heat" (lecture given in 1847), reprinted together with papers by Mayer and Helmholtz in S. G. Brush, *Kinetic Theory*, Vol. 1, *The Nature of Gases and of Heat*, New York: Pergamon Press, 1965

E. C. Kemble, *Physical Science*, Chapters 13 and 14

## SOURCES, INTERPRETATIONS, AND REFERENCE WORKS

A. C. Bell, *Christiaan Huygens and the Development of Science in the 17th Century*, London: Arnold, 1947

S. C. Brown, *Benjamin Thompson—Count Rumford: Count Rumford on the Nature of Heat*, New York: Pergamon Press, 1967. Includes extracts from his writings.

T. M. Brown, "Resource Letter EEC-1 on the Evolution of Energy Concepts from Galileo to Helmholtz," *American Journal of Physics*, Vol. 33, no. 10 (October 1965)

S. G. Brush, "The wave theory of heat: A forgotten stage in the transition from the caloric theory to thermodynamics," *British Journal for the History of Science*, Vol. 5, pages 145–167 (1970)

D. S. L. Cardwell, *From Watt to Clausius: The Rise of Thermodynamics in the Early Industrial Age*, Ithaca, N.Y.: Cornell University Press, 1971

R. Dugas, *A History of Mechanics*, New York: Central Book Company, 1955

R. Dugas, *Mechanics in the Seventeenth Century*, Chapter XIV

Y. Elkana, "Helmholtz' 'Kraft': An illustration of concepts in flux," *Historical Studies in the Physical Sciences*, Vol. 2, pages 263–298 (1970)

Y. Elkana, "The conservation of energy; a case of simultaneous discovery?" *Archives Internationales d'Histoire des Sciences*, Vol. 23, pages 31–60 (1970)

E. Hiebert, *Historical Roots of the Principle of Conservation of Energy*, Madison, Wisconsin: State Historical Society of Wisconsin, 1962

A. Koslow, *The Changeless Order*, New York: Braziller, 1967. Extracts from writings of Leibniz, Joule, Helmholtz, and others on conservation laws.

T. S. Kuhn, "Energy conservation as an example of simultaneous discovery," pages 321–356 in *Critical Problems in the History of Science* (edited by M. Clagett), Madison, Wisconsin: University of Wisconsin Press, 1959; reprinted in *The Sociology of Science* (edited by B. Barber and W. Hirsch), New York: The Free Press of Glencoe, 1962

Ernst Mach, *The Science of Mechanics*, pages 433–440; *The History and Root of the Principle of Conservation of Energy*, translated from German, Chicago: Open Court Publishing Co., 1911; "On the principle of the conservation of energy," in his *Popular Scientific Lectures*

D. Roller, "The early development of the concepts of temperature and heat: the rise and decline of the caloric theory," pages 117–214 in *Harvard Case Histories in Experimental Science* (edited by J. B. Conant), Cambridge, Mass.: Harvard University Press, 1948; also available as a separate pamphlet

D. W. Theobald, *The Concept of Energy*, New York: Barnes and Noble, 1966

# Chapter 18

# THE LAW OF DISSIPATION OF ENERGY

In the last three chapters we have seen how the conservation laws of mass, momentum, and energy came to be established. With certain modifications, such as the possibility of interconversions between mass and energy, to be discussed in Chapter 30, these laws are still considered valid today. Taken as a whole, they seem merely to confirm an opinion common in the seventeenth century: that the universe is like a perfect machine whose parts never wear out, and which never runs down. In such a "world machine," nothing really changes in the long run; the pieces of matter that constitute our universe can only perform simple or complex cycles of motion, repeated indefinitely. Moreover, Newton's laws of motion make no distinction between past and future: Any sequence of motions predicted by the laws can equally well be run backward.

This idea of the world as a machine (or perhaps as a mechanistic device analogous to a clock) does not in fact correspond to the general scientific viewpoint about the physical world today, nor to the facts, if one looks carefully. As the British physicist William Thomson pointed out in 1874, Newton's laws alone do not explain why a possible state of affairs cannot in general be constructed by simply reversing the velocities of all the atoms in a system:

"If the motion of every particle of matter in the universe were precisely reversed at any instant, the course of nature would be simply reversed for ever after. The bursting bubble of foam at the foot of a waterfall would reunite and descend into the water; the thermal motions would reconcentrate their energy, and throw the mass up the fall in drops reforming into a close column of ascending water. Heat which had been generated by the friction of solids and dissipated by conduction, and radiation with absorption, would come again to the place of contact, and throw the moving body back against the force to which it had previously yielded. Boulders would recover from the mud the materials required to rebuild them into their previous jagged forms, and would become reunited to the mountain peak from which they had formerly broken away. And if also the materialistic hypothesis of life were true, living creatures would grow backwards, with conscious knowledge of the future, but no memory of the past, and would become again unborn. But the real phenomena of life infinitely transcend human science; and speculation regarding consequences of their imagined reversal is utterly unprofitable."*

The gradual decline of the world-machine idea in modern science has been a long and complicated process which is not yet completed, and we shall not be able to tell the full story in this book (it involves, among other things, the rise of evolutionary biology). But we can in this chapter at least indicate one principle that was added to the system of physical laws in the nineteenth century and which throws a different light on the idea of conservation: the *principle of dissipation of energy*, sometimes known as the generalized *second law of thermodynamics*.

## 18.1 NEWTON'S REJECTION OF THE "NEWTONIAN WORLD MACHINE"

It is one of the ironies of the history of science that the one man whose work by the nineteenth century had done most to establish the idea that the world is like a machine or clock was the most eager to reject that idea. Newton pointed out in his *Opticks* (Query 31) the consequences of adopting a principle of vector momentum in a world in which collisions are not always elastic (see Section 16.2):

"By reason of the ... weakness of elasticity in solids, motion is much more apt to be lost than got, and is always upon the decay. For bodies which are either absolutely hard, or so soft as to be void of elasticity, will not rebound from one another. Impenetrability makes them only stop. If two equal bodies meet directly in a vacuum, they will by the laws of motion stop where they meet, and lose all their motion, and remain in rest, unless they be elastic, and receive new motion from their spring.

---

* William Thomson, *Proceedings of the Royal Society of Edinburgh,* Vol. 8, page 325 (1874).

If they have so much elasticity as suffices to make them rebound with a quarter, or half, or three quarters of the force with which they come together, they will lose three quarters, or half, or a quarter of their motion.... Seeing therefore the variety of motion which we find in the world is always decreasing, there is a necessity of conserving and recruiting it by active principles, such as are the cause of gravity, by which planets and comets keep their motion in their orbs, and bodies acquire great motion in falling; and the cause of fermentation, by which the heart and blood of animals are kept in perpetual motion and heat; the inward parts of the earth are constantly warm'd, and in some places grow very hot; bodies burn and shine, mountains take fire, the caverns of the earth are blown up, and the sun continues violently hot and lucid, and warms all things by his light. For we meet with very little motion in the world, besides what is owing to these active principles. And if it were not for these principles, the bodies of the earth, planets, comets, sun, and all things in them, would grow cold and freeze, and become inactive masses; and all putrefaction, generation, vegetation and life would cease, and the planets and comets would not remain in their orbs."

Newton seems to be arguing here that the continual dissipation of motion in the world, which would occur in spite of the law of conservation of momentum, somehow is or must be counteracted by the effects of "active principles." It is tempting to interpret the mysterious phrase "active principles" as a suggestion of the role of *energy* in the universe; but the tenor and context of Newton's argument do not support the hypothesis that he was willing to let the continued operation of his system depend simply on the conservation of energy. Indeed, in an earlier edition of the *Opticks* there appeared the phrase "what is owing either to these active principles or to the dictates of a will" which was modified in the 1717 edition (perhaps as a result of the criticisms mentioned below) to read "what is owing to these active principles," as in the quotation above.

A further indication of Newton's reluctance to accept that his laws of motion would guarantee the existence of an eternal world machine is given by another passage from the same work:

"For while comets move in very eccentric orbs in all manner of positions, blind fate could never make the planets move one and the same way in orbs concentric, some inconsiderable irregularities excepted which may have arisen from the mutual actions of comets and planets upon one another, and which will be apt to increase, till this system wants a reformation."

What would a "reformation" be? A miraculous act of God, putting the comets and planets back into their proper orbits from which they had strayed? Any such suggestion of a need for divine intervention would cast doubt on the adequacy of the laws of motion to maintain the world machine in perfect working order. Newton was challenged directly on this point by Leibniz, the German philosopher-scientist who had previously developed the theory of *vis viva* (Section 17.1) and who had invented the calculus independently of Newton. In a letter (soon made public) to Caroline, Princess of Wales, in 1715, Leibniz began by complaining that Newton and his friend, the philosopher John Locke, were undermining belief in God and the soul. He continued:

"Sir Isaac Newton, and his followers, have also a very odd opinion concerning the work of God. According to their doctrine, God almighty needs to wind up his watch from time to time; otherwise it would cease to move. He had not, it seems, sufficient foresight to make it a perpetual motion. Nay, the machine of God's making, is so imperfect, according to these gentlemen, that he is obliged to clean it now and then by an extraordinary concourse, and even to mend it, as a clockmaker mends his work; who must consequently be so much the more unskilful a workman, as he is

oftened obliged to mend his work and set it right. According to my opinion, the same force and vigour [energy] remains always in the world, and only passes from one part to another, agreeably to the laws of nature, and the beautiful pre-established order . . . ."*

Rather than replying directly, Newton dictated his answers to a young theologian, Samuel Clarke, who had previously translated Newton's *Opticks* into Latin. Speaking through Clarke, he firmly rejected the idea of a world machine that can continue to run forever without divine intervention, along with Leibniz's suggestion that the law of conservation of energy is sufficient to account for all natural phenomena. He pointed out that the clockmaker, whose highest achievement would be to make a clock that runs forever without any need for adjustment, does not himself create the forces and laws of motion that enable the clock to run, but must take them as given. But since God, according to Newton, "is Himself the author and continual preserver of [the world's] original forces or moving powers," it is

". . . not a diminution, but the true glory of His workmanship, that nothing is done without his continual government and inspection. The notion of the world's being a great machine, going on without the interposition of God, as a clock continues to go without the assistance of a clockmaker, is the notion of materialism and fate, and tends (under pretence of making God a *supramundane* intelligence) to exclude providence and God's government in reality out of the world."

Thus the theory of the clockwork universe, far from being complimentary to God, as Robert Boyle, Leibniz, and others had claimed, was likely to lead instead to atheism. Skeptics, Newton predicted, could easily push the creation of the universe—the first and last act of God—further and further

---

*  H. G. Alexander, *The Leibniz-Clarke Correspondence*, Manchester: Manchester University Press (1956), pp. 11–12.

backward in time and might even conclude eventually that the universe has *always* existed as it does now.

As Newton's system of the world was further developed by European scientists in the eighteenth and early nineteenth centuries (especially Euler and the French mathematicians Lagrange, Laplace, and Poisson), it appeared that the instability which Newton had predicted would result from mutual gravitational interactions of comets and planets would not occur; these interactions would simply cause the average distances of the planets from the sun to oscillate periodically within finite limits. Thus, as Newton had feared, the need for divine intervention or guidance to keep the system running smoothly seemed to disappear. According to legend, when Napoleon met Laplace, after glancing through Laplace's masterpiece, *Celestial Mechanics*, Napoleon remarked that he found no mention of God at all in the book; to which the author replied, "Sir, I have no need of that hypothesis!" The idea of the world machine had triumphed—at least for the time being; and it seemed to be established that Newton's laws by themselves guaranteed the stability of the solar system over indefinitely long periods of time, both past and future.

## 18.2 THE PROBLEM OF THE COOLING OF THE EARTH

Although the successes of Newtonian planetary theory seemed to lead convincingly to the idea of the world as a machine, speculations about the earth's interior in the eighteenth and early nineteenth centuries gave rise to another viewpoint. We must now venture at least briefly into geophysics, an important though sometimes neglected part of physics.

Many writers in earlier centuries had suggested that the earth was originally formed in a hot, molten state, and has subsequently cooled down, solidifying at least on the outside. Leibniz endorsed

this theory in 1693, arguing that many rocks found on the surface of the earth appear to have been formed by the action of fires which may still be burning inside the earth. The great French naturalist Buffon (1707–1788) suggested that the discovery of ivory tusks of elephants in polar regions, where these animals cannot now survive, supports the idea that the earth's surface was once warmer than it is now. Buffon also conducted a series of experiments on the cooling of heated spheres of iron and other substances, in order to estimate the time required for the earth to cool down to its present temperature from the postulated molten state; his final result was about 75,000 years. While this figure seems extremely low in the light of twentieth-century knowledge, it represented a shockingly long expanse of time to his contemporaries, who believed that the chronology of the Bible fixed the creation of the universe at a time only a few thousand years ago.

From the theory of the earth's central heat and gradual refrigeration, it was but a short step to the conjecture that all bodies in the universe are cooling off and will eventually become too cold to support life. This step seems to have been first taken explicitly by the French astronomer Jean-Sylvain Bailly (1736–1793) in his writings on the history of astronomy. According to Bailly, all the planets must have an internal heat and are now at some particular stage of cooling; Jupiter, for example, is still too hot for life to arise for several thousand more years; the moon, on the other hand, is already too cold. But Bailly predicts that all bodies must eventually reach a final state of equilibrium in which all motion ceases.

James Hutton (1726–1797), one of the founders of geological science in Britain, accepted the hypothesis that the inside of the earth is now much hotter than the surface as a convenient explanation for volcanos and hot springs, but did not believe that there had been any cooling during past epochs. In his *Theory of the Earth* (1795), he wrote that "subterraneous fire had existed previous to, and ever since, the formation of this earth," but also "that it exists in all its vigour at this day." Hutton proposed a cyclic view of the earth's history, with periods of erosion and denudation leading to destruction of mountains and even entire continents, followed by consolidation of sediments and uplifting of new continents (powered by the subterranean fires). Hutton's disciple John Playfair explained this cyclic viewpoint in a widely read book, *Illustrations of the Huttonian Theory of the Earth* (1802):

"How often these vicissitudes of decay and renovation have been repeated, is not for us to determine: they constitute a series, of which, as the author of this theory has remarked, we neither see the beginning nor the end; a circumstance that accords well with what is known concerning other parts of the economy of the world. In the continuation of the different species of animals and vegetables that inhabit the earth, we discern neither a beginning nor an end; and, in the planetary motions, where geometry has carried the eye so far both into the future and the past, we discover no mark, either of the commencement or the termination of the present order. It is unreasonable, indeed, to suppose, that such marks should anywhere exist. The Author of nature has not given laws to the universe, which, like the institutions of men, carry in themselves the elements of their own destruction. He has not permitted, in his works, any symptoms of infancy or of old age, or any sign by which we may estimate either their future or their past duration. He may put an end, as He no doubt gave a beginning, to the present system, at some determinate period; but we may safely conclude, that this great *catastrophe* will not be brought about by any of the laws now existing, and that it is not indicated by any thing which we perceive."

In a supplementary note, Playfair cited the mathematical investigations of Lagrange and Laplace which have shown that "all the variations in our [solar] system are periodical; that they are confined within certain limits; and consist of alternate diminution and increase . . . . The system is thus endowed with a stability, which can resist the lapse of unlimited duration; it can only perish by an external cause, and by the introduction of laws, of which at present no vestige can be traced." The Huttonian system itself has the same stability; both systems, remarks Playfair, are just like machines designed by a superior intelligence.

The principle that geological phenomena are to be explained by using only those laws and physical processes that can *now* be observed to operate—as opposed to postulating catastrophic events in the past, such as the Biblical Flood—was known as *Uniformitarianism*. This principle was used by Sir Charles Lyell (1797–1875) in his great synthesis of geological knowledge, first published in 1830, and provided a basis for most geological speculation up to the twentieth century. Although uniformitarianism was an essential tool of the movement to make geology a science independent of theological dogma, some nineteenth-century scientists criticized it as being itself a dogma incompatible with the laws of physics. In particular, Lyell refused to accept the theory that the earth had previously been molten and has been cooling by losing heat; that would imply that some geological processes had been much more intense in the distant past than now.

The theory of the cooling of the earth had in the meantime been put on a quantitative basis by the labors of the French mathematician Jean Baptiste Joseph Fourier (1768–1830). Motivated originally by the problem of terrestrial temperatures, Fourier had developed a theory of heat conduction in solids, together with a powerful method for solving the equations which arose in the theory. Both the general theory of heat conduction and the method of solving its equations (by expressing an arbitrary mathematical function as the sum of an infinite series of trigonometric functions) have had enormous impact on modern theoretical physics. Although they cannot be discussed here, one significant fact must be remarked: Fourier's heat conduction equation, unlike Newton's laws of motion, is *irreversible* with respect to time. *It postulates that wherever temperature differences exist, they tend to be evened out by the flow of heat from high temperature to low.* Conversely, the existence of such temperature differences now can be used to calculate the even greater differences of temperature that existed in the past.

Fourier used his theory to study the thermal history the earth may have had, taking account of the periodic variations in the heat received from the sun at various points on the surface, combined with a slowly diminishing internal heat of the earth. By using measurements of the rate of increase of temperature as one descends into the interior of the earth, in mines, together with estimates of the thermal conductivity of rocks, Fourier was able to infer the earth's internal temperatures at earlier times. He concluded that the theory of a hot interior, gradually cooling down, is compatible with available data, although in recent geological epochs the temperature of the interior has not had any significant effect on the temperature of the earth's surface.

A more comprehensive theory was developed by William Hopkins, a British geophysicist who was also known as a brilliant teacher of mathematics at Cambridge University. Hopkins pointed out in 1839 that when a hot liquid sphere cools down, it does not necessarily solidify first at the outside, even though the outer surface is most directly exposed to the lower temperature of the surrounding space. For a sphere the size of the earth, the weight of the outer layers produces enormous pressures on the inside, and these pressures may produce solidification even at temperatures far above the normal melting point of the substance. In 1842, Hopkins published an elaborate mathematical study of the effect a partially fluid earth could have on the moon's motion, from which he concluded that a fairly large portion of the earth's interior must be solid. He proposed that the central core of the earth is solid, and that a liquid region lies between this and a fairly thick crust. According to Hopkins' theory, changes in the surface climate have been produced by geographical changes rather than by cooling of the interior. This idea, which had previously been proposed by Lyell, was reinforced by the evidence collected by the Swiss-American naturalist Louis Agassiz (1807–1873), showing that the surface had been much *colder* during certain epochs in the past than at present, and that ice had covered large parts of the continents.

Hopkins emphasized the point that if the earth's surface can warm up after an Ice Age, the older theories of the cooling of the earth cannot provide adequate explanations for most geological phenomena. But at the same time, while reviewing the situation in his presidential address to the Geological Society of London in 1852, Hopkins reiterated the principle that over very long periods of time the cooling of the earth *is* important, and imposes an overall irreversibility or progressive change on terrestrial phenomena:

"I assume the truth of the simple proposition, that if a mass of matter, such, for instance, as the earth with its waters and its atmosphere, be placed in a space of which the temperature is lower than its own, it will necessarily lose a portion of its heat by radiation, until its temperature ultimately approximates to that of the circumambient space, unless this reduction of temperature be prevented by the continued generation of heat."

Moreover, the earth cannot be saved from this fate by postulating an influx of radiation from the sun and stars to balance the heat it loses, for all heat in the universe must be subject to the general tendency to flow from hot to cold. For this reason, Hopkins insisted, "I am unable in any manner to recognize the seal and impress of eternity stamped on the physical universe"; the universe must be governed by this overall law of equalization of temperature, which makes it tend irreversibly toward an ultimate limit, the cold of outer space.

## 18.3 THE SECOND LAW OF THERMODYNAMICS AND THE DISSIPATION OF ENERGY

Hopkins' enunciation of the principle of irreversible heat flow in geological processes did not attract much attention outside the audience of geologists to which he addressed it, but as it happened an even more general principle of irreversibility was proclaimed in the same year, 1852, by William Thomson (1824–1907) (later to become Lord Kelvin).* To understand the basis for Thomson's principle, we must go back to the origin of thermodynamics in the analysis of steam engines.

In 1824, a young French engineer, Sadi Carnot, published a short book entitled *Reflections on the Motive Power of Fire*, in which he raised the question: What is the maximum *duty* of an engine? (The *duty* of an engine was defined as the amount of mechanical work it could perform, using a given amount of fuel; it is related to the concept of *efficiency*.) Carnot used the caloric theory of heat in his analysis of steam engines; he assumed that heat is not actually converted into work, but that the flow of heat from a hot body to a cold body can be used to *do* work, just as the fall of water from a high level to a low level can be used to do work. Although the caloric theory upon which his results were based was later rejected by other scientists (and even by Carnot himself, in notes that remained unpublished until 1878), his conclusions are still valid. He found that there *is* a maximum value for the duty. There is a fixed upper limit on the amount of work that can be obtained from a given amount of heat by using an engine, and this limit can never be exceeded regardless of what substance—steam, air, or anything else—is used in the engine.

Even more significant than the existence of this maximum limit is Carnot's conclusion that all real

---

* Thomson was one of several outstanding theoretical physicists who had been tutored in mathematics by Hopkins at Cambridge; his interest in geophysical applications may have been reinforced by this association.

engines fail to attain that upper limit in practice. The reason is that whenever a difference of temperature exists between two bodies, or two parts of the same body, there is a *possibility* of doing work by allowing heat to expand a gas as the heat flows from one body to the other. But if heat flows by itself from a hot body to a cold body, and we do not design our engine properly, we will lose the chance of doing work that might have been done.

Carnot's analysis of steam engines shows that the process of equalization of temperature by the flow of heat from hot bodies to cold bodies is always taking place, not only in engines but in all of nature, and that this equalization represents a loss of the possibility of doing mechanical work. This is what we mean when we say that energy is "dissipated"— *the total amount of energy in a closed system (or in the world) is always the same, but energy tends to transform itself into less useful forms.*

After the discovery of the conservation law for energy (Chapter 17), Carnot's conclusions about steam engines were incorporated into the new theories of heat, and became known as the second law of thermodynamics. This law has been stated in various ways, all of which are roughly equivalent, and express the idea that *the tendency of heat to flow from hot to cold makes it impossible to obtain the maximum amount of mechanical work from a given amount of heat.*

When William Thomson in 1852 asserted a "universal tendency in nature to the dissipation of mechanical energy," he relied primarily on Carnot's principle. But he also drew the conclusion that "within a finite time past the earth must have been, and within a finite period of time to come the earth must again be, unfit for the habitation of man . . . ." Thomson was especially interested in the theory that the earth has been very hot in the past and will be very cold in the future, for he was one of the first scientists in Britain to take up Fourier's theory of heat conduction and its geophysical applications.

In the 1860s Thomson came to the conclusion that a period of 100 million to 200 million years might have been required for the earth, assumed initially to be at a uniform temperature of 7000° to 10,000°F, to have reached its present state. Before that, the temperature at the surface would have been above the melting points of all known rocks. At the same time, he estimated that the sun could not have been illuminating the earth for much more than a few hundred million years, assuming that its energy supply was limited to sources known in the nineteenth century (primarily gravitational contraction).

But these conclusions seemed to be inconsistent with statements of Uniformitarian geologists that processes such as erosion had been continuing at the same rate for hundreds of millions of years, and that the history of the earth is essentially cyclic with no long-term changes (see Section 18.2). Thomson therefore argued that the basic principles of Uniformitarian geology were contrary to established laws of physics, in particular the second law of thermodynamics and Fourier's theory of heat conduction.

Thomson's attack on the geologists was also indirectly an attack on Charles Darwin's theory of evolution, for in his *Origin of Species* (1859) Darwin had conjectured, in support of his assumption that sufficiently long times have been available for natural selection to operate, that certain geological processes might have been going on for as long as 300 million years. Insofar as such remarks made it appear that the theory of evolution depended on the continuous existence of conditions favorable to life for hundreds of millions of years, Darwin had unwittingly given his opponents a new argument: Such long periods of time were not considered meaningful or acceptable to some physicists, such as Thomson.

The debate between William Thomson (supported by other physicists) and the Uniformitarian geologists continued throughout the rest of the nineteenth century. The *problem of the age of the earth*, as it came to be called, seemed to call forth a bitter conflict between geology and physics, with each side asserting that its own evidence should

not be ignored simply because it seemed to be incompatible with the other side's. Because of the prestige of physics, resulting from its past successes, the geologists were forced to revise their estimates of the rate of geological processes in order to fit the shorter time scale that seemed to be required by Thomson's calculations. Supporters of Darwinian evolution tried to deflect Thomson's attack by pointing out that the validity of evolutionary theory did not depend on any particular time scale; since there was as yet no definite measurement of the rate of evolutionary change, even 10 million years might be sufficient.

At the beginning of the twentieth century the situation changed radically, for a reason that neither geologists nor physicists could have anticipated—the discovery of radioactivity. After Marie and Pierre Curie isolated the new element, radium, from samples of rocks, it was realized that the heat generated by radioactivity inside the earth must be taken into account in any calculations of the rate of cooling, and might easily be sufficient to compensate for most or all of the loss of heat by conduction through the crust. Thomson's conclusions would thus be invalid because he had incorrectly assumed that there are no sources of heat inside the earth.

An even more direct refutation of his results came from a more detailed analysis of the process of radioactive decay (see Chapter 27). Measurements of the relative proportions of lead, helium, radium, and uranium in rocks could be used to estimate the "ages" (time elapsed since solidification) of those rocks. The first results of this radioactive dating technique suggested that the earth's solid crust was nearly 3 billion years old: a figure ten times as great as Darwin's 300 million years, which Thomson had called much too high!

Evolution, cooling of the earth, and radioactivity are all processes which fall outside the neat framework of Newtonian physics. They are *progressive* or *irreversible*, but not *cyclic*. The failure of Thomson's theory does not by any means prove that physical theory cannot deal with irreversible processes—it is easy enough to fix his analysis by inserting a heat-source term in Fourier's heat-conduction equation. But it does suggest that the physicist who is accustomed to working with laboratory experiments and mathematical analysis may have something to learn from scientists who are more experienced in observing and interpreting natural phenomena that take place over large spans of time and space. Astronomy and the earth sciences are not to be regarded merely as fields in which already-established physical laws may be applied, but rather as possible sources for new physical discoveries and concepts.

## 18.4  ENTROPY AND THE HEAT DEATH

Although Thomson's quantitative estimates of the age of the earth turned out to be much too small, his qualitative principle of dissipation of energy was not challenged. Hermann von Helmholtz, in a lecture delivered in 1854, pointed out that Thomson's principle implied the cooling of the entire universe. All energy would eventually be transformed into heat at a uniform temperature, and all natural processes would cease; "the universe from that time forward would be condemned to a state of eternal rest." Thus was made explicit the concept of the *heat death* of the universe, as foreseen nearly a century earlier by Bailly, but now apparently based much more firmly on the new science of thermodynamics.

Another way of stating the principle of dissipation of energy was suggested by the German physicist Rudolf Clausius in 1865. Clausius had previously introduced a new concept to which he now gave the name *entropy*,* defined in terms of the heat transferred from one body to another. More precisely he defined the *change* in entropy, $\Delta S_1$, resulting from a transfer of heat $H$ from a body 1 at

---

* From the Greek words ἐνέργεια (energy) and τροπή (transformation). The concept had been used since 1854 under the cumbersome name, "equivalence-value of a transformation."

temperature $T_1$ as

$$\Delta S_1 = -H/T_1,$$

where $T_1$ is the absolute temperature (see Section 19.3). When the same amount of heat is received by another body, 2, that is kept at temperature $T_2$, the change in entropy is

$$\Delta S_2 = +H/T_2.$$

If $T_2$ is less than $T_1$, the total change in entropy, $\Delta S = \Delta S_1 + S_2$, will be *positive*. The change in entropy would be zero only if $T_1 = T_2$. (The total amount of entropy in a body cannot be measured directly, but is usually defined in such a way that $S = 0$ at absolute zero temperature, $T = 0$.)

The above indicates that the total entropy of the system always increases whenever heat flows from a hot body to a cold body. It also increases whenever mechanical energy is changed into internal (thermal) energy, as in inelastic collisions and frictional processes, since here one body absorbs heat but no other body loses it.

A more general definition of entropy was proposed by the Austrian physicist Ludwig Boltzmann in 1878. According to Boltzmann's definition, entropy depends on the probabilities of molecular arrangements, and can change even when there is no heat flow, e.g., if a system becomes more random or disordered (when you scramble an egg, or mix two different pure substances).

The principle of dissipation of energy (generalized second law of thermodynamics) can now be stated very simply: *The entropy of an isolated system always tends to increase.* In particular, this means that heat does not flow by itself from cold bodies to hot bodies; a ball dropped on the floor will not bounce back higher than its original position by converting heat into mechanical energy; and an egg will not unscramble itself. All these (and many other) events would not be violating any of the principles of Newtonian mechanics, if they took place—but they would decrease the entropy of an isolated system and are thus forbidden by the second law of thermodynamics.

Clausius stated the two laws of thermodynamics in the form:

1) The energy of the universe is constant.

2) The entropy of the universe tends toward a maximum.

These laws have been verified directly only for terrestrial phenomena, and to extend them to the entire universe requires a rather bold application of Newton's rules of reasoning (Section 9.3). Indeed, within the last few decades, theories of the evolution of the universe have been proposed which would violate both laws. Nevertheless, they are so useful that they will probably continue to be accepted as working hypotheses until definitely disproved.

The second law implies that *time* is not a neutral mathematical variable in the same sense as the space coordinates $(x, y, z)$; it has a definite *direction*—what has been called "time's arrow"—pointing from past to future. This statement is rescued from the status of a ponderous triviality only by the recognition that no such directionality is associated with Newton's laws of motion. As William Thomson's statement quoted at the beginning of this chapter vividly illustrates, it is difficult to imagine a world in which the second law is *not* valid. In such a world nothing would forbid such reversals as the ones he describes.

The second law gained considerable notoriety in the last part of the nineteenth century; the concept of an eventual heat death for the universe exercised a certain morbid attraction on popular writers, caught up in the *fin de siècle* mood of pessimism that swept some parts of European and American society. Since increase of entropy means more randomness and disorder, perhaps that was an explanation for social disintegration and environmental degradation!

The American historian Henry Adams tried to put all these ideas together in a series of essays on the application of thermodynamics to human history. He quoted Thomson's statement of the principle of dissipation of energy, remarking that "to the vulgar and ignorant historian it meant only that the ash-heap was constantly increasing in size." If historians were to concern themselves, as Adams thought they should, with the future as well as the past, then they could hardly ignore the latest result of physics, which implied (according to him) that human society along with the physical universe must end in degradation and death, even though this prophecy was distasteful to the evolutionists, who preached nothing but eternal progress. Adams complained that evolution, together with Lyell's geological doctrine of Uniformitarianism, had already started to take the place of religious dogma, and "in a literary point of view the Victorian epoch rested largely—perhaps chiefly—on the faith that society had but to follow where science led . . . in order to attain perfection." But Adams could find numerous examples in anthropological studies and the popular press to confirm the notion that the human race was going from bad to worse, and would soon be extinct.*

A more balanced interpretation of the second law will have to be postponed until Chapter 22, but it should be evident already why the writer C. P. Snow, in his famous book, *The Two Cultures and the Scientific Revolution*, remarked that for a humanist to be completely ignorant of the second law of thermodynamics would be as shocking as for a scientist to be completely ignorant of the works of Shakespeare.

## RECOMMENDED FOR FURTHER READING

H. Butterfield, *The Origins of Modern Science*, Chapter 12

R. Feynman, *The Character of Physical Law*, Chapter 5

E. C. Kemble, *Physical Science*, Chapter 15

---

* Henry Adams, *The Degradation of the Democratic Dogma* (essays first published 1894–1909).

## SOURCES, INTERPRETATIONS, AND REFERENCE WORKS

F. D. Adams, *The Birth and Development of the Geological Sciences*, Dover paperback reprint of the 1938 edition; Chapters VII–XI

Henry Adams, *The Degradation of the Democratic Dogma*, New York: Capricorn Books (Putnam's), 1958, paperback reprint of the 1919 edition. Three essays on the application of thermodynamics to history, and a long introduction by Henry's brother Brooks on the background of the Adams family.

H. G. Alexander (editor), *The Leibniz-Clarke correspondence, with extracts from Newton's Principia and Opticks*, New York: Philosophical Library, 1956

M. K. Barnett, "Sadi Carnot and the Second Law of Thermodynamics," *Osiris*, Vol. 13, pages 327–357 (1958)

S. G. Brush, "Thermodynamics and History," *The Graduate Journal*, Vol. 7, pages 477–565 (1967)

Sadi Carnot, *Reflections on the Motive Power of Fire, and Other Papers on the Second Law of Thermodynamics by E. Clapeyron and R. Clausius* (edited by E. Mendoza), New York: Dover Publications, 1960; paperback

B. Gal-Or, "The crisis about the origin of irreversibility and time anisotropy," *Science*, Vol. 176, pages 11–17 (1972)

F. C. Haber, *The Age of the World from Moses to Darwin*, Baltimore: Johns Hopkins University Press, 1959

E. Hiebert, "The uses and abuses of thermodynamics in religion," *Daedalus*, Fall 1966

F. O. Koenig, "On the history of science and of the second law of thermodynamics," pages 57–111 in *Men and Moments in the History of Science* (edited by H. M. Evans), Seattle: University of Washington Press, 1959

D. Kubrin, "Newton and the cyclical cosmos: Providence and the mechanical philosophy," *Journal of the History of Ideas*, Vol. 28, pages 325–346 (1967)

S. Toulmin and J. Goodfield, *The Discovery of Time*, New York: Harper, 1965; paperback

PART F

# ORIGINS OF THE ATOMIC THEORY IN PHYSICS AND CHEMISTRY

# ORIGINS OF THE ATOMIC THEORY IN PHYSICS AND CHEMISTRY

The story of the development and the gradual acceptance into science of the atomic view of matter is astonishing in many ways—in its origins, in the length and vigor of the debate, in the diversity of men and fields involved, in the often quite unexpected concurrence of separate arguments, and in the ever-growing flood of consequences that has finally swept over all branches of science.

The account in this section will carry the development to about the end of the nineteenth century. In retrospect it is clear that the model of the atom at that point had grown out of three separate types of questions: What is the physical structure of matter, particularly of gases? What is the nature of heat? What is the basis of chemical phenomena? Although at first glance these three topics may seem completely dissociated, the answers were obtained simultaneously, in a *joint* conceptualization of the nature of heat, of gases, and of chemical reactions, in terms of one quantitative atomic scheme. In fact, the atomic view and the conservation of energy were very closely linked concepts, and the fight for both involved the same set of protagonists. As in a fugue, the separate themes introduce themselves, approach one another in the development, superpose, grow apart, and in the end again coalesce.

The concepts and derivations we shall encounter in this story are not always simple, but there exists hardly a better example on this level to show a conceptual scheme evolving from the labors of generations—a tapestry spun out of widely scattered experimental observations, shrewd assumptions, simplifying approximations, detected errors, penetrating intuitions, and mathematical manipulations. And there are three other important reasons why the closer study of this topic may be fruitful. First, it will give us a historically important and generally useful model of the elementary structure of matter. From this model, and from the outburst of successful speculations which it generated, there arose much of the physics and chemistry of the period immediately preceding the twentieth century—and its failures were the stimuli for much in the new physical science of our own time. Second, not only is the kinetic theory a connecting link between the Newtonian and the contemporary approaches to physics, but it leads from *macrocosmic* physics to the world of *sub-microscopic* phenomena. And finally, this topic will introduce us, though of necessity rather sketchily, to the important type of physical law which, instead of *certainty* and the prediction of individual events, deals with *probability* and the discussion of a multitude of events.

# Chapter 19

# THE PHYSICS OF GASES

## 19.1 THE NATURE OF GASES— EARLY CONCEPTS

From the time of the earliest recorded philosophical speculations to this day, the mind of man has been haunted by some paradoxical propositions: On the one side we can with our hands cut or subdivide gross matter—a rock or a quantity of water—into smaller and smaller parts, until the crudeness of our tools or the deficiency of our sight, *but never the material itself*, calls a halt to our experiments. Consequently, we might conclude that matter is inherently infinitely subdivisible. But on the other hand, our imagination encounters difficulty when we demand that it present to us matter as a collection of truly infinitely small parts, and this view makes for logical problems as well. So where our senses tend to one view of matter, our reason persuades us of another.

The great compromise was formulated about 2400 years ago by the Greek philosophers Leucippus and Democritus, and later extended by Epicurus. It is of course the atomistic idea. Our senses and our reason would both rest satisfied if we assumed that matter is indeed divisible far beyond immediate experience, but that there does exist an ultimate substructure of infinitely hard, uncuttable,

indivisible particles, which we might provisionally name *corpuscles* or *atoms*.*

The Roman poet Lucretius, whose *De Rerum Natura* was cited before in connection with the origins of the law of conservation of mass, exemplified the teaching of those now largely lost Greek original works, and gave a summary of this atomistic science that "marks the climax of Roman scientific thought" (Sarton)—although we have noted, of course, that this work is not to be regarded as primarily a science text.

His leading theme is "All nature then, as it exists by itself, is founded on two things: there are bodies and there is void in which these bodies are placed and through which they move about . . . ." Ordinary matter is composed of these two realities: solid, everlasting particles, and, on the other hand, the void, or what we might loosely call the vacuum. Note that by their very definition these atoms are not directly perceptible, and until about 150 years ago there were very few fruitful and unambiguous consequences that could be drawn from the atomic hypothesis to serve as experimental confirmations of this challenging speculative idea. Even at the

---

* From the Greek *atomos*, indivisible.

turn of this century, when such confirmation was at hand, there were important scientists who still rejected the concept of atoms as too convenient a fable, unworthy of serious consideration by "hard-boiled" investigators.

It is therefore not surprising that the structure of matter was for a long time the subject of inconclusive discussions, whether the atomistic view was rejected (as by Plato, Aristotle, and their diverse followers) or in some form accepted (by Galileo, Gassendi, Bacon, Descartes, Newton, Leibniz, and many of their contemporaries). We might speculate that the idea of small particles became more convincing as a consequence of Newton's work, in which, as we have seen, the postulates leading to the law of universal gravitation provided that the *particles* of matter are the agencies of mutual attraction. Of course, these are not necessarily identical with ultimately *indivisible* particles or atoms. Yet the surpassing success of the treatment of gravitation in terms of the mechanics of particles could turn the thoughts again to the idea that the mechanics of small discrete bodies is the key to all phenomena. Newton himself wrote in the introduction to the *Principia*:

"Then from these [gravitational] forces, by other propositions which are also mathematical, I deduce the motions of the planets, the comets, the moon, and the sea. I wish we could derive the rest from the phenomena of nature by the same kind of reasoning from mechanical principles; for I am induced by many reasons to suspect that they all may depend upon certain forces by which the little parts of bodies...are either mutually impelled to one another and crowd together according to regular figures, or are repelled and recede from one another; which forces being unknown, philosophers have hitherto attempted the search of nature in vain, but I hope the principles here laid down will have brought some light either to that or to a truer method of philosophy."

Later, in the *Opticks* (published 1704), we find this visionary, yet still unformalized opinion on atomism:

"All these things being considered, it seems probable to me that God in the Beginning formed Matter in solid, massy, hard, impenetrable, movable Particles, of such Sizes and Figures, and with such other Properties, and in such Proportion to space, as most conduced to the end for which he formed them; and as these primitive Particles being Solids, are incomparably harder than any porous Bodies compounded of them; even so very hard as never to wear or break in pieces; no ordinary Power being able to divide what God himself made one in the first Creation.... And therefore that Nature may be lasting, the Changes of corporeal Things are to be placed only in the various Separations and new Associations, and Motions of these permanent Particles; compound Bodies being apt to break, not in the midst of solid Particles, but where those Particles are laid together...

"God is able to create Particles of Matter of several Sizes and Figures and in general Proportion to the space they occupy, and perhaps of different Densities and Forces.... Now, by the help of these Principles, all material Things seem to have been composed of the hard and solid Particles above mentioned—variously associated in the first Creation by the Counsel of an intelligent Agent. For it became him who created them to set them in order. And if he did so, it's unphilosophical to seek for any other Origin of the World, or to pretend that it might arise out of a Chaos by the mere Laws of Nature; though being once form'd, it may continue by those Laws for many Ages."

Part of Newton's preoccupation here is theological, and for a most excellent historic reason that we must not pass over if we wish to understand fully the development of the atomistic view. Ever since its early inception, atomism has generally been regarded as atheistic; as the excerpt from Lucretius (Section 15.1) indicated, the atomist openly professed to give an explanation of matter and events *not* in terms of impenetrable designs of an ever-present Creator and Ruler, but by means of the interplay and structure among material bodies. In the seventeenth century, philosophers such as Pierre Gassendi had attempted to remove the taint of atheism by proposing that atoms are not self-animated, but are inert pieces of matter which required a divine agency to set them into motion at the creation of the world. In the "mechanical philosophy," as it was advanced especially in England by Robert Boyle, God was assigned the role of designer and creator of the world machine which, once set in motion, could run from then on by His pleasure without continual active intervention; all observable events should then be ultimately explainable by the configurations, sizes, and shapes of the eternal particles or by the mathematical laws of force and motion ruling among them.

As we have seen in the previous chapter, Newton himself did not accept the extreme version of the world-machine theory which would deny that God now takes any active part in the physical world. Yet the above quotation from the *Opticks* shows that in formulating the principles of his own system of the world, Newton was primarily concerned with specifying the results of God's initial creative acts, so that in practice his system could easily be interpreted as a world machine.

Some of Newton's other remarks on the properties of atoms did give support to those who were dissatisfied with the mechanistic philosophy advocated by Descartes (Section 6.1). The Cartesians rejected the hypothesis of "action at a distance," preferring to believe that all apparent long-range forces, such as gravity, could be explained by contact actions propagated through an intervening material medium (the ether). Newton, while rejecting in principle the possibility of true action at a distance (see his letter to Bentley, quoted in Section 11.10), did explain many phenomena in terms of attractive and repulsive forces acting

between atoms. These remarks encouraged scientists like Boscovich and Priestley in the eighteenth century to develop an atomic theory of matter that relied more on the atoms' attractive and repulsive forces than on their hardness and impenetrability. The concept of an atom as a point center of attractive and repulsive forces survived into the twentieth century, along with the alternative concept of the atom as a billiard ball that could interact with other billiard balls only by contact.

## 19.2 AIR PRESSURE

In the meantime the atomistic view of matter was extended from an entirely different direction—the research on gases.

Galileo, in his *Two New Sciences* (1638), said that a suction pump cannot lift water more than about 10.5 meters. This fact was presumably well known by the time Galileo wrote his book; pumps were already being used to obtain drinking water from wells, and to remove water from flooded mines, so their limitations must have become evident to many workmen. One important consequence of the limited ability of pumps to lift water was that some other method was needed to pump water out of deep mines, and this need provided the initial stimulus for the development of steam engines that allowed water to be conveyed out in buckets. Another consequence was that physicists in the seventeenth century became curious to discover the reason why the suction pump worked at all, as well as why there should be a limit on its ability to raise water.

If you remove air from a container and create a vacuum, there is a tendency for things to be sucked in. The Aristotelian philosophers explained this fact by saying that Nature abhors a vacuum. It is unnatural, they said, for space not to be filled with matter, and therefore matter will move so as to fill up any empty space. This is an example of a teleological explanation, one based on an "ultimate cause" (another one would be: "Rain falls because crops need water"). One of the basic goals of the

Evacuated spaces

*h*

**Fig. 19.1** Torricelli barometer (two forms). The mercury column in each vertical glass tube is balanced by the pressure of the atmosphere on the free mercury surface in the trough. Usually *h* is about 760 mm at sea level.

new mechanical philosophy of the seventeenth century was to eliminate such teleological explanations, and to explain phenomena instead in terms of immediate physical causes. The Aristotelian theory seemed especially weak in this particular case: Even if one accepts the assertion that Nature abhors a vacuum, one finds it difficult to explain why this abhorrence is just sufficient to raise water by 10.5 m, no more, no less!

Evangelista Torricelli, once briefly the pupil of Galileo, realized that the atmosphere exerts a pressure at the surface of the earth, and suspected that this air pressure might be sufficient to account for the phenomena previously attributed to Nature's abhorrence of a vacuum. He made the fortunate guess that mercury, a liquid nearly 14 times as dense as water, might be more convenient than water for laboratory experiments; indeed, it turned out that the same vacuum pump that could raise a column of water 10.5 m could lift a column of mercury only 0.76 m.

The pump itself is not even necessary to hold up a column of liquid in the simple instrument that came to be known as the Torricelli *barometer*. One takes a straight glass tube somewhat more than 0.76 m long, open at one end and sealed at the other; a bowl; and enough mercury to fill the tube

and the bowl. Fill the tube to the brim, close the open end with your finger, turn it upside down in the bowl, and remove your finger, and you will find that the mercury runs down into the bowl, but only until the difference between the level in the tube and the level in the bowl is about 0.76 m. The result is shown in Fig. 19.1 (the left tube). There is now found to be an evacuated space at the top of the tube above the mercury. Torricelli then repeated the same procedure with a tube which had a large bulb at the closed end, thereby making available an evacuated space of larger size in which small objects could be fastened.

The announcement of Torricelli's experiments in 1643 prompted other scientists (Otto von Guericke and Robert Boyle were among the first) to develop better vacuum pumps for experimenting on vacuum phenomena and the properties of gases at low pressures.* With a pump and a glass jar, one could prepare an *experimental space* in which to discern the effect of light, sound, and magnetic or electric forces within an atmosphere rarefied in varying degrees. One could attempt to discover, by the difference of weight of a hollow vessel before and after "exsuction," the weight of the gas within. Conversely, one could also pump gases into a vessel at higher pressures.

According to Torricelli and other followers of the mechanical philosophy, the force that holds the column of mercury up in the tube is simply the pressure of the atmosphere, transmitted through the mercury in the bowl. The air pushes down on the surface of the mercury, and since there is practically no gas in the space at the top of the tube, the mercury rises until its weight is sufficient

---

* These pumps were relatively crude and could not produce what we would now call a vacuum. For that matter, the complete removal of every trace of gas from a closed vessel is impossible today also. But considerable advances in the construction of vacuum pumps in the second half of the nineteenth century made possible evacuation good enough to allow, on the one hand, the discovery of subatomic particles, and on the other hand the development of the incandescent electric light bulb.

to balance the force exerted by the air outside on the top surface of the pool of mercury.

Note that the height of the mercury in the tube does *not* depend on the diameter of the tube or the size of the bowl. This might at first seem rather strange, since the force exerted by the mercury depends on how much of it there is—that is, on its total volume, not just its height. Similarly, the force of the air outside will be greater on a large bowl than on a small one. Nevertheless, the mercury always rises to the same height whether the tube is thick or thin, and whether the bowl is large or small. This fact is explained if we recognize that it is the balance of *pressure*, not *force*, that determines the height of the mercury column (and the equilibrium of fluids in general).

It is important to distinguish clearly between pressure and force. Pressure is defined as the amount of force acting perpendicular to a surface, divided by the area of the surface:

$$P = F/A. \qquad (19.1)$$

A large force may produce only a small pressure if it is spread over a large enough area; for example, you can walk on snow without sinking in, if you wear large snowshoes. On the other hand, a small force may produce a very large pressure if it is concentrated in a small area; when spike heels were used on ladies' shoes some years ago, many wooden floors and carpets were damaged because the *pressure* at the place where the heel touches the floor—a very small area—can be greater than that exerted by an elephant's foot!

It is instructive to calculate the numerical value of the pressure exerted by the atmosphere, to see how large it is and why it does not depend on the diameter of the tube in the barometer. The column of mercury stands 0.76 m high (the average height of the barometer column at sea level at 0°C). If this column has a cross-sectional area of 1 m², the volume of the column is 0.76 m³. A cubic meter has a mass of 13,600 kg. The mass of the mercury in the column is therefore

$$m = 0.76 \text{ m}^3 \times 13,600 \text{ (kg/m}^3) = 10336 \text{ kg.}$$

The force exerted by this amount of mercury is

$$F = mg = 10336 \text{ kg} \times 9.8 \text{ m/sec}^2$$

$$= 1.013 \times 10^5 \text{ newtons.}$$

Since the pressure is defined as $F/A$, and $A = 1$ m², we find that

atmospheric pressure = $1.013 \times 10^5$ newtons/m².

Note that we had to multiply by the cross-sectional area to calculate the force, and then divide by this same area to calculate the pressure, so that the final result is independent of the value assumed for the area.

The actual height of the mercury column in a barometer is not precisely 0.76 m; it varies from one day to the next, and from one place to another, being itself an index of changing climatic conditions. It is therefore sufficiently precise, and more convenient in practice, to *define* the pressure of the atmosphere at the earth's surface as $10^5$ newtons/m², and this quantity has been given a special name, the *bar*. Weather maps usually give local atmospheric pressures in terms of *millibars* (1 bar = 1000 millibars).

Another unit of pressure frequently used in scientific work is the *torr*, named in honor of Torricelli. A pressure of one torr corresponds to a barometer reading of 1 mm of mercury. Since atmospheric pressure corresponds to 0.76 m or 760 mm of mercury, the toor is $\frac{1}{760}$ atm.

It was hard for many people to believe that the atmosphere really exerts such a large pressure—in common units about 15 pounds per square inch. A famous experiment, conducted by Otto von Guericke in Magdeburg, Germany, in 1654, helped to dramatize the magnitude of atmospheric pressure. Two large hollow bronze hemispheres were fitted carefully edge to edge, and the air was removed from the inside of this space by a vacuum pump. A team of eight horses was harnessed to each hemisphere and the two teams were driven in opposite directions; they were just barely able to pull the hemispheres

**Fig. 19.2**  Robert Boyle (1627–1691). Note in the background one of his pumps, set up to evacuate a glass globe.

apart against the force with which the external atmospheric pressure pressed them together.

The properties of air and of atmospheric pressure were thoroughly investigated by Robert Boyle, who heard about Guericke's experiments in 1657. Using an improved vacuum pump built for him by Robert Hooke, Boyle was able to obtain a large evacuated space in which several new experiments could be done; for example, he found that small animals became unconscious and eventually died when deprived of air. In one experiment, Boyle placed a Torricelli barometer inside a closed container and then removed the air from the container. The level of the mercury in the tube fell as the air was removed from the container, until finally it reached the same level as the mercury in the bowl. Boyle realized that this result proved it was the

pressure of the atmosphere outside which had been holding up the mercury in the tube, rather than the vacuum itself exerting some pull on the top of the tube.

Soon after Boyle published his experiments in 1660, his conclusions were challenged by Franciscus Linus, a Jesuit scientist and professor of mathematics and Hebrew at the University of Liège, in Belgium. Linus asserted that the apparently empty space above the mercury really contains an invisible cord or membrane, which he called the *funiculus* (a Latin word meaning "small rope"). According to Linus, when air is stretched or rarefied, the funiculus exerts a violent attractive force on all surrounding objects, and it is this attraction which pulls the mercury up the tube. He would argue that if you put your finger over the end of the tube from a vacuum pump (or a modern vacuum cleaner), you can actually feel the funiculus pulling in the flesh of your finger. (Try it!)

The funiculus theory sounds like a fantastic idea which no scientist would take seriously nowadays. But this kind of pseudo-mechanical explanation of physical phenomena was quite popular in the early days of modern science. Boyle did not simply ignore Linus, perhaps realizing that the new idea of air pressure needed further justification before it would be generally acceptable. Along with his refutation of the funiculus hypothesis, Boyle published in 1662 some quantitative measurements of the relation between pressure and volume of air, in support of what is now called *Boyle's law*, to be discussed in the next section. While it is for his pressure–volume law that Boyle is chiefly famous, we should not forget that in his own time his work in establishing the qualitative importance of air pressure was more significant.

## 19.3  THE GENERAL GAS LAW

The compressibility of air, a fact realized generally for some time, was put on a quantitative basis through Boyle's celebrated experiments, from which it appeared that for a given mass of air trapped in a vessel at a constant temperature, any decrease of volume raises the gas pressure proportionally; conversely, any increase in pressure decreases the volume.* In modern terminology, we write this result

$$P \cdot V = \text{constant} \qquad \text{(at constant temperature)}$$
(19.2)

where $P$ refers to the actual pressure of the gas considered (e.g., in newtons/m$^2$) and $V$ to the volume of the gas (e.g., in m$^3$).

Following Boyle's work, continued effort was made to discover the effect of *changes of temperature* on the pressure or the volume of a gas. In the eighteenth century, Guillaume Amontons and Jacques Charles suggested that if a constant pressure is maintained in a sample of gas, its volume would increase proportionately to the rise in temperature ($\Delta V \propto \Delta T$ at constant $P$). This law was definitely established by extensive experiments on a number of gases conducted by the French chemist Joseph Louis Gay-Lussac, around 1800, and confirmed shortly afterward in England by John Dalton. For brevity we shall call it Gay-Lussac's

---

* The complicated history of the discovery of Boyle's law has only recently been disentangled by historians. It appears that the law was first proposed by two other British scientists, Henry Power and Richard Towneley, on the basis of their experiments begun in 1653. They did not publish their results immediately, but after Boyle's first experiments on air pressure had been published in 1660, Power sent a paper describing the results of their joint work to his friend, William Croone, in London. The title of the paper was "Additional experiments made at Towneley Hall, in the years 1660 and 1661, by the advice and assistance of that Heroick and Worthy Gentleman, Richard Towneley." But Power neglected to put his own name on the paper. Croone sent the paper to Boyle, forgetting to mention that Power was the author. Boyle was very careful to give proper credit for the information he had received, and in his monograph of 1662 replying to Linus, he stated that he had not realized that the simple relation $PV = \text{constant}$ applied to his own data until Richard Towneley pointed it out. Later scientists, who read Boyle's works carelessly or not at all, assumed that Boyle had made the discovery all by himself.

law, after the French scientist who seems to have had the greatest share in establishing its validity, recognizing however that, just as in the case of Boyle's law, several scientists made important contributions by proposing and testing it.

To visualize the meaning of Gay-Lussac's law in an actual experiment, think of a quantity of gas contained in a very thin, easily stretched balloon. The pressure inside and outside is always the same, and constant at one atmosphere. As we now proceed to heat the surroundings ($\Delta T$) the balloon expands ($\Delta V$). What is so very startling is the fact that the fractional increase in volume per degree rise in temperature is exactly the same for all gases, even though their chemical constitutions are different. To appreciate the unexpectedness of this finding, recall that on heating a solid or a liquid one ordinarily also obtains a length or volume expansion which is proportional to the temperature change, but there is a great difference in the relative amount of volume expansion per degree of temperature between different solids or liquids (about 0.01% for ice, 0.0025% for quartz, 0.02% for mercury and for water, and 0.15% for acetone).

But under normal conditions, almost all gases expand roughly $\frac{1}{273}$ (that is, 0.37%) of their volume when heated from 0°C to 1°C. The same volume change is also observed at much lower temperatures, provided that the pressure of the gas is sufficiently low. (If a gas is continually cooled at a constant pressure, it will eventually condense to a liquid, and then, of course, cease to obey Gay-Lussac's law.) We can thus define an *ideal gas* as a gas which obeys both Boyle's law and Gay-Lussac's law with complete precision; the behavior of real gases approaches this as a limit at low pressures.

Assume that an ideal gas could be cooled to 273° below the ice point (i.e., to −273°C); then since by Gay-Lussac's law—if it holds over such a large range—it should contract $\frac{1}{273}$ of its volume at 0°C for each degree of cooling, it would eventually have no volume at all. While this condition cannot be realized experimentally, it suggests that we should

**Figure 19.3**

$-273°C$ or $0°K$    $0°C$ or $273°K$

recognize the special temperature $-273°C$ (more precisely, $-273.15°C$) as the *absolute zero of temperature*, in the sense that no lower temperature can *ever be reached.*\* Consequently, we can simplify the law $\Delta V \propto \Delta T$ (at constant $P$) by writing $V \propto T$ at constant $P$ *if temperature $T$ is measured on a new scale of temperature*, i.e., the absolute scale (in degrees absolute, $°K$), where $0°K = -273°C$. The letter K was chosen in honor of William Thomson, Lord Kelvin, who first proposed this scale in 1848, and showed that it is consistent with the second law of thermodynamics. (Recall that the definition of entropy, given by Clausius in 1865, uses this temperature scale; see Section 18.4.)

Evidently the rate of exchange between these two scales is

temperature (in $°K$) = temperature (in $°C$) + 273,

a simple enough relationship. For example, $0°C$ is equivalent to $273°K$; room temperature ($20°C$) is $293°K$; water boils at $373°K$; and a *one-degree interval* on the absolute scale is just as large as a one-degree interval on the centigrade scale.

---

**Problem 19.1.** Figure 19.3 is a hypothetical plot of experimental data for a sample of gas at constant pressure; here the volume $V$ is plotted against the temperature $T$. The graph illustrates $\Delta V/\Delta T =$

---

\*    According to the third law of thermodynamics, proposed by Walther Nernst in 1906, all thermal properties approach zero values as the absolute zero of temperature is approached, and as a consequence it is never possible to reach this temperature.

constant (Gay-Lussac's law). Show by geometrical reasoning that if Gay-Lussac's law holds, then also $V \propto T$ (at constant $P$) provided $T$ is measured on the absolute scale. Show that the proportionality $V \propto T$ is *not* true if $T$ is measured in $°C$.

---

What interests us most, however, is the obvious next step: combining the two laws into one. If for each gas, by Boyle's law, $V \propto 1/P$, and by Gay-Lussac's law, $V \propto T$, then $V \propto T/P$. This implies that the quotient $PV/T$ is a constant (usually called $r$), whose value depends only on the sample. Thus we can write:

$$PV = rT. \qquad (19.3)$$

We may call the last equation, which is of great usefulness, the *ideal gas law*; it relates in a simple way the three variables of interest in any given sample of gas, and is approximately valid for all real gases under ordinary conditions.

---

**Problem 19.2.** Find $r$ in Eq. (19.3) for a sample of hydrogen gas weighing 2 g and filling 22.4 l at $0°C$ and 1 atmosphere pressure (i.e., 1 bar, as defined on page 301). Then determine the volume of this sample at room temperature ($20°C$) and a *gauge pressure* of 100 atmospheres. [*Note:* Gauge pressure is defined as the pressure *above* atmospheric.]

---

At this point we may call attention to an important temperature: the *triple point* of water, the unique temperature at which all three phases (gas, liquid, and solid) can coexist. The triple point was found to be at $273.16 \pm 0.01$ $°K$. At the Tenth General Conference on Weights and Measures in 1954, it was decided to establish a new absolute temperature scale based on two fixed points: absolute zero and the triple point of water, which is now *defined* to have the exact value $T = 273.16000$ $°K$. The new scale agrees with the old one for nearly all practical purposes, but has the advantage that the fixed points are no longer quite so arbitrary.

(Since the melting and boiling points of a substance vary with pressure, one had to specify a particular pressure, namely the average pressure of the atmosphere at the earth's surface. The triple point, however, occurs at a definite pressure, 4.58 torr.)

## 19.4 TWO GAS MODELS

One is immediately tempted to interpret these observations and the experimental gas law which summarizes them in terms of some model or structure. Even qualitatively, the great compressibility of all gases, the "spring of the air," seems to call for some explanation. In 1660 Boyle himself, while favoring one model "without peremptorily asserting it," presented two opposing explanations by models, both atomistic, for he and his time had begun to accept atomism as part of the rising materialistic world view. Expressed in his words "for all I can clearly discern, whatsoever is performed in the material world, is really done by particular bodies, acting according to the laws of motion."

The two views which Boyle presents may be called respectively the *static* and the *kinetic* views. If air consists of contiguous particles at rest, the corpuscles must themselves be compressible, rather like little springs or, as Torricelli had preferred, like pieces of wool. If, on the other hand, the corpuscles of the gas do not touch at all times, they need be neither variable in size themselves nor maintained by a static assembly, but then they must be in violent agitation, being whirled through all available space within a turbulent though "subtle" fluid. We recall that the subtle fluid theory, then recently proposed by Descartes to account for the motion of planets, had captured the imagination throughout Europe; and indeed, it seemed to explain qualitatively the behavior of gases well enough.*

---

* Boyle attempted to find experimentally this subtle fluid, separated from the particles on which it supposedly acted. This was the start of a 250-year futile search for *direct* evidence of ethereal fluids.

Having named the first of these the static model of a gas, and the second the kinetic model, we must make a brief comment on each. As to the first, a springlike compressibility of static particles cannot easily account for one very conspicuous property of gases—their ability to expand infinitely in all directions, unless of course we allow the improbable, namely, that each atom itself can grow infinitely on expanding. For a pile of ordinary springs opposes compression, but does not disperse continuously when put into the open. Therefore some followers of the static view found it necessary to consider that gas corpuscles do not change in size but can repel one another at a distance. Evidently some special new agency of repulsion, some new force among gas atoms, would have to be invented, because the well-known mutual gravitational force between large bodies, the logical candidate for intercorpuscular forces also, unfortunately was always found to be a force of attraction, as is evidently also the nature of those forces which hold together the particles in *solids* and *liquids*.

The distinguished authority of Newton seems to lend some aid to the static picture, for in the *Principia* he had shown, by way of a mathematical demonstration, that *if* a sample of gas were made of mutually repulsive particles, and *if* the force of repulsion between any two particles were *inversely proportional* to the distance between their centers, then the gas pressure in this sample would increase with a small decrease of volume, just as found by Boyle's law. Newton did not mean to prove that gases really do correspond to this model; on the contrary, he writes, "But whether elastic fluids do really consist of particles so repelling each other, is a physical question," i.e., one which has to be settled by appeal to experience. And indeed, this postulated type of force would cause gases to exhibit different behavior for different *shapes* of the container, which does not correspond to anything observed. Nevertheless, Newton's was the first *quantitative* deduction obtained for the behavior of gases on the basis of any atomistic hypothesis, and it appears to have been quite impressive.

**Fig. 19.4**  The pressure on the piston exerted by the weight *P* is balanced by the impact of high-speed gas particles.  (From Daniel Bernoulli's *Hydrodynamica,* 1738)

The second or kinetic view of the gas might well have been formulated without recourse to an imponderable fluid (if it had not been for the inescapable, almost hypnotic hold of that idea), for it had been the opinion of Greek atomists, as reported by Aristotle and Lucretius, that particles, particularly those not joined in specific form, are "ever passing about in the great void," just as minute specks of dust can be seen dancing irregularly in a sunbeam that pierces a dark room.*

In fact, the Swiss physicist Daniel Bernoulli in 1738 did publish a prophetic, quantitative development of a Lucretian type of kinetic model for gases. Bernoulli thought of the "corpuscles" of the gas as so minute as to be "practically infinite in number" under ordinary conditions, even in a small container.  In their rapid motion hither and thither these corpuscles collide with one another and also with the rigid walls of the closed vessel.  But the collisions can be assumed to be perfectly elastic; therefore, as we would put it, the kinetic energy of the particles is conserved and the motion can

---

*    If it be objected that such incessant motion of corpuscles *per se* is impossible, then the eternal turbulence of Descartes' vortices is open to the same objection.

continue undiminished.  The pressure which the gas is expected to exert against all sides of the container is thus caused by the incessant impact of millions of high-speed particles—hence the name *impact theory* of gas pressure.

Let us say that the gas-filled container used here is cylindrical, and that the top end can be made to slide in like a piston (see Fig. 19.4).  If the volume is slowly decreased, the corpuscles are more crowded in the progressively smaller space and the number of collisions per second with the walls is larger, i.e., the pressure should become greater, as actually observed.  Bernoulli even calculates the magnitude of this expected increase and finds that it corresponds to Boyle's experimental law.  Then he continued:

"The elasticity of air [the pressure in the container] is not only increased by the condensation [decrease of volume] but by heat supplied to it, and since it is admitted that heat may be considered as an increasing internal motion of the particles, it follows that if the elasticity of the air of which the volume does not change is increased [by heating the sample of gas], this indicates a most intense motion in the particles of air; which fits well with our hypothesis . . . ."

Lastly, the recognition that $PV/T$ is constant for a sample of gas is, in a general way, implied in the final summary.

This model, this line of reasoning, and this result, are all quite similar to the work which more than a century later finally clarified simultaneously the main problems of the nature of gases, heat, and chemistry.  But at the time, Bernoulli's work was generally neglected. That lack of general attention, that long interval, are both remarkable, and are not entirely explainable by pointing out that gases were not yet clearly understood by chemists, that density measurements were lacking which would have been needed to obtain the numerical values for the average speed of the particles, and so forth.

The other model, the Boyle–Newton theory based on repulsive forces between gas particles, proved to be more congenial to the eighteenth-century scientists who adopted the caloric theory of heat (Section 17.7). The repulsive force was now attributed to atmospheres of caloric fluid surrounding each atom. Heating a gas meant pouring in more caloric, expanding the atmosphere, and thereby intensifying the repulsive force. Although the caloric theory did not completely exclude the notion that heat might be associated with the motion of particles, it did seem to be incompatible with Bernoulli's hypothesis that heat is *nothing but* the motion of particles. The discovery of latent heat by Joseph Black seemed to indicate that a simple proportionality between heat and temperature could not be generally valid, since a finite amount of heat must be added to a liquid in order to change it to a gas without any change of temperature; this was one of several phenomena that were thought to be more easily explicable by the caloric theory than by the kinetic theory. Again, Bernoulli's theory, by postulating free motion of the atoms through empty space, ignored the ether which was thought to pervade all space; whereas caloric could easily be identified as a special kind of ether. A modified kinetic theory such as that proposed in 1727 by another Swiss scientist, the famous mathematician Leonhard Euler, would have been a better compromise, for it yielded some of Bernoulli's results on the basis of a model of atomic particles *rotating* in an ethereal fluid. But Euler's theory attracted no more attention than Bernoulli's in the eighteenth century, and was much less like the theory eventually established in the nineteenth century.

Perhaps we may compare Bernoulli's astonishing insight to some quick, isolated, premature stab from a general holding position, far through enemy lines, bringing no strategically valuable consequences until in time the whole main force has moved ahead to link up with this almost forgotten outpost. For Bernoulli in effect had made in his thinking two enormous jumps, for which the great mass of scientists was not ready: first, the direct equivalence of heat and internal molecular motion, *ignoring* any interactions with the ether; and second, the idea that a well-defined numerical relationship, such as Boyle's simple law, could be deduced from the chaotic picture of randomly moving particles. We must postpone our account of the final victory of these ideas to a later chapter, turning now to examine some other ideas about atomic properties which were being developed in the meantime.

## RECOMMENDED FOR FURTHER READING

Robert Boyle, "The spring of the air," and extracts from writings of Newton and Bernoulli reprinted in S. G. Brush, *Kinetic Theory*, Vol. I

J. B. Conant, "Robert Boyle's experiments in pneumatics," pages 1–63 in *Harvard Case Histories in Experimental Science* (edited by J. B. Conant), Cambridge, Mass.: Harvard University Press, 1948. Also available as a separate pamphlet.

Marie Boas Hall, "Robert Boyle," *Scientific American*, August 1967, pages 97–102

E. C. Kemble, *Physical Science*, Chapter 12

## SOURCES, INTERPRETATIONS, AND REFERENCE WORKS

I. B. Cohen, "Newton, Hooke, and 'Boyle's Law' (Discovered by Power and Towneley)," *Nature*, Vol. 204, page 618 (1964)

R. Dugas, *Mechanics in the Seventeenth Century*, Chapter VIII

M. B. Hall, *Nature and Nature's Laws*, pages 184–215. Extracts from writings of Torricelli, Pecquet, and Boyle.

W. S. James, "The discovery of the gas laws," *Science Progress*, Vol. 23, pages 261–279 (1928), and Vol. 24, pages 57–71 (1929)

L. T. More, *The Life and Works of the Honourable Robert Boyle*, London and New York: Oxford University Press, 1944

C. Truesdell, "Early kinetic theories of gases," pages 272–304 in his *Essays in the History of Mechanics*, New York: Springer-Verlag, 1968

C. Webster, "The discovery of Boyle's law, and the concept of the elasticity of air in the seventeenth century," *Archive for History of Exact Sciences*, Vol. 2, pages 441–502 (1965)

# Chapter 20

# THE ATOMIC THEORY
# OF CHEMISTRY

## 20.1 CHEMICAL ELEMENTS AND ATOMS

When in the preceding chapters we used the words *atom, corpuscle, particle,* and the like, it was not to be inferred that the writers of antiquity, or even of the seventeenth and eighteenth centuries, had in mind what we now mean by *atom*. The general idea of a "smallest particle" was a component of various conceptual schemes in physics and chemistry, usually mechanical models which postulated small indivisible bodies by which certain of the properties of matter could be explained. But, although theorists might assign particular properties such as shape or elasticity to their atoms, these properties were not susceptible to direct measurement or observation and could differ widely from one theory to the next.

As we have seen, the atomistic hypotheses up to the nineteenth century had not given rise to an unambiguous and precise theory either of matter or of heat. Yet, between the publication of Robert Boyle's *Sceptical Chymist* in 1661 and Lavoisier's death in 1794, there had developed a fundamental concept which was destined to give a new, vital, and precise meaning to the speculative atomistic hypothesis. This concept was the *chemical element*. Boyle had tried to demolish the older view, deriving from the Aristotelians and the followers of the sixteenth-century chemist Paracelsus, that the

elements are essentially different kinds of matter, which are always found mixed together, never pure. Instead, Boyle proposed that chemistry must be built on an identification and knowledge of those chemical substances which are found in nature and which cannot be separated into different components by any known methods. By the end of the eighteenth century, chemists had come to accept this operational definition of *element*: a substance which cannot be decomposed into other substances by chemical or physical means available at the present time. Whether these elements would in the future turn out to be really compounds of yet more basic substances was a question that could safely be left to metaphysicians, or to future scientists.

Lavoisier himself, in the preface to the *Elements of Chemistry* (1789), wrote this revealing and, in part, quite modern analysis of the concept "element" and its relation to the atomistic hypothesis of his predecessors:

"All that can be said upon the number and nature of elements is, in my opinion, confined to discussions entirely of a metaphysical nature. The subject only furnishes us with indefinite problems, which may be solved in a thousand different ways, not one of which, in all probability, is consistent with nature. I shall therefore only add upon this subject, that if, by the term *elements*, we mean to express those simple and indivisible atoms of which matter is composed, it is extremely probable we know nothing at all about them; but, if we apply the term *elements*, or *principles of bodies*, to express our idea of the last point which [chemical] analysis is capable of reaching, we must admit, as elements, all the substances into which we are capable, by any means, to reduce bodies by decomposition. Not that we are entitled to affirm, that these substances we consider as simple may not be compounded of two, or even of a greater number of principles; but, since these principles cannot be separated, or rather since we have not hitherto discovered the means of separating them, they act with regard to us as simple substances, and we ought never to suppose them compounded until experiment and observation has proved them to be so."

In addition to the few elements known to the ancients (e.g., metals, sulfur), a growing number of new ones had been identified (about one-third of our present list) although, of course, many recalcitrant compounds still posed as elements until electrolysis and other powerful means became available to separate their components. Not so very surprisingly, caloric (heat considered as a substance) also was generally listed among the elements.

The establishment of a relation between the two concepts *chemical element* and *atom* is now, in retrospect, perhaps all too obvious. Atoms of a chemical element would be the smallest physical particles that can participate in the identifying chemical reactions or physical tests. The thought urges itself on our minds to postulate, until evidence to the contrary appears, that each atom of an element is identical to every other atom of that same element ("like manufactured articles," as the scientists of industrialized nineteenth-century Britain used to say), although we may be prepared to allow that different elements may turn out to have

physically quite different atoms (as indeed is the case). It is also reasonable to expect that an element in the gaseous state, say oxygen or hydrogen, consists of individual atoms, and that a chemical compound, say water vapor, might be imagined as constituted of particles, each containing two or more atoms of different elements. In particular, it being agreed to make initially the simplest assumption, the smallest corpuscles of water vapor (we might at first call them *atoms* of water) should be regarded as made of a close union of an atom each of hydrogen and oxygen, the two gases which yield water when mixed and exploded.

This, in brief, outlines the thoughts of the Englishman John Dalton (1766–1844), who, in his work from 1800 on, effectively performed the marriage of the ideas of chemical element and the ancient atomistic hypothesis. Dalton was largely self-taught. A retiring person, son of a hand-loom weaver, he supported himself poorly as a teacher (as early as his twelfth year) and general tutor in Manchester. Although he had a strong drive and a rich imagination, particularly along the line of mechanical models and clear mental pictures, his supreme gift was his astonishing physical intuition, which allowed him to arrive at correct conclusions despite being only "a coarse experimenter," as Humphry Davy (Section 17.7) called him.[*]

## 20.2 DALTON'S MODEL OF GASES

Like Kepler and Newton, whom he resembled in many ways, Dalton published on a great variety of subjects; but it appears that his main work originated in an interest in the *physical* structure of gases. He reports that through his interest in

---

[*]   For an account of Dalton's work which was one of the main sources in the preparation of this chapter, see Leonard K. Nash, "The atomic molecular theory," Case 4 in the *Harvard Case Histories in Experimental Science* (edited by J. B. Conant); Cambridge: Harvard University Press (1950).

meteorology he had been led to wonder why the earth's atmosphere, being composed of a mixture of several gases of different density (e.g., nitrogen, oxygen, and water vapor), should be so homogeneous—that is, why samples taken at rather widely differing altitudes should have the same proportion of chemical elements, even though nitrogen gas, being lighter per unit volume in the pure state than oxygen gas, might be expected to "float" on top of the latter, somewhat like oil on water. (Fortunately perhaps for the advancement of atomic theory, it was not yet possible to gather samples of air from regions higher than a few miles above the earth's surface; there *is* a noticeable difference in composition at higher altitudes.)

Count Rumford would have had a good answer to the puzzle; he might have said, as he did in an analogous case of mixed liquids, that by thermal agitation, by their "peculiar and continual motion," all the individual particles of the different kinds of gases must diffuse among one another and thus mix thoroughly. However, Dalton could never accept this kinetic view of matter for a moment. He adopted the static model of a gas, being (wrongly) convinced that the great Newton had *proved* it in the brief passage from the *Principia* to which we referred in Section 19.4. "Newton had demonstrated," Dalton once wrote, "from the phenomena of condensation and rarefaction that elastic fluids are constituted of particles, which repel one another by forces which increase in proportion as the distance of their centers diminishes: in other words, the forces are reciprocally as the distances. This deduction will stand as long as the Laws of elastic fluids continue to be what they are."

Dalton was naturally forced to reject the kinetic view of heat, along with the kinetic view of gases, and to adopt the caloric theory which was then, of course, still the general conceptual scheme for heat phenomena. The only satisfactory solution to the problem of homogeneity of gas mixtures based on a static model was, to Dalton's mind, the following set of assumptions and deductions, some of them taken from the work of predecessors:

a) *Each particle of a gas is surrounded by an atmosphere of caloric.* As Dalton put it in his main work, *A New System of Chemical Philosophy*:

"A vessel full of any pure elastic fluid presents to the imagination a picture like one full of small shot. The globules are all of the same size; but the particles of the fluid differ from those of the shot, in that they are constituted of an exceedingly small central atom of solid matter, which is surrounded by an atmosphere of heat, of greater density next the atom, but gradually growing rarer according to some power of the distance."

b) *The particles are essentially at rest, their shells of caloric touching.* At the top of Fig. 20.1, taken from Dalton's notebook, is shown the case of two neighboring gas atoms. Throughout, Dalton accepts the notion that the only means for the communication of forces among atoms is direct contact. He is as uncomfortable with the notion of action at a distance as Newton's contemporaries had been.

c) *The total diameter of each particle,* counting in the caloric atmosphere or shell, *differs from one substance to the next.*

"...the following may be adopted as a maxim, till some reason appears to the contrary: namely,— That every species of pure elastic fluid has its particles globular and all of a size; but that no two species agree in the size of their particles, the pressure and temperature being the same."

This point proved crucially important in the later development. Dalton thought he had deduced this maxim unambiguously from experimental evidences of the following type. One of the products formed when oxygen and nitrogen gas are combined is nitric oxide or, as Dalton called it, "nitrous gas." On other grounds, Dalton had decided that in this gas one nitrogen atom combines always with just one oxygen atom. However, in his own experi-

**Fig. 20.1** From Dalton's notebook. At the top: two elementary atoms, as they would exist in a gaseous *mixture.* The straight lines indicate the region of caloric around each central mass. Below: a molecule of a *compound,* consisting of two atoms in close contact, surrounded by a common, nearly spherical atmosphere of caloric. (As reproduced by Roscoe and Harden)

ments, it seemed that the *ratio by volume* of oxygen gas to nitrogen gas needed in the reaction was not 1:1, which would have indicated equal volumes for both types of atoms, but more nearly about 0.8:1, pointing to the conclusion that nitrogen atoms were somewhat larger than oxygen atoms. (In actual fact, the experimental results here are easily misleading because several different compounds of nitrogen and oxygen may be formed concurrently.)

The same set of experiments seemed to furnish Dalton with another confirmation of his "maxim." The nitrous gas so formed took about double the volume of either of the component elements. In terms of a gas whose particles all touch, this seemed to indicate clearly that the particles or (compound) atoms of nitrous gas take about twice the volume of either oxygen or nitrogen particles.

Experiments on water vapor seemed to bear out the same type of conclusion: "When two measures of hydrogen and one of oxygen gas are mixed, and fired by the electric spark, the whole is converted into steam, and if the pressure be great, this steam becomes water. It is most probable then that there is the same number of particles in two measures of hydrogen as in one of oxygen . . ."—that is, hydrogen atoms are about twice as large as those of oxygen.

d) Turning back to the initial problem, Dalton concluded on the basis of some qualitative speculations that there could be no stratification of the air into its elements *because the contiguous particles of several sizes would tend to push one another away* until stability is reached in a homogeneous mixture.

e) Therefore, he reasoned, the constituents of the atmosphere must have been mixed with one another in the disequilibrium during their original encounter, and since then have remained homogeneous.

This suggestive "pile of shot" picture of the atmosphere prompted Dalton to consider next the number and relative weights of atoms, particularly "of all chemical elementary principles which enter into any sort of combination one with another." But before we turn to look at the enormously fruitful results that followed this decision, let us in passing note the amazing fact that each and every one of the points (a) to (e) above is, from the present point of view and often even in terms of internal consistency, *wrong*; that furthermore the premise itself, Newton's supposed proof of the static model, is Dalton's misinterpretation.

This is a fine illustration of the position we have argued in Part D, namely that science is not to be regarded as the inevitably successful result of following a clear, step-by-step method. On the other hand, neither can we conclude that science is a blind man's game, that we must grope from error to error, making a few valid discoveries only by chance, by sheer luck, or simply as by-products of a prodigious amount of largely fruitless labor being done everywhere and throughout the centuries.

**Problem 20.1.** Regarding as unquestioned the stimulating influence which Dalton's general atomic theory had on chemistry (the details will be presented in the following pages) write a brief analysis of Dalton's creative process in relation to the dual aspect of science, $S_1$ and $S_2$ (Sections 13.1 to 13.7).

## 20.3 PROPERTIES OF DALTON'S CHEMICAL ATOM

But to return to the growth of the atomic theory of chemistry: Dalton's great opus, *A New System of Chemical Philosophy*, from which most of the previous quotations have been taken, was first published in two parts, in 1808 and 1810. From that work and from his other writings, we can abstract the several principles which reflect the basic ideas in his conceptual scheme.

a) *Matter consists of indivisible atoms.*

"Matter, though divisible in an *extreme degree*, is nevertheless not *infinitely* divisible. That is, there must be some point beyond which we cannot go in the division of matter. The existence of these ultimate particles of matter can scarcely be doubted, though they are probably much too small ever to be exhibited by microscopic improvements. I have chosen the word *atom* to signify these ultimate particles . . . ."

b) *Atoms are unchangeable.* The atoms of different elements "never can be metamorphosed, one into another, by any power we can control," as the failure of centuries of alchemy had made clear. By continual *test* the transmutation of elements was found to be impossible—*a postulate of impotency* arrived at in a manner exactly analogous to the rise of the law of conservation of energy from the long failure to find a perpetuum mobile.

c) *Compounds are made of molecules.* Chemical compounds are formed by the combination of atoms of two or more elements into "compound atoms" or, as we now prefer to name the smallest particle of a compound, *molecules*.

"I call an ultimate particle of *carbonic acid* a *compound atom*. Now, though this atom may be divided, yet it ceases to be carbonic acid, being resolved by such division into charcoal and oxygen. Hence I conceive there is no inconsistency in speaking of compound atoms, and that my meaning cannot be misunderstood."

Although we may be tempted to think of the particles of compounds, the molecule, as made of two or more atoms, so to speak *side by side*, and although Dalton's ideographs, as we shall see at once, at first glance give the same impression, he did not have this conception. Rather, he thought that the compound atom was essentially round or "globular," the centers of the combining atoms being "retained in physical contact by a strong affinity, and supposed to be surrounded by a common atmosphere of heat." (See Fig. 20.1.) In the specific example of the formation of water, he asked that it should "be supposed that each particle of hydrogen attaches itself to a particle of oxygen, and the two particles so united form *one*, from which the repulsive energy emanates [by a redistribution of the combined caloric]: then the new elastic fluid may perfectly conform to Newton's Law . . . ."

d) *All atoms or molecules of a pure substance are identically alike.*

"The ultimate particles of all homogeneous bodies are perfectly alike in weight, figure, etc. In other words, every particle of hydrogen is like every other particle of hydrogen . . . ."

We should here realize that Dalton was now simplifying even beyond the older atomists, who usually had felt it necessary to allow different sized atoms for the same element, just as pebbles made of the same material might have different shapes and sizes.

e) *In chemical reactions, atoms are only rearranged, not created or destroyed.*

"Chemical analysis and synthesis go no farther than to the separation of particles one from another, and to their reunion. No new creation or destruction of matter is within the reach of chemical agency. We might as well attempt to introduce a new planet into the solar system, or to annihilate one already in existence, as to create or destroy a particle of hydrogen. All the changes we can produce, consist in separating particles that are in a state of cohesion or combination, and joining those that were previously at a distance."

Here at last we have a strikingly simple physical picture to explain the law of conservation of mass which Lavoisier had postulated and experimentally demonstrated decades before.

## 20.4 DALTON'S SYMBOLS FOR REPRESENTING ATOMS

We have already noted a passage in which the combination of hydrogen and oxygen into water vapor is pictured. To picture to himself this and other chemical reactions, Dalton found it extremely helpful to make use of simple but ingenious symbols (Fig. 20.2). Whereas even the earliest alchemists had represented different substances by different symbols, Dalton's ideographs represented *individual atoms*. But note well that in Fig. 20.2 some of the first 20 symbols do not represent the atoms of elements but the molecules of compounds as

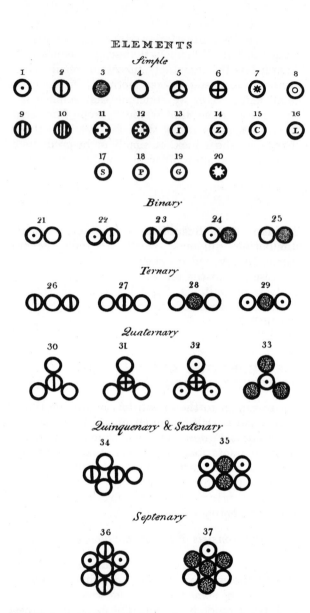

**Fig. 20.2** Dalton's symbols for elementary and compound atoms, taken from his work, *A New System of Chemical Philosophy*, Part I. Dalton writes, "This plate contains the arbitrary marks or signs chosen to represent the several chemical elements or ultimate particles." The modern names for the first few atoms are: 1, hydrogen; 2, nitrogen; 3, carbon; 4, oxygen; 5, phosphorus; 6, sulfur.

Dalton understood them. For example, symbols 9 and 10, which stand for soda and potash, were known to Dalton from Davy's work of the same year (1808) to be "metallic oxides" (actually metallic hydroxides).

Because we shall make some use of these historic symbols, let us briefly note those for the compounds. For example, 21 depicts water, which Dalton considered to be a binary compound made of one atom of oxygen and one of hydrogen; 22, "an atom of ammonia, composed of one of azote [the then current name for nitrogen] and one of hydrogen"; 23 represents "an atom of nitrous gas [now called nitric oxide] composed of one of azote and one of oxygen"; 24 was Dalton's sign for what we now call methane; and 25 for carbon monoxide, then called "carbonic oxide." Among the ternary compounds 26 is nitrous oxide, still called thus, although the name laughing gas is more familiar; 27 is a compound atom of "nitric acid," which we would name a molecule of nitrogen dioxide; 28 represents "carbonic acid," now called carbon dioxide. In the same fashion, Dalton pictured to himself other compounds which he knew to contain more than three elementary atoms, namely quaternary, quinquenary, etc.

Nowadays, of course, we are used to other symbols, those introduced by the Swedish chemist Berzelius in 1819. Letters are used to represent the substance. Hydrogen, oxygen, nitrogen, carbon, and chlorine are respectively H, O, N, C, and Cl (the full list of all elements appears in Appendix III), and compounds are now rendered by the joining of the respective symbols of elements. For example, hydrochloric acid would be represented by HCl, and carbon dioxide by $CO_2$. A chemical reaction, say the combustion of carbon (the formation of carbon dioxide), might then be given in one of these ways, the first being Dalton's, the last being the present convention:

$$\oplus + \bigcirc + \bigcirc \rightarrow \bigcirc \oplus \bigcirc$$

or

$$C + O + O \rightarrow CO_2$$

or

$$C + O_2 \rightarrow CO_2.$$

## 20.5 THE LAW OF DEFINITE PROPORTIONS

The accumulation of more and more accurate analytic data during the prior half century, and particularly the vigorous dispute between the French chemists Proust and Berthollet, had made increasingly acceptable what we now call the *law of definite proportions* of combining weights: No matter in which locality a pure substance—say a metallic oxide, a drop of some liquid, or a complex gas—is found, no matter whether it is artificially prepared or taken from a natural source, the proportions by weight of the constituent elements in every case are always the same. For water, this ratio by weight between hydrogen and oxygen gas is 1:8 (approximate modern value; Dalton's was first 1:6 and later 1:7).*

Now, combining the simple pictorial view of atoms and the law of definite proportions by weights of combining elements, Dalton was forced to the hypothesis that in a chemical reaction there is always a fixed or definite proportion by *number* of combining atoms. To use a very crude analogy, if it were found on shelling peas that the weight of

---

* We must stop a moment to realize that like the law of conservation of mass, which was the fruit of a fierce struggle between Lavoisier and his followers in their fight against the phlogistonists, so also was the establishment of this law much speeded by the vigor of the dispute between able antagonists. Thereby were also established at the time several other most valuable results, e.g., clearer distinctions between compounds and mixtures, and a large amount of data that provided the raw material for further theorizing. From our dynamic view of science we can well understand this new illustration of the evolutionary mechanism in the struggle of ideas.

Again, fortunately for the development of atomic theory, the weight of experimental evidence as of 1800 was in favor of Proust's theory of definite proportions; only much later was it established that there do exist *some* compounds having variable composition.

the total quantity of peas is a *constant fraction* of the weight of the empty pods, no matter how many pods are included or where the harvest is gathered, we should be quick to propose that the number of peas is the same in each pod, even if we do not know what that number actually is. To put Dalton's hypothesis simply, *in forming molecules during chemical reactions, the numbers of combining atoms of the different elements form simple, definite ratios.*

This part of Dalton's atomic theory denied a then widespread and distinguished opinion concerning chemical reactions, namely, that particularly in the case of two gases the compound is formed by a solution of one in the other, rather as salt and water form salt water. Dalton, a little querulously, gives several arguments "advanced to shew the absurdity of this doctrine...." Instead of regarding water as made of an indefinite or variable combination of hydrogen and oxygen particles, Dalton conceived of each molecule of water vapor as made the same way, namely, of one atom of each element. But there's the rub; on the basis of what has developed so far, Dalton has, of course, no way of knowing *just how many* atoms of each element do in fact combine in one water molecule; as far as we can determine, the ratio might not be 1:1 but perhaps 2:1 or, for that matter, 23:61. The relative volumes of combining gases do not offer a clue, since Dalton has already assumed the atoms to have different and, of course, unknown sizes. All that can be said is this: The relative combining weights between hydrogen and oxygen being 1:8, and *assuming* that the combination proceeds by joining a single atom of each element *for every single molecule*, then we may conclude that the relative atomic weights of hydrogen and oxygen are as 1:8 also. On the other hand, if you assume that the molecular formula is not HO but $H_2O$ (that is, $H + H + O \rightarrow H_2O$ or $H_2 + O \rightarrow H_2O$), then the relative atomic weights would be 1:16; and similarly for other formulas. The missing link to an understanding of chemical reactions, therefore, is the knowledge of some simple molecular formulas.

## 20.6 DALTON'S RULE OF SIMPLICITY

Dalton was finally driven to decide on an *assumption*, one which appears to typify the advantages and dangers of any hypothesis, his rule of greatest simplicity of the most common stable compounds:

"When only one combination of two bodies [elements] can be obtained, it must be assumed to be a binary one, unless some other cause appears to the contrary.... When two combinations are observed, they must be presumed to be a binary and a ternary...."

and so forth for more than two combinations. Specifically *water*, then the only known combination of hydrogen and oxygen, must be a binary compound (HO or ⊙◯).* In the same manner, the only two known combinations of carbon and oxygen were identified as follows: the lighter of the gases (*carbonic oxide*) as CO, the heavier (*carbonic acid*) as $CO_2$. The common compound of nitrogen and hydrogen, ammonia, was regarded as NH (we now know it to be $NH_3$). Ethylene and methane were identified as CH and $CH_2$ (whereas now they would be written $C_2H_4$ and $CH_4$).

One brief look at any modern table of chemical formulas will at once convince us of the fallaciousness of this maxim. For example, the only iron-iodine compound, iron iodide, is not binary but has the formula $FeI_2$; and the simplest iron chloride is not FeCl but $FeCl_2$. The iron phosphides are not FeP, $FeP_2$, and $Fe_2P$, as Dalton might have expected, but FeP, $Fe_2P$, and $Fe_3P$. The common oxides of iron are now recognized to be FeO, $Fe_2O_3$, and $Fe_3O_4$. However, the analytic data in most cases (we might almost say luckily) were too ambiguous to point out a large-scale failure of his axiom, and in some cases it gave correct results (e.g., for the carbon oxides and for the nitrogen oxides). Above all, Dalton's devotion to the time-honored idea of "the regularity and simplicity" in the laws of nature did make possible a start, and

---

*    Hydrogen peroxide ($H_2O_2$) had not yet been identified.

a fruitful one, toward the identification of molecular formulas. Where some early guess of a molecular formula was wrong, it could be corrected through internal consistency checks; we recall again the admonition that truth can come out of error more easily than out of confusion.

## 20.7 THE EARLY ACHIEVEMENTS OF DALTON'S THEORY

We might discuss a variety of subtle consequences of Dalton's work: that his pictorial conceptualizations and his ideographs must have produced a deeper feeling for the reality of atomic processes at the time; that his identification of chemical atoms helped in the final differentiation between the concepts of pure chemical compounds and mixtures; and other details of this sort. But above all, we must now summarize briefly the initial main achievements of Dalton's theory—its power to gather known facts and laws, and to provide a simple, workable conceptual scheme; its fruitfulness in suggesting new relations and stimulating research; and its helpfulness in the solution of problems in contemporary science. Elements of each of these functions may be found in the following examples.

a) As we saw, Dalton's chemical atom is the concept that gives physical meaning to the law of conservation of mass, and that integrates it with the law of definite proportions. The latter, although essentially established at the time by empirical and exhaustive test, was not yet explicitly and generally appreciated. Those experimental results were now given physical meaning and were supported by the inherent plausibility of Dalton's picture; in this sense we may say that Dalton did for the embattled law of definite proportions what Newton did for Kepler's established laws of planetary motion. Other established chemical results could be similarly accommodated (e.g., the law of equivalent proportions).

b) Perhaps the best support of the atomic theory, by common consent at the time, was an extraordinarily powerful insight that followed directly and naturally from Dalton's conceptual scheme: *When one substance unites with another to form more than one compound and so forms more than one proportion* (for example, 2:1 and 3:2), *these different proportions bear a simple ratio to one another* (for example,

$$\frac{2:1}{3:2} = 4:3).$$

This is the *law of multiple proportion*. We see at once that by necessity it is fully implied in Dalton's assumptions. For example, if we picture *carbonic oxide* and *carbonic acid* in Dalton's manner,

○◉ (carbonic oxide)        ○◉○ (carbonic acid),

it is clear that we should find associated with a given quantity of carbon twice as much oxygen by weight in the second case compared with the first; or, to put it into the language of the law, the respective weight proportion of oxygen to carbon in the second substance, *divided* by the proportion of oxygen to carbon in the first, should be as 2:1, a simple ratio indeed. As to the experimental check of this prediction, the necessary data for combining weights had, significantly, been long available in about the following form:

In carbonic acid, 73% oxygen and 27% carbon (by weight).
In carbonic oxide, 57% oxygen and 43% carbon.

If we reduce these approximate percentages to proportions, using ratios of small integers, we obtain:

In carbonic acid, ratio of oxygen to carbon is about 8:3.
In carbonic oxide, ratio of oxygen to carbon is about 4:3.

Therefore the ratio between these proportions = $(8:3)/(4:3) = 2$ (or 2:1).*

Similar confirmations of the law of multiple proportions could be obtained by an analogous treatment of other data from previously published scientific work. We might say that the facts had been before everyone's eyes, but the simplicity they contained remained hidden because there was previously no good reason to look for it. As has often been said on these pages, one cannot see much without the aid of a hypothesis to organize the perceptions; in a sense, the old saying "seeing is believing" should be supplemented by another, "believing is seeing."

As might have been expected, this discovery of the law of multiple proportions stimulated a great deal of new research. Given any one proportion of combining weights for two substances, one could now plan for and predict the analysis or synthesis of other compounds of the same elements. The results were impressive, although we now know that the postulate of "simple" ratios is again too limited and that *large* whole numbers may be involved. For example, in the complex structure of the organic compounds, then still mostly inaccessible, each molecule may contain several dozens of atoms of a few substances, so that the ratio of carbon-to-hydrogen proportions is not such a simple one for two substances such as $C_{23}H_{22}O_6$ and $C_{22}H_{42}O_3$.

---

**Problem 20.2.** In 1810, Dalton gave the approximate results listed in the following table for the relative densities of five gases (his terminology),

---

* The deduction does not proceed unambiguously in the opposite direction, however. From the fact that the ratio is 2:1 we could conclude either that the two molecular formulas are CO and $CO_2$, or that they are $C_2O$ and CO. Dalton did not give any explicit rule for resolving this ambiguity in all such cases, but used various additional kinds of evidence to determine molecular formulas. See the discussion by A. J. Bernatowicz, *Journal of Chemical Education*, August 1970.

each made of molecules involving only nitrogen and oxygen. He also concluded that the lightest of these gases was made of molecules containing one atom each of N and O. (a) Decide on the molecular formula for each gas on the basis of this information. (b) Confirm the law of multiple proportions for these five gases.

| Substance | Relative density | % by weight | |
|---|---|---|---|
| | | Nitrogen | Oxygen |
| Nitrous gas | 12.1 | 42.1 | 57.9 |
| Nitrous oxide | 17.2 | 59.3 | 40.7 |
| Nitric acid | 19.1 | 26.7 | 73.3 |
| Oxynitric acid | 26.1 | 19.5 | 80.5 |
| Nitrous acid | 31.2 | 32.7 | 67.3 |

c) A third achievement of the atomic theory of chemistry, one without which modern physics and chemistry would be unimaginable, was the start toward the *determination of atomic weights*. In Dalton's words, "Now it is one great object of this work, to shew the importance and advantage of ascertaining the relative weights of the ultimate particles, both of simple and compound bodies..." —an undertaking previously neither possible nor, indeed, even meaningful. Note that Dalton only claimed to determine the *relative* weights of the atoms of different substances; no reliable method was known for determining absolute weights until 1865, when Loschmidt used the results of kinetic theory for this purpose. (We now know that the atoms of elements have masses of the order of $10^{-24}$ g each.)

Historians have disputed for some time the exact way in which Dalton first came to compute his atomic weights. The most recent view is that he was attempting to develop a mechanical theory of the solubility of gases in water, and found that it was important to determine the sizes and weights of the particles of different gases. Entries in his laboratory notebook show that he had calculated a table of atomic and molecular weights by September, 1803, although the table was not published until 1805, and a systematic explanation of the method of constructing it did not appear until 1808 in the *New System*. Our foregoing discussion has hinted strongly how the relative weights of atoms could be obtained: from the *known combining weights* of the elements of a substance, and from the *assumed molecular formulas*. For example, from the data given in Problem 20.2 it would appear, as you may check in five ways, that the relative weights of the atoms of nitrogen and oxygen could be taken to be nearly 6:8.2 (from better experiments we now have the values of 7:8, or more precisely, 14.008:16.000).

We may use other analytic data in the very same way: The combining weights of hydrogen and oxygen in Dalton's synthesis of water being as 1:6 (later 1:7, and by present measurements nearly 1:8), and the formula for water having been assumed to be HO ($\odot\bigcirc$), the relative atomic weights were also at first supposed to be as 1 to 6. By making a compromise between the various results of this type, Dalton could assemble a table of most probable values of relative atomic weights (although the words "relative combining *masses*" would have been more appropriate). By referring all values to the arbitrary value of unity for hydrogen, Dalton assembled tables for most elements and compounds then available.* One of the earlier tables, published in 1808, gave the following list of relative atomic weights: hydrogen 1, nitrogen 5, carbon 5, oxygen 7, phosphorus 9, sulfur 13, soda (a "compound atom," namely NaOH, sodium hydroxide) 28, and so forth.

It soon became obvious to Dalton that in using the rule of greatest simplicity, inconsistencies were bound to occur in the determination of atomic weights based on various compounds. In the specific case of water he said as early as 1810, "After all, it must be allowed to be possible that

water may be a ternary compound" ($H_2O$ or $HO_2$ instead of HO). In the former case the (modern) combining weights of 1:8 for the two gases would indicate that the relative atomic weight of oxygen is 16 if hydrogen is 1; in the latter case, if H = 1, O = 4. It is conceivable that by allowing such flexibility in the assumption of molecular formulas, and by eliminating contradictions, one might eventually have arrived at a set of consistent atomic weights (the process is illustrated in the next problem). But just then there appeared from France an experimental discovery and from Italy a theoretical insight which at once aided this task, and which together were destined to replace the gradually less and less defendable Daltonian rule of greatest simplicity, putting atomic chemistry on a new basis.

---

**Problem 20.3.** In the table below is given a set of approximate *combining weights* for five substances. For each, Dalton's principle of greatest simplicity would demand a *binary* structure, although in fact most of them are more complex. Analyze these data and obtain, without recourse to any other aids and exclusively from the consistency of these data, (a) a set of the most consistent molecular formulas for all these substances, and (b) a set of the most consistent atomic weights for all the elements involved.

*Then* consult a collection of constants for inorganic compounds (as for example in the *Handbook of Chemistry and Physics*) and find the correct formulas and the current names for these compounds.

| Sub-stance | Combining weights in grams | | | |
| | Iron | Sulfur | Hydro-gen | Carbon |
|---|---|---|---|---|
| A | 14 | | | 9 |
| B | | 16 | 1 | |
| C | 7 | 8 | | |
| D | | 16 | | 3 |
| E | | 48 | 1 | 6 |

---

* Modern tables use carbon as the reference substance, so that hydrogen now has an atomic weight of about 1.008; see Appendix III.

[*Hint:* Begin by writing for each of the *first four* substances several possible molecular formulas, and the relative atomic weights for each of these cases as they would follow from the given combining weights. Next, select the one most consistent set of atomic weights, and thereby you have decided on the most probable molecular formulas also. *Then*, as a test of your scheme, turn to the last substance and find whether it can be accommodated.]

## 20.8 GAY-LUSSAC'S LAW OF COMBINING VOLUMES OF REACTING GASES

In 1808 Joseph Louis Gay-Lussac (see Section 19.3), a master of experimental chemistry, had noted that Cavendish's experiment on the explosion of hydrogen and oxygen gas to yield water vapor pointed to a beautiful simplicity. On repeating the experiment most carefully, he found that the *combining volumes* of hydrogen and oxygen are as 2:1, within 0.1%. Now Gay-Lussac remembered that gases as a group exhibit simple and regular physical laws, i.e., the ideal gas laws, which are approximately valid for all gases, and he speculated that therefore the combination of gases (by volume) in general might be guided by a set of simple rules. From his analysis of previously published work and his own careful research he concluded, "It appears to me that gases always combine in the simplest proportions when they act on one another; and we have seen in reality in all the preceding examples that the ratio of combinations is 1 to 1, 1 to 2, or 1 to 3." (To this we should now add other simple ratios by volume, such as 2:3.) Furthermore, he noted that if the compound resulting from the chemical reaction remains a gas or a vapor at the temperature of the experiment, the volume of the compound is also simply related to the volumes of the participating elements. We might write results of this kind as follows:

1 vol nitrogen gas + 3 vol hydrogen gas
                    = 2 vol ammonia vapor,

2 vol hydrogen gas + 1 vol oxygen gas
                   = 2 vol water vapor,

1 vol nitrogen gas + 1 vol oxygen gas
                   = 2 vol nitric oxide gas,

1 vol hydrogen gas + 1 vol chlorine gas
                   = 2 vol hydrochloric acid gas.

---

**Problem 20.4.** Write out a concise statement of Gay-Lussac's law of combining volumes.

---

This marvelous simplicity would appeal to us at first sight, but for many reasons Dalton rejected these results utterly and to the very end. We recall that his model of gases, in order to explain the homogeneity of the atmosphere, required that the atoms of different elements be of different sizes. To accept the possibility of such general simplicity, even occasionally of *equal* combining volumes, seemed to him tantamount to accepting the notion that equal gas volumes contain equal numbers of atoms, or at any rate multiples thereof, that the atoms therefore might be of equal sizes, and that consequently he would have to give up both his static model and what he supposed to be Newton's "immutable Law of elastic fluids."

Dalton found his arguments confirmed by the fact that his own (faulty) volumetric experiments contradicted Gay-Lussac's results, and that these, of course, were admittedly not *exactly* whole numbers (we recall that scientific measurements can never be made without some uncertainties; and in these delicate tests the likelihood of error was often considerable, though in fact Gay-Lussac is generally recognized to have been a much more accurate experimenter than Dalton). Therefore Gay-Lussac had had to round off his figures, and to Dalton this discrepancy was not a negligible experimental error, but the most significant truth. This dispute summarizes for us brilliantly and compactly some typical features of the life of science: the search for simplicity, the fascination with

integers, the inexactitude of measurement, the role
of a theory to guide and interpret (or misinterpret)
experimental results; indeed, we are here reminded
a little of the dispute between Galileo and the
scholastics regarding the interpretation of free-
falling motion.

Quite apart from Dalton's objections, Gay-
Lussac's results do not at this point lend themselves
directly to an interpretation by Dalton's atomic
theory of gases. To make the two dovetail required
an amendment of the theory, after which, however,
the atomic hypothesis emerged stronger than ever,
in fact, approximately in its present state.

## 20.9 AVOGADRO'S MODEL OF GASES

The amendment was not long in coming; it was the
work of the Italian physicist Lorenzo Romano
Amedeo Carlo Avogadro di Quaregua e di Cerreto
(1776–1856), usually known as Amedeo Avogadro
(Fig. 20.3). Avogadro's obscurely phrased but
penetrating publication of 1811 presents a model of
gases which may be analyzed in five separate parts:

a) As had been often and persistently suggested
by Gay-Lussac and others, Dalton's picture of a gas
in which the particles are in contact is now replaced by another (it does not matter whether a
static or a kinetic model) in which the particles
have only a very thin shell of caloric and are quite
small compared with the distances between them.
The volume actually taken up by the atoms is
consequently very much smaller than in Dalton's
model, in fact, a negligible fraction of the total
volume of any sample. It is a return, if you wish,
to Lucretius' "atom and void" picture of the gas.
With this assumption we must, of course, also give
up the hope of obtaining any information on
*relative atomic dimensions* on the basis of combining volumes. But this problem of dimensions,
so important to Dalton, immediately becomes quite
insignificant in this formulation.

b) Contrary to initial considerations of simplicity,
or to the firm faith of Dalton, or, in fact, to every

**Fig. 20.3**   Amedeo Avogadro (1776–1856).

reasonable expectation at the time, the ultimate
particles of gaseous *elements* are conceived to be
not necessarily just single solitary atoms, but
instead may be thought to be made of two or more
chemically alike atoms "united by attraction to
form a single whole." Calling the smallest such
unit a *molecule*, we may then imagine that the
molecules of element $X$ in gaseous form all consist
of two atoms each (a diatomic gas), or in the case
of some element $Y$ of three atoms (triatomic), or in
gas $Z$ of single atoms (monatomic), and so on. We
still bear in mind that since the molecules of an
element are made up of, at most, a few atoms they
are negligibly small compared with the intermolecular distances.

c) In a *gaseous compound*, e.g., water vapor, the
molecules evidently each consist of definite numbers of chemically unlike atoms. But when two
gases combine, as in the formation of water from

hydrogen and oxygen gas, we must not just *assume* that one or more molecules of hydrogen and one or more of oxygen simply associate into a new or larger unit; we must examine the possibility that such a larger group forms only for a brief moment and then at once breaks up into two or more equal parts. We may represent the various possibilities symbolically as follows:

$$H + O \rightarrow HO \tag{1}$$

$$2H + O \rightarrow H_2O \tag{2}$$

$$H_2 + O \rightarrow H_2O \tag{3}$$

$$2H + O_2 \rightarrow [H_2O_2] \rightarrow 2HO \tag{4}$$

$$H_2 + O_2 \rightarrow [H_2O_2] \rightarrow 2HO \tag{5}$$

$$2H_2 + O_2 \rightarrow [H_4O_2] \rightarrow 2H_2O \tag{6}$$

$$2H_3 + O_2 \rightarrow [H_6O_2] \rightarrow 2H_3O \tag{7}$$

$$2H_4 + O_4 \rightarrow [H_8O_4] \rightarrow 2H_4O_2 \tag{8}$$

et cetera.

d) But the most important part of Avogadro's proposal is that, at equal temperatures and pressures (conditions largely obeyed in Gay-Lussac's experiments), *equal volumes of all gases, whether elements or compounds, or even mixtures, contain equal numbers of molecules.* This statement is generally known as *Avogadro's hypothesis* or *Avogadro's law.* Avogadro was not the first scientist to propose it, but he did succeed in incorporating it into a coherent and convincing theory.

To turn to specific cases, let us first represent Gay-Lussac's experimental results for water in symbolic form:

Each of these squares represents a unit volume (say 1 liter). In accord with Avogadro's postulates, we must draw into each square the same number of molecules; for simplicity, let this be just one molecule. If with Dalton we now assume that hydrogen and oxygen gases are monatomic, and if the reaction proceeds according to Eq. (1) above, we should at once find it impossible, without splitting the very atom of oxygen, to place into each of the two volumes of water vapor one whole molecule of the same material. We must also reject Eqs. (2) and (3) for a similar reason. If we really believe in Avogadro's hypothesis deeply enough to entertain the possibility that one of the other equations represents the true reaction, we may next turn to the assumption that hydrogen is indeed monatomic but that oxygen is diatomic; then Eq. (4) applies, and filling in our squares,

a reaction that is evidently in harmony both with the expectations of Avogadro's hypothesis and Gay-Lussac's experimental results.

But alas! Equations (6) and (7), and many others that may be imagined, would *also* be consistent with these two factors. Evidently there is an unlimited number of such possibilities; in abandoning Dalton's too simple reaction H + O = HO as wrong, we have discovered too many alternatives that may all be right. However, we can at once exclude some of these by considering how these gases behave in other reactions. From Gay-Lussac's experiments we know of the following volume relation in the production of ammonia vapor:

3 vol hydrogen + 1 vol nitrogen = 2 vol ammonia.

This result shows us at once that hydrogen cannot be monatomic if each unit volume of ammonia is to have the same number of molecules as each unit volume of hydrogen, since that would entail the impossible conclusion that a molecule of ammonia contains 1½ atoms of hydrogen. But the experimental result harmonizes with Avogadro's hypothesis *if hydrogen is diatomic* and nitrogen is also diatomic; for in this case we may write:

$$3H_2 + N_2 \rightarrow [N_2H_6] \rightarrow 2NH_3, \text{ or}$$

| Hydrogen gas | Nitrogen gas | | Ammonia vapor |
|:---:|:---:|:---:|:---:|
| 3 vol | 1 vol | | 2 vol |

Having now ruled out that hydrogen gas is monatomic, we have automatically denied Eq. (4) for the production of water vapor. Equation (5) is impossible because it conflicts with Gay-Lussac's results on the combining volumes (demonstrate). So Eq. (6) emerges as the simplest one that may still be correct:

$$2H_2 + O_2 \rightarrow [H_4O_2] \rightarrow 2H_2O, \text{ or}$$

| Hydrogen gas | Oxygen gas | | Water vapor |
|:---:|:---:|:---:|:---:|
| 2 vol | 1 vol | | 2 vol |

e) At this point Avogadro himself, quite properly, assumes also a "rule of simplicity"; that is, he regards as the correct process the one that is consistent with other chemical reactions, is simplest, *and* is in harmony with Gay-Lussac's results and his own postulate. Thereby he chooses Eq. (6) over (7), (8), and all the others that may be constructed along these lines. But this is—and was—perhaps a bit disappointing; for all along we had hoped to do without any arbitrary rules of this kind. However, there is this difference: Dalton's rule, by assigning molecular formulas based on fairly incomplete data, had led at once to conflicts, whereas Avogadro's rule of simplicity may be used as a final criterion of choice *after* all the known conflicts have been considered.

---

**Problem 20.5.** Summarize the experimental and theoretical steps which lead to the result that the formula for the molecules of water vapor is $H_2O$, and not HO as Dalton thought originally. Is $H_2O$ the only possibility on the basis of these evidences? If not, why might it be generally accepted as the correct formula?

**Problem 20.6.** A frequently used system for generating hydrogen on a large scale in a portable plant involves the following reaction: A unit volume of alcohol treated with 1 vol steam produces 1 vol carbon dioxide and 3 vol hydrogen gas. Prove that consequently the molecular formula of the alcohol is $CH_3OH$ (wood alcohol).

**Problem 20.7.** Show that the reaction $N_2 + O_2 \rightarrow 2NO$ is consistent both with Gay-Lussac's volumetric experiment on nitric oxide and with Avogadro's hypothesis. Why are we sure that this reaction cannot proceed according to the equation $N + O \rightarrow NO$? If the production of nitrous oxide ($N_2O$) is given by the equation $2N_2 + O_2 \rightarrow 2N_2O$, what must be the volumetric relations in this reaction?

**Problem 20.8.** Accepting Avogadro's hypothesis, show that the following experimental result is harmonious with the proposition that hydrogen and chlorine gas are both diatomic, but is in contradiction to the early idea that hydrogen and chlorine gas are monatomic:

1 vol hydrogen gas + 1 vol chlorine gas

$$= 2 \text{ vol hydrochloric acid vapor.}$$

## 20.10 AN EVALUATION OF AVOGADRO'S THEORY

In summary, let us note the following features of Avogadro's accomplishment.

a) Unlike Dalton, he accepted Gay-Lussac's results as correct, and showed that the work of both investigators was mutually consistent, within experimental error.

b) He replaced the static model where atoms are in contact and fill up all space, with another model in which the molecules (possibly containing two or more atoms) are quite small compared to the spaces between molecules. This assumption, which was needed in order to explain why a gas of large molecules might occupy the same space as an equal number of small molecules, turned out to be essential also for the kinetic theory of gases developed later on in the century.

c) In postulating molecules, Avogadro could offer no plausible explanation for the proposition that chemically identical atoms in the gas of some elements cling together in twos or threes, etc.; indeed, this proposition directly contradicted the prevalent idea that atoms in a gas exert *repulsive* forces on each other (Section 19.4). If, however, forces of attraction could exist between like atoms, why, it was asked, do these forces not cause the whole bulk of the gas to condense? On such grounds Dalton had, in 1809, rejected as absurd the notion of gaseous elements other than monatomic. Nor did others take more kindly to the idea. Chemists after 1811 began to become more and more convinced that the stability of molecules such as those of water could be explained by strong positive or negative charges attached to the atoms of different elements; therefore it was difficult to visualize that

two hydrogen atoms, which on this model evidently would present each other with equal charges, should nevertheless be firmly bound together in a molecule. As a matter of fact, it has been possible to answer such questions as why atoms of some elements do form diatomic, triatomic, etc., molecules only within the last few decades, with the help of the quantum theory (Chapter 29).

d) At the outset there had been only four elements known in gaseous form, namely, nitrogen, oxygen, hydrogen, and chlorine, and for chemical reactions involving these, Avogadro's hypothesis worked well, provided diatomic molecules for each were assumed. Avogadro's theory might therefore have been accepted if it had simply required the postulate that all elements are diatomic in the gaseous state, even without a plausible explanation of why that should be so. But things were not as simple as that. Other elements were studied in gaseous form at higher temperatures, e.g., mercury, phosphorus, arsenic, and sulfur; their chemical and physical behavior could be accounted for on the basis of Avogadro's hypothesis only by assuming that the molecules of these gases were *not* all diatomic, but that, instead, the molecular formulas were $Hg$, $P_4$, $As_4$, and, depending on the temperature, $S$, $S_2$, $S_6$, and $S_8$.

Why this diversity? It seemed to many that Avogadro, while trying to achieve simplicity, had produced confusion. Actually, an independent check on the number of atoms in a molecule of gaseous material eventually bore out Avogadro's hypothesis fully, but that was made possible only much later by the kinetic theory of gases, to be discussed in Chapter 22.

e) There seemed to be little plausibility attached to the idea—and less evidence for it—that during the reaction of gases, intermediate unstable molecules were formed which at once broke into two or more equal fragments. Although we know today that the intermediate stages in such reactions are far more complex, it is still true that Avogadro's remarkable intuition was closer to the facts than the conception of his contemporaries.

f) Avogadro's hypothesis that all gases at the same pressure and temperature have the same number of molecules in a given volume did not allow him to discover the *exact value* of that number, other than the vague suggestion that it would be generally very large. Experiments in many separate branches of physical science since that time have indirectly produced values, and the evidences all agree fairly well on about $2.69 \times 10^{19}$ molecules per $cm^3$ of gas at 0°C and 1 atm pressure. Larger volumes would contain proportionately larger numbers, for example, 22.4 liters at that pressure and temperature would contain $6.02 \times 10^{23}$ molecules—a number which will prove of importance in the sequel. Although Avogadro's hypothesis and these values are correct only for ideal gases, they apply well enough for real gases which are not too dense or too near the point of liquefaction.

g) At the time, two not inconsiderable obstacles to the immediate acceptance of Avogadro's work were the obscure manner of his presentation and the large admixture of other daring—but erroneous —conclusions.

## 20.11 CHEMISTRY AFTER AVOGADRO: THE CONCEPT OF VALENCE

Some of these points may make plausible why Avogadro's work lay neglected for almost half a century, during which time chemistry, initially spurred on by Dalton's contribution, progressed vigorously. But this very activity had an inevitable result: It unearthed more and more evident data testifying to an insufficiency in Dalton's specula-tions. Consequently, while Avogadro's contribution remained in obscurity, other possible amendments to Dalton's conceptual schemes were strongly debated. They all, however, suffered from funda-mental flaws, e.g., the persistent assumption that

the atoms in a gas are contiguous. The discourage-ment of unavoidable contradictions seems to have produced in the 1840s a gradual loss of faith in the whole atomic theory of chemistry.

The solution to the dilemma came after 1858 when the Italian chemist Stanislao Cannizzaro called for a return to Avogadro's work and showed, partly through his own experiments, that it pro-vided indeed a rational basis for chemistry. In fact, this was the final stage of the revolution of chem-istry—thus were achieved the fundamental notions on which modern chemistry is built. But if Can-nizzaro was successful, his persuasiveness had a powerful ally. By that time chemists had become more prepared to accept the ancient suggestion, the kinetic model of a gas, for in the meantime, a revolution had begun in physics—the abandonment of the caloric fluid, and the growing acceptance of the kinetic model of gases in that field.

Dalton's atomic theory suggests a very simple interpretation of the empirical law of definite pro-portions. Carbon monoxide, for example, always consists of molecules made of one carbon atom and one oxygen atom each; each molecule of ammonia invariably consists of one nitrogen atom in close connection with three hydrogen atoms; and sim-ilarly for each molecule of any other compound. However, contemplation of this gratifying reg-ularity soon brings up perplexing questions: Why do atoms have such fixed combining powers with one another? What determines the fact that hydrogen and chlorine always make HCl and not $HCl_3$ or $H_5Cl_2$? Why, given H and O, can we find either $H_2O$ or $H_2O_2$, but not $H_2O_3$; or, given C and O, can we find CO or $CO_2$ but not $C_2O$?

Familiarity with the composition of many molecules might lead to a picturization of hydrogen atoms as having one "hand" or "hook" each by which to engage a similar "hook" on other atoms, as indicated by the formulas HCl and $H_2O$. Thus one hook should be assigned to the chlorine atom, two to oxygen atoms (see Fig. 20.4). The difference between the structure of $H_2O$ and that of $H_2O_2$ is then directly visualizable. The nonexistence of

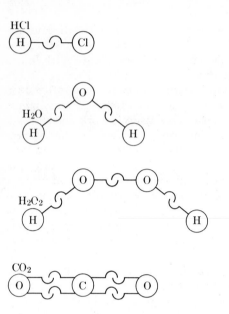

**Fig. 20.4**  Schematic representation of bonding in molecules.  It is not to be imagined that all molecules are actually strung out in one line as represented here.

$H_2O_3$ might be interpreted as the inability of these atoms to form stable molecules when the hooks or bonds are not all engaged.   The formula $CO_2$ implies that the carbon atom has four such coupling mechanisms to engage the two of each oxygen atom; but in CO it would seem that two hooks of carbon are left unoccupied, after all, and ready to snatch more oxygen—a simple picture that co-incides with the fact that CO is such a powerful reducing agent (and so dangerous in a closed space in which there may be competition by living organisms for the available oxygen).

Ultimately this type of analogy is found to be incomplete and dangerous.   The question of com-bining powers of atoms had to be answered in terms of the structure of the atoms themselves. Until that type of answer could be given, however, the general approach indicated in Fig. 20.4 formal-ized the subject in a powerful way, even though it

remained at first mainly a descriptive device.  Thus we encounter, as of 1852, the concept of *valence* (or valency) that became indispensable for the description and understanding of chemical pro-cesses.*  The valence of an element, speaking very roughly, is the number of "hooks" which an atom can put out to enter into chemical partnership with other atoms or with stable groups of atoms.   The simplest way to determine the valence of an element is to look up the chemical formula of a substance involving only this element and hydrogen, which we regard *a priori* as having a valence of 1 (mono-valent).   For example, the formulas $H_2O$, $NH_3$, $CH_4$ indicate valence numbers of O = 2 (divalent), N = 3 (trivalent), C = 4 (quadrivalent).   The for-mula $H_2SO_4$ shows that the sulfate group ($SO_4$) has the valence of 2 (divalent).   Next we may find further valences by noting the number of hydrogen atoms an element can *displace* in a molecule, e.g., in CaO, Ca takes the place of 2 hydrogen atoms in $H_2O$, hence Ca has the valence 2.   The occurrence of HF (hydrofluoric acid) and of $OsF_8$ (osmium octafluoride) shows that osmium can have a valence of 8, the highest valence number that exists. Similarly, from NaCl, $CaCl_2$, $AlCl_3$, $SnCl_4$, $PCl_5$, and $WCl_6$ we derive valences for Na = 1, Ca = 2, Al = 3, Sn = 4, P = 5, and W = 6.

This process of enumeration, however, leads to difficulties.   Carbon is quadrivalent and oxygen is divalent by the above test.   Yet we just saw that they do form CO.   There the total number of avail-able "hooks" is not engaged, and the valences of the parts of the molecule do not balance.  We must now say either that two of the four valence bonds of C are not used in CO, or that the valence of carbon is effectively 2 in CO but 4 in $CO_2$.  In the long run the latter mode of expression is preferable, although it leads to the disappointing realization that carbon and many other elements can take on different valences depending on what molecule

---

*    Modern chemistry books have begun to abandon the term *valence* and use instead some phrase such as *combining capacity*.

Ethane
(C₂H₆)

Ethyl alcohol
(C₂H₅OH)

Benzene
(C₆H₆)

Glucose
(C₆H₁₂O₆)

Fructose
(C₆H₁₂O₆)

**Fig. 20.5** Structural formulas for some organic compounds.

they help to form. The oxides of nitrogen ($N_2O$, $NO$, $N_2O_3$, $NO_2$, $N_2O_5$) indicate that nitrogen can take on the valence numbers 1, 2, 3, 4, and 5, respectively ($O = 2$ throughout).

By the same argument, the valence of iron atoms is 2 in $FeO$, $FeCl_2$, and $Fe(NO_3)_2$, but is 3 in $Fe_2O_3$, $FeCl_3$, and $Fe(NO_3)_3$. Uranium can have five different valences, chlorine six, rhenium seven. Historically, this element of variability brought on a great deal of debate throughout the last half of the nineteenth century.*

Of the many valuable consequences of the valence concept, one is particularly noteworthy at this point. The device of valence bonds, in essence the same as the hooks used in Fig. 20.4, supplemented Dalton's pictorial representation of the atoms themselves and invited speculation on

the structural arrangement in the molecule, particularly for the complex organic substances where the mind needed such a crutch badly. This was all the more urgent because there had been discovered sets of *isomers*,† i.e., substances with precisely the same chemical formula but with very different physical and chemical properties, and the only hope for explaining the differences was the hypothesis of different placement of the atoms within the molecule. Hence the rise of the graphic structural formulas for representing chemical compounds such as those in Fig. 20.5 for $C_2H_6$ (ethane), $C_2H_6O$ (ethyl alcohol, better written $C_2H_5OH$), the famous case of $C_6H_6$ (benzene; note the closed ring), and the isomers $C_6H_{12}O_6$, one being glucose (also called dextrose and grape sugar) and the other fructose (or levulose, fruit sugar).

But in the 1860s, when such representation first became widely used among chemists, there

---

*    The concept of valence brought other troubles also. For instance, a compound sometimes has an anomalous valence number, as in $Fe_3O_4$. It is probable that this is a mixture of two valence states, $FeO$ and $Fe_2O_3$.

---

†    From the Greek *isos* + *meros*, composed of equal parts.

was hesitation in regarding them initially as any-thing more than working schemes without neces-sary relation to the actual three-dimensional and essentially inaccessible structure of molecules. In truth, although the structural formula is indis-pensable as a rough sketch of nature's plan for the molecule, the correct disposition of atoms and stable groups of atoms in the molecule of a com-pound is generally difficult to determine even with our present powerful tools of research. And even if the structure of all compounds were known with the greatest accuracy, this would still give us no clue to the initial puzzle: Why do atoms combine in such limited and specific compounds, and what is the physical mechanism that holds the components of a molecule together? These questions, like so many related ones, were answered only much later after the physical structure of the atom became clear. In the meantime, the concept of valence was a keystone in the construction of another inspired scheme (to be taken up in the next chapter) that contributed to the advance of chemistry—almost as much as the atomic theory itself.

**Problem 20.9.** Evaluate Dalton's atomic theory of chemistry from the point of view of our criteria for theories in Chapter 3.

**Problem 20.10.** Discuss the growth of the atom-istic hypothesis to this point in the light of Chap-ter 12.

## 20.12   MOLECULAR WEIGHTS*

Given a more rational procedure for obtaining molecular formulas, we may at once expect an extension and correction of the table of relative atomic weights. For gases the method is extremely simple. If, for example, we compare (at the same temperature and pressure) the *relative weights of*

---

* This section completes the discussion of Dalton's atomic theory, but may be omitted without loss of continuity.

**Table 20.1.**   Relative Molecular Weights†

| Gas or vapor | Relative molecular weight ($M$) | Molecular formula |
|---|---|---|
| Elements: | | |
| Hydrogen | 2.016 | $H_2$ |
| Helium | 4.003 | He |
| Nitrogen | 28.01 | $N_2$ |
| Oxygen | 32.00 | $O_2$ |
| Fluorine | 38.00 | $F_2$ |
| Argon | 39.95 | A |
| Chlorine | 70.90 | $Cl_2$ |
| Phosphorus | 123.90 | $P_4$ |
| Mercury | 200.59 | Hg |
| Compounds: | | |
| Water vapor | 18.02 | $H_2O$ |
| Carbon monoxide | 28.01 | CO |
| Carbon dioxide | 44.01 | $CO_2$ |
| Uranium hexafluoride | 352.02 | $UF_6$ |
| Mixture: | | |
| Air | 29 (approximate average) | $N_2$ (78%), $O_2$ (21%), $CO_2$ (0.03–0.04%), also water vapor, rare gases, etc. |

† In a less strict sense of the word the term molecular weight is used even for gaseous mixtures (as for air), or for the molecules of a solid or liquid where the molecules may or may not exist in groups, and where molecular formulas, consequently, hardly have the same meaning, as in the case of gases.

*equal volumes* of several gases, we would find by experiment the following relative masses: 2.016 for hydrogen, 28.01 for nitrogen, and about 32.00 for oxygen. If we adopt the physical scale of atomic weights based on an isotope of carbon (see below, Section 27.1), we find that oxygen still has a relative mass of 32.00 to four significant figures. On this basis we may assemble a table of relative molecular weights of some gas molecules—for elements as well as compounds (Table 20.1).

The *relative atomic weights* are, of course, obtained at once by dividing the molecular weight of the element (cf. Table 20.1) by the number of atoms per molecule; the required molecular formulas are obtained from the nature of chemical reactions, particularly from the volumetric relations of gases, as has been demonstrated. A complete list of modern values for all elements will be found in the table of elements (Appendix IV).

**Problem 20.11.** What should be the relative molecular weight of wood alcohol (methyl alcohol, $CH_3OH$)? Ethyl alcohol ($C_2H_5OH$)? Diethyl ether [$(C_2H_5)_2O$]? Quinine ($C_{20}H_{24}N_2O_2$)? Glucose ($C_6H_{12}O_2$)? Clay [$Al_2O_3 \cdot 2(SiO_2) \cdot 2(H_2O)$]?

**Problem 20.12.** Did our chemical evidences exclude the possibility that hydrogen is triatomic? What type of experiment would you look for in order to prove or disprove the hypothesis that hydrogen is triatomic? If that hypothesis should be found to be correct, what do the experiments cited in this chapter tell us about the molecular structure of the other gases? How would the molecular weights given above be modified?

In Problem 20.3 you obtained exclusively from the internal consistency of data on combining weights the probable molecular formulas of five compounds. Note that of these, two were solids, one liquid, and two gaseous. Later we saw that the work of Gay-Lussac and Avogadro supplemented and much simplified such calculations, at least for gaseous reactions. These approaches evidently made possible a complete determination of all atomic weights. Now we must make more explicit the relationship between these three important factors in any reaction, namely, the combining weights, the formula for the reaction, and the atomic and molecular weights of the substances involved. *Given any two of these, the third can be*

*calculated.* These two examples will help to illustrate this point.

**Example 1.** In the reduction of the black iron ore, hematite, iron is obtained by burning carbon with the ore in the following reaction:

$$Fe_2O_3 + 3C \rightarrow 2Fe + 3CO.$$

Given this reaction equation, and the atomic weights (Fe = 55.85, O = 16.00, C = 12.01), find the combining weights.

*Solution:* The following must represent the proportions by weight of the substances involved:

$$
\begin{array}{cc}
\text{Hematite} & \text{Carbon} \\
\underbrace{[(55.85) \times 2 + (16.00) \times 3]}_{159.70} + \underbrace{[3 \times 12.01]}_{36.03}
\end{array}
$$

$$
\begin{array}{cc}
\text{Iron} & \text{Carbon monoxide} \\
\rightarrow \underbrace{[2 \times 55.85]}_{111.70} + \underbrace{[3 \times (12.01 + 16.00)]}_{84.03} \\
\rightarrow \quad 111.70 \quad + \quad 84.03
\end{array}
$$

This schedule informs us that if we hope to obtain about 112 g of iron, we must supply the furnace with about 160 g ore, 36 g pure coal (or equivalent), and must take precautions to dispose of 84 g of carbon monoxide gas.

**Problem 20.13.** How many tons of carbon must we supply if we expect to obtain 80 tons of iron?

**Problem 20.14.** Depending on the temperature of the furnace, another possible reaction is

$$Fe_2O_3 + 3CO \rightarrow 2Fe + 3CO_2.$$

What are the relative weights of the participants in this reaction?

**Example 2.** When 304 g of a certain oxide of chromium are ignited with 108 g aluminum (Al, at. wt. about 27), there is formed an aluminum oxide and 208 g of the pure metal chromium (Cr, at. wt.

about 52). From these data, find the equation for this reaction.

*Solution:* First of all, there must be $(304 + 108) - 208$ g of aluminum oxide, i.e., 204 g. Now we do not know the molecular formulas, but they can be written for the time being as follows,

$$Cr_xO_y + zAl \rightarrow Al_2O_y + xCr,$$

where $x$, $y$, and $z$ may each stand for a whole integer (1, 2, 3, ...), for evidently the equation, if it refers to combinations of atoms, must contain as many atoms of each element on one side as on the other; that is, the equation must *balance*.

We may now regard the relative combining weights as indicative of the relative molecular weights, and say

| Oxide of chromium | Alumi- num | Aluminum oxide | Chro- mium |
|---|---|---|---|
| $[(52x) + (16)y]$ | $[z27]$ | $[(27)z + (16)y]$ | $[x52]$ |
| 304 | 108 | 204 | 208 |

$$z \cdot Al = 108 \quad \text{or} \quad z(27) = 108; \quad z = \frac{108}{27} = 4,$$

$$x \cdot Cr = 208 \quad \text{or} \quad x(52) = 208; \quad x = \frac{208}{52} = 4.$$

Therefore

$$Cr_xO_y = Cr_4O_y = [(4 \times 52) + (y \times 16)] = 304;$$

$$y = \frac{304 - 208}{16} = 6.$$

Our reaction equation is then

$$Cr_4O_6 + 4Al \rightarrow Al_4O_6 + 4Cr.$$

More simply, $2Cr_2O_3 + 4Al \rightarrow 2Al_2O_3 + 4Cr$, and dividing through by 2, finally

$$Cr_2O_3 + 2Al \rightarrow Al_2O_3 + 2Cr.$$

---

**Problem 20.15.** At one time, aluminum was produced by reducing aluminum chloride with metallic sodium (Na, at. wt. about 23), yielding sodium chloride and aluminum (Al, at. wt. about 27). If 34.5 g sodium are needed to obtain 13.5 g aluminum from 66.8 g aluminum chloride, what is the simplest reaction formula? [*Note:* To obtain $x$, $y$, and $z$ in integral form, one may have to multiply all these given quantities by some one number.]

---

Finally, we should mention a fruitful concept that makes these calculations very much more simple: the *gram-molecular-weight* or "mole." By definition, one gram-molecular-weight, or 1 mole, of any substance is an amount whose mass is numerically equal to its relative molecular weight. Thus 1 mole of oxygen is nothing but 32.000 g of oxygen; 1 mole of hydrogen = 2.016 g of hydrogen; 3 moles of mercury = 601.77 g of mercury (i.e., $3 \times 200.59$; note that for monatomic substances the term gram-molecular-weight is often replaced by "gram-*atomic*-weight"); $n$ moles of water = $n \times 18.016$ g of water. Conversely, if a substance is known to have the molecular weight $M$, then $y$ g of that substance correspond to $y/M$ moles.

It stands to reason that *all pure substances must have the same number of molecules in one mole*, no matter what the state, temperature, or pressure. The proof is simple: The total mass of a substance ($m$) is the mass per molecule ($m_0$) × number of molecules ($N_m$), or $m = m_0N_m$. If the mass is taken to be 1 mole, it is numerically equal to $M$, the molecular weight; then $M = m_0N_m$. If we compare two substances with molecular weights $M'$ and $M''$, we may write

$$\left. \begin{aligned} M' &= m_0'N_m' \\ M'' &= m_0''N_m'' \end{aligned} \right\}.$$

Taking the ratio,

$$\frac{M'}{M''} = \frac{m_0'}{m_0''} \cdot \frac{N_m'}{N_m''}.$$

But *by definition* the ratio $m_0'/m_0''$ is the relative molecular weight of one substance with respect to the other, or $M'/M'' = m_0'/m_0''$. Thus $N_m' = N_m''$, as was to be proved. Furthermore, the actual number of molecules per mole, by many indirect tests, is $6.02 \times 10^{23}$ (called *Avogadro's number*, symbolized by $N$).

We recall that at $0°C$ and 1 atm pressure, all pure gases contain $2.69 \times 10^{19}$ molecules per $cm^3$. It is now evident that under these standard conditions, $6.02 \times 10^{23}$ molecules would occupy a volume of $(6.02 \times 10^{23})/(2.69 \times 10^{19}) = 22.4 \times 10^3$ $cm^3 = 22.4$ liters. This brings us to a common form of stating Avogadro's hypothesis: *At $0°C$ and 1 atm pressure, one mole of any pure gas, containing $6.02 \times 10^{23}$ molecules, takes a volume of 22.4 liters.*

A useful and immediate application of the concept *mole* springs from this argument: If we know that in a certain reaction every molecule of $A$ joins with $x$ molecules of $B$, then we must provide for each *mole* of $A$ exactly $x$ *moles* of $B$ (or corresponding fractions) to assure a complete reaction.

**Example 3.** Of a substance $X$ we are told that it is formed when *equal numbers* of moles of two elements combine. When a quantity of 40 g of $X$ is analyzed, it is found to contain about 35.5 g chlorine and 4.5 g of an unidentified element $Y$; the stated amount of chlorine is $\frac{1}{2}$ mole of the chlorine gas $Cl_2$, because

$$n = \frac{m}{M} = \frac{35.5\text{ g}}{71\text{ g/mole}} = \frac{1}{2}\text{ mole};$$

therefore the 4.5 g of $Y$ represent $\frac{1}{2}$ mole of $Y$ also. This makes the molecular weight of $Y \doteq 9$. If $Y$ is monatomic, this is also its atomic weight; if diatomic, the atomic weight is 4.5; if triatomic, 3; etc. A glance at the table of atomic weights shows

that, in fact, only one of these choices exists—namely, beryllium (Be) with an atomic weight 9 (or more accurately, 9.01). Thus we have identified the substance $X$ as $BeCl_2$.

---

**Problem 20.16.** How many moles of ammonia is 1 g of that vapor? How many molecules are there in this quantity? How many *atoms*? What would be its volume at $0°C$ and 1 atm? at $100°C$ and 5 atm? (Use the general gas law.) How many moles of hydrogen does this vapor contain?

**Problem 20.17.** The reaction representing the rusting of iron may be written

$$3Fe + 4H_2O \rightarrow Fe_3O_4 + 4H_2.$$

How many moles of hydrogen are liberated for each gram of iron completely rusted?

**Problem 20.18.** Knowing Avogadro's number, the molecular weights of certain gases, and their molecular formulas, calculate the "absolute" mass (in grams) of a single atom of hydrogen, of oxygen, and of chlorine.

---

### RECOMMENDED FOR FURTHER READING

H. A. Boorse and L. Motz (editors), *The World of the Atom*, New York: Basic Books, 1966. Extracts and discussion of writings of Dalton, Gay-Lussac, Avogadro, and Prout, pages 139–191. (Or see the similar extracts in Hurd and Kipling's book below.)

H. Butterfield, *The Origins of Modern Science*, Chapter 11

D. L. Hurd and J. J. Kipling, *The Origins and Growth of Physical Science*, Vol. 2, pages 19–36, 46–80. Extracts from writings of Dalton, Gay-Lussac, and Avogadro.

L. K. Nash, "The atomic-molecular theory," pages 215–321 in *Harvard Case Histories in Experimental Science* (edited by J. B. Conant), Cambridge, Mass.:

Harvard University Press, 1948. Also available as a separate pamphlet.

E. C. Patterson, *John Dalton and the Atomic Theory*, Garden City, N.Y.: Anchor Books, 1970; paperback

## SOURCES, INTERPRETATIONS, AND REFERENCE WORKS

A. J. Bernatowicz, "Dalton's rule of simplicity," *Journal of Chemical Education*, Vol. 47, pages 577–579 (1970)

D. S. L. Cardwell (editor), *John Dalton and the Progress of Science*, New York: Barnes & Noble, 1968. Papers presented to a conference held to mark the bicentenary of Dalton's birth.

Ida Freund, *The Study of Chemical Composition*, Chapters III–XIII

F. Greenaway, *John Dalton and the Atom*, Ithaca: Cornell University Press, 1967

H. Guerlac, "Some Daltonian Doubts," *Isis*, Vol. 52, pages 544–554 (1961)

A. J. Ihde, *The Development of Modern Chemistry*, New York: Harper & Row, 1964, Chapters 4 and 6

D. M. Knight, *Atoms and Elements, a Study of Theories of Matter in England in the Nineteenth Century*, London: Hutchinson, 1967

L. K. Nash, "The origins of Dalton's chemical atomic theory," *Isis*, Vol. 47, pages 101–116 (1956)

W. G. Palmer, *A History of the Concept of Valency to 1930*, New York: Cambridge University Press, 1965

J. R. Partington, *A History of Chemistry*, Vol. IV, London: Macmillan, 1964

A. L. Smyth, *John Dalton 1766–1844: A Bibliography of Works by and About Him*, Manchester, England: Manchester University Press, 1966

A. W. Thackray, "The emergence of Dalton's chemical atomic theory," *British Journal for the History of Science*, Vol. 3, pages 1–23 (1966); "Documents relating to the origins of Dalton's chemical atomic theory," *Memoirs of the Manchester Literary and Philosophical Society*, Vol. 108, no. 2, pages 1–22 (1966); "The origin of Dalton's chemical atomic theory: Daltonian doubts resolved," *Isis*, Vol. 57, pages 35–55 (1966)

# Chapter 21

# THE PERIODIC TABLE
# OF ELEMENTS

## 21.1 THE SEARCH FOR REGULARITY IN THE LIST OF ELEMENTS

As the early list of chemical elements became longer, and as their physical and chemical properties became increasingly better established, there arose rather naturally the urge to find some relation among the elements that would bring them into a logical sequence—a situation analogous to the Keplerian quest for principles that would relate the separate planets.

At first glance the task would seem hopeless. The most immediately striking thing is that each element seems to have unique properties that differentiate it from each of the other elements. Most are solids under ordinary circumstances, but many are gases and some are liquids; their relative abundances on earth have widely different values, and so have their physical characteristics—boiling point, density, and the like. Nevertheless, there are some clues for speculation. In 1815–1816 the English physician William Prout had suggested that the atoms of all elements were but the condensations of different numbers of hydrogen atoms, which themselves were the true fundamental particles. Consequently the atomic weights could be expected to be whole multiples of the atomic weight of hydrogen; and indeed the early determination of atomic weights by Dalton did allow this hypothesis.

However, more accurate measurements were soon forthcoming, particularly those of the great chemists Berzelius, Dumas, and Stas, all demonstrating real deviations from Prout's rule. A glance at a modern and accurate table of elements shows that only a few of the atomic weights are even close to being exact multiples of the atomic weight of hydrogen. Thus Prout's hypothesis, despite its direct appeal, had to yield to the contrary evidences of quantitative analysis. (It should be noted, however, that much of this quantitative analysis was performed just for the purpose of testing Prout's hypothesis, which must therefore be judged as having been useful to science regardless of the fact that it was ultimately refuted.)

But at the same time, these growing results favored another trend of thought toward bringing order into the apparently haphazard list of elements, namely, the attempts to connect the atomic weights of elements with their chemical properties. It had long been known that dispersed throughout the list of elements are *families* of elements of similar chemical behavior. An example of such a family is that of the *halogens*—a group that includes the elements fluorine (F), chlorine (Cl), bromine (Br), and iodine (I); their respective atomic weights are approximately 19, 35.5, 80, 127. Although they exhibit some striking dissimilarities (e.g., the first two are gases, the third ordinarily a liquid, the last a volatile solid), these four elements have much in common. They all combine with many metals to form white, crystalline salts (*halogen* means "salt-former"). The boiling points change progressively as we shift attention from the first to the last one (F, $-187°C$; Cl, $-35°C$; Br, $59°C$; I, $184°C$), and similarly for their melting points. The solubility in water progressively decreases, and so does the chemical activity, as manifested by the violence of chemical reactions (speed of reaction, amount of energy liberated).

This list of properties relating the halogens by continuous changes is to be supplemented by a second list of properties which are the same for all these elements. Halogens form compounds of similar formulas in violent reaction with metals (e.g., NaF, NaCl, NaBr, NaI; or $AlF_3$, $AlCl_3$, $AlBr_3$, $AlI_3$); that is, they have the same valence in analogous reactions. With hydrogen, all four elements form simple compounds that dissolve in water and form acids. All four form a diatomic vapor under ordinary conditions.

Another family, referred to as the *alkali metals,* is formed by lithium (Li, approximate atomic weight 7), sodium (Na, 23), potassium (K, 39), rubidium (Rb, 85.5), and cesium (Cs, 133). Here too, there is a long list of chemical and physical properties which are common to all five elements or which change in the same sense with increasing atomic weight.

In the restless search for some quantitative connection between the elements, there appeared as early as 1829 the discovery of the German chemist Johann Wolfgang Döbereiner, who found that within several chemical families the atomic weight of one element had a value equal, or nearly so, to the average of its two immediate neighbors; thus 23 for Na is the average of 7 for Li and 39 for K; and in the triad of Cl, Br, and I, 80 for Br is roughly the mean of 35.5 (Cl) and 127 (I). But no physical meaning could be attached to such numerical regularities (we are at once reminded of Bode's law). In fact, the

underlying explanation of such relations turned out to be extremely complex (involving the stability of nuclei) and has not yet been fully given. However, the search for numerical regularities of this kind proved popular, as always, and was carried on with vigor. It is another example where research that proceeded for a time on the wrong track was nevertheless directly fruitful, for through this effort it became more evident that the list of elements did indeed have a meaningful general structure, although not according to such atomic weight groups.

From 1860 on, about the time when Cannizzaro's agitation on behalf of Avogadro's hypothesis began to succeed, it was possible in principle to fix with very good accuracy the atomic weights of almost all elements and the molecular weights of compounds. Now one could place the elements into a sequence based entirely on a property of the single atom itself, namely, its atomic weight. Such a sequence reveals a curious fact: one may discover a regular spacing between the elements belonging to one family, a recurrence of numerical position quite apart from Döbereiner's atomic weight triads. In the words of the English chemist Newlands, one of several to seek such a regularity of spacing, "the eighth element, starting from a given one, is a kind of repetition of the first, like the eighth note in an octave of music."

Newlands' attempt was, however, doomed to failure. Although it proved to be correct that chemically similar elements repeat periodically, and although it was possible to perceive several "octaves" in Newlands' arrangement, his analogies more often than not had to be rather far-fetched, for he did not foresee that the periodicity in the end would not be regular throughout the list of elements, and he had not considered the necessity of leaving spaces for still undiscovered elements. Here you might object that prior to the discovery of the new elements one could not possibly leave gaps for them in the proper places. However, it was exactly this achievement which is at the core of the work of Mendeléeff.

## 21.2 THE EARLY PERIODIC TABLE OF ELEMENTS

In the publications of the great Russian chemist Dmitri Ivanovitch Mendeléeff (1834–1907) from 1869 on—and, starting a little later, of the German chemist Julius Lothar Meyer—we find the culmination of the six decades of quantitative investigations and speculative inductions since Dalton's atomic theory first became generally known. A few of the most striking aspects of Mendeléeff's contribution can be given in his own words, even though a great deal of descriptive material between excerpts must here be omitted. Thus his initial paper on the subject (1869) gives us the following account of the background and motivation of his work:

"During the course of the development of our science, the systematic arrangement of elements has undergone manifold and repeated changes. The most frequent classification of elements into metals and non-metals is based upon physical differences as they are observed for many simple bodies, as well as upon differences in character of the oxides and of the compounds corresponding to them. However, what at first acquaintance with the subject-matter appeared to be completely clear, lost its importance entirely when more detailed knowledge was obtained. Ever since it became known that an element such as phosphorus could appear in non-metallic as well as in metallic form, it became impossible to found a classification on physical differences. . . . In recent times the majority of chemists is inclined to achieve a correct ordering of the elements on the basis of their valency. There are many uncertainties involved in such an attempt . . .

"Thus, there does not exist yet a single universal principle which can withstand criticism, that could serve as guide to judge the relative properties of elements and that would permit their arrangement in a more or less strict system. Only with respect to some groups of elements there are no doubts that they form a whole and represent a

Fig. 21.1  Dmitri Ivanovitch Mendeléeff (1834–1907).

natural series of similar manifestations of matter. Such groups are: the halogens, the alkaline earth metals, the nitrogen group, partially also the sulfur group, the companions of platinum, the companions of cerium, and a few others . . . . The discovery of rubidium, cesium and thallium gave rise to the reflection that our knowledge of the elements is very limited; and the attempt to construct a system appears to be premature as long as there exists no hypothetical basis which could serve as the foundation of a strict system.

"The investigations regarding the simple relations of atomic weights have caused many, in particular Dumas, Pettenkofer, Sokolow and others, to point out the numerical relations between the atomic weights of those elements which form a group; but so far as I know, they have not led to a systematic arrangement of all known elements.... When I undertook to write a handbook of chemistry entitled *Foundations of Chemistry*, I had to make a decision in favor of some system of elements in order not to be guided in their classification by accidental, or instinctive, reasons but by some exact, definite principle. In what has been said above we have seen the nearly complete absence of numerical relations in the construction of systems of elements; every system, however, that is based upon exactly observed numbers is to be preferred of course to other systems not based upon numbers because then only little margin is left to arbitrariness. The numerical data available regarding elements are limited at this time. Even if the physical properties of some of them have been determined accurately, this is true only of a very small number of elements. Properties such as the optical, and even the electrical or magnetic ones, cannot serve as a basis for the system naturally, since one and the same body, according to the state in which is happens to be at the moment, may show enormous differences in this regard. With respect to this fact, it is sufficient to remember graphite and diamond, ordinary and red phosphorus.... However, everybody does understand that in all changes of properties of elements, *something* remains unchanged, and that when elements go into compounds this material something represents the [common] characteristics of compounds the given element can form. In this regard only a numerical value is known, and this is the atomic weight appropriate to the element. The magnitude of the atomic weight, according to the actual, essential nature of the concept, is a quantity which does not refer to the momentary state of an element but belongs to a material part of it, a part which it has in common with the free element and with all its compounds.... For this reason I have endeavored to found the system upon the quantity of the atomic weight.

"The first attempt I undertook in this direction was the following: I selected the bodies with the smallest atomic weight and ordered them according to the magnitude of their atomic weights. Thereby it appeared that there exists a periodicity of properties and that even according to valence, one element follows the other in the order of an arithmetical sequence.

$$
\begin{array}{llll}
\text{Li} &= 7 & \text{Be} = 9.4 & \text{B} = 11 \\
\text{Na} &= 23 & \text{Mg} = 24 & \text{Al} = 27.4 \\
\text{K} &= 39 & \text{Ca} = 40 & \ldots \\
\\
\text{C} &= 12 & \text{N} = 14 & \text{O} = 16 & \text{F} = 19 \\
\text{Si} &= 28 & \text{P} = 31 & \text{S} = 32 & \text{Cl} = 35.3 \\
\text{Ti} &= 50 & \text{V} = 51'' & \text{(et cetera)}
\end{array}
$$

Note that in this arrangement hydrogen, the first and lightest of the elements, is left out, for it has rather unique properties; that helium, together with the other elements of the family of *rare gases*, had not yet been discovered; and that Mendeléeff's values for atomic weights sometimes differ significantly from the more recent values. He writes the first seven elements from lithium to fluorine in sequence of increasing atomic weight, then writes the next seven, from sodium to chlorine, below the first line. The periodicity of chemical behavior is already obvious. In the first vertical column are the first two alkali metals, in the seventh column the first two halogens, and each of the columns between contains two chemically similar elements. Furthermore, the main valences for these columns are 1 for the first, 2 for the second, 3 for the next, then 4, 3, 2, and 1. Reading from left to right, any element (let us use a general symbol R) in each column may form higher oxides or hydrides according to the following scheme.

| Column | Higher oxide or hydride |
|--------|-------------------------|
| 1st | $R_2O$ |
| 2nd | $R_2O_2 (= RO)$ |
| 3rd | $R_2O_3$ |
| 4th | $R_2O_4 (= RO_2)$ |
|     | $H_4R$ |
| 5th | $H_3R$ |
| 6th | $H_2R$ |
| 7th | $HR$ |

When Mendeléeff now adds to the two rows of elements the third, K comes to stand directly below Li and Na, and of course we know that K is indeed a member of the same family with the same valence of unity. Next comes Ca (calcium), divalent like Be and Mg above it. Next should come the next heaviest element then known, which was titanium (Ti), and previous workers trying to develop such tables had generally done this. In Newlands' table, which almost entirely coincides with Mendeléeff's to this point, another element was placed in its stead whose atomic weight actually was much too high. This point offers us accurate insight into the dominant thoughts of Mendeléeff. He recognizes that the chemical properties of Ti suggest that it appear under C and Si in the fourth vertical column, not in the third column. If the classification is to be complete and meaningful, there ought to exist a hitherto unsuspected element with an atomic weight somewhere between that of Ca (40) and Ti (50, new value 47.9), and with a valence of 3. This amounts to a definite prediction, and later in the same work Mendeléeff finds other cases of this sort among the remaining elements.

The whole scheme to this point may be under-

stood in terms of a very crude analogy. It is as if a librarian were to put all his books in one heap, weigh them individually, and then place them on a set of shelves according to increasing weight—and find that on each shelf the first book is on art, the second on philosophy, the third on science, the fourth on economics, and so on. Our librarian may not understand in the least what the underlying explanation for this astonishing regularity is, but if he now discovers on one of these shelves a sequence art-science-economics, he will perhaps be very tempted indeed to leave a gap between the two books on art and science and look about for a missing philosophy book of the correct weight to fill the gap.

Mendeléeff had no illusions that he understood the reason for this symmetry of arrangement, but he saw clearly that his work would eventually lead to a physical explanation, and that in the meantime "new interest will be awakened for the determination of atomic weights, for the discovery of new elements, and for finding new analogies among the elements." At a later time he added, "just as without knowing the cause of gravitation it is possible to make use of the law of gravitation, so for the aims of chemistry it is possible to take advantage of the laws discovered by chemistry without being able to explain their causes." The initial paper continues:

"I shall now present one of the many systems of elements which are based upon the atomic weight. They form but one attempt to represent the results which can be achieved in this direction. I am quite conscious of the fact that this attempt is not final, but it appears to me to express quite clearly already the applicability of my proposed principle to all elements whose atomic weight is determined with some reliability. Above all I was interested to find out a general system of elements."

Table 21.1 gives Mendeléeff's arrangement in the version published in 1872. We note that Mendeléeff

**Table 21.1.** Periodic classification of the elements (Mendeléeff, 1872)*

| Group→ | I | II | III | IV | V | VI | VII | VIII |
|---|---|---|---|---|---|---|---|---|
| Higher oxides and hydrides | $R_2O$ —— | $RO$ —— | $R_2O_3$ —— | $RO_2$ $H_4R$ | $R_2O_5$ $H_3R$ | $RO_3$ $H_2R$ | $R_2O_7$ $HR$ | $RO_4$ —— |
| 1 | H(1) | | | | | | | |
| 2 | Li(7) | Be(9.4) | B(11) | C(12) | N(14) | O(16) | F(19) | |
| 3 | Na(23) | Mg(24) | Al(27.3) | Si(28) | P(31) | S(32) | Cl(35.5) | |
| 4 | K(39) | Ca(40) | —(44) | Ti(48) | V(51) | Cr(52) | Mn(55) | Fe(56, Co(59), Ni(59), Cu(63) |
| 5 | [Cu(63)] | Zn(65) | —(68) | —(72) | As(75) | Se(78) | Br(80) | |
| 6 | Rb(85) | Sr(87) | ?Yt(88) | Zr(90) | Nb(94) | Mo(96) | —(100) | Ru(104), Rh(104), Pd(106), Ag(108) |
| 7 | [Ag(108)] | Cd(112) | In(113) | Sn(118) | Sb(122) | Te(125) | I(127) | |
| 8 | Cs(133) | Ba(137) | ?Di(138) | ?Ce(140) | —— | —— | —— | |
| 9 | —— | | | | —— | —— | —— | |
| 10 | —— | —— | ?Er(178) | ?La(180) | Ta(182) | W(184) | —— | Os(195), Ir(197), Pt(198, Au(199) |
| 11 | [Au(199)] | Hg(200) | Tl(204) | Pb(207) | Bi(208) | —— | —— | |
| 12 | —— | —— | —— | Th(231) | —— | U(240) | | |

*Series* runs vertically alongside the series numbers 1–12.

* Atomic weights are shown in parentheses. Horizontal lines in the boxes indicate "missing" elements which Mendeléeff expected to be discovered in the future.

distributes the 63 elements then known (with 5 in doubt) in 12 horizontal lines or *series*, starting with the lightest element, hydrogen, at the top left, and ending with uranium at the bottom right. The others are written in order of increasing atomic weight, but are so placed that each element is in the nearest vertical column or *group* of chemically similar elements. Thus the halogens are in group VII, group VIII contains only ductile metals, groups I and II contain light and low melting-point metals, and the specific family of alkali metals is in group I. The more detailed arguments for each particular placement, as Mendeléeff's extensive discussion shows, are often exceedingly delicate, guided by his profound and often almost intuitive knowledge of chemical nature.

**Fig. 21.2** Atomic volumes for the first portion of the Table of Elements. (After Lothar Meyer, but using modern values)

The arrangement does, however, leave many gaps: "Vacant places occur for elements which, perhaps, shall be discovered in the course of time." And several of the elements toward the end of this table are in doubt: "The higher atomic weights belong to elements which are rarely encountered in nature, which do not form large deposits, and which therefore have been studied relatively little.... With respect to the position of some elements that are rarely encountered in nature, there exists, quite understandably, complete uncertainty."

But the main part of the table is fairly complete, and another feature of this scheme reveals itself: "For a true comprehension of the matter it is very important to see that all aspects of the distribution of the elements according to their atomic weights essentially express one and the same fundamental dependence—*periodic properties.*" By this is meant that in addition to the gradual change in physical and chemical properties within each *vertical* group or family, there is also a periodic change of such properties among the elements if they are scanned in a horizontal sequence. "The properties of the elements as well as the forms and properties of their compounds are in periodic dependence on, or (to express ourselves algebraically) form a periodic function of, the atomic weights of the elements." This *periodic law* is the heart of the matter, a generalization of Newlands' suggestion (with whose work Mendeléeff was initially not acquainted). We can best illustrate it as Lothar Meyer did, by drawing a curve of the value taken on by some physical quantity as a function of atomic weight, for example, the *atomic volume* of each element (the ratio of atomic weight to density, the mass per unit volume, in the liquid or solid phase; Fig. 21.2). Note the irregular spacing of the "waves" yet the simultaneous periodicity of the position of analogous elements on the curve, entered here specifically for the halogens and for the alkali metals.

**Problem 21.1** Use the data given in the *Handbook of Chemistry and Physics* or a similar source, and plot the melting points of the elements against the atomic weights for the elements from hydrogen to barium (Ba) inclusive. Then discuss the periodicities that appear on the graph, e.g., the positions of the two families of elements mentioned thus far.

## 21.3 CONSEQUENCES OF THE PERIODIC LAW

More detailed examination of the periodic scheme yields some additional information.

a) If an element R enters into chemical combinations with oxygen, the composition of one of its higher oxygen compounds is determined by the *group* to which R belongs, i.e., the first group gives $R_2O$, the second $R_2O_2$ (or RO), the next $R_2O_3$, and so forth; similarly for combinations with hydrogen as indicated along the top of Table 21.1.

b) A distinction is to be made between the even and the odd series, indicated in Table 21.1 by a slight shift in the printing of the symbols to the right or left. Chemical and physical similarities are generally closer among all the elements in one group that belong to the even series only or the odd series only. Thus in group VI, S, Se, and Te are nonmetals, whereas Cr, Mo, and W are heavy metals.

c) Each element is a member of a vertical (family) sequence and a horizontal (periodic) sequence. Thus any one property of an element is related to the rest of the table somewhat like the piece in the middle of a jigsaw puzzle or the letter in a crossword puzzle—and like it, can be induced from what is known about the neighboring regions. Here, to continue with arguments taken from Mendeléeff's textbook, is one example:

"If in a certain group there occur elements $R_1$, $R_2$, $R_3$, and if in that series which contains one of these elements, for instance $R_2$, and element $Q_2$ precedes it and an element $T_2$ succeeds it, then the properties of $R_2$ are determined by the mean of the properties of $R_1$, $R_3$, $Q_2$, and $T_2$. Thus, for instance, the atomic weight of $R_2 = \frac{1}{4}(R_1 + R_3 + Q_2 + T_2)$. For example, selenium occurs in the same group as sulfur, S = 32, and tellurium, Te = 127, and, in the 5th series, As = 75 stands before it and Br = 80 after it. Hence the atomic weight of selenium should be $\frac{1}{4}(32 + 127 + 75 + 80) = 78.5$, which is near to that generally accepted, Se = 79, in which there is a possible error in the first decimal, so that 78.5 may be nearer the actual figure."

We see at once what this leads to:

"... the periodic dependence of the properties on the atomic weights of the elements gives a *new means for determining*, by means of the equivalent weight, *the atomic weight* or atomicity of imperfectly investigated but known elements, for which no other means could as yet be applied for determining the true atomic weight. At the time [1869] when the periodic law was first proposed there were several such elements. It thus became possible to learn their true atomic weights, and these were verified by later researches. Among the elements thus concerned were indium, uranium, cerium, yttrium, and others .... Thus a true law of nature anticipates facts, foretells magnitudes, gives a hold on nature, and leads to improvements in the methods of research, etc."

d) But there is another, much more important consequence:

"[The] properties of selenium may also be determined in this manner; for example, arsenic forms $H_3As$, bromine gives HBr, and it is evident that selenium, which stands between them, should form $H_2Se$, with properties intermediate between those of $H_3As$ and HBr. Even the physical properties of selenium and its compounds, not to speak of their composition, being determined by the group in which it occurs, may be foreseen with a close approach to reality from the properties of sulfur, tellurium, arsenic, and bromine. *In this manner it is possible to foretell the properties of elements still unknown*, especially when it is surrounded by well-known elements. For instance, in the position IV, 5 [that is, in the IVth group and 5th series] an element is still wanting. These unknown elements may be named after the preceding known element of the same group by adding to the first syllable the prefix *eka-*, which means *one* in Sanskrit. The element IV, 5, follows after IV, 3, and this latter position being occupied by silicon, I named this formerly unknown element ekasilicon and its symbol Es. The following are the properties which this element should have on the basis of the known properties of silicon, tin, zinc, and arsenic. Its atomic weight is nearly 72, it forms a higher oxide $EsO_2$, a lower oxide EsO, compounds of the general form $EsX_4$, and chemically unstable lower compounds of the form $EsX_2$. Es gives volatile organometallic compounds; for instance, $Es(CH_3)_4$, $Es(CH_3)_3Cl$, and $Es(C_2H_5)_4$, which boil at about 160°, etc.; also a volatile and liquid chloride, $EsCl_4$, boiling at about 90° and of density about 1.9. $EsO_2$ will be the anhydride of a feeble colloidal acid, metallic Es will be rather easily obtainable from the oxides and from $K_2EsF_6$ by reduction, $EsS_2$ will resemble $SnS_2$ and $SiS_2$, and will probably be soluble in ammonium sulphide; the density of Es will be about 5.5, and $EsO_2$ will have a density of about 4.7, etc . . . ."

This summary of his own prediction in 1871 shows fully the genius and daring of the man, the power of his scheme. And in 1887, C. Winckler of Freiberg did indeed discover a metal, now called germanium (Ge), having an atomic weight of about $72\frac{1}{2}$, a density of $5\frac{1}{2}$, an oxide $GeO_2$ of density 4.7, a derivative $Ge(C_2H_5)_4$ that boiled at 160°C, a liquid chloride $GeCl_4$ with a boiling point at 83°C and density 1.9, and so forth. In a similar fashion Mendeléeff had described the probable specifications of the elements at group III, series 4 and 5, later found to have the predicted properties (gallium and scandium).

Even though not every aspect of Mendeléeff's work offered such marvelous agreement, these are indeed stunning results (again reminiscent of the discovery of the asteroids in the general orbits predicted for them by Bode's law). Numerical predictions of this sort are the very essence of science, and the chemical profession soon cast off its initial reluctance to embrace the main concepts of a periodic classification of elements.

## 21.4  THE MODERN PERIODIC TABLE

It remains to tell briefly of some rather important rearrangements that had to be made subsequently in the periodic table. Table 21.2 is one of several modern forms, and comparison with Mendeléeff's table reveals these specific differences:

a) Hydrogen is placed uniquely so as not to be directly associated with any single family; it does in fact have a chemical relationship with both the halogens and the alkali metals.

b) The placement of the symbol in each rectangle is arranged so that elements along one vertical line belong to one chemical family. The elements have been given consecutive numbers, the so-called *atomic numbers,* to indicate their positions in the periodic table. Several elements have been assigned more correct atomic weights and new places.

c) Many new elements have been entered, and there are now no more additions expected within the table. In the 57th place belongs not one element but a whole group of 14 chemically almost indistinguishable elements, the *rare earths,* written out for convenience in a separate row and given atomic numbers from 57 to 71 inclusive. These elements were largely unknown in Mendeléeff's time, and indeed they somewhat disturb the symmetry of the table. Similarly, the elements of the actinide series may best be placed into the one position marked 89.

d) The lower end of the table has been pulled together. The horizontal rows fall into 7 periods, the first three short, the following three long ones (2 series each).

e) In 1894 the startling discovery was made that the atmosphere contained about 1% of a hitherto unnoticed gas, later called *argon* (A), which had escaped previous detection by chemists because of its complete unwillingness to enter into any chemical combination with other elements. Now other such gaseous elements were discovered, forming a family of six inert gases (He, Ne, A, Kr, Xe, Rn).

**Table 21.2.** Periodic table of elements*

| Group→ | | I | II | III | IV | V | VI | VII | VIII | | | O |
|---|---|---|---|---|---|---|---|---|---|---|---|---|
| Period | Series | | | | | | | | | | | |
| 1 | 1 | 1 H 1.0080 | | | | | | | | | | 2 He 4.003 |
| 2 | 2 | 3 Li 6.940 | 4 Be 9.013 | 5 B 10.82 | 6 C 12.011 | 7 N 14.008 | 8 O 16.0000 | 9 F 19.00 | | | | 10 Ne 20.183 |
| 3 | 3 | 11 Na 22.991 | 12 Mg 24.32 | 13 Al 26.98 | 14 Si 28.09 | 15 P 30.975 | 16 S 32.066 | 17 Cl 35.457 | | | | 18 A 39.944 |
| 4 | 4 | 19 K 39.100 | 20 Ca 40.08 | 21 Sc 44.96 | 22 Ti 47.90 | 23 V 50.95 | 24 Cr 52.01 | 25 Mn 54.94 | 26 Fe 55.85 | 27 Co 58.94 | 28 Ni 58.71 | |
| | 5 | 29 Cu 63.54 | 30 Zn 65.38 | 31 Ga 69.72 | 32 Ge 72.60 | 33 As 74.91 | 34 Se 78.96 | 35 Br 79.916 | | | | 36 Kr 83.80 |
| 5 | 6 | 37 Rb 85.48 | 38 Sr 87.63 | 39 Y 88.92 | 40 Zr 91.22 | 41 Nb 92.91 | 42 Mo 95.95 | 43 Tc [99] | 44 Ru 101.1 | 45 Rh 102.91 | 46 Pd 106.4 | |
| | 7 | 47 Ag 107.880 | 48 Cd 112.41 | 49 In 114.82 | 50 Sn 118.70 | 51 Sb 121.76 | 52 Te 127.61 | 53 I 126.91 | | | | 54 Xe 131.30 |
| 6 | 8 | 55 Cs 132.91 | 56 Ba 137.36 | 57–71 Lanthanide series* | 72 Hf 178.50 | 73 Ta 180.95 | 74 W 183.86 | 75 Re 186.22 | 76 Os 190.2 | 77 Ir 192.2 | 78 Pt 195.09 | |
| | 9 | 79 Au 197.0 | 80 Hg 200.61 | 81 Tl 204.39 | 82 Pb 207.21 | 83 Bi 209.00 | 84 Po 210 | 85 At [210] | | | | 86 Rn 222 |
| 7 | 10 | 87 Fr [223] | 88 Ra 226.05 | 89– Actinide series** | | | | | | | | |

| | | | | | | | | | | | | | | | |
|---|---|---|---|---|---|---|---|---|---|---|---|---|---|---|---|
| *Lanthanide series: | 57 La 138.92 | 58 Ce 140.13 | 59 Pr 140.92 | 60 Nd 144.27 | 61 Pm [147] | 62 Sm 150.35 | 63 Eu 152.0 | 64 Gd 157.26 | 65 Tb 158.93 | 66 Dy 162.51 | 67 Ho 164.94 | 68 Er 167.27 | 69 Tm 168.94 | 70 Yb 173.04 | 71 Lu 174.99 |
| **Actinide series: | 89 Ac 227 | 90 Th 232.05 | 91 Pa 231 | 92 U 238.07 | 93 Np [237] | 94 Pu [242] | 95 Am [243] | 96 Cm [245] | 97 Bk [249] | 98 Cf [249] | 99 E [253] | 100 Fm [255] | 101 Md [256] | 102 No | 103 |

*   Atomic weights are the most recent (1961) adopted by the International Union of Pure and Applied Chemistry. The numbers are all relative to the atomic weight of the principal isotope of carbon, $^{12}C$, defined as exactly 12.00000. For artificially produced elements, the approximate atomic weight of the most stable isotope is given in brackets. The full names of the elements are given in Appendix III.

They all could be accommodated together in the periodic scheme by simply adding one vertical column, either interlaced with group VIII or, better, marked as a separate group with group number zero, in accord with the zero valence of each member. This harmonious fit of a whole group of unexpected additions was a considerable triumph for the scheme—too much of a triumph, indeed, for the neat conclusion that these elements have zero valence and hence form no chemical compounds seems to have prevented chemists until quite recently from looking for inert-gas compounds, some of which have in fact now been found to exist.

f) Mendeléeff's basic scheme does break down in a few places. Note that A (argon) and K (potassium) are placed respectively in the eighteenth and nineteenth positions as demanded by their chemical properties, whereas on the basis of their respective atomic weights alone (A = 39.944 and K = 39.096) one should really transpose the two. Other inversions of this kind can be found in the table, e.g., for elements 52 (Te = 127.61) and 53 (I = 126.92). Mendeléeff had confidently expected that the atomic weight for Te would on redetermination be found to be lower than that for I. The necessity of an inversion would have been to him not an inconvenience but a catastrophe. In connection with just such a point, he said:

"The laws of nature admit of no exception . . . . It was necessary to do one or the other—either to consider the periodic law as completely true, and as forming a new instrument in chemical research, or to refute it. Acknowledging the method of experiment to be the only true one, I myself verified what I could, and gave everyone the possibility of proving or confirming the law, and did not think, like L. Meyer, when writing about the periodic law that 'it would be rash to change the accepted atomic weights on the basis of so uncertain a starting point.' In my opinion, the basis offered by the periodic law had to be verified or refuted, and experiment in every case verified it."

Clearly, he overestimated the *necessity* of the periodic law in every detail, particularly as it had not yet received a physical explanation; and although the anomalous inversions in the sequence of elements are real, their existence did not invalidate the scheme. As so often, the complexity of nature is here greater than any one generalization can accommodate. Yet one must understand that it was just this utter, enthusiastic belief in the existence of a simple scheme that carried the undaunted Mendeléeff to his far-reaching conclusions.

The usefulness of the periodic law was not confined to the redetermination of some atomic weights and the prediction of properties of new elements. It had provided an inner order in the list of elements that revealed several previously unsuspected analogies among elements. The whole study of inorganic chemistry was thereby revitalized. And perhaps most important of all, just as the Keplerian discoveries of simple harmonies and mathematical

order in planetary motions posed the tantalizing problem of how to account for it by a fundamental physical law, so was science now challenged to provide an explanation for the observed regularities of the periodic table in terms of a physical model of the atom. This task, and its solution in our century, so reminiscent of Newton's solution in the analogous case, is to be recounted in Part G. The atom of the last century, however, by and large remained incapable of supplying a deeper insight into the fundamental processes of chemistry. On the other hand, it did achieve startling successes in the hands of the nineteenth-century physicists. And we now turn to pick up that part of the story in the development of the atomic theory.

## ADDITIONAL PROBLEMS

**Problem 21.2.** In the 1897 English edition of his book *Principles of Chemistry,* we find Mendeléeff's own summary of his original work on the periodic table. Discuss each of the eight points, choosing further illustrations from the material in this chapter:

"The substance of this paper is embraced in the following conclusions: (1) The elements, if arranged according to their atomic weights, exhibit an evident *periodicity* of properties. (2) Elements which are similar as regards their chemical properties have atomic weights which are either of nearly the same value (platinum, iridium, osmium) or which increase regularly (e.g. potassium, rubidium, cesium). (3) The arrangement of the elements or of groups of elements in the order of their atomic weights corresponds with their so-called *valencies.* (4) The elements, which are the most widely distributed in nature, have *small* atomic weights, and all the elements of small atomic weight are characterised by sharply defined properties. They are therefore typical elements. (5) The *magnitude* of the atomic weight determines the character of an element. (6) The discovery of many yet unknown elements may be expected. For instance, elements analogous to aluminum and silicon, whose atomic weights would be between 65 and 75. (7) The atomic weight of an element may sometimes be corrected by aid of a knowledge of those of the adjacent elements. Thus the combining weight of tellurium must lie between 123 and 126, and cannot be 128. (8) Certain characteristic properties of the elements can be foretold from their atomic weights. The entire periodic law is included in these lines."

**Problem 21.3.** In the *Handbook of Chemistry and Physics* there is printed, below one of the periodic tables, a plot of valence numbers for the elements versus atomic number. Neglect the *negative* valence numbers and plot (to element 65) a graph of *maximum* valences observed versus atomic *weight*. What periodicity is found? Is there any physical or chemical significance to this periodicity? Does there have to be any?

## RECOMMENDED FOR FURTHER READING

H. A. Boorse and L. Motz, *The World of the Atom*, New York: Basic Books, 1966. See pages 298–312 for an extract from Mendeléeff's writings, and a biographical sketch; or the extract in D. L. Hurd and J. J. Kipling, *The Origins and Growth of Physical Science*, Vol. 2, pages 81–100.

## SOURCES, INTERPRETATIONS, AND REFERENCE WORKS

H. Cassebaum and G. B. Kauffman, "The periodic system of the chemical elements: The search for its discoverer," *Isis*, vol. 62, pages 314–327 (1971)

Ida Freund, *The Study of Chemical Composition*, Chapter XVI

A. J. Ihde, *The Development of Modern Chemistry*, New York: Harper & Row, 1964, Chapter 9

D. M. Knight, ed., *Classical Scientific Papers— Chemistry* (second series), New York: American Elsevier, 1971. A collection of original sources pertaining to Prout's hypothesis and the arrangement of the elements

D. Q. Posin, *Mendeleev: The story of a Great Scientist*, New York: McGraw-Hill Book Co., 1948

J. W. von Spronsen, *The Periodic System of Chemical Elements*, New York: American Elsevier, 1969

# Chapter 22

# THE KINETIC-MOLECULAR THEORY OF GASES

## 22.1 INTRODUCTION

Having watched the establishment of the chemical atom through the labors of Dalton and Avogadro, we finally come, full circle, back to the topic treated by Democritus and Leucippus, the atomicity of physical matter and the nature of heat. In this chapter we shall see how, within a few years after the generalized law of conservation of energy and the equivalence of heat and mechanical work had been established (as described in Chapter 17), Bernoulli's gas model (Chapter 19) had been revived and transformed into a powerful theory of the thermal and mechanical properties of gases, using the atomic hypothesis.

One of the first scientists to point the way toward this new theory was an old acquaintance from Chapter 17, James Prescott Joule of Manchester, who showed in 1847 that a kinetic theory of heat, joined to an atomic theory of matter, can account, at least *qualitatively*, quite well for a wide variety of physical phenomena. The mathematical work needed to permit quantitative tests of the theory against experimental data was quickly accomplished by Rudolph Clausius in Germany, James Clerk Maxwell in Britain, and Ludwig Boltzmann in Austria, starting in 1857. In 1873, J. D. van der Waals in Holland showed that the kinetic theory could explain, at least approxi-

mately, not only the properties of the gaseous state of matter but also the transition between gas and liquid; and by the beginning of the twentieth century, J. Willard Gibbs in the United States had developed a general method of *statistical mechanics* which could be applied to all three states of matter. When the assumptions of statistical mechanics were modified by the development of quantum theory and new knowledge about the internal structure of the atom (Part H), the theory was found to be capable of providing a satisfactory explanation of the magnetic and electrical properties of matter as well as thermal and mechanical properties. Thus, as if to compensate for its early neglect of Bernoulli's idea, physics has now applied it to almost every conceivable phenomenon.

We saw in Chapter 20 that in the years before 1847, Dalton's atomic theory of chemistry had raised the atomistic idea itself from the qualitative level of speculation; the densities and relative molecular weights of gases were becoming known with some accuracy; Rumford and Davy had called the caloric theory in question; the wave theory of heat had popularized the notion that heat is related to vibrations of a medium; Mayer had discussed the idea (in 1842 and 1845) that heat and mechanical energy both obeyed an overall law of conservation of energy; Joule's laborious experiments had independently established Mayer's inspired postulate, and had, in fact, yielded a value for the mechanical equivalent of heat. Yet during almost all of this time it was believed that atoms do not move freely through space, even in a gas, but, if they move at all, may only vibrate around fixed positions, caught in a medium composed of caloric or ether.

Although a few scientists before 1847 expressed views favorable to the theory that atoms can move quite freely in a gas, only one of them was able to develop such a kinetic theory into a serious challenge to the prevailing orthodoxy: John Herapath (1790–1868), a British scientist and editor of a railway journal. Without knowing of Bernoulli's work, Herapath proposed a similar theory, but developed it much further. He obtained from his model the fundamental relation between the pressure ($P$) of a gas and the speed ($v$) of the particles of which it is supposedly composed:

$$P = \tfrac{1}{3}\rho v^2 \qquad (22.1)$$

where $\rho$ is the density (mass/volume) of the gas. (This equation will be derived in Section 22.4.) Since the pressure and density of a gas can be determined directly by experiment, this equation can be used to compute the speed the particles would have to have; Herapath did this, and found a result that is fairly close to the speed of sound in the gas (about 330 m/sec for air).

Herapath's calculation of the speed of an air molecule (first published in 1836) can be considered an important event in the history of science, but it was ignored by most of his contemporaries. Herapath's earlier papers on the kinetic theory, written in 1820, had been rejected for publication by the Royal Society of London, and despite a long and bitter battle (including letters to the editor of *The Times* of London), Herapath had not succeeded in getting any recognition for his theory. Only Joule saw the value of Herapath's theory, but Joule's own attempt to establish the kinetic theory (based on Herapath's work and his new experiments) did not attract much attention either for another decade.

In the 1850s the kinetic theory was again revived, and this time it was quickly established on firm foundations. For an account of the basic assumptions and some of the main results of the kinetic theory, we can do no better than quote passages (of which almost every word still corresponds to our current ideas) from the fundamental paper by Rudolf Clausius—the paper "On the nature of the motion which we call heat" (1857), which initiated the modern phase of development of the theory. Although Clausius had adopted the kinetic theory some years before in connection with his studies on thermodynamics, he did not publish a systematic exposition until August Krönig, a German chemist, had announced the kinetic hypothesis, in a very rough form, in a short note published in 1856. Clausius had learned indirectly of Joule's work by 1857, but was not able to obtain a copy of the Englishman's paper until a year later.* He did not yet know that both Krönig and Joule had been anticipated by Bernoulli and Herapath.

"Krönig assumes that the molecules of a gas do not oscillate about definite positions of equilibrium, but that they move with constant velocity in straight lines until they strike against other molecules, or against some surface which is to them impermeable. I share this view completely, and I also believe that the expansive force of the gas arises from this motion. On the other hand, I am of the opinion that this is not the only motion present.

"In the first place, the hypothesis of a rotatory as well as a progressive motion of the molecules at once suggests itself; for at every impact of two bodies, unless the same happens to be central and

---

* Joule had published his paper in the *Memoirs of the Literary and Philosophical Society of Manchester*; it might seem an odd place to look for important contributions to physical science, but in fact many of John Dalton's papers had appeared in the same journal four decades earlier.

rectilineal, a rotatory as well as a translatory motion ensues.

"I am also of opinion that vibrations take place within the several masses in a state of progressive motion. Such vibrations are conceivable in several ways. Even if we limit ourselves to the consideration of the atomic masses solely, and regard these as absolutely rigid, it is still possible that a molecule, which consists of several atoms, may not also constitute an absolutely rigid mass, but that within it the several atoms are to a certain extent moveable, and thus capable of oscillating with respect to each other."

Although Clausius could not provide a direct theoretical calculation of the amount of energy contained in these molecular vibrations and rotations, he argues that an equilibrium must be established, as a result of collisions among the molecules, so that eventually a fixed proportion of the total energy will be found in rotation and vibration, and the remainder in the translatory motion of the molecules as a whole.

According to the kinetic theory, he held, "the pressure of a gas against a fixed surface is caused by the molecules in great number continually striking against and rebounding from the same," and this pressure is, for constant volume, proportional to the kinetic energy of translatory motion of the molecules (Eq. 22.1).

"On the other hand, from Gay-Lussac's law we know that, under constant volume, the pressure of a perfect gas increases in the same ratio as the temperature calculated from $-273°C$, which we call the absolute temperature. Hence, according to the above, it follows that the absolute temperature is proportional to the kinetic energy* of the translatory motion of the molecules. But as, according to a former remark, the several motions in one and the

**Fig. 22.1**  Rudolph Clausius (1822–1888). (Courtesy of the Deutsche Museum, Munich)

---

*     We have replaced the term *vis viva*, still in use in 1857, by its modern equivalent (except for a factor of $\frac{1}{2}$), kinetic energy.

same gas bear a constant relation to each other, it is evident that the kinetic energy of the translatory motion forms a constant proportion of the total energy, so that the absolute temperature is also proportional to the total kinetic energy in the gas."

From this model it would follow that (as Clausius had already suggested in an earlier paper) the specific heat of a gas should be independent of temperature, and that the two specific heats (at constant volume and at constant pressure) differ by a constant term. Both of these consequences are in fact found to correspond to the properties of real gases over a large range of temperatures.

Although Clausius could perform detailed calculations only for a model corresponding to the gaseous state of matter—in which molecules are presumed to move freely through a nearly empty space with only occasional collisions—he was willing to propose a plausible explanation of the difference between gaseous, liquid, and solid states, based on a qualitative kinetic-molecular viewpoint. In the *solid* state, he suggested that the molecules simply vibrate around fixed equilibrium positions and rotate around their centers of gravity. In the *liquid* state, the molecules no longer have definite positions of equilibrium, but can move from one place to another, though always acted on by the forces exerted by neighboring molecules. In the *gaseous* state, however, the molecules finally escape the influence of their neighbors and fly off in straight lines; if two molecules collide, they simply rebound in accordance with the laws of elastic collisions.

From this same qualitative picture, Clausius developed an explanation of changes of state. Thus, *evaporation* of a liquid could be explained by assuming that even though the average speed of a molecule may not be sufficient to carry it beyond the range of the attractive forces of its neighbors, a few exceptionally fast molecules may be able to escape from a liquid surface, even at temperatures below the boiling point. Inherent in this explanation is the notion that, even at a single temperature,

molecules have a range of different speeds; this idea was soon to be developed with great effectiveness by Maxwell (see below, Section 22.6).

This summary has only hinted at the richness of conclusions and the generality of usefulness of the kinetic-molecular theory which, through the further work of such men as Maxwell, Boltzmann, van der Waals, and Gibbs, eventually became a major triumph of nineteenth-century physical science. For our purposes it will be well to leave the strictly historical road, and to summarize in modern language some qualitative consequences before we turn to the quantitative side of the concept that heat is a mode of motion of submicroscopic particles.

## 22.2 SOME QUALITATIVE SUCCESSES OF THE KINETIC-MOLECULAR THEORY

a) Any work done on a substance (e.g., by friction) may be converted into kinetic energy of motion or potential energy between the atoms or molecules. In gases, the increase in kinetic energy of the particles means greater random thermal motion; in solids, the energy may be stored in increased vibration of the crowded atoms; and in a liquid, both thermal motion and thermal vibration play their parts. Heat conduction from a hot to a cold body is thus to be thought of as the transfer of energy of motion from one set of molecules to the other at the interface, the surface of contact. Expansion of a solid or liquid on heating is then the consequence of increased agitation of individual molecules, and of their resulting greater separation through more powerful mutual collisions. Therefore, if a glass-in-mercury thermometer is inserted in a vessel containing a gas, and if now work is done on the gas (by compression or friction), the thermometer indicates a rise to a higher temperature reading on the scale because (i) the work done on the gas serves to increase the kinetic energy of the gas molecules, which in turn means (ii) stronger, more frequent collisions of these molecules with the mercury-containing glass bulb, (iii) a transfer of some of this energy to the mercury, and (iv) ultimately an expansion of the mercury column as the increased agitation of mercury molecules results in their greater mutual separation.

b) Our own sensations of heat or cold by touching are not so simply explained by this or any other physical theory, but evidently we can neither feel nor see directly the individual agitation of the molecules, for their dimensions are a thousand times smaller than the wavelength of light or the sensitive receptors of our skin. Yet, if we consider a small dust particle or bacterium suspended in still air or on the surface of a liquid and bombarded from all sides by swarms of invisible but highly agitated molecules of the surrounding gas or liquid, we understand why under high-power magnification the particle is seen to dance in perpetual, irregular motion, as though alive.

In fact, when first observed in suspensions of organic particles in the eighteenth century, this phenomenon was suspected of being evidence of "vitality"; but in 1827 the English botanist Robert Brown demonstrated that similar motions could be seen in suspensions of almost any kind of particle of a certain size, whether organic or inorganic. Thus what came to be known as "Brownian movement" was attributed to the effects of impacts of invisible molecules on the particles, though the development of a satisfactory quantitative theory had to wait until Einstein's work of 1905. (See Fig. 22.2.)

c) The summary of Clausius' paper mentioned another qualitative success of the kinetic-molecular theory: an explanation of the difference between the three states of matter—solid, liquid, and gaseous—in terms of the relative agitation of the molecules. (Essentially the same explanation had been given earlier by Davy and by Joule.) The same fundamental ideas also help to make clear several other simple phenomena, for example, (i) the diffusion of one gas through another (one type of thermally agitated molecules working their way through a space filled with another type of molecules), or (ii)

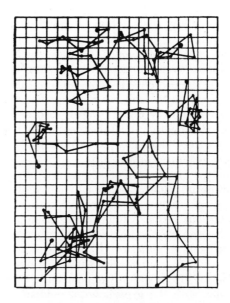

**Fig. 22.2** Brownian movement. The positions of three grains on water, as seen through the microscope at 30-sec intervals. (After J. Perrin)

the immediate expansion of a gas to fill any vessel no matter how large (since the mutual attractive forces between gas molecules are supposedly negligible and the velocities of random motion very high), and (iii) the very existence of pressure exerted by a gas against its container (understandable as the cooperative effect of billions of collisions per second by all the agitated gas molecules with the walls of the container).

## 22.3 MODEL OF A GAS, AND ASSUMPTIONS IN THE KINETIC THEORY

Turning now to quantitative achievements of the kinetic-molecular theory (and dropping "molecular" from the phrase, in conformity with custom), we shall concentrate mainly on its applications to *gases*, this being most illustrative, mathematically easiest, and, in fact, physically most developed at the present time. We shall follow a simplified derivation, in which our problem reduces to this: Let us construct a plausible *model of a gas* which will yield as a logical consequence (thus "explain") the known laws of the behavior of gases, and which if possible will lead us to look for new verifiable effects. The model and the underlying hypotheses leading to the model together form the *theory* which is useful to the extent that it explains previously known laws and generates new results.

In constructing the model of a gas, we now carefully examine all clues. Some may be well-known experimental facts about matter, others may be plausible but hitherto unchecked assumptions, and still others may be bold new hypotheses which are justifiable *a posteriori* if they lead to verifiable results. Let us summarize the ten major facts or assumptions from which we must build up our initial model, with the experimental evidence for each, where meaningful.

a) *The general gas law*. The experimental facts about a given sample of a relatively rarefied and stable gas are, as we saw, well summarized in the equation $PV/T = r$.*

b) *Molecular hypothesis*. One of the fundamental assumptions of our gas picture is, of course, the idea that gases are made of tiny submicroscopic particles—atoms or groups of atoms called molecules. The best experimental evidence originally was chemical—the success of Dalton's atomic theory, particularly in dealing with the laws of combining proportions. For our purposes we shall at first discuss only pure single gases made of identical building blocks (one type of molecule or one type of atom).

---

* At extreme conditions of pressure and temperature this simple equation is no longer sufficiently accurate, but can be replaced by a better approximation which is theoretically nicely defensible. Ideal, perfect, or true gases, by definition, are gases that behave accurately according to $PV/T =$ constant—which fortunately applies to most actual gases far from such extreme conditions.

c) *The equation of state*. Another experimentally supported concept, which we may either build into our model at the start or else must demand as one of its derivable consequences, is Avogadro's hypothesis, which, simply stated, requires that all true gases at a given pressure, volume, and temperature have the same number of molecules. Having defined Avogadro's number $N$ as the number of molecules per mole of substance ($6.02 \times 10^{23}$ molecules per mole of any substance), we can reformulate the general gas law, which reads

$$\frac{PV}{T} = r \qquad (22.2)$$

and which has the inconvenient character that the value of $r$ has to be determined experimentally for every particular sample of every gas under investigation ($r$ changes both with the mass and with the chemical nature of the gas). Specifically, experiments with samples of the same gas but with different total masses $m$ would reveal that

$$\frac{PV}{T} \propto m.$$

Furthermore, choosing gas samples of equal mass $m$ but with different chemical constitution (i.e., different molecular weight $M$) would yield the relation

$$\frac{PV}{T} \propto \frac{1}{M}.$$

There being no other factors that need to be taken into account, we can write

$$\frac{PV}{T} \propto \frac{m}{M} \quad \text{or} \quad \frac{PV}{T} = \frac{m}{M} \times \text{(a constant } R\text{)}, \qquad (22.3)$$

where $R$ is indeed a factor which is equal for every sample of every gas, and therefore may be called the *universal gas constant*. You will also note that $(m/M)$, being the ratio of the mass of a sample and its molecular weight, is nothing else than $n$, the number of moles of the gas sample at hand. Therefore we now write

$$PV = nRT. \qquad (22.4)$$

This most important equation, which is essentially a summary of experimentally observed facts, is the so-called *equation of state* for (perfect, ideal) gases. Since $R$ is a *universal constant*, the equation $R = PV/nT$ should give us a numerical value for $R$ if we know any one set of values of $P$, $V$, $n$, and $T$ which all gases take on simultaneously. And indeed we do: At $T = 273°$K and $P = 1.013 \times 10^5$ newtons/m², all gases have, according to Avogadro's hypothesis, the same volume per mole ($V/n = 2.24 \times 10^4$ cm³/mole). Thus $R = 8.31$ (joules/mole)/degree (check the value and the units).

The new form of the gas law has enormous practical advantages over Eq. (22.2). Now we can predict the physical behavior of a quantity of gas if only we know its chemical composition and its mass! This is evident in the following example, which could not have been solved directly before Eq. (22.4) was formulated.

**Example 1.** What pressure will 10 g of helium exert if contained in a 500 cm³ cylinder at room temperature (21°C)?

*Solution:* The relative molecular weight of He is approximately 4 (see Table 21.1). One mole would be 4 g, $x$ g corresponds to $x/4$ moles, therefore $n = 10/4 = 2.5$ moles.

$$P = \frac{nRT}{V}$$

$$= \frac{[2.5 \text{ moles}][8.31 \text{ (joules/mole)/degree}] \times [294°\text{K}]}{500 \text{ cm}^3}$$

$$= 1.22 \times 10^8 \text{ newtons/m}^2,$$

which is more than 1000 atm (and therefore probably too much pressure for this container).

**Problem 22.1.** (a) A 50 cm³ cylinder is safe only up to 100 atm. Find the temperature at which 3 g of oxygen gas in the cylinder will burst the cylinder. (b) Determine the number of molecules of oxygen that must be present in a vessel of 10 cm³ capacity at 20°C to produce a pressure of 50 atm.

To repeat: $PV = nRT$ is an experimentally verified equation for ideal gases, and therefore it must either appear axiomatically in our gas model, or else must follow as a consequence of the model.

d) *Size of molecules.* The model of the gas, which we are piecing together, must include the provision that the individual molecules are extremely small, perhaps of the order of $10^{-10}$ to $10^{-9}$ m. The experimental evidences for this point are numerous, varied, and conclusive. For example, vapors and gases occupy volumes that may be thousands of times larger than the liquids from which they can be obtained and, conversely, gases may be compressed to minute fractions of their original volume. They diffuse through invisibly small pores in unglazed porcelain. They combine in exact accord with the laws of chemical proportions even in extremely minute quantities. Even as clumsy and large molecules as those of oil (glycerin trioleate, $C_{57}H_{104}O_6$) cannot exceed $10^{-9}$ m in diameter because, by measurement, that is the approximate thickness of oil films spread very thinly on water.

We conclude therefore that ordinarily we may neglect the actual volume of all molecules in comparison with the total volume which the gas occupies by virtue of its unceasing motion, and indeed we might at the start consider them as point masses, or at any rate as very tiny, smooth spheres.

e) *Number of molecules.* By the same evidence as in (d) we must suppose that there are enormous swarms of molecules in every cubic centimeter of ordinary gases; in fact, Avogadro's number expresses this point quantitatively. Even in the best vacuum ordinarily obtained in the laboratory (about $10^{-10}$ atm) there are at ordinary temperatures still about $10^9$ molecules/cm$^3$.

f) *Mean free path.* As the enormous expansion of a gas obtained from a liquid indicated, the average distances between molecules are correspondingly huge compared with their own size. On the average, a molecule may travel for distances hundreds or thousands of times its own dimensions before encountering another molecule. These distances, measured indirectly, are of the order of $10^{-7}$ m for ordinary densities, and are referred to as *mean free path.*

g) *Forces between molecules.* Although the mean separation of molecules may be large by atomic dimensions, it is so small an absolute quantity that we may wonder about the mutual forces between gas molecules (which we guess to be strong in the liquid and solid states because of the large latent heats).

First we may suspect conventional gravitational forces, but they are trivial, as a quick calculation will show. The masses of, say, oxygen molecules, are calculated from Avogadro's postulate. Since 32 kg of $O_2$ contain more than $6 \times 10^{26}$ molecules, each molecule weighs about $32/(6 \times 10^{26})$ or about $5 \times 10^{-26}$ kg. Two such molecules at about $10^{-7}$ m apart exert a mutual gravitational force of

$$F = G \frac{m_1 m_2}{d^2}$$

or

$$F = (6.67 \times 10^{-11}) \times \frac{(5 \times 10^{-26})^2}{(10^{-7})^2} \text{ newtons,}$$

about $2 \times 10^{-47}$ newtons, evidently a completely negligible quantity. (So also is the weight of each molecule by itself for all our purposes.)

But there may be among these particles forces other than gravitational—perhaps electrical ones. Although we may not readily imagine exactly what or how strong they are, we can make an experiment to see whether or not they exist. If they do, then a gas under pressure which is allowed to expand into

a big *evacuated* vessel, on separating from its fellow molecules, will invoke such intermolecular forces to do work. If the forces are attractions, then work is done by the molecules on separating and since, according to the fundamental assumptions, the required energy for this process can come only from heat energy, the gas ought to cool itself on expanding into a vacuum. Conversely, if the intermolecular forces are repulsions, the total mass of the gas should heat up on expansion. The experimental fact observed by Gay-Lussac (see Section 17.8), is, however, that on expansion into a vacuum only a very small amount of net cooling or heating occurs in the whole mass of the actual gases,* hence no significant intermolecular forces exist except during actual collisions, which presumably take very little time and on the average occupy a negligible span for each molecule compared with its whole history. Incidentally, you will recall that this is exactly the experiment which most directly refutes the old *static* model of mutually repelling gas molecules.

Here we may digress for a moment: even if it be true that the molecules do not exert significant forces on one another while in gaseous form and widely separated, we must still expect that when they come closer together, as in liquids and solids, such intermolecular forces become important. We must postulate that slight increases in the intermolecular separation distances call into action a certain amount of mutual *attraction* in the case of liquids and solids, for otherwise we cannot explain *cohesion*, the resistance offered to efforts tending to separate the material in solid or liquid form into parts. There also must be the converse effect, mutual *repulsion* on decreasing the intermolecular distance, otherwise we cannot account for the large resistance of liquids and solids to the slightest compressions. One may assemble all this information about intermolecular forces in one qualitative graph of distance vs. force (where *positive* forces represent repulsion among molecules, *negative* ones attraction; see Fig. 22.3).

_____

* For an excerpt of Joule's experiment on this point, see Magie, *Source Book in Physics*, pages 172–174.

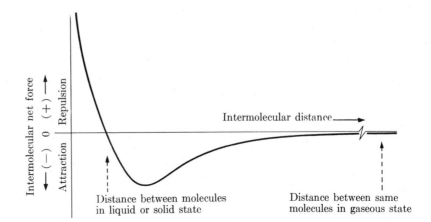

**Fig. 22.3** Qualitative graph of the force between two molecules versus the intermolecular distance.

---

**Problem 22.2.** Justify the indicated position in Fig. 22.3 for the intermolecular distances in solids and gases.

---

h) *Thermal agitation.* Strongly implied in the developing model, in the fundamental assumptions of the kinetic theory, and in the observations on Brownian movement and diffusion, is the requirement of *rapid* motion of the gas molecules at usual and high temperature. The picture of the gas then expands to the view of moving particles, passing through relatively "long" stretches of straight motion, and colliding briefly but many million times per second with other molecules or with the walls of the container (where each collision contributes to the maintenance of pressure). So brief are these mutual collisions, so random the motions, that we shall in our analysis neglect the effect of such encounters entirely. We shall also see that thermal agitation, as implied by the observation of Brownian movement, must be omnipresent. All assemblies of atoms (if above 0°K) exhibit such agitation, not only in gases and liquids but even in solids, where it is expressed by a small-scale vibration among neighboring atoms.

i) *Nature of collisions; Newton's laws apply.* Now that we have allowed the molecules to undergo collisions, we must consider the mechanical laws governing such events. Naturally, we turn to Newton's mechanics, although we have no direct evidence that microscopic worlds obey these laws, which had developed from experiments with large-scale objects. But although we may have to keep our eyes open to retract this dangerous extrapolation if facts demand it, we are more or less forced to start with the assumption that Newtonian laws hold among molecules, for we have no *other* mechanics available to us at this point. In his pioneering work on this subject, James Clerk Maxwell discusses the relation between the atomistic model and the forces of interaction as follows.*

"Instead of saying that the particles are hard, spherical, and elastic, we may if we please say that the particles are centers of force, of which the action is insensible except at a certain small distance, when it suddenly appears as a repulsive force of very great intensity. It is evident that either assumption will lead to the same results."

j) *Conservation of kinetic energy.* We must also assume that the molecular collisions with one another and with the container are perfectly elastic, i.e., that on the average the kinetic energy (KE)

---

\*    *Philosophical Magazine*, Vol. 19, pages 19–32 (1860).

of a molecule before the collision is the same as immediately afterward; only *during* the brief collision will the KE pass into potential energy of elastic distortion, presumably just as in the case of an elastic ball bouncing off a wall.

The reasons for this simplifying assumption are evident: If in the long run the collisions were not perfectly elastic, the KE of each molecule would gradually disappear even in a gas filling a heat-insulated container completely left to itself; with a diminishing total KE goes a reduced pressure, until eventually the speeds are so low that the feeble gravitational and other attractive forces induce liquefaction or solidification of the gas.

Nothing like that is observed at all. Nor can we at this point readily imagine what could happen to KE lost in a hypothetical inelastic collision of a molecule. Of course some energy could be—and sometimes is—somehow stored as potential energy, as in a change of position or increased vibration of the parts of a molecule with respect to one another or even in rotation of the molecule. But this is not an unlimited reservoir for energy, and in any case we should try out the simpler hypothesis of perfectly elastic behavior.* The previously mentioned suggestion that the molecules be thought of as tiny, round, smooth spheres is simply an expression of this requirement of conservation of kinetic energy.

## 22.4 THE DERIVATION OF THE PRESSURE FORMULA

We now have a fine mental picture or *model* of a typical ideal gas. It is overpoweringly tempting to visualize it in terms of a three-dimensional assembly of agitated microscopic billiard balls. Therefore a warning is in order to be prepared for a breakdown of the analogy where it is carried beyond the evidence, and for new assumptions to be introduced where necessary, and to be ready for surprising consequences of our model, for behaviors quite beyond anything an assembly of billiard balls would ordinarily dream of doing.

As an initial test of the validity of our model we now choose any well-established *quantitative* law of gases and try to deduce it from the model. Following the usual choice, we decide on the general gas law [in its modern form, Eq. (22.4)], $PV = nRT$. If we can derive this equation, we should be completely satisfied. But actually we cannot do this directly, and instead we shall obtain from our analysis another equation, of the form $PV = XYZ$. (where $XYZ$ stands for some factors such as the mass and velocity of the molecules). The proof that our model is valid then depends on showing that $XYZ = nRT$, or that the equation $PV = XYZ$ actually yields experimentally verifiable consequences. This we shall do, with exciting secondary results.

To derive a relationship between the pressure, volume, and other variables of a gas, consider a strong and rigid cubical container with sides of length $l$,† enclosing the molecules of a pure ideal gas (Fig. 22.4). Each molecule has the same mass, $m_0$, and moves rapidly back and forth in the box, frequently colliding with the six walls, in accordance with the gas model which we have agreed on.

a) First imagine that there is only a single molecule rattling around in the box. For simplicity, we also assume that its speed $v$ (and therefore its kinetic energy) does not change numerically except during actual collisions, which, however, take only a negligibly short time. Every time the molecule

---

*    On the other hand, no large-scale body, not even a sphere of the best steel, shows perfectly elastic collisions; there are always some losses of mechanical energy through internal friction and a consequent rise in temperature through increased internal molecular agitation. But what could be the meaning of frictional heat losses in the collisions of the molecules themselves? Note that we have here a good example of the breakdown of a concept gained from large-scale experience when applied to the realm of the submicroscopic.

†    If this first simplifying assumption seems too restrictive, consider that any real noncubical container may be thought of as a set of many such very tiny cubical subspaces.

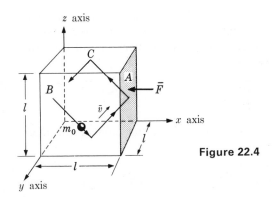

**Figure 22.4**

collides with one flat and rigidly fixed side of the box, say $A$ (shaded in Fig. 22.4), it is reflected again with the same speed; evidently the component of the velocity perpendicular to side $A$ is simply reversed, and the other two components of the velocity are not affected at all by that collision.

This situation reminds us of the result of Example 2, Section 17.1: If a very light object undergoes an elastic collision with a very massive object, it will simply bounce off with the same speed in the opposite direction. More generally, if the surface of the massive object is parallel to the $zy$ plane and the light object strikes it (e.g., at $A$) obliquely with a velocity vector $v$ having components $v_x$, $v_y$, $v_z$, then the component $v_x$ will be reversed by the collision, but $v_y$ and $v_z$ will remain unchanged. In this case the very light object is a molecule, and the very massive object is one wall of the cubical container. It is important to note that the kinetic energy of the molecule is conserved in these collisions, as long as the wall remains fixed and the speed of the molecules does not change. There is a continual transfer of momentum between the molecules and any wall with which it collides, and we shall calculate that at once for one wall. (But for a rigid box taken as a whole, the actions at the different walls will in time cancel out on the average as the molecule rattles around in the box.)

b) If $v_x$ is the velocity component of the molecule in the $x$ direction, and $t$ is the time which the molecule needs to travel the $l$ cm from side $B$ (opposite $A$) to side $A$, then

$$v_x = \frac{l}{t} \quad \text{and} \quad t = \frac{l}{v_x}.$$

Then $\tau$, the time interval needed for a complete round trip from one collision with side $A$ to the next collision with that side, is $2t$. Thus we write:

Time elapsed per collision (with side $A$, for one molecule) $\equiv \tau = 2l/v_x$.

If $\tau$ is the number of seconds per one collision, the reciprocal quantity $1/\tau$ is *the number of collisions per second,* given by $1/\tau = v_x/2l$.

c) During each of the $v_x/2l$ collisions occurring every second between the molecule of mass $m_0$ and the rigid side $A$, the velocity component $v_x$ is changed from positive (to the right in Fig. 22.4) to negative (to the left), a total change of $2v_x$. The change of *momentum* of the molecule (being the mass times the change of velocity) $= 2\,m_0 v_x$ for *each collision*; but during *one second*, the sum of all such momentum changes is

$$(2m_0 v_x) \times \left(\frac{v_x}{2l}\right),$$

that is, (momentum change/collision) × (collisions/sec). Of course,

$$(2m_0 v_x) \times \left(\frac{v_x}{2l}\right) = \left(\frac{m_0 v_x^2}{l}\right),$$

and this still applies for only a single molecule.

But let us now consider not just one but the actual number, $N_m$, of molecules in this box. Their total change of momentum per second at side $A$ is

$$N_m \left(\frac{m_0 v_x^2}{l}\right),$$

if they all ordinarily have the same speed. If they do not (and it would be too much of a limitation to put in such an improbable restriction) we need do no more at this time than write $\overline{v_x^2}$ instead of $v_x^2$ in the last expression, where $\overline{v_x^2}$ means the *mean* (straight average) value of the sum of the square of the $x$ components of velocity for all $N_m$ molecules present; that is,

$$\overline{v_x^2} \equiv \frac{[(v_x)_1^2 + (v_x)_2^2 + (v_x)_3^2 + \cdots + (v_x)_{N_m}^2]}{N_m} \quad (22.5)$$

This presupposes nothing about the way actual values of $v_x$ are distributed among the $N_m$ molecules, a question we may investigate later.

Thus we now write: The total change of momentum per second of all $N_m$ molecules at side $A$ is equal to

$$\frac{N_m m_0 \overline{v_x^2}}{l}.$$

(Watch how this expression is growing and changing in meaning in the development.)

d) By Newton's second law, the change of momentum = *average net force $\overline{F}$* exerted, multiplied by the *time* during which the change of momentum occurs. Consequently, the change of momentum *per second* experienced by the molecules is given by $\overline{F}$ numerically; or, to put it differently, the average force exerted on the molecules by the side $A$ as they rebound is numerically given by

$$\overline{F} = \frac{N_m m_0 \overline{v_x^2}}{l}. \quad (22.6)$$

Since the hail of molecules against the wall is pretty thick, this force is what might be observed steadily if the wall or side $A$ were a piston held in place (as indicated in Fig. 22.4) by means of a compression-type spring balance against the blows of the molecules. By the law of action and reaction it is, of course, also numerically equal to the average (perpendicular) force with which the molecules push on side $A$.

e) We are not after the force exerted on the wall, but after the *pressure*. We recall that pressure

$P \doteq$ perpendicular force/area. If we divide both sides of Eq. (22.5) by $l^2$, which is the area of side $A$ (and is equally the area of every one of the other five sides), we now have

$$\frac{\overline{F}}{l^2} (\equiv P \text{ on side } A) = \frac{N_m m_0 \overline{v_x^2}}{l^3}.$$

But $l^3$ is the volume $V$ of the cubical container, hence

$$P = \frac{N_m m_0 \overline{v_x^2}}{V} \quad \text{or} \quad PV = N_m m_0 \overline{v_x^2}. \quad (22.7)$$

This is a very promising development in our effort to derive the relationship between $P$, $V$, and other measurable quantities; all that remains to be done is to prove that the right side of Eq. (22.7) is equivalent to $nRT$. But that will not be the easiest part of our work!

f) First of all, the expression $\overline{v_x^2}$ in Eq. (22.7) can be replaced by $v^2/3$; the argument is rather simple. From Eq. (22.7),

$$\overline{v_x^2} = \frac{PV}{N_m m_0},$$

where $P$ is the pressure on face $A$, and $V$, $N_m$, and $m_0$ are constants for this box and this gas sample. If we had chosen to pay attention to one of the other faces, say the top face $C$ which stands perpendicular to the $z$ axis, we should have obtained

$$\overline{v_z^2} = \frac{PV}{N_m m_0}.$$

The pressure $P$ at $C$ is no different from that on $A$ or on any other face, and $V$, $N_m$, and $m_0$ are still the same for each case; therefore

$$\overline{v_x^2} = \overline{v_y^2} = \overline{v_z^2},$$

that is, we assume that there is *no preference* among the molecules for motion along any one of the three axes. By Pythagoras' theorem, extended to a vector $v$ in three dimensions, $v^2 = v_x^2 + v_y^2 + v_z^2$.

Therefore we may write $\overline{v^2} = \overline{v_x^2} + \overline{v_y^2} + \overline{v_z^2}$, or in our case, $\overline{v^2} = \overline{3v_x^2} = \overline{3v_y^2} = \overline{3v_z^2}$. Therefore $\overline{v_x^2} = \frac{1}{3} \cdot \overline{v^2}$, and Eq. (22.7) can be rewritten:

$$PV = \tfrac{1}{3} N_m m_0 \overline{v^2}. \qquad (22.8)$$

(This is clearly equivalent to Eq. (22.1), since $\rho = N_m m_0 / V$.)

Note that $\tfrac{1}{2} m_0 \overline{v^2}$ would be the average kinetic energy of translation ($\overline{KE}_{trans}$) *per molecule*, and $N_m(\tfrac{1}{2} m_0 \overline{v^2})$ is the *total* $KE_{trans}$ for all $N_m$ molecules in the sample of gas in our box. Therefore it becomes significant to recast Eq. (22.8).

$$PV = \tfrac{2}{3} N_m \frac{m_0 \overline{v^2}}{2} = \tfrac{2}{3}(\text{total } KE_{trans} \text{ of gas}). \quad (22.9)$$

In words, the product of the pressure and the volume of a gas is two-thirds of the total kinetic energy of translation of the molecules.

This is an astonishing consequence of our model. In a qualitative way we supposed all the time that the pressure of the gas against the container was due to the incessant blows of highly agitated gas molecules, but we had no idea that so simple and beautiful a relation between the pressure and the energy of the gas molecules could exist. Equation (22.9) is the so-called pressure equation for an ideal gas. We may be chagrined that we did not obtain $PV = nRT$ by our calculations; but we have the promised relation $PV = XYZ$, and so the validity of our model can still be tested by examining whether the proposition $XYZ = nRT$ yields verifiable consequences. Moreover, if the factor $\overline{v^2}$ depends only on the temperature of the gas (to be proved), then Eq. (22.8) agrees with Boyle's law: $PV = $ constant at given $T$. This is, of course, most gratifying.

## 22.5 CONSEQUENCES AND VERIFICATION OF THE KINETIC THEORY

a) First let us see whether Eq. (22.9) by itself is reasonable from the point of view of *direct experimental verification.*

It follows from Eq. (22.8) or (22.9) that $\overline{v^2} = (3PV/N_m m_0)$. But $N_m m_0 = $ total mass of gas actually present, and $(N_m m_0/V) = $ mass density $\rho$ of the gas (kg/m³).

Therefore $\overline{v^2} = 3P/\rho$ and $\sqrt{\overline{v^2}} = \sqrt{3P/\rho}$. The expression $\sqrt{\overline{v^2}}$, for which we shall use the symbol $c$, is the square root of the mean (straight average) of the squares of the various velocities—in brief, the *root-mean-square* value of the molecular velocities (or rms velocity). It is mathematically not the same thing as a straight mean value $\bar{v}$, but, in fact, we find that the value of $\sqrt{\overline{v^2}}$ differs only a few per cent from the straight average* in actual gases, or $c \doteq \bar{v}$.

Now the prediction following from our model is that the rms velocity $c$, or nearly enough the average velocity $\bar{v}$ of the molecules, should be $= \sqrt{3P/\rho}$. Substituting representative values for air at one atmosphere pressure ($1.01 \times 10^5$ newtons/m²) and 0°C (where $\rho = 1.29$ kg/m³) we find that the rms velocity should be

$$\sqrt{\frac{3 \times 1.01 \times 10^5}{1.29}} = 4.85 \times 10^2 \text{ m/sec},$$

nearly 500 m/sec, or about $\tfrac{1}{3}$ mi/sec—comparable to the speed of a slow bullet!† (We may now confirm Herapath's conclusions that the average velocity of the molecules of water vapor, say at 100°C, is in excess of 1 mi/sec.) This is an enormous speed compared with the motion of everyday objects, but it is of the same order as the speed of propagation of

---

* To appreciate the difference between $\bar{v}$ and $c$, note that if there were only three molecules with speeds $v_1 = 5$ units, $v_2 = 10$, and $v_3 = 15$,

$$\bar{v} = \frac{5 + 10 + 15}{3} = 10 \text{ units},$$

$$c \equiv \sqrt{\overline{v^2}} = \sqrt{\frac{5^2 + 10^2 + 15^2}{3}} = 10.8 \text{ units}.$$

This is a difference of 8%.

† Bernoulli might have carried his own work to this point if he had had good measurements of gas densities.

*sound* at that pressure and temperature (331 m/sec). The rms velocities of molecules in other gaseous materials at that pressure and temperature have, by calculation, similarly high values, ranging from 1800 m/sec for hydrogen gas to 1900 m/sec for mercury vapor, and in each case the velocity of sound is not much less. But this is just what we should expect in terms of our model of a gas, for the obvious way to think of the propagation of sound waves is to visualize it as a directional motion of the molecules superposed on their random chaotic motion, so that the energy of the sound wave is carried as kinetic energy from one gas molecule to its next near neighbors with which it collides. The molecules themselves, despite their high speeds, do not move very far during any appreciable interval, being contained within a relatively small volume by multiple collisions, around a billion per second, with their neighbors—but the energy of the sound wave *is* communicated from one to the next molecule with that high speed.* As to the numerical values involved here, we do not really expect that the speed of sound should be exactly as great as $c$ (why not?); in fact, the numerical discrepancy is exactly predictable from the complete theory of sound propagation.

---

**Problem 22.3.** Dalton, who first of all was a meteorologist, discovered experimentally a law concerning gas pressures, which we call Dalton's law of partial pressures, and which may be stated thus: If several (e.g., three) gases are mixed in one container at a given temperature, then the total pressure $P$ exerted by all of them together is the sum of the individual pressure (for example,

---

* Incidentally, this picture of molecules being restrained by mutual collisions despite their individual large speeds explains why the scent of a bottle of perfume (being nothing but a swarm of vapor molecules from that liquid) takes a relatively long time to travel from one corner of a room to the other; this realization disposed of one of the initial strong objections to the kinetic theory.

$P = P_1 + P_2 + P_3$) which each would exert if it were alone in that container at that temperature. Prove that our kinetic theory model is in harmony with this law.
[*Hint:* Accept the general gas law as applicable even to gas mixtures. From Eq. (22.9), $P = (\frac{2}{3})$ total $KE_{tranr}/V$, and note that kinetic energies add by simple scalar addition.]

---

b) Returning to the main line of the argument, the fact that the theoretical result of our model ($PV = \frac{1}{3}N_m m_0 \overline{v^2}$) must coincide with the experimental facts about gases ($PV = nRT$), we proceed to look for derivable and *verifiable* consequences of the implication that the right sides of both equations are equal, that

$$nRT \overset{?}{=} \tfrac{1}{3}N_m m_0 \overline{v^2}. \qquad (22.10)$$

This can be rewritten:

$$T \overset{?}{=} \frac{2}{3}\left(\frac{N_m}{nR}\right)\left(\frac{m_0\overline{v^2}}{2}\right). \qquad (22.11)$$

That is, *the temperature of a gas (in °K) is proportional to the mean kinetic energy of translation per molecule for the gas.* It is a most important result, and a very fruitful interpretation of the concept of temperature.

Furthermore, note that the factor $\frac{2}{3}(N_m/nR)$ is a universal constant for *all* gases. The proof is as follows:

$$\frac{N_m}{n} = \frac{\text{number of molecules in given sample}}{\text{moles of gas in sample}}.$$

But the number of molecules per mole is, by definition, Avogadro's number, $N$, a universal constant. Therefore

$$\frac{2}{3}\left(\frac{N_m}{nR}\right) = \frac{2}{3}\frac{N}{R}.$$

But as we noted above, $R$ is also a universal constant; and so it is convenient and customary to write the last expression as

$$\frac{2}{3}\frac{1}{k}, \quad \text{where}$$

$$k = \frac{R}{N} = \frac{8.31 \text{ (joules/mole)/degree}}{6.02 \times 10^{23} \text{ molecules/mole}}$$

$$= 1.38 \times 10^{-23} \text{ (joules/degree)/molecule.}$$

(The universal constant $k$ so defined is of greatest significance in many branches of physics, and is called *Boltzmann's constant*, named in honor of the great nineteenth-century Austrian physicist whose work paved the way to many advances in this field.) We can now rewrite Eq. (22.11):

$$T = \left(\frac{2}{3k}\right)\frac{m_0\overline{v^2}}{2}$$

$$= \text{(constant)} \times (\overline{\text{KE}}_{\text{trans}} \text{ per molecule).} \quad (22.12)$$

In words (since $k$ is truly a universal constant), *the temperature of a gas is directly proportional to the mean kinetic energy of translation per molecule for any gas*. This, then, is the consequence of the proposal that Eq. (22.10) is correct.

c) Consider one of the many consequences of this last paragraph. At the same temperature $T$, for two different types of gases (with values of $m_1$ and $v_1$ for the one, $m_2$ and $v_2$ for the other) our theory requires that

$$T = \frac{2}{3k}\frac{m_1\overline{v_1^2}}{2} \quad \text{and} \quad T = \frac{2}{3k}\frac{m_2\overline{v_2^2}}{2},$$

or

$$\frac{m_1\overline{v_1^2}}{2} = \frac{m_2\overline{v_2^2}}{2}. \quad (22.13)$$

Hence, at a given temperature, the molecules of any gas whatever have the same $\overline{\text{KE}}_{\text{trans}}$ as any other gas, or

$$\frac{\overline{v_1^2}}{\overline{v_2^2}} = \frac{m_2}{m_1} \quad \text{and} \quad \frac{\sqrt{\overline{v_1^2}}}{\sqrt{\overline{v_2^2}}} \equiv \frac{c_1}{c_2} = \sqrt{\frac{m_2}{m_1}}. \quad (22.14)$$

In words, *the ratio of rms speeds of molecules of two gases is equal to the square root of the inverse ratio of their masses.*

This conclusion is susceptible to immediate experimental test. If we take two different gases at equal temperatures, each in its own tight rigid container at equal pressures, and then open in each container an about equally tiny hole toward a vacuum for the molecules to escape, the gas with the lighter molecular mass, having a relatively higher rms speed, should escape faster than the heavier gas with lower rms speed. This experiment in *effusion* is in principle quite possible, and although the experiment itself obviously would be difficult to perform, the result does, in fact, prove Eq. (22.14) to be correct.

---

**Problem 22.4.** The mass of each molecule of hydrogen gas is to the mass of a molecule of nitrogen gas as 2:28 (approximately). What are the relative speeds of effusion for two such gases under equal conditions of temperature, etc.?

---

Another proof of Eq. (22.14) is to be found in the *diffusion* of two different gases trapped in one container with porous walls toward a vacuum; the lighter gas will again escape faster than the heavier one. This had been established empirically by Thomas Graham in 1830, and now was explainable in terms of a theory.

This process of diffusion is the basis of one type of large-scale separation of the light and heavy isotopes of uranium, the fateful metal. Specifically, uranium (chemical symbol U), as found in nature, consists primarily of an intimate mixture of two chemically equivalent but not equally heavy types of atoms, the isotope U-238 (i.e., relative molecular weight approximately 238, making up 99.3% of natural pure uranium) and the more potent but rarer isotope U-235 (molecular weight approximately 235, abundance about 0.7%). Small traces of a third type of uranium atom, U-234, are inevitably also present, but are negligible now. The historic problem of "enriching" natural uranium to make it useful for nuclear reactors or weapons, i.e., of increasing the relative abundance of U-235 with the help of the kinetic theory, was first discussed in the famous *Smyth Report* of 1945.* For the complete understanding of the following extract, note that the ratio $c_1/c_2$ in Eq. (22.14) corresponds also to the ratio of the *number* of molecules with these respective speeds which will have found their way through a porous barrier of a vessel after a short time interval. This ratio $c_1/c_2$ is therefore also called *the ideal separation factor*, symbolized by $\alpha$. The Smyth Report gives these details (pages 115–116):

"As long ago as 1896 Lord Rayleigh showed that a mixture of two gases of different atomic weight could be partly separated by allowing some of it to diffuse through a porous barrier into an evacuated space. Because of their higher average speed the molecules of the light gas diffuse through the barrier faster, so that the gas which has passed through the barrier (i.e., the 'diffusate') is enriched in the lighter constituent, and the residual gas (which has not passed through the barrier) is impoverished in the lighter constituent. The gas most highly enriched in the lighter constituent is the so-called 'instantaneous diffusate'; it is the part that diffuses before the impoverishment of the residue has become appreciable. If the diffusion process is continued until nearly all the gas has passed through the barrier, the average enrichment of the diffusate naturally diminishes.... On the assumption that the diffusion rates are inversely proportional to the square roots of the molecular weights, the separation factor for the instantaneous diffu-

sate, called the 'ideal separation factor' $\alpha$, is given by

$$\alpha = \sqrt{\frac{m_2}{m_1}},$$

where $m_1$ is the molecular weight of the lighter gas and $m_2$ that of the heavier. Applying this formula to the case of uranium will illustrate the magnitude of the separation problem. Since uranium itself is not a gas, some gaseous compound of uranium must be used. The only one obviously suitable is uranium hexafluoride, $UF_6$, .... Since fluorine has only one isotope,* the two important uranium hexafluorides are $U^{235}F_6$ and $U^{238}F_6$; their [approximate] molecular weights are 349 and 352 [respectively]. Thus if a small fraction of a quantity of uranium hexafluoride is allowed to diffuse through a porous barrier, the diffusate will be enriched in $U^{235}F_6$ by a factor

$$\alpha = \sqrt{\frac{352}{349}} = 1.0043."$$

Note that the enrichment is only 0.43% over the previous proportions.

Some of the difficulties of this process of separation are evident, not the least being the need for the final chemical extraction of uranium from uranium hexafluoride. Another serious point is that this separation factor $\alpha$ is far too low to permit use of the feebly enriched uranium hexafluoride after one such process. Therefore the first diffusate is made to go through another porous barrier which again discriminates against U-238 in favor of the prized U-235. Since $\alpha$ remains about the same, each single stage changes the relative abundance of U-235 by only very little. In Smyth's words:

"To separate the uranium isotopes, many successive diffusion stages (i.e., a cascade) must be used.... [Studies] have shown that the best flow

---

*    H. D. Smyth, *A General Account of the Development of Methods of Using Atomic Energy for Military Purposes Under the Auspices of the United States Government, 1940–1945*, Washington, D.C.: U.S. Government Printing Office, August 1945.

---

*    Of relative molecular weight 19.

**Fig. 22.5** Diffusion plant of the Clinton Works (now Atomic Energy Commission) at Oak Ridge, Tennessee, for the separation of uranium isotopes.

arrangement for the successive stages is that in which half the gas pumped into each stage diffuses through the barrier, the other (impoverished) half being returned to the feed of the next lower stage . . . . If one desires to produce 99 percent pure $U^{235}F_6$, and if one uses a cascade in which each stage has a reasonable over-all enrichment factor, then it turns out that roughly 4000 stages are required."

The construction by 1945 of many acres of such diffusion chambers in a Tennessee valley (Clinton Engineering Works, Oak Ridge, Tenn.; Fig. 22.5) is described in the report. In a certain sense, the first atomic bomb which used U-235 was then a gigantic confirmation of the kinetic theory that led to Eq. (22.14) and this diffusion process.

Yet a third proof of Eq. (22.14), on a rather more modest scale, is the closer observation of Brownian movement. If we watch the thermal agitation of a single tiny dust particle (mass $m_1$, rms velocity $c_1$), we note that its erratic dance movement never reaches the excessive speeds which we know the surrounding air molecules possess. In fact, Eq. (22.14) assures us that the rms speed $c_1$ of the dust particle is equal to $c_2 \sqrt{m_2/m_1}$; and since $m_2$, the mass of the air molecules, is many million times smaller than $m_1$, the mass of our dust particle, we cannot be surprised that $c_1$ is so small.*

Bodies larger than dust particles and the pollen of flowers are, of course, also randomly bombarded by the surrounding atmosphere, but because their masses $m_1$ are larger still, their $c_1$ is correspondingly smaller and their Brownian movement becomes unnoticeable. Yet very delicately suspended light mirrors, as used in accurate instruments, do show the slight random jitter of Brownian movement.

## 22.6 THE DISTRIBUTION OF MOLECULAR VELOCITIES

Consider another attack on Eq. (22.12),

$$T = \frac{2}{3k} \cdot \frac{m_0 \overline{v^2}}{2}.$$

For any given gas we can measure $T$, we know its mass per molecule $m_0$,† and we have calculated $k$ before; thus

$$c = \sqrt{\overline{v^2}} = \sqrt{3kT/m_0}.$$

If we now were able to measure $\sqrt{\overline{v^2}}$ directly, we could check Eq. (22.12) at once. But how are we to make a measurement of the incredible speed of in-

visible molecules? To prepare for the answer, we must make a short excursion.

One of the most significant ways in which the development of the kinetic-molecular theory has influenced modern science is by showing that regular, predictable behavior on the macroscopic level can result from irregular, random behavior on the microscopic level. By promoting a "statistical view of nature," Maxwell and Boltzmann helped to prepare the way for twentieth-century quantum theory, which was eventually to replace the Newtonian idea of the world as a machine.

At the beginning of the nineteenth century, the deterministic aspect of Newtonian mechanics had been epitomized in the following well-known passage of the French mathematician, Pierre Simon, Marquis de Laplace (1749–1827):

"We ought then to regard the present state of the universe as the effect of its previous states and as the cause of the one which is to follow. Given for one instant a mind which could comprehend all the forces by which nature is animated and the respective situation of the beings who compose it— a mind sufficiently vast to submit these data to analysis—it would embrace in the same formula the movements of the greatest bodies of the universe and those of the lightest atom; for it, nothing would be uncertain and the future, as the past would be present to its eyes."

Ironically this assertion was made in Laplace's *Philosophical Essay on Probabilities* (1814), serving as a preface to a major and influential contribution to statistical theory. From the labors of Laplace and his contemporaries Poisson and Gauss emerged a powerful set of tools for dealing with events and processes that are either *random* or, though deterministic, so complicated that statistical assumptions are more convenient.

One of the most useful results of nineteenth-century statistical theory was the *law of errors* or *normal distribution*. According to the law of errors, if any quantity can vary randomly around a certain average value, then the curve showing the fre-

---

* Actually the direction of motion of the dust particle changes so rapidly that it is extremely difficult to observe $c_1$ directly. It was for this reason that attempts to base a quantitative theory of Brownian movement on Eq. (22.14) in the nineteenth century proved futile.

† The mass $m_0$ in grams of a gas molecule is given by

$$m_0 = \frac{\text{mass in grams per mole of gas}}{\text{number of molecules per mole}}$$

$$= \frac{\text{number of grams/mole}}{6.02 \times 10^{23} \text{ molecules/mole}}$$

by Avogadro's law. For nitrogen, for example, this reduces to

$$m_0 \doteq \frac{28}{6.02 \times 10^{23}} = 4.65 \times 10^{-23} \text{ g/molecule.}$$

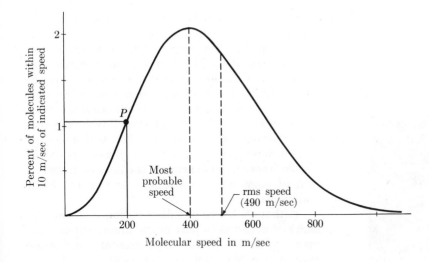

Fig. 22.6 Distribution of speeds for nitrogen molecules at 0°C.

quencies of deviations from the average value will be the familiar *bell-shaped curve,* with a peak in the region of the average value and a sharp decline on either side. Statisticians found that the law of errors applies in many cases involving large numbers of people; for example, in the distribution of heights and weights. Students are familiar with the law because it is supposed to apply usually to the distribution of scores on tests; teachers sometimes use the bell-shaped curve in deciding which scores should correspond to the various letter-grades, hence the expression "grading on the curve."

The suggestion that speeds of molecules in a gas might also be distributed according to a similar curve was made by Maxwell in his 1860 paper on kinetic theory. Even though the motions of the individual molecules were generally thought to be determined by Newton's laws, it seemed to be impossible to determine them directly—or, even if that could be done, to use the information in calculating the properties of gases. But Maxwell realized that in physics, as in human affairs, we often have to draw conclusions even when we lack most of the relevant details. An insurance company must set its rates, not of course knowing when any par-

ticular person seeking insurance is going to die, but relying on mortality tables that give an estimate of the probable life span of people at various ages. Similarly, we can calculate the average pressure of a gas of many molecules using the kinetic theory, even though we do not know the motions of each individual molecule in the gas.

Maxwell proposed that molecular speeds are distributed in a way similar to the law of errors. The curve representing the magnitudes of the speeds (Fig. 22.6) is actually unsymmetrical, since there is a minimum value (zero) but no maximum. For the same reason, the rms speed (Eq. 22.14) is greater than the most probable speed (the higher speeds have relatively greater weight in the calculation of the root-mean-square).

We shall not need to use the precise mathematical formula for Maxwell's distribution* or give a rigorous derivation of it. The crucial point

* According to his (now well proved) distribution, the relative probability for a molecular speed $v$ is proportional to $v^2 e^{-3mv^2/2kT}$, where $e = 2.718$. When the exponent of $e$ is a large negative number, corresponding to kinetic energies much greater than the average ($3kT/2$), the probability is extremely small but not zero.

**Fig. 22.7** James Clerk Maxwell (1831–1879). (Courtesy of Trinity College Library, Cambridge University)

to note here is simply that the collisions of molecules with each other do *not* tend to make all the speeds equal. You might think that, if a slow molecule collides with a fast one, their speeds after the collision would be closer to the average of the speeds before the collision. Some early theorists were led to think so by this line of reasoning, but it is fallacious. The fact is that even if two molecules with identical speeds collide, they will generally have different speeds after the collision, as can be shown from an application of the laws of conservation of momentum and kinetic energy to elastic collisions.

Although Maxwell's distribution law soon became established as part of the kinetic theory, for many years no direct experimental verification seemed possible. But, exactly because a scientific problem looks most paradoxical and difficult—how can one measure the speeds of molecules which are themselves invisible?—good minds will ponder it and frequently produce most ingenious solutions. An especially interesting technique, developed by Otto Stern and others in the 1920s, requires that atoms of a metal be evaporated in all directions from a hot wire of known temperature. The wire runs down the center of a close-fitting hollow cylinder which has a thin slit that can be opened and shut like a fast shutter. When the shutter is briefly uncovered, a short, thin, well-defined "squirt" of atoms escapes into the evacuated space beyond the cylinder; as they continue undisturbed in a straight line, the fast atoms outdistance the slower ones, thus spreading into a procession what started out to be a thickly packed crowd of atoms.

If now all atoms are allowed to impinge on a screen or slide of glass, they will quickly condense on it and form a thin coating of metal at a well-defined spot. But in order to separate on the screen the fast atoms from the slow ones, the screen is pulled across the path of the atoms as they fall on it; thus the coating becomes a lengthy smear instead of one spot, with the fastest atoms falling at the head of the deposit, the slowest ones at the tail end.

This is rather like shooting off a load of buckshot against a moving train at the very moment when the first car passes by. The speediest pellets will indeed hit the first car, and the slowest might hit the last. From the distribution of hits along the side of the whole train, you could analyze the distribution of velocities of all pellets.

One cannot simply and directly count how many atoms fall on any part of the moving plate, but from the geometry and known motion of the slide in this arrangement, we can calculate the absolute speeds (in m/sec) needed by an atom to reach a given portion of the screen. By mounting the slide and the shutter on the inside of two concentric rotating drums with the oven at the center (see Fig. 22.8), one can accurately repeat the procedure every time that these drums make one revolution, and gather many such layers of spread-out "squirts" of atoms on the same plate. Then the relative opaqueness, the measurable *density* of the total smear tells us directly what *fraction* of atoms had what speeds.

The very first thing that a qualitative examination of such a smear reveals is the presence in the stream of atoms of all speeds, from practically zero to practically infinitely large. The percentage of atoms (i.e., the relative density of the smear) at the extreme values is almost negligibly small but rises progressively as we shift our attention to the more central region of the deposit. For silver atoms evaporated from a filament kept at a temperature several hundred degrees below the actual melting point of the metal, the maximum density of the silver coating in the original experiments fell on that portion of the slide which corresponds to a speed of about $5.5 \times 10^3$ m/sec. This speed we may call the *most probable speed*, but it is not necessarily the same as the rms speed; according to Maxwell's theoretical work, the most probable speed should be definitely lower than the rms speed (Fig. 22.6). To determine the latter, we divide the deposit on the slide into small regions and measure more exactly what fraction or percentage of the atoms had speeds of $0–1 \times 10^3$ m/sec, $1–2 \times 10^3$ m/sec, $2–3 \times 10^3$ m/sec, etc. We might find that for the

**Fig. 22.8** Schematic representation of apparatus for measuring the velocity distribution of gas molecules.

given vapor and temperature, within the error of measurement, the following *distribution* of velocities exists.*

| Velocity range in $10^3$ m/sec | 0–1 | 1–2 | 2–3 | 3–4 | 4–5 | 5–6 |
|---|---|---|---|---|---|---|
| | 6–7 | 7–8 | 8–9 | 9–10 | 10–11 | 11–12 | 12–13 | above 13 |
| % of atoms in range | | 0.5 | 3 | 7 | 11 | 15 | 16.5 |
| | 14 | 11.5 | 9 | 7 | 3 | 2 | 1 | 0.5 |

This may be represented by a plot, as in Fig. 22.9. We see at once from the figure that (a) the observed speeds are bunched strongly around the maximum, the "most probable" value, dropping off to around 20% of the maximum at twice the most probable speed, and (b) that the distribution of speeds is quite lopsided toward the higher values

---

* These values actually apply only to truly random speeds such as exist in a gas or vapor in a closed box. The atoms in the *beam* described above have a slightly different distribution because the shutter arrangement automatically gives preference to one type of atoms—those with high velocity components toward the slit. The values in the table are thus really deduced indirectly from the actual density measurements, but this need not concern us here.

Fig. 22.9  Velocity distribution of atoms in silver vapor.

Fig. 22.10  Plot of speed distribution of the molecules of one gas at two different temperatures.

of speed. Even the straight average speed $\bar{v}$ is higher than the most probable speed—as you may ascertain by calculation.

In addition to confirming the shape of the velocity-distribution curve predicted by Maxwell, Stern's experiment also showed that the rms speed coincides with the theoretical value calculated from kinetic theory (see Eq. 22.12):

$$\sqrt{\overline{v^2}} = \sqrt{\frac{3kT}{m_0}}.$$

Here is the final result. If we carry through the calculation of the rms speed based on these measurements (how is this to be done?), our experiment yields a value of about $6.5 \times 10^3$ m/sec; on the other hand, the calculated value of $\sqrt{\overline{v^2}}$, using the

derived equation for this gas and for this temperature, gives us the *same value* within the expected limits of error!

Before we turn to another achievement of kinetic theory, let us briefly reformulate our results. For the sake of more accurate computations, we may plot Fig. 22.8 again for thinner strips and smaller velocity intervals, until the stepwise graph merges into a smooth line as in Fig. 22.6, drawn here for nitrogen gas at 0°C.

The meaning of any point on the curve, say $P$, is now the following: In this gas at this temperature only about 1.1% of all molecules have speeds in a $10^3$ cm/sec interval centered around $2 \times 10^4$ cm/sec (i.e., have speeds of $2 \times 10^4$ cm/sec $\pm 0.5 \times 10^3$ cm/sec). Figure 22.10 represents the speed distribution in a given gas at two different temperatures.

A useful concept arising from this discussion is *probability*. We may say that the probability is 0.011 that any randomly selected nitrogen molecule at 0°C will happen to have, within $\pm 5$ m/sec, a speed of 200 m/sec (see Fig. 22.5). As we use the words here, "probability of 0.011" means therefore that previous observations of the several speeds of many molecules make it safe to predict that of the next randomly selected 1000 molecules, about 11 will have the specified speed range. If we look at the speed of only 100 molecules, perhaps one will have the desired speed, but the smaller our sample becomes, the less predictable are the results. For a *single* molecule we cannot predict any definite speed, except perhaps to offer the platitude that the probability is 1.00 (100%) that its speed lies somewhere between zero and infinity; in fact, this whole concept of probability does not apply to such a singular situation.

**Problem 22.5.** From Fig. 22.6, find the probabilities corresponding within the given range to the most probable speed, to the rms speeds, and to twice and three times the most probable speeds.

Note that although the probabilities are smaller and smaller at extreme (high and low) speeds, they are not totally zero; that is, a few precocious molecules could be found at all times (for temperatures above absolute zero) which will show enormous speeds and consequently enormous energies, whereas at any moment a few others will seem practically to stand still. Similar distribution curves apply to liquids and even to solids, and they explain why some molecules, even at low temperatures, will be able to escape ("evaporate") from the surface, as Clausius had suggested in his qualitative theory (Section 22.2).

**Problem 22.6.** Equation (22.12) can be written as $\frac{3}{2}kT = \frac{1}{2}m_0\overline{v^2}$. Thus the $\overline{KE}_{trans}$ *per molecule* of any gas is $\frac{3}{2}kT$ and, for any gas,

total $KE_{trans}$ *per mole* $= (\frac{3}{2}kT)N = \frac{3}{2}RT$

a) Calculate this energy (in joules) present in one mole of water vapor (superheated steam) at 200°C; also find the amount of energy given up (to the receding piston) by 9 g of expanding steam in a steam engine cylinder, if the steam cools itself

during an adiabatic expansion to 100°C (all heat losses to the surroundings neglected).

b) Calculate also the amount of energy that must be removed and might be usefully available if 1 mole of ordinary air were cooled from room temperature to about −200°C, where these gases liquefy. Why don't we ordinarily utilize this energy to do useful work?

## 22.7 ADDITIONAL RESULTS AND VERIFICATIONS OF THE KINETIC THEORY

a) An important implication of Eq. (22.12) is that the $\overline{KE}_{trans}$ of a perfect gas decreases linearly as the temperature (in °K) is lowered toward zero; this gives physical significance to *absolute zero temperature* as the temperature at which—to put a subtle point very crudely—the random thermal motion of gas molecules has ceased. (This does not mean that *all* motion has ceased.)

b) Again on the qualitative side, one of the very important purposes of the kinetic theory was to explain why a gas in a thermally insulated container (adiabatic process) rises in (absolute) temperature in proportion to the work done on it by reducing its volume, e.g., by means of a piston pushing in one side of the container. In fact, it was this type of observation which first served as a foundation for the comprehensive statement of the law of conservation of energy. The work done on the gas, or the energy given to it by the advancing piston, is equal to the increase of translational kinetic energy picked up by the gas molecules during collision with the moving piston. But by Eq. (22.12), any change in $\overline{KE}_{trans}$ is reflected in a proportional change in the temperature: $\Delta\overline{KE}_{trans} \propto \Delta T$; hence the proportional rise in temperature.

The same argument in reverse explains why an (idealized) gas cools itself when it is allowed to expand—not into a vacuum, in which case it does no work and, by experiment, suffers no total temperature drop, but when it expands against a receding piston, or the inside walls of a balloon, or against another gas at lower pressure, or some "obstacle" of that sort. A familiar example is presented by a warm mass of air rising to a higher region, expanding against the more rarefied atmosphere, and so cooling itself to a temperature where the moisture brought along in the air is forced to condense into a rain or snow cloud.

Turning briefly again to the kinetic theory as applied to liquids, we now realize why evaporation from the surface cools the body of the liquid: since only the fastest molecules overcome the surface attraction and escape by evaporation, the average KE of the remaining ones drops correspondingly; however, a drop in $\overline{KE}_{trans}$ is, in fact, equivalent to a drop in $T$.

c) An important result of the kinetic theory, on which we can only touch briefly, was that it permitted a fair estimate of the *size* of individual molecules. Recall that Dalton tried to estimate the relative sizes of atoms on the basis of the mistaken model of a static gas in which the atoms touched. In the kinetic theory we have been assuming throughout that the volume of the molecules is negligible, seemingly destroying all hope of estimating sizes. However, there are several phenomena which can be treated by our model without making that assumption, specifically, (1) the pressure-volume behavior of denser gases, and vapors, where the molecular volume becomes of importance and where consequently Boyle's law breaks down; (2) the speed with which one gas diffuses through another, where evidently the size of the atoms will determine how often they collide and how easily and quickly they will pass by one another; (3) the speed with which heat can be conducted through a gas, another instance where the size of atoms will determine their progress despite collisions; and (4) the measurable viscous drag which a gas exerts on a larger object moving through it, and which again brings the factor of molecular size into the calculations. Suffice it to say that each of these problems involved enormous theoretical difficulties, that some are still the subject of fundamental in-

vestigation, and that all lines of evidence, from the 1860s on, converged on the result that the "diameters" of simple gas molecules are of the order of $10^{-10}$ to $10^{-9}$ m.*

d) Because it was Dalton's question of why the atmosphere's constituent gases are so thoroughly mixed which started him on his road to the atomic theory, we should surely give a brief look at the solution now given by kinetic theory to this problem. You recall that Dalton believed the mixing to be a result of a brief initial period of disequilibrium among unequally large, mutually repelling atoms and that he was wrong on several counts. The kinetic theory picture of gases predicts that two or more gases, intermixed like the atmosphere, should diffuse one through the other by virtue of the thermal movement of the molecules. However, and quite ironically, the heavier gas (namely, oxygen) is indeed expected to be found relatively somewhat more frequently per 100 molecules at lower levels than at levels higher in the stratosphere (why?). While this does not correspond at all to the marked stratification between a heavy and a light liquid—say water and oil—it is nevertheless just the type of phenomenon which Dalton thought does not happen. And there is a further twist: Whereas the expectations of the kinetic theory on that subject are fulfilled in a suitable quantity of gas under ideal conditions, our own atmosphere is so disturbed by wind currents that there *is* a fairly even mixing of all gases throughout the region accessible to direct and indirect measurements; so that it is after all a disequilibrium in the gas which explains the observations, but a continuous one, of a type not looked for in the Daltonian context.

---

*    Evidently we must be on guard against conceiving of too rigorous a model for molecules. The "diameter" may refer only to some average distance of closest approach among molecules. For a qualitative account of the determinations of size and mean free path of molecules, see T. W. Chalmers, *Historic Researches*, London: Morgan Brothers (1949), page 156 ff. A fine technical account is in T. G. Cowling's *Molecules in Motion*, London: Hutchinson, 1950.

e) A startling set of observations to which the kinetic theory lends direct explanation concerns the *escape of the atmosphere* from planets and satellites. As everyone knows, the moon and the smaller planets, with their smaller gravitational pull, have long ago lost their atmospheres—if they ever had any—and our earth slowly is doing the same, although we may be sweeping up enough interplanetary gas and vapor from meteorites to balance the loss. The process of escape is evident: The finite weight of the atmosphere "packs" it more densely near the earth, but the density decreases rapidly with height. The distance traveled by each molecule between collisions becomes correspondingly greater—at sea level it is only $10^{-7}$ m, but at a height of 1000 km it may be many miles. The paths of molecules will then be ordinary trajectories, but if their "initial speeds" are large enough (more than $10^4$ m/sec at that height above our earth) they may shoot too far out to be pulled back by the diminishing gravitational field. One glance at Fig. 22.6 shows that for nitrogen at 0°C there are only very few molecules indeed having such speeds. However, there are always some molecules at all speeds, and the temperatures at that height being much greater (perhaps 1000°C) the rate of escape for the common atmosphere is a little larger than would first appear, although still practically negligible. Finally, the smaller the molecular weight the larger the proportion of high speed molecules at any given temperature, which means, in the extreme, that a hydrogen molecule is likely to escape from our earth only about 1000 years after it is liberated (in chemical decomposition).

## 22.8  SPECIFIC HEATS OF GASES

We now come to one of the most striking and intriguing applications of the kinetic theory—the attempt to relate the internal motions of molecules to measurable thermal properties of gases.

To find the specific heat of a gas (see Section 17.8), we may take 1 g of the gas, put it in a rigid, heat-insulated container, and feed it a measured amount of heat energy (in calories) until its temperature rises by one degree (on the centigrade *or* the absolute scale—in both, one degree corresponds to the same *interval*). The specific heat* of the gas $s_v$ = calories needed per gram of gas, per °C of temperature rise. This experiment would yield widely different values for different gases, 0.75 (cal/g)/degree for helium, 2.41 for hydrogen, 0.15 for carbon dioxide, etc. Instead of using 1 g, we might have used 1 mole of the gas, defining thereby another but related concept, the *molar* specific heat, $C_v$, that is, the number of (calories/*mole*)/degree. $C_v$ would correspondingly differ from the values of $s_v$; evidently $C_v = s_v \times$ number of grams per mole of gas = $s_v \times$ relative molecular weight of the gas.

In either case, our model of the gas tells us that for every calorie we supply the total $KE_{trans}$ of the molecules will rise by 4.18 joules (law of conservation of energy); for $Q$ calories of heat energy supplied, $KE_{trans}$ increases by $\Delta KE_{trans} = QJ$ ergs (where $J$ is 4.18 joules/calorie, the mechanical equivalent of heat).

Consider 1 mole of gas, about to receive $Q$ calories. By Eq. (22.12), the total $KE_{trans}$ per mole of this gas = $\frac{3}{2}RT$. Since the temperature and the total KE are linearly proportional, $\Delta KE_{trans} = \frac{3}{2}R \, \Delta T$; let us write this $\Delta KE_{trans}/\Delta T = \frac{3}{2}R$. From the law of conservation of energy, $\Delta KE_{trans} = Q \cdot J$; thus $QJ/\Delta T = \frac{3}{2}R$, and therefore we write

$$\frac{Q}{\Delta T} = 3 \, \frac{R}{2J}. \qquad (22.15)$$

The left-hand side is the heat energy (cal) supplied to this one mole per degree rise of temperature—but that is exactly the definition of $C_v$ for a gas.

$$C_v = 3 \, \frac{R}{2J}. \qquad (22.16)$$

---

* At constant volume, care being taken that none of the heat energy supplied is wasted in work done by an expansion of the gas.

We note with astonishment that by our theory the molar specific heat of all gases should be the same, $R$ and $J$ being universal constants! Specifically, substituting values for $R$ and $J$, we get

$$C_v = 3 \times \frac{8.31}{2 \times 4.18} = 3 \text{ (cal/mole)/degree}$$
$$\text{(very nearly).}$$

We immediately rush to determine $C_v$ experimentally—and indeed, for helium, argon, and mercury vapor, $C_v$ is 3.02, 2.95, and 2.94 respectively, and the values for the other monatomic gases also confirm our theory within tolerable limits.

But only the *mon*atomic gases! Diatomic ones such as hydrogen, nitrogen, oxygen, carbon monoxide, and hydrogen chloride have experimental values of $C_v$ near 5 (cal/mole)/degree instead of the predicted value of 3. For these it is as though our theory must somehow be modified to yield

$$C_v \text{ (diatomic gases)} = 5 \frac{R}{2J}.$$

In short, diatomic molecules seem to have a way of absorbing (receiving and storing) more energy per degree temperature rise than do monatomic molecules. Every calorie supplied to molecules made of only a single atom each did turn up as translational energy, as expected; but every calorie supplied to a diatomic molecule seems to split 5 ways— only 3 parts go into translational energy and occasion the observed temperature rise, and the remaining 2 parts somehow are stored away without affecting the $\text{KE}_{\text{trans}}$ or the temperature. What is the physical explanation of this?

This is simply a breakdown of our model. Joule, who brought his work to this point, was stumped by the discrepancy in the specific heats of polyatomic gases, the more so as in his day no monatomic gases had yet been generally identified. In fact, his initial thought was that the experimental values of $C_v$ then available might be in error; and when new determinations only widened the gap between prediction and experimental results, he

seems to have been so discouraged as to leave the whole subject of kinetic theory. He had not made any provision for a molecule to receive other energy than translational kinetic energy, and up till now there was no reason to revise this simple idea. Now, however, we search for a plausible mechanism to account for the unsuspected heat capacity of diatomic molecules. And it might occur to us (this was done first by Clausius in 1857) that even if monatomic gases seem to fill the role of agitated point masses or small spheres without *rotational* energy, diatomic molecules might well be thought of as dumbbell-like objects, as two small spherical masses a short distance apart, an arrangement which has a much larger rotational inertia than either of the individual "point masses," and which therefore is much more capable of being put into rotation and of containing a significant amount of rotational kinetic energy ($\frac{1}{2}I\omega^2$) at plausible speeds of rotation.

Figure 22.11 shows a very crude schematic presentation of a diatomic molecule and its three possible types of rotation, distinguished from one another by the positions of the axis of rotation— analogous to our method of differentiating between the three possible types of translational motions, i.e., motion along the $x$, $y$, or $z$ axis.

---

**Problem 22.7.** Compare the rotational inertia of a hydrogen molecule ($H_2$) conceived (a) as a small sphere, as in our first model, and (b) as a dumbbell molecule with two equal (spherical) hydrogen atoms at about one atomic diameter apart. [*Note:* $I$ depends on the axis of rotation (see Fig. 22.11). A dumbbell molecule has a different value in (a) than in (c).] Numerical values for $I$ are obtainable if we make the assumption, based on other experiments, that the diameter of a single H atom is *about* $10^{-10}$ m, and that all but about $\frac{1}{2000}$ part of the total mass of the atom is concentrated at its center, the nucleus, whose radius is perhaps a $10^{-4}$ part of the atom's total radius.

---

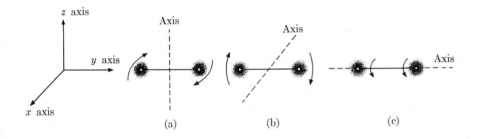

**Fig. 22.11** Three independent ways in which a diatomic molecule can rotate.

It is very convenient to invent the term *degree of freedom of motion* to denote each one of the independent ways in which a molecule can receive energy. For example, a monatomic gas molecule which does *not* rotate, as we initially pictured it, can acquire kinetic energy only by virtue of its velocity along the $x$ axis or $y$ axis or $z$ axis. These three motions are independent components, in the sense that we could imagine that one component is increased or decreased without the others being affected. In other words, the molecule has three degrees of freedom. If, however, we consider a particle made of two or more atoms, capable of rotational as well as translational motion, it has three degrees of freedom for translation *plus* three more for the three possible independent ways of rotating (about the three different axes shown in Fig. 20.10), or six degrees of freedom altogether.

However, we must be prepared to face the possibility that a molecule for some reason may not make use of all the conceivable degrees of freedom to which it is entitled by simple geometry. Thus we have already implicitly denied a monatomic molecule the use of its own three geometrically quite conceivable degrees of freedom of rotation or spin by saying that its rotational inertia is too small to make them effective, or that it is a point mass—a rather dubious trick, to be re-examined later, and, in fact, a veiled way of saying that in order to explain the experimental facts of specific heat in terms of our model we need not and must not allow

rotation for monatomic gas molecules in our model. Similarly, if rotation is neglected for a single atom, it should also be neglected for that one degree of freedom of rotation of a diatomic molecule which corresponds to motion about the lengthwise axis [Fig. 22.11(c)]; therefore molecules of $H_2$, $N_2$, $O_2$, CO, HCl, etc. should have only five degrees of freedom (three for translation, two for rotation), or two degrees more than monatomic molecules.*

Here, then, we have an explanation why a mole of diatomic gas absorbs more energy per degree of temperature rise than does a mole of monatomic gas. Of any and every unit of heat energy supplied to the former, only $\frac{3}{5}$ goes into translational motion [$\Delta KE_{trans}$, which by Eq. (22.12) causes a proportional rise in temperature], and the remaining $\frac{2}{5}$ of the supplied energy is stored in the two active modes of rotation ($\Delta KE_{rotation}$, which does not directly influence the rise in temperature). Therefore, to raise the temperature of a mole of diatomic gas by one degree, we must give it as much energy as that required for one mole of monatomic gas *plus* almost as much again ($\frac{2}{5}$ of the total) to satisfy the two active rotational degrees of freedom. So we understand that $C_v$ (monatomic gas) $= 3(R/2J)$, whereas $C_v$ (diatomic gas) $= 5(R/2J)$.

---

* The actual existence of rotational motion of polyatomic molecules, which affects neither the pressure nor the temperature, has been confirmed independently by direct experimentation on Brownian movement and on the light emitted by rotating molecules.

Reinspection of these two equations now reveals some interesting facts. If we were dealing with polyatomic gases whose molecules were so complex as to have significant rotational inertias about all three axes of symmetry, we should expect the third degree of freedom of rotation also to play its part. With a total of six active degrees of freedom, we expect $C_v$ (polyatomic) $= 2 \times C_v$ (monatomic) $= 6(R/2J)$, or about 6(cal/mole)/degree. In actual fact, many triatomic gases *do* have values of $C_v$ of about 6 (5.9 for $CO_2$ at low temperatures, 6.1 for water vapor at 100°C). Other polyatomic gases have even higher values, indicating that further hitherto neglected degrees of freedom begin to be important, namely, those representing *vibration* between the components of the molecule. For example, the complex molecules of ethyl ether ($C_4H_{10}O$) have a molar specific heat $C_v$ of about 30. Thus the idea of degrees of freedom of motion superposed on our initial model of the gas allows us to interpret the experimental facts of specific heat very satisfactorily. Within reasonable limits, we may write

$$C_v = \delta \left( \frac{R}{2J} \right), \qquad (22.17)$$

where $\delta =$ the number of active degrees of freedom of motion for the specified gas molecules.

This notion can be expanded to explain the observed specific heats of solids in terms of kinetic theory. But note now the astonishing fact that the specific heat of a gas has just given us a strong clue as to the molecular structure of gases, the very information Avogadro had lacked to protect his theory from the worst doubts. For example, the fact that $C_v$ for hydrogen gas is ordinarily 5(cal/mole)/degree shows that it must be diatomic—not monatomic as Dalton had thought. Similarly, $C_v$ for mercury vapor is about 3(cal/mole)/degree—clear evidence for its monatomic structure even though the other (chemical) evidences may be ambiguous or difficult to obtain.

Implied in the last few paragraphs was an important idea, the *equipartition of energy* among all active degrees of freedom in a gas; we have assumed that any and every unit of heat energy supplied to a mole of gas is distributed in equally large shares to each active degree of freedom. In terms of individual molecules this means that at a given temperature each gas molecule has, on the average, energy in amount equal to $\frac{1}{2}kT$ for each of its active degrees of freedom (about $1.88 \times 10^{-21}$ joule at 0°C). Hence a typical average monatomic molecule (with $\delta = 3$) possesses a total energy of $\frac{3}{2}kT$ per molecule (translational energy only, and thus equal to $\frac{1}{2}m\overline{v^2}$), diatomic ones each have $\frac{5}{2}kT$, polyatomic ones have even more. If any external energy $\Delta E$ is supplied to the gas, each active degree of freedom absorbs an equal share of $\Delta E$. This principle of equipartition, which we have introduced to explain the discrepancy in specific heats, is derivable more rigorously from the experimental facts, and is of foremost importance in the study of gases.

Finally, we should examine the points where our kinetic theory frankly breaks down. We expected such a possibility from the beginning, because our gas model was admittedly very simplified, and the introduction of the idea of degrees of freedom and of equipartition of energy may not have had enough sophistication in the long run. And not only did we expect the inevitable deficiency which appears sooner or later in every theory, but we eagerly look for it as a clue to the necessary modifications and perhaps as a key to the strange laws governing the world of atoms. One-third of the excitement in the study of physics lies in constructing a theory, another third in confirming the theory, and yet another in pushing beyond the theory.

The outstanding defect in our theory can be summarized in the following paradox: the principle of equipartition of energy requires that equal amounts of energy exist at a given temperature in each degree of freedom—except in those (the inactive) degrees of freedom where *no* energy at all resides, e.g., in the geometrically possible rotational

**Fig. 22.12** Variation of specific heat $C_v$ of molecular hydrogen gas. (To condense the graph, the temperature of the gas is plotted along the abscissa on a logarithmic scale.)

degree for diatomic molecules about the lengthwise axis. We are forced to ask, *exactly what determines whether a degree of freedom should be active or inactive?*

This question becomes even more intriguing if we note a discovery (made in 1912) concerning the amazing behavior of hydrogen gas, $H_2$. At ordinary temperatures its value of $C_v$ is a little below 5(cal/mole)/degree, as becomes a diatomic gas. But at lower temperatures, $C_v$ for this gas falls to about 3(cal/mole)/degree (at about 100°K), the value appropriate for a monatomic gas (see Fig. 22.12)! We must conclude that the two rotational degrees of freedom have "frozen in." Nothing in our theory has prepared us for such shocking behavior.* And what is worse, when $H_2$ gas or others such as $O_2$ are heated to extremely high temperatures, $C_v$ increases to over 5(cal/mole)/degree (at about 2000°C), as though other degrees of freedom begin to play a part not usual for diatomic molecules at ordinary temperatures.

---

\* If, on cooling, the two nuclei of $H_2$ came very close together, we might perhaps accept this as an explanation; but the molecular dimensions change only very little during cooling, as we find from other lines of investigation.

The fact of the matter is that here we come into entirely new and unexpected territory; this will be our link between the "classical" concept of the atom as formed in kinetic theory and the "modern" concept which the next part of the text will develop. But we may indeed rejoice in how well our simple mechanical model now seems to serve us in so many contexts (with that exception of specific heats). So successful was the "billiard-ball" picture of the atom that it seemed that any revision should prove only minor. That hope, however, was in vain —the revision was accompanied by a fundamental revolution of physics itself. This is the topic to which we turn in later chapters.

## 22.9 THE PROBLEM OF IRREVERSIBILITY IN KINETIC THEORY: MAXWELL'S DEMON

In Chapter 18 we discussed the tendency toward irreversible heat flow in connection with the second law of thermodynamics. In deriving the general gas law from the kinetic theory in this chapter we saw that it was necessary to assume a similar tendency toward equalization of temperatures of the wall of the container and the gas molecules inside, as read on two thermometers properly placed. Hence we accepted the observed fact that in these cases phenomena were not "symmetrical," in the sense that when a movie is taken of the temperature readings on the two thermometers, it corresponds to physically realizable phenomena only when run forward, and not when run backward. Yet in the rest of our discussion of kinetic theory we have accepted the validity of Newtonian mechanics, which implies complete reversibility. That is, we assumed symmetry between positive and negative time directions in physical processes in the sense that a movie taken of the perfectly elastic collision of two billiard balls shows physically realizable phenomena whether it is run forward or backward. This apparent inconsistency between the basic principles of mechanics which we have assumed to apply to the molecular level, on the one hand and the undoubted fact that observable natural processes are irreversible, on the other is called the *reversibility paradox*. It was pointed out forcefully by William Thomson in 1874, in the paper from which we quoted at the beginning of Chapter 18, and again by Josef Loschmidt in 1876.

A fairly satisfactory solution of the reversibility paradox was proposed by Maxwell, Thomson, and (much more elaborately) by Ludwig Boltzmann. It involves two recognitions: the surprising admission that the principle of dissipation of energy (second law of thermodynamics) is *not* an inviolable law of physics—it may occasionally, and for short time spans, be violated—and that in most ordinary circumstances the probability of such violations is extremely small.

Maxwell proposed a thought experiment to show how the second law of thermodynamics could be violated by an imaginary person who could observe individual molecules and, with virtually no expenditure of energy, could sort them out, thereby causing heat to flow from hot to cold. Suppose a container of gas is divided into two parts, A and B, by a diaphragm (solid partition). Initially the gas in A is hotter than the gas in B. This means that the molecules in A have greater *average* speeds than those in B. However, since the speeds are distributed according to Maxwell's distribution law, a few molecules in A have speeds less than the average in B, and a few molecules in B have speeds greater than the average in A.

Maxwell pointed out that with such an arrangement, and without expending energy, there would be a possibility of making heat flow "uphill" from a cold gas (as in b, Fig. 22.13) to a hot gas (side B in Fig. 22.13), using this overlapping of the distributions for gases at different temperatures: "Now conceive a finite being," Maxwell suggested, "who knows the paths and velocities of all the molecules by simple inspection but who can do no work except open and close a hole in the diaphragm by means of a slide without mass." (If the slide or shutter has no mass, no work will be needed to move it.) Let this "finite being"—which has come to be known as *Maxwell's Demon*—observe the molecules in A, and when he sees one coming whose speed is less than the average speed of the molecules in B, let him open the hole and let it go into B. Now the average speed of the molecules in B will be even lower than it was before. Next, let him watch for a molecule of B, whose speed is greater than the average speed in A, and when it comes to the hole let him draw the slide and let it go into A. Now the average speed in A will be even higher than it was before. Maxwell concludes:

"Then the number of molecules in A and B are the same as at first, but the energy in A is increased and that in B diminished, that is, the hot system has got hotter and the cold colder and yet no work has been done, only the intelligence of a very observant and neat-fingered being has been employed."

Thus Maxwell's thought experiment shows that if there were any way to sort out individual molecules by inspection, the principle of dissipation of energy could be violated.

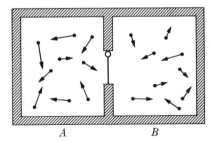

(a) Initially the average KE of molecules is greater in $A$.

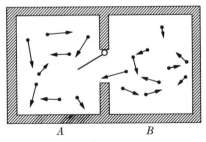

(b) The shutter is opened in such a way that only fast molecules are allowed to go from $B$ to $A$.

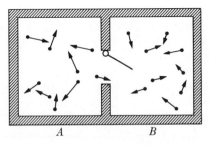

(c) Also, only slow molecules are allowed to go from $A$ to $B$.

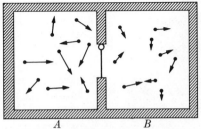

(d) As this continues, the average KE in $A$ increases and the average KE in $B$ decreases.

**Fig. 22.13** Operation of Maxwell's Demon.

The fact that Maxwell's Demon operates by creating *order* on the molecular level—putting the fast molecules in one box and the slow ones in another—suggests conversely that the usual processes of dissipation of energy involve increasing the amount of *disorder* or *randomness* of a system. Boltzmann made this idea explicit by developing a precise definition of entropy in terms of the probabilities of molecular arrangements. He found that if the entropy of a state is defined as

$$S = k \log W,$$

where $W$ is the number of possible molecular arrangements corresponding to the state, then the average entropy has the same properties as the entropy defined by Clausius (see Section 18.4). Disordered states of a system are those which can have a large number of molecular arrangements, so that $W$ and $\log W$ are both large, as in an ordinary gas; ordered states can have only a small number of possible arrangements (e.g., for a solid at absolute zero temperature, if there were only one possible molecular arrangement, $W = 1$, the entropy would be zero).

Boltzmann argued that if one were to list all the possible arrangements the molecules of a gas in a container can assume (for example, by making many "snapshots"), nearly every arrangement would have to be considered "disordered." In only a few cases, for example, would all the molecules be in one corner of an otherwise empty container, or all the fast molecules be on one side and all the slow ones on the other side. It is to be expected that if we start from an ordered arrangement of molecules in a gas, the arrangement will in time become less ordered; random arrangements have the advantage

**Fig. 22.14**  Ludwig Boltzmann (1844–1906).

of high probability. Similarly, if we put a hot body in contact with a cold one, it is almost certain that after a short time both will have nearly the same temperature, simply because there are many more possible arrangements in which fast and slow molecules are mixed together, than arrangements in which most of the fast molecules are in one place and most of the slow molecules are in another place.

To illustrate Boltzmann's argument, consider what happens if one is shuffling a pack of cards. If there were only two cards, $A$ and $B$ (analogous to two colliding particles), one of all possible arrangements (namely either $AB$ or $BA$) is as likely to occur as the other. But if we use many cards, the case looks different; most of the possible arrangements of the cards that occur are more or less disordered, and ordered arrangements come only very rarely out of a shuffled deck. If we *start* with an ordered arrangement—for example, the cards sorted by rank or suit—then shuffling would almost certainly lead to a more disordered arrangement. High probability is on the side of randomness. It does occasionally happen, and may be expected, that a player is dealt 13 spades, even if one uses a well-shuffled deck. But it is *almost* certain that this will not happen in real life.

According to Boltzmann's view, it is almost certain that disorder (entropy) will increase in any physical process among molecules. The second law is therefore a statistical law that applies to collections of many molecules, and it has no meaning when applied to one individual molecule. Since it is a statistical law, however, there is a remote possibility that a noticeably large fluctuation may occur in which energy is concentrated rather than dissipated. For example, the molecules in a glass of water are usually moving randomly in all directions; but it is not entirely impossible that at one moment they might all just happen to move in the same direction. The water would then jump out of the glass. (If you held the glass in your hand at that moment you would expect that the glass moves downward at the same time, since momentum still must be conserved.) In such a case a disordered motion has spontaneously turned into an ordered motion; entropy has decreased instead of increased, and the second law of thermodynamics (if it is regarded as an *absolute* law of physics) has been violated. Such large fluctuations seem extremely unlikely; yet if they can occur at all we must recognize that the second law has limits which do not appear to be shared by the other fundamental laws of physics.

## 22.10 THE RECURRENCE PARADOX

In their attempts to prove the stability of the "Newtonian world machine," eighteenth-century mathematical physicists had arrived at the conclusion that the gravitational perturbations of planetary orbits would not produce a continual change in the same direction on any one planet, but would have a cyclic effect. Thus the earth would not, on the one hand, gradually get closer to the sun over a long period of time, or, on the other hand, gradually get farther away; instead, its orbit would oscillate between definite inner and outer limits. This result had been used by Playfair to justify Hutton's geological theories (Section 18.2), but in the 1860s William Thomson cast some doubts on it by pointing out that the calculations were only approximate. In 1889 the French mathematician Henri Poincaré attacked the problem again using more accurate methods, and proved that *any* mechanical system subject to Newton's laws, whether the forces are gravitational or not, must be cyclic in the sense that it will eventually return to any initial configuration (set of positions and velocities of all the particles). However, the proof was subject to the condition that the system is restricted to a finite space as well as having fixed total energy.

Poincaré recognized that his *recurrence theorem*, if applied on the molecular level, would contradict the principle of dissipation of energy, since a gas enclosed in a container would eventually return to any initial ordered state if the recurrence

theorem held on this level. Also, on the cosmic level, the theorem implied that while the universe might undergo a heat death as all temperature differences disappeared, it would ultimately come alive again. Poincaré wrote:

"A bounded world, governed only by the laws of mechanics, will always pass through a state very close to its initial state. On the other hand, according to accepted experimental laws (if one attributes absolute validity to them, and if one is willing to press their consequences to the extreme), the universe tends toward a certain final state, from which it will never depart. In this final state, which will be a kind of death, all bodies will be at rest at the same temperature . . . .

". . . The kinetic theories can extricate themselves from this contradiction. The world, according to them, tends at first toward a state where it remains for a long time without apparent change; and this is consistent with experience; but it does not remain that way forever; . . . it merely stays there for an enormously long time, a time which is longer the more numerous are the molecules. This state will not be the final death of the universe, but a sort of slumber, from which it will awake after millions of centuries.

"According to this theory, to see heat pass from a cold body to a warm one, it will not be necessary to have the acute vision, the intelligence, and the dexterity of Maxwell's Demon; it will suffice to have a little patience."

Though Poincaré was willing to accept the possibility of a violation of the second law after a very long time, others were less tolerant. In 1896, Ernst Zermelo (at that time a student of Max Planck) published a paper attacking not only the kinetic theory but the mechanistic conception of the world in general, on the grounds that it contradicted the second law. Boltzmann replied, repeating his earlier explanations of the statistical nature of irreversibility, and pointing out that the recurrences predicted by Poincaré's theorem would occur so far in the distant future that they could have no practical effect on the applicability of the second law to ordinary processes. When these arguments failed to satisfy Zermelo, Boltzmann (half seriously) proposed the following hypothesis: The history of the universe is really cyclic, so that the energy of all its molecules must eventually be reconcentrated in order to start the next cycle. During this process of reconcentration, all natural processes will go backward, as described in the passage by William Thomson (Chapter 18). However, the human sense of time depends on natural processes going on in our own brains. If these processes are reversed, our sense of time will also be reversed. Therefore we could never actually observe "time going backward" since we would be going backward along with time!*

In the dispute between Boltzmann and Zermelo it was eventually seen that both sides were partly right and partly wrong. Zermelo and other critics of atomic theory such as Ernst Mach were correct in their belief that a complete microscopic description of matter cannot be based only on Newton's laws of mechanics. Gases are not collections of little billiard balls without internal structure, as Maxwell himself had concluded as a result of the specific heats discrepancy. But Boltzmann was right in his belief in the usefulness of the molecular model; the kinetic theory is correct except only for those properties involving the detailed structure of atoms and molecules.

In 1905, Albert Einstein pointed out that the fluctuations predicted by kinetic theory need not be so rare as both Poincaré and Boltzmann had thought; they should be able to produce an effect which can be observed and measured quantitatively: what had long been known as the *Brownian movement* of small particles suspended in fluids.

Subsequent studies inspired by Einstein's theory, in particular by the French physicist Jean Perrin, verified the quantitative predictions deduced from kinetic theory, and at the same time provided another method for determining molecular sizes. This new success reinforced the emerging statistical view of nature, but left open the question of whether the basic laws of nature are irreversible at the atomic level. This question is still a subject of lively interest among physicists today.

## RECOMMENDED FOR FURTHER READING

Ludwig Boltzmann, "On the relation of a general mechanical theorem to the second law of thermodynamics," English translation in S. G. Brush, *Kinetic Theory*, Vol. 2, New York: Pergamon, 1966

H. A. Boorse and L. Motz, *The World of the Atom*, New York: Basic Books, 1960. See pages 195–274 for extracts from the writings of Herapath, Brown, Waterston, Joule, and Maxwell.

Rudolf Clausius, "The nature of the motion which we call heat," English translation reprinted in S. G. Brush, *Kinetic Theory*, Vol. 1, New York: Pergamon, 1965

T. G. Cowling, *Molecules in Motion*, London: Hutchinson, 1950; Harper Torchbook paperback reprint. A comprehensive but elementary exposition of kinetic theory.

E. C. Kemble, *Physical Science*, Chapter 16

H. Poincaré, "Mechanism and experience," English translation in Brush, *Kinetic Theory*, Vol. 2

*Project Physics Reader*, Vol. 3. Articles by Kenneth W. Ford, George Gamow, Robert M. Coates, Jacob Bronowski, and James R. Newman, dealing with statistical concepts in kinetic theory, irreversibility, Maxwell's Demon, and the life of Maxwell. See also, in the same volume, "The Great Molecular Theory of Gases" by Eric M. Rogers, for a lively discussion of some further applications of the theory; and "Randomness and the twentieth century," by Alfred M. Bork.

---

* Boltzmann's theory has been revived in some recent cosmological discussions of the *big-bang* hypothesis. The present phase of expansion of the universe might be followed by a contraction (during which subjective time directions would be reversed), leading to a collapse, and then to another big bang, *ad infinitum.*

## SOURCES, INTERPRETATIONS, AND REFERENCE WORKS

Rudolf Arnheim, *Entropy and Art, an essay on order and disorder*, Berkeley: University of California Press, 1971

Ludwig Boltzmann, *Lectures on Gas Theory*, English translation of the 1896–98 edition by S. G. Brush, Berkeley: University of California Press, 1964

S. G. Brush, "The development of the kinetic theory of gases," *Annals of Science*, Vol. 13, pages 188–198, 273–282 (1957), Vol. 14, pages 185–196, 243–255 (1958); *American Journal of Physics*, Vol. 29, pages 593–605 (1961), Vol. 30, pages 269–281 (1962). *Kinetic Theory*, New York: Pergamon Press, 3 Vols., 1965–72, includes reprints and translations of papers of Clausius, Maxwell, Boltzmann, William Thomson, Poincaré, Zermelo, and others. "James Clerk Maxwell and the Kinetic Theory of Gases: A Review Based on Recent Historical Studies," *American Journal of Physics*, Vol. 39, pages 631–640 (1971)

A. S. Eddington, *The Nature of the Physical World*, Cambridge, Eng.: Cambridge University Press, 1928, Chapter IV

Paul and Tatiana Ehrenfest, *The Conceptual Foundations of the Statistical Approach in Mechanics*, translated by M. J. Moravcsik from an encyclopedia article published in 1912, Ithaca, New York: Cornell University Press, 1959

M. Eliade, *Cosmos and History: the Myth of the Eternal Return*, translated from French by W. R. Trask, New York: Harper Torchbook paperback reprint, 1959

E. W. Garber, "Clausius and Maxwell's kinetic theory of gases," *Historical Studies in the Physical Sciences*, Vol. 2, pages 299–319 (1970)

M. J. Klein, "Maxwell, his Demon, and the second law of thermodynamics," *American Scientist*, Vol. 58, pages 84–97 (1970)

R. Schlegel, *Time and the Physical World*, New York: Dover Pubs., 1968

PART G

# LIGHT AND ELECTRO-MAGNETISM

# LIGHT AND ELECTRO-MAGNETISM

To most scientists of the late nineteenth century, the physical world, despite all its superficial complexity, appeared to be endowed with a truly superb unity. The mechanics of Galileo and Newton explained the motion of all bodies—whether everyday objects or planets. Furthermore, the kinetic-molecular theory had triumphantly extended the reign of this mechanics into submicroscopic worlds by interpreting most phenomena of heat as the motion of atoms. Sound, too, had become a branch of mechanics, having been analyzed as a vibrating disturbance traveling among the molecules, according to those same classical laws of motion.

Earlier there had been hope that the phenomena of light, electricity, and magnetism could also be explained mechanically, either by postulating special kinds of particles with appropriate force laws, or as influences propagated through space by the vibrations of an ethereal medium. But as theorists shifted from the former to the latter type of explanation, they found that while their equations became more and more accurate in describing and predicting the phenomena, the physical basis of those equations was becoming less and less comprehensible. The great triumph of Maxwell's electromagnetic theory of light, which finally unified an amazing range of phenomena previously considered unrelated (if they were known at all), was achieved only at the cost of erecting a mechanical model so complex and artificial that it was unanimously discarded as implausible.

What we have to describe in this part, then, is the growth of a dinosaur which gobbles up so much food that it finally collapses under its own weight. But that is not quite an accurate metaphor; as we shall see in Part H, the skeleton of electromagnetic theory is still strong enough to provide the backbone for modern quantum mechanics and relativity.

# Chapter 23

# THE WAVE THEORY
# OF LIGHT

In this chapter we shall not attempt a comprehensive survey of optics, but aim to introduce the chief ideas and experiments that shaped the development of the wave theory of light. The main theme will be a question that was long debated in the history of the development of optics: Do the laws of mechanics, presented earlier in this book, apply also to the phenomena of light?

## 23.1 THEORIES OF REFRACTION AND THE SPEED OF LIGHT

We begin with René Descartes' theory of refraction, not because it is now considered correct but because, as the first quantitative model to be published, it had a considerable influence on the ideas of Newton and other scientists. Descartes begins his *Optics* (1637) by asserting that the propagation of light is similar to the transmission of a mechanical impulse through a stick which a blind man uses to guide himself, poking it at various objects. (This is a traditional analogy which can be found in scientific writings as early as the sixth century A.D.) Light is thus associated with motion in a medium, more precisely with a disturbance propagated quickly by mechanical means from one place to another (rather than, say, a bulletlike stream of little corpuscles). Descartes definitely rejects the possibility that "something material passes from

the objects to our eyes to make us see colors and light."

In order to explain properties of light such as reflection and refraction, Descartes proposes what now seems to be a rather artificial mechanical model. He compares the behavior of light when it strikes a surface (e.g., the interface between two media such as water and glass) to that of a tennis ball. If a tennis ball strikes a perfectly flat surface it is reflected elastically in such a way that the angle of incidence equals the angle of reflection (Fig. 23.1). (You can convince yourself that this result follows from the law of conservation of momentum, assuming that the force acting on the object during the collision is perpendicular to the surface.)

To find an explanation for the refraction of light, Descartes now considers the case in which a tennis ball hits a cloth "which is so weakly and loosely woven that this ball has the force to rupture it and to pass completely through it, while losing only a part of its speed." If the component of the ball's velocity perpendicular to the cloth is diminished but the component parallel to the cloth remains unchanged, the ball is deflected *away from the normal* (see Fig. 23.2).

At first it appears that Descartes is going to identify this behavior with that of a light ray going

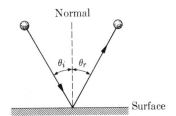

**Fig. 23.1** "Reflection" of a ball showing equal angles of incidence and reflection.

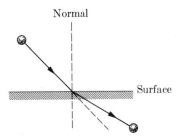

**Fig. 23.2** Refraction (ball deflected away from normal).

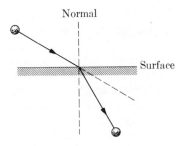

**Fig. 23.3** Refraction (ball deflected toward normal).

from a rare to a denser medium, for example from air into water, the natural assumption being that the ray is slowed down as it crosses such an interface. But, as Descartes is well aware, that would lead to error, for it is well known from common experience that when light rays pass from air into water they are refracted *toward the normal.*

The other possibility which Descartes considers is that the tennis ball, when it arrives at the interface, is hit by a racket in the direction perpendicular to the surface, increasing this component of its velocity but leaving the parallel component unchanged. The path is then bent *toward the normal.* This is in accord with the observed behavior of light rays going from air into water (see Fig. 23.3); of course it implies that light travels faster in water than in air, but that could not be tested until about 200 years later.

The major success of Descartes' model is that, once its premises are accepted, it accounts for the quantitative law of refraction discovered by the Dutch mathematician Willebrord Snell, around 1620.* But Descartes must now accept the consequence of his model that light rays travel faster in dense media than in rare ones, whereas he had previously asserted that light is always propagated instantaneously (that is, at a speed $v = \infty$). This inconsistency is typical of the paradoxes that have

afflicted theories of light since the seventeenth century. But it must also be remembered that Descartes did not take too seriously his picture of light rays as tennis balls moving at different speeds. It was more in the nature of a useful analogy. As he had written to a friend in 1634, he believed that light always does travel at infinitely high speed, that it "reaches our eyes from the luminous object in an instant . . . for me this was so

---

*    The law states that the sine of the angle of incidence is proportional to the sine of the angle of refraction (both angles being measured from the normal). Descartes was actually the first to publish the law; it is not known whether he learned it from Snell, to whom he makes no reference.

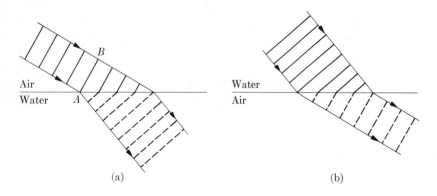

**Fig. 23.4**  (a) Refraction of an impulse on passing into a medium in which its speed is lower.  (b) Refraction of an impulse on passing into a medium in which its speed is higher.

certain, that if it could be proved false, I should be ready to confess that I know absolutely nothing in philosophy."

Newton took Descartes' model somewhat more literally as a theory of light.  He concluded that if moving particles are accelerated or decelerated by forces acting at the interface between two media, then one, at least provisionally, explains refraction by assuming that light is actually composed of particles.  Other properties of light seemed to favor a similar interpretation, for example, the fact that sharp shadows can be formed by solid obstacles placed in a beam of light.  If light were composed of pulses in a medium, he reasoned, then it should be able to bend easily around corners, as does sound.  (Though Newton was familiar with the diffraction of light, he considered it a less important phenomenon which could be explained by a particle theory of light if forces were assumed to act on the particles as they pass by the surface of objects.)

Newton also thought that an impulse theory could not account for *polarization*, a property of light unknown to Descartes.  In 1669, the Danish scientist Erasmus Bartholinus discovered that crystals of Iceland spar had the curious property of splitting a ray of light into two rays.  One of these rays can be split again if it strikes another spar crystal, but only for certain orientations of the second crystal; the ray seems to have a directional character.  (This directional character of light is now familiar to anyone who has played with the lenses of polarizing sunglasses.)  To Newton it was clear that this behavior could be explained only by assuming that the ray has "sides" so that its properties depend on its orientation with respect to the axis (direction of propagation) of the ray.  This would be easy enough to understand if the ray is a stream of rectangular particles, but rather more difficult if light is a wave disturbance in a medium.  As Newton remarked in his book *Opticks*,

"For Pressions or Motions, propagated from a shining body through a uniform medium, must be on all sides alike; whereas by those Experiments it appears, that the Rays of Light have different Properties in their different sides."

In spite of Newton's criticisms, other seventeenth-century scientists such as Robert Hooke and Christian Huygens continued to think of light in terms of impulses in a medium.  This was not yet the "wave theory" in the modern sense, because the *periodic* nature of the pulses had not yet been recognized (see the next section); ironically it was Newton who suggested that light might have to be somehow assigned also periodic properties in order to account for the phenomena of colors, even though such properties do not fit into his own particle theory very satisfactorily.

It was soon discovered that the pulse theory of light could after all account for refraction in a more plausible way than the particle theory—if one assumed that pulses are slowed down on entering a denser medium. This explanation was first published in 1644 by Thomas Hobbes, the political philosopher, and was presented in a more systematic and comprehensible manner by Huygens in his *Treatise on Light* (1690). The chief idea is to consider what happens to the *wave front*, an imaginary surface representing the disturbance or pulse, which is usually perpendicular to the direction of propagation of light, i.e., the direction of the ray. As shown in Fig. 23.4, when a pulse strikes an interface at an oblique angle, one part of the wave front enters the new medium and travels at its new speed, while the other part is still moving in the original medium. If the speed is smaller in the second medium, the ray is bent toward the normal to the interface, as one would expect when light passes from a rare to a dense medium; if the speed is greater in the second medium, the ray is bent away from the normal.

To decide between the Newtonian particle theory and the pulse theory of light, a "crucial experiment" is called for: Measure the speed of light in air and in water or glass, to see which is greater. But before this experiment could be done, it was necessary to settle the more fundamental question: Is the speed of light finite or infinite? Galileo discussed this problem in his *Two New Sciences*; he points out that everyday experiences might lead us to conclude that the propagation of light is instantaneous. But these experiences, when analyzed more closely, really show only that light travels much faster than sound. For example, "When we see a piece of artillery fired, at a great distance, the flash reaches our eyes without lapse of time; but the sound reaches the ear only after a noticeable interval." But how do we really know that the light moved "without lapse of time" unless we have some accurate way of measuring whatever lapse of time there might be? Galileo could not find any satisfactory way to measure the extremely short time interval required for light to travel any terrestrial distance. He concluded that the speed

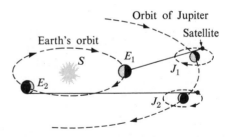

**Fig. 23.5** Romer's explanation of the delay in the observed time of an eclipse of one of Jupiter's satellites. It takes light longer to travel the distance $J_2E_2$ than the distance $J_1E_1$.

of light is probably not infinite, but was not able to estimate a definite value for it.

The first definite evidence that light moves at a finite speed was found by a Danish astronomer, Ole Rømer. In September 1676, Rømer announced to the Academy of Sciences in Paris that the eclipse of a satellite of Jupiter, which was expected to occur at 45 seconds after 5:25 A.M. on the 9th November, would be exactly 10 minutes late. On November 9, 1676, astronomers at the Royal Observatory in Paris, though skeptical of Rømer's mysterious prediction, made careful observations of the eclipse and reported that it occurred at 45 seconds after 5:35 A.M., just as Rømer had predicted.

Two weeks later, Rømer revealed the theoretical basis of his prediction to the baffled astronomers at the Academy of Sciences. He explained that the delay in the eclipse was simply due to the fact that light from Jupiter takes a longer or shorter time to reach the earth, depending on the relative positions of Jupiter and the earth in their orbits. (See Fig. 23.5.) By comparing observed eclipse times at various points in the earth's orbit, he estimated that it would take 22 minutes for light to cross the earth's orbit.

Shortly thereafter, Huygens used Rømer's data to make the first calculation of the speed of light.

Combining Rømer's value of 22 minutes for light to cross the earth's orbit with his own estimate of the diameter of the earth's orbit based on the latest astronomical observations, he obtained a value for the speed of light *in vacuo* which (in modern units) is about $2 \times 10^8$ m/sec. Though this is only about two-thirds of the presently accepted value of the speed of light,* $2.998 \times 10^8$ m/sec, it was a considerable achievement of the time to find a value that good. The correct value of the speed of light is now considered one of the most important constants, and is usually denoted by the letter $c$.

The significance of Rømer's work in the context of seventeenth-century science was not so much that it led to a particular value of the speed of light, but rather that it established that the propagation of light in free space is *not* instantaneous, but takes a finite time. Yet the fact that it only takes a few minutes for light to cover the enormous distance from the sun to the earth indicated the considerable technical difficulties involved in measuring the speed directly in a terrestrial laboratory. This feat was accomplished in the middle of the nineteenth century; not until that time could it be shown that light does indeed travel faster in air than in water. Here was the long-sought confirmation of the wave theory; but when it came, most physicists had already abandoned the particle theory and accepted the wave theory for other reasons.

**Fig. 23.6**  Propagation of a transverse wave along a rope.

## 23.2 THE PROPAGATION OF PERIODIC WAVES

In order to understand the concept of periodicity, which is essential to the modern wave theory of light, let us consider the vibrations of a string that is clamped at one end. We take hold of the free end,

as in Fig. 23.6, and move it rhythmically up and down in simple harmonic motion. As one particle of string is connected with the next, the motion is communicated from one to the next, but each lags behind its neighbor's motion, being coupled to it by elastic forces rather than by a rigid connection. We may visualize what happens in terms of the series of successive "snapshots" of Fig. 23.6.

Note how the motion of particle $A$, forced upon it by the hand, is communicated to $B, C, D \ldots$ and so on down the line. In particular, the maximum displacement, or *amplitude* of vibration, is reached by each particle in turn. So we see a "crest" being

---

*    The discrepancy is mainly due to the fact that as we now know light actually takes only about 16 min to cross the earth's orbit; Huygens' value for the diameter of the orbit was fairly close to the modern value, about $3 \times 10^{11}$ m.

generated, and after it a "valley"; and as both travel down the string to the right, they are followed by another crest, another valley, and so forth. In short, soon a complete *periodic wave* travels along the string, bringing motion from one end to the other, and with it, of course, energy. Note parenthetically that the wave travels here horizontally while the particles themselves move up and down. Technically, this mutually crosswise motion makes the wave a *transverse* one. If the motion of the particles is back and forth along the direction of motion of the wave itself, the wave would be called *longitudinal*.

The speed of propagation of the wave $v_w$, which we may here take to be the speed with which a crest travels, is of course the ratio of the distance $s$ which a crest travels on the string to the time $t$ needed to negotiate that distance. We can evaluate $v_w$ as follows: If we look at the last of the snapshots in Fig. 23.6, we note that particles $B$ and $J$ are both doing the same thing, i.e., they are both at their maximum displacement upward and about to move down again. The two are "in phase" with each other; and the distance between successive points in phase is called the *wavelength* (symbol $\lambda$, the Greek letter lambda). It is clear at a glance how long the crest at $B$ will take to cover distance $\lambda$ from $B$ to $J$. Not until particle $J$ has completed one full cycle of motion down and up will the crest arrive there. The time needed per cycle is, by definition, the observable period of vibration, $T$. Therefore

$$v_w = \frac{\lambda}{T}. \qquad (23.1)$$

If we recall that the frequency of vibration $n$ is the reciprocal of $T$, we can write instead

$$v_w = \lambda n, \qquad (23.2)$$

a most useful relationship between velocity, wavelength, and frequency, applicable not only to waves in a rope, but to waves of any kind, including light waves.

Fig. 23.7 Propagation of a sound wave in air. The displacement of the layers of gas molecules is greatly exaggerated.

**Problem 23.1.** Sound waves are longitudinal instead of transverse, i.e., the particles in the path of the wave oscillate in the same direction as the motion of the wave itself. Consequently, the sound wave, instead of crests and valleys, produces successive condensations and rarefactions. Figure 23.7 shows layers of molecules, at first equally spaced, then acted on by a progressing wave, i.e., by a piston or diaphragm which imparts its own motion to the gas in front of it. Copy this series and complete it in the same manner as Fig. 23.6 (13 lines altogether). Point out condensations (crowding of the planes) and rarefactions. Then mark off the distance corresponding to one wavelength.

Another property of vibrating strings should at least be mentioned here because it will turn out to help us later in the development of ideas in quantum mechanics (Chapter 29). Suppose a string of length $L$ is fixed at both ends. If it is plucked in the middle, it is able to vibrate as shown in Fig. 23.8(a). There

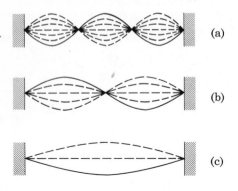

**Fig. 23.8** Fundamental vibration and first two overtones of a string fastened at both ends.

is a *node* (point of zero displacement) at each end. If properly plucked or otherwise agitated, the same string can also take on other vibratory patterns, of which the first two are shown in Figs. 23.8(b) and (c). These vibrations are examples of *stationary waves* or *standing waves*. We can understand them if we think of them as the result of two overlapping *traveling* waves, of the same frequency and amplitude, going through the medium (the string) in opposite directions; for example, when the string is plucked in the middle, two waves go off along it to the right and the left, each with velocity $v_w$, and are reflected again and again at the two ends, crisscrossing on the same string. At some moments the displacements due to each of the two waves combine vigorously (interfere constructively); at other moments they cancel (interfere destructively), and for an instant the string is completely horizontal.

These standing wave patterns, however, can happen only when the waves' velocity ($v_w$), the frequency of each of the waves ($f$), and the length of the string $L$ have the relation

$$f = v_w/2L \qquad \text{(for Fig. 23.8a)}$$

or

$$f = 2v_w/2L \qquad \text{(for Fig. 23.8b)}$$

or

$$f = 3v_w/2L \qquad \text{(for Fig. 23.8c)}$$

or, in general,

$$f = hv_w/2L, \qquad \text{where } h = 1 \text{ or } 2 \text{ or } 3 \dots$$

The proof is as follows. The oppositely traveling waves cancel exactly (produce nodes) at some point $x_1$ at which the instantaneous displacement of one wave is, say, $+y$, and that of the other is $-y$.[*] The next node is at $x_2$, where the first wave has displacement $-y$ and the second has displacement $+y$. Then there is a node at $x_3$, where the first wave again has displacement $+y$ and the second $-y$. But now we have advanced a full wavelength in going from $x_1$ to $x_3$. In other words, the distance between successive nodes is *half* a wavelength, $\lambda/2$.

The wave patterns at three successive moments of time are shown in Fig. 23.9. It can be seen from these diagrams that the nodes always occur at the same points $x_1, x_2, x_3, \dots$ even though the individual displacements that cancel to give these nodes differ from one moment to the next.

Since the two fixed ends of the string must be nodes, the total length of the string $L$ must be an integer multiple of the distance $\lambda/2$ between nodes,

$$L = \lambda/2, \; 2\lambda/2, \; 3\lambda/2, \dots \qquad (23.3)$$

Or, if we consider $L$ fixed, the possible wavelengths for such standing waves are

$$\lambda = 2L, \; 2L/2, \; 2L/3, \dots$$

---

[*]   At a later time the first has displacement $y + \Delta y$, and the second displacement $-y - \Delta y$, so they still cancel; the changes in the displacements of the two waves must be equal and opposite, since they were stated to have the same frequency and amplitude.

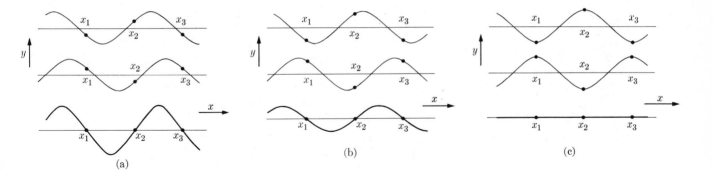

**Fig. 23.9** Interference of two similar waves traveling in opposite directions, producing standing waves. Three parts, (a), (b), and (c), represent successive instants of time; in each part the first and second patterns (light lines) are the individual waves, and the third (bottom) pattern (heavy line) is their superposition.

as shown in Fig. 23.8. The corresponding frequencies are

$$f = v_w/2L,\ 2v_w/2L,\ 3v_w/2L,\ \ldots \qquad (23.4)$$

as stated above.

The frequencies following the fundamental frequency $v_w/2L$ are called *overtones*. The difference in tone quality of two musical string instruments playing the same note (same fundamental frequency) is due to the fact that overtones are always present to some degree along with the fundamental, but are produced with different intensities in different instruments.

## 23.3 THE WAVE THEORY OF YOUNG AND FRESNEL

From the time of Newton until the early years of the nineteenth century, the particle theory of light was favored by most physicists, largely because of the prestige of Newton—though the theory was not as dogmatically entertained by the master as by his later disciples. The English scientist Thomas

Young and his younger French colleague Augustin Fresnel fought to put the wave theory of light on a sound theoretical and experimental basis during the years 1800–1820. They exploited a key feature that distinguishes waves from particles: When two or more waves are superimposed, in the same medium (as we have already seen in discussing standing waves) they may either reinforce or cancel out the displacement of particles at various places in the medium. This property is known as *interference*; it can be either *constructive interference* (when the waves reinforce) or *destructive interference* (when the waves cancel). Streams of particles, when they join, do not seem to have such behavior in all our normal observations.

The fact that interference effects can be demonstrated in optical experiments helped to establish the wave theory of light. The first of two famous experiments was Young's *double-slit experiment*, represented in Fig. 23.10. The interference pattern of alternating bright and dark regions—produced when a beam of light is passed through two slits or holes, and the spreading, separate parts then

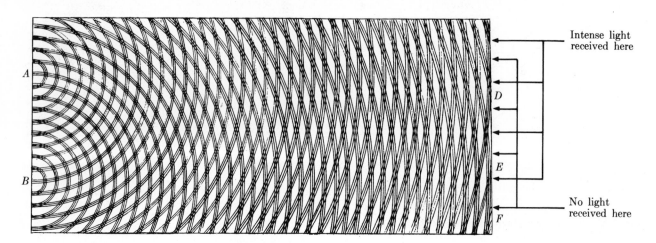

**Fig. 23.10** Thomas Young's original drawing, showing interference effects expected when waves from slits *A* and *B* overlap. (Place the eye near the left edge and sight at a grazing angle along the diagram.) At the screen on the right, no light is received where the waves from *A* constantly interfere destructively with the waves from *B*. Note that at those places the wave crests or fronts (shown in circular arcs) from *A* fall just between those from *B*.

recombine—is easily explained by the wave theory, but seemed quite mysterious from the particle viewpoint.

The second experiment was actually proposed as a test of Fresnel's theory by a supporter of the particle theory, the French mathematical physicist Simon Poisson. Poisson was one of the judges appointed by the French Academy of Sciences to examine Fresnel's paper, submitted for a prize in 1818. Using Fresnel's mathematical equations for the diffraction of light waves around obstacles, Poisson showed that if these equations really did describe the behavior of light, a very peculiar thing ought to happen when a small round disk is placed in a beam of light. A screen placed behind the disk should have a bright spot in the center of the shadow, because diffraction of the light waves spilling all around the edge of the disk should lead

to constructive interference at the center of the shadow. According to Newton's particle theory, there could not be any such bright spot, and it had never before been observed. Poisson fully expected that this absurd prediction would lead to the refutation of the wave theory, and challenged Fresnel to do the experiment.

Fresnel accepted the challenge and arranged for this prediction to be tested by experiment. The result: There *was* a bright spot in the center of the shadow!

When the significance of the Young double-slit experiment and the Poisson bright spot (Fig. 23.11) was realized, support for the particle theory of light began to crumble away. By 1850, when Fizeau and Foucault found that the speed of light in water is indeed less than the speed in air (as predicted by the wave theory), the validity of the Young-Fresnel

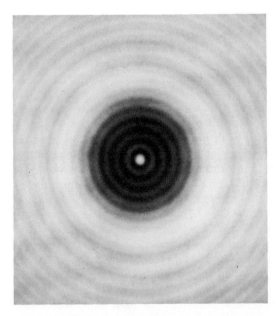

**Fig. 23.11** The Poisson bright spot. (Photograph of the diffraction pattern of a small circular obstacle, showing the bright spot predicted by Poisson from Fresnel's wave theory.)

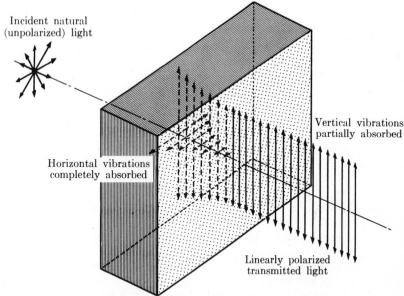

**Fig. 23.12** An unpolarized beam of light (vibrations in all directions in a plane perpendicular to the direction of travel) passes through a crystal such as tourmaline, which removes the horizontal component of the vibrations, leaving only vertical vibrations.

theory was generally accepted, and physicists had already begun to work out the mathematical consequences of the theory and its application to all the different properties of light.

One property of light that had to be explained was *polarization*. Newton, as we noted earlier, had argued that a particle theory was needed to explain the fact that light rays seem to have directional properties because impulses propagated through a medium "must be on all sides alike." But that is only true of *longitudinal* waves, such as sound. Young and Fresnel pointed out that the phenomena of polarization could be explained by assuming that light waves are *transverse*; thus the vibrations could be vertical (as in Fig. 23.6) or horizontal, or in any other direction in a plane perpendicular to the direction of propagation. Ordinary light from the sun or light bulbs is a mixture of waves vibrating

in all these directions, and hence is said to be *unpolarized,* but on reflection or refraction by certain substances the waves polarized in some directions are selectively removed, leaving a beam of light polarized, as shown, for example, in Fig. 23.12.

But the assumption that light consists of transverse waves brought new difficulties, for the only kind of transverse waves known to Young and Fresnel were waves that travel in a *solid* medium. In order to transmit transverse waves, the medium must offer some resistance to "twisting" motions; it must have a tendency to return to its original *shape* when deformed. This is true of a solid but not, in general, of liquids or gases. It therefore seemed necessary to assume that the "luminiferous" (light-bearing) ether—the medium that was thought to occupy all otherwise "empty" space, that penetrated all matter, and whose vibrations were identified at the time with light waves—is a solid!

It was also known that ordinarily the speed of the waves is equal to $\sqrt{\varepsilon/\rho}$, where $\varepsilon$ is a measure of the elasticity of the medium and $\rho$ its density; this formula was confirmed by experiments on the speed of sound in solids, liquids, and gases, and therefore was assumed to apply also to light waves. But for a speed as great as that of light, the medium of propagation, the ether, would have to show simultaneously a very high elasticity and an extremely low density. In order to explain the properties of light, the nineteenth-century physicists were therefore compelled to assume that all space is filled with a highly elastic but rarefied solid. How could one believe that such an ether really exists, when it was well known that the planets move freely through space just as if they were going through a vacuum that offered no resistance at all?

Scientists of today, who more than their ancestors have grown up without feeling the need to account for phenomena by mechanical models, would find it harder to conceive of a medium with such contradictory properties than to think of the propagation of light through space without the benefit of an intermediate material medium. But it should be clear that in the nineteenth century, instead of discouraging the proponents of ether models, such paradoxes only increased their efforts to find better models.

The wave theory also held out the hope of explaining phenomena other than light by considering vibrations of different wavelengths. Newton had suggested in 1672 that the different colors of the spectrum, formed by refracting a beam of white light through a prism, might be imagined to correspond to ether vibrations of different magnitude. (At that time, Newton had not committed himself to the particle theory of light.) Thomas Young was able to measure the wavelengths corresponding to the colors of the spectrum, and found that they ranged from about $7 \times 10^{-7}$ m for red to about $4 \times 10^{-7}$ m for violet. But Sir William Herschel had discovered in 1800 that there always existed in the spectrum of hot solids—along with the visible light—some radiation of longer wavelengths (beyond the red, a region named the *infrared*). The suggestion that this "invisible light" was identical with radiant heat was proved when heat rays were shown to exhibit all the characteristics of light—refraction, interference, polarization, and so forth. Already there is a simplification here: Newton had thought that radiant heat might need a separate ether for its propagation; now the luminiferous ether could serve a double purpose. And, as was noted in Section 17.7, once the wave nature of *light* was accepted, it was only natural to suppose that heat also has a wave nature, and can be explained in terms of motions of matter. Thus the wave theory of light seemed to provide an essential link in the chain binding together all the forces of nature, including later (as will be seen in Chapter 25), electricity and magnetism.

## RECOMMENDED FOR FURTHER READING

P. S. Epstein, "Centennial of the undulatory theory of light," *Science*, Vol. 63, pages 387–393 (1926). A dramatic account of Fresnel's discoveries which marked a turning point in the history of optics.

D. L. Hurd and J. J. Kipling, *The Origins and Growth of Physical Science*, Vol. 2, pages 149–181. Extracts from the writings of Newton and Huygens on light.

*Project Physics Reader*, Vol. 4, extracts from writings of A. A. Michelson, W. A. Shurcliff, and S. S. Ballard

## SOURCES, INTERPRETATIONS, AND REFERENCE WORKS

A. C. Bell, *Christiaan Huygens and the Development of Science in the 17th Century*

J. G. Burke, "Descartes on the refraction and the

velocity of light," *American Journal of Physics*, Vol. 34, pages 390–400 (1966)

I. B. Cohen, "Roemer and the first determination of the velocity of light," *Isis*, Vol. 31, pages 327–379 (1940). Includes facsimile reprints of Roemer's papers.

I. B. Cohen, "The first explanation of interference," *American Journal of Physics*, Vol. 8, pages 99–106 (1940)

René Descartes, *Discourse on method, Optics, Geometry and Meteorology*

Christian Huygens, *Treatise on Light*, translated from French by S. P. Thompson, London: Macmillan, 1912; Dover paperback reprint

Vasco Ronchi, *The Nature of Light*, translated by V. Barocas, Cambridge, Mass.: Harvard University Press, 1970

A. I. Sabra, *Theories of light from Descartes to Newton*, London: Oldbourne Book Company, 1967; New York: American Elsevier

J. F. Scott, *Scientific Works of René Descartes*

E. Sparberg, "Misinterpretation of theories of light," *American Journal of Physics*, Vol. 34, pages 377–389 (1966). On the problem of choosing between the theories of Newton and Huygens.

F. R. Tangherlini, "On Snell's law and the gravitational deflection of light," *American Journal of Physics*, Vol. 36, pages 1001–1004 (1968)

E. T. Whittaker, *A History of the Theories of Aether and Electricity*, London: Nelson, revised edition, 1951; Vol. I, Chapters I, IV and V

# Chapter 24

# ELECTROSTATICS

## 24.1 INTRODUCTION

The physical atom of the nineteenth century was discussed largely in terms of mechanical force, momentum, kinetic energy, elastic collision, and the like. But as the century drew to a close, there was discovered the electron, that carrier of a fixed negative charge, endowed with a definite mass, a particle much smaller than the atom and undeniably a part of its structure. Then radioactivity brought out more charged particles that had to be somehow accommodated in the atom, re-enforcing long-standing demands from other branches of physics that electric phenomena must be associated with atoms directly. Finally, the twentieth century gave us a model that stresses concepts like charge, electric potential, and electric fields at least as prominently as its predecessor had stressed mechanical concepts. Therefore, as we now turn toward the modern atomic theory, it is not only fitting but necessary to undertake first of all a rather detailed study of the basic ideas of electricity.

In reviewing the development of electrostatics we note for the first time a major contribution by an American scientist in the early formulation of a physical theory. Benjamin Franklin (1706–1790) was a well-known statesman and writer as well as a scientist. He was greatly interested in the phe-

nomena of electricity; his famous kite experiment and invention of the lightning rod gained him wide recognition. His other inventions include the Franklin stove, bifocal glasses, and the glass harmonica. While in Philadelphia, Franklin organized a debating club that developed into the American Philosophical Society (one of the oldest scientific societies still in existence), and he helped to establish an academy which later became the University of Pennsylvania.

## 24.2 ELECTRIFICATION BY FRICTION

The word *electricity* is derived from the Greek word for amber, *electron* (ἤλεκτρον), for it was in ancient Greece that the observation seems to have been first made that after amber has been rubbed with cloth it *attracts* small pieces of lint, as we now know, by electrostatic forces. Not until the seventeenth century were forces of repulsion observed in such simple experiments as with bodies electrically charged by rubbing. We still use Franklin's terminology to describe the main observations:

a) When two different bodies such as a glass rod and a silk cloth are rubbed against each other, they will thereafter attract each other with a new, non-gravitational force, because they have developed electric charges of opposite sign. By arbitrary convention we call the charges on the glass positive (+), those on the silk negative (−). Note well that the observed attraction defines two things: that charges are present, and that they are of different kinds.

b) Two glass rods, rubbed in turn with a silk cloth, both become charged, but repel each other. We note that positive charges repel one another.

c) When a rod of resin (or amber, rubber, etc.) is rubbed against a new silk cloth, it also becomes charged, as seen by the appearance of a force of attraction between the resin and silk. But this charged resin rod also attracts the previously charged glass rod, wherefore we decide to assign a minus sign to charges on the resin.

d) Finally, two resin rods charged in the same way repel, and we conclude that negative charges repel one another.*

Further experiments with similar combinations reveal no case in which the cloth is repelled by the material which it has freshly charged, nor any cases in which a charged body is attracted to (or repelled from) both + and − charges. All these observations can therefore be summarized by the following hypotheses (and the deduction from the above experiments is a useful exercise in clear thinking):

i) Two neutral bodies can charge each other by friction, but always with opposite charges.
ii) There exist only two kinds of charges which interact, *positive* and *negative*. Positive charges appear on glass when it is rubbed with silk (definition of + charge), negative ones appear on resin when it is rubbed with silk or, more effectively, with fur (definition of − charge).

---

* The postulate that negative charges repel each other was not part of Franklin's original theory but was added by the German physicist Franz Aepinus in his systematic treatise on electricity published in 1759.

If suitable care is taken, any object can be charged by friction with a suitable cloth, and the charges thereby produced turn out to be + or −, but never of a third kind.

---

**Problem 24.1.** What is the action of a silk cloth which has charged a glass rod, on another silk cloth which charged a rubber rod? Can you predict what will happen if you rub silk against silk? Silk against fur?

**Problem 24.2.** Although they are not found in nature, some kinds of charges other than + and − are logically possible. Describe what properties they would have, i.e., how you would *know* when charges different from both + and − are present.

## 24.3 LAW OF CONSERVATION OF CHARGE

Even before we define charges quantitatively, we may suspect a law, to be confirmed later more rigorously, of *conservation of charge*, i.e., a law stating that electric charges can be produced only in pairs of equally strong positive and negative charges. The preliminary experimental proof, due also to Franklin, is a simple observation: Wrap an uncharged cloth around an uncharged glass rod. Rub the rod and separate cloth and rod; individually each exerts, as expected, strong electric forces of repulsion or attraction on some third object, perhaps a charged leaf of tinfoil. But now wrap the charged cloth again around the charged rod; the electric forces exerted on some third object disappear. Our conclusion is that the effect of the opposite charges cancels when they are brought near each other, therefore cloth and rod charged each other equally strongly though with opposite signs. No *net* charge (excess of + or of −) was created by mutual friction. As usual, such a law of conservation is the deepest justification for the power of the concept "electric charge" itself.

## 24.4 MODERN MODEL FOR ELECTRIFICATION

Simple experiments like these led to some reasonable and fruitful early theories concerning the nature of electric charges.  Our modern model cannot be presented as a logical development without reference to many more subtle and quantitative recent experiments, but let us simplify and speed up our discussion of electrostatics by stating the modern concepts now and justifying them later.

All materials—solid, liquid or gaseous—are atomic in structure, as we convinced ourselves some time ago in discussing the successes of the kinetic-molecular theory.  Each atom can ordinarily be considered as a neutral, uncharged object, but actually it is made of a strongly positively charged nucleus (the small central kernel of the atom which represents most of its mass within a radius of about $10^{-14}$ m), surrounded at a relatively large distance (about $10^{-10}$ m) by negatively charged electrons (from 1 to more than 100 around each nucleus, depending on the particular element).  Electrons are all alike, and very light compared with the nuclei.

The neutrality of the ordinary atom is the consequence of an exact balance between the positive charges of its nucleus and the negative charges of its extranuclear electrons.  When an electron is somehow torn from its parent atom, the remaining atom lacks one negative electron-charge, i.e., it is a *positive ion.*\*  The separated electron meanwhile represents a free negative charge; it may in time attach itself to a positive ion and so restore the neutrality of that atom, or it may attach itself to a neutral atom and so transform it into a *negative ion.*

In solids, the atoms at the surface are at rest except for thermal vibrations.  Their electrons are held near the nuclei by electrostatic attraction, but in some substances it is possible to interfere with this attraction (even by simple surface contact with

\*    Further discussion of ions will be presented in Section 24.13.

another material) and to snatch some surface electrons, leaving a positively charged surface on one material and carrying away an equally strong negative charge on the other.  This is our picture of electrification by intimate contact, as by friction, although we are not prepared to say just how or why some materials, like glass, joyfully allow their electrons to be carried off, whereas others, like amber and resin, tend to grab extra electrons very copiously.

Here then is a simple model to explain the experimental facts of electrification by friction.  Furthermore, it accounts for the observation that there are only two kinds of charge, and it will offer an adequate explanation of all other observed phenomena in electrostatics, e.g., the fact that in principle a practically limitless amount of electric charges can be separated by friction, understandable now in terms of the huge quantity of atoms available on every small portion of surface.

Note that the charges of ions and electrons are not something *apart from* those bodies.  For example, it is not possible to discharge an electron and leave a small neutral mass; the whole electron *itself* is a carrier of negative charge.  It is best to realize right here that there is no easy picture for visualizing the concept "charge."  An electron has a definite mass (this statement undoubtedly and unfortunately invokes a picture of a tiny, hard, round sphere) and the electron also carries a constant negative charge (this statement adds nothing to our picture except that we must endow that little sphere with the ability to repel other electrons and to be attracted by positive ions).  In fact, the concept "electric charge" has no meaning apart from some of these nongravitational forces which bodies sometimes exert on one another.  The following quotation illustrates the conceptual difficulties in electricity very well.

"Some readers may expect me at this stage to tell them what electricity 'really is.' The fact is that I have already said what it is. It is not a thing like St. Paul's Cathedral; *it is a way in which things behave.*  When we have told how things behave

Figure 24.1

when they are electrified, and under what circumstances they are electrified, we have told all there is to tell. When I say that an electron has a certain amount of negative electricity, I mean merely that it behaves in a certain way. Electricity is not like red paint, a substance which can be put on to the electron and taken off again, it is merely a convenient name for certain physical laws." (Bertrand Russell, *ABC of Atoms*, E. P. Dutton, 1923)

## 24.5  INSULATORS AND CONDUCTORS

Experimentally we can distinguish between two types of materials: (a) those that retain an electric charge wherever on the surface it was placed initially, and (b) those that allow equal charges placed on their surfaces to repel one another, to "flow" and distribute those charges all over the body no matter where they were placed initially. The first behavior characterizes *insulators* (or non-conductors, dielectrics) such as glass, fur, silk,

paper, resin, etc. The second type of materials, of course, are *conductors* such as metals and graphite.

We can also differentiate between conductors and insulators by the relative resistance which they offer to an electric current. In good conductors like metals, some of the electrons more loosely connected to their atoms quite easily drift away from their parent atoms when some external electric force acts on them; hence they are called *conduction electrons*, and while they move they constitute an *electron current* through the conductor (or along its surface). In a good dielectric, however, the electrons are held more tightly by the parent atoms and are not free to flow away when acted on by external electric forces of attraction or repulsion.

Many of the experiments leading to these conclusions can be presented in concise diagrammatic form. Let us agree to indicate by a + or − the sign of the excess of + or − charges at any point on a solid. Then Fig. 24.1 tells the following simple story.

a) When it is rubbed vigorously, a piece of fur loses electrons to a resin or hard rubber rod; the fur is charged positively, the rod negatively.

b) The fur is withdrawn, the rod remains charged.

c) When a conducting metal sphere on an insulating stand is brought near, the conduction electrons are repelled to one side, leaving the sphere differently charged at the two sides; this separation of charges on a body by external electric forces is called *induction*. The rod and sphere now attract each other, because the rod is nearer to the unlike charges on the left part of the sphere than to the like charges at the right.

d) We touch the right side of the metal sphere with our finger or with a wire connected to a large conducting body such as the (moist) earth (*grounding*). At once almost all the "free" negative charges at the right flow off, still repelled by the charged rod and by one another, and eager to distribute themselves as far away from the rod and from one another as possible. The positive charges at the left are *bound* charges, that is, they cannot be neutralized by grounding at that side as long as the charged rod holds them in spell. (Explain why.)

e) Just withdrawing the finger or ground connections at the right leaves that side still neutral, the left still positively charged. If the rubber rod were now also withdrawn, the + charges, now free, would freely distribute themselves all over the whole metal sphere which is now *charged by induction*.

f) Touching the rod against the sphere at one point also changes practically nothing: none of the excess electrons on the nonconducting rod except those by chance on the point of contact can flow to annihilate the excess + charges on the sphere. Only if we twist and turn the rod, so that in time most of the rod has made contact with the sphere, is each + charge on the sphere able to pick up its neutralizing electron from the rod.

**Fig. 24.2**   Effect of induction on a nonconductor.

---

**Problem 24.3.** What would have happened if step (c) had been omitted? (Draw the sequence.) Step (f) would then be called *charging by contact*. How does the sign of this charge on the sphere differ in the two cases? What would happen if in step (f) the negative charge on the rod were very large compared with the positive charge on the sphere?

**Problem 24.4.** Repeat this series of pictures to illustrate the same sequence, starting, however, with a glass rod and a silk cloth.

**Problem 24.5.** A metal rod held by means of an insulating handle may develop negative charges when rubbed with cloth. Redraw the sequence in Fig. 24.1 using a metal rod instead of a nonconductor.

---

Another interesting problem is presented by the action of an uncharged *non*conductor in the presence of a nearby charged body. There is, of course, no flow of electrons on or in a true nonconductor, but we may imagine that the negative parts of each atom or molecule are pushed a little away from an external negative charge without being entirely severed from the positive parts; this corresponds to a distortion rather than a separation in the atoms or molecules of the material. Although such a small shift is negligible if it affects only a single atom, the total effect of billions of atoms is to *induce* charges, actually bound surface charges, on the two ends of the dielectric (see Fig. 24.2). As soon as the external charge is withdrawn, the induced surface charges usually also disappear; but until it is withdrawn the two bodies in Fig. 24.2 attract each other. We see therefore that a charged body attracts a previously neutral one, whether they are both conductors or not; and this explains the observation of the Greeks that a rubbed piece of amber attracts small pieces of lint.

---

**Problem 24.6.** Redraw the sequence in Fig. 24.1, using a *non*conducting sphere.

---

The exact degree of mobility of charges in different bodies depends a great deal on the experimental circumstances, and no really sharp division is possible. Cool, dry gas acts like a nonconductor, but very hot gases are good conductors. Perfectly pure water does not carry electric currents, but tap water containing certain impurities is a fairly good conductor and becomes an excellent one if a pinch of salt or any other substance (electrolyte) is added that separates in the solvent to form ions. The human skin is a fair conductor compared with clean glass, a fair nonconductor compared with metals.

The situation in conducting liquids (electrolytes) is more complex than in solids. The moving charges there are positive and negative ions produced when molecules of such substances as salts or acids become separated into oppositely charged parts on dissolving in the liquid. In heated or otherwise agitated gases, the currents consist of electrons as well as positive and negative ions.

## 24.6  THE ELECTROSCOPE

In the discussion centering around Fig. 24.1 you may well have asked, How do we know what the charges are at every step? We certainly cannot tell the sign or even presence of an electric charge by just looking at the object.*

One possible device for detecting the presence of a charge might be a tiny glass sphere at the end of

---

* Or by weighing, for the weight of the displaced electrons is inconceivably small compared with the bodies involved.

**Fig. 24.3**  The leaf electroscope.

a thin, long nonconductor. If we rub the sphere gently with silk we place on it a small positive charge. This sphere is then useful for exploring the neighborhood of another charged body: A strong repulsion indicates the presence of a like (positive) charge, and strong attraction the opposite.

However, we should soon find that we need a much better instrument—the leaf electroscope (Fig. 24.3). The metal rod $B$ which is separated from the protecting case $D$ by means of a good insulator $C$ is terminated at the top by a metal knob and at the bottom by very light leaves of gold or aluminum, $A$. If a charge (+ or −) is "placed" on the knob and allowed to distribute itself over the stem and leaves, then the charged leaves will fly apart by mutual repulsion. The same effect is seen if the neutral knob, instead of being charged

(a)        (b)        (c)        (d)        (e)

Ground

**Fig. 24.4**  Charging an electroscope by induction.

by contact, is simply being approached by another charged body; then, by induction, there is a separation of charges in the knob and one type of charge is "driven" to the leaves, the other remains on the knob.  Again the leaves deflect.  In general, the deflection or divergence of the leaves indicates a presence of charge in the leaves, and we assume that the divergence of the leaves, which is a measure of the repulsion between them, also is a qualitative measure of the *quantity of charge*, to be defined more rigorously at once.

---

**Problem 24.7.**  Figure 24.4 shows without text how an electroscope is charged *by induction*.  Supply text for each picture; then repeat this series of pictures for the case in which an initially positively charged rod is used.  [*Note:* Only the sign, not the quantity, of the *excess* charges is indicated.]

---

A good electroscope, once charged, holds its charge extremely well, as seen by the steady deflection of the leaves; of course, the insulation at $C$ must be carefully chosen, and all traces of conducting films of moisture or grease must be absent.  If, however, the air surrounding the knob is heated or is made conducting in some other way, the charge leaks off and the leaves collapse.  From the rate of collapse one may judge the strength of a source of x-rays or of radioactive emission, for both types of high-energy radiation *ionize* the gases through which they pass.  Hence electroscopes are important instruments in modern research, although the details of construction may differ widely from Fig. 24.3.  For example, one type of rugged electroscope, as small and slim as a fountain pen, is nowadays

carried in the pockets of many research workers who must expose themselves to nuclear radiations; the electroscope is charged at the beginning of the day, and afterward a periodic check of its deflection reveals whether the carrier has had more than the experimentally known safe dosage of high-energy radiation.

Electroscopes of some form have played prominent roles in some of the most important discoveries of modern physics.  One of the first clues of *cosmic rays* came when it was found that, in time, even the very best and the most carefully guarded electroscopes steadily lose some of their charge; the residual ionization of the air around the knob or leaves persisted even if the electroscope was shielded by thick layers of brick or metal.

---

**Problem 24.8.**  Explain the following simple phenomena or devices in terms of the discussion of electrostatics: Walking across a carpet is followed by a shock when one's hand touches a metal object.  Violent motion of moist air or raindrops is followed by lightning.  Gasoline trucks often drag chains, and cars are brushed by metal strips as they cross a toll bridge.

---

## 24.7 COULOMB'S LAW OF ELECTROSTATICS

Place some electric charges on two very small metal spheres; the relative amounts of charge, by measurement with the can arrangement, are $q_1$ and $q_2$ in magnitude (for example, $q_1 = -2$ units, $q_2 = +2$ units).  Both spheres—so small and so far apart that we may call them *point* charges—are suspended in a vacuum, and are so far away from other materials that the observed forces on the charged spheres are in effect only the electric force $F$ of $q_1$ on $q_2$ and the reaction of $q_2$ on $q_1$. *By experiment,* $F \propto 1/r^2$, where $r$ is the distance between the two point charges. Also, $F \propto q_1$ ($q_1$ may be increased to $-3$, $-4$, $-5$,... units by recharging and referring to the electroscope calibration); and $F \propto q_2$. In short, $F \propto q_1q_2/r^2$, quite analogous to the law of gravitation. Introducing a constant of proportionality, $k$, we can write:

$$F = k_e \frac{q_1q_2}{r^2}. \qquad (24.1)$$

The accurate measurements needed to establish Eq. (24.1) were published by Charles Augustin de Coulomb in France in 1785; this expression for the electric force between two point charges is therefore called *Coulomb's law,** and the unit of electric charge is called the *coulomb*. This unit is officially defined (by international agreement) in terms of the unit of electric current, the *ampere*: the coulomb is the amount of charge that flows past a point in a wire when the current is equal to one ampere. The ampere, in turn, is defined in terms of the force between two current-carrying wires. Since we cannot go into the details of electrical measurements here, we must simply state the result: The constant $k_e$ in Coulomb's law (Eq. 24.1) is very nearly equal to nine billion newton-meters squared per coulomb squared ($9 \times 10^9$ newton-m²/coul²).† This means that two objects, each with a *net* charge of one coulomb, separated by a distance of one meter, would exert forces on each other of nine billion newtons. As can be seen from this number, a coulomb is an enormous amount of charge; most of the familiar phenomena of static electricity involve charges of only about a millionth of a coulomb.

**Problem 24.9.** (a) How do we know when $q_1$ and $q_2$ are each 1 coul? (b) From accurate experiments it is known that each electron carries about $1.6 \times 10^{-19}$ coul of negative charge. The neutral hydrogen atom has only one such electron, about $0.5 \times 10^{-10}$ m from its positively charged central nucleus. What is the electric force holding the electron to the nucleus? (c) If the mass of an electron is approximately $10^{-30}$ kg and of a hydrogen nucleus is about $2 \times 10^{-27}$ kg, what is the magnitude of the gravitational force compared with the electrostatic force? (d) How many electrons make up $-1$ coul of charge? (e) How many electrons must be placed on each of two spheres 0.1 m apart to cause each sphere to exert a force of 0.001 newton on the other?

By experiment, it has also been found that if air instead of a vacuum is placed between $q_1$ and $q_2$, the force these charges experience is sensibly unchanged. However, if the space between and around the two charges is completely filled with another homogeneous isotropic dielectric such as paraffin or pure water, the force experienced by each charge is reduced by a factor of about $\frac{1}{2}$ and $\frac{1}{81}$, respectively. This is not because Coulomb's law is there invalid, but because in the presence of $q_1$ and $q_2$ the charges that are always present in the dielectric separate and in turn exert forces of their own on $q_1$ and $q_2$ that tend to counteract the direct effects of $q_1$ on $q_2$ and $q_2$ on $q_1$. However, this is a complex situation, and we shall usually have little occasion to talk about any cases except that of charges *in vacuo*, for which Eq. (24.1) describes accurately the total force actually observed.

**Problem 24.10.** Plot a graph of relative force on a point charge $q'$ vs. distance from a second charge $q$ if there are no other charges present.

---

* The accuracy of the statement $F \propto 1/r^2$ in Coulomb's law has since been confirmed indirectly to within 1 part in one billion.

† More precisely, $k_e = 10^{-7} c^2 = 8.9874 \times 10^9$ newton-m²/coul². The reason why $k_e$ is related to the speed of light $(c)$ will be explained in Section 25.3.

To find the electrostatic net force on a point charge $q$ in the neighborhood of a large charged body, we must add vectorially the several forces exerted individually on $q$ by each of the point charges distributed on the large body. Evidently this is tedious, if not impossible, in a general case with the mathematical methods at our command, but there are two special cases which come up so often in our later studies that we might now at least look at the obtainable results.

a) The net force $F$ on a single small point charge $q$ in air or vacuum *outside* a uniformly charged *spherical* body is given by $F = k_e q \cdot Q / r^2$, where $Q$ is the net charge on that spherical body, and $r$ is the distance between $q$ and the center of the sphere. Note the similarity between this conclusion and Newton's results for the gravitational attraction exerted by a spherical mass on a small point mass. There, too, we heard that the effect of a (homogeneous) spherical body is the same as if the bulk of the sphere were shrunk to a point at its center.

b) The net force on a single small point charge is constant in direction and magnitude anywhere inside the air gap between two oppositely charged large parallel *plates*, an arrangement called a condenser or capacitor. (Note that the force on a charge is *not* constant along the line between two opposite *point* charges; cf. Problem 24.10.)

---

**Problem 24.11.** Compare Newton's law of universal gravitation as applied to two small point masses, and Coulomb's law of electrostatics applied to two point charges; consider the similarities and outstanding differences. What conclusions do you draw from the repeated appearance of the inverse-square law in the various branches of physics? (Do not assume that a dogmatic answer to this question exists.)

---

## 24.8  THE ELECTROSTATIC FIELD

If we attach a small body to the end of a thin non-conducting stick and give it a positive charge, say +1 coulomb, the body (a test point charge) will experience a force $F$ when placed anywhere in the vicinity of other electric charges. We may use this probe to explore every point in space, and to find the electric force (in magnitude and direction) at every point. We could populate this whole space mentally with arrows to a given scale to show the situation at every point at a glance. In Fig. 24.5 a few such electric force vectors have been represented in the neighborhood of simple charge distributions, namely (a) around a positively charged sphere, (b) around two oppositely charged spheres. Obviously the force vectors become negligibly small far away from these charges, but even in their immediate neighborhood we realize that there must be an infinity of such vectors, one for each point of space. It is inconvenient that these force vectors depend on two factors, i.e., on both the charge $Q$ that sets up the action and the charge $q$ that is acted upon. Therefore another very useful concept has been defined which, by a standardization of the test charge used to explore the action of $Q$, is independent of $q$. This is called the *electric intensity* at $P$ or *electric field strength* at $P$, and is directly defined as the electrostatic force per positive unit charge placed at that point $P$. The letter $E$ is usually assigned to this concept. By definition,

$$E \text{ at } P = \frac{\text{net force } \boldsymbol{F} \text{ on positive charge } q \text{ at } P}{q},$$

(24.2)

usually expressed in units of newtons/coul. The arrows in Fig. 24.5, if they referred to forces *per unit charge*, now represent directly the vector $\boldsymbol{E}$ at each selected point. *The totality of such E-vectors in a given region we call "the electrostatic field" in that region.*

This is a loose definition of one of the most fruitful concepts of physics (gravitational field, magnetic field, etc., are similarly conceived).

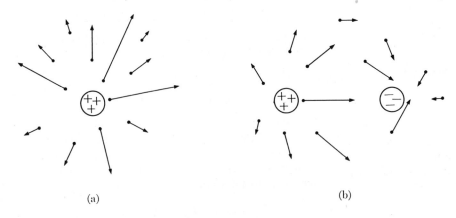

**Fig. 24.5** Force vectors on a small positive test charge
($q$ coulomb) at selected points near fixed charges.

Note three important characteristics of the
concept "electric intensity."

1) The *magnitude* of $E$ at a point $P$ may be determined by measuring the force $F$ on a test
charge of, say, +15 coul placed at $P$, and
dividing this force $F$ by 15 coul, since $E$ is the
*ratio* of force by charge. If 30 coul had been
chosen for the test charge, $F$ would have been
twice as large, but the ratio $E$ at that point
comes to the same value no matter how large
the test charge.* The magnitude of $E$ is determined by the charges that *set up* the field,
and not by the small test charges used to explore
its strength. In fact, that is the main usefulness
of the concept.

2) The *direction* of $E$ at $P$, by convention, is the
same as the direction of the force on a *positive*
test charge at $P$. We might equally well have
chosen the opposite convention, but once we
agree on the present one we must stick to it
consistently.

3) The *units* of $E$ are the units of force/charge,
usually newtons/coulomb.

**Example 1.** (a) The force $F_P$ on a +4 coulomb
test charge at a point $P$, 10 cm from a small charged
sphere, is 12 newtons toward the sphere, by test.
Find the electric intensity $E_P$ at $P$.

*Solution.* $E_P = 12$ newtons/(+4 coul) = 3 new-
tons/coul (toward sphere).

b) Now find the total charge $Q$ on the sphere.

*Solution.*

$$F_P = 12 \text{ newtons} = \frac{k_e Q \cdot (+4 \text{ coul})}{100 \text{ cm}^2};$$

therefore

$$Q = k_e \left( \frac{12}{4} \times 100 \right) \text{ coul},$$

and $Q$ is *negative* since it attracted a + test charge.
The defining equation

$$E = \frac{\text{force on positive test charge}}{\text{magnitude of test charge}}$$

---

*    But the test charge still must not be so large that it
induces perceptible charge rearrangements in its
neighborhood.

Figure 24.6

becomes particularly simple if the field-producing charge $Q$ is a point charge at a distance $r$ from the location of test charge $q$, or if $Q$ is uniformly distributed on any sphere whose center is $r$ cm from $q$. In either case (see Fig. 24.6),

$$E_P = \frac{F_P}{q} = \frac{\left(k_e \frac{Q \cdot q}{r^2}\right)}{q} = k_e \frac{Q}{r^2}. \qquad (24.3)$$

But for all other cases, the value of $E$ at point $P$ must be found by experiment from the defining equation (force/charge), or else by adding vectorially the contributions to $E$ at $P$ of each point charge $Q$ at its particular distance $r$, each contribution calculated by Eq. (24.3). We have here neither the tools nor the need for carrying out this mathematically advanced exercise.

---

**Problem 24.12.** At a distance of 1 m from the center of a uniformly charged large sphere the electric intensity is 4 newtons/coul. What is the force on a point charge of $10^{-9}$ coul at that distance from the sphere, and what is the total charge in coulombs on the sphere?

---

The field concept can be used to describe other kinds of forces as well. The simplest case is the gravitational field, for which (by analogy with Eq. 24.2) we may define a gravitational field strength as follows:

gravitational field strength $f_G$ at a point $P$ =

$$\frac{\text{gravitational force } F_G \text{ on mass } m \text{ at } P}{m}.$$

Thus the magnitude of the gravitational field strength at a distance $R$ from a point mass $m_2$ would be, according to Eq. (11.6),

$$f_G = G \frac{m_2}{R^2}.$$

Conversely, the gravitational force $F_G$ acting on a mass $m_1$ at point $P$ can be expressed in terms of $m_1$ and the value of the field strength at $P$, without reference to the mass or masses which may have created that field in the first place:

$$F_G = m_1 f_G.$$

Note that $f_G$ is in every sense equal to the concept which we previously identified as the gravitational acceleration $g$.

## 24.9  LINES OF FORCE

To visualize the electric field around a charged body we must try mentally to supply each and every point in space with an arrow indicating the force per unit charge at that point; this is obviously a theoretically feasible but practically impossible task. As an aid to our limited powers of imagination we may picture the field in another way.

We know that if we bring a small positive test charge $q$ into an electric field created by some other charges, $q$ will experience repulsive forces from + charges, attractive forces to − charges. Figure 24.7 shows the paths of a small positive test charge $q$ *released* at various points in the neighborhood of some other charges. All forces except electric ones are neglected, and the inertia of the body carrying $q$ is also negligible. Therefore the direction of travel (indicated by the arrow on each line) follows the direction of the *force* acting on $q$ at every point along the path, and we may call these lines the *lines of force*. At every point along each line, the force acting on $q$ is tangential to the line, but the line only tells us the direction of the force vector on the positive charge, not its magnitude.

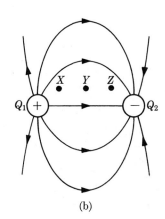

(a)                                    (b)

**Fig. 24.7** Lines of force (a) in the neighborhood of a single positive charge, (b) around equally large point charges of opposite sign.

**Problem 24.13.** Invent a modification of this idea to present at one glance both direction and magnitude at every point along the path.

Clearly it is possible in every case to draw infinitely many different lines of force around a charged body. Every point in space lies on a line of its own; for example, you can readily imagine (and draw now) a curve from $Q_1$ to $Q_2$ through point $X$ in Fig. 24.7(b). But even without filling in all possible lines, we get a good feeling for the state of affairs by looking at only a few such lines. As Fig. 24.7(a) informs us, a crowding of the lines of force means numerically a relatively stronger field (near the charged body), a larger spread means numerically a relatively smaller field strength $E$. Thus point $X$ is in a stronger field in Fig. 24.7(b) than point $Y$; and in this direct relation, so pictorial and helpful to our imagination, lies much of the usefulness of this concept of lines of force.

**Problem 24.14.** (a) Note that a line of force, starting at a positive charge, either goes to infinity or terminates on a nearby negative charge. Why?

(b) Can lines of force intersect each other? (c) Draw some lines of force (i) around two positive charges placed like $Q_1$ and $Q_2$ in Fig. 24.7(b), (ii) between a parallel-plate condenser charged oppositely on the two plates. (d) Is the electric field around a charged body *real*? Are the lines of force real?

**Problem 24.15.** It is a fact that if a conductor carries electric charges which have found their *equilibrium* position (steady state), the lines of force join the conductor perpendicularly at every point on the surface. Explain why. [*Hint:* Consider what would happen to the charges on the surface if the lines of force were oblique instead of normal.]

## 24.10 ELECTRIC POTENTIAL DIFFERENCE— QUALITATIVE DISCUSSION

We turn now to another concept which is of fundamental importance in electricity—electric potential difference, for which the symbol $V$ will be used.

The word *potential* reminds us at once of potential energy, defined in mechanics. In fact, the new concept also refers to work done by virtue of a change of position (of charged bodies), although it is now work done by or against electric forces rather than gravitational or elastic ones. But the analogy is quite deceptive in many respects and is better not invoked at all. For a definition we rather choose the following.

The potential difference $V_{AB}$ between any two points $A$ and $B$ is the ratio obtained when the work done in moving a positive electric charge $q$ from $B$ to $A$ is divided by the magnitude of the charge $q$; $V_{AB}$ = work done per unit charge moved, in joules/coul. (By definition, 1 joule/coul = 1 volt.)

For example, refer again to Fig. 24.7(b). When we attempt to carry a positive charge $q$ from $Y$ to $X$, our muscles must do work against the repulsion from $Q_1$ on the left and the attraction to $Q_2$ on the right. The net work done per unit charge can be found experimentally (suggest a way), and turns out to be independent of the exact path chosen between $Y$ and $X$, just as the net work done in raising a weight depends on the total displacement rather than on the exact path taken.

By convention, we say that $X$ is at a more positive potential than $Y$ or $Z$ because our muscles or some other nonelectric forces must do work on a positive charge to carry it from $Y$ or $Z$ to $X$. On the other hand, $Z$ is at a more negative potential with respect to $Y$ or $X$ because on carrying a positive charge to $Z$ from $Y$ or $X$ all the work is done by the electric field instead of our muscles. In brief, the relative positive and negative potential at a point is recognized by the behavior of a positive charge which, "on its own," tends to go *away* from a point at relatively *positive* potential and toward one at relatively *negative* potential.

While in this manner the words "potential difference between $A$ and $B$" are easily given operational significance, the frequently encountered phrase "the potential at point $A$" is not so clear, because no mention is made in it of the starting or reference point. *By convention,* the *implied* reference point is some infinitely distant uncharged body or, in practice, for example, the surface of the earth (*electrical ground*) at some distance below point $A$. Therefore if we hear "the potential is

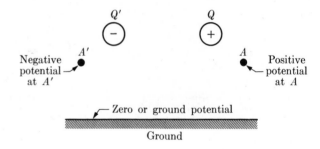

**Figure 24.8**

positive at point $A$," we know that a positive test charge brought to point $A$ will experience a net force of repulsion and will try to escape to ground if a conductor is stretched from point $A$ to ground (see Fig. 24.8). Similarly, a point $A'$ is at a negative potential if a positive charge is readily attracted from ground toward that point $A'$. The numerical value of the potential at a point $A$ (no second reference point given) is then simply the potential difference between $A$ and any point on the ground, or the work done per unit charge on taking a positive charge from ground to point $A$. For all our purposes, then, we may always replace the words "potential at a point" by "potential difference between that point and ground."

---

**Problem 24.16.** What is the operational meaning of the phrase "a point $P$ has a negative potential of 1 joule/coul"?

**Problem 24.17.** Decide whether the potential is $+$, $-$, or zero at points $X$, $Y$, and $Z$ in Fig. 24.7(b), providing $Q_1 = Q_2$ numerically but is of opposite sign.

**Problem 24.18.** At electrostatic equilibrium (no current), the potential difference between any two points on the surface of a conductor is zero. Explain why.

**Problem 24.19.** In Fig. 24.8 there will also be points between $Q$ and $Q'$ where the potential is zero. Indicate a few. [*Hint:* The potential at any point $P$ with respect to ground is zero if no work is done moving a charge from ground to $P$. Furthermore, no work is done moving always perpendicular to lines of force; why?]

## 24.11  POTENTIAL DIFFERENCE— QUANTITATIVE DISCUSSION

So far we have developed only the qualitative and the experimental aspects of the concept of potential difference; now we ask the inevitable question, How can one *calculate* and predict the potential difference between two points?

Consider the simplest possible case, the potential difference $V_{AB}$ between points $A$ and $B$, in the vicinity of a charged sphere $Q$ and away from all other charges (Fig. 24.9). Let us assume the charge $Q$ is positive. Point $A$ is therefore at a higher positive potential than $B$, since one has to supply work to transport a small positive charge $q$ from $B$ to $A$ against the repulsion of $Q$. $V_{AB}$ is this work per unit charge; and if the force on $q$ between $B$ and $A$ were constant (say $F$) the work in this simple case would be $F \cdot (r_B - r_A)$, $V_{AB}$ being consequently $F(r_B - r_A)/q$. But the force on $q$ is not at all constant during its transit from $B$ to $A$. In vacuum or air the force changes from $k_e Qq/r_B^2$ at $B$ to $k_e Qq/r_A^2$ at $A$, and not linearly either (where $k_e$ is the electrostatic force constant). Frankly, we do not know how to calculate the exact work done under such circumstances without calculus, but we can make an instructive approximate calculation of $V_{AB}$ which will point to the simple result

$$V_{AB} = k_e \frac{Q}{r_A} - k_e \frac{Q}{r_B}. \qquad (24.4)$$

*Derivation.* To find an approximate value for the work done on $q$ from $B$ to $A$, we mentally subdivide the path $BA$ into very small parts, each of

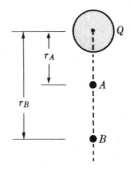

**Fig. 24.9** The difference in potential between points $A$ and $B$ near a uniformly charged sphere $Q$ is $V_{AB} = k_e Q/r_A - k_e Q/r_B$.

length $\Delta s$. The force does not change much during a small displacement $\Delta s$. For example, on going from $B$ at distance $r_B$ to a near point at distance $r_1$ from $Q$ (whence $\Delta s = r_B - r_1$), the average force is nearly $k_e Qq/r_B^2$ or also $k_e Qq/r_1^2$ or $k_e Qq/r_B r_1$; the smaller $\Delta s$, the more nearly true this is. But the work $\Delta W_1$ done during this small displacement $\Delta s$ is simply the average force × displacement (forces and displacements being here parallel):

$$\Delta W_1 = \frac{k_e Qq}{r_B r_1} \Delta s = \frac{k_e Qq}{r_B r_1} (r_B - r_1)$$

$$= k_e Qq \left( \frac{1}{r_1} - \frac{1}{r_B} \right).$$

On proceeding toward $A$ over the next small interval to a near point at distance $r_2$ from $Q$, that is, through a distance $(r_1 - r_2)$, the work done is

$$\Delta W_2 = k_e Qq \left( \frac{1}{r_2} - \frac{1}{r_1} \right),$$

similarly obtained. The total work on $q$ from $B$ to $A$, the sum of such contributions for $n$ small intervals, is

$$W_{\text{total}} = \Delta W_1 + \Delta W_2 + \cdots + \Delta W_n$$

$$= k_e Qq \left(\frac{1}{r_1} - \frac{1}{r_B}\right) + k_e Qq \left(\frac{1}{r_2} - \frac{1}{r_1}\right) + \cdots$$

$$+ k_e Qq \left(\frac{1}{r_A} - \frac{1}{r_n}\right)$$

$$= k_e Qq \left(\frac{1}{r_1} - \frac{1}{r_B} + \frac{1}{r_2} - \frac{1}{r_1} + \cdots + \frac{1}{r_A} - \frac{1}{r_n}\right)$$

$$= k_e Qq \left(\frac{1}{r_A} - \frac{1}{r_B}\right),$$

all other terms canceling. By definition,

$$V_{AB} \equiv \frac{W_{\text{total}}}{q} = \frac{k_e Q}{r_A} - \frac{k_e Q}{r_B}; \qquad \text{Q.E.D.}$$

This last result applies only to the region (in air or vacuum) around a single point charge $Q$, or to a single sphere on which $Q$ charges are uniformly distributed. An important consequence of Eq. (24.4) becomes apparent if we let $r_B$ become very large. As $r_B \to \infty$, $V_{AB}$ now means the potential difference of point $A$ with respect to an infinitely distant uncharged body (e.g., to ground); $V_{AB}$ is now "the potential at a point $A$ which is $r_A$ m from a single point charge $Q$ or a uniformly charged sphere." In this case, since $1/r_B = 1/\infty = 0$, let us write simply $V_A$ instead of $V_{AB}$, and we have

$$V_A = \frac{k_e Q}{r_A}. \qquad (24.5)$$

---

**Problem 24.20.** In Problem 24.9 we considered the force at $0.5 \times 10^{-10}$ m from a nucleus of a hydrogen atom. Now find the potential at that distance. Also find the potential difference between two points $0.5 \times 10^{-10}$ m and $2 \times 10^{-10}$ m from the nucleus, respectively.

**Problem 24.21.** Find the potential at a point halfway between $Q_1$ and $Q_2$ in Fig. 24.7(b) if $Q_1 = +3$ coul, $Q_2 = -5$ coul, and the distance between them is 10 cm.

## 24.12 USES OF THE CONCEPT OF POTENTIAL

There are several interesting features of Eq. (24.5).

a) If the sign of the charge $Q$ is used along with its numerical value, $V_A$ automatically appears with its correct $+$ or $-$ sign.

b) Since the units of $k_e$ are newton-m²/coul², the units of $V_A$ are newton-m/coul; this is consistent with our previous statement that potential difference has units of joules/coul, since 1 joule = 1 newton-m.

c) Like the concept of work itself, potential is a scalar; therefore simple algebraic addition gives the total potential at any point in the vicinity of more than a single point charge. Hence a point equally far from equally strong opposite charges, like point $Y$ in Fig. 24.7(b), is at zero potential with respect to ground (but the *force* is *not* zero there! Is this a contradiction?).

d) In our derivation there was no restriction on how close to the charged sphere the point $A$ may be for the equation $V_A = k_e Q/r_A$ to hold. Actually, point $A$ may even be right on the surface of the sphere in which case the potential of the sphere, $V_s$, is the ratio of its charge $Q$ to its radius $r_s$, i.e., $V_s = k_e Q/r_s$. Thus 1 coulomb of charge deposited on a 2-cm sphere ($r_s = 1$ cm) raises the sphere to $(9 \times 10^9 \text{ newton-m}^2/\text{coul}^2)(1 \text{ coul})/(0.01 \text{ m}) = 9 \times 10^{11}$ joule/coul $= 9 \times 10^{11}$ volts potential. It is clear that relatively small bodies can easily achieve large potentials if supplied with enough charge; a rubber or glass rod can quickly be given enough charge by friction to develop several thousand volts on a dry day.

The usefulness of the concept *electric potential* may be further illustrated by two other instances:

e) We can now quickly calculate the potential energy of a point charge $q$ at distance $r_A$ from a point charge $Q$. The potential at that distance is $V_A$; by definition this is the potential energy of a charged body *per unit charge*. Therefore the total potential energy of $q$ at $A$ is the product $V_A \times q$.

**Fig. 24.10** The electric force $F_E$ on a small charged body $q$ between oppositely charged plates.

If, for example, $Q$ refers to the nucleus of a hydrogen atom and $q$ refers to its electron at $r_A = 0.5 \times 10^{-10}$ m, the potential energy of the electron is

$$PE = \frac{k_e Q}{r_A} \times q$$

$$= (9 \times 10^9 \text{ newton-m}^2/\text{coul}^2)(1.6 \times 10^{-19} \text{ coul})$$
$$\times (-1.6 \times 10^{-19} \text{ coul})/0.5 \times 10^{-10} \text{ m}$$

$$= -4.6 \times 10^{-18} \text{ joules.}$$

(What is the meaning of the negative sign?)

f) We have found before that the electric force $F_E$ on a charge between two equally and oppositely charged plates is constant. As a consequence, the electric intensity $E$ in such a gap is also constant. The work that must be done *per unit charge* to transport a small positive test charge $q$ from the negative plate a distance $s$ straight across the gap to the positive plate (Fig. 24.10) is, by definition, $V$ (its value is read off most conveniently from the label of the battery that might be used to provide the potential difference). In this situation, the work that must be done on $q$ against the electric forces is work per unit charge $\times$ total charge, or

$$F_E \times s = V \times q.$$

Or transposing, and noting that $F_E/q = E$,

$$E = V/s. \qquad (24.6)$$

If $V$ is in joules/coul ($= $ volts) and $s$ is in meters, $E$ is in newtons/coul $=$ volts/meter.

The importance of Eq. (24.6) is that since $s$ is easily measured and $V$ is known from the battery used, $E$ can be readily calculated at all points of the space between the plates.

**Problem 24.22.** Two parallel horizontal plates are 2 cm apart in vacuum, and are connected to a supply providing a 10,000-volt potential difference. An electrically charged oil drop, which from other measurements appears to have a mass of $10^{-4}$ g, is seen to be suspended motionlessly between the plates. What is the electric force on that oil drop? What is its charge $q$? What charge must exist on the drop to account for its vertical motion in the electric field with an acceleration of 2 cm/sec² up? 10 cm/sec² down?

**Problem 24.23.** A Van de Graaff generator is essentially a metallic sphere on an insulating stand, charged to a high electric potential, which is used to repel and accelerate charged atomic particles along an evacuated tube to reach extremely high speeds; when they crash into a "target" they will have an energy suitable for "atom smashing." If we wish to use in this accelerator a sphere 10 ft in diameter at 2,000,000 volts, how much charge must be supplied? (Watch the units.) What might be the practical limits to reaching much higher potentials with such a device?

**Problem 24.24.** If an electron is repelled from a point near the surface of this same Van de Graaff generator and is allowed to accelerate freely, what will be its final kinetic energy as it hits a target far from the generator? What will be its final speed? (Mass of electron $= 9.11 \times 10^{-28}$ g, charge $= -1.6 \times 10^{-19}$ coul.) What was the potential energy of the electron just as it began to leave the surface of the sphere?

**Problem 24.25.** How much kinetic energy (in joules) does an electron gain when it freely passes through a potential difference of 1 volt? [This quantity of energy has been given the unfortunate name *1 electron volt* (1 eV), an important unit of energy in atomic physics.]

## 24.13  ELECTROCHEMISTRY

The history of electricity shows that the fundamental notions of charge, conductor, insulator,
induction, and the like were acquired only slowly
and sometimes after lengthy struggles. It was not
until about 1800, the year of Alessandro Volta's
discovery of the voltaic cell, that the subject of
electricity really opened itself to widespread and
fruitful investigations touching on the structure of
matter. One of the first consequences was the discovery of the spectacular decomposing effect of
current on certain liquids. Electric current proved
to be a tool capable of breaking up into their
elements dissolved or molten materials whose molecules previously had resisted separation by chemical or physical means. It was by this process,
called electrolysis, that Davy in 1807 broke down
potash and soda, until then considered to be
elements, and obtained from them the metals
potassium and sodium.

While the study of electric currents led physicists to the subject of electromagnetism, which is
part of our next consideration, further research on
electrolysis itself led chemists to the atomic hypothesis, by another door than Dalton's. In a systemati-
zation and extension of Davy's methods, Michael
Faraday proved that the passage of current through
a liquid (e.g., a diluted acid or a salt solution) would
dissociate the liquid at the two electrodes in a
regular and accurately repeatable manner. The
mass of an element liberated from the liquid, either
in the form of gas bubbles (as for hydrogen or
chlorine) or as metal deposited at or near the electrode (e.g., copper), is proportional both to the
molecular weight of the material freed and to the
quantity of charge transferred, but inversely proportional to the valence of the material liberated.
For example, charge passing through a molten
sample of common salt (NaCl) caused the sodium
metal to collect at the negative electrode (the
*cathode*, Fig. 24.11), and liberated chlorine gas
bubbles at the positive electrode (*anode*). The mass
of metallic sodium is found to be about $24 \times 10^{-5}$ g

**Fig. 24.11** Migration of ions to electrodes during electrolysis.

per coulomb of charge so transferred, and the mass
of chlorine gas about $12 \times 10^{-14}$ g. It follows that
in order to obtain 1 gram-atomic weight of Na
(namely, 23 g), one needs 23 g/($24 \times 10^{-5}$ g/coul) or
96,500 coul; at the same time, the transfer of this
quantity of charge (occasioned, say, by a 26.8-
ampere current lasting one full hour) also releases
1 gram-atomic weight of chlorine (35.5 g) at the
other electrode. The transfer of twice that charge
yields twice the respective amounts of these mono-
valent substances. If we now electrolyze water,
96,500 coul will release 1 g of hydrogen but only 8 g
of oxygen, for oxygen has the valence 2, and the
experimental law of electrolysis teaches that a
quantity of charge capable of releasing 1 gram-
atomic weight of a *mono*valent element will produce only $\frac{1}{2}$ gram-atomic weight of a *di*valent element. In equation form, these experimental facts
(essentially Faraday's law of electrolysis in modern
terms) can be summarized as follows:

Mass of material liberated at one electrode (g)

$$= \frac{\text{charge passed (coul)}}{96,500 \text{ (coul per g-at. wt)}} \times \frac{\text{at. wt of material}}{\text{valence of material}}.$$

(24.6)

Incidentally, the recurring quantity of 96,500
coulombs is given the convenient name *one faraday*
of electric charge, so that we may say simply: in
electrolysis, 1 faraday releases 1 gram-atomic
weight of any monovalent material, $\frac{1}{2}$ gram-atomic
weight of divalent materials, etc., or, in terms of
concrete examples, 1.008 g hydrogen, 31.78 g copper,
40.59 g antimony.

**Problem 24.26.** The element barium (Ba, at. wt 137.36, valence 2) was also discovered by Davy by electrolysis. If you repeat an experiment of this kind and for one hour pass a current of one ampere through molten barium chloride ($BaCl_2$), how much barium and how much chlorine do you expect to collect? [*Note:* 1 ampere = 1 coulomb/sec.]

## 24.14  ATOMICITY OF CHARGE

Now that the hard-won concepts of atom and electron are so familiar to physical scientists, it is easy to visualize a mechanism that will explain the main phenomena of electrolysis. In such simple cases as the breakdown of molten sodium chloride one can think of the liquid as made of separate atoms of sodium and chlorine. But since Na migrates to the cathode and Cl to the anode, we must account for the influence of electric forces on the atoms; each Na atom after the breakup of the molecule must somehow be positively charged, and each Cl atom must be negatively charged. The passage of electricity through the liquid is then accomplished by the movement of *charged* atoms, or *ions* (from the Greek word for *to wander, to go*).

At this point enters a theoretical picture of the molecule: we are led to the assumption that both Na and Cl, in the pure state of the element sodium and the element chlorine, are ordinary electrically neutral atoms by virtue of a perfect balance between the positive and negative components that make up the atom. Specifically, the negatively charged parts of the atom are the electrons, each having the same definite mass $m_e$ and charge $e$. Now when one neutral atom of Na and one of Cl together form one molecule of NaCl in the solid salt, one of the electrons from the sodium atom attaches itself to the chlorine atom (later we must explain why this should happen), so that on separation during electrolysis the Na ion lacks one electron and the Cl ion has one electron more than it does when in its neutral state. These ions can be symbolized by $Na^+$

and $Cl^-$. Similarly, a molecule of copper sulphate, $CuSO_4$, splits during electrolytic dissociation into $Cu^{++}$ and $SO_4^{--}$ ions by the transfer of two electrons.

An explanation of Faraday's law of electrolysis as summarized by Eq. (24.6) suggests itself. Every time one $Cl^-$ ion reaches the anode, it gives up its excess electron to that electrode, becomes neutral, and rises eventually in a bubble with other such chlorine atoms. At about the same time, an $Na^+$ ion picks up one of the electrons at the cathode and settles out as neutral sodium metal; we may imagine that this neutralizing electron was made available when the $Cl^-$ ion gave up its excess electron to the circuit on the other electrode. Thus for each molecule of NaCl electrolyzed into its elements, the charge corresponding to *one* electron moves through the external circuit (wires, battery, and current meters).

This hypothesis leads at once to an estimate of the value of the charge $e$ on one electron: We know that the passage of 96,500 coul of charge results in the electrolysis of one mole or $6.02 \times 10^{23}$ ($N$) atoms of a monovalent element. At the rate of transfer of one electron per atom, we have

96,500 coul/mole = ($N$, the number of atoms/mole)

$\times$ ($e$, charge in coul/electron)

or

$$e = 96,500/N, \qquad (24.7)$$

from which we can calculate that the charge per electron is $1.6 \times 10^{-19}$ coul. The confirmation of this value for $e$ in several independent and entirely different experiments (e.g., Millikan's oil-drop experiment, spectroscopic data) gives us confidence that the whole physical picture leading to this point is plausible. Or conversely, Eq. (24.7) together with separate determination of $e$ enables us to compute a value for Avogadro's number $N$, a value that is indeed in agreement with the results of other experiments.

This concordance of numerical results obtained by widely different methods shows us again the interconnectedness of physical science. From isolated observable facts and limited hypotheses we fashion, as it were, a net or web, and the constants of nature—the constant of universal gravitation, the values of $N$ and $e$, the atomic weights, and the like—are at the points of intersection of different strands, the knots that hold together the whole structure.

The two ideas of importance introduced here are, first, the atomicity of matter and, second, the "atomicity" of electric charges (i.e., charges in liquids are transferred by ions bearing $1, 2, 3, \ldots$ electrons more or less than when the atoms are neutral). A third idea is also implied, namely, that the electron is ordinarily a part of the atom. But all three postulates were far from evident when Faraday phrased his generalizations on electrolysis in 1833. Atomicity of matter soon became acceptable on other grounds (through chemistry and kinetic theory), but the existence of discrete charges called electrons was first proposed in definite form only in 1874 by G. Johnstone Stoney, and was not completely acceptable for more than 20 years more, when precise measurements of the mass and charge on the electron became available through the study of beams of electrons in cathode-ray tubes. The part that electrons play in atoms was then the next major puzzle; its solution, which is the substance of Chapters 28 and 29, formed one of the most exciting portions in the recent history of physical science.

## RECOMMENDED FOR FURTHER READING

D. L. Anderson, *The Discovery of the Electron*, Princeton, N.J.: D. Van Nostrand, 1964; Momentum Book paperback

D. L. Hurd and J. J. Kipling, *The Origins and Growth of Physical Science*, Vol. 2, pages 191–216, extracts from the writings of Franklin, Coulomb, and Volta; pages 275–285, Faraday on electrolysis.

D. Roller and D. H. D. Roller, "The development of the concept of electric charge: electricity from the Greeks to Coulomb," in *Harvard Case Histories in Experimental Science* (edited by J. B. Conant), Cambridge, Mass.: Harvard University Press, 1948, pages 541–639; also available as a separate pamphlet

## SOURCES, INTERPRETATIONS, AND REFERENCE WORKS

I. B. Cohen, "Conservation and the concept of electric charge: an aspect of philosophy in relation to physics in the nineteenth century," in *Critical Problems in the History of Science* (edited by M. Clagett), Madison, Wisconsin: University of Wisconsin Press, 1959

I. B. Cohen, *Franklin and Newton*, Philadelphia: American Philosophical Society, 1956

Benjamin Franklin, *Experiments and Observations on Electricity* (edited by I. B. Cohen), Cambridge, Mass.: Harvard University Press, 1941

C. S. Gillmor, *Coulomb and the Evolution of Physics and Engineering in Eighteenth-Century France.* Princeton, N.J.: Princeton University Press, 1971

Mary Hesse, *Forces and Fields*, Chapter VIII

W. J. King, "The quantification of the concepts of electric charge and electric current," *The Natural Philosopher*, Vol. 2, pages 107–127 (1963)

Joseph Priestley, *The History and Present State of Electricity, with Original Experiments.* New York: Johnson Reprint Corporation, 1966

W. C. Walker, "The detection and estimation of electric charges in the eighteenth century," *Annals of Science*, Vol. 1, pages 66–100 (1936)

E. T. Whittaker, *A History of the Theories of Aether and Electricity*, Vol. I, Chapters II and III, New York: Humanities Press, 1961

# Chapter 25

# ELECTROMAGNETIC FIELD THEORY

## 25.1 INTRODUCTION

The two previous chapters have been devoted to subjects that so far may appear to be entirely separate: light and electricity. We must now introduce yet a third apparently distinct topic: magnetism. By exploring the relations between electricity and magnetism, we shall arrive at a general theory (Maxwell's theory of electromagnetic fields) which turns out, perhaps unexpectedly, to provide an explanation for the propagation of light. Just as the concept of *energy* provided a unifying link for mechanical and thermal phenomena (Chapter 17), the concept of *field* brought electricity, magnetism, gravity, and light into a common framework of physical theory. And the two concepts themselves are closely related, for, as we shall see, a field can be regarded as a form of energy in space. If it could be established that "space" really contained (consisted of) an ether having mechanical properties (so that all energy is really mechanical energy, traveling in a medium, the ether, as Maxwell and other nineteenth-century physicists believed), then this finding would have to be declared the ultimate triumph of Newtonian mechanics, and the evolution of concepts and theories in physical science would have reached the goal of a complete mechanical explanation of nature. That indeed was the judgment of a few physicists at the end of

the nineteenth century; it led, for example, to A. A. Michelson's famous statement in 1894 that there were no more fundamental discoveries left to be made, but only increasingly refined determinations of the physical constants to six or seven decimal places.* But not everyone was so sanguine at that time; other leading scientists like Lord Kelvin perceived "clouds over the dynamical theory of heat and light" which foreshadowed the coming revolution in physical theory that will be the subject of the last chapters of this book.

## 25.2 CURRENTS AND MAGNETS

For most phenomena of electrostatics described in the preceding chapter there exist similar counterpart phenomena in *magnetostatics*, the principal difference being that magnetic poles, unlike electric charges, are always found in *pairs*, called north and south poles. (Single poles or *monopoles* might exist, but at the present time search for them has been fruitless.) The obvious similarity between electricity

---

\*    The circumstances and significance of this statement have been discussed in several letters to the editor in *Physics Today*, April 1968, page 56; August 1968, page 9; January 1969, page 9.

and magnetism suggested long ago that there should be some interaction between electric charges and magnetic poles, but no such interaction could be demonstrated experimentally until the early part of the nineteenth century.

The first concrete evidence of an interaction between electricity and magnetism came in 1820, when the Danish physicist Hans Christian Oersted (1777–1851) performed a series of experiments of momentous consequences. The original inspiration for these experiments, according to Oersted's own account, was the metaphysical conviction of the unity of all forces in nature, which he derived from the German "nature philosophers," in particular Friedrich Schelling. Although nature philosophy was generally characterized by mystical, anti-mathematical speculation of a type which has usually been of little benefit to science, in this case it did encourage the trend toward unification which had some outstanding successes in the nineteenth century. As we insisted in Section 12.3, there is no single method which all scientists can be expected to follow; the route to some of the greatest discoveries in science has been amazingly unorthodox.

In his famous experiment, Oersted placed a magnetic compass needle directly beneath a long horizontal, electrically conducting wire. He had placed the wire along the earth's magnetic north-south line, so that the magnetic needle was aligned parallel to the wire. When the wire was connected to the terminals of a battery, the compass needle swung to an east-west orientation—perpendicular to the wire! Although it had been shown much earlier that electric charge at rest does not affect a magnet, now it was clear that charge in motion (a current) does exert a strange kind of "sideways" force on a magnet placed near the conducting wire.

Oersted's results were the first instance in which a force was observed that did not act along a line connecting the sources of the force. The force that the current-carrying wire exerts on a magnetic pole is not along the line from the wire to the pole: The force on the pole is *perpendicular* to any such line. The magnetic needle is not attracted to, or repelled by, the current; it is *twisted* sidewise by forces acting on its poles.

The unusual way in which a compass needle is affected by an electric current helps to explain why it took so long before anyone found this interaction between electricity and magnetism. To begin with, there is no effect between static electric charges and magnetic poles. Also, steady electric currents had not been readily available for laboratory experiments until Volta developed the battery in 1800. But even if one has currents and compass needles available, one does not observe the effect unless the compass is placed in the right position so that the needle can respond to a force that seems to act in a direction *around* the current rather than *toward* it. We are dealing here with what had been a psychological obstacle to discovery: the presupposition that all forces are like gravitational and electrostatic forces, acting directly along the line between pieces of matter or charge. Although no such presupposition is explicitly stated in Newton's writings, later scientists imbued with Newtonian methods and ideas seem to have unconsciously accepted it.

The announcement of Oersted's discovery produced a sensation in scientific circles in Europe and America. The excitement was caused not simply by the fundamental nature of his work, showing a relation between two hitherto separate physical phenomena, but perhaps even more by an interest in technological applications of electric currents. It was already suspected that electricity might offer an effective way of transmitting and using the large amounts of energy available from steam engines. If a current can exert a force on a magnet, as in Oersted's experiment, one could expect (by Newton's third law, if for no other reason) that a magnet should also exert a force on a current, and it does not take too much imagination to speculate that a magnet might somehow *produce* a current. If this could be done significantly more cheaply than by Volta's battery (which consumed large quantities of expensive metals to yield only small amounts of current), one would find oneself at the threshold of the Electrical Age. And indeed it turned out that only one more bold spirit was needed to open that door.

In 1821 the editor of the British journal *Annals of Philosophy* asked a young man, Michael Faraday, to undertake a historical survey of the experiments and theories of electromagnetism, to summarize the large burst of activity in this field inspired by Oersted's discovery in the previous year. Faraday, who was at that time a laboratory assistant to the well-known chemist Humphry Davy, did not yet have a reputation in science, but he was eager to learn all he could. Faraday agreed to accept the assignment, but soon found that he could not limit himself to merely reporting what others had done; he felt he had to repeat the experiments in his own laboratory. Not being satisfied with the theoretical explanations proposed by other physicists, Faraday started to work out his own theories and plans for further experiments. Before long he had launched a series of researches in electricity that was to make him (despite his lack of formal training in science) one of the most famous physicists of his time.

Faraday's first discovery in electromagnetism was made (according to the careful records he kept) on September 3, 1821. Repeating Oersted's experiment by holding a compass needle at various places around a current-carrying wire, Faraday realized that the force exerted by the current on the magnet is *circular* in nature. As he expressed it a few years later, the wire is surrounded by an infinite set of concentric circular *lines of force*, so that a magnetic pole which is free to move experiences a push in a circular path around a fixed conducting wire. The collection of these lines of force is called the *magnetic field* of the current, a term introduced by Faraday. (See Fig. 25.1; the symbol $H$ is used for the magnetic field vector at every point around the current $I$. $H$ is everywhere tangential to the line of force.)

Faraday immediately constructed an "electromagnetic rotator" based on this idea, whereby a bar magnet pivoted at one end could rotate around the wire along the line of force acting on the freely

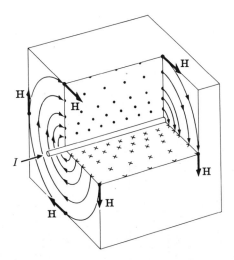

**Fig. 25.1** Magnetic field around a long straight conductor.

moving pole. He used mercury to complete the electric circuit (see Fig. 25.2, left side). He also designed an arrangement in which the magnet was fixed and the current-carrying wire rotated around it, thereby showing that, in accordance with Newton's third law, the magnet does exert a force on the wire (Fig. 25.2, right side). As in many other cases, Faraday was guided by the idea that for every effect of electricity on magnetism, there must be a converse effect of magnetism on electricity— though it was not always so obvious what form the converse effect would take!

Armed with his lines-of-force picture for understanding electric and magnetic fields, Faraday joined in the search for a way of producing currents by magnetism. Scattered through his diary in the years after 1824 are many descriptions of such experiments. Each report ended with a note: "exhibited no action" or "no effect."

Finally, in 1831, came the breakthrough. Like many discoveries which have been preceded by a

**Fig. 25.2** Two versions of Faraday's electromagnetic rotator. In each, the cup is filled with mercury so that a large electric current can be passed between the top wire and the base. At the left, the south end of a bar magnet is fixed, and the north end is free to revolve along one of the circular magnetic lines of force surrounding the current. At the right, the rod carrying the current revolves around the firmly fastened bar magnet.

period of preliminary research and discussion among scientists, this one was made almost simultaneously by two scientists working independently in different countries. Faraday was not quite the first to produce electricity from magnetism; that was actually done first by an American scientist, Joseph Henry. Henry was at the time teaching mathematics and philosophy at an academy in Albany, New York. (Shortly thereafter he was appointed professor of natural philosophy at Princeton; he later served as the first secretary of the Smithsonian Institution in Washington, D.C.) Unfortunately for American science, teachers at the Albany Academy were expected to spend all their

time on teaching and administrative duties, with no time left for research. Henry had hardly any opportunity to follow up his discovery, which he made during a one-month summer vacation. He was not able to publish his work until a year later, and in the meantime Faraday had made a similar discovery and published his results.

Faraday is known as the discoverer of *electromagnetic induction* (production of a current by magnetism) not simply because he established official priority by first publication, but primarily because he conducted exhaustive investigations into all aspects of the subject. His earlier experiments and his thoughts about lines of force had suggested to him the possibility that a current in one wire ought to be able to induce a current in a nearby wire, perhaps through the action of the magnetic lines of force in the space around the first current.   But again, presuppositions based on earlier ideas delayed the discovery by restricting the range of possible combinations.   Oersted had shown that a *steady* electric current produced a *steady* (constant) magnetic effect around the circuit carrying the current.   By arguing from "symmetry," one could easily expect that a steady electric current could somehow be generated if a wire were placed near or around a magnet, although a very strong magnet might be needed to show an effect. Or a steady current might be produced in one circuit if a very large current was flowing in another circuit nearby.   Faraday tried all these possibilities, with no success.

At last, on August 29, 1831, Faraday found an effect.   He had put two wires, *A* and *B*, near each other; current was sent through one of them (*A*) from a battery.   He noticed that a current did appear in the other wire (*B*), but only while the current from the battery in the first wire *started or stopped.* The current "induced" in wire *B* lasted only for a moment, just while contact was being made with the battery.   But as long as there was a *steady* current in wire *A*, there was no current in wire *B*. When the current in wire *A* was stopped, again there was a momentary current induced in wire *B*.

To summarize Faraday's result: A current can induce another current only while it is *changing.* A steady current in one wire cannot induce a current in another wire.

Faraday was not satisfied with merely observing and reporting this result, which he had reached by accident (i.e., without having looked for this particular form of the long-suspected effect).   Guided by his concept of lines of force, he tried to find out what were the *essential factors* involved in electromagnetic induction.

In Faraday's first successful experiment, wire *A* had not been a straight segment, but was wrapped around one side of an iron ring (the *primary coil,*

**Fig. 25.3**  Faraday's electromagnetic induction experiment.

shown as *A* in Fig. 25.3) and could be connected to a battery.   The other wire (the *secondary coil B* in Fig. 25.3) was wrapped around the other side of the ring and connected to a current-indicating device (as shown in Fig. 25.3, but a compass needle placed near wire *B* would serve the same purpose).   Faraday reasoned that the change in the current in the primary coil would cause the lines of magnetic force throughout the iron ring to change, and this change in turn in some manner served to induce a current in the secondary coil.   But if this was the correct explanation of induction, Faraday asked himself, should it not be possible to produce the same effect in another way?   In particular:

1) Is the iron ring really necessary to produce the induction effect, or does it merely intensify an effect that would occur anyway?

2) Is a fixed primary coil and a switched current in *A* really necessary, or could current be induced in *B* merely by changing in some other way the magnetic field in which it is bathed, such as by doing entirely without wire *A* and, instead, moving a bar magnet relative to wire *B*?

Faraday answered these questions almost immediately by further experiments.   First, he showed that the iron ring was not necessary; starting a current in one coil of wire would induce a momentary current in a nearby second coil, separated from the first only by air (or even empty space). Second, he found that when a bar magnet was inserted into the end of a coil of wire, a current was induced in the coil at the instant of insertion.

Having done these and many such experiments, Faraday stated his general principle of electromagnetic induction: Changing lines of magnetic force can cause a current in a wire. The "change" in the lines of force can be produced either by (a) a magnet moving relative to the wire, or (b) a change in the current going in a second wire. Using Faraday's term *field*, we can say that a current is induced in a circuit when variations are set up in a magnetic field around the circuit; such a variation may be caused either by the relative motion of the wire and field (for example, when a magnet near a fixed wire is pushed back and forth by means of a steam engine), or by any change in the intensity of the field.

Let us note at least briefly the enormous consequences for all of society that can spring from such "disinterested" research in basic (or "pure") science. Faraday's rotators were the first motors. From these toys sprang the development of larger, more effective *motors* over the next half century, all of which work on essentially the same original principle. From Faraday's experiment on induced current followed the whole set of *generators* of electric current, each again working essentially on the same principle as his original faint and clumsy current generator. Together, the generator and motor made possible a transformation of society through the "electrification" of most tasks that require transmission and use of energy or of information. To give just one example, once electric motors could be made for effective use in streetcars, subways, and elevators, the size of cities (in all three dimensions) could quickly expand—for good or ill. The effective generation of electricity and the invention of the electric light bulb also helped to change the quality of life, for example, by bringing cheap and safe light into the farmer's house.

**Fig. 25.4**  James Clerk Maxwell (1831–1879).

## 25.3 ELECTROMAGNETIC WAVES AND ETHER

We now turn to the work done in the 1860s by the great British scientist James Clerk Maxwell. We have already met Maxwell through his contributions to the kinetic theory of gases (Chapter 22); Fig. 25.4 is a portrait of him at a later period.

The importance of Maxwell's electromagnetic theory for physical science has often been said to approach that of Newton's work, but we must here be content with the briefest summary. Maxwell's theory of the propagation of light waves is now referred to as the *classical theory*. He first considered what happens when an electric current oscillates along a straight piece of wire, or circulates in a loop of wire. His theoretical conclusion was that electrical energy is lost, that it is radiated

**Fig. 25.5** Electric ($E$) and magnetic ($H$) field intensity vectors at the same point $P$ and at eight consecutive moments.

away from the electric current and spreads out, wavelike, in all directions. By this is meant that any electric charges in a distant obstacle in the path of the radiation, say in a second wire, are set into oscillation at the same frequency as that of the original source, the first wire; any magnetic poles present in the obstacle also undergo similar vibrations, and the obstacle itself will feel a slight mechanical pressure. Technically we can put all this into different language: A vibrating or circulating electric charge (called a transmitter or oscillator) sets up in the region all around it a fluctuating electric and magnetic field; when other electric charges or magnetic poles are introduced into this fluctuating electromagnetic field (as in an antenna or receiver) they are acted on by electric and magnetic forces that change with the same periodicity as the transmitter's original oscillations.

Figure 25.5 shows the electric and magnetic field intensity vectors at one point $P$ in the electromagnetic field, at eight successive moments. If you concentrate your attention on the behavior of the electric field vector exclusively—it is the only one that we shall pay much attention to—a useful analogy suggests itself: If some receiver, perhaps a metal wire containing electrons that are free to move, is placed at point $P$, the fluctuating electric field acts on these charges in the same manner as a wave passing along a rope agitates the particles of one section of the rope. In fact, it is perfectly correct to say that an electromagnetic *wave* is passing by this point $P$ in space, and to refer to the energy responsible for the motion of the charges as electromagnetic energy or electromagnetic radiation.

Maxwell's theory also predicted that metallic objects would reflect a beam of incident electro-

magnetic energy like a mirror; that on entering other obstacles, such as a sheet of glass, the path of the beams would be bent; and most startlingly, that this radiation must travel through vacuum or air with the speed of about $3 \times 10^8$ m/sec, a value equal to that for the previously known speed of propagation of light.*

These and similar predictions of Maxwell's theory all agree at one point: The radiation expected on theoretical grounds from vibrating or rotating currents should be found to behave in every way *like light*. Even the fact that light exerts a pressure had long been suggested by the observation that the material in the tail of comets always points away from the sun at every position along the comet's path (see Section 16.7), although this phenomenon is now thought to be in large part due to a stream of particles emitted by the sun, in addition to light pressure.

The analogy between the known behavior of light and the expected (but not yet proved)

---

*    In Maxwell's theory, the ratio of the electrostatic force constant $k_e$ to the electromagnetic force constant $k_m$ is shown to be equal to the square of the velocity of propagation of electromagnetic waves. The constant $k_m$ arises in the law of force for two parallel currents $I_1$ and $I_2$ at distance $R$, established by Ampère: $F = 2k_m I_1 I_2 L / R_{12}$. (This is the force on length $L$ of the wire.) The Eleventh General Conference on Weights and Measures, in 1960, decided to adopt the arbitrary value $k_m = 10^{-7}$. This choice serves to define the *ampere* as that current which, when flowing in two parallel conductors separated by a distance of one meter, results in a force on each conductor of $2 \times 10^{-7}$ newtons per meter of length of each conductor. Since, according to Maxwell's (now established) theory, $k_e/k_m = c^2$, we obtain the value $k_e = 10^{-7}c^2$ in the appropriate units of newton-m²/coul², as stated in Section 24.7.

behavior of Maxwell's theoretically predicted electromagnetic waves was indeed tantalizing. At the time Maxwell was developing his electromagnetic theory, the theory of light as a mechanical wave propagating in some ether was generally accepted (Chapter 23). Furthermore, in order to account for the polarization of light, people had assumed that light waves are transverse rather than longitudinal, and it was thought that transverse waves could be propagated only in a solid ether. As can be seen from Figs. 25.5 and 25.6, the electromagnetic waves whose existence Maxwell predicted are also transverse, since the "disturbance" being propagated—the pattern of changing electric and magnetic field vectors—would be perpendicular to the direction of propagation of the wave. Different directions of polarization of light could correspond simply to different orientations of the $E$ and $H$ vectors in the plane perpendicular to the direction of propagation.

The question naturally arose whether the propagation of disturbances in electric and magnetic fields in space would not also depend on the action of a mechanical medium, an ether. It seemed that the mind could not contemplate waves without visualizing a specific medium. In fact, Maxwell's own thinking and his derivations specifically depended on his assumption of certain rather complex mechanical models for the ether. To this day there are echoes of this conceptual struggle, which we recognize as having its equivalent in physics since the time of Aristotle.

The merger of the two ethers postulated for radiant heat and for light respectively into *one* (Section 17.7) had been the first step in the direction of unification, of making one ether do the job of several. Maxwell took a second, big step: He proposed that the propagation medium for magnetic and electric effects take over the function of transmitting light and heat waves also. (We shall see in the next section that experiments later proved that Maxwell's assumption was right—visible light *is* merely one species of propagating electromagnetic disturbances.) In the long run, however, this simplification had a very unexpected result: When

the time came to abandon the ether concept, it was easier to do so by giving up one general ether than it would have been to disprove and dismiss several ether concepts separately.

That time came in 1905. The one feature all hopeful ether models had in common was that they provided a medium for the propagation of light which existed apart from the source and the observer, just as air is a medium for sound transmission. To pursue this analogy further, consider how one measures the speed of sound: One might station a loudspeaker and a microphone many meters apart and time the progress of a short pulse of sound along that distance. At a temperature of 0°C, the speed in air is ordinarily 330 m/sec. But if this experiment were made on an open railway flatcar, or in some other "laboratory" that moves with respect to the medium (air), then the speed measurement for sound would be affected by our own motion. For example, if the sound is directed to a microphone at the rear end of the flatcar, it will get there sooner if the car moves forward than if the car stands still. Similarly, the measured velocity of light should vary depending on the motion of the "laboratory" through the ether and the placement of the equipment relative to the motion. Our earth itself is a fast-moving "platform" as it revolves about the sun and rotates on its own axis; light moving along a meter stick that points to the east is caught up in the relative eastward motion of the ether through which the earth moves, and arrives at the other end sooner than if it had to travel the opposite way or along a north-south line.

This expected difference in the speed of light, which should yield the velocity of the earth through the ether or, equally well, the relative velocity of the ether past the earth, was essentially what the American experimenters A. A. Michelson and E. W. Morley attempted to find in 1887. On the basis of their technique one could have expected a clear and convincing positive result—and yet, no difference of speeds was observed! Here was a new postulate of impotency: One cannot measure the relative velocity of the ether.* After a period of adjustment and speculation, with varying degrees of success, this realization was incorporated into a more far-reaching postulate, one of the two basic pillars of Einstein's relativity theory (Chapter 30) as announced in 1905: The measured velocity of light in space is constant, no matter what the motion of the observer or of the source may be. On the one hand, this extends the previous discussion on the relativity principle (Section 8.5) to include optical in addition to mechanical experiments in the statement, "All laws of physics observed in one coordinate system are equally valid in any other coordinate system moving with a constant velocity relative to the first." And on the other hand, the acceptance of the postulate of constancy of light speed as a fundamental principle relieves us of the need to account for the propagation of light by an ether model. For if the principle is right, as would be found by checking experimentally all consequences derivable from it, then we are better off to accept the principle as a basic point of departure than we should be if we continued the dubious quest for an ether free from all internal contradictions and in accord with all experiments. The chief point is that one must allow that a useful principle, even one unfruitful of convenient mental images, is to be preferred over a clear model which is not consistent logically and not in accord with empirical facts.

This point reminds us of the fate of the caloric, the medium whose function was to provide a visualizable mechanical model for the processes of heat transfer. It was challenged by Rumford and Davy on the grounds that it failed to exhibit the behavior expected of material media—specifically, that it had no measurable weight, and could, in friction, be created without limit and without drawing from the materials out of which it arose. The comparison between the fate of the caloric and of the ether is, however, faulty in one detail. The caloric was

---

*    That the earth was not simply "dragging" along the ether in its vicinity was made clear in other experiments.

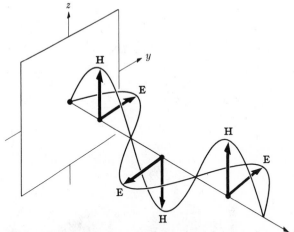

**Fig. 25.6** Instantaneous pattern of electric and magnetic field vectors in an electromagnetic wave. The entire pattern moves through space in the positive x direction.

eventually replaced by a scheme built on another visualizable model, the kinetic theory picture of the atom, whereas the ether was replaced by a principle, and by the habit of speaking of "the field" as the arena for physical processes that occur *in vacuo*, in the absence of ponderable matter. If we wish, we may conceive of light for the present as progressing through space by the propagation of electric (and magnetic) field vectors (see Fig. 25.6 for a "snapshot" of a region through which light travels, with some field vectors drawn in). The distance between successive instantaneous maxima is one wavelength $\lambda$, and the frequency of the wave $f$ is the number of crests passing a given point per second. The velocity of light $c$ is given by the usual relation (equivalent to Eq. 23.2), $c = \lambda f$. Our elimination of the ether from this discussion does not invalidate the rest of the Maxwellian theory of electromagnetic waves.

## 25.4 HERTZ'S EXPERIMENTS

If Maxwell's theory of light as an electromagnetic radiation emitted by the periodic fluctuation of

**Fig. 25.7** Heinrich Hertz (1857–1894).

**Fig. 25.8** Spark gaps in Hertz's oscillator.

currents is correct, a heated object, e.g., the metal filament of an incandescent lamp, sends out light by virtue of the agitation on the filament of sub-microscopic electric charges such as one expects to be distributed throughout neutral bodies in balancing amounts.

We cannot decide at this point whether these charges are parts of the atoms in the emitter or separate entities such as electrons or positive particles; so permit us to call them for the moment, in a general way, *oscillators*. Somehow these oscillators are able to emit electromagnetic energy that our eye interprets as light, the frequency $v$ of the light waves being equal to the periodic frequency $f$ of the oscillators.

This "classical" interpretation of light emission might be more acceptable to us if we could refer to direct experimental confirmation. One very obvious test might be to produce an oscillating current in a

wire at a frequency equal to that of, say, green light, and watch whether green light is actually released into the surrounding region. Even before the days of radio and microwaves it was easy enough to produce electric oscillations in wires, but the frequencies achieved were relatively low in Maxwell's day, and even today the higher frequencies of electric oscillation in the most modern equipment can hardly be pushed above $10^{12}$ sec$^{-1}$, far below the frequency of visible light ($\sim 10^{15}$ sec$^{-1}$). Consequently, a direct test seemed out of the question, and partly for this reason Maxwell's work did not attract a great deal of attention for two decades.

There is, however, another point of attack. The gifted German physicist Heinrich Hertz (1857–1894) (Fig. 25.7) undertook in 1887–1888 to show that at least the main premise of Maxwell's view was correct, namely, that an oscillating electric current (he attained a frequency of about $10^9$ sec$^{-1}$) does indeed send out electromagnetic waves that have all the characteristics of light except, of course, visibility. Hertz's proof succeeded so brilliantly that hardly one important physicist remained unconvinced of the equivalence of light and electromagnetic radiation.

In essence, Hertz's experimental method was simple. Two rods, each about 5 in. long and terminated at one end by a small polished metal sphere, were fixed along a line so that a small air gap separated the spheres (Fig. 25.8). A pair of wires led from the spheres to a device supplying short pulses of very high potential difference, and during those intervals brought to the two spheres large amounts of electric charge of opposite signs, until the point was reached at which a spark jumped across. The air in the gap remained conducting for a short time and so provided a path for these charges as they oscillated from one rod to the other until electrical equilibrium was achieved. Then, the air being restored to nonconductivity, the stage was set for the next "voltage" pulse and another spell of oscillations by the same processes.*

If Maxwell was right, this arrangement should, during each spark, give off electromagnetic waves of the same frequency as the oscillations, up to about $5 \times 10^8$ sec$^{-1}$ by Hertz's estimate, corresponding to a wavelength of about 60 cm. Furthermore, this radiation should spread out from the rods and be detectable by its production of a fluctuating current in a wire some distance away, as has been pointed out.

Hertz's first triumph was to observe this very effect, even when he placed the receiving wire (antenna) many yards from the sending oscillator. This new invisible electromagnetic radiation, still sometimes called Hertzian waves, is of course identical with our now familiar radio waves.

---

* Let us take notice of a short and somewhat mystifying remark of Hertz's which will assume strange importance later, namely, that "it is essential that the pole surfaces of the spark gap should be frequently repolished" to assure proper operation of the spark.

† The demonstration of Hertzian waves proved to be not only the intended triumph of Maxwell's theory but also the unintended beginning of radio and television. It is not amiss to contemplate briefly that strange fruit on the tree of science. Consider an ordinary television set from all the following points of view—symbolically, as it were: how the first hint of radio waves originated in the mathematical speculations of Maxwell, and through the ingenious experiments of Hertz came to the notice of the young Italian Marconi. Consider how even these three men, in the cooperatively cumulative nature of their work and the differences of their nationalities, religion, and training, exemplify the interdependence of man, the basis of the humanistic ideal. Then consider the meaning of television as a huge industry in the economics of the nation, and how society uses this tool, sometimes for mass education or genuine entertainment, more often for mass propaganda, mass tranquilization, and mass vulgarity.

Hertz showed next that the radiation could be made to go through all the characteristic behavior of light: reflection, focusing into a beam, refraction, interference, polarization, and so forth. One interesting test, the direct determination of the velocity of this radiation, was beyond the instrumentation available to Hertz. Experimentation conducted after 1895 confirmed the assumption that the speed was the same as that of light.†

One conclusion was soon almost universally accepted even before detailed additional evidence came in: Since Hertzian waves and light waves behave so much alike, light too must be caused by the rapid oscillatory motion of charged particles, perhaps those in the atoms of the light emitter. Here then was the model for light emission. And in terms of this picture, a great many details became clear and could be integrated into one scheme—the mechanism for transmission of light through crystals and other (transparent) objects, the reflection from metal surfaces, and so on. One particularly impressive, even sensational, confirmation of the theory was provided by P. Zeeman, who discovered (1896) that a glowing gas sends out light at changed frequencies if the emitting material is placed in a strong magnetic field. This effect could be explained for many cases by a straightforward adoption of the fundamental Maxwellian idea that light energy is released by the vibration of charged particles, since the observed change in light frequency brought about by the field is equal to that calculated from Maxwell's theory.

## RECOMMENDED FOR FURTHER READING

D. L. Hurd and J. J. Kipling, *The Origins and Growth of Physical Science*, Vol. 2, pages 217–266. Extracts from the writings of Ampère, Faraday, Maxwell, and Hertz.

D. K. C. MacDonald, *Faraday, Maxwell, and Kelvin*, Garden City, N.Y.: Anchor Books, 1964; paperback

P. and E. Morrison, "Heinrich Hertz," *Scientific American*, December 1957, pages 98–106.

*Project Physics Reader*, Vol. 4, articles by Maxwell, Newman, Einstein and Infeld, and Feynman, Leighton, and Sands

## SOURCES, INTERPRETATIONS, AND REFERENCE WORKS

J. Bromberg, "Maxwell's displacement current and his theory of light," *Archive for History of Exact Sciences*, Vol. 4, pages 218–234 (1967)

B. Dibner, *Oersted and the Discovery of Electromagnetism,* Norwalk, Conn.: Burndy Library, 1961

Albert Einstein, "Maxwell's influence on the development of the conception of physical reality," and other articles in *James Clerk Maxwell: A Commemoration Volume*, London: Cambridge University Press, 1931

M. Hesse, *Forces and Fields,* Chapter VIII

K. F. Schaffner, *Nineteenth-Century Aether Theories*, New York: Pergamon Press, 1972

W. T. Scott, "Resource Letter FC-1 on the evolution of the electromagnetic field concept," *American Journal of Physics*, Vol. 31, no. 11 (November 1963)

R. S. Shankland, "Michelson-Morley experiment," *American Journal of Physics*, Vol. 32, pages 16–35 (1964)

T. K. Simpson, "Some observations on Maxwell's *Treatise on Electricity and Magnetism*," *Studies in History and Philosophy of Science*, Vol. 1, pages 249–263 (1970)

R. C. Stauffer, "Speculation and experiment in the background of Oersted's discovery of electromagnetism," *Isis*, Vol. 48, pages 33–50 (1957)

C. Süsskind, "Observations of electromagnetic-wave radiation before Hertz," *Isis*, Vol. 55, pages 32–42 (1964)

L. S. Swenson, Jr., *The Ethereal Aether, A History of the Michelson-Morley-Miller Aether-Drift Experiments, 1880–1930,*   Austin: University of Texas Press, 1972

R. A. R. Tricker, *Early Electrodynamics, the First Law of Circulation*, New York: Pergamon Press, 1965; *The Contributions of Faraday and Maxwell to Electrical Science*, New York: Pergamon Press, 1966. Both books include historical and scientific commentary and extracts from original sources.

E. T. Whittaker, *A History of the Theories of Aether and Electricity*, Vol. I, Chapters VI-X and XIII

L. P. Williams, *Michael Faraday*, New York: Basic Books, 1965; Simon and Schuster paperback

# Chapter 26

# THE QUANTUM
# THEORY OF LIGHT

## 26.1 CONTINUOUS EMISSION SPECTRA

The study of light leads into modern physics by two separate paths. The problem of the earth's motion through the hypothetical ether has been mentioned in Section 25.3; we shall return to that topic when we discuss Einstein's relativity theory in Chapter 30. The other path now leads us, by a somewhat roundabout route, to the quantum theory.

Our newly gained—though as yet rather vague—model of the emitter of light, the submicroscopic electric oscillator, raises our hopes for the solution of some long-standing puzzles about radiation: Can we not now begin to *explain the colors (frequencies) and distribution of light\* energies* sent out by different sources of light?

Before we can grapple with this problem we must turn to a brief account of the large amount of experimental knowledge available by the turn of the twentieth century. Newton had shown, in the "celebrated experiment," that light from the sun could be resolved by a prism into a spectrum (Newton's word) of various colors; and just such a "rainbow" spectrum had since been obtained by examining the light from *all* glowing solids (filament, carbon in arc) and glowing liquids (molten metals). With only a little idealization, these emission spectra all have three features in common:

a) They are *continuous* spectra, that is, there are no gaps, no "color" bands missing from the spectrum, even if measurements of the radiation are carried into the infrared or ultraviolet region.

b) All glowing solids and liquids, *no matter what their chemical nature*, send out light with about the same color balance if they are at the same temperature. A red-glowing coal has that color not because there are no other colors but red being radiated from it, but because of all the visible radiation that in the "red" region is most intense, and an iron poker put into the coals, after reaching the same temperature, glows with the same light.

This must be put more precisely. The wavelength of each color (and, by means of the relation $c = \lambda \nu$, the frequency also) can be measured with great accuracy by means of a spectroscope.†

---

\* The identity of the character of ultraviolet rays, visible light, infrared (heat) rays, and longer wavelength (Hertzian) waves having been established, we shall from now on refer to all these as "light."

† Although not a prism but a diffraction grating is used for the determination of absolute wavelengths. See bibliography for detailed discussion.

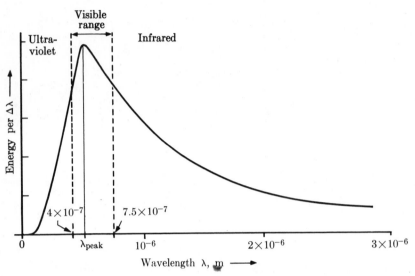

**Fig. 26.1** Distribution of energy in light emitted from a glowing body at 60,000°K.

Approximate ranges in wavelength for different colors are given below. Note that wavelengths are usually not given in centimeters but in units of $10^{-10}$ m, called 1 *angstrom* or 1 Å, in honor of the Swedish spectroscopist A. J. Ångström.

**Table 26.1**. Wavelengths of colors in the spectrum

| Color | Range of $\lambda$ in Å ($10^{-10}$ m) |
| --- | --- |
| Red | 6100 to about 7500 |
| Orange | 5900 to 6100 |
| Yellow | 5700 to 5900 |
| Green | 5000 to 5700 |
| Blue | 4500 to 5000 |
| Violet | about 4000 to 4500 |
| (Infrared) | Longer than 7500 |
| (Ultraviolet) | Shorter than 4000 |

The exact amount of energy contributed by the components of different wavelengths in a beam of light can now also be measured with precision, although originally the details of experimentation were extremely delicate and difficult. For the infrared ranges, the energy can be detected conveniently by sensitive heat-measuring devices (thermopiles, bolometers), and in the visible and ultraviolet ranges by photographic exposures (so used since about 1840), photocells, and the like.

A typical distribution curve of radiated energy vs. wavelength, specifically for a temperature corresponding to that on the surface of the sun (6000°K) is seen in Fig. 26.1. In this case there is relatively little energy at very short wavelengths. The peak of the curve falls into the green part of the visible region, and the largest share of the total energy emitted is in the infrared; our eye perceives only the particular mixture of "visible" radiation, and compounds it to give the impression of white light. The chief points of interest here are that such a curve can be drawn for every temperature of the emitter and that all emitters of the same temperature, regardless of chemical composition, yield about the same curve.*

---

\* To be more accurate, the radiation has the distribution shown for emitters that, when cool, are perfect absorbers of light. These are called *blackbodies*, but we can here disregard deviations from blackbody radiation.

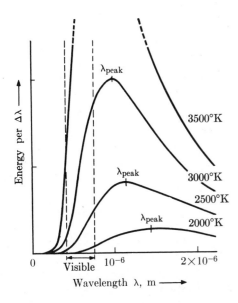

**Fig. 26.2** Distribution of energy in light radiated from a glowing solid at different temperatures.

For example, $\lambda_{peak}$ for the sun is, by experiment, about $5.5 \times 10^{-7}$ m. It follows that $T = 5300°$K. But this figure is a little too low, partly because some of the energy radiated by the sun, particularly at short wavelengths, is absorbed in the atmosphere before reaching our instruments, so that the true $\lambda_{peak}$ is lower; and also because the sun is not the ideal radiator (so-called *blackbody radiator*) for which Eq. (26.1) is designed. Allowing for these difficulties, one is led to a value of about $6000°$K for the sun's surface.

Provided Wien's displacement law holds even in extreme temperature ranges, we may at once make some surprising calculations: The radiation from distant stars can be analyzed to find where $\lambda_{peak}$ falls. From this we may calculate approximate stellar surface temperatures. For the "hot" stars, whose $\lambda_{peak}$ is small and lies in the blue end of the spectrum or even beyond, the temperatures are correspondingly higher than that of the sun (for example, $14{,}000°$K for Vega). The cooler, reddish stars have larger values for $\lambda_{peak}$, and the calculated values of $T$ are consequently lower (for example, $2500°$K for Antares).

c) The third feature which all hot solid or liquid emitters have in common is that their energy distribution curve shifts with changing temperature $T$, as shown in Fig. 26.2 for four temperatures. With increasing $T$, more energy is radiated in each wavelength region, and the peak in the curve moves toward shorter wavelengths.

The exact data were summarized by the German physicist Wilhelm Wien* in 1893 in the so-called "displacement law," which in part states that for a perfect emitter of continuous spectra, the product of $\lambda$ peak (in cm) and its temperature (in °K) is a constant with the empirical value of 0.2897 cm-°K:

$$\lambda_{peak} \text{ (cm)} \times T(°K) \doteq 0.0029 \text{ m-°K}. \quad (26.1)$$

---

* Wien received a Nobel prize in 1911 for his work on the theory of radiation.

**Problem 26.1.** What is the wavelength of the most intense radiation for Vega, for Antares, and for a body at room temperature? Are there qualifications to your answer?

**Problem 26.2.** The total area under the curve for intensity of radiation vs. wavelength is proportional to the total energy radiated throughout the spectrum. Trace the curves in Fig. 26.2 and find an approximate law connecting the total energy radiated and the temperature of the emitter. (In the ideal case the relation is total energy $\propto T^4$.)

**Problem 26.3.** On the basis of our model of light emitters, would you expect to find a distribution curve such as shown in Fig. 26.1 or instead that all energy is sent out at *one* frequency? Should chemical and physical consistency influence the light emitted? What might temperature have to do with the distribution of intensity vs. wavelength? (Defend your conclusions, but be frank about vague guesses.)

The displacement law could actually be derived theoretically from the classical (Maxwellian) theory of light emission, and that was indeed another impressive victory! But at that very same point we encounter serious trouble. Surely we should also be able to explain the exact shape of the characteristic emission curve as a whole by derivation from the basic model of emission. This would involve first a calculation of the fraction of individual oscillators for each range of emitted frequencies, then the addition of their individual contributions to find the over-all pattern.

*But all such attempts proved fruitless.* Specifically, Fig. 26.3 compares the actual emission curve with the prediction made by two modifications of the classical theory that differed from each other by some of the assumptions introduced in the classical model.

Neither modification yields the correct curve. We recall instances of similar dilemmas in the history of science. The most memorable case is perhaps Kepler's attempt to derive from the Copernican theory a path for the planet Mars that would fit the observations. He failed, just as we now find ourselves unable to fit theory and fact. His solution was not to forget about the fairly small discrepancy —that is permissible only if one can account for it exactly by computation of the secondary effects at work, or at least tentatively by a consistent plausibility argument, as Galileo accounted for small deviations from the law of free fall by citing the effects of air friction. In Kepler's case, the small discrepancy provided a real challenge, the point of departure for a profound change of outlook in the whole field. You will recall that Kepler turned from the unfulfilled theoretical prediction to find a mathematical formulation of the actual (elliptical) planetary path, and to adopt that as the starting

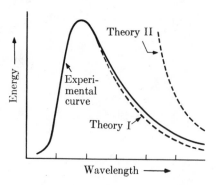

**Fig. 26.3** Comparison of experimental emission curve with predictions from two classically based theories.

point of a modification of the Copernican heliocentric theory. This led him next to the great discovery of the three mathematical laws of planetary motion. Finally, a good physical interpretation of Kepler's laws was given by Newton in terms of the gravitational forces, thereby "explaining" Kepler's empirical findings.

A very similar turn of events will soon be revealed in our coming discussion of the emission problem: We shall see that the obstinate mismatch of the curves in Fig. 26.3 prompted first a mathematical formulation of the actually observed curve; then came a physical interpretation of the mathematical formula which modified the old theory of light emission as drastically as Newton's work improved on the Copernican concept of the universe. For the moment, however, we must look at greater length at more of the experimental results that sparked the revolution in physical science, the quantum theory.

## 26.2 PLANCK'S EMPIRICAL EMISSION FORMULA

On October 19, 1900, the German physicist Max Planck (1858–1947) (Fig. 26.4) disclosed that he had found for continuous emission spectra from ideal (blackbody) radiators an equation somewhat analogous to the one Balmer had given for the line emission spectrum of hydrogen; that is, Planck had discovered the one mathematical expression for representing the experimental energy-distribution curves at every temperature (Fig. 26.2). One way of formulating his result is this: If we designate by the symbol $E_\lambda$ the radiation energy emitted per second, per unit area of emitter surface, per unit wavelength interval, then

$$E_\lambda = \frac{C_1 \lambda^{-5}}{e^{C_2/\lambda T} - 1},  \qquad (26.2)$$

where $C_1$ and $C_2$ are constants, and $e$ is a number (about 2.718).

But Planck, one of the greatest physicists of his time, was not at all satisfied with this accomplishment. After all, one can generally find a formula to represent any family of curves. Finding Eq. (26.2) was to Planck no more of an explanation of the physical mechanism responsible for the phenomenon of emission than was Bode's law an explanation of the construction of the solar system (Section 11.8). Planck was, in fact, then working on the problem of deriving the energy distribution curve by assuming that light was emitted from a Maxwell-Hertz type of oscillator. In his autobiographical notes he relates that the problem appealed to him deeply because of the feature which we have had cause to marvel at—that the continuous spectrum from (ideal) emitters is entirely independent of their chemical properties. We are reminded of the law of universal gravitation, which also was found to hold regardless of the chemical nature of the participating masses. Here again there seems to be a universal law at work. In Planck's own words, this fact concerning emission "represents something absolute, and since I had always regarded the search for the absolute as the loftiest goal of all scientific activity, I eagerly set to work." We note parenthetically the philosophic kinship between Planck and Newton; the tone here expressed distinctly forms a counterpoint in contemporary philosophy of science.

As had all others who struggled with the emis-

**Fig. 26.4** Max Planck (1858–1947).

sion problem, Planck failed at first, but at any rate his empirical formula of intensity vs. wavelength was a startling success. The very night of Planck's public announcement in a scientific meeting, a colleague extended the measurements beyond the range and accuracy previously obtained, and confirmed a complete agreement between the precise facts and Planck's formula. Spurred on by this success, Planck once more set to work to provide a conceptual scheme from which his formula would follow as a necessary consequence—"until after a few weeks of the most strenuous labor of my life the darkness lifted and a new, unimagined prospect began to dawn." December 14, 1900, when Planck presented his solution, marked the effective birthday of contemporary physics.

**Fig. 26.5** The energy content which a single Planckian oscillator may have changes discontinuously from one level to the next.

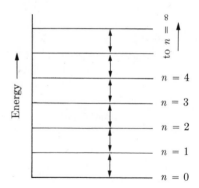

**Fig. 26.6** Energy level scheme for one oscillator of frequency $f$. For any level, the energy is $nhf$; transitions are permitted between neighboring levels, as indicated.

## 26.3  THE QUANTUM HYPOTHESIS

Hertz's work had convinced Planck that radiation must be explained by the action of submicroscopic electric oscillators.*  The wall of a glowing solid emitter may thus be imagined as populated by such oscillators (perhaps the vibrating electrons, but this need not be specified), all at different frequencies.  We know how this oscillator may send off energy.  But how does such an oscillator gain energy?  One mechanism might be the incessant collisions with neighbors as heat is supplied to the wall to keep it emitting.  And how is light radiated? By sending out electromagnetic waves during the vibration, in varying amounts of energy and at the frequency of the particular oscillator.  But just here Planck finds he must make a radical change in the classical picture: to fit observed facts, one must *not* assume, as seemed a matter of course until then, that an oscillator of some frequency $f$ can send out any part of its total amount of energy, no matter how small.  On the contrary, the energy content of the oscillator, its kinetic and potential energy at any moment, can only be an integral multiple of the quantity $hf$, where $f$ is the frequency of its vibration and $h$ is a universal constant, now known as Planck's constant, whose value remained to be found by experiment.  For example, the total energy at any one instant may be zero, $hf$, $2hf$, ... $nhf$

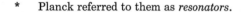

\* Planck referred to them as *resonators*.

($n$ = any integer), but never $1.85\,hf$. That is to say, the energy of the oscillator is *quantized*, limited to integral multiples of one given amount or *quantum of energy*, $hf$. For if the energy changes are in steps of $hf$, the total energy $E$ can be considered given by $E = nhf$ ($n = 0, 1, 2, \dots$ any integer).  If an ordinary pendulum, on losing its energy by friction during vibration, behaved according to such a rule, its amplitudes of vibration would not fall off gradually, but would decrease in a stepwise manner, each step representing an energy interval proportional in magnitude to the frequency of vibration or a multiple of that frequency.

A more concise rendition of the fundamental ideas of quantum theory can be given in terms of the concept *energy levels*: If the frequency $f$ of an oscillator is constant despite changes in amplitude, then we might represent its energy content over a short time interval by a graph like that in Fig. 26.5. Each of these steps or changes of levels is occasioned by absorption or emission of energy corresponding to a single quantum of magnitude $hf$. If we now draw all the different energy levels which the oscillator may reach at one time or another (Fig. 26.6), we obtain a ladder with equal steps. At any moment, the oscillator, as far as its energy content is concerned, is "located" on one or another of these energy levels, if it is not just then changing levels. Planck knew nothing about the physical nature of the oscillator, or about its actual behavior in time; therefore the energy level diagram is the best representation of oscillators, for it avoids premature or erroneous pictorializations of the physical situation within the emitter.

To this point we have been looking only at a single oscillator of frequency $f$. Now we extend our view to encompass all the uncountable oscillators of different frequencies on the surface of the glowing body. This body as a whole, in Planck's view, emitted the continuous spectrum of electromagnetic (light) waves by virtue of containing simple oscillators of all frequencies, each emitting light of a frequency $\nu$ identical with its own frequency of vibration $f$. The energy $E$ of each oscillator being quantized according to $E = nhf$, emission of energy can occur only during a sudden change in the amplitude of oscillation, for this amplitude is directly related to the energy content.

---

**Problem 26.4.** A small object of mass $m$ at the end of a spring oscillates about an equilibrium position with simple harmonic motion, an initial amplitude $x_1$, and zero initial speed. Recall (Section 17.6) that the potential energy of the system in that state of extension or compression is $\frac{1}{2}\kappa x_1^2$; here $\kappa$ is a constant indicating the stiffness of the spring, that is, $\kappa$ = the *force constant*, the amount of force (newtons) required per m extension. If in a particular case $m = 3$ kg, $\kappa = 100$ newtons/m, $x_1 = 0.4$ m, find answers to the following: (a) What is the potential energy of the oscillator at the extreme ends of the vibration? When it passes through equilibrium position? What is the kinetic energy in these two cases? What is the total energy at all times? (b) Assume that the period of vibration is given by $T = 2\pi\sqrt{m/\kappa}$. What is the frequency $f$ of vibration for this particular spring? (c) If the total energy of this *hypothetical* oscillator can only change in quanta of $hf$, where by experiment $h = 2.18$ joule ·

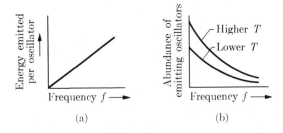

**Figure 26.7**

sec [not the true value of Planck's constant], compute its energy if it were to gain a quantum, or lose one quantum. What is the amplitude of vibration in each of these two cases? (d) Draw a graph for the first few energy levels corresponding to Fig. 26.6.

---

To Planck's fundamental postulate that the electric oscillator can change its energy only between neighboring energy levels $E''$ and $E'$ according to $\Delta E = E'' - E' = nhf$, we must add the requirement that at any moment the likelihood is much greater that the emitting body contains more elementary oscillators at low energies (low $f$) than at high energies. From assumptions such as these, fully and quantitatively developed in a brilliant series of publications, Planck was able to deduce the emission formula [Eq. (26.2)] which he had previously found empirically, and he derived also the other well-established characteristics of continuous emission spectra, as for example Wien's displacement law.

Although the details of the argument are on a most advanced level of mathematical physics, we may, in a qualitative way, make Planck's work plausible by means of the following curves. Figure 26.7(a) shows the energy emitted by individual oscillators during a drop in energy level as a function of the frequency of vibration or of emitted light; it is a straight-line graph, according to the assumption $\Delta E = hf$, and so shows that the high-frequency radiation carries away the largest energy during emission, whereas low-frequency emission does not deplete the emitter of much energy. On the other hand, the *number* of high-frequency oscillators present in an emitter is postulated to be much smaller than the number of low-frequency oscillators; see Fig. 26.7(b), where the relative abundance is shown in a schematic way for two temperatures; a rise in the temperature of the emitter, of course, brings into play more oscillators at all frequencies.

Now if the assumptions in these two graphs be granted, we can at once make a prediction of the distribution curve for the *total* energy by multiplying the ordinates (vertical coordinates) of these two graphs at each frequency to obtain a new graph. The new ordinate is (energy emitted per oscillator) × (number of oscillators at given frequency), i.e., total energy radiated at (or within a small frequency interval around) a given frequency; the abscissa (horizontal coordinate) is still frequency. When we make such a graph, two things become apparent. The maximum intensity of radiation must be in the range between the small and high frequencies, and for higher temperatures the peak of intensity shifts to the right, toward higher frequencies, i.e., shorter wavelength, exactly as expected by Wien's law. In short, the graph derived from Planck's quantum hypothesis exhibits the main features of the corresponding graph (Fig. 26.2) obtained by experiment.

---

**Problem 26.5.** (a) Copy Fig. 26.7(a) and (b), superimpose on each a grid of vertical and horizontal lines for reference purposes, and construct the total energy distribution graph (for both temperatures). (b) Note that in Fig. 26.2, with which our new graph is to be compared, the abscissa is not $f$ but $\lambda$. Therefore draw a second graph from the data for the first with the same quantity as ordinate but $\lambda$ as abscissa (recall $\lambda = c/f$). Compare its shape with Fig. 26.2.

---

From his fundamental assumptions, Planck deduced the following precise mathematical relation between the energy radiated ($E_\lambda$), the wavelength or frequency of the radiation ($\lambda$ or $v$) at which the energy $E_\lambda$ is measured, and the temperature ($T$) of the whole (ideal, blackbody) emitter:

$$E_\lambda = \frac{8\pi\lambda^{-4}hv}{e^{hv/kT} - 1},\tag{26.3}$$

where $k$ = Boltzmann's constant (see Section 22.4), and $h$ is the constant of proportionality that relates the smallest possible energy change of an electric oscillator, namely the quantum of energy, to its frequency of oscillation $f$. Although we cannot follow Planck's mathematical derivation in detail, Eq. (26.3) will seem more plausible if we accept that, on the basis of the quantum hypothesis, (a) the average energy per degree of freedom of the emitter was not $kT$, as in the classical theory based on the principle of equipartition,* but is given rather by the expression

$$\frac{hv}{e^{hv/kT} - 1},$$

and that (b) the number of degrees of freedom of the emitter at wavelength $\lambda$ is $8\pi\lambda^{-4}$. Multiplying the last two expressions, we obtain Eq. (26.3).

By inspection, we can reach at once two conclusions:

a) The theoretically derived Eq. (26.3) is indeed perfectly equivalent to the empirical Eq. (26.2), for if we replace $v$ by $c/\lambda$, where $c$ = velocity of light, we can identify $C_1$ with $8\pi ch$ and $C_2$ with $ch/k$, the quantities $c$, $h$, and $k$ being universal constants.

b) In the last statement it is implied that we may use the relations there to calculate $h$. Given the measurements of $E_\lambda$, $\lambda$, and $T$, and knowing

---

* Each vibrating oscillator has, in general, potential and kinetic energy, and $\frac{1}{2}kT$ is assigned for each of these two modes of carrying energy.

the factors $c$ and $k$, there remains only $h$ as unknown. On solving, we find

$$h = 6.626 \times 10^{-34} \text{ joule-sec.}$$

Planck's own calculations of $h$ at the time gave him a value correct to within 1% of the present well-confirmed figure. It is one of the most fundamental constants in atomic physics, of the same importance as $G$, of the law of universal gravitation, is in astrophysics, or as $N$ is in chemistry. The quantity $h$, initially evaluated as discussed above, made its appearance thereafter in many other connections, so that there are now several independent methods of calculating and confirming its value.

Before discussing the role of Planck's emission formula in the development of quantum theory, we may note that Eq. (26.3) has found important applications in astrophysics, the most recent being the discovery that the universe is filled with residual radiation produced by the "primeval fireball" whose explosion is believed to have begun the present epoch of cosmic history. The frequency distribution of this background radiation has been found to agree with Planck's formula if the temperature is set equal to 2° or 3°K.

---

**Problem 26.6.** An atomic oscillator sends out light at 7000 Å (red). At what frequency does the oscillator vibrate? What is the quantum of energy for this oscillator? What is it for another oscillator that emits violet light (at 4000 Å)? *Note well that the quantum of energy depends on the frequency of the oscillator*; energy is not quantized, like electrical charge, in uniformly large bundles under all conditions.

**Problem 26.7.** An atomic oscillator sends out light at 5000 Å. Draw some portion of its energy level diagram (to a stated scale) according to the quantum hypothesis. What is the magnitude between successive steps on the diagram?

**Problem 26.8.** In Problem 26.4, the value of the quantity $h$ was chosen quite arbitrarily. In actual fact, all oscillating systems, even springs, are considered to obey the rule that the smallest change in energy can proceed only according to the equation $\Delta E = hf$, where $h$ = Planck's constant, $f$ = frequency of vibration. Given the same spring as before, with the same mass $m$, stiffness $\kappa$, and initial amplitude $x_1$, calculate now the actual change of amplitude when the energy changes by one quantum. Why does the decay or increase of the oscillatory motion of springs, pendulums, and the like generally seem to us continuous when it is really quantized?

---

We might think that these successes must have gratified Planck deeply. However, he later confessed that his own postulate of energy quantization, to which he had been driven by the facts of emission, always remained to him deeply disturbing. He had indeed attained his goal of finding an "absolute" as nearly as man can—that is, he had found a universal and basic process—but at the cost of introducing a new hypothesis that did not seem to be in harmony either with Newtonian mechanics or with the wave theory of light. Furthermore, like most physicists of the time, he was still fundamentally convinced that natural processes are continuous; as Newton had expressed it, *natura non saltus facit.* Indeed, both the ancient faith in a sequence of cause and effect and even the usefulness of the mathematical calculus itself, seemed to depend on the proposition that natural phenomena do not proceed by jumps.

Not unexpectedly, the quantum hypothesis was regarded as rather too radical and unconvincing by almost all physicists at the time. Planck relates that he himself spent years trying to save science from discontinuous energy levels while the quantum idea "obstinately withstood all attempts at fitting, in a suitable form, into the framework of the classical theory." From a broader point of view, the general initial hesitation to plunge into quantum theory is only symptomatic of a useful law of inertia in the growth of science. One had to scan carefully the full implications of a step into such strange territory.

Moreover, Planck's initial treatment was not free from ambiguities that had to be resolved. There was, for example, a subtle paradox in Planck's reasoning. On the one hand, he jealously held on to the Maxwellian wave theory of light, considering it unquestionable that the light energy given off by a quantized oscillator spreads continuously over an ever-expanding set of wave fronts; and on the other hand, he repudiated the Maxwellian theory by implying that the subatomic oscillators would radiate energy *not* while oscillating with a certain amplitude, but only while suddenly decreasing that amplitude. Lastly, Planck's work did not contain a physical picture or model for explaining the nature of the "oscillator" or for solving such puzzles as the difference between continuous and line emission spectra (the latter will be discussed in Chapter 28).

If one chose, one could regard Planck's scheme as a formalism which at best left the "real" questions of physical mechanisms unanswered. The general acceptance of Planck's work depended therefore not just on experimental verifications, but on further demonstrations of the usefulness, even the necessity, of the quantum hypothesis in scientific thought. For some years the quantum hypothesis was uneasily suspended between doubt and neglect. Then Einstein extended and applied Planck's conception in rather the same manner that Avogadro developed Dalton's, and thereafter the doubts slowly began to fall away. Because Einstein's specific contribution to this topic depended on the clarification of a long-standing puzzle, the photoelectric effect, we shall first trace out the relevant facts.

## 26.4 THE PHOTOELECTRIC EFFECT

Ironically, the first facts leading to the photoelectric effect, and through that eventually to the recognition that the classical theory of light had to be

**Fig. 26.8** A photoelectric current is set up through the circuit when ultraviolet light falls on the electrode at the left.

fundamentally revised, were by-products of the very same research that seemed at the time to furnish the most convincing proof for the classical theory—namely, Hertz's experiments on the propagation of electric waves. As we noted, Hertz mentioned that an electric spark would jump more readily between the metal spheres if they were freshly polished. It soon became clear, through the work of various contributors, that the ultraviolet component of light—as from a spark itself—had the effect of drawing or expelling negative charges from the clean surface of metals, and that it was these charges that aided in maintaining a suitable transient current between the spheres in Hertz's equipment. Indeed, some metals, notably the alkali metals, exhibited the same effect even with visible light. The air around the metal was not involved; on the contrary, when two clean metal plates (Fig. 26.8) are sealed in a transparent, well-evacuated quartz tube and are connected to a battery which charges one of the plates negatively, the latter, if illuminated, copiously emits negative charges that are then attracted to the positive plate and recirculated by way of the battery. The current so obtained clearly deserves the name *photoelectric current*.

The exact nature of those negative charges was not clarified until 1897. At that time, principally through the work of J. J. Thomson, electricity was found to be carried in fixed quantities of charge by discrete particles much smaller than the lightest atoms. This identification of electrons was first made from a study of the current through an almost completely evacuated tube when a sufficiently high potential difference was applied between two electrodes. Although the electron concept was eventually found to be essential in explaining every phenomenon involving electric charges, the free electron was initially a radical hypothesis, in serious contradiction to the classical view of the structure of matter, for it had previously not been necessary to entertain the notion that any material particle smaller than the atom should exist by itself. J. J. Thomson himself wrote:

"At first there were very few who believed in the existence of these bodies smaller than atoms [i.e., the electrons]. I was even told long afterwards by a distinguished physicist who had been present at my lecture at the Royal Institute that he thought I had been 'pulling their legs.' I was not surprised at this, as I had myself come to this explanation of my experiments with great reluctance, and it was only after I was convinced that the experiments left no escape from it that I published my belief in the existence of bodies smaller than atoms."

However, the electron quickly explained and coordinated a wide variety of phenomena. The flow of electrons accounted not only for currents in cathode-ray tubes and photoelectric cells, but also in solid metallic conductors; in a completely evacuated tube, a heated metal wire (filament) "boils off" electrons which are attracted by a positively charged plate nearby and can be recirculated by a battery connected between plate and filament (*thermionic* tube, or diode as used in radio circuits); the overabundance or deficiency of one or more electrons would turn an atom into a negatively or positively charged ion, and so help to account for the facts of electrolysis; the electrons associated with a neutral atom, one in which the positive and negative charges balance, could be identified more definitely with the oscillators responsible for the emission of light on the classical

**Fig. 26.9** To overcome the retarding potential $V_r$, the photoelectron must have a sufficiently large initial kinetic energy.

(and later on the quantum) model of radiation.

More detailed studies of the photoelectric effect itself revealed the following experimental facts.

a) When, in the apparatus shown in Fig. 26.8, the light incident on the left-hand plate is of one given frequency ("monochromatic" light), the number of electrons ("photoelectrons") emitted per second per unit area of plate, as measured by the current through the tube and the wires, increases with the intensity of the incident light. This is not unexpected, for the energy of the incident light must somehow be absorbed by the electrons as they leave the cathode; in this conversion from light energy to the energy in an electric current, an increase in the former should be attended by a corresponding increase in the latter.

b) When photoelectrons of mass $m_e$ are "knocked out" from the metal by the incident light, they must leave with some initial kinetic energy $\frac{1}{2}m_e v^2$. To find this energy experimentally, we simply remove the battery connections in Fig. 26.8 which put an accelerating potential between the electrodes, and apply instead a potential difference in the opposite sense, a *retarding voltage* $V_r$ (see Fig. 26.9). Under these conditions only those electrons will reach the right side whose initial kinetic energy $\frac{1}{2}m_e v^2$ is as large as or larger than $V_r e$. In such an experiment we may increase $V_r$ until no photoelectric current flows at all. At that point the "retarding voltage" has become a "stopping" potential, $V_s$; that is, even those electrons emitted by the action of the light with the highest speed $v_{max}$ just fail to reach the plate; in equation form,

$$V_s e = \tfrac{1}{2} m_e v_{max}. \qquad (26.4)$$

We should expect to find that the maximum kinetic energy of photoelectrons increases in some way with the intensity of the incident light, just as high ocean waves will not only move more of the pebbles on the shore, but also give them greater speeds. However, the experiments showed that *the maximum kinetic energy of photoelectrons is not a function of the intensity of incident light of a given frequency*. This was unexpected. It implied that an electron in the cathode can draw from the incident light energy up to a fixed maximum only, rejecting the remaining energy if the light waves are very intense, and, if the light is very feeble, presumably accumulating that maximum energy before leaving the surface.

---

**Problem 26.9.** In 1902 P. Lenard determined that photoelectrons released from a metal on which the ultraviolet portion of sunlight was falling needed a stopping potential of 4.3 volts. What was the maximum speed of these electrons? (Be careful to use consistent units.)

---

c) It is natural to inquire next whether the $KE_{max}$ of photoelectrons depends on the frequency $v$ of the incident light, and experiments showed, though at first only qualitatively, that the KE does in fact increase with frequency. But it was not a case of simple proportionality: For each metal there was some frequency of light $v_0$ below which *no photoelectric emission could be obtained, no matter how strong the incident radiation*. This *threshold frequency*, $v_0$, for zinc is about $8.5 \times 10^{14}$ sec$^{-1}$ ($\lambda_0$ is about 3500 Å), and for metallic sodium is around $5 \times 10^{14}$ sec$^{-1}$ ($\lambda_0$ below 6000 Å). This observation raised the same type of conceptual difficulties as noted in (b) above concerning the mechanism by which the electron absorbs energy from the light.

d) Finally, it was also observed that whereas no amount of illumination below the threshold frequency would yield any photoelectrons from a given metal, even the feeblest intensity of light of a higher frequency than $v_0$ sufficed to obtain photoelectric currents *without any delay*. Modern experiments have confirmed that the time interval between the arrival of light and the departure of electrons from the surface is less than $3 \times 10^{-9}$ sec. That, however, is in flat contradiction to our expectations: We have already measured the energy photoelectrons may have, and we can calculate how long we should have to let light fall on a metal surface in a typical experiment before that much energy is available to the electron. One result of such a calculation is that an electron would have to wait for some seconds before it could accumulate the necessary energy from a weak beam of light—always assuming, as we do in the Maxwellian picture of light propagation, that the energy of a light wave is uniformly distributed over the whole wave front.

## 26.5  EINSTEIN'S PHOTON THEORY

The resolution of these problems came in another 1905 paper of Einstein,[*] entitled simply, "On a heuristic point of view concerning the generation and transformation of light."[†] Einstein (Fig. 26.10)

---

[*]    For a splendid biography of Einstein, see P. Frank, *Einstein, His Life and Times*, New York: Knopf, 1947.

[†]    *Heuristic* = useful as an aid in the advancement of a conceptual scheme, but not necessarily in the final correct form. An example of a heuristic device is the use of analogies. It is curious to note that Newton, in communicating to the Royal Society in 1675 a corpuscular view of light, entitled his paper "An Hypothesis Explaining the Properties of Light," and wrote that this hypothesis was also a heuristic one: "... though I shall not assume either this or any other hypothesis, not thinking it necessary to concern myself whether the properties of light discovered by men be explained by this or any other hypothesis capable of explaining them; yet while I am describing this, I shall sometimes, to avoid circumlocution ... speak of it as if I assumed it."

begins by paying tribute to the wide usefulness of Maxwell's theory of light; it "has proved itself excellently suited for the description of purely optical phenomena [reflection, refraction, interference, polarization, etc.], and will probably never be replaced by another theory." But this classical theory of light was designed and tested above all with reference to the *propagation* of light, its progress through space measured over relatively long time intervals. Therefore, he continues, one should not be surprised if contradictions with experience arise when the investigation turns to problems in which everything is contingent on a *momentary* interaction between light and matter. Two prominent examples of the latter case in which troubles have developed are the origin or emission of light from an oscillator and the transformation of light energy into the kinetic energy of photoelectrons.

Now Einstein proposes the following conceptual scheme: *The energy of light is not distributed evenly over the whole wave front, as the classical picture assumed, but rather is concentrated or localized in discrete small regions* or, if we may say so, in "lumps" or "bundles". A mental image is difficult to come by, but this analogy might help: If the same statement could be applied to the energy of water waves, we would expect that an incoming wave could move a floating cork up and down (i.e., give the cork energy) only at certain spots along the wave front; between those spots, the passing wave would leave the cork undisturbed.

One can see the roots of Einstein's idea in Planck's work. We remember that the Planckian oscillator can lose energy only in quanta of $hf$, and that the light thereby emitted has the frequency $v = f$. But in introducing a new concept of light emission Planck did not suggest in the least that a change is also necessary in the classical view concerning the nature and propagation of light waves after the moment of emission; the Maxwellian picture of expanding electromagnetic waves with evenly and continuously distributed energy remained effective. However, this raised conceptual problems of the following kind: If a Planckian

**Fig. 26.10**  Albert Einstein in 1912.

oscillator can change its energy only in steps of $hf$, how can it absorb energy from light in which the energy is thinly distributed? If not by a slow continuous increase in oscillation, which is prohibited by the quantum theory, how can the oscillator store up the stream of energy given to it by incident light until there is enough for a jump to a higher energy level? Even if a storage mechanism could be imagined, the fact of instantaneous emission of photoelectrons [item (d) in Section 26.4] shows that no such storage actually occurs.

Roughly speaking, Einstein therefore extended the quantum concept to the energy in the emitted light. The quantum of energy ($hf$ ergs) lost by the emitter does not spread out over the expanding wave front, but *remains intact as a quantum of light with energy hv ergs, where v of light = f of oscillator.* (To the quantum of light the name *photon* was later given.) Each wave front of light emitted by a glowing solid, or the sun, etc., must therefore be imagined to be studded with photons, like little dots on a balloon, traveling with the wave at the speed of light; and when the wave expands as it progresses, the light becomes weaker because the distance between neighboring photons increases. Ordinarily the light beams we observe in experiments on reflection, refraction, polarization, and the like make use of beams with substantial cross section; and then we are dealing with enormous numbers of photons on each wave front. Therefore the individuality of those photons is masked, just as an ordinary stone appears to our eyes to be made of continuous matter rather than of individual atoms. But in the absorption and transformation of light energy by an individual submicroscopic oscillator, as in the photoelectric effect, the finer "structure" of the light wave becomes important. That is why the classical theory of light remains completely valid and most appropriate for dealing with the ordinary problems of optics, whereas the quantum theory of light must be used to understand the interaction between light and individual atoms.

An application of these fundamental ideas by Einstein yielded at once a quantitative explanation of the photoelectric effect. A quantum of light, a photon, falling on an atom may or may not give it enough energy—all in one "bundle"—to loosen an electron;* if the photon's energy is enough, the electron can free itself by doing the work $W$ that must be done against the force of electric attraction binding it to the atoms of the cathode; if the photon's energy is less than $W$, the electron will not be emitted. This at once explains why emission is instantaneous if it takes place at all [item (d)].

---

* Note that no picture is supplied for imagining how the photon energy is converted to the energy of the electron.

But now the meaning of the threshold frequency $v_0$ is clear also: To break the electron from an atom at the surface, the photon energy $hv$ must be at least equal to $W$, or

$$hv_0 = W. \qquad (26.5)$$

Furthermore, we expect chemically different atoms to have different internal structure, so that $W$, and consequently also $v_0$, should vary from one element to another, as indeed it does.

For light of higher frequency than $v_0$ each photon has an energy $hv$ larger than $W$, so that energy is left over to let the photoelectron leave with some initial speed. Because light can penetrate the topmost layers of atoms, the photoelectric effect may take place below the surface. If the photoelectron then has to pass through layers of atoms before reaching the surface of the metal plate, it may lose some of its speed; but some electrons suffer no such losses and come off with the full speed $v_{max}$, directly toward the anode. For these we write

$$hv = \tfrac{1}{2}m_e v_{max}^2 + W. \qquad (26.6)$$

This is Einstein's equation for the photoelectric effect, an application, on the level of individual atoms, of the law of conservation of energy to the quantum theory of photoelectricity. It contains the algebraic expression of the experimental fact noted in Section 26.4(b) and (c) that the $KE_{max}$ of photoelectrons increases with the frequency $v$, but not with the intensity of the incident light. On the other hand, a larger intensity means more photons and larger *numbers* of photoelectrons per second, in accordance with observation (a). Thus Eq. (26.6) summarizes all the observations in *one* mathematical law.

But again, it did more. We must realize that at the time of Einstein's proposal, the data on hand were incomplete and often only qualitative. Precise and reproducible measurements of the photoelectric effect are still most difficult; for example, roughness or small impurities on the metal surfaces greatly affect the results for $v_0$ and $v_{max}$. Therefore Einstein's equation was essentially still a prediction rather than a summary of detailed quantitative facts. It was the fulfillment of these predictions in later experiments which gave Einstein's work such importance and changed the photon from a "heuristic point of view" to a fundamental part of contemporary scientific thought.*

In Eq. (26.6), the constant $W$ is often called the work function of the metal and $h$ is, of course, Planck's constant. The equation indicates how both quantities can be found by experiment, for if we were able—by careful experimentation with two different frequencies of light $v_1$ and $v_2$—to ascertain in terms of the necessary stopping potentials $V_{s1}$ and $V_{s2}$ what the term $\frac{1}{2}m_e v_{max}$ is for the photoelectrons in each case [see Eq. (26.4)], then we could write

$$h v_1 = V_{s1} e + W, \qquad h v_2 = V_{s2} e + W.$$

Therefore

$$h = (V_{s1} - V_{s2})e/(v_1 - v_2). \qquad (26.7)$$

After $h$ is obtained from Eq. (26.7), the constant $W$ can be quickly calculated from Eq. (26.3) or (26.6). And of course here is a chance to check the value for $h$ from Eq. (26.3), i.e., from Planck's theory of energy quantization in the radiating oscillator, a theory supplemented but by no means invalidated by Einstein's work. The credit for making the necessary precise measurement under extremely difficult experimental conditions belongs to the same man who had previously measured the electronic charge $e$, namely the American physicist Robert A. Millikan.† The result, obtained in 1916, was a coincidence of the values of $h$ by these two very different methods, within the experimental error.

---

\* Nominally, it was mainly for this work on the quantum theory of light rather than for his relativity theory that Einstein was awarded a Nobel prize in 1921.

† Recipient of the 1923 Nobel prize in physics.

**Problem 26.10.** In a typical experiment, $10^{-6}$ joules of light energy from a mercury arc are allowed to fall on one $cm^2$ of surface of iron during each second. Clean iron reflects perhaps 98% of the light, but even of the remaining energy only about 3% falls into the spectral region above the threshold frequency of the metal. (a) How much energy is available for the photoelectric effect? (b) If all the energy in the effective spectral region were of wavelength 2500 Å, how many photoelectrons should be emitted per second? (c) What is the current in the photoelectric tube in amperes? (d) If the threshold frequency is $1.1 \times 10^{15}$ $sec^{-1}$, what is the work function $W$ for this metal? (e) What is the stopping potential (in volts) for photoelectrons emitted by light at 2500 Å?

**Problem 26.11.** Sketch the graph you expect to obtain when plotting $v$, the frequency of light incident on a metal surface (abscissa), vs. $\frac{1}{2}m_e v_{max}^2$ or $V_s$ for the photoelectrons (ordinate). Should the line be straight? What is the physical meaning of the intercept on the horizontal axis? What is the equation for the line? What is the meaning of the slope of the line?

## 26.6 THE PHOTON–WAVE DILEMMA

Einstein's photon model of light was soon found fruitful in explaining other phenomena, e.g., photochemistry and anomalous specific heats, two topics to which Einstein himself greatly contributed. As has been aptly said:

"It is striking how thoroughly the theory affects our judgment of the importance of one field or another. Strictly from the experimental point of view, the photoelectric effect seems to be a rather remote corner of physics with technical applications of no great significance. However, Einstein's theory makes it clear that the photoelectric effect gives evidence of the nature of light, one of the greatest problems of physics."*

By showing that the energy of light was quantized, Einstein contributed directly toward the acceptance of Planck's original quantum theory, then still without general support. But, a little like Dalton, who repudiated Gay-Lussac and Avogadro, Planck himself was far from pleased with Einstein's photon. By its acceptance, he wrote in 1910, "the theory of light would be thrown back by centuries" to the time when the followers of Newton and Huygens fought one another on the issue of corpuscular vs. wave theory. All the fruits of Maxwell's great work would be lost by accepting a quantization of energy in the wave front—and all that "for the sake of a few still rather dubious speculations."†

Indeed, the inconvenience of having to deal with photons is very real. They represent bundles of energy without having ordinary *rest mass*; in this they differ from the Newtonian *corpuscles* of light, leaving only a faint analogy between them, even though it is customary to refer to Einstein's photon theory as the *corpuscular* or *particulate* theory of light. But our mind tends to insist on a good picture, and it requires great self-control to visualize a quantum of energy without bringing in some matter to which to attach it. It was a little easier to think of Maxwell's light energy spread evenly through the "field" along a wave front. Then, too, there are other questions: How large in volume and cross-sectional area is the "spot" on the wave front

where the photon is located? What could be the meaning of "wavelength" and "frequency" of light, which determine the energy content of the photon by $E = h\nu$, if the photon is, so to speak, only a dot on the wave front, and not part of a whole wave train? By what mechanism does the wave determine the path of the photon in such wave phenomena as interference and polarization? How does an electron absorb a photon?

In the past, one partial answer to such questions has been to hold concurrently two separate views on light—the wave and the photon model—and to apply one or the other as required by the problem. The more recent answer, which allows, in principle, a solution of every problem with physical meaning in this field, involves combining both views, and assuming the photons distributed over the wave front in a statistical way, i.e., not individually localized at a particular point. But at our level the more practical solution is, first of all, to realize that some of these questions, while possibly very disturbing, are asked on the basis of a mechanical view of atomic phenomena that may stem from an erroneous transfer of experience with large bodies obeying simple Newtonian laws. Thus the "size" of the photon is not a concept that we should expect to have the same meaning as the size of marbles and projectiles. Furthermore, photons (and atomic particles as well) differ from water waves, pebbles, and other large-scale entities in that one cannot make various experiments on the *same* subatomic entity. One can localize and measure and weigh a stone, find its velocity, etc., and all the while it is the same unchanged stone. But a photon, after it has been sent into a Geiger counter or a photographic emulsion, is no more; two photons on which we impress different experimental conditions—for example, in searching for wave and for corpuscular properties—are not, strictly speaking, the same entities.

As the physicist Max Born said:

"The ultimate origin of the difficulty lies in the fact (or philosophical principle) that we are com-

---

* From *Introduction to Atomic Physics*, by O. Oldenberg (1949).

† It is amusing to note that the work of Thomas Young in 1801, which introduced Huygens' wave theory and became the basis for Maxwell's view, was generally attacked at the time by the proponents of the current corpuscular theory on the basis that the wave theory, as one critic put it, "can have no other effect than to check the progress of science and renew all those wild phantoms of the imagination which . . . Newton put to flight from her temple." (Henry Brougham in the *Edinburgh Review*, Vol. 1, page 451, January 1803.)

pelled to use the words of common language when we wish to describe a phenomenon, not by logical or mathematical analysis, but by a picture appealing to the imagination. Common language has grown by everyday experience and can never surpass these limits. Classical physics has restricted itself to the use of concepts of this kind; by analyzing visible motions it has developed two ways of representing them by elementary processes: moving particles and waves. There is no other way of giving a pictorial description of motions—we have to apply it even in the region of atomic processes, where classical physics breaks down.

"Every process can be interpreted either in terms of corpuscles or in terms of waves, but on the other hand it is beyond our power to produce proof that it is actually corpuscles or waves with which we are dealing, for we cannot simultaneously determine all the other properties which are distinctive of a corpuscle or of a wave, as the case may be. We can therefore say that the wave and corpuscular descriptions are only to be regarded as complementary ways of viewing one and the same objective process, a process which only in definite limiting cases admits of complete pictorial interpretation . . . ."*

However, no matter how uncomfortable the wave-photon duality, or how great the disparity between concept and intuition at this stage, the photon theory stands on its proven power to explain, to predict, to stimulate further discoveries. We may finally illustrate this power by brief references to a few examples.

## 26.7 APPLICATIONS OF THE PHOTON CONCEPT

a) *Fluorescence.* A material is fluorescent if it absorbs light and immediately reradiates it. It had long been noticed that in general the frequency of

---

* From *Atomic Physics*, by Max Born, New York: Hafner, 4th edition (1946), page 92.

**Table 26.1**   Spectrum of electromagnetic radiation

| Name of radiation | Approximate range of wavelengths |
|---|---|
| Radio waves | a few meters and up |
| Microwaves | a few mm to meters |
| Infrared waves | 7500 Å to 0.01 cm |
| Visible light | 4000 to 7500 Å |
| Ultraviolet light | 100 to 4000 Å |
| X-rays | 0.1 Å to 500 Å |
| Gamma-rays | less than 0.5 Å |

such reradiated light is equal to or less than the original incident light. This is at once plausible when we imagine the following processes. The incident photon of energy $h\nu$ raises the atomic oscillator to a new level of higher energy than before. There is a general tendency of all oscillators to remain only briefly at high energy levels and to spend their time mostly at the lowest levels; thus the oscillator quickly reradiates the energy, though it may do so in two or more stages. If the absorbed photon is again given up in one process, we have resonance radiation (see Section 28.2), but if the original energy $h\nu$ is to be divided, the frequency $\nu_r$ of the reradiated light must be lower.

---

**Problem 26.12.** What does the above account reveal concerning the relationship between the frequency $f$ of the oscillator and $\nu$ of incident, absorbed light?

**Problem 26.13.** In some cases light is reradiated at a *higher* frequency. How can the model account for this?

---

b) *X-ray photons.* In 1895, W. C. Röntgen discovered that a beam of cathode rays (soon after identified as high-speed electrons) could create a new type of radiation when allowed to fall upon an obstacle such as the glass of the tube itself. In time,

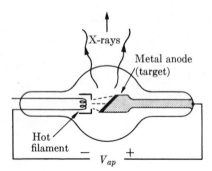

**Fig. 26.11** X-rays are generated when a stream of high-speed electrons falls on a target.

these so-called x-rays were shown to be high-frequency electromagnetic radiation of the same fundamental nature as radio waves, gamma-rays, and all other forms of "light" (see Table 26.1).

Nowadays x-rays are generated by the bombardment of a metal target forming the anode in a hot-filament cathode-ray tube. In a sense this is the reverse of the photoelectric effect; now photons are "knocked out" from the substance by incident electrons. We may provisionally imagine that the energy of the incident electron is first absorbed by the atomic oscillators, and then is reemitted as a photon when the oscillator returns to its original energy level. Therefore we expect that the energy of the x-ray photon is equal to that of the electrons in the beam, or

$$h\nu = \tfrac{1}{2}m_e v^2. \qquad (26.8)$$

In practice, we find that Eq. (26.8) gives the highest frequency of x-rays obtainable for a given electron beam. Furthermore, when such an x-ray in turn is allowed to fall on a clean metal surface, it causes the emission of photoelectrons of about the same velocity as that of the original beam (the work function being very small compared with the energies involved in x-rays).

**Problem 26.14.** Figure 26.11 shows schematically an apparatus for producing x-rays. If $V_{ap} = 50{,}000$ volts, (a) what is the speed of the electrons incident on the anode, and (b) what is the energy of the x-ray photons emitted? [*Note*: this figure gives only the maximum energy, for the electron may not give all its energy to an oscillator.]

### 26.8 QUANTIZATION IN SCIENCE

The enormous change in the natural sciences during the hundred years following 1808 can be correlated with a strangely similar discovery in each of such widely different fields as biology, chemistry, and physics—the discovery that there exists a structure, an elementary discrete quantization as it were, in the material on which each field builds. The case just discussed, that of energy quantization in systems that exhibit vibratory motion or periodic fluctuation (as in a Planck oscillator or in the electromagnetic field), is but the last of a series of such developments.

We recall that Dalton showed in the 1800s that chemical matter is not to be regarded as an infinitely divisible continuum, but that it consists of atoms which maintain their integrity in all chemical transactions. In biology, the cell theory of Schleiden (1838) and Schwann revealed that all living matter consists of individual cells and their products. Gregor Mendel's work (1865) led to the immensely fruitful idea that the material governing heredity contains a structure of definite particles, or genes, which may be handed on from one generation to the next without change.

Meanwhile the physics of heat, electricity, magnetism, and light—that is, that part of physics which the eighteenth century had visualized largely in terms of the action of imponderable fluids—was being rephrased in a similar way. Sensible heat was identified with the motion of discrete atoms. Electrical charge was found to be carried by the electron (whose motion would serve to set up magnetic fields, thereby reducing two topics to one). Finally, the energy of light and of atomic oscillators was found to be quantized also. In short, it was as if these new views in the various branches of science stemmed from a similarly directed change of the mental model used to comprehend fundamental quantities and processes, a change in the guiding idea from a *continuum* to a well-defined particle or *quantum*. In part, one can explain such changes by noting that the development of better instrumentation and techniques opened up a new level of observation. But this is largely arguing *a posteriori*, for such technical developments frequently were stimulated by the proposal of the idea which they later helped to fortify. One must allow the possibility of a change of tone and mood in the whole field, of a general conceptual movement not related to any single cause—yet another expression of the organic and dynamic nature of science.

## RECOMMENDED FOR FURTHER READING

H. A. Boorse and L. Motz, *The World of the Atom*, pages 462–501, 533–557. Extracts from the writings of Planck and Einstein, and biographical sketches.

M. J. Klein, "Thermodynamics and quanta in Planck's work," *Physics Today*, November 1966, pages 23–32.

Max Planck, "A scientific autobiography," translated from German by F. Gaynor, in *Scientific Autobiography and other papers*, New York: Philosophical Library, 1949

## SOURCES, INTERPRETATIONS, AND REFERENCE WORKS

B. Dibner, *Wilhelm Conrad Röntgen and the Discovery of X-rays*, New York: Watts, 1968

D. ter Haar, *The Old Quantum Theory*, New York: Pergamon Press, 1967: Part 1, Chapters I and II; Part 2, reprints of papers by Planck and Einstein, translated from German

A. Hermann, *The Genesis of Quantum Theory*, Cambridge, Mass.: MIT Press, 1971

M. Jammer, *The Conceptual Development of Quantum Mechanics*, New York: McGraw-Hill, 1966, Chapter 1.

M. J. Klein, "Max Planck and the beginnings of the quantum theory," *Archive for History of Exact Sciences*, Vol. 1, pages 459–479 (1962); "Einstein's first paper on quanta," *The Natural Philosopher*, Vol. 2, pages 57–86 (1963)

Max Planck, *A Survey of Physical Theory*, translated from German by R. Jones and D. H. Williams, Dover paperback reprint. A collection of lectures and essays.

E. T. Whittaker, *A History of the Theories of Aether and Electricity*, Vol. I, Chapter XII

## GENERAL BOOKS ON THE HISTORY OF MODERN PHYSICS (PRIMARILY TWENTIETH CENTURY)

A. d'Abro, *The Rise of the New Physics*, Dover paperback reprint of the 1939 edition; two volumes

R. J. Blin-Stoyle *et al.*, *Turning Points in Physics*, Amsterdam: North-Holland, 1959; Harper Torchbook paperback reprint

M. Capek, *Philosophical Impact of Contemporary Physics*, Princeton, N.J.: Van Nostrand, 1961

B. Cline, *Men Who Made a New Physics: Physicists and the Quantum Theory*, New York: Crowell, 1965; Signet Science Library paperback

Albert Einstein and Leopold Infeld, *The Evolution of Physics*, New York: Simon and Schuster, paperback reprint of the 1938 edition

G. P. Thomson, *The Inspiration of Science*, New York: Oxford University Press, 1961; Anchor Books paperback reprint

E. T. Whittaker, *A History of the Theories of Aether and Electricity*, Volume II

PART H

# THE ATOM AND THE UNIVERSE IN MODERN PHYSICS

# THE ATOM AND THE UNIVERSE IN MODERN PHYSICS

The first years of the twentieth century saw the publication of a handful of short papers by Max Planck and Albert Einstein which, in retrospect, appear to have been among the chief events to usher in a revolution in the fundamental concepts of physics. As we saw in Part G, the groundwork for what might be called the *second scientific revolution* (the first being the one initiated by Copernicus and essentially completed by Newton) had already been laid in the nineteenth century through studies of light and electromagnetic radiation, although scientists did not yet fully realize how the classical concepts were being undermined. Planck's quantum hypothesis seemed at first no more than a somewhat arbitrary modification of Maxwell's electromagnetic theory, and Planck himself devoted considerable effort to trying to fit it into the framework of classical theory. Similarly, when the Michelson-Morley experiment appeared to indicate that the earth's motion through the ether could never be observed, G. F. FitzGerald and H. A. Lorentz suggested as an *ad hoc* hypothesis that bodies moving through the ether might simply contract in length, in such a way that a measuring device would fail to indicate any motion.

It was Einstein who first turned from all *ad hoc* attempts to patch up the old, failing theories and insisted on the need for new physical laws, e.g., to explain the properties of radiation. His work was soon paralleled in related fields by that of Bohr, Rutherford, and many others in developing detailed theories of atomic structure. Out of the mathematical formulations of Schrödinger and Heisenberg came a new model of the atom, less easily visualizable in mechanical terms than the classical models, but far more effective. The revised model gave us the new, larger unity. That is, the contemporary theory of the atom is the meeting ground of modern science, where such fields as physics, chemistry, biology, astronomy, and engineering often join or overlap. The growth of science became virtually explosive, the harvest in pure and applied knowledge unparalleled, the promise for future developments ever more challenging.

In the meantime, as most other physicists concentrated on understanding the microstructure of matter, Einstein and a few other physicists, astronomers, and mathematicians were generalizing his relativity theory to deal with models of the structure and evolution of the entire universe. In returning to the subject of astronomy, with which we started this book, we find that we cannot tell the "end of the story" at all, for today's scientific journals and newspapers continue to announce discoveries which show that the astronomical universe is even more remarkable and manifold than we had so recently thought.

# Chapter 27

# RADIOACTIVITY AND
# THE NUCLEAR ATOM

## 27.1 ISOTOPES

The account to this point has given us an atom that has some well-defined and some still mysterious features. Around 1910, just about one century after Dalton's announcement of the atomic theory of chemistry, one could be fairly certain of a few facts. For example, the size of atoms was known, though only very roughly, from kinetic theory; and although the internal structure of the atom was not known, it was certain that some of the components had to be charged, for electric currents of all kinds needed the electron, and all theories of light agreed on electric oscillators in the atom. Also, it had been shown by Lenard in 1895 that a stream of cathode rays (electrons) would easily penetrate foils or sheets of material and, although very thin, such foils still represented many thousand layers of atoms, so that the fact of penetration led one to think of the atom no longer as a ball-like solid, but as a porous, open structure.

A new set of astonishing facts came on the scene after the discovery of radioactivity in 1896 by Henri Becquerel. Atoms of the heavier elements such as uranium, thorium, and radium were found to disintegrate spontaneously by the emission of either a high-speed $\alpha$-particle (with a positive charge numerically equal to that of two electrons and a structure that was later found to be that of the

helium nucleus), or a high-speed $\beta$-ray (a particle with the same rest mass and negative charge as any ordinary electron). In many elements, the emission of either particle was attended by that of a $\gamma$-ray, a very penetrating type of short-wavelength electromagnetic radiation.

The most central facts concerning natural radioactivity, which will serve as background for the following discussion, are not difficult to summarize in modern terms. Above all, an element (e.g., uranium) as prepared by a chemist is not the simplest starting point of investigation for radioactivity, because such a sample consists generally of a mixture of isotopes (atoms of the same chemical behavior but differing from one another in atomic weight*) in natural radioactivity, and in the reaction to nuclear bombardment.

All our references to isotopes so far have emphasized their differences in weight, and this is the difference on which was based the first separation of isotopes by J. J. Thomson in 1912. If the gas or

---

* This is contrary to the most fundamental assumptions of Dalton and all chemical thinking in the nineteenth century. Does this discovery invalidate the work of the chemists?

**Fig. 27.1** The path of a beam of negative ions is bent in a strong electric field. For given velocity and charge, light ions fall at $P_2$, heavy ones at $P_1$, nearer $O$.

vapor from one element is ionized so that the individual atoms carry a charge, a beam of such ions can be deflected by an electric field, a magnetic field, or both applied simultaneously. But the amount of deflection depends on the mass of the ions. A brief hint of the operation of an isotope separator (*mass spectrograph*) working on such a principle will suffice.

Figure 27.1 represents an ion beam emerging with initial speed $v$ from the box at the left into the gap between plates $A$ and $B$, held at a potential difference $V_{AB}$. If each ion has the same (negative) charge $q$, each experiences a constant force $F_{net} = (V_{AB}q)/d$ and an acceleration $a = F_{net}/m$ while traversing the distance $l$. Just as in the case of a projectile in a constant gravitational field and with horizontal initial velocity, the displacement $s_y$ of the ion during the time of transit $t = l/v$ under the action of force $F_{net}$ is

$$s_y = \tfrac{1}{2}at^2 = \frac{1}{2}\frac{F_{net}}{m}t^2 = \frac{1}{2}\frac{V_{AB}q}{dm}\frac{l^2}{v^2}. \quad (27.1)$$

It follows from Eq. (27.1) that for a beam of ions with equal charge $q$ and equal initial speed $v$, the displacement in transit between the plates is inversely proportional to the mass $m$ of the particular ion, all other factors ($V_{AB}$, $q$, $l$, $d$, and $v$) being constant for all ions by hypothesis. In continuing their flight, the "heaviest" ions therefore hit a detector (e.g.,

a photographic plate) at the spot $P_1$ near spot $O$ in Fig. 27.1, whereas the lighter ions may come to $P_2$. Conversely, from the points of incidence of the beam on the plate and from the construction of the apparatus, the mass $m$ of the ion may be determined.

In practice, it is hard to make this method work precisely; for example, the equality of $v$ and $q$ for all ions is by no means easily obtained. The actual mass spectograph is therefore designed on much more refined principles, but the final results are not different in kind from those given above.

---

**Problem 27.1.** On the principles given, design a mass spectograph to separate the three isotopes of uranium whose masses are approximately 238, 235, and 234 amu, where 1 amu $= 1.66 \times 10^{-24}$ g. Assume a singly ionized beam ($q = 1$ electronic charge) of initial speed $v = 10^6$ m/sec. You are free to choose $V_{AB}$, $l$, etc. What is the separation between the points on the detector (in cm) where these three isotopes arrive? If there appear traces of the arrival of some ions just halfway between the points of incidence for U$^{238}$ and U$^{235}$, what is the mass of these unknown ions? What assumptions underlie the last answer? How can we check them?

---

Now we are in a position to summarize some of the findings concerning isotopes. The most striking fact is probably the great variety. Some elements have only a single isotope (and, strangely, their placement numbers in the periodic table of elements, counting from hydrogen = 1 to uranium = 92, tend to be *odd*); examples are fluorine, sodium, aluminum, phosphorus. Other elements have a great number of natural isotopes; six or seven are not uncommon.\* Second, with the exception of

---

\*     Many more can usually be made artificially by nuclear bombardment or as fission products, but these are never stable.

some elements that have radioactive isotopes, the ratios of abundances for the isotopes are generally quite constant in a given element, no matter where it was obtained.* For example, neon gas, whose isotopes were in fact the first to be separated, has three isotopes of approximate mass 20, 21, and 22 amu, in respective amounts of 90.9%, 0.26%, and 8.82% of a chemically pure sample from any natural source. Oxygen, the gas which the chemist has selected to give him a standard of relative atomic weight (16.0000 exactly), also has three isotopes, $O^{16}$ (99.76%), $O^{17}$ (0.04%), and $O^{18}$ (0.20%).

Third, the mass of an isotope, measured relative to $O^{16}$, can be obtained with very great precision (six or even seven significant figures are now usually given) and invariably is very close to a whole number. This is, of course, the justification for writing the mass number, i.e., the superscript in $H^1$ or $U^{238}$, as a whole number. The mass number, which may best be defined as the true isotopic mass (in amu) rounded off to the nearest whole integer, is in fact always within less than 1% of the true isotopic mass, as Table 27.1 shows.

It is evident from the table that the isotopes of an element are generally not too far removed from one another in mass. The table also introduces a very convenient symbolism, namely, the subscript at the left of each element symbol, the *atomic number*, which refers to the place of the element in the chemist's periodic table. But at the same time the physicist has not strictly adhered to the chemist's convention in the calibration of the isotopic masses. Instead of standardizing on the atomic weight of the element oxygen, i.e., of the mixture of its three isotopes in their usual proportions, as the chemist does, the physicist needs a standard of his own for isotopic mass measurements, and has chosen to call the isotopic mass of $C^{12} \equiv 12.0000$ amu

**Table 27.1**  Isotopes of a few elements

| Element | Natural isotopes | Isotopic mass, amu | Abundance, % |
|---|---|---|---|
| Hydrogen | $_1H^1$ | 1.00783 | 99.985 |
|  | $_1H^2$ | 2.0141 | 0.015 |
| Helium | $_2He^3$ | 3.01603 | $\sim 10^{-6}$ |
|  | $_2He^4$ | 4.00260 | $\sim 100$ |
| Oxygen | $_8O^{16}$ | 15.99492 | 99.76 |
|  | $_8O^{17}$ | 16.99913 | 0.04 |
|  | $_8O^{18}$ | 17.99916 | 0.20 |
| Neon | $_{10}Ne^{20}$ | 19.99244 | 90.9 |
|  | $_{10}Ne^{21}$ | 20.99385 | 0.26 |
|  | $_{10}Ne^{22}$ | 21.99138 | 8.82 |
| Uranium | $_{92}U^{234}$ | 234.0409 | 0.006 |
|  | $_{92}U^{235}$ | 235.0439 | 0.72 |
|  | $_{92}U^{238}$ | 238.0508 | 99.27 |

exactly. By this convention, the weighted average of the oxygen isotopes as found in nature is 15.9994 amu on the physical scale as compared with the *chemical atomic mass* of 16.0000, by definition, on the chemical scale. The ratio between these figures is 0.99996, which is the conversion factor to use in going from one scale to the other. But actually we shall make little use of the conversion factor, and instead shall use the physical scale for discussions concerning isotopes and nuclei, and the chemical scale for problems in the chemical behavior of elements and compounds, as indeed we have tacitly done so far.

---

**Problem 27.2.** Using the data in Table 27.1, (a) calculate the average atomic weight of the element uranium on the physical scale, and (b) convert the result to the chemical scale.

---

\* Because chemical processes do not separate isotopes, and because the abundances are constant, the chemist generally does not need to concern himself at all with the concept.

**Problem 27.3.** The isotopes of ordinary argon are $_{18}A^{36}$ (35.9675 amu, 0.34%), $_{18}A^{38}$ (37.9627 amu, 0.06%), $_{18}A^{40}$ (39.9624 amu, 99.60%). It is by no means clear at present why the abundances are distributed as they are. Calculate (approximately) what the molecular weight of argon would be if the abundances of $_{18}A^{38}$ and $_{18}A^{40}$ had been exchanged. Consult the periodic table, and note how this would have straightened out one of the perplexing inversions.

**Problem 27.4.** Ordinary hydrogen gas consists largely of $_1H^1$ atoms (in molecules of $H_2$), but with some traces of the $_1H^2$ isotope (called *heavy hydrogen* or *deuterium\**) as indicated in Table 27.1. (a) Explain how the abundance of the latter could be increased by a physical process. (b) In the formation of water molecules, various isotopic combinations are possible; list them and give the molecular weight of each on the physical scale. (c) What is the molecular weight, on the chemical scale, of "heavy water," i.e., of water formed by the combination of $_1H^2$ atoms with the usual mixture of oxygen isotopes?

**Problem 27.5.** Note the doubly astonishing facts that the isotopic masses are all nearly *whole* numbers, but that they all do deviate a little from exact whole numbers. What may be the physical reasons for these two observations?

## 27.2 RADIOACTIVE HALF LIFE

After the recognition of the existence of isotopes, the next important step toward an understanding of radioactivity is the concept of *half life*. Ernest Rutherford, the man who originally noted the distinction between $\alpha$- and $\beta$-rays and who named them, suggested in 1902–1903, together with Frederick Soddy, that each of the radioactive ele-

---

\*    From the Greek *deuteros*, second.

**Fig. 27.2** Radioactive decay curve. After a time interval *T*, one-half the initial number of atoms has survived without radioactive decay.

ments has its own characteristic rate of emission of $\alpha$- or $\beta$-rays. If we make an experiment with any number ($N_0$) of atoms of a specific radioactive isotope (radioisotope), the number of atoms $N_t$ left after a time interval $t$ without having yet experienced radioactive decay follows a curve similar to Fig. 27.2. There eventually elapses a time interval $T$ when just half the original atoms ($N_0/2$) survive without having yet emitted a ray. Rutherford called $T$ the half life of the particular isotope, since an atom that has emitted no longer belongs to the type of isotopes under observation.

The principal advantage of the concept lies in the *experimental* fact that the half life can be defined at any point in the total existence of the radioisotope; no matter how old the sample is now, in another $T$ seconds half the present sample will still survive. Note that this is very different for a population of, say, humans instead of radioactive atoms. If we select $N_0$ babies, half that number may survive the 60th birthday, but of the $N_0/2$ oldsters none is likely to celebrate a 120th birthday. However, a radioisotope sample with a half life of 60 years will still have $N_0/4$ atoms intact after 120 years, $N_0/8$ after 180 years, and so forth.

**Table 27.2** Half lives of some isotopes

| Radioisotope | Half life $T$ | Type of particle emitted | Emission energy, MeV* |
|---|---|---|---|
| $_{92}U^{234}$ | $2.48 + 10^5$ yr | $\alpha$ | 4.77 |
| $_{92}U^{235}$ | $6.92 + 10^8$ yr | $\alpha$ | 4.50 |
| $_{92}U^{238}$ | $4.51 + 10^9$ yr | $\alpha$ | 4.20 |
| $_{88}Ra^{223}$ | 11.2 days | $\alpha$ | 5.71 |
| $_{88}Ra^{224}$ | 3.64 days | $\alpha$ | 5.68 |
| $_{88}Ra^{226}$ | $1.62 + 10^3$ yr | $\alpha$ | 4.78 |
| $_{88}Ra^{228}$ | 6.7 yr | $\beta$ | 0.04 |

There are several ways of writing an expression relating $N_0$, $N_t$, $T$, and the elapsed time $t$ of observation in a specific experiment (compare Section 12.8). We shall adopt the following form:

$$N_t = N_0 e^{-0.693t/T}. \qquad (27.2)$$

This is also the equation for the curve in Fig. 27.2. For a very active radioisotope, the half life $T$ as determined by experiment is short, the exponent is large, and the curve of $N_t$ vs. $t$ drops rapidly to smaller values; conversely for slowly decaying substances.

Some representative experimental values for $T$ of naturally occurring isotopes of uranium and of radium are assembled in Table 27.2. Note the variety of half-life values. There exist, for other elements, some larger and many far smaller values than those shown here; it is possible to speculate that radioactive elements, present in great quantities at the time of some hypothetical original creation of all elements, have since decayed so rapidly that no measurable traces are now left. On the other hand, other elements decay so slowly that only the most refined experiments could reveal that they do not have a constant, unchanging activity.

We also realize that a chemically pure sample of natural uranium, for example, would, by the simultaneous emission of its three isotopes, com-plicate the data quite seriously. Sometimes, as in the case of radium, the emission of one species is different in kind from the rest ($\beta$ instead of $\alpha$), so that, because of the much greater penetrating power of one radiation than the other, one can "shield out" the rest. Otherwise one can sometimes differentiate between the components by the difference of energies of similar rays, as the last column shows. The energy of $\alpha$-rays, given here in $10^6$ electron volts (MeV),* varies a little from one isotope to the other, and this may help to identify them individually, as by the difference of path length through air. Lastly, the decay curve for a mixture of isotopes differs from that given in Fig. 27.2 for a single isotope, and a resolution of the total curve into its components can reveal the activity of the separate elements.

**Problem 27.6.** You are given 1 g of radium $_{88}Ra^{226}$, the long-lived and useful one among the radium isotopes. (a) On the basis of Avogadro's hypothesis, how many atoms of radium are there in the initial sample? (b) Make a plot of $N_t$ versus time for a period of about 8000 years. How many atoms of $_{88}Ra^{226}$ are left at that time? How much total mass has been lost because of this particular decay? (c) What is the speed of the $\alpha$-particles of $Ra^{226}$?

**Problem 27.7.** For an illustration of the last point, consider Fig. 27.3, which sketches the $N_t$ versus $t$ graph for three isotopes $X^a$, $X^b$, and $X^c$, and for equal values of $N_0$ in each case. If these isotopes occur in the initial ratios of 6:3:2, derive graphic-

---

\* See Section 24.12: 1 electron volt = energy gained by an electron in crossing a 1-volt potential difference = $1.60 \times 10^{-19}$ joule; this unit of energy may be used for all particles. Multiples of the electron volt are frequently encountered: 1 keV = 1000 eV (k for the prefix "kilo" in the metric system); 1 MeV = 1,000,000 eV (million electron volts); 1 BeV or 1 GeV = 1,000,000,000 eV (billion electron volts; but unfortunately this is ambiguous in international circles because the British billion is $10^{12}$, so the term GeV is preferable; the G stands for the metric prefix "giga").

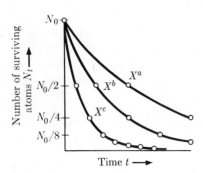

**Figure 27.3**

ally the shape of the composite decay curve. (Note that the number of undecayed atoms $N_t$ present at any time $t$ compared to the initial number $N_0$ is inferred from the relative strength of activity of the sample, e.g., from the number of particles emitted at $t$ and counted by a Geiger counter or measured by the ionization of the gas in an electroscope. Therefore in an actual experiment one does not plot $N_t$ vs. $t$, but the equivalent information, namely, the number of disintegrations or amount of ionization vs. $t$.)

**Problem 27.8.** Justify the statement that one can compare $N_0$ with $N_t$ after $t$ sec for a sample by comparing the number of particles emitted at the start and after $t$ sec.

## 27.3  RADIOACTIVE SERIES

The previous two sections have seriously upset our original idea, derived from chemistry and kinetic theory, of the atom as a stable unit, uniquely given for each element. It is no wonder that the concepts of radioactive decay and of isotopes were initially difficult to assimilate. But by early 1913, both these concepts had been fitted together by several work-

ers, principally by Soddy, in a truly beautiful theory that illuminated all the different aspects of the complex picture. One must accept, on *physical* evidence, that generally the product of radioactive decay is itself radioactive. Thus $Ra^{226}$ decays, with the emission of an $\alpha$-particle, into a substance that is itself an $\alpha$-particle emitter, although of much shorter half life and much greater energy of emission. This in turn means a new decay product, which is again radioactive in a very individual way. But on *chemical* evidence we must also accept that each of these decay products behaves chemically in a different manner than its immediate parent and its daughter product. For instance, the metal radium ($Ra^{226}$) is a decay product of thorium ($Th^{230}$) and in turn gives rise to a rare gas, radon ($Rn^{222}$). The great wealth of chemical and physical facts was summarized by recognizing that each radioisotope falls on one of three long chains, or *radioactive series*, stretching through the last rows of the periodic table. These natural series are named after an element at or near the head of each, the *uranium*, the *thorium*, and the *actinium series*.

Table 27.3 gives the first of the chains by showing in some detail the inescapable life history of each atom of $_{92}U^{238}$. After a time $T$ of $4.5 \times 10^9$ years the chances are equal that it will have decayed. Having lost an $\alpha$-particle, the atom is now about 4 amu lighter, hence its mass number is 234. Although we may name it Uranium $X_1$, the chemist will prove to us that the new atom is thorium, and therefore belongs in the ninetieth place in the periodic table (atomic number 90). This atom of $_{90}Th^{234}$ in turn emits a $\beta$-ray, which does not change the mass number materially, but which, again on chemical evidence, leaves us with an atom of protactinium, $_{91}Pa^{234}$.

The rest of the chain shows that eventually the atom is for a time radium, then radon, and eventually stable lead ($_{82}Pb^{206}$), although on the way it occasionally reverses its progress through the periodic table to become temporarily another isotope of an element it had previously represented. Each decay is accomplished by the emission of an

Table 27.3  The uranium series*

| Energy (MeV) | | 4.20 | 0.19 | 2.32 | 4.77 | 4.68 | 4.78 | 5.49 | 6.0 |
|---|---|---|---|---|---|---|---|---|---|
| Emission | | $\alpha$ | $\beta, \gamma$ | $\beta, \gamma$ | $\alpha$ | $\alpha, \gamma$ | $\alpha, \gamma$ | $\alpha$ | $\alpha$ |
| | $_{92}U^{238}$ | $_{90}Th^{234}$ | $_{91}Pa^{234}$ | $_{92}U^{234}$ | $_{90}Th^{230}$ | $_{88}Ra^{226}$ | $_{86}Rn^{222}$ | $_{84}Po^{218}$ | |
| Alternative symbol | UI | $UX_1$ | $UX_2$ | UII | Io | — | — | RaA | |
| Half life | $4.5 + 10^9$ yr | 24.1 days | 1.18 min | $2.5 + 10^5$ yr | $8 + 10^4$ yr | 1620 yr | 3.82 days | 3.05 min | |

| Energy (MeV) | | 0.7 | 3.17, 1.8 | 7.68 | 0.02 | 1.16 | 5.3 |
|---|---|---|---|---|---|---|---|
| Emission | | $\beta, \gamma$ | $\beta, \gamma$ | $\alpha$ | $\beta, \gamma$ | $\beta$ | $\alpha, \gamma$ |
| | $_{82}Pb^{214}$ | $_{83}Bi^{214}$ | $_{84}Po^{214}$ | $_{82}Pb^{210}$ | $_{83}Bi^{210}$ | $_{84}Po^{210}$ | $_{82}Pb^{206}$ |
| Alternative symbol | RaB | RaC | RaC' | RaD | RaE | RaF | RaG |
| Half life | 26.8 min | 19.7 min | $1.69 + 10^{-4}$ sec | 19.4 yr | 5 days | 1.38 days | (stable lead) |

---

*    There are in this and all other series points at which a small fraction of the total radioactive decay proceeds in a branch line with different characteristics.  These have been omitted for simplicity.

$\alpha$- or $\beta$-particle of stated energy, usually associated with a $\gamma$-ray.

Examination of this and the other two series discloses a regularity of behaviour which has the name *displacement law of radioactivity: When an atom undergoes $\alpha$-decay, its mass number decreases by 4 and its atomic number decreases by 2; when an atom undergoes $\beta$-decay, however, its mass number does not change, and its atomic number increases by 1.* Mark well that the full law was a generalization from experimental facts and, like the existence of isotopes or of radioactive decay as such, was not comprehensible in terms of any model of the atom as derived from the physical sciences of the preceding era.

**Problem 27.9.** The actinium series begins with $_{92}U^{235}$; each atom sends out in succession the following particles: $\alpha, \beta, \alpha, \beta, \alpha, \alpha, \alpha, \alpha, \beta, \beta, \alpha$. From this information and by consulting the periodic table, write out an account of the actinium series on the model of Table 27.3, complete with the necessary symbols and the figures for atomic mass and atomic number.

**Problem 27.10.** From plutonium ($_{94}Pu^{241}$), artificially prepared by bombarding uranium in a nuclear reactor, stems one of the many artificial radioactive series ("neptunium series"). The first members of the chain are $_{94}Pu^{241}$, $_{95}Am^{241}$, $_{93}Np^{237}$, $_{91}Pa^{233}$, $_{92}U^{233}$, $_{90}Th^{229}$. Unlike all three natural series, this one does not end in a stable isotope of lead, but in $_{83}Bi^{209}$. For the first six members, draw the series as in the previous problem or Table 27.3.

**Problem 27.11.** The third natural series originates from thorium, $_{90}Th^{232}$. On the basis of the following information, replace the symbols here used and write the complete series.   Indicate alternative possibilities where they can exist:

$$_{90}\text{Th}^{232} \xrightarrow[?]{\alpha} \text{MsTh}_1? \xrightarrow[?]{?} {}_{89}\text{MsTh}_2? \xrightarrow[?]{\beta} \text{RaTh}? \xrightarrow{\alpha}$$

$$\rightarrow \text{ThX}? \xrightarrow[?]{?} \text{Tn}? \xrightarrow[?]{?} {}_{84}\text{ThA}? \xrightarrow[?]{?} \text{ThB}? \xrightarrow{\beta}$$

$$\rightarrow \text{ThC}? \xrightarrow[?]{?} {}_{84}\text{ThC}^1? \xrightarrow[?]{?} \text{ThD}^{208} \text{ (stable)}$$

Describe briefly what experiments might have been made to determine the series as given above. (Note that this example indicates the type of interpolation that was initially often required.)

**Problem 27.12.** Make a plot on graph paper of atomic number vs. mass number for the elements in the uranium series.   Enter each element as a labeled dot, and draw a straight arrow from each element to its immediate decay product. This gives one of the useful and widely current representations of radioactive series.

## 27.4 RUTHERFORD'S NUCLEAR MODEL

We return now to the state of affairs at about 1910. The chemical and physical facts of radioactivity presented a stupendous challenge. The existence of radio elements, of $\alpha$-, $\beta$-, and $\gamma$-radiation, of separate and measurable half lives was well established. The idea of isotopes, arising both from the deflection of ion beams and from the first glimpses of radioactive series, was just beginning to be voiced. Although the need to accommodate such findings in any one model of the atom seemed to complicate the picture beyond the power of imagination, the solution was hidden in the problem itself after all. The swift $\alpha$- and $\beta$-rays, the by-products, as it were,

of radioactivity, provided exactly the tool needed to take the next important step, the means for probing the perplexing interior of the atom.

Soon after Lenard's experiment on the transmission of cathode rays through thin foils, several attempts were made to construct an atomic model which would incorporate the porosity of the atom as indicated by such experiments, as well as the presence of electrons revealed by the facts of electricity and light emission.   It was variously suggested that each atom was made of myriads of massive fragments, each having a positive and a negative charge of equal amounts (Lenard's *dynamids*, 1903), or that it was a homogeneous sphere with thinly distributed mass and positive electricity, inside which there were negatively charged electrons either placed like seeds in a pumpkin or moving in circular orbits (J. J. Thomson, 1898, 1904).   The attempt was made to arrange the electron positions or orbits in a stable fashion so that the mutual electric forces would not lead one to expect a dispersal or collapse of the whole structure (Thomson, Nagaoka, 1904).   But the persistent and fatal shortcoming of such models was that no way existed of using them to explain the old puzzle of line emission spectra. Even if one wished to identify the source of emission of light with a small vibration of the electrons to either side of their equilibrium position or orbit, yet the known, complex spectra did not follow.

An entirely new basis for all speculations along such lines was provided in 1909–1911 by Ernest Rutherford (1871–1937), the New Zealander who had already shown an unmatched virtuosity in experimental physics by his work on radioactive emission in Canada, and who had moved in 1907 to Manchester University in England, where he headed a most productive research laboratory.* Rutherford had noticed that a stream of $\alpha$-particles, when it passed through a thin film of mica or metal, was somewhat broadened (*scattered*) by the material

_____

*    Nobel prize in chemistry, 1908.

through which it had passed.  This is interesting, for the amount of scattering might give a hint concerning the disposition of the atom's mass, then generally thought to be fairly evenly distributed, as in Thomson's ill-defined sphere of positive electricity.  But just how Rutherford was led to the discovery that almost all the mass of the atom exists in a small central *nucleus* has been best summarized by Rutherford himself:

"...I would like to use this example to show how you often stumble upon facts by accident.  In the early days I had observed the scattering of α-particles, and Dr. Geiger in my laboratory had examined it in detail.  He found, in thin pieces of heavy metal, that the scattering was usually small, of the order of one degree.  One day Geiger came to me and said, 'Don't you think that young Marsden, whom I am training in radioactive methods, ought to begin a small research?'  Now I had thought that, too, so I said, 'Why not let him see if any α-particles can be scattered through a large angle?'  I may tell you in confidence that I did not believe that they would be, since we knew that the α-particle was a very fast, massive particle, with a great deal of energy, and you could show that if the scattering was due to the accumulated effect of a number of small scatterings the chance of an α-particle's being scattered backwards was very small.  Then I remember two or three days later Geiger coming to me in great excitement and saying, 'We have been able to get some of the α-particles coming backwards ....'  It was quite the most incredible event that has ever happened to me in my life.  It was almost as incredible as if you fired a 15-inch shell at a piece of tissue paper and it came back and hit you.  On consideration, I realized that this scattering backwards must be the result of a single collision, and when I made calculations I saw that it was impossible to get anything of that order of magnitude unless you took a system in which the greater part of the mass of the atom was concentrated in a minute nucleus.  It was then that I had the idea of an atom with a minute massive center carrying a

**Fig. 27.4**  Lord Rutherford (1871–1937).

charge.  I worked out mathematically what laws the scattering should obey, and I found that the number of particles scattered through a given angle should be proportional to the thickness of the scattering foil, the square of the nuclear charge, and inversely proportional to the fourth power of the velocity.  These deductions were later verified by Geiger and Marsden in a series of beautiful experiments."*

This, then, was the birth of the nuclear atom— a conception almost immediately accepted into science despite the initial uncertainties of detail.

---

*    From the essay (1936) "The Development of the Theory of Atomic Structure," by Lord Rutherford, in *Background to Modern Science*, New York: Macmillan (1940).

Because it gave the picture of a central concentration of a (positively charged) mass about which the light electrons can swarm at relatively great distances, the whole problem of atomic theory was immediately simplified. For now one could think in separate terms about the very different phenomena for which the atom had to be made responsible; the extranuclear electrons would be concerned with chemical behavior, electricity, the emission of light, etc., whereas the nucleus would serve for the phenomena of radioactivity, the explanation of isotopes, and the like.

Now we turn to the quantitative details of the work of Rutherford and his co-workers, H. Geiger and E. Marsden. The projectiles used were α-rays with speeds of about $2 \times 10^7$ m/sec, fired in a narrow beam against metallic films, e.g., a gold foil only $4 \times 10^{-7}$ m thick. On the basis of the approximate atomic dimensions deduced from kinetic theory, one may expect such a film to represent about $10^3$ layers of atoms. Most of the α-particles do go right through the foil without noticeable deviation, and therefore we have additional evidence for the proposition that the atoms are *not* "solid". They must contain electrons, but these can hardly be held responsible for changing the path of those α-particles that are scattered so drastically as to come almost straight back. At each encounter with an electron, the α-particle can experience only a very minute change in momentum because of the great disparity between the respective masses. The α-particle has a mass of $m_\alpha \doteq 6.85 \times 10^{-24}$ g, whereas the mass of the electron, $m_e$ is only $9.11 \times 10^{-28}$ g, i.e., about $\frac{1}{7500}$ as large; even if we assume a head-on collision, the forward component of the velocity of the α-particle changes only very little each time—to be precise, about 0.01%. The probability is vanishingly small that an α-particle, in thousands of successive encounters with electrons, will experience small deviations from its original direction that are so predominantly in one sense that the net result is an actual turning back. The alternative explanation of the observed scattering is, therefore, that an

α-particle may meet a single object in the foil that is heavy enough to turn it back in one encounter, just as a bullet ricochets back from a boulder but not from a loose pile of sand particles.

---

**Problem. 27.13.** An α-particle of mass $m_\alpha$ with initial speed $v_{\alpha1}$ approaches an electron with mass $m_e$, the latter initially at rest. After the head-on (perfectly elastic) encounter, the α-particle continues with speed $v_{\alpha2}$ and the electron moves with speed $v_e$. (a) Write the two equations for conservation of momentum and of energy for this event, and prove that

$$v_e = \left(\frac{2m_\alpha}{m_\alpha + m_e}\right) \cdot v_{\alpha1}, \qquad v_{\alpha2} = \left(\frac{m_\alpha - m_e}{m_\alpha + m_e}\right) v_{\alpha1}.$$

(b) Calculate the actual change in speed and in momentum for an α-particle in one encounter (in percentage).

---

In the paragraphs above we have assumed that an α-particle interacts with an electron as though this were a mechanical collision between marbles, but of course we should consider now that both are oppositely charged particles. The mechanism of interaction is then primarily not an elastic collision but rather a mutual attraction, the same effect by which the sun causes a comet that is passing close by to turn around after half circling the sun. To be precise, the comet's path is hyperbolic if the comet does not remain in the solar system but escapes again from the gravitational attraction between it and the sun. But as for the momentum and energy changes, those are the same no matter whether the mechanism of interaction is elastic or electric. The fact of occasional large-angle scattering of α-particles is not explainable by multiple encounters with electrons even if we consider that the force of interaction is electric attraction.

On the other hand, visualize the hypothetical behavior of an α-particle if it encounters a small heavy nucleus of positively charged mass. Figure

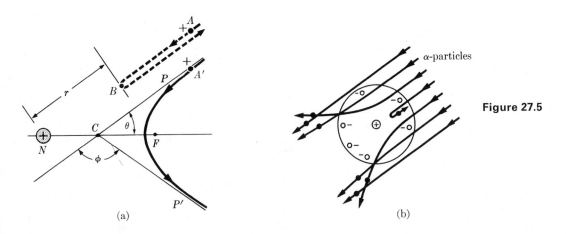

Figure 27.5

(a)                                             (b)

27.5(a) is based on a figure in Rutherford's publication "The scattering of α- and β-particles by matter and the structure of the atom," in the British scientific journal, *Philosophical Magazine and Journal of Science* ("Phil. Mag.") in 1911, which may be said to have laid the foundation for the modern theory of atomic structure. The α-particle at A approaches a nucleus at N, is repelled by the (Coulomb's law) force between the positively charged bodies, and follows a path PP' that must be a hyperbola.*

Now one may inquire how large the angle of deviation or scattering from the original direction should be for a parallel beam of α-particles approaching a layer of distributed nuclei. Particle A', for example, is heading directly toward N, approaches only up to point B, and then returns directly back; most other particles do not pass so

close to any nuclei and so do not experience such serious deflections [Fig. 27.5(b)].

Rutherford considered the effect on α-particle scattering of all important factors such as charge on the nuclei, speed of incident particle $v_\alpha$, and thickness of film $t$, and derived the following predictions. If scattering occurs by single encounters with small positive nuclei, the fraction of α-particles scattered by angle $\phi$ should be directly proportional to (a) $[\sin(\phi/2)]^{-4}$, (b) the foil thickness $t$, (c) the magnitude of the nuclear charge, and (d) $(v_\alpha)^{-4}$. Predictions (a), (b), and (d) could be checked quickly; in fact, the previous year Geiger had published scattering measurements which were most favorable, and a more complete and fully convincing check was given in 1913 by Geiger and Marsden.

Figure 27.6 represents schematically the apparatus used by Geiger and Marsden. B is a lead box from which an α-emitter (radon) sends out a beam through a small diaphragm; the foil F scatters the rays in varying degrees through different angles $\phi$. The number of scattered particles at each angle is found by letting them fall on a zinc-sulphide screen S, where each incidence causes a momentary flash of light (scintillation) observable and countable through the microscope M; S is fixed to M and can be rotated in this plane up to $\phi = 150°$.

---

*    On the basis of Newton's *Principia* it had long been known that if a body A moves toward another, N, against a force of repulsion which is inversely proportional to the square of the distance between them, as is the case with these positively charged particles, then the path taken by A is the same as if, instead of N, there were a similar but oppositely charged body at F, the focus of the curve. This defines a hyperbola with center at C.

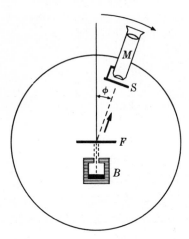

**Fig. 27.6**  Diagram of apparatus for verifying Rutherford's theory of alpha-particle scattering.

**Problem 27.14.** A beam of α-particles of equal speed is directed at a metal foil; about 100 are observed to penetrate the foil each minute without any scattering. Plot on the basis of Rutherford's prediction a graph of number of scintillations observed per minute vs. angle $\phi$ of scattering.

Although the observation at different angles $\phi$, with foils of different $t$, and with α-particle beams of different $v_\alpha$, confirmed three of Rutherford's four predictions, there appeared no direct way of checking on the effect of the nuclear charge on scattering. However, one could turn the argument around, assume that the experiments had sufficiently proved the general supposition, and then calculate the nuclear charge from the data. This was done for carbon (in a sheet of paraffin wax), and for metal foils from aluminum to gold, and it turned out that for these materials the nuclear charge, expressed

in multiples of the numerical electronic charge $e$ ($1.6 \times 10^{-19}$ coul) was roughly $\frac{1}{2}$ (at. wt) $\times e$. This would be $6e$ for carbon, $13e$ or $14e$ for aluminum, $78e$ or $79e$ for gold, and so on.

These experiments therefore provided simultaneously a vindication of Rutherford's hypothesis, and an important piece of new information, for if the nucleus carries a charge equal to ($\frac{1}{2}$ at. wt) units of electronic charges, there must be equally many electrons around it to provide for the electric neutrality of the whole atom. In fact, previous experiments on the scattering of x-rays and $\beta$-particles had given the number of electrons about equal to ($\frac{1}{2}$ at. wt) also. It was pointed out soon afterward by A. van den Broek in 1913 that the important comparison to make was not between nuclear charge and half the atomic weight, but between the former and *atomic number* ($Z$). It is an astonishing fact that for most substances, $Z = \frac{1}{2}$ at. wt within narrow limits; for example, $Z$ for carbon is 6, the at. wt is 12.010. Thus the experimentally founded suggestion that the number of positive charges on the nucleus or, equally well, of electrons around the nucleus, be simply equal to the atomic number $Z$ at once made the picture of the atom more precise and complete. A hydrogen atom ($Z = 1$) in its neutral state has one electron moving around its positively charged nucleus, a helium atom ($Z = 2$) has 2 electrons, and so on up to uranium ($Z = 92$) with 92 extranuclear electrons. This simple scheme was made more plausible when in ion beams in the mass spectrograph there were found singly ionized hydrogen ($H^+$) and even doubly ionized helium ($He^{++}$), but never $H^{++}$ or $He^{+++}$, evidently because H atoms had only a single electron to lose, and He atoms only two.

A second important consequence from scattering experiments concerned the size of the nucleus, which Rutherford estimated by an ingenious argument. Since the rest of his deductions depended on the validity of Coulomb's law of electric forces between the nucleus and α-particle, and since these deductions (as that of the variation of scattering with angle) had been borne out, we may assume

Coulomb's law to hold during the scattering process. Now surely it would not hold if the projectile were to come so close to the nucleus as to deform it elastically or touch it in any way. Therefore the closest distance of mutual approach between the centers of the target nucleus and the $\alpha$-particle must be less than the sum of their radii. The actual distance of nearest approach can, however, be calculated, for in such a case the $\alpha$-particle must be heading straight toward the nucleus [see path $A'B$, Fig. 27.5(a)] and penetrates to point $B$ at a distance $NB \equiv r$ from the nucleus, where the kinetic energy, by the work against mutual repulsion, is all converted to potential energy; after this the projectile turns back like a stone at the top of its trajectory. At $B$, the point of momentary rest, $KE_{init} = PE$ or, numerically,

$$\tfrac{1}{2}m_\alpha v_\alpha^2 = \frac{k_e(Ze) \cdot (2e)}{r},$$

where $v_\alpha$ is the initial speed of the $\alpha$-rays, $(2e)$ its charge, and $(Ze)$ the charge on the target nucleus $N$. [Here we have assumed that the potential energy of the $\alpha$-particle at distance $r$ is given by Eq. (24.5) for point charges.]

For a gold foil bombarded with $\alpha$-particles of about $2 \times 10^9$ cm/sec speed, the distance $r$ can be calculated to be about $3 \times 10^{-12}$ cm (see Problem 27.15). Compared to the overall size of the atom ($\sim 10^{-10}$ m), the nuclear size must therefore be of the order of $10^{-14}$ or $10^{-15}$ m, that is, less than $\frac{1}{10,000}$ as large! The mass of electrons being almost negligible compared with that of even the lightest atom as a whole, most of the mass of the atom is thus concentrated within a distance that bears to the total atomic radius a relation similar to that between the sun's radius and the radius of the total solar system measured to Pluto's orbit. One may assume that it is the presence of moving electrons around the nucleus at the relatively large distance of $10^{-10}$ m which determines the "size" of the atom. Evidently this scheme presents us with an atom that is mostly empty space, as indeed it must be to account for the relative ease of penetration of $\alpha$- and $\beta$-radiation through thousands of atoms, as in a metallic foil or in the gas of a cloud chamber.

---

**Problem 27.15.** Determine the distance $r$ of nearest approach between an $\alpha$-particle with $v_\alpha = 2 \times 10^7$ m/sec, and a nucleus of a gold atom. (This calculation was made by Rutherford in his 1911 paper, but he considered $Z \doteq \frac{1}{2}$ at. wt.)

---

Immediately, the developing atom raises three issues: (a) What is the structure of the nucleus? (b) What keeps the nucleus in general from exploding by the mutual repulsion of its components if it is made of densely packed charges of equal sign? (c) How are the electrons arranged around the nucleus?

To the first of these questions, Rutherford initially could do no more than to suggest that at least in some elements the charge and mass of the nucleus can be explained by assuming a grouping of a number of helium nuclei. "This may be only a coincidence, but it is certainly suggestive in view of the expulsion of helium atoms [$\alpha$-particle] carrying two unit charges from radioactive matter."

---

**Problem 27.16.** According to this suggestion, how many helium nuclei would make up one nucleus of carbon, of nitrogen, of oxygen? Does the scheme work for all elements in the periodic table?

---

While therefore the first question was actually unsolvable at the time, the second one concerning nuclear stability was even more puzzling, and had to be dismissed by Rutherford as follows: "The question of the stability of the atoms proposed need not be considered at this stage, for this will obviously depend on the minute structure of the atom, and on the motion of the constituent charged

parts." Rutherford's later work on artificial trans-
mutations opened the way to a better answer to
both problems, as we shall see at once.

Finally, the third question raised above was
destined to be solved soon by Bohr, another co-
worker of Rutherford's. At the time, Rutherford
simply noted that the Japanese physicist Nagaoka
had speculated in 1904 on a model of the atom in
which a central nucleus was pictured as surrounded by electrical charges revolving in rings
similar to those seen around Saturn.

---

**Problem 27.17.** If in a gold foil the atoms are so
close as to touch one another, with a distance of
about $3 \times 10^{-10}$ m between their nuclei, and if the
radii of the nuclei of the atoms and of the α-particles
are for this rough calculation each assumed to be
about $3 \times 10^{-14}$ m, calculate the probability of an
α-particle approaching a nucleus virtually head-on
in any one layer of metal atoms. [*Hint*: Divide the
diameter of a gold atom into small sections, each
equal to the diameter of the α-particle; then assume
that the chances for the α-particle to pass through
any section are evenly distributed.]

## 27.5 MOSELEY'S X-RAY SPECTRA

In the year 1913, when Geiger and Marsden's
scattering experiments completely verified the
main features of the theory of the nuclear atom,
there came yet another profound contribution from
Rutherford's laboratory. There the young physicist H. G. J. Moseley (1887–1915) (Fig. 27.7) had
been investigating the wavelengths of x-rays emitted when a stream of high-speed electrons (cathode
rays) impinges on metallic targets (see Fig. 26.11).
A plot of the intensity of x-rays vs. wavelength
gives a graph rather like Fig. 27.8. It reminds one
somewhat of a continuous emission spectrum of
light from a hot solid, but with a line spectrum
superimposed on it. Indeed, we might analyze the
graph in two parts. The continuous spectrum of so-

**Fig. 27.7** H. G. J. Moseley (1887–1915). (From the
collection of the late Professor George Sarton)

called *general radiation*, which serves as "background" for the sharply defined spikes, is generated
by the conversion of part of the kinetic energy of
cathode rays into photons as they pass through the
atoms of the target. Those electrons that give up
all their energy in one interaction with the target
atoms give rise to the highest frequency x-ray
photons by the relation $h\nu = \frac{1}{2}m_e v^2$ (Eq. 26.8), but
as $\lambda = c/\nu$, the shortest wavelength in Fig. 27.8 is
$\lambda_{\min} = hc/\frac{1}{2}m_e v^2$, and is well defined for a given
electron beam.

[27.5]

**Figure 27.8**

Table 27.4 Wavelengths of prominent characteristic K lines (in $10^{-8}$ cm)

| Element | At. number | At. wt | $\lambda$ of $K_\alpha$ | $\lambda$ of $K_\beta$ |
|---|---|---|---|---|
| Manganese (Mn) | 25 | 54.93 | 2.10 | 1.91 |
| Iron        (Fe) | 26 | 55.85 | 1.93 | 1.75 |
| Cobalt      (Co) | 27 | 58.94 | 1.79 | 1.62 |
| Nickel      (Ni) | 28 | 58.69 | 1.66 | 1.50 |
| Copper      (Cu) | 29 | 63.54 | 1.54 | 1.39 |

and $K_\beta$ lines of five neighboring metallic elements in the periodic table; in each row, the wavelength of $K_\alpha$ or $K_\beta$ is shorter than in the preceding row.

At once, several conclusions could be drawn from this discovery of regularity.

a) From his initial data on the characteristic radiation for nine elements from calcium to copper, and from his further work in 1914, which embraced elements between aluminum ($Z = 13$) and gold ($Z = 79$), Moseley found that the frequency $v$ of the $K_\alpha$ line could generally be given by an empirical equation that may be written

$$v \text{ (sec}^{-1}) = 2.48 \times 10^{15}(Z - 1)^2. \qquad (27.3)$$

(An analogous equation was given for $L_\alpha$ lines.) Such an equation has immediate uses. For example, although the metal scandium ($Z = 21$) was not then available as a target material, the frequency of $K_\alpha$ could be calculated with considerable confidence. When that line was later measured, the agreement was, in fact, found to be good. This is reminiscent of the story connected with Balmer's formula for the ultraviolet portion of the hydrogen spectrum, and indeed, both formulas must find a similar type of physical explanation in the final theory of the atom.

**Problem 27.18.** (a) What voltage must be applied to the x-ray tube to obtain a continuous spectrum with $\lambda_{min} = 0.5$ Å? (Recall that, in consistent units, $\frac{1}{2}mv_e^2 = Ve$, where $V$ = potential difference between electrodes, $e$ = charge on electron.) (b) How would you use these relations to measure the ratio $h/e$ with precision? (c) Show that $\lambda_{min} = 12,400/V$, where $V$ is in volts.

Moseley, however, was not concerned with the continuous spectrum (which was not fully understood at the time). His work concentrated on the other part of the total x-ray spectrum, the "spikes" or lines or *characteristic radiation*. For each of his targets, the result was similar to Fig. 27.8; there were several such lines, usually in two groups named K and L, although later, better equipment brought out more groups (M, N, . . .) and, in each group, usually many more lines. Nevertheless, even the initial data revealed an astonishing simplicity. The lines in the characteristic x-ray spectrum progressively shifted to shorter wavelengths as the *atomic number* of the target material increased. Table 27.4 compares the prominent $K_\alpha$

**Problem 27.19.** The prominent $K_\alpha$ line of scandium has a wavelength of $3.03 \times 10^{-10}$ m. Check to what extent Moseley's empirical formula holds for this line.

b) The predictive value of Moseley's formula is, of course, most gratifying. So much confidence could later be placed in it that the identification of a newly discovered element depended on its having characteristic x-ray lines that fitted properly between those of its immediate neighbors in the periodic table, and conversely, wherever a great gap occurred between the wavelengths of $K_\alpha$ for neighboring elements, there, and only there, could one suspect the presence of a yet unknown element.

Yet Moseley had a rather different point in mind in originally undertaking his research, a problem which he brilliantly illuminated. It must be remembered that in Moseley's time the concept of *atomic number* was not yet of importance in physics; the idea was in the air that it corresponded to the number of positive charges on the nucleus, as Rutherford said later, but it had not been announced. In fact, it was Moseley who seems to have first used the term atomic number explicitly. It requires only one glance at Table 27.4 to see that the progressive regularity in the K lines of elements depends on the atomic number and not on the atomic weight. Cobalt and nickel represent an inversion in the periodic table, yet the K lines follow regularly. (The same point was established for the other inversions.)

Here is comfort for the chemist who was forced to make inversions on the basis of chemical behavior and macroscopic physical properties, and here is also proof that whatever be the cause of the characteristic x-ray spectrum, it is connected with the atomic number, i.e., with the number of positive charges on the nucleus. This is quite reasonable, for we cannot help but guess, pending a full explanation, that x-rays are generated when the penetrating cathode ray sets the electrons near the nucleus into such motion as will cause the emission of high-energy photons. But the behavior of an electron near the nucleus surely must be governed, above all, by the field strength in that region owing to the charges on the nucleus. Moreover, as Moseley pointed out, "The very close similarity between the x-ray spectra of the different elements shows that these radiations originate inside the atom, and have no direct connection with the complicated light spectra and chemical properties governed by the structure of its surface."

c) That x-rays are purely the property of atoms was also made clear by noting that if the target in the x-ray tube were made of brass, an alloy of copper and zinc, the undisturbed characteristic radiations of both Cu and Zn atoms were observed simultaneously.

To summarize the central points in Moseley's own words, "We have here a proof that there is in the atom a fundamental quantity, which increases by regular steps as we pass from one element to the next. This quantity can only be the charge on the central positive nucleus, of the existence of which we already have definite proof." A great deal of research soon made it quite certain that this charge, measured in units of electronic charge but with a positive sign, was in every case exactly equal to the atomic number. Moreover, here we have at last been given one property that changes in a regular way from one element in the periodic table to the next. The chemical properties, such as valence, and the physical behavior, such as melting points, do not vary in continuous gradation but in definite periods, with sudden breaks between one and the next. Even the atomic weight progresses unevenly, as the inversions prove. Some other features, such as line-emission spectra, establish no simple relation whatever between neighboring elements. Only the progressively increasing charge on the nucleus gives us a truly secure foundation for the arrangement of elements in a unique order.

## 27.6 FURTHER CONCEPTS OF NUCLEAR STRUCTURE

Because this chapter is to be in the main an account of the theory of nuclear structure, we may briefly deviate from the historical development to summarize a few of the more recent concepts of nuclear physics. The emission from natural radioactive elements of $\alpha$- and $\beta$-particles suggested naturally that a model of the nucleus start with these building blocks. But the very first element, hydrogen, offered difficulty, having but a single positive nuclear charge and about one-fourth the mass of an $\alpha$-particle.* The nucleus of the usual atom of hydrogen was accessible for observation, either in the form of $H^+$, the ionized atom whose single electron had been removed in the violence of an electric discharge, or as the result of bombardment of hydrogen gas with high-speed $\alpha$-particles, where occasionally a $_1H^1$ nucleus would be seen to speed through the cloud chamber or to a scintillation screen after having experienced a collision. The name *proton* was generally adopted for the particle in 1920. The previous year, Rutherford had reported that such particles appeared to be emitted when nitrogen is bombarded by $\alpha$-particles: "... we must conclude that the nitrogen atom is disintegrated under the intense force developed in a close collision with a swift $\alpha$-particle, and that the hydrogen atom which is liberated formed a constituent part of the nitrogen nucleus." He added prophetically, "The results on the whole suggest that, if $\alpha$-particles —or similar projectiles—of still greater energy were available for experiment, we might expect to break down the nuclear structure of many of the lighter atoms."

Now the problem of nuclear structure seemed much nearer to solution. Each nucleus carries a number of protons equal to its atomic number $Z$. Each proton has one unit charge and a mass of about 1 amu, so that this scheme so far accounts for the proper positive charge but, in general, for only half the mass; we may therefore imagine the nucleus to contain sufficient neutral pairs of pro-

---

* The heavy isotope of natural hydrogen, which has about half the mass of an $\alpha$-particle, was not known until 1931, and of course cannot be explained as a combination of an $\alpha$- and a $\beta$-particle.

tons and (negative) electrons to make up the difference between the observed atomic weight and the atomic number. Thus a helium nucleus would be composed of 4 protons and 2 nuclear electrons, to give a charge of $+2$ and a mass of about 4 (the electrons having only 1/1836 of the mass of a proton, therefore not adding significantly to the total nuclear mass). Similarly, a nucleus of $_{92}U^{238}$ would contain 92 un-neutralized protons and $(238 - 92)$ neutral proton-electron pairs.

This hypothesis has two appealing features. It places in the nucleus electrons that are known to come from there in radioactive decay as $\beta$-radiation, and it allows us to understand at last why an element, after $\alpha$- or $\beta$-decay, is found in a different position in the periodic table. Since the atomic number is given by the number of positive nuclear charges, the loss of two positive units by $\alpha$-emission should move the element down two places, and the gain of one positive unit through the loss of a (negative) nuclear electron in $\beta$-emission should move the element up by one place—all just as observed.

The hypothetical combination proton-electron was given the appropriate name *neutron* in the early 1920s, and a wide search was made to discover it experimentally in a free state, e.g., as a product of inelastic collision of fast $\alpha$-particles with other nuclei. None of the experiments planned specifically for this purpose succeeded, but in 1930 it was reported that energetic $\alpha$-particles, when stopped by light elements (lithium, beryllium, boron) could create a radiation so penetrating that they were considered to be high-energy $\gamma$-rays. Indeed, the energies obtained were greater than of any then known $\gamma$-radiation. In a historic experiment, the French physicists Frédéric Joliot and his wife Irène Curie (daughter of the discoverers of radium) noted in 1932 that this new radiation could in turn eject protons of very high speed from hydrogen-containing materials such as paraffin wax. Immediately after the publication of this result, the British physicist James Chadwick, who had been an assistant of Rutherford's, showed that the protons could not have been ejected by a $\gamma$-radiation unless the latter not only differed in behavior from all other known electromagnetic radiations but also, if one wished to account for the great speed given to the protons, the $\gamma$-radiation would somehow have to leave the parent atoms with an energy 10 times greater than had been given up by the $\alpha$-particles that initially created it. In his paper "The Exist ence of a Neutron," Chadwick then goes on to prove that "the experimental results were very difficult to explain on the hypothesis that the beryllium radiation [the rays created when $\alpha$-particles fall on Be] was a quantum [$\gamma$, photon] radiation, but followed immediately if it were supposed that the radiation consisted of particles of mass nearly equal to that of a proton and with no net charge, or neutrons." Having no charge, these neutrons are non-ionizing, therefore do not lose their energy easily in passing through matter, can penetrate thick sheets of lead, and can approach a charged nucleus directly without experiencing a force of repulsion. In a head-on collision with a hydrogen nucleus, of approximately equal mass, the neutron's energy would be virtually completely transmitted to the proton.

At once it became clear that the neutron, together with the proton, would account completely for the nuclear structure of all elements. But one no longer thinks of the neutron as a particle composed of 1 proton and 1 electron; instead, the neutron is regarded as a fundamental particle that may, under certain conditions, decay into a proton, an electron ($\beta$-ray) and a neutrino.

Figure 27.9 represents schematically the structure of several nuclei ($n$ = neutron, $p$ = proton). For $_1\text{H}^2$, the "heavy" isotope of hydrogen, or *deuterium*, one neutron and one proton are required, a combination referred to as a deuteron. The helium nucleus or $\alpha$-particle comprises two neutrons and two protons. Similarly for the other nuclei.

---

**Problem 27.20.** (a) Draw similar representations for $_8\text{O}^{16}$, $_8\text{O}^{17}$, $_8\text{O}^{18}$, $_{15}\text{P}^{31}$. (b) How many protons

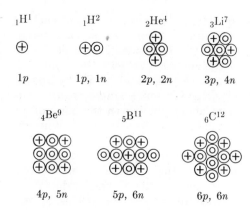

**Fig. 27.9** Purely pictorial representation of some nuclei; $\odot$ = neutrons, $\oplus$ = protons.

and neutrons are there in each of the natural isotopes of uranium?

---

We may conclude the account of the nuclear atom with a brief summary of some typical concepts that are useful in contemporary research. Most important is probably the question of the stability of the nucleus, the problem that Rutherford originally wisely by-passed. Mechanisms have been invented that account adequately for the main facts of stability and for the general facts of natural and induced radioactive decay, although by no means in the framework of the classical physics which we have largely relied on so far in describing the nuclear model. To us one most intriguing point is that the neutron and proton, having about equal mass, very close to 1 amu, necessarily create nuclei that have almost integral masses on the amu scale.

Thus we are led back to Prout's hypothesis (Section 21.1), although in a manner he did not envisage at all, and we now understand the previously most mysterious fact that all isotopic masses are nearly whole integers. On the other hand, we are quite ready to expect that the isotopic masses are not all *exactly* integral multiples of the mass of $_1\text{H}^1$, because the neutron has a slightly larger mass than the proton (1.00866 compared with 1.00728 amu).

But even when we take account of the slight difference between the mass of the proton and that of the neutron, we cannot quantitatively explain the masses of isotopes. For example, we find that the nucleus of a helium atom has a smaller mass (4.0026 amu) than the sum of the individual particles in their free states (2 protons = $2 \times 1.00728$, 2 neutrons = $2 \times 1.00866$, together 4.0319 amu), a difference $\Delta m$ of about 0.0293 amu. What has happened to this small but measurable amount of mass in the process of forming the helium nucleus from its constituents? The answer, which will emerge from Einstein's theory of relativity (Chapter 30), is that this missing mass corresponds to energy ($4.5 \times 10^{-12}$ joule or 28 MeV) lost or given up during the formation of the atom out of its parts. Only if this much energy is supplied to the atom can the helium nucleus break up again into its constituents. The absence of this much energy binds the protons and neutrons together in the nucleus, just as the absence of extra kinetic energy keeps our moon from breaking away from its orbit around the earth.

But a detailed discussion of nuclear structure will better be suspended temporarily, until we introduce the radical changes in physical theory that led to the generalization and combination of the laws of conservation of mass and energy. Now that we have reached a model of the nucleus that is actually as close to pictorialization as is permissible in a subject largely given over to mathematical rather than physical models, we must turn to another set of questions: What is the structure of the swarm of electrons that we have tentatively postulated to exist around the nucleus? What are the laws governing their behavior, and to what extent can they account for our remaining puzzles of chemistry and physics, e.g., valences and emission spectra? When these questions have been answered, we shall indeed be in possession of the key to a unified understanding of many of the major questions of physical science.

## RECOMMENDED FOR FURTHER READING

E. N. da C. Andrade, *Rutherford and the Nature of the Atom*, Garden City, N.Y.: Doubleday Anchor Books, 1964; paperback

H. A. Boorse and L. Motz, *The World of the Atom*, pages 385–407, 427–461, 641–646, 693–733, 779–824. Extracts from the writings of Roentgen, Becquerel, the Curies, Rutherford, etc., and biographies.

B. L. Cline, *Men who Made a New Physics* [published earlier as *The Questioners*], New York: New American Library, Signet paperback, 1969, Chaps. 1 and 2

D. L. Hurd and J. J. Kipling, *The Origins and Growth of Physical Science*, Vol. 2, pages 319–338, 363–401. Extracts from the writings of Röntgen, Becquerel, the Curies, Rutherford and Soddy.

*Project Physics Reader*, Vol. 6. Almost every one of the articles reprinted here, exploring aspects of the development of nuclear physics and its social impact, is relevant to the subject matter of the present chapter.

C. Weiner, "1932—Moving into the new physics," *Physics Today*, Vol. 25, No. 5, pages 40–49 (May 1972)

## SOURCES, INTERPRETATIONS, AND REFERENCE WORKS

D. ter Haar, *The Old Quantum Theory*, Part 1, Ch. III, and Part 2, reprint of Rutherford's paper

B. Jaffe, *Moseley and the Numbering of the Elements*, Garden City, N.Y.: Doubleday, 1971

A. Romer, *The Discovery of Radioactivity and Transmutation*, New York: Dover Publications, 1964; *Radiochemistry and the Discovery of Isotopes*, New York: Dover Publications, 1970. Both books contain a selection of reprints and translations of original papers, with commentary.

S. Wright, *Classical Scientific Papers—Physics*, New York: American Elsevier, 1965

# Chapter 28

# BOHR'S MODEL OF
# THE ATOM

## 28.1 LINE EMISSION SPECTRA

As we saw in Chapter 26, it was the study of continuous emission spectra that led Max Planck to formulate his quantum hypothesis in 1900. We were concerned only with the spectra from glowing solids or liquids, and incidentally also from gases at those extreme conditions of density and temperature found in our sun and other stars, but not in our laboratories. Such spectra were found to depend only on the *temperature* of the source, not on its chemical composition. And of course it was precisely this *universal* character of emission spectra that attracted Planck's attention and made him suspect that a fundamental law of physics must be involved. Having identified, at least provisionally, the general features of this law (Section 26.3) and having seen how it was successful in providing an interpretation of the photoelectric effect (Sections 26.4 and 26.5), we must now see how it might apply to the more specific problems of atomic structure. For this purpose we turn to another type of spectrum.

It had long been known that light is emitted by gases and vapors when "excited" by the passage of an electric spark or arc or, as in the case of neon lights, by a continuous electric current established through a thin tube filled with some of the gaseous

material (Fig. 28.1), or when volatile matter is put into a nonluminous flame. Furthermore, the light so emitted, when resolved into its components by the prism spectroscope, was found to have a spectrum startlingly and fundamentally different from the continuous emission spectra of glowing solids and liquids: Gases and vapors have *line emission spectra*, that is, instead of a solid band of changing "rainbow" colors they show light only at some well-defined wavelengths, with dark gaps between neighbors. As seen through the spectroscope or photographed on a plate in a spectrograph, the spectrum therefore appears as a set of lines, usually irregularly spaced, some very bright, others less so.

The second point of difference with respect to continuous emission spectra is that the line emission spectra are markedly different for different radiating elements. Each substance has its own characteristic pattern of wavelengths throughout the whole observable region (Fig. 28.2). The unaided eye synthesizes these separate lines and recognizes the mixture of colors as reddish for neon, pale blue for nitrogen, etc. Some materials reveal a most complex emission spectrum, others are far simpler; iron vapor shows about six thousand bright lines, whereas sodium has only two strong,

-Large potential difference-

**Fig. 28.1**  Electric discharge tube.  Current from one sealed-in electrode to the other causes the gas to emit light.

**Fig. 28.2**  Sections of the line emission spectra of mercury, helium, and neon.  (Redrawn from photographic record)

close yellow lines in the visible range.  The great variety of patterns and the individual spacing of lines seemed quite unaccountable; why should gases emit line spectra at all, and why should closely related elements present such differing line patterns?  All this was destined to remain most puzzling for nearly three generations.  But at any rate, each material could be identified from its unique line emission spectrum as surely as criminals can be traced by means of their characteristic fingerprints.

Rapidly, a whole art of spectroscopy arose; more and more powerful prism and grating spectroscopes became available to chart the exact wavelengths of these patterns.  As early as the 1830s, the suggestion was made that the presence, identity, and abundance of materials in ores, etc., might be determined from the type and intensity of the emission spectra.  And in fact, the physicist G. R. Kirchhoff and the chemist R. W. Bunsen at the University of Heidelberg, who were the foremost pioneers in the development of spectroscopy, discovered two elements, rubidium and cesium, in the vapor of a mineral water by noting hitherto uncharted emission lines.  It was the first of a brilliant series of such discoveries, and the origin of a technique that has given us the speedy chemical analysis of small samples by routine spectroscopy, a vital tool in today's research laboratory and industry alike.  Another spectacular achievement that caught the imagination of all Europe was the spectroscopic analysis of meteorite material.  When the emission spectra of vaporized meteorites were compared, it was established that these contained only well-known elements present on our earth, a direct confirmation of the unity of terrestrial and celestial science that would have gladdened the hearts of Galileo and Newton.

## 28.2  ABSORPTION LINE SPECTRA

In 1814, the German optician Joseph Fraunhofer studied a peculiarity in the appearance of the continuous spectrum of sunlight.  When light was passed through a narrow slit and its visible spectrum very carefully examined with good prism systems, the continuity of colors appeared to be

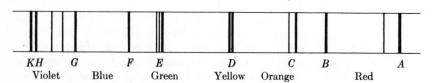

KH     G        F      E            D           C        B              A
Violet      Blue       Green      Yellow   Orange       Red

**Fig. 28.3** Fraunhofer lines in the sun's continuous spectrum; only some of the most prominent lines are shown.

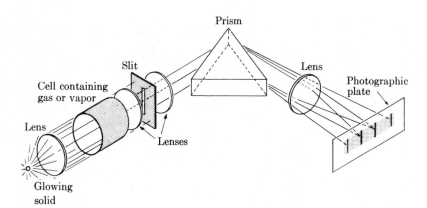

**Fig. 28.4** Diagram of spectrometer set up for recording an absorption spectrum of a vapor.

disrupted by a series of fine, steady, irregularly spaced *dark lines* (Fig. 28.3). Fraunhofer counted more than 700 such lines (now, with better instruments, we know of more than 15,000) and assigned the letters *A, B, C,* . . . to the most prominent ones. He also observed many such lines in the spectra of some bright stars.*

---

\* It is significant that Fraunhofer's interest in these dark lines was initially a practical one; they identified and labeled portions of the solar spectrum, and so provided reference points useful for comparing the degree of dispersion of colors attainable with different types of glass. Here, as so often in the history of science, the skills of the craftsman and the needs of technology eventually bring to the notice of scientists a new natural phenomenon and techniques for dealing with it; and from such beginnings, in the hands of a gifted scientist, profound and unexpected contributions may follow.

The key observation toward a better understanding of both the absorption and the emission spectra came in 1859. By that time it was generally known that the emission line from the heated vapor of sodium metal had the same wavelength as the very prominent dark line† in the solar spectrum to which Fraunhofer had assigned the letter *D*.

What Kirchhoff now found looks simple in retrospect. In essence the argument is this: When the light from a pure, glowing solid is ordinarily analyzed, it does not show dark lines crossing the continuous emission spectrum, but if (as shown in Fig. 28.4) that same light on its way to the prism is

---

† There are two lines, with wavelengths of 5889.953 Å and 5895.923 Å, but only modern instruments of great precision can resolve them that well, and so we shall speak of *a* sodium line at about 5890 Å.

Absorption
spectrum

Emission
spectrum

**Fig. 28.5** Absorption spectrum of sodium compared with the line emission spectrum of the same substance.

←— Ultraviolet —→|←— Visible —→|←— Infrared —→

now allowed to pass through a suitably large volume of gas or a cloud of vapor, e.g., vaporized sodium. metal, then the spectrum does exhibit one or more dark lines of exactly the same wavelength and appearance as do those in the Fraunhofer observation. Each vapor was found to have a characteristic line absorption pattern (Fig. 28.5); evidently each strongly absorbs certain characteristic frequencies from the passing "white" light. And, most interesting of all, the wavelengths of the *absorption* lines in each case correspond exactly to some lines in the catalog of *emission* patterns for the particular gaseous substance.

We may therefore attempt a preliminary interpretation of the dark lines in solar and other stellar spectra: The missing lines correspond to light that was withdrawn or somehow scattered from the total beam on passage through the (relatively cooler) atmosphere surrounding the star and, to some extent, through the atmosphere of our earth. This view leads us to regard the Fraunhofer lines as evidences of the presence of specific elements in the atmosphere of a star. We see at once the enormous potentialities of this theory for astrophysical research.

---

**Problem 28.1.** How would you decide, on an experimental basis, which absorption lines in the solar spectrum are due to absorption near the sun's atmosphere rather than to absorption by gases and vapors in our own atmosphere?

**Problem 28.2.** How might one decide from spectroscopic observations whether the planets shine by their own light or by reflected light from the sun?

---

Among the consequences following from this explanation of Fraunhofer lines, three are of particular interest to us now:

a) Each single Fraunhofer line ought to be assignable to the action of some specific gas. Therefore, when with better techniques a new line was discerned in the green part of the solar spectrum which corresponded to none of the lines known to be emitted or absorbed by terrestrial material, it was plausible to ascribe it to the action of a gas in the atmosphere of the sun not yet discovered on earth. The appropriate name *helium* was chosen and, indeed, many years later a gas with such an absorption line was found, first as a constituent of many minerals, later also in natural gases and in our own atmosphere.

b) After absorbing its characteristic colors from white light, a gas may reradiate it, and generally at the same frequency. This conclusion was original with Kirchhoff. We may understand the essence of his argument quickly in terms of an analogy: A beam of sound containing many frequencies is sent past a mounted and silent tuning fork tuned to 440 cycles/sec. Any sound of that particular frequency does not simply pass by the tuning fork as do all the other frequencies; instead, being "in resonance" with the instrument, its energy is absorbed as it sets the prongs vibrating. A distant listener along a line between sound source and tuning fork might therefore notice that this one component reaches him with less than the full energy. However, the tuning fork now vibrates, and sends out again the sound energy it received— but in all directions, not just in the direction in which the absorbed sound was traveling. The proportion of sound energy reradiated *toward* the observer is very small compared with the total energy absorbed. The conclusions from the analogy are, first of all, that a dark line in an absorption spectrum appears dark only because of the relatively much greater intensity of the "background" of unaffected radiation at either side; *some* light is always also reradiated in the original direction of propagation. Second, we expect to see some light reradiated toward us from a volume of gas while it is absorbing light, even if we view it at 90° to the direction of propagation of the original radiation; for example, we expect from the edges of the absorbing layer of the sun's atmosphere a bright line spectrum corresponding exactly to a *reversed* Fraunhofer spectrum. This should happen during those few moments while the main disk of the sun itself is just hidden in a total eclipse, and this deduction was first and spectacularly fulfilled during an observation of this kind in 1870.

c) It is not difficult to visualize the basic mechanism for this *resonance* interpretation of absorption in terms of the classical theory of light as it developed soon after Kirchhoff's work. We recall that an "oscillator" corresponds to an electric charge vibrating or revolving at the same frequency as the light radiation thereby emitted. When such an oscillator is inactive because no energy is being supplied to it (by heating or by an electric discharge), it nevertheless remains "tuned" to its particular frequency, ready to receive energy from a passing light ray of that same frequency, and once in full vibration, it must give up (reradiate) energy as fast as it is supplied.

These points, which followed previously from the tuning fork analogy, are in general accord with the Maxwellian scheme. But if the classical theory could furnish no explanation for the distribution of intensities in the relatively simple and uniform case of continuous emission spectra, the explanation of bright line emission spectra seemed infinitely more hopeless. Here are some of the conceptual difficulties:

One might argue that the atoms of different elements may well have characteristic structures of their own, so that each elementary atom by itself would emit at the frequency of its particular oscillatory vibration. But the spectra from solids or liquids or very hot and dense gases do *not* show any differences from one element to another, and this can only mean that in those physical states the nearness and incessant mutual collision among atoms blurs the individual vibration characteristics, just as pendulums or springs can no longer vibrate with their free frequencies when connected together in large groups. This picture qualitatively accounts for the difference between the continuous and the line emission spectra, but there is this serious flaw: Even in the rarefied gas in a glowing discharge tube, atoms are continually colliding billions of times per second, according to our kinetic theory model of gases, and usually with enormous speeds; yet we do get from them definite line spectra. What could account for the complete blurring of spectra in one case and the lack of interaction among colliding "oscillators" in the other case?

Quite apart from this last difficulty, how are we to account for the exact line pattern itself? In some of the more involved spectra, one might have to visualize an internal structure of "oscillators" in the atom which would make a grand piano exceedingly simple by comparison (to use an analogy then current). Equally disappointing was that the line spectra, although definite for each element, still were apparently without order. They showed no progressive change from one element to the next; as noted, even chemically related elements often exhibited strikingly different arrangements of lines.

A further complexity was introduced by the observation that the number of absorption lines in every case depended on the violence of the electric discharge; that further, although every absorption line indeed corresponded to some emission line for that same gas, yet there were always many lines in the emission spectrum without equivalents in absorption spectra. This point conflicted with the idea of resonance radiation, for surely a musical instrument, to return to our analogy, should freely absorb every sound to which it is "tuned" and which, on being played, it can emit.

Another most serious difficulty was that each emission line corresponds to a definite state of vibration for the atomic oscillator. According to the kinetic theory of gases, these internal motions of the atom ought to share the energy of the entire molecule, thereby increasing the theoretical value of the specific heat of the gas. But the theoretical value even without these internal motions was already higher than the experimental value! Maxwell and other physicists were well aware that any attempt to give an atomic interpretation of spectral lines would intensify the difficulty of the kinetic theory in explaining specific heats (Section 22.3), but they could not find a way to resolve the problem within the framework of Newtonian mechanics.

Because the field of experimental spectroscopy after the work of Kirchhoff and Bunsen grew so vigorously, the avalanche of data tended to point up the chaotic features with more and more discouraging certainty. True, there were a few hints of order, by themselves also unexplainable: The general appearance of the spectra of alkali metals showed *some* features of similarity when the regions beyond the visible were included; the wavelength difference between neighboring lines for some parts of the sodium and the zinc spectrum was constant, and groupings of two or three lines tended to recur in some other spectra; but not much else. This was a period of almost obsessed searching for some hint of a numerical relation among the tantalizing lines, for some mathematical key to decode the message that surely had to be behind the appearances. It is as if we watched again the pupils of Plato trying to reduce the offensive disorderliness of planetary

motion to a series of circular motions, or Kepler poring endlessly over his data to find relationship between the radii of planetary orbits and the times of revolution. And, most prominently, we recall the chemists from Döbereiner to Mendeléeff trying to find order in the list of chemical elements. We are here very close to a vital process in the evolution of the sciences.

## 28.3  BALMER'S FORMULA

As sometimes happens, the first important break in this problem came from a fairly obscure corner. In 1885 a Swiss schoolteacher, Johann Jakob Balmer (1825–1898) published a paper entitled simply "Notice concerning the spectral lines of hydrogen." His starting point was Ångström's fine measurements of the wavelengths of the four quite well-known visible hydrogen emission lines, commonly named $H_\alpha$ (red), $H_\beta$ (green), $H_\gamma$ (blue), $H_\delta$ (violet). Quite probably by a straight cut-and-try method, Balmer hit on a formula that would relate these four wavelengths, and which we may write as

$$\lambda \text{ (m)} = C \times \left( \frac{n^2}{n^2 - 2^2} \right), \qquad (28.1)$$

where $C$ is a constant quantity, equal to $3645.6 \times 10^{-10}$ m, and $n$ is a whole number. Specifically, $n = 3$ if $\lambda$ for $H_\alpha$ is to be found, $n = 4$ for $H_\beta$, $n = 5$ for $H_\gamma$, and $n = 6$ for $H_\delta$. Table 28.1 shows the excellent agreement (to within 0.02%) between

Table 28.1  Data from Balmer's paper

| Name of line | Wavelength $\lambda$, Å | |
|---|---|---|
| | Experimental | From Balmer's formula |
| $H_\alpha$ | 6562.10 | 6562.08 ($n = 3$) |
| $H_\beta$ | 4860.74 | 4860.8  ($n = 4$) |
| $H_\gamma$ | 4340.1 | 4340.0  ($n = 5$) |
| $H_\delta$ | 4101.2 | 4101.3  ($n = 6$) |

Series↑  Ultra- →|←— Visible ——→  H$_\gamma$      H$_\beta$                    H$_\alpha$
limit    violet

**Fig. 28.6**  Balmer series of hydrogen emission spectrum.

Balmer's empirical formula and Ångström's measurements on which it was based.*

Balmer speculated next what the wavelength of a fifth line might be if it existed. Using $n = 7$, he finds $\lambda$ of H$_\varepsilon$ = 3969.65 Å, from Eq. (28.1), and he writes, "I knew nothing of such a fifth line, which must lie within the visible part of the spectrum...and I was compelled to assume that the temperature relations [in the glowing gas] were not favorable to the development of this line or that the formula was not generally applicable." But then, he continues, a colleague mentioned that more lines had been found in the violet and ultraviolet, in the analysis both of emission from discharge tubes and of stellar (absorption) spectra. The fifth line did indeed exist, and had the predicted wavelength! In a postscript Balmer relates that just then he received the measured wavelengths for nine more lines in the ultraviolet region, and they all fitted his predictions within better than 0.1%. His formula seemed completely vindicated; he was equally correct in his prediction that all higher lines must crowd more and more together as $n \to \infty$, for according to his formula, the shortest wavelength ($\lambda$ at $n = \infty$) is 3645.6 Å, the so-called *convergence limit*.

---

*    With the sublime self-confidence of a numerologist, Balmer comments at this point: "A striking evidence for the great scientific skill and care with which Ångström must have gone to work."

Increasingly improved techniques in spectroscopy soon made it possible to record more and more of the ultraviolet region, as for example by photographing the emission spectrum of the hot gases in the sun's chromosphere during an eclipse; the 35 consecutive "Balmer" lines so recorded were all in good agreement with his formula. The convergence limit came to 3645.81 Å. Figure 28.6 represents a Balmer spectrum on a plate sensitive into the ultraviolet region.

---

**Problem 28.3.**  (a) Check Balmer's calculation for the fifth line, H$_\varepsilon$, and for the convergence limit. (b) How many hydrogen lines should there be in the range between 3650.0 Å and 3660.0 Å?

---

But there was more. Balmer thought that on purely speculative grounds one might expect several complete new series of hydrogen lines, for which

$$\lambda = C \times \left( \frac{n^2}{n^2 - 3^2} \right), \qquad \lambda = C \times \left( \frac{n^2}{n^2 - 4^2} \right)$$

and so on; and he also expressed the hope that this formula might guide the way to discoveries of series relationships among other spectra than that of hydrogen. It soon became evident that Balmer's suggestions were quite correct, although there was as yet no hint of a *physical* basis on which to arrive at this formula.

It is now customary to write Balmer's formula in a perfectly equivalent but more suggestive way, which we shall adopt from now on:

$$\frac{1}{\lambda} = R \left( \frac{1}{2^2} - \frac{1}{n^2} \right), \qquad (28.2)$$

where the value for the constant $R$ is, according to modern measurements, $1.09678 \times 10^7$/m. The letter $R$ has been chosen to honor the Swedish spectroscopist J. R. Rydberg (*Rydberg constant, R*).

---

**Problem 28.4.** Derive Eq. (28.2) from (28.1), and show that $R = 4/C$.

---

Now we can restate Balmer's prediction of other series of hydrogen lines, by writing speculatively

$$\frac{1}{\lambda} = R \left( \frac{1}{3^2} - \frac{1}{n^2} \right), \qquad \frac{1}{\lambda} = R \left( \frac{1}{4^2} - \frac{1}{n^2} \right),$$

etc., perhaps also

$$\frac{1}{\lambda} = R \left( \frac{1}{1^2} - \frac{1}{n^2} \right).$$

In general, all these possible series can be summarized in one formula, namely,

$$\frac{1}{\lambda} = R \left[ \frac{1}{(n')^2} - \frac{1}{(n'')^2} \right], \qquad (28.3)$$

where $n'$ is a fixed integer $(1, 2, 3, \ldots)$ that distinguishes one series from the other (for the Balmer series, $n' = 2$), and where $n''$ also represents integers (namely, $n'' = n' + 1, n' + 2, n' + 3 \ldots$), but those integers that distinguish one line from the next in a given series, so that for $H_\alpha$ and $H_\beta$, $n''$ is 3 and 4, respectively.

And indeed, in 1908, Paschen in Germany found two new hydrogen lines in the infrared whose

**Table 28.2**

| Name of series | Date of discovery | Values in Eq. (28.3) |
|---|---|---|
| Lyman series | 1906–1914 | $n' = 1,$ $n'' = 2, 3, 4, \ldots$ |
| Balmer series | 1885 | $n' = 2,$ $n'' = 3, 4, 5, \ldots$ |
| Paschen series | 1908 | $n' = 3,$ $n'' = 4, 5, 6, \ldots$ |
| Brackett series | 1922 | $n' = 4,$ $n'' = 5, 6, 7, \ldots$ |
| Pfund series | 1924 | $n' = 5,$ $n'' = 6, 7, 8, \ldots$ |

wavelengths were given by the first of the speculative formulas or, in terms of Eq. (28.3), by $n' = 3$ and $n'' = 4$ and 5. Many other lines in this series have been identified since.

To the Balmer series and the Paschen series there have gradually been added others, as by the improvement of experimental apparatus and techniques new spectral regions could be explored. In the above list, the name of the series corresponds to its first observer, except, of course, that the *existence* of Balmer's lines was common knowledge at the time he formulated the series expression.

---

**Problem 28.5.** Find the wavelength of the third line (always counting up from the longest wavelength) in the Lyman series, and the convergence limit for that series.

**Problem 28.6.** Draw a long line and divide it into 20 equal intervals, about 1 cm apart. Let this represent a wavelength scale from 0 to $80,000 \times 10^{-10}$ m. On this scale enter the first, second, third, and last line of each of the five series. Do they overlap? Of how many series could one *conceive* of between $\lambda = 0$ and $\lambda = \infty$? How many lines *in toto*?

The rest of Balmer's initial suggestions began to be fulfilled even sooner than those concerning other series for hydrogen emission; his formula did indeed prove suggestive in the search for relationships among the spectra of other gases. In 1896 E. C. Pickering observed a series of hitherto unknown lines in the absorption spectra of a star that virtually coincided with that of the Balmer series for hydrogen, except that there was an additional line between every couple of Balmer-type lines. The whole set of wavelengths fitted the relation

$$\frac{1}{\lambda} = R \left[ \frac{1}{(n'/2)^2} - \frac{1}{(n''/2)^2} \right], \qquad (28.4)$$

where $n' = 4$ and $n'' = 5, 6, 7 \ldots$ (later, other series of this type were found, e.g., one with $n' = 2$ and one with $n' = 3$). The thought was voiced at first that this was evidence for some celestial modification of common hydrogen. The true solution to this intriguing puzzle was one of the earliest triumphs of the new model of the atom, to which the historical development is here gradually leading us; it is such a good story that we shall save it for the last section of this chapter.

While Balmer's formula itself was not found to serve directly in the description of other spectra, it inspired formulas of similar mathematical form that were very useful in discerning some order in portions of a good many complex spectra. The Rydberg constant itself also reappeared in such empirical formulas. It became more and more clear that somehow the same physical mechanism had to be at work behind the variety of spectroscopic observations, although simultaneously the exciting search threatened to bog down under the sheer weight of data. But as we know now, it might almost have been equally good for the progress in this field if for a generation after Balmer no line spectrum analysis had been attempted, because the key pieces for the construction of the necessary conceptual scheme had to come first from an entirely different direction. And we now turn to follow that path.

## 28.4 NIELS BOHR AND THE PROBLEM OF ATOMIC STRUCTURE

The rest of this chapter is an account of the original work in 1913 of a young Danish physicist, and its consequences for the physical sciences. Niels Bohr (page 449), born in 1885, the year of Balmer's publication concerning the hydrogen spectrum, had just received his Ph.D. degree from the University of Copenhagen in 1911 when he joined J. J. Thomson in Cambridge as visiting researcher. After a few months he left to join Rutherford's group at Manchester. The work at that laboratory was just bringing proof of the theory of nuclear atoms; at the same time, the rest of the atom, the configuration of the electrons about the nucleus, was still a puzzle. There were only the more or less general suggestions that the electrons moved about the nucleus in closed orbits, a scheme rather analogous to the planetary system. Indeed, Thomson recently had been able to show that the system would be stable if one assumed the electrons to be arranged in rings, with two electrons on the innermost ring, up to 11 in a second ring, and so on.

But the hypotheses had two fatal weaknesses. First, although they provided electrons to account for the *presence* of the phenomena of electric currents, ionization, photoelectricity, the neutrality of the ordinary atom, x-ray scattering, and so forth, they were unable to produce quantitative explanations for the details of any of these processes, nor could they explain even qualitatively the presence of line emission spectra. Second, the assumption that the electrons revolved about the nucleus was necessary for explaining why the electrons did not follow the force of electric attraction by falling into the nucleus, but this created immediately a seemingly insuperable problem. According to the theory of light emission, any vibrating or revolving charge emits electromagnetic energy, so that each revolving electron should continually spiral closer and closer to the nucleus with ever-increasing speed while emitting light of steadily increasing frequency. As Rutherford said later: "I was perfectly aware when I put forward the theory of the nuclear atom that according to classical theory the electron ought to fall into the nucleus . . . ." Of course, both the emission of continuous spectra for ordinary (gaseous) elements and the annihilation of the electrons were predictions in catastrophic conflict with actual fact.

It was at that point that Bohr entered the picture. In essence, he joined the new idea of the nuclear atom with another great recent conceptual scheme, the quantum theory, which had just begun to make its way into scientific thought in Britain. As Arthur Eddington wrote in 1936; "Let us go back to 1912. At that time quantum theory was a German invention which had scarcely penetrated to England at all. There were rumors that Jeans had gone to a conference on the continent and been converted; Lindemann, I believe, was expert on it; I cannot think of anyone else."

## 28.5 ENERGY LEVELS IN HYDROGEN ATOMS

Bohr's original paper was published in the *Phil. Mag.* in July 1913 under the title "On the Constitution of Atoms and Molecules"; other papers followed at short intervals. We cannot here follow Bohr's own line of thought exactly, partly because of its advanced mathematical level, partly because, as even Max Planck himself once exclaimed, Bohr's work is written "in a style that is by no means simple." Nevertheless, the main arguments are quite accessible to us.

Bohr begins by reviewing Rutherford's atomic model and the difficulties it raised concerning electronic orbits and light radiation. Then appears the new theme. Classical electrodynamics, which had originally been confirmed by Hertz's experiments with large circuits creating radio waves, is inadequate to describe the behavior of systems of atomic size. Essentially this is the lesson to be drawn from the work of Planck and Einstein on the quantum theory of light, although they had not

analyzed the detailed motion of the atomic "oscillator." The objection to Rutherford's theory that the electron revolving about the nucleus continually loses energy had already been met by Planck, who had proved that the energy in the oscillator does *not* change except discontinuously, in steps of $hf$, where $h$ = Planck's constant and $f$ = frequency of rotation or vibration.

In this light, consider an atom of hydrogen. It has a nucleus with one positive charge, circled by a single electron at a distance known from kinetic theory to be of the order of $10^{-10}$ m. Bohr's first postulate therefore is that while the electron is in a given orbit, which for the time being we may imagine to be circular, *it does not radiate energy*. To radiate a photon, or to absorb energy, the electron must do so while decreasing or increasing its orbit, i.e., during the transition from one *stable* (i.e., nonradiating) orbit to another. Thus to every level on the energy level diagram of the oscillator there corresponds a definite stable orbit, and a sudden transition from one orbit to the next is accompanied by a change of energy level, i.e., from some energy $E''$ to $E'$, so that $E'' - E' = \Delta E$, which by Einstein's photon theory is equal to the energy of the photon emitted:

$$E'' - E' = \Delta E = h\nu. \qquad (28.5)$$

Obviously there are now several questions, above all: What are the actual energies in the different energy levels? Planck's theory only supplied this hint: The energy levels in any one oscillator are all equally spaced, and throughout any sample of radiating atoms there are distributed oscillators of every conceivable type of vibration. These assumptions, as we have seen, led Planck to the correct derivation of the emission curves for glowing solids and liquids, but they had proved powerless for the interpretation of the line emission from gases.

Now Bohr quite boldly denies that for radiating atoms in an ordinary gas the same process holds as in solids, liquids, or dense gases. He postulates that

**Fig. 28.7** A provisional energy-level scheme to account for emission on first four Balmer lines.

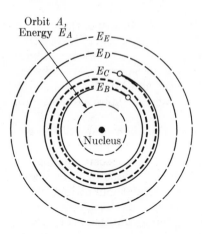

**Fig. 28.8** Provisional model of five stable orbits for the electron around the hydrogen nucleus. The electron is spiraling from orbit C to orbit B.

in this case *each atomic emitter of hydrogen has identically the same energy level scheme.* Furthermore, these energy levels cannot be evenly spaced; if they were, all hydrogen atoms would send out photons of only one frequency.

We seem to see a way of proceeding. We know the frequencies of light which hydrogen gas emits, therefore we know the energies of the photons. This gives us directly the spacing between energy levels, and so we may construct empirically the energy level diagram. For example, to account for the first four lines in the visible spectrum ($\lambda =$ $6563 \times 10^{-10}$ m, $4861 \times 10^{-10}$, $4340 \times 10^{-10}$, $4102 \times 10^{-10}$), we calculate the energy of these photons ($h\nu \sim 3.02 \times 10^{-19}$ joule, $4.07 \times 10^{-19}$, $4.57 \times 10^{-19}$, $4.84 \times 10^{-19}$), then build up an energy "ladder" which incorporates these four intervals (Fig. 28.7). Now the emission of the line $H_\delta$ ($\lambda =$ $4340 \times 10^{-10}$ m) corresponds to a change of energy level from $E_D$ to $E_C$, and similarly for the others. In principle, we can extend such a ladder to include all observed emission lines.

There occurs a first doubt. Both Planck's and Einstein's work on the quantum theory of light assumed axiomatically that the frequency $\nu$ of the emitted light is equal to the frequency $f$ of oscillation or rotation in the atom. It follows that an electron, while changing from one energy level to the next, must be rotating at a fixed orbital frequency $f = \nu$. This is at first sight not inconceivable, if an electron might suddenly spiral from one stable, nonradiating orbit into the next, say to one nearer the nucleus, while retaining its period and its frequency of revolution (Fig. 28.8). But on the one hand this raises the troublesome question of how the electron can thereafter change to a new orbital frequency when it may subsequently have to negotiate the interval from $E_B$ to $E_A$ to emit a photon of higher frequency, and on the other hand, one can prove quickly that the electron *cannot* keep a constant frequency or period while it spirals from one stable orbit to the next over any considerable distance. The net force acting at any instant on the revolving electron, a centripetal Coulomb's law force of electric attraction directed to the central nucleus, as seen in Fig. 28.9, is simply given (numerically) by

$$F_c = \frac{m_e v^2}{r} = \frac{k_e e \cdot e}{r^2}, \qquad (28.6)$$

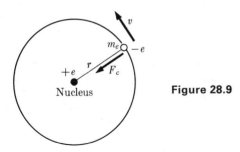

**Figure 28.9**

where $m_e$ = mass of electron, $e$ = numerical charge on the electron and on the hydrogen nucleus, and $r$ = distance between electron and nucleus. But the speed of the electron is $v = 2\pi fr$, and although this was defined for circular motion, it should hold well enough for motion along portions of a close spiral. Substituting and rearranging terms in Eq. (28.6) yields

$$\frac{1}{f^2} \equiv T^2 = \left(\frac{4\pi^2 m_e}{k_e e^2}\right) r^3. \qquad (28.7)$$

In short, $T^2 \propto r^3$, nothing less than Kepler's third law! The conclusion is that the period (and frequency) of the electron's orbital revolution does indeed change with the distance from the nucleus; therefore radiation cannot be emitted at constant frequency in the transition between stable orbits. The whole scheme collapses.

There is, however, a daring way of saving the fundamental idea. In the discussion so far, it was assumed that ordinary laws of mechanics and electricity hold during the transition of the electron between energy levels. This led us to a contradiction with experience. Therefore Bohr states a second postulate: *Ordinary mechanics and electricity* [e.g., Eqs. (28.6) and (28.7)] *might be used to discuss the motion of the electron in a stable and nonradiating orbit, but "the passing of the systems*

---

*    "Stationary" is Bohr's term for what we have called "stable."

*between different stationary\* states cannot be treated on that basis."* On the contrary, rather than try any further to visualize what happens during the transition, he asks us to realize that the classical laws were not at all designed to accommodate such intra-atomic processes, that a different mode of thought is required for that region of phenomena, and that the most straightforward approach is to accept a third postulate: During the transition between energy levels, the electron, in a way that cannot be pictured, emits a photon whose frequency $v$ is given by the Planck-Einstein formula ($\Delta E = E'' - E' = hv$, therefore $v = \Delta E/h$), *regardless of the frequency f of orbital revolution of the electron.*

That $f$ is not equal to $v$ was so completely foreign to all physical thinking that Bohr felt compelled to append this simple remark, "The [last] assumption is in obvious contrast to the ordinary ideas of electrodynamics, but appears to be necessary in order to account for experimental facts." In its impact on the contemporary scientist, it was a statement quite equivalent to Galileo's assertion about three centuries before that all bodies fall at equal rates of descent no matter what their weight. At a later time, Bohr re-emphasized the philosophy of sound scientific pragmatism; classical physics, when applied to the problem of electron stability and line emission, again and again has led to the wrong answers; evidently, the tool is not suited to the problem. "We stand here almost entirely on virgin ground, and upon introducing new assumptions we need only take care not to get into contradiction with experiment. Time will have to show to what extent this can be avoided, but the safest way is, of course, to make as few assumptions as possible."

To summarize, the three radical postulates do give us a scheme for explaining the emission of line spectra by a hydrogen atom (or for that matter, by any other system). We need only to examine all known emission lines for that element, calculate the photon energies, and construct an energy ladder with intervals corresponding to these energies. For each step on that ladder the electron is considered to move in a stable orbit. When the atom absorbs energy from the outside (by the collision with atoms in a hot gas, with particles in electric discharge, with photons, etc.) the electron moves to a new stable orbit with greater energy. When the electron returns to the previous orbit, the energy difference is given up as a photon.

Such a scheme, though it might be perfectly self-consistent and congruent with all known experimental data, would still be unsatisfactory, for it would leave the origin of the "energy ladder" itself completely unexplained. Before 1913 there had been several attempts to find relationships between the frequencies of spectral lines and the "allowed" freqeuncies of vibration of various mechanical models of the atom, corresponding to the fundamental and overtone frequencies of a string (Section 23.2). All such attempts had failed, in part because (as we have just seen) the frequency $v$ of the emitted light cannot be identified with the frequency $f$ of the electron's motion in a particular orbit.

Nevertheless, Bohr suspected that there was probably some definite relation between the single frequency $v$ of the emitted light and the two different frequencies of the initial and final orbits, and that this relation involved the energies $E'$ and $E''$ of these orbits. He saw that what was needed was, first of all, a formula expressing the frequency of rotation in an orbit, $f$, in terms of the energy of that orbit, $E$. Such a formula could easily be obtained from classical theory: The energy of the electron, orbiting as indicated in Fig. 28.9, is simply the sum of its kinetic energy and its potential energy (see Section 24.11):

$$E = \tfrac{1}{2}m_e v^2 - k_e e^2/r. \qquad (28.8)$$

Since, according to Eq. (28.6),

$$\tfrac{1}{2}m_e v^2 = \tfrac{1}{2}k_e e^2/r,$$

we can write the equation for the energy of the electron in its orbit of radius $r$ as

$$E = -\tfrac{1}{2}K_e e^2/r. \qquad (28.9a)$$

Rearranging, we obtain

$$r = -\tfrac{1}{2}k_e e^2/E. \qquad (28.9b)$$

On substituting this formula for $r$ into Eq. (28.7) and rearranging, we find the desired relation between orbital frequency and energy:

$$f = \frac{\sqrt{2}(-E)^{3/2}}{\pi k_e e^2 \sqrt{m}}. \qquad (28.10)$$

(Note that the energy $E$ is negative for electrons bound in orbits around nuclei, so $-E$ will be a positive quantity.)

So far Bohr has not gone beyond classical physics, although in choosing to express frequency in terms of energy he has obviously been guided by Planck's hypothesis. But now he takes a bold step on the basis of little more than a hunch; he assumes that if the electron is initially at rest at a very large distance from the nucleus, so that (by Eqs. 28.9a and 28.10) its energy and frequency of revolution are both essentially zero, and if, when it is captured or bound by the atom, it drops into a final state with frequency and energy related by Eq. (28.10), *the frequency $v$ of the emitted light is simply the average of the initial and final frequencies of orbital revolution.* If the initial frequency is zero and the final frequency, say, $f$, the average is just $\tfrac{1}{2}f$. In a paragraph that marks the crucial breakthrough in the development of the modern theory of atomic structure, Bohr writes:

"Let us now assume that, during the binding of the electron, a homogeneous radiation is emitted of a frequency $v$, equal to half the frequency of revolution of the electron in its final orbit; then, from Planck's theory, we might expect that the amount of energy emitted by the process considered is equal to $nhv$, where $h$ is Planck's constant and $n$ an entire [integer] number. If we assume that the radiation emitted is homogeneous, the second assumption concerning the frequency suggests itself, since the frequency of revolution of the electron at the beginning of the emission is zero."

Note that $n$ means the *number of quanta* of energy $hv$ that are emitted, according to Planck's hypothesis, when an electron moves from rest at an infinite distance to an orbit with frequency $f$ and energy $E$. Bohr proposes to set

$$E = \tfrac{1}{2}nhf.$$

because he is asserting that the frequency of the emitted radiation is $v = \tfrac{1}{2}f$. Substituting this expression for $E$ into Eq. (28.10), we can determine both $v$ and $E$ in terms of the charge and mass of the electron, Planck's constant, and the integer $n$:

$$v = \frac{2\pi^2 m_e e^4 k_e^2}{n^3 h^3}, \qquad (28.11)$$

$$E = \frac{-2\pi^2 e^4 m_e k_e^2}{n^2 h^2}. \qquad (28.12)$$

When Bohr substituted numerical values into these equations, he found that—provided that $n$ was set equal to 1—he could obtain almost exactly the experimental values for the "ionization energy"* and for the spectroscopic frequency corresponding to the capture of a free electron into the lowest energy level of a hydrogen atom. But, even more remarkable, he found that he could explain in a very simple way *all* the spectral series of hydrogen if he *changed the definition of n.* Instead of using $n$ to represent the number of quanta emitted when an electron "falls from infinity" into a given state of energy $E$, Bohr proposed that $n$ is the number of the allowed stable orbit the electron can

be in, starting from the *orbit closest to* the nucleus ($n = 1$). The various possible orbits correspond to $n = 1, 2, 3, \ldots$ up to infinity; $n$ is now called a *quantum number*, and is used to label the orbits, and Eq. (28.12) can be used to compute the energy of the electron in each of the allowed orbits.

To understand what these orbits are like, it is helpful to go back and substitute Eq. (28.12) for $E$ into Eq. (28.9b) to get radius $r$ of the possible orbits. *It is then found that the radius of the orbit with quantum number n is proportional to* $n^2$:

$$r = \frac{n^2 h^2}{4\pi^2 m_e e^2 k_e}. \qquad (28.13)$$

Orbits with small values of $n$ are close to the nucleus and (by Eq. 28.12) have large (negative) binding energies $E$; conversely, orbits with large values of $n$ are far away from the nucleus and have small binding energy (since $E$ is inversely proportional to $n^2$). When $n = 1$, $r$ is smallest, and has by calculation the value

$$r = 0.53 \times 10^{-10} \text{ m}.$$

But this is just the order of magnitude expected for the dimensions of the (normal, or unexcited, or ordinary) atom (see Loschmidt's estimate from kinetic theory, Section 22.3), and this success provides further evidence in favor of Bohr's model.

Bohr proposed that the frequency of the photon emitted when the electron jumps from a distant orbit with quantum number $n''$ to an orbit nearer

---

* When we substitute numerical values into Eq. (28.12), with $n = 1$, we find $E = -2.18 \times 10^{-18}$ joule. To put it another way, the electron must be given $2.18 \times 10^{-18}$ joule of energy to allow it to move (against the force of attraction to the nucleus) from the first orbit to infinity, or to a large enough distance from the nucleus so that the remaining energy is virtually zero. But this state of affairs, the loss of the electron, corresponds to the ionization of the hydrogen atom, and therefore Bohr's theory predicts that one must supply $2.18 \times 10^{-18}$ joule to a hydrogen atom to cause its ionization. However, the actual energy requirement for ionizing atoms can be measured experimentally by accelerating a beam of

electrons into a tube filled with a little hydrogen gas. If the electrons in the beam are sufficiently energetic, during a collision they impart to the orbital electrons in hydrogen atoms enough energy to allow the orbital electrons to leave their parent atoms entirely, which in turn causes the gas to become suddenly ionized and highly conducting. It takes an accelerating voltage of 13.6 volts for the electron beam to induce a sudden change in conduction through or ionization in the gas tube. But with that accelerating voltage, each projectile, having the electronic charge $e$, carries an energy of $13.6 \times 1.6 \times 10^{-19} = 21.7 \times 10^{-19}$ joules, in excellent agreement with the prediction from Bohr's theory.

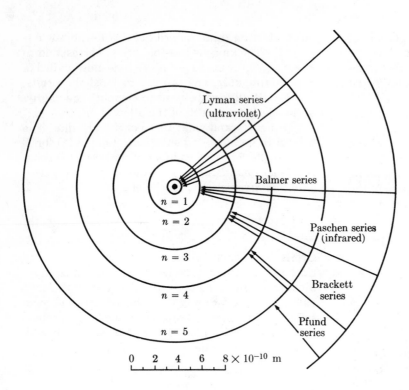

n = 1

n = 2

n = 3

n = 4

n = 5

Lyman series
(ultraviolet)

Balmer series

Paschen series
(infrared)

Brackett
series

Pfund
series

0    2    4    6    8 × 10⁻¹⁰ m

**Fig. 28.10** Final scheme of transitions between stable Bohr orbits in hydrogen atom (not drawn to scale).

the nucleus with quantum number $n'$ can be computed from Planck's postulate (Eq. 28.5), if one simply uses these quantum numbers to compute the energies $E''$ and $E'$ from Eq. (28.12):

$$v = \frac{E'' - E'}{h} = \left[ -\frac{2\pi^2 m_e e^4 k_e^2}{(n'')^2 h^3} \right] - \left[ -\frac{2\pi^2 m_e e^4 k_e^2}{(n')^2 h^3} \right]$$

$$= \frac{2\pi^2 m_e e^4 k_e^2}{h^3} \left[ \frac{1}{(n')^2} - \frac{1}{(n'')^2} \right].$$

In terms of the more directly measurable wavelength $\lambda$, this becomes

$$\frac{1}{\lambda} = \frac{2\pi^2 m_e e^4 k_e^2}{h^3 c} \left[ \frac{1}{(n')^2} - \frac{1}{(n'')^2} \right]. \quad (28.14)$$

We recognize at once that this *derived* expression is indeed identical with the *empirical* formula for the hydrogen spectra (Eq. 28.3),

$$\frac{1}{\lambda} = R \left[ \frac{1}{(n')^2} - \frac{1}{(n'')^2} \right], \quad (28.15)$$

provided that (a) the combination of constants $2\pi^2 m_e e^4 k_e^2/h^3 c$ is equal to $1.09678 \times 10^7$/m, Rydberg's empirical constant $R$ (which indeed turns out, on calculation, to be the case within 0.01%*), and (b) that we may interpret the quantum numbers $n'$ and $n''$ according to the following scheme.

---

*    Using modern values. In early 1913, the discrepancy was 6%, yet the agreement was very impressive.

Fig. 28.11 Energy-level diagram for hydrogen (final form).

When $n' = 2$ and $n'' = 3, 4, 5, \ldots$, which gives us the lines of the Balmer series from the empirical Eq. (28.15), the ground state for all transitions is the second smallest orbit (see Fig. 28.10). When $n' = 3$ and $n'' = 4, 5, 6, \ldots$ (the condition that is met for the Paschen series), the ground state is the third orbit; and so on for the other series.

For example, $n' = 1$, $n'' = 2, 3, 4, \ldots$ refers to the Lyman series, which was not fully identified until 1914, but which in 1913 was generally ex-

pected to exist in the ultraviolet. Figures 28.10 and 28.11 show where in the orbital scheme and on the energy level diagram the Lyman series fits in with the others. The radii and energy levels for this series can be calculated, independent of spectroscopic data, from Eqs. (28.13) and (28.12) by assigning $n = 1$ (the ground state for the spectral line series is here the natural state for the *unexcited* atom). Note well that nothing of the kind was possible in our initial, unsuccessful attempt to construct energy level diagrams and orbital schemes from the available experimental data of line emission. Lacking lines from the Lyman series, we could not at that time avoid the (incorrect) assumption that the ground state for transitions connected with Balmer series emission was the lowest natural state; only with an independent postulate can one hope to fix the radii and energy levels before all the lines are known.

Thus we have come to the initial successes of Bohr's model. It predicts the size of the unexcited atom, and the prediction is of the correct order of magnitude. It yields an expression for the wavelengths of all lines that are known to come from radiating hydrogen atoms, and this expression coincides exactly with one that summarizes the experimental facts of line emission, including lines not known at the time the theory was formulated. It accounts for Rydberg's empirical constant in terms of physical quantities. It provides us with a visualizable (although therefore perhaps dangerous) system, and establishes a physical order among the events accompanied by emission. The model introduces quantum theory into the atom and thereby on the one hand gives a physical basis for Planck's idea that the energy in the atom is quantized, and on the other hand removes the problem of the stability of electron orbits from the classical theory, which could not provide the solution.

Bohr did not arrive at his theory of electron orbits by a process of logical reasoning from a few basic postulates, although the theory was later formulated in this way. Nor did he make his discovery by induction from experimental data, although he was guided by knowing the general type of phenomenon that he wanted to explain. The critical steps—assuming that the frequency of emitted radiation is the average of the frequencies of revolution in the two orbits, and then redefining the quantum number $n$—were leaps in the dark which could only be justified by their subsequent success. Yet the discovery was not due to chance in the sense that anyone could have made it by guesswork; it was only after familiarizing himself with various aspects of the problem and trying out several approaches that, as Bohr reported in the paragraph quoted, the right assumption "suggested itself" to him.

**Problem 28.7.** Verify by calculation the agreement between Rydberg's constant and the expression in Eq. (28.14). (For data refer to Appendix I.)

**Problem 28.8.** With proper techniques, each of these four lines may be found in the hydrogen spectrum: $\lambda = 1215.7$ Å, $\lambda = 3970.1$ Å, $\lambda = 9500$ Å, $\lambda = 74,000$ Å. Try to identify in each case to which series the line belongs, and fix the number of the orbits (or the value of the quantum numbers) involved in the particular transition.

## 28.6 FURTHER DEVELOPMENTS

Although the derivation of the empirical hydrogen spectrum formula was a superb feat in itself, Bohr also extended his argument from the very beginning to encompass a large variety of problems—rather like Newton, who followed his explanation of Kepler's empirical laws with a discussion of the inequalities in the moon's motions, the problem of comets, of the tides, and the like. In fact, it is the hallmark of great conceptual schemes that their applicability far exceeds the boundaries of initial intent, as has been amply illustrated in the preceding cases.

Bohr's original paper of 1913 discussed a number of other applications. For example, E. C. Pickering in 1896–1897 had described a series of lines in stellar spectra which do not agree with the frequencies given by Eq. (28.15). In 1912, A. Fowler also found these lines in electric discharges through mixtures of hydrogen and helium. Their wavelength is given by

$$\frac{1}{\lambda} = R \left( \frac{1}{(n'/2)^2} - \frac{1}{(n''/2)^2} \right). \qquad (28.4)$$

However, Bohr continues, "we can account naturally for these lines if we ascribe them to helium [specifically, single ionized helium, He⁺]. A neutral atom of the latter element consists, according to Rutherford's theory, of a positive nucleus of charge $2e$ and two electrons." Now note that if one of the electrons is stripped off through the violence either of the electric discharge or of mutual atomic collisions in hot stars, the remaining helium ion is just like a hydrogen atom *except* for a doubly positive central charge. The light emitted by transitions of the single remaining electron can be calculated at once.

---

**Problem 28.9.** Starting with Eq. (28.6), revise the derivation leading to Eq. (28.14) for the case of an electron of charge $e$ in an orbit around a nucleus of charge $2e$. Show that the result may be written in the form

$$\frac{1}{\lambda} = \frac{2\pi^2 m_e e^4 k_e^2}{h^3 c}\left[\frac{1}{(n'/2)^2} - \frac{1}{(n''/2)^2}\right]. \quad (28.16)$$

But Eq. (28.16) coincides with the observations expressed in Eq. (28.4), since it has been proved that Rydberg's constant $R$ is equivalent within small limits to the factor outside the bracket! Bohr says: "If we put $n' = 3$, and let $n''$ vary [that is, $n'' = 4$, $5, \ldots$], we get a series which includes 2 of the [lines] observed by Fowler .... If we put $n' = 4$, we get the series observed by Pickering in the spectrum of $\zeta$ Puppis. Every second of the lines in this series is identical with a line in the Balmer series of the hydrogen spectrum." This is indeed an exciting interpretation of an old puzzle.

---

It is clear that Eq. (28.16) differs from Eq. (28.14) only by the factor $(2)^2$. If an atom with $Z$ positive nuclear charges were stripped of all but one electron, Eq. (28.16) should apply to its spectrum except that $(2)^2$ is replaced by $Z^2$. And indeed, it is possible to produce such violent discharges that not only helium but each of the light elements up to the eighth, oxygen, is suitably ionized. The lines emitted from such gases are in full agreement with Bohr's theory.

---

**Problem 28.10.** The vapor of lithium ($Z = 3$) may be doubly ionized (Li⁺⁺). Derive the equation analogous to Eq. (28.16) for the line spectrum from Li⁺⁺, and draw an energy level diagram and a diagram of the first few orbits analogous to Figs. 28.10 and 28.11.

---

We must now see whether Bohr's model accounts for absorption of light as well as it did for emission. Two facts have been noted in the discussion of line absorption spectra: (i) that light with a continuous spectrum, on passing through a gas, may lose frequencies that correspond exactly to those the gas would itself emit in a line emission spectrum, and (ii) that the absorption spectrum of a gas is, however, usually less rich than its emission spectrum. In this specific case, monatomic hydrogen gas generally absorbs only lines corresponding to the Lyman spectrum, as we now know; it has to be brought to an excited state if it is to absorb the lines corresponding to the Balmer spectrum, as Bohr pointed out. If we make the assumption, so consistent with the whole model, that an orbital electron can absorb a photon only if the energy so absorbed brings it out exactly into another permitted orbit, the whole set of observations becomes clear. Ordinarily, hydrogen atoms are in the most stable, the "natural," state corresponding to $n = 1$. The energy differences between that level and all others correspond to lines in the Lyman spectrum; photons with less energy, as all those in the Balmer spectrum, cannot be absorbed because they are unable to make the electron go even from the first innermost orbit to the second one. Only if the gas is initially in an excited state will sufficient atoms be present in the particular ground state of the Balmer series ($n = 2$) to absorb Balmer series frequencies.

On the same model, another puzzle may be solved. It had been noted that above a certain frequency of incident light there is no longer *line* absorption, but *continuous* absorption of all frequencies. Evidently a photon that corresponds to the highest frequency line in any series is one that, if absorbed, produces complete ionization of the hydrogen atom by carrying the electron to the level $n = \infty$. That electron is then no longer bound to the atom, but is free; but for free electrons there is no energy quantization. They can accept and carry (as translational kinetic energy) any amount of energy whatever. Evidently, when a photon with more energy than needed to free the electron is absorbed, part of the energy is used to ionize the atom, and the rest is given to the electron as kinetic energy.

We realize now that we have actually been speaking about the photoelectric effect as applied to gases: Only photons above a threshold frequency $\nu_0$ will cause photoelectric emission, and $\nu_0$ is $\Delta E/h$, where $\Delta E$ is the difference of energy levels between the ground state of the illuminated gas and the state for which $n = \infty$. Bohr says, "Obviously, we get in this way the same expression for the kinetic energy of an electron ejected from an atom by photoelectric effect as that deduced by Einstein, that is $KE = h\nu - W \ldots$"

**Problem 28.11.** A sample of monatomic hydrogen gas is illuminated from a source emitting only light of wavelength $\lambda = 1.026 \times 10^{-7}$ m (second Lyman line). If Bohr's picture of absorption and radiation is correct, and if the hydrogen atoms later re-emit the energy that they absorb, what *two* wavelengths of light do we expect to be re-radiated from the sample?

**Problem 28.12.** Calculate the work function $W$ for the photoelectric effect in unexcited monatomic hydrogen gas.

**Problem 28.13.** As Bohr was aware, Rutherford had noted in 1912 that fast $\beta$-rays sent through a gas "lose energy in distinct finite quanta." Explain this observation in terms of Bohr's model.

**Problem 28.14.** According to the kinetic-molecular theory of gases, the average kinetic energy per gas molecule is given by $\overline{KE}_{trans} = \frac{3}{2}kT$, where $k$ = Boltzmann's constant, $1.3806 \times 10^{-23}$ joule/deg, and $T$ = temperature in °K [see Eq. 22.11]. If the orbital electron can be thrown into a higher energy level by absorbing the kinetic energy during an inelastic collision between two atoms, calculate the temperature of the gas (for monatomic hydrogen) at which most gas atoms are in an energy level where $n = 2$ and consequently can contribute to the prominent absorption of lines with the Balmer series frequencies. [*Note*: Even at a lower temperature, some fraction of the total number of atoms is sufficiently excited to aid in this type of absorption. Why?]

To this point, Bohr's concern was only with single-electron systems, e.g., hydrogen and ionized helium. After he had explained such physical properties of hydrogen atoms as dimensions, emission spectra, and ionization energies with brilliant success, the next problem was obviously atoms whose nuclei are surrounded by more than one electron. Bohr made a first attack on this problem in his 1913 paper, and returned to it again several times during the next decade. Other scientists, attracted by the initial success of Bohr's model for hydrogen, soon joined in this work. By 1920 there had been developed a *shell model* of atomic structure, which was able to account at least qualitatively for the chemical properties of the elements.

As the x-ray experiments of Moseley (Section 27.5) had shown, the atomic number of an element may be identified with the electric charge carried by the nucleus in Rutherford's model. It is also equal to the number of electrons circulating in orbits around the nucleus of the neutral atom. These orbits are (with slight modifications) the same ones available to the electron in a hydrogen atom, but it was found that certain apparently arbitrary restrictions must be imposed on the number of electrons that can occupy a given orbit. The innermost shell (corresponding to $n = 1$ in Fig. 28.10) is called the K shell, and has room for a maximum of two electrons. Helium's two electrons "fill" or "close" the shell, so that lithium's third electron must start a new one, the L shell. (Refer to Table 21.2 for atomic numbers of the elements.) The L shell is assumed to have room for eight electrons, so that neon, with a total of ten electrons, fills completely the K and L shells. The next element is sodium, whose eleventh electron places itself in the third or M shell; in this way the atomic structures of the rest of the elements can be built up.

The chemical properties of the elements are explained by postulating that an atom tends to gain or lose electrons in such a way as to achieve a closed-shell structure. The rare gases (He, Ne, A, etc.), which already have such a structure, do not enter into chemical reactions at all, with few exceptions. The alkali metals (Li, Na, . . . ) have one extra electron in the outermost shell; this electron is easily given up. The halogens (F, Cl, . . . ), on the other hand, are lacking one electron, and thus tend to acquire one whenever possible. More generally the *valence* of an element (Section 20.11) can be identified with the number of electrons needed to be gained or lost in order to form a closed shell.

In summary, not only the main physical properties but also the chemical behavior of the elements seems to be governed in principle by the structure of the electron cloud around the nucleus. But the configuration itself seems determined largely by the atomic number, i.e., the *number* of electrons present, and this recognition, a triumph of old Pythagorean and neo-Platonic tendencies in modern science, brought this remarkable reaction from Bohr himself:

"This interpretation of the atomic number may be said to signify an important step toward the resolution of a problem which for a long time has been one of the boldest dreams of natural science, namely, to build up an understanding of the regularities of nature upon the consideration of pure numbers."

The list of successes of Bohr's fundamental approach has by no means been exhausted in this recital. We might continue, as is done in the books recommended at the end of this chapter, to apply the theory and its subsequent modifications to solving such puzzles as the precise wavelengths of x-ray spectra (e.g., Moseley's unexplained $K_\beta$ line), the difference in the appearance of line spectra under different experimental conditions, the relative intensities of some of the observed lines, and much besides.

But the fact is that Bohr's conceptual scheme, although today still indispensable and basic to an initial understanding of both physics and chemistry, has by now been overtaken by the relentless and swift progress of modern physical science. Several fundamental transformations became necessary to account for the large wealth of phenomena not included even in such a sweeping synthesis as Bohr's. In essence, these transformations centered on abandoning the last vestiges of visualizability of single events in the electron cloud; the rather comfortable and intuitively meaningful orbits, shells, and "jumps" of electrons had to be given up as essentially meaningless from an operational point of view, and, in fact, misleading for some purposes of prediction. The contemporary atom, which still retains the now familiar concept of energy levels, quite intentionally no longer presents any simple physical picture to guide our imagination. It is a mathematical model, a set of postulates and equations—the ultimate triumph of ascetic thought. As Bohr himself said at the very beginning of this phase of the development (1925):

"To the physicists it will at first seem deplorable that in atomic problems we have apparently met with such a limitation of our usual means of visualization. This regret will, however, have to give way to thankfulness that mathematics in this field, too, presents us with the tools to prepare the way for further progress."

---

**Problem 28.15.** Show that the kinetic energy of an electron in Bohr's model is (a) proportional to the total energy $E$, (b) inversely proportional to $n^2$. Hence show that (c) the velocity of the electron in the orbit model is inversely proportional to $n$. This result will be used in the next chapter.

## RECOMMENDED FOR FURTHER READING

H. A. Boorse and L. Motz, *The World of the Atom*, pages 363–368 and 734–765. Extracts from the writings of Balmer and Bohr, and biographical sketches.

B. L. Cline, *Men who Made a New Physics*, Chaps. 6 and 7

G. Gamow, *Thirty Years that Shook Physics*, Garden City, N.Y.; Doubleday, 1966, Chapters II and III. A lively account, well illustrated, based on the author's own recollections as a participant and observer of the development of quantum theory. Not all details are, however, historically accurate.

R. Moore, *Niels Bohr: The Man, His Science, and the World They Changed*, New York: Knopf, 1966

## SOURCES, INTERPRETATIONS, AND REFERENCE WORKS

C. E. Behrens, "Atomic theory from 1904 to 1913," *American Journal of Physics*, Vol. 11, pages 60–66 (1943)

D. ter Haar, *The Old Quantum Theory*, Part 1, Chapters IV and V, and Part 2; reprints of papers by Bohr and Pauli

J. L. Heilbron and T. S. Kuhn, "The genesis of the Bohr atom," *Historical Studies in the Physical Sciences*, Vol. 1, pages 211–290 (1969)

A. Hermann, *The Genesis of Quantum Theory*, Cambridge, Mass.: M.I.T. Press, 1971

W. R. Hindmarsh, *Atomic Spectra*, New York: Pergamon Press, 1967. Includes reprints of papers by Balmer, Rydberg, Bohr, and others.

T. Hirosige and S. Nisio, "The genesis of the Bohr atom model and Planck's theory of radiation," *Japanese Studies in the History of Science*, No. 9, pages 35–47 (1970)

J. MacLean, "On harmonic ratios in spectra," *Annals of Science*, Vol. 28, pages 121–137 (1972). Some precursors of Balmer.

W. McGucken, *Nineteenth-Century Spectroscopy*, Baltimore: The Johns Hopkins Press, 1969

M. Jammer, *The Conceptual Development of Quantum Mechanics*, New York: McGraw-Hill, 1966, Chapters 2 and 3

R. McCormmach, "The atomic theory of John William Nicholson," *Archive for History of Exact Sciences*, Vol. 3, pages 160–184 (1966)

S. Rozental (editor), *Niels Bohr: His Life and Work as Seen by his Friends and Colleagues*, New York: Wiley, 1967; paperback

# Chapter 29

# QUANTUM MECHANICS

## 29.1 A CRISIS IN THE FOUNDATIONS OF PHYSICS

With this chapter we come to the end of Newtonian mechanics as a fundamental basis for physical theory, even though almost all its practical applications remain valid at the present time. We have already seen the beginnings of one theory destined to replace Newtonian mechanics, the quantum hypothesis for light and Bohr's extension of it to a model of the atom; these results were soon to be incorporated into a unified quantum mechanics. In the last chapter of this book we shall discuss the other new theory, Einstein's theory of relativity. The complete amalgamation of these two theories into a consistent general system, which could play the same role as Newton's laws or Aristotle's natural philosophy, still lies largely in the future. It is partly for this reason, and partly because of the considerable mathematical complexity of quantum mechanics and relativity, that we have had to devote to them less space than their present importance might warrant.

To set the stage for the introduction of quantum mechanics, let us recall the situation in physics around 1920. First, Planck's quantum hypothesis had been generally accepted as a basis for atomic theory, and the photon concept had been given strong support by Robert A. Millikan's experi-mental confirmation of Einstein's theory of the photoelectric effect. Yet, despite the early success of Bohr's theory of the hydrogen atom, no one had found a satisfactory way to extend the theory to multi-electron atoms and molecules without introducing arbitrary additional hypotheses tailored to fit each particular case. Furthermore, it seemed necessary to use principles of Newtonian mechanics and Maxwellian electromagnetism in some parts of the theory, while rejecting them in others.

Although the photoelectric effect, as interpreted by Einstein, seemed to offer definite proof that light energy comes in discrete units, this did not by any means persuade physicists that the wave theory of light should be abandoned in favor of a particle theory. The experiments on interference and diffraction of light, and indeed all the properties of electromagnetic radiation discovered in the nineteenth century, seemed to require wave properties. Even the quantum theory, which assigned each "particle" of light an energy proportional to its "frequency" ($E = h\nu$), depended on a wave property, periodicity. Thus one was led to the paradoxical conclusion that light behaves in some respects like particles *and* in other respects like waves.

Another puzzling aspect of the physical theories developed in the first two decades of the twentieth century was their failure to prescribe definite *mechanisms* on the atomic level.    Thus Bohr's theory specified the initial and final states of an atomic radiation process, but did not describe just how the electron got from one orbit to another. The theory of radioactive decay was even less definite: It simply predicted the *probability* that a nucleus would emit radiation, but did not state when or how any particular nucleus would decay.    It appeared that the *determinism* or *causality* characteristic of the Newtonian world machine—the predictability, at least in principle, of the detailed course of events of any atomic process, given the initial masses, velocities, and forces—was slipping away.    Just as Maxwell and Boltzmann had shown that the second law of thermodynamics was only statistically valid as a description of observable processes, so some physicists began to suspect that Newton's laws of motion and the laws of conservation of energy and momentum might, in the atomic world, be true only "on the average," but not necessarily at every instant of time.

But if neither the wave theory nor the particle theory of light could be proved exclusively correct, and if the classical laws of mechanics and electromagnetism could no longer be relied on to describe atomic processes, clearly something new was needed, some fundamental law from which all these partially or statistically valid theories could be derived.    And perhaps even the basic Newtonian presuppositions about the nature and purpose of theories would have to be modified.    As they gradually recognized the seriousness of these problems, physicists early in the 1920s began to talk (much more urgently than they usually do) about a "crisis in the foundations of physics."

## 29.2 THE WAVE NATURE OF MATTER

It is a curious fact that radical conceptual changes in a science are often initiated by men who did not receive their initial professional training in that science.    The person who first approaches the problems of a discipline with the mature perspective of an outsider, rather than being indoctrinated previously with the established methods and attitudes of the discipline, can sometimes point to unorthodox, though (when seen in retrospect) remarkably simple solutions to those problems.    As we saw in Chapter 17, for example, the generalized law of conservation of energy was introduced into physics from engineering, physiology, and philosophy, though its value was accepted readily enough by physicists.

Louis de Broglie (b. 1892) received his first degree in history at the University of Paris in 1910, intending to go into the civil service.    But he became intrigued by scientific problems as a result of discussions with his brother, the physicist Maurice de Broglie, and by reading the popular scientific expositions of the mathematician Henri Poincaré.*    Although he managed to acquire sufficient competence in using the standard tools of theoretical physics to satisfy his professors, his real interest was in the more fundamental problems of the nature of space, time, matter, and energy.

In his doctoral dissertation of 1924 (based in part on papers published a year earlier), de Broglie proposed a sweeping symmetry for physics: Just as photons behave like particles as well as like waves, so electrons should behave like waves as well as like particles.    In particular, an electron of mass $m$ moving with velocity $v$ will have a "wavelength" $\lambda$ given by the simple formula

$$\lambda = \frac{h}{mv}, \qquad (29.1)$$

where $h$ is Planck's constant.

Although de Broglie's hypothesis was based originally on theoretical arguments, it was soon

---

*    These books are still delightful reading; they are available in English under the titles *The Value of Science*, *Science and Method*, *Science and Hypothesis*, and *Mathematics and Science: Last Essays* (all Dover paperback reprints).

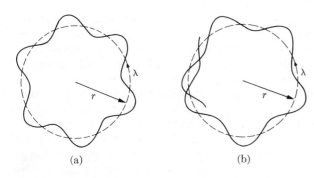

Fig. 29.1   Only certain wavelength will "fit" around a circle, e.g., in (a) not in (b).

realized that it could be checked rather directly by experiments of a kind which had already been performed by C. J. Davisson in the United States. In these experiments, electrons emitted from an electrode in a vacuum tube were allowed to strike a metal surface, and the scattering of the electron beam at various angles was measured. Walter Elsasser in Germany pointed out that Davisson's results could be interpreted as showing a *diffraction* of the beam of electrons as they passed through rows of atoms in the metal crystal, and that other experiments on electron scattering by C. Ramsauer could be attributed to *interference* of waves that somehow determined the path of the electrons. In both cases the apparent wavelength associated with the electrons agreed with the values calculated from de Broglie's equation (29.1). In 1926, Davisson together with L. H. Germer carried out further experiments which confirmed de Broglie's hypothesis, and in 1927 G. P. Thomson* demonstrated the phenomenon of electron diffraction by still another method. Later experiments showed conclusively that not only electrons but protons, neutrons,

* The son of J. J. Thomson, who had received the Nobel Prize in physics (1906) for his discovery (1897) of the electron and "proof" that it is a particle (Section 26.4). In 1937, Davisson and G. P. Thomson were jointly awarded the Nobel Prize in physics for demonstrating the wave nature of the electron.

heavier nuclei, and indeed all "material particles" share this wavelike behavior.

---

**Problem 29.1.** From Eq. (29.1) calculate the de Broglie wavelength $\lambda$ for an electron with kinetic energy equal to 50 electron volts. In order to show diffraction effects, a wave disturbance must go through an opening whose linear dimension is about the same order of magnitude as its wavelength, or smaller. Show that the space between rows of atoms in a crystal could be a suitable opening.

**Problem 29.2.** Estimate the de Broglie wavelength for a baseball moving at 100 m/sec and for an automobile moving at 50 m/sec. On the basis of these results, suggest why the wave nature of matter was not discovered until the twentieth century.

---

In his 1924 memoir, de Broglie pointed out that the quantization of allowed orbits in the Bohr model of the hydrogen atom can also be deduced from his formula for the wavelength of an electron, if one makes the natural assumption that the circumference of the orbit must be just large enough to contain an integral number of wavelengths, so that a continuous waveform results, as in Fig. 29.1. The assumption, applied to a circular orbit of radius $r$, would mean that

$$2\pi r = n\lambda = \frac{nh}{mv},$$

where $n$ is an integer. According to Bohr's theory, the kinetic energy ($\frac{1}{2}mv^2$) of the electron is proportional to its total energy $E$, which varies inversely as the square of the quantum number $n$ (Eq. 28.12). Therefore, if the mass $m$ is fixed, the velocity is inversely proportional to $n$ (see Problem 28.15), and the radius is proportional to $n^2$, in agreement with Eq. (28.13).

De Broglie's hypothesis suggests that the reason why atomic energy levels are quantized is the same as the reason why the frequencies of over-tones of a vibrating string are "quantized" (Section 23.2): The waves must exactly fill up a certain space. Hence a theory of atomic properties might be constructed by analogy with the theory of vibrations of mechanical systems—reminiscent of the old program of Pythagoras and Kepler, who based many of the laws of nature on the laws that govern the operation of musical instruments (Sections 1.1 and 4.4).

Another year passed before this clue was followed up by an Austrian physicist, Erwin Schrödinger (1887–1961).* Stimulated by the favorable remarks of Einstein and of Peter Debye on de Broglie's theory, Schrodinger developed a mathematical theory of atomic properties in which the quantization of energy levels was based on the allowed values, or *eigenvalues*,† of the electron wavelengths. His theory, known as "wave mechanics," was published early in 1926. It was soon recognized to be a satisfactory generalization of Bohr's theory; it provided a set of postulates which determined uniquely a fundamental equation governing the behavior of the electrons in any given atom or molecule (see below, Eq. 29.3). In the simple cases in which Schrödinger's equation can be solved exactly or almost exactly, it was found to predict the correct values of spectral frequencies and the intensities of spectral lines, as well as all other *observable* properties of the system. (For most atoms and molecules the equation has not been solved exactly because of its extreme complexity when many electrons are involved.)

The major defect (if it can be called that) of Schrödinger's theory, like de Broglie's theory from

---

* Schrödinger was born in Vienna and received his Ph.D. there in 1910. It should be noted that he was quite familiar with the theory of sound, and published a paper on the acoustics of the atmosphere in 1917.

† The German prefix *eigen* means "own," "individual," "proper," or "characteristic." The complete German word *eigenwert* has been half-translated into scientific English as "eigenvalue."

which it evolved, is that it fails to explain just what is vibrating or "waving"; it is definitely *not* the electron, conceived as a point particle analogous to a billiard ball, because that would still imply the classical conception of a particle localized at a particular spatial position and having a particular velocity, both of which change in a deterministic way from one instant to the next. Instead, the electron itself *is* in some (not visualizable) sense the wave. If the electron is in a definite *state* or energy level in the atom (corresponding to one of the allowed orbits in Bohr's theory), then it would seem to occupy simultaneously the entire orbit and to have simultaneously all values of the velocity vector corresponding to the different possible directions of motion in a circular orbit at constant speed. Even this statement is unduly restrictive, for the theory can be interpreted to mean that the electron is "spread out" over a region of space around the Bohr orbit and has a large range of different speeds (see below, Section 29.3). This is a natural description of a wave disturbance in a continuous medium, but it seems difficult to reconcile it with our earlier conception of the electron as a particle whose mass, charge, and velocity can be experimentally determined (Section 26.4). This is just as difficult, indeed, as it is to think of an electromagnetic wave as a vibration of some ethereal medium (Section 23.3) and at the same time as a particle having a definite momentum (Section 16.7) and energy (Section 26.5)!

In the case of electromagnetic waves, we could at least say that the vibrating or oscillating quantities were electric and magnetic field intensities (Section 25.3). For Schrödinger's electron waves, we can say that the vibrating quantity is a *wave function*, usually denoted by $\psi$ (Greek letter psi). The word "function" is used because, like the electric and magnetic fields $E$ and $H$, $\psi$ is a mathematical function of space and time variables, i.e., its value depends on $x$, $y$, $z$, and $t$. (For an electron in a stationary state corresponding to a Bohr orbit, $\psi$ does not depend on $t$.) However, *unlike* $E$ and $H$, the values of $\psi$ may be *complex numbers* (numbers which depend on the imaginary unit $\sqrt{-1}$), and therefore it defies any simple physical interpretation. Partly for this reason, we suggest that the following remarks be construed as a general overview, not to be studied in detail, rather than a clear pedagogic structure (see also the readings suggested at the end of the chapter).

According to Schrödinger's theory, one must first construct the wave equation for a physical system by converting the expression for the kinetic energy of each particle into a mathematical operation on the wave function $\psi$. This operation is similar to finding the acceleration of a particle whose position $s$ is expressed as a function of the time $t$ (see Sections 6.4, 6.5, and 10.3) except that the independent variable $t$ is replaced by each of the three space variables $x$, $y$, $z$ in turn (corresponding to the three velocity components). The operations can be expressed verbally as "find the rate of change with respect to $x$ of the rate of change with respect to $x$ of" the function $\psi$. A further difference is that $\psi$ is *not yet known* as a function of $x$, $y$, $z$, so these operations cannot yet be performed, but are only written down symbolically, as "kinetic energy operations on $\psi$," $KE_{op}\psi$.

In order to determine the stationary states of an atomic or molecular system, Schrödinger's theory requires us to add this kinetic energy operation to a potential energy operation $PE_{op}$, where $PE_{op}$ is defined simply as multiplying $\psi$ by the algebraic expression for the total potential energy of the system, PE. (PE usually consists simply of a sum of terms of the form $e_i e_j / r_{ij}$, where $e_i$ and $e_j$ are the charges of two interacting electrons or nuclei and $r_{ij}$ is the distance between them.) The total operation is called the *Hamiltonian* operator, $H$:

$$H = KE_{op} + PE_{op} \qquad (29.2)$$

---

\* After William Rowan Hamilton, the nineteenth-century Irish mathematician who first explored the analogies between optical and mechanical theories.

The fundamental equation of Schrödinger is then simply

$$H\psi = E\psi, \qquad (29.3)$$

where $E$ is a numerical constant which is equal to the energy of the system.

The Schrödinger equation, Eq. (29.3), must now be solved for the unknown wave function $\psi$ and the unknown energy $E$. In general there are several possible $\psi$ functions which are solutions of this equation; each has its own eigenvalue $E$. Thus *the fact that an atom has a discrete set of possible states and a discrete set of corresponding energy values emerges automatically as a mathematical property of the equation; the set of rules for constructing the Schrödinger equation replaces the previous postulates about allowed orbits in the Bohr theory.* These rules apply to all conceivable systems, whereas in Bohr's theory the postulates had to be constructed by trial and error for each case, with only the physicist's intuition and experience to guide him.

## 29.3 KNOWLEDGE AND REALITY IN QUANTUM MECHANICS

Although the wave function $\psi$ appears to play a central role in Schrödinger's wave mechanics, it is according to some physicists only a superfluous auxiliary construct. The rationale for this viewpoint derives from the success of another theory published a few months before Schrödinger's: Werner Heisenberg's theory, now known as *matrix mechanics*. Heisenberg, a young German physicist working in Max Born's institute at Göttingen, constructed an alternative set of rules for calculating the frequencies and intensities of spectral lines, using only relations between observable quantities. With the collaboration of Born and Pascual Jordan, Heisenberg soon developed his method into a complete mathematical theory of atomic processes which could yield the same results as Schrödinger's theory.

As Max Jammer has remarked, in his comprehensive study of the history of this subject, "It is hard to find in the history of physics two theories designed to cover the same range of experience, which differ more radically than these two"— Schrödinger's wave mechanics and Heisenberg's matrix mechanics.* The former emphasized the *continuity* of physical processes and the wavelike behavior of the electron—almost visualizable, despite the obscure meaning of the wave function. The latter proceeded from the *discontinuity* of physical processes, suggested by the observed discreteness of spectral lines, and regarded the electron as a particle, though without assigning it any definite space–time description in the classical sense. Nevertheless it was quickly discovered that the two theories are mathematically equivalent— the proof of this fact being given by Schrödinger and independently by Carl Eckart (at the California Institute of Technology) in 1926.

The development of physics since 1926 has not undermined in any significant way the validity of the Heisenberg–Schrödinger theories, considered now as two alternative formulations of a single theory, *quantum mechanics.* The philosophical viewpoints about the nature of the physical world and our knowledge of it, deriving from quantum mechanics, are of considerable interest, the more so as this is still an area of vigorous research. For example, quantum mechanics has revived the old question: Do physical objects have an existence and properties entirely *independent* of the human observer?

The early success of quantum mechanics led physicists to try to interpret its concepts in terms of the classical categories applicable to particles. The first such attempt was made by Max Born in 1926. He proposed that the electron is really a particle, and that the wave function simply represents the *probability* that it is located at a particular

point in space.* The use of the word "probability" in Born's interpretation is not the same as in classical physics, for example in the kinetic theory of gases (Chapter 22), in which it was assumed that the molecules have definite positions and velocities which are determined at each instant of time by Newton's laws and could in principle be calculated exactly.

In the kinetic theory, statistical methods were introduced as a matter of convenience in doing calculations: It seemed impossible to measure all the positions and velocities of $10^{22}$ molecules or more, or to use this information even if it were available. "Probability" referred to the mode of description employed by the scientist and did not affect the idea that the properties of an individual molecule are considered exactly determinable.

But in Born's view, the positions and velocities of each subatomic particle are basically random; they cannot be said to have any definite values, even in principle, but have only certain probabilities of exhibiting particular values. Born argued that the contrary assumption (adopted at first by Schrödinger), namely, that the electron is really a wave spread out in space with a charge density given by the wave function, is inadmissible. For if the electron is not bound in an atom, its wave function eventually will spread out over an infinite space; yet experiment indicates that all the electron is actually present at some particular place. This result is still consistent with the idea that the electron has a certain probability of being at any given place; performing the experiment to find its position converts this probability into a certainty—it

---

* Max Jammer, *The Conceptual Development of Quantum Mechanics*, page 271. New York: McGraw-Hill, 1966.

---

* More precisely, since the value of the wave function is a complex number (and only the square of a complex number has entirely real components), it is the square of the absolute value of the wave function which is proportional to the probability that the electron has the position $x, y, z$. If the complex number $\psi$ is expressed as a sum of real and imaginary parts, $\psi = a + \sqrt{-1}\, b$, then the square of its absolute value is defined as $|\psi|^2 = a^2 + b^2$, which is a real number.

either *is* or *is not* there, but it cannot be "partly there." It is already evident that any attempt to pin down a definite property of the electron involves an act of measurement.

Max Born was awarded the Nobel Prize for physics in 1954, primarily for his statistical interpretation of quantum mechanics. The fact that the award came so long after the original publication (1926) indicates that this interpretation was not at once regarded as an indubitable, basic "discovery," but that it gradually came to be accepted as a fruitful viewpoint in physical theory. Insofar as many scientists do accept the notion that natural processes are fundamentally random rather than deterministic, we have to admit that the "Newtonian world machine," or the mechanistic viewpoint advocated by Descartes and Boyle in the seventeenth century, has been abandoned.

Another way of characterizing the consequences of quantum mechanics is through Heisenberg's *indeterminacy principle*, often called the uncertainty principle. In 1927, Heisenberg proposed as a postulate that there are certain pairs of physical properties of a particle which cannot simultaneously be measured to an arbitrarily high degree of accuracy. The more accurately we try to measure one property in this pair, the less accurate will be our measurement of the other one. For example, if we try to determine both the position ($x$) and momentum ($p$) of an electron, then the product of the average errors or uncertainties of measurement (symbolized by $\delta$) must be at least as great as $h/2\pi$:

$$(\delta x) \cdot (\delta p) \geqq \frac{h}{2\pi}. \tag{29.4}$$

Although the principle applies to the position and velocity of any object, it is a significant limitation only for atomic or subatomic particles, since Planck's constant $h$ is such a small number.

**Problem 29.3.** (a) An electron is known to have a speed of $100 \pm 1$ m/sec. What will be the uncertain-

ty in its measured position? (b) A rock of mass 100 g is known to have a speed of $100 \pm 1$ m/sec. What will be the measured uncertainty in its position?

**Problem 29.4.** For an electron in a Bohr orbit in a hydrogen atom, the uncertainty in position at any moment may be said to be equal to the radius of the orbit (its distance from the center being within $\pm r$), and the uncertainty in velocity at that moment may be said to equal the numerical value of the speed (the magnitude of the velocity component in any direction fluctuates between $+v$ and $-v$ as it goes around a circular orbit). Write down the condition that the de Broglie wavelength of the electron must be no greater than the circumference of the orbit; then write down the indeterminacy principle for position and momentum of this electron; then show that these two statements are equivalent.

---

The indeterminacy principle can be deduced from quantum mechanics, but its validity was also established by considering various possible experiments which might be designed to measure both $x$ and $p$. Such experiments were analyzed in some detail in a famous discussion between Bohr and Einstein.* The conclusion was that any measurement involves an *interaction* between the observer (or his apparatus) and the object being observed. For example, one could measure the position of an electron by bouncing photons off it, in the same way that we "see" ordinary objects; but in order to do this very accurately it would be necessary to use very short-wavelength photons—otherwise they would simply be diffracted around the electron rather than reflected. But a short-wavelength photon has a large momentum and would therefore give the electron a strong kick when it hit it (see

---

\* See Bohr's account in the volume *Albert Einstein, Philosopher–Scientist* (edited by P. A. Schilpp), Library of Living Philosophers, 1949, pages 199–241.

Section 16.7), changing the electron's momentum in an unpredictable way, thereby making it impossible for us to determine $p$ accurately at the same time.

Heisenberg's principle also applies to the energy $E$ of a particle and the time interval $t$ during which it has that energy. It is impossible to determine both $E$ and $t$ simultaneously with unlimited accuracy, because of the restriction

$$(\delta E) \cdot (\delta t) \geqq \frac{h}{2\pi} .$$

Heisenberg's principle might be interpreted as merely a restriction on how much we can *know* about the electron, taking into account the limitations of existing experimental methods, without thereby rejecting the belief that the electron really does have a definite position and momentum. The term "*uncertainty* principle" would then be appropriate, with its implication that the principle applies to the observer's knowledge rather than to nature itself.

There are two objections to that view, however. The first is the feeling of many physicists that science should be concerned only with concepts that have operational definitions (Section 12.4). The case of the ether (Sections 23.3 and 25.3) had shown that considerable time and effort can be wasted if one insists on constructing theories based on hypothetical entities whose properties cannot be measured experimentally. A second reason for being skeptical about the assumption that an electron really has a definite position and momentum is that no one has yet succeeded in constructing a satisfactory theory based on this assumption, despite numerous attempts.

Heisenberg did not suggest that the terms "position" and "momentum" be banished from physics, for they are needed to describe the results of experiments. Even the term "path" or "orbit" of an electron is still admissible as long as we take the attitude that "the path comes into existence only when we observe it."* This remark appears to some philosophers to align quantum mechanics with the traditional philosophy of subjective idealism, according to which the real world consists only of the perceptions of an observer, and physical objects have no objective existence or properties apart from such human observations. Taken together with Born's assertion that atomic quantities are inherently random, Heisenberg's viewpoint would seem to deny the possibility of finding out anything certain about a physical system taken by itself, since the observer or the measuring apparatus is an essential part of any physical system we may choose to study. (For example, the photon which was used to "see" the electron in the earlier experiment becomes part of the system being studied.)

A somewhat similar situation is quite familiar in the behavioral sciences: In almost any psychological experiment, the subject knows that his behavior is being observed, and may not behave in the same way that he would if no psychologist were present. But at least in that case one feels confident that there *is* such a thing as "behavior when the psychologist is not present," even if we do not observe it, and we probably believe that if the psychologist were sufficiently clever he could carry out his experiment without the knowledge of the subject if his scientific and moral code of ethics did not forbid it.

A better analogy might be the intelligence test, which purports to measure an inherent property of the individual, but which is now recognised to be so strongly dependent on the circumstances of testing and on cultural bias in fashioning the test that it is doubtful whether "intelligence" really has an objective existence, except in the opera-

---

* W. Heisenberg, *Zeitschrift für Physik*, Vol. 43, page 185 (1927), as quoted by Jammer, *op. cit.*, page 329. According to this viewpoint, Schrödinger's wave function is also a somewhat dubious concept; it cannot be given an operational definition, though it may be quite useful in doing calculations with the theory.

tional sense of ability to perform *on a particular test* ("intelligence is what the intelligence test tests").

Quantum mechanics is rescued from the quicksands of subjectivism by the fact that it makes perfectly definite predictions about the properties of physical systems, and these properties can be measured with the same results (including the measurement uncertainty) by all observers. If one prefers, one does not *have* to abandon the idea that there is a real world "out there," and even the fact that the measuring apparatus is strictly speaking inseparable from the things we are measuring does not in practice make very much difference.

Moreover, in spite of the fact that our knowledge of the position and momentum of electrons, photons, and all other particles is restricted to probabilities, the wave function which determines these probabilities is itself completely determined by the theory, and changes with time in a very definite, "causal" way: If $\psi$ is given at a particular time and the Hamiltonian operator of the entire system is known (Eq. 29.2), then $\psi$ can be computed at any other time. (The "entire system" includes of course any measuring device that may interact with the system.) Thus as Heisenberg himself pointed out, "probability" itself acquires a certain "intermediate reality," not unlike that of the *potentia* (possibility or tendency for an event to take place) in Aristotle's philosophy. An isolated system which is evolving in this way but not being observed is, according to Heisenberg, "potential but not actual in character"; its states are "objective, but not real."[*]

Although quantum mechanics has stimulated much discussion by philosophers, it would be inaccurate to claim that the adoption of this theory in physics requires the acceptance of either of the traditional doctrines in philosophy, idealism or realism. The significance of quantum mechanics is rather that it forces us to take an attitude toward the descriptive categories when doing atomic or subatomic physics that is different from the attitude we employ in talking about the physical world in other cases.[†]

One expression of the new attitude is the *complementarity principle*, proposed by Niels Bohr in 1927. According to this principle, two modes of description of a given system may appear to be mutually exclusive, yet both are necessary for complete understanding of the system. Thus, on the one hand, we may wish to emphasize *causality* by pointing to the fact that there is a well-determined evolution of the state of the system (defined by its wave function), as mentioned in the last paragraph. Yet this description is meaningful only if we refrain from making any observations of space and time variables, since the process of making such observations would disturb the state in an unpredictable way and destroy the causality. On the other hand, we might prefer to emphasize the spatiotemporal description and sacrifice the causality. Bohr held that each description gives a partial glimpse of the total "truth" about the system taken as a whole.

Bohr accepted the indeterminacy principle as a useful quantitative limitation on the extent to which complementary (and therefore partial) descriptions may overlap—that is, an estimate of the price that must be paid for trying to use both descriptions at once. But he did not follow Born and Heisenberg in their insistence that the "wave nature of matter" meant nothing more than the inherently random behavior of entities that are really particles. Instead, Bohr argued that the wave and particle theories are examples of complementary modes of description, each valid by itself though (in terms of Newtonian physics) incompatible with each other.

---

[*]    W. Heisenberg, page 27 in *Niels Bohr and the Development of Physics* (edited by W. Pauli *et al.*), London: Pergamon Press, 1955.

---

[†]    This attitude can itself be considered a philosophical doctrine, often called "instrumentalism."

tarity principle is a major com-
now called the *Copenhagen*
ntum mechanics. This inter-
that the physical world has just
properties that are revealed by experiments,
including the wave and particle aspects, and that
the theory can only deal with the results of observations, not with a hypothetical "underlying
reality" that may or may not lie beneath the
appearances. Any attempt to go further than this,
to specify more precisely the microscopic details
of the structure and evolution of an atomic system,
will inevitably encounter nothing but randomness
and indeterminacy.

The Copenhagen interpretation met, and still
meets, with strong objections from a minority of
leading physicists, including both de Broglie and
Einstein. There is understandably a deep-seated
resistance to abandoning the traditional conception of physical reality, and much appeal in Einstein's famous critique of inherent randomness:
"God does not play dice." But so far no satisfactory
alternative interpretation of quantum mechanics
(nor any experimental refutation of the theory
itself) has been found.*

## RECOMMENDED FOR FURTHER READING

D. Bohm, *Causality and Chance in Modern Physics*,
London: Routledge and Kegan Paul, 1957; Harper
Torchbook reprint. A lucid account of the development of mechanism in physical theory, ending with
an argument that quantum mechanics should not
be considered a complete theory with ultimate
validity.

Niels Bohr, "Discussion with Einstein on epistemological problems in atomic physics," pages 199–

---

* For a survey and critique of earlier proposed alternatives to the Copenhagen interpretation, see Heisenberg's
article, *op. cit.* More recent discussions may be found in
the publications of Ballentine, Bohm, DeWitt, and
Schlegel listed below.

241 in *Albert Einstein, Philosopher-Scientist* (edited
by P. A. Schilpp), New York: Library of Living
Philosophers, 1949. See also Einstein's comments,
pages 665–683 of the same book.

H. A. Boorse and L. Motz, *The World of the Atom*,
pages 1021–1122. Extracts from the writings of
Hamilton, de Broglie, Schrödinger, Born, and
Heisenberg, and biographical sketches.

B. Cline, *Men Who Made a New Physics*, New York:
Crowell, 1965; Signet Science Library Paperback;
Chapters 8–11

R. Feynman, *The Character of Physical Law*,
Chapters 6 and 7

Werner Heisenberg, "The development of the
interpretation of the quantum theory," pages 12–
29 in *Niels Bohr and the Development of Physics*
(edited by W. Pauli *et al.*), London: Pergamon Press,
and New York: McGraw-Hill, 1955; *Physics and
Philosophy*, New York: Harper and Row, paperback
reprint of the 1958 edition

*Project Physics Reader*, Vol. 5, pages 121–192.
Extracts from writings of C. P. Snow, Dirac, Infeld,
and Schrödinger.

Erwin Schrödinger, *Science, Theory, and Man*,
Dover paperback reprint of the 1935 edition

S. Toulmin (editor), *Physical Reality*, New York:
Harper and Row, 1970; paperback. Essays by
Planck, Mach, Einstein, Bohr and others on the
philosophical interpretation of physical theories.

## SOURCES, INTERPRETATIONS, AND REFERENCE WORKS

L. A. Ballentine, "The statistical interpretation of
quantum mechanics," *Reviews of Modern Physics*,
Vol. 42, pages 358–381 (1970). Argues, primarily on
technical grounds, for an alternative to the Copenhagen interpretation.

Louis de Broglie, *The Current Interpretation of*

*Wave Mechanics, a critical study*, New York: American Elsevier, 1964

W. H. Cropper, *The Quantum Physicists and an Introduction to their Physics*, New York: Oxford University Press, 1970, Chapters 1–4

B. S. DeWitt, "Quantum mechanics and reality," *Physics Today*, September 1970, pages 30–35.

P. Forman, "Weimar culture, causality, and quantum theory, 1918–1927. Adaptation by German physicists and mathematicians to a hostile intellectual environment," *Historical Studies in the Physical Sciences*, Vol. 3, pages 1–115 (1971)

J. L. Heilbron, "Quantum historiography and the Archive for History of Quantum Physics," *History of Science*, Vol. 7, pages 90–111 (1968)

Werner Heisenberg, *The Physical Principles of the Quantum Theory*, translated from German by C. Eckart and F. C. Hoyt, Dover paperback reprint of the 1930 edition

G. Holton, "The roots of complementarity," *Daedalus*, Vol. 99, pages 1015–1055 (1970)

M. Jammer, *The Conceptual Development of Quantum Mechanics*, New York: McGraw-Hill, 1966, Chapters 4–9

M. J. Klein, "Einstein and the wave-particle duality," *The Natural Philosopher*, Vol. 3, pages 1–49 (1964)

A. Kompaneyets, *Basic Concepts in Quantum Mechanics*, New York: Reinhold, 1966. A good elementary text.

G. Ludwig, *Wave Mechanics*, New York: Pergamon Press, 1968. Includes extracts from papers by de Broglie, Schrödinger, Heisenberg, Born, and Jordan.

K. Przibram (editor), *Letters on Wave Mechanics*: *Schrödinger, Planck, Einstein, Lorentz*, translated from German by M. J. Klein, New York: Philosophical Library, 1967

V. V. Raman and P. Forman, "Why was it Schrödinger who developed de Broglie's ideas?" *His-torical Studies in the Physical Sciences*, Vol. 1, pages 291–314 (1969)

R. Schlegel, "Statistical explanation in physics: the Copenhagen interpretation," *Synthese*, Vol. 21, pages 65–82 (1970). Claims that a recent laser experiment with single photons "implies a subjectivism on the level of individual quantum-level occurrences."

B. L. van der Waerden, *Sources of Quantum Mechanics*, Amsterdam: North-Holland, 1967; Dover paperback reprint

# Chapter 30

# EINSTEIN'S THEORY OF RELATIVITY

## 30.1 BIOGRAPHICAL SKETCH OF ALBERT EINSTEIN

Before turning to a brief account of the theory of relativity, let us sketch the career of its chief developer. Albert Einstein was born on March 14, 1879, in Ulm, a middle-sized city in the Swabian part of Bavaria in southwestern Germany. A year later his family moved to Munich, the political and intellectual center of southern Germany. His father, Hermann Einstein, operated (not too successfully) a small electrochemical factory. His mother, born Pauline Koch, had artistic interests and instilled in her son a love for music. An uncle who lived nearby and ran the technical side of the business aroused Albert's first interest in mathematics by teaching him some algebra.

Einstein was not a child prodigy; on the contrary, he was late in learning to talk, and did not show much proficiency at learning his lessons in school. He disliked any kind of regimentation or pressure toward conformity, whether from his teachers or his classmates.

His family was Jewish, but "entirely irreligious," as Einstein later described it; young Einstein did not receive much religious instruction at home, and his parents sent him to a Catholic elementary school, in part because it was closer and more convenient. He became, on his own, deeply interested in the Jewish religion. But, according to his autobiographical notes, at age 12,

"...through the reading of popular scientific books I soon reached the conviction that much in the stories of the Bible could not be true. The consequence was a positively fanatic [orgy of] freethinking coupled with the impression that youth is intentionally being deceived by the state through lies; it was a crushing impression. Suspicion against every kind of authority grew out of this experience, a skeptical attitude towards the convictions which were alive in any specific social environment—an attitude which has never again left me, even though later on, because of a better insight into the causal connections, it lost some of its original poignancy."[*]

When Albert was 15 his father's business failed. Hermann Einstein decided to leave Munich and try again in Italy. Albert was left behind to finish his secondary school education in Munich, while living in a boarding house. He was unhappy there, and soon left to join his parents. His exit apparently

---

[*] Paul Arthur Schilpp, editor, *Albert Einstein, Philosopher–Scientist*, New York: Harper, 1949, page 5.

did not sadden the Munich school—in fact, one of his teachers had suggested he leave on the grounds that his mere presence in the class "destroyed the respect of the students." Apparently Einstein's obvious, honestly expressed distaste for the mechanical drill of lessons and for military discipline, combined with the fact that he was nevertheless exceedingly profficient in mathematics and physics, was too much for the teacher to take.

In Italy, his father's business did not go too well; it was now clear that young Einstein would soon have to support himself. He decided to study electrical engineering at the Swiss Federal Polytechnic University in Zurich, Switzerland. Though only 16, and two years younger than most of the candidates, he presented himself for the entrance examination—and failed (in biology and French). To prepare for another try, he went to a Swiss high school in Aarau, and for the first time found himself in a congenial, democratically run school. There he seems to have flourished, and he recalled later that it was at Aarau that he had his first ideas leading to the relativity theory. By the time he graduated from the Polytechnic, his interests had settled on theoretical physics (with the intention to become a teacher), but he had to teach himself such advanced subjects as Maxwell's electromagnetic theory, which had not yet become part of the standard curriculum.

After graduation, Einstein (still an alien) was unable to find a satisfactory teaching position in Switzerland. Finally the father of one of his friends and former classmates helped to get him a job at the patent office in Bern. Soon after taking this position, Einstein married Mileva Maritsch, a fellow student at Zurich who had come from Hungary. She shared Einstein's interest in physics, but was rather reserved and uncommunicative. Two sons were born of the marriage.

Although it might seem an unusual occupation for a theoretical physicist, the position as patent examiner did have certain advantages for Einstein at this stage in his career. The work was not dull; he was continually having to analyze and criticize ideas for new inventions. Moreover, unlike most teaching positions available to young physicists, the patent office job gave Einstein enough free time to do his own reading, thinking, and research. The complete absence of contact with professional physicists during this period was perhaps a blessing in disguise, for it permitted Einstein to develop his rather unorthodox approach to the problems of physics.

In 1905, Einstein published three papers, each of which was eventually recognized as the work of genius. The first, applying the kinetic-molecular theory to Brownian movement, was mentioned in Section 22.7. The second, on the photon hypothesis and the photoelectric effect, was discussed in Section 26.5. The third and most revolutionary, on what was later called the special relativity theory, will be treated in the following sections. In a separate brief note, published shortly afterward, Einstein pointed out that as a consequence of his theory of relativity, the mass $m$ of an object should be related to its energy content $E$ through the formula $E = mc^2$ (see below, Section 30.8).

By 1909, these and other publications had brought him so much attention in the scientific world that he was appointed Professor Extraordinary (equivalent to Associate Professor) at the University of Zurich, and a year later he was called to the chair of Theoretical Physics at the German University in Prague, Czechoslovakia. In 1912 he returned to Zurich as Professor of Theoretical Physics at the Polytechnic School from which he had graduated, and began to develop his general theory of relativity (Section 30.10). The next year, he was offered what was at the time the highest position to which a European scientist could aspire: professor at the University of Berlin, research director at the newly founded Kaiser Wilhelm Institute there, and member of the Prussian Academy of Science.

The decision to go to Berlin was not an easy one for Einstein. As a pacifist and social democrat, Einstein did not like the political and social atmosphere in Berlin. But in other respects the position was irresistible: Among Einstein's colleagues during his stay in Berlin (1913–1933) were many of Europe's leading physicists—Max Planck, Walther Nernst, Max von Laue, James Franck, Gustav Hertz, Lise Meitner, and Erwin Schrödinger.

One of Einstein's most spectacular triumphs came just after the end of World War I. He predicted, on the basis of his general theory of relativity, that light from a star should be deflected by the gravitational force of the sun, and calculated that the deflection should be a maximum of 1.75 sec of arc for a beam of light passing very close to the sun. This prediction could be tested accurately only during a solar eclipse, and the observations would have to be made from certain special places on the earth's surface where a *total* eclipse would occur.

Einstein's prediction became known in England shortly before the end of World War I, and British astronomers decided to make an expedition to test it during the eclipse expected on March 29, 1919. It would be necessary to travel to northern Brazil and to an island off the coast of West Africa for optimum viewing conditions, and such expeditions would have been rather dangerous while England and Germany were at war. But after the war ended on November 11, 1918, the British had time to plan and carry out the project; the observed result coincided with the expected value within the estimated experimental error. Since there was no other known theory that predicted this result, it could be quickly celebrated as a victory for Einstein's theory.

One can readily imagine the impact on world opinion of a British-sponsored expedition confirming a "German" theory, just a few months after the British and their allies had defeated the Germans in a lengthy and costly war. After the bitterness of combat, the acclaim of this work was an example of international reconciliation through scientific cooperation. And of course the rapturous press reports about the scientific ideas being verified—the "weight of light" and the "curvature of space"—helped spread the idea that Einstein

had wrought a revolution that overthrew the old order and opened a new and mysterious door to the deepest secrets of the universe.

Einstein was now, and for more decades to come, the most famous scientist in the world, and inevitably became the target of appeals for support of worthy personal and international causes (as well as for attacks by crackpots). He decided to lend his prestige to the support of three undertakings outside science itself which have, each in their own way, had a major impact on twentieth-century history. The first was support of the Zionist movement that led eventually to the establishment of the state of Israel; in this he was largely motivated by the tragic persistence of hidden and explicit antisemitism which he saw all around him. The second was one letter he wrote to President F. D. Roosevelt to suggest establishing the atomic energy project in the United States; it led eventually to the construction of nuclear weapons, initially intended to preempt the Germans who, in a war to conquer the world, had embarked on a project of their own to produce such weapons. The third cause was peace, from his manifesto against German militarism in World War I to his support of the international "Pugwash" movement of scientists, started in the 1950s, on behalf of international arms control.

In the early 1920s, when Einstein began to identify himself publicly with the Jews, he attracted the criticism of the political movement that soon developed into the Nazi party in Germany. This meant that from these quarters relativity theory itself was violently attacked on political and racial grounds; it was supposed to be an example of "Jewish" or "non-Nordic" physics, or even of "Bolshevik" physics. Some previously respectable German physicists joined in these attacks, in particular the Nobel Prize winner Philipp Lenard, whose measurements on the photoelectric effect in 1902 had provided data for Einstein's early formulation of the quantum theory.

In early 1933, when Hitler came to power and the racial purges began in Germany, Einstein was just returning from one of his frequent visits to the United States. When he landed in Belgium, he decided that it would be impossible for him to go on to Berlin so long as the Nazis were in charge. (Indeed, they had already confiscated his property in Berlin.) He therefore accepted an offer to join the newly founded Institute for Advanced Study at Princeton, New Jersey, and this became his home until his death in 1955.

## 30.2 THE FITZGERALD-LORENTZ CONTRACTION

As will be seen in the next section, Einstein's primary concern in his 1905 paper on relativity was to show that Maxwell's electromagnetic theory could be understood in a way that would be both physically accurate and logically consistent for objects and observers moving at high speeds relative to each other. But this reformulation required him to make a revolutionary critique of the traditional concepts of space and time, and subsequently of mass and energy. At the heart of this critique was the adoption of a set of transformation equations (now known as the Lorentz transformation) by means of which the values of physical quantities for a phenomenon as observed or measured from one frame of reference could be translated into the values for the same phenomenon as observed or measured from another frame of reference that is moving at a constant speed with respect to the first. As it happens, the mathematical form of these equations had been deduced by other physicists before Einstein, but had been given a different physical interpretation by them, one that was more in line with traditional ideas. Instead of introducing right away both the equations and their physical interpretation as Einstein presented them in 1905, it will be pedagogically easier, and is more usual, to present first those earlier ideas (on which Einstein himself did not rely).

In Section 25.3 we mentioned Michelson's and Morley's beautiful experiment of 1887, designed to detect the motion of the earth relative to the ether.

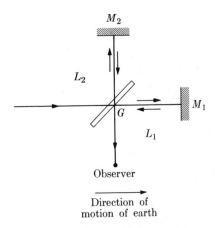

**Fig. 30.1**   The Michelson-Morley experiment (simplified sketch).

The "failure" of this experiment gave physicists such as H. A. Lorentz in the 1890s the first clue leading to the new transformation equations. In the experiment, shown schematically in Fig. 30.1, a beam of light is divided by a partly reflecting glass plate $G$ into two beams at right angles, one of which travels along a path of length $L_1$, the other along a path of length $L_2$. Each is reflected back by a mirror ($M_1$ or $M_2$) at the end of its path; the two beams are then recombined at $G$, and are seen by an observer at $O$. If the times required to travel the two separate routes are different, then the beams will be "out of phase" when they recombine, and interference will result. Michelson had designed a very sensitive interferometer which could detect extremely small phase differences in this way.

Suppose first that path $L_1$ is parallel to the direction of the earth's velocity in space, $v$. Michelson and Morley assumed that the speed of light traveling along this path would be compounded of the speed $c$ with respect to the ether and the earth's speed $v$. As the light goes from $G$ to $M_1$, $M_1$ is receding from it, so the trip should

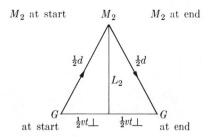

$M_2$ at start      $M_2$      $M_2$ at end

$\frac{1}{2}d$      $\frac{1}{2}d$

$L_2$

$G$      $G$
at start    $\frac{1}{2}vt_\perp$    $\frac{1}{2}vt_\perp$    at end

**Fig. 30.2**  Path of light from $G$ to $M_2$ to $G$ as seen by someone fixed in the "ether."

take *longer* than if the earth were at rest in the ether; but the return trip would take a shorter time since the plate $G$ is moving forward to meet the light that is coming back from $M_1$. Thus the time required for the first part (from $G$ to $M_1$) should be $L_1/(c - v)$, and the time for the second part (from $M_1$ to $G$) should be $L_1/(c + v)$. The total time required for the round trip along the path parallel to the earth's motion would therefore be

$$t_\parallel = \frac{L_1}{c - v} + \frac{L_1}{c + v}$$

$$= \left(\frac{2L_1}{c}\right) \frac{1}{1 - v^2/c^2}$$

$$= \frac{2L_1}{c(1 - v^2/c^2)}. \qquad (30.1)$$

The calculation of the time needed by the other beam, the one that goes along the path perpendicular to the earth's motion ($GM_2G$), is a little harder, and Michelson got it wrong the first time he tried it (in 1881). He thought that the time required to travel at right angles to the earth's motion would be unaffected by this motion, and would simply be equal to $2L_2/c$. However, one must see the situation as if one were fixed in the "ether" (Fig. 30.2): The central plate $G$ has moved a distance

equal to $vt_\perp$ during the time $t_\perp$ that it takes the beam of light to travel the total distance $d$ (from $G$ to $M_2$ and back to $G$). This distance $d$ is not simply $2L_2$, but is related to it by the Pythagorean theorem (see Fig. 30.2):

$$(\tfrac{1}{2}d)^2 = (\tfrac{1}{2}vt_\perp)^2 + L_2^2$$

or

$$d = \sqrt{v^2 t_\perp^2 + 4L_2^2}.$$

Hence the time $t_\perp$ has to be found by solving the equation

$$t_\perp = d/c.$$

On substituting for $d$ as above and solving, we find

$$t_\perp = \frac{2L_2}{c\sqrt{1 - v^2/c^2}}. \qquad (30.2)$$

If the apparatus could be constructed so perfectly that $L_1$ were precisely equal to $L_2$, then we could simply measure the difference between $t_\parallel$ and $t_\perp$ as given in Eqs. (30.1) and (30.2), and from this difference we could determine $v$. However, it cannot be assumed that these lengths can be made precisely the same within a small fraction of one wavelength of light, and any small discrepancy might well have a greater effect on the time difference than the earth's motion through the ether.

Michelson and Morley avoided this difficulty in an ingenious way: First they set $L_1$ and $L_2$ to give destructive interference between the two beams; then they rotated the whole interferometer through an angle of 90°, so that the arm $L_2$ was turned to a position parallel to $v$ and the arm $L_1$ was perpendicular to $v$. While it was rotating, they watched whether the new alignments of the arms of the interferometer with respect to the ether changed the interference pattern. An effect was clearly expected, as we can now show. In this new arrangement (indicated by a superscript $r$, for "rotated," above the symbols $t_\parallel$ and $t_\perp$), the same reasoning used above leads to the formulas

$$t_\parallel^{(r)} = \frac{2L_2}{c(1 - v^2/c^2)}, \quad (30.3)$$

$$t_\perp^{(r)} = \frac{2L_1}{c\sqrt{1 - v^2/c^2}}. \quad (30.4)$$

For example, study the differences among the last four equations. Note that if $L_1$ and $L_2$ are adjusted to assure equality of $t_\parallel$ and $t_\perp$ in Eqs. (30.1) and (30.2), $t_\parallel^{(r)}$ and $t_\perp^{(r)}$ will differ from each other; that is, if $(t_\parallel - t_\perp)$ is zero, $(t_\parallel^{(r)}$ and $t_\perp^{(r)})$ is not zero. In fact, the sum of the time differences in the two arrangements depends on the sum of the two lengths, $L_1 + L_2$:

$$\Delta t \equiv (t_\parallel - t_\perp) + (t_\parallel^{(r)} - t_\perp^{(r)})$$

$$= \frac{2(L_1 + L_2)}{c} \left[ \frac{1}{1 - v^2/c^2} - \frac{1}{\sqrt{1 - v^2/c^2}} \right].$$

$$(30.5)$$

(Note that we assumed throughout that the lengths of the arms, $L_1$ and $L_2$, respectively, do not change while we rotate the apparatus.)

---

**Problem 30.1.** (a) Use commonly available data to show that the speed of the earth in its orbit is approximately 30 km/sec. (b) In Michelson's original apparatus, the lengths $L_1$ and $L_2$ were each about 120 cm. Assuming that at some time of the year the earth is moving at a speed of 30 km/sec relative to the ether, calculate $\Delta t$, the difference between time intervals (as defined in Eq. 30.5), which one would expect to observe in this case. (c) If the wavelength of the light used in the experiment is 6000 Å, what fraction of a cycle of a light wave would the value of $\Delta t$ calculated in part (b) correspond to?

The value of $\Delta t$ expected to be observed in the Michelson-Morley experiment, as estimated in Problem 30.1, was not large, but it was substantially greater than the experimental error. Thus, when Michelson and Morley did the experiment and found virtually no time difference at all, they were most surprised and uncomfortable.

One explanation that offered itself was that one of the basic assumptions employed in the derivation of Eq. (30.5) was incorrect. In 1889 the Irish physicist G. F. FitzGerald suggested that the negative result of the Michelson-Morley experiment might be explained by simply assuming that the arm of the interferometer contracts by a factor $\sqrt{1 - v^2/c^2}$ in the direction of motion through the ether at a speed $v$. Thus an arm measuring 1 m when perpendicular to $v$ will really be only $\sqrt{1 - v^2/c^2}$ meters long when it is placed parallel to $v$. Thus, by multiplying $L_1$ by this factor in the formula for $t_\parallel^{(1)}$, Eq. (30.1), we would obtain a corrected time interval,

$$t_\parallel^{*(1)} = \frac{[2L_1\sqrt{1 - v^2/c^2}]}{c(1 - v^2/c^2)} = \frac{2L_1}{c\sqrt{1 - v^2/c^2}}. \quad (30.1^*)$$

If for the same reason the same correction is applied to $L_2$ in the second arrangement, Eq. (30.3) (but the length of the other arm, now perpendicular to the earth's motion through the ether, is restored to its full value $L_1$), we then find that the sum of the time differences is exactly zero,

$$\Delta t^* = (t_\parallel^{*(1)} - t_\perp^{(1)}) + (t_\parallel^{*(2)} - t_\perp^{(2)}) = 0. \quad (30.5^*)$$

This coincides with observation, and so tends to "explain" it.

FitzGerald tried to justify his contraction assumption (or hypothesis) by pointing out that intermolecular forces are probably electrical in nature. If (as in Maxwell's theory) electrical forces were propagated through the ether, then it was not unreasonable to suppose that they would be affected by motion through it; so the contraction hypothesis might be made more plausible by suggesting the existence of an increase in intermolecular attractive forces which drew the molecules closer together.

The contraction hypothesis was also proposed a few years later by the Dutch physicist H. A. Lorentz (who probably thought of it independently, since FitzGerald's publication of 1889 was not widely known). Lorentz developed a general theory of electrodynamics, in which he assumed that the lengths of all objects in a system moving with respect to the ether shrink by a factor $\sqrt{1 - v^2/c^2}$ (the correction factor still applied only to the component in the direction of motion). Furthermore, Lorentz proposed that the values of *mass* and of *time* intervals in the moving system are *dilated* by a similar factor: They must be multiplied by a factor $1/\sqrt{1 - v^2/c^2}$, which means a significant *increase* unless $v$ is very small compared to $c$. (The time dilation would not affect the interpretation of the Michelson-Morley experiment since, unlike the length contraction, it would be the same both along the perpendicular and the parallel path.)

Although the FitzGerald-Lorentz contraction hypothesis suggested that strange things happen when objects move at speeds close to the speed of light, it did not challenge traditional concepts of space and time. It was still assumed that certain phenomena depend on motion through an ether, and that one can distinguish in some absolute sense between a system that is moving and one that is not. Even the idea of time dilation did not contradict the presupposition that there is an absolute time scale for a system at rest in the ether, and that this time scale is more fundamental than any "local time" appropriate to a moving body and observed on clocks transported along with the moving body.

---

**Problem 30.2.** List *all* the assumptions made in deriving Eq. (30.5).

## 30.3  EINSTEIN'S FORMULATION (1905)

Einstein's first paper on relativity, published in 1905, appears to have emerged from a prolonged period of reflection on some aspects of Maxwell's electromagnetic theory rather than from any special concern on his part with the results of the Michelson-Morley experiment. At the age of 16, as he reported later in his "Autobiographical Notes," he had discovered the seminal paradox:

"If I pursue a beam of light with the velocity $c$ (velocity of light in a vacuum), I should observe such a beam of light as a spatially oscillatory electromagnetic field at rest. However, there seems to be no such thing, whether on the basis of experience or according to Maxwell's equations. From the very beginning it appeared to me intuitively clear that, judged from the standpoint of such an observer, everything would have to happen according to the same laws as for an observer who, relative to the earth, was at rest. For how, otherwise, should the first observer know, i.e., be able to determine, that he is in a state of uniform motion?"*

Thus the possibility of reducing the relative speed of light to zero for some observer seemed to be an absurdity; yet there was nothing in mechanics as then understood to exclude such a possibility.

In the opening paragraph of his 1905 paper, Einstein pointed out an inconsistency in the conventional interpretation of electromagnetic theory:

"It is known that Maxwell's electrodynamics—as usually understood at the present time—when applied to moving bodies, leads to asymmetries which do not appear to be inherent in the phenomena. Take, for example, the reciprocal electro-

---

* Quoted from page 53 in *Albert Einstein, Philosopher–Scientist* (edited by P. A. Schilpp). New York: Library of Living Philosophers, 1949.

dynamic action of a magnet and a conductor. The observable phenomenon [namely the direction and magnitude of the induced current] here depends only on the relative motion of the conductor and the magnet, whereas the customary view draws a sharp distinction between the two cases in which either the one or the other of these bodies is in motion."

For example, when the magnet is in motion and the conductor at rest, one explains the current in this way: There arises in the neighborhood of the magnet an electric field with a certain definite energy, producing a current at the places where parts of the conductor are situated. But in the second case, if the magnet is stationary and the conductor in motion, one does not say that an electric field arises in the neighborhood of the magnet. Rather, one says that in the conductor there appears an electromotive force; there is no corresponding energy to it, but the electromotive force gives rise—assuming equality of relative motion of conductor and magnet in the two cases discussed—to an electric current of the same path and intensity as that produced by the electric field in the first case. To put it differently: To arrive at the *same* current, one has to use two very different sets of ideas and sets of equations in the two cases. Our physics seems to be more complicated than Nature.

Einstein then proceeds to locate the difficulty in the assumption of the existence of an absolute space. Influenced by his reading of the critiques of Newtonian science by the eighteenth-century Scottish philosopher David Hume and the nineteenth-century Austrian physicist Ernst Mach, he decided to reject this time-honored assumption, and to build physics on a new assumption: "The same laws of electrodynamics and optics will be valid for all frames of reference for which the equations of mechanics hold good." Those frames of reference are, of course, the "inertial" frames of reference—to put it simply, those systems which are moving at constant velocity with respect to the fixed stars,* and within which Newton's laws of motion apply. Einstein continues briskly:

"We will raise this conjecture (the purport of which will hereafter be called the 'Principle of Relativity') to the status of a postulate, and also introduce another postulate, which is only apparently irreconcilable with the former, namely, that light is always propagated in empty space with a definite velocity $c$ which is independent of the state of motion of the emitting body. These two postulates suffice for the attainment of a simple and consistent theory of the electrodynamics of moving bodies based on Maxwell's theory for stationary bodies. The introduction of a 'luminiferous ether' will prove to be superfluous inasmuch as the view here to be developed will not require an 'absolute stationary space' provided with special properties, nor assign a velocity-vector to a point of the empty space in which electromagnetic processes take place."

The first postulate means that the same law of nature (for example Maxwell's equation) holds in an "inertial" frame of reference that is at rest with respect to a fixed star as in one that moves at, say, nearly the velocity of light.

## 30.4 GALILEAN TRANSFORMATION EQUATIONS

Einstein's first major goal (in what is now known as the "special theory of relativity," to distinguish it from the later, more general theory of relativity that applies to rotating and other accelerating

---

*   A terrestrial laboratory is not quite an inertial frame, since the earth is rotating around its axis. But for most purposes we can ignore or easily correct for such discrepancies. And once we have identified a single inertial frame, we can immediately identify an infinite number of others, namely all those frames moving at constant velocity with respect to the first one.

**Fig. 30.3** Frame of reference $K'$ is moving at constant velocity **v** in the $x$ direction relative to $K$. At time $t = 0$ on the clock of $O$, when it is also $t' = 0$ on the clock of $O'$, the origin of the two coordinate systems momentarily coincided. When these two origins had moved a distance $x = vt$ apart, an event $E$ happened (e.g., an explosion); $O$ says it happened at time $t$ and position $x$ (with $y = 0$, $z = 0$); $O'$ says it happened at time $t'$ and position $x'$ (with $y' = 0$, $z' = 0$). In classical physics, we naturally assume that $x' = x - vt$ and that $t' = t$, in accord with experience.

frames of reference) was to determine the way in which the result of measurements of space and time intervals for the same phenomenon differ when made from two inertial frames of reference that are moving at constant velocity with respect to each other. Einstein himself used the example of measurements made from a railroad train moving along an embankment and from the fixed station embankment.*

An observer $O$ standing in the station on the embankment sees an explosion down along the track, and measures the distances to that spot and the time when it happened in his own frame of reference (let us call it $K$), using the variables $x$, $y$, $z$, and $t$. Another observer $O'$, riding in the train (moving at a velocity $v$ in the $x$ direction, relative to the embankment) and observing *the*

---

* In the following discussion we shall be using the approach of Einstein's book *Relativity, The Special and General Theory*, first published in German in 1917. This exposition, available in English translation in a paperback edition (Crown Publishers, New York), is one of the best for beginning students.

*same event*, would measure distances and times with respect to his own frame of reference (let us call it $K'$), using variables which we may call $x'$, $y'$, $z'$, and $t'$. In the physics of Galileo and Newton, it would naturally be assumed that if at an initial time $t = t' = 0$ the origins of the two coordinate systems coincided (when the values of $x$, $y$, $z$ and $x'$, $y'$, $z'$ all equal zero for the *origins* of the co-ordinate systems), then at any later time an event which the observer on the embankment reports to occur at time $t'$ and at position $(x, y, z)$ would be reported to occur at time $t'$ and at position $(x', y', z')$ by the observer on the train (see Fig. 30.3). Moreover,

$$x' = x - vt$$
$$y' = y$$
$$z' = z \qquad (30.6)$$
$$t' = t$$

Einstein called this set of equations the "Galilean transformation equations." They are, so to speak, a dictionary by which an observer in $K$ can translate into his own language a report coming to him from the observer in $K'$, and vice versa.

Equation 30.6 is best understood by referring to Fig. 30.3: For any given point in space, the value assigned to its position coordinate by measuring in $K'$ along the $x'$ coordinate is *smaller* than the corresponding value measured in $K$ along the $x$ coordinate—and the difference is simply the distance that the frame $K'$ has traveled relative to $K$, namely $vt$. The $y$ and $z$ coordinates are not affected by the relative motion of $K$ and $K'$ (since their motion is along the $x$ and $x'$ axis only). Of course, it is assumed also in all of classical physics that there is a single universal time variable $t$ which is the same for all observers, hence $t = t'$.

---

**Problem 30.3.** Prove that it follows from Eq. (30.6) that $x = x' + vt'$. This is called the inverse transformation equation, and it is useful for predicting what $O$ will measure when all we have is the data taken by $O'$.

**Problem 30.4.** The length of a rigid stick resting in $K$ and pointing along the $x$ axis can be identified by the symbols $L$ or $\Delta x$, where $\Delta x \equiv x_2 - x_1$. The symbols $x_2$ and $x_1$ refer of course to the coordinates of each end of the stick, measured along the $x$ axis. Use Eq. (30.6) to show that when the length of the same stick is measured from $K'$ (and is symbolized by $\Delta x'$), it is found that $\Delta x' = \Delta x$. This coincides with the intuitively obvious point that lengths in classical physics are invariant with respect to transformation.

[*Note:* The length of an object at rest in one's own reference frame is called the "proper length." The above examples are not cases of measurement of "proper length"—but in classical physics the values obtained would be the same.]

**Problem 30.5.** Derive from the transformation equation (30.6) that when the speed of an object is $u'$ as determined in $K'$, the speed $u$ for the same object as determined in $K$ is given by the equation $u' + v$. [For example, a man walking toward the locomotive inside a train ($K'$) is going at 3 mi/hr ($u'$) with respect to the train, and the train itself is going at 80 mi/hr ($v$) with respect to the embankment. From the embankment the measured speed of the man ($u$) is 83 mi/hr.]

Then use Eq. (30.6), and show conversely that the equation $u' = u - v$ gives the speed $u'$ (measured from $K'$) of a man who is walking on the embankment with speed $u$ with respect to the embankment.

[*Note:* The equation $u = u' + v$ (or the equivalent equation $u' = u - v$) is called the classical addition theorem, and is in accord with ordinary experience and classical physics.]

**Problem 30.6.** Nothing is more important in problems of relativity than the proper use of words and concepts—e.g., to distinguish between position

($x$) and length (which is a distance interval, $\Delta x$), instantaneous time ($t$) and time interval ($\Delta t$), measurement (made with instruments in one frame of reference) of events or motion that can also be observed from another frame of reference. A related need is that of taking an operational view of measurements, e.g., to be quite clear how a given measurement would be carried out. For this reason, do the following problem.

A newspaper editor wishes to know the exact length (in feet) at a given instant (say, exactly at noon) of a parade of people that has been moving down a straight road at an even pace. Describe exactly and in detail how you would arrange to measure this length (of a "moving object") without stopping the parade.

[*Hints:* The best way will be to hire a good number of helpers, spaced out at known or measurable positions along the route. Give them synchronized watches, so that they can ascertain where the head or tail of the procession is at the same agreed-upon instant of time.]

**Problem 30.7.** In the above example, we discovered graphically that a measurement of the length of a moving object involves the use of watches, i.e., the concept of time! (a) Why is this not the case in making a measurement of "proper length"? (b) In the above example, assume now that the helpers after all had watches that were not well synchronized; the watches of those near the head of the parade showed 12:00 noon while the watches of those placed near the tail still read 11:58 a.m. What effect would this lack of synchrony have on the measurement of the length of the parade?

**Problem 30.8.** In the light of the last problem, how would you arrange to ensure that a row of helpers, spaced along a line, did have synchronized clocks? Describe your procedure in detail.

[*Hint:* Any two clocks at different places are synchronized in any inertial system if a light wave, spreading out (preferably in vacuum) from a point exactly midway between the clocks, reaches the

**Fig. 30.4** The observer $O$ on the embankment has before him two mirrors, tilted so that he can at the same time watch light coming from both $A$ and $B$. Also $AO = OB$. Lightning bolts strike at $A$ and $B$ simultaneously, as judged by $O$ later (when light from $A$ and $B$ comes to his eyes at the same time). When lightning struck, $O'$ on the moving train was closest and opposite to $O$, $A$ opposite $A'$, and $B$ opposite $B'$. What will $O'$ see, and when?

two clock faces when both show the same time reading. To ensure that the two readings are the same not just by accident, and that the clocks do not have different rates, we might well do the experiment again a bit later to see if the clocks both show the same new time reading.

An alternative way of checking whether two clocks are synchronized is by sending from each, at the same time, a beam of light *toward* a common point that is exactly midway between the clocks. An observer at the midpoint receives on his retina both light beams at the same instant, if and only if the two beams are each really sent out, say, at 12:00 noon as shown on both clocks, i.e., if these were synchronized clocks. Again, to guard against an accidentally equal setting, one can repeat the experiment later. Use this alternative way to describe a method to double check that the editor's assistants all have synchronized clocks.]

**Problem 30.9.** Note that the definition of synchronization of clocks at different places involves light waves. Why could it not be done in just the same way, but using sound waves instead of light? How is it that we are allowed to use the method with light beams to synchronize clocks in *any* inertial system (even one moving at high speed past us)?

The Galilean transformation equations seem so obvious that they were rarely stated explicitly before 1905, least of all by Galileo himself; nevertheless they are an essential feature of Newtonian physics. But Einstein discovered that if we are to accept the two postulates, (1) that all laws of nature must be the same in $K$ and $K'$, and (2) that the speed of light is the same for an observer in

$K$ and in $K'$, then we must discard the intuitively clear Galilean transformation equations and replace them by others that may go counter to our intuitions. In particular, we shall have to accept the apparently strange notion that distance intervals $\Delta x$ and time intervals $\Delta t$ measured in $K'$ are different from the values for $\Delta x'$ and $\Delta t'$ measured in $K'$ for the same entities.

## 30.5 THE RELATIVITY OF SIMULTANEITY

To demonstrate the need for this change, Einstein poses the following famous (and pretty) *gedanken* experiment: Suppose it is asserted that two events happen "simultaneously" at two different places—for example, that two bolts of lightning strike "simultaneously" at the two places marked $A$ and $B$ in Fig. 30.4. How can we verify this assertion? We might ask an observer $O$, placed on the fixed embankment, who happened to be at a spot halfway between the charred spots $A$ and $B$ (that is, $AO = OB$). Let us assume that he had, in anticipation, set up before him an apparatus (for example, two mirrors at $O$) which enabled him to look at the same time at light from both $A$ and $B$. If he perceived the arrival at $O$ of two flashes, reaching his eye at the same time from two places which he knows to be equally distant (i.e., from where lightning had struck at $A$ and $B$), then he can say that the bolts struck there earlier and *simultaneously*.

Now suppose that there was also an observer $O'$, with similar mirrors, sitting on a moving train, and that $O'$ just happened to be abreast of (coincided with) $O$ precisely when the bolts of lightning struck at $A$ and $B$, as judged from the embankment. Since the train, with $O'$ on it, is moving relative to the embankment at a velocity $v$ the observer at $O'$ will have moved away from $O$ by the time the two flashes reach $O'$. In fact $O'$ is moving toward $B$ and away from $A$, so that he sees the flash coming from $B$ *before* that flash reaches the "stationary" observer $O$ on the embankment at $M$. But he sees the other flash, the one coming from $A$, *after* the observer at $O$ sees it, since $O'$ is moving away from $A$ and the beam of light from $A$ takes a little extra time to catch up with him. Thus the observer on the train sees the flash from $B$ before he sees the flash from $A$. However, the lightning bolts will have burned or charred a spot $A'$ and $B'$ on each end of the train as these ends were passing by and coincided with $A$ and $B$ at the instant when the bolts struck there. As $O'$ can check by measuring the distance along the train to $A'$ and $B'$ in each direction, $A'O' = B'O'$. In short, $O'$ will say that the two bolts struck at equally distant points, but that the light from these points reached him at different moments (first from $B$, then from $A$). Therefore $O'$ will say that the bolts *could not have struck A and B simultaneously*.

[*Note:* It is implied in this argument that $O'$ assumes that the speed of light is the same for light coming from $A$ and from $B$. But this is precisely what the second principle of relativity allows, or rather commands, him to do: He is in an inertial system, and every observer in every inertial system can and must assume that the speed of light as measured in his own frame of reference is constant in all directions in space, and equal to $c$. (In fact, of course, we make this assumption safely all the time in our "fixed" laboratory—without worrying about the fact that our spaceship, earth, is in motion with respect to the sun and other stars.)]

In short, $O'$ judges that the two lightning bolts had not hit simultaneously. This is different from $O$ who judges that the same two events *were* simultaneous. From this example, Einstein concludes that the concept of "simultaneity" is relative with respect to the coordinate system in which the determination is made. The concept has no

"absolute" physical significance for two events occurring at different places. Two events may be simultaneous for an observer in one frame of reference, but not simultaneous for an observer in a frame that is moving relative to the first.

The discussion so far is not easy to accept at first reading. Somehow one has the intuitive feeling that if the observer $O'$ on the train were to measure the speed of the light of a flash coming toward him from the point $B$ on the embankment, he would find that the speed was greater than the speed of the light of a flash coming toward him from the point $A$. Indeed this is what we would expect if we were thinking about *sound* waves from $A$ and $B$. And the same would apply if the beams of light from $A$ and $B$ were ripples in an ether stationary with respect to the embankment; in such a case the speed of a beam as measured by $O$ in his frame of reference $K$ would be $c$, but the speed of the same beam relative to the observer $O'$ on the train, when the beam is coming to him from point $B$, would be $c + v$—the beam's speed relative to him would be greater than $c$ because they are approaching each other. Conversely, if light moved in a fixed ether, the beam of light from $A$ would be expected to have a speed $c - v$ relative to $O'$, since he is fleeing from $A$.

But all these conclusions contradict Einstein's second postulate, that the speed of light is the same (namely $c$) for all observers in all inertial frames. And it also contradicts a variety of experiments, none of which had found any evidence for different values of the speed of light from different directions. (The null result of the Michelson-Morley experiment was just one of the experiments of which Einstein may have known.)

As soon as we accept the second postulate and give up "absolute simultaneity," it follows, as Einstein next shows, that we must adopt a new set of transformation equations, *different* from the Galilean transformation equations (Eq. 30.6). To find the new equations, he starts again from the assumption that for a given set of events the values of $x'$ in $K'$ must be related to the values of $x$ as determined in $K$ in such a way that the measured speed of light will come out to be equal to $c$ in both frames. Thus, if a light signal starts from some point where $x = 0$ in frame $K$ at a time $t = 0$ (as also determined in frame $K$), then at a later time $t$ the light signal will have reached a position $x$, determined by the equation

$$x = ct$$

or                                                                                    (30.7)

$$x - ct = 0.$$

In frame $K'$, whose origin initially coincides with $K$ (when $t = t' = 0$) but has been moving at a relative velocity $v$ (as in Fig. 30.3), the second postulate assures us that the same light signal has a speed $c' = c$. After some time $t'$, it will have reached a point at a distance $x'$ in $K'$, where $x'$ is related to $c$ and $t'$ by the equation

$$x' = ct'$$

or                                                                                    (30.8)

$$x' - ct' = 0.$$

Note that Eqs. (30.7) and (30.8) are *general relations* between the variables $x$ and $t$, on the one hand, and $x'$ and $t'$ on the other hand; we have not yet said that any particular time $t'$ and position $x'$ determined in one frame is to be associated with any particular position $x$ and time $t$ in the other. But now we impose the additional requirement that *the existence of a light signal at a particular time and place is an event which we must be able to describe consistently, by the appropriate equations, in either of the two frames of reference*, i.e., by Eq. (30.7) as well as by (30.8).

## 30.6 THE RELATIVISTIC (LORENTZ) TRANSFORMATION EQUATIONS

We must now make a little detour of an algebraic kind.* One way to express the requirement at the

---

* On the first reading it may be advisable to skip the material from here to Eq. (30.16).

end of the last paragraph mathematically is to write down the most general equation which is satisfied if, and only if, Eqs. (30.7) and (30.8) are both true for all possible values of $x$, $x'$, $t$, and $t'$, namely:

$$(x' - ct') = \lambda(x - ct), \tag{30.9}$$

where $\lambda$ is an as-yet-undetermined constant, independent of $x$, $x'$, $t$, and $t'$. [*Note:* It is not enough to write $(x' - ct') = (x - ct)$ just because each parenthesis by itself is zero. It is true that $0 = 0$; but so is $0 = 3 \cdot 0$ or $0 = \frac{1}{3} \cdot 0$.]

We can obtain another equation similar to (30.9) by applying the same requirements to a light signal sent along the $x$ axis in the opposite (negative) direction. For such a signal $-x = ct$, or $x = -ct$; similarly, $x' = -ct'$; and the two equations are compatible if

$$(x' + ct') = \mu(x + ct), \tag{30.10}$$

where $\mu$ is another as-yet-undetermined constant.

When we solve Eqs. (30.9) and (30.10) together for $x'$ and $t'$, we obtain the relations

$$x' = ax - bct, \qquad ct' = act - bx, \tag{30.11}$$

where

$$a \equiv \frac{\lambda + \mu}{2} \quad \text{and} \quad b \equiv \frac{\lambda - \mu}{2}.$$

The new constants $a$ and $b$, which determine the relations between time and space measurements made in the two frames of reference $K$ and $K'$, depend on the relative velocity $v$ of the frames. To see this, and to calculate $a$ and $b$ in terms of $v$, we can begin by picking a particular point that is fixed in one frame, say a point at $x' = 0$, that is, at the origin of $K'$, and see how it moves with respect to (when seen from) the other frame. When we put $x' = 0$ into the first of Eqs. (30.11), we find

$$x = \frac{bc}{a} t.$$

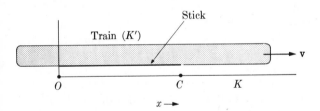

**Fig. 30.5** Both $O$ and $C$ are fixed in the frame of reference $K$. They are collaborators who measure the length of the stick that is moving with the train, and they each have synchronized, identical watches. When $t = 0$ (or noon) on both watches, $O$ notes that one end of the moving stick is just moving past $O$, and $C$ notes that the other end is just moving past $C$. They therefore agree that the distance $OC$ along the $x$-axis is the length of the moving stick as measured in $K$.

This equation describes a point moving at a constant velocity $v$. But by definition, $v = x/t$; therefore $v$ is also given by

$$v = \frac{bc}{a}, \qquad (30.12)$$

when measured relative to position $x = 0$, the origin of frame $K$.

It is already evident from the appearance of of the factor $a$ in the first of Eqs. (30.11) that distances—for example, the spatial dimension of any particular object—have different values as measured in the two frames of reference. Thus consider an observer $O$ in $K$ who looks at a stick lying in and moving with frame $K'$—that is, at a moving, rigid object. Assume that the length as measured by an observer in $K'$ who is at rest with respect to that object is reported (over a radio, say) to be exactly 1 m (that is, $\Delta x' = 1$ m); indeed, perhaps it is a standard meter stick. Yet, $O$ himself would perceive its length, measured in $K$, to be different by a factor $1/a$. We assume that $K$ makes this measurement at the instant $t = 0$ (according to his clocks in frame $K$) when the origins of the two frames coincide. One way of doing it might be to stand at point $O$ (Fig. 30.5), and to note that the left end of the moving stick, fastened to the side of the train, goes past point $O$ just as his watch

says $t = 0$ or noon. He also has a collaborator $C$ further along the embankment, with a watch going at the same rate and otherwise identical and synchronized. $C$ was instructed to find where on the $x$ axis the other end of the stick is just when $t = 0$ (or noon) by $C$'s watch. In this way, $O$ and $C$ find the length, as measured in $K$, of the moving stick.

They find that the distance interval of the stick which by definition is $\Delta x' = 1$ m in frame $K'$ is seen as

$$\Delta x = \frac{1}{a} \text{ meters}$$

in frame $K$. [*Note:* We get the same result when we set $t$ equal to zero in the first of Eqs. (30.11).] The stick when observed from $K$ is shorter than when observed in its own frame $K'$, the frame with respect to which it is at rest.

A more astonishing result follows at once. Remember that both inertial frames, $K$ and $K'$, are *equivalent*—the laws of nature are postulated to be the same in each, and each observer, $O$ and $O'$, has the same right to regard himself as "really at rest" and the other as moving with respect to his system. It follows that if observers in $K'$, on the train, are looking at a stick at rest in $K$ (perhaps at the meter stick used by the men on the embankment to make their measurement), they would notice that it is also shorter, and in precisely the same ratio. The whole situation is symmetrical. For the observer in frame $K'$, the time at which he and his own collaborator make their observations is $t' = 0$ on their own identical and synchronized watches; and if we substitute this value into the second of Eqs. (30.11), we find*

$$x' = a \left( 1 - \frac{v^2}{c^2} \right) x. \qquad (30.13)$$

---

\* From the first of Eqs. (30.11), $t = (ax - x')/bc$, and from the second, after setting $t' = 0$, we have $act = bx$. Therefore $bx = ac(ax - x')/bc = (a^2x/b) - (ax'/b)$. Solving this for $x'$, we find $x' = a[1 - (b^2/a^2)]x$. From Eq. (30.12), we have $(b/a) = (v/c)$, and this substitution brings us to Eq. (30.13).

Thus if the length of a stick in frame $K$ is 1 m (hence one end is at $x = 0$, the other at $x = 1$ m) the observer in frame $K'$ will find the length of the same object to be $a[1 - (v^2/c^2)]$ meters (i.e., one end is at $x' = 0$ and the other at $x' = a[1 - (v^2/c^2)]$ meters), and so $\Delta x' = a(1 - v^2/c^2)$ meters.

Since $O$ and $O'$ see each other's meter sticks shorter in exactly the same way, it must be true that $\Delta x$ (measured in $K$) in the first measurement is numerically equal to $\Delta x'$ (measured in $K'$) in the second. Hence

$$\frac{1}{a} = a\left(1 - \frac{v^2}{c^2}\right).$$

On solving this equation for $a$, we find

$$a = \frac{1}{\sqrt{1 - (v^2/c^2)}}. \tag{30.14}$$

We can now find $b$ by using Eq. (30.12):

$$b = \frac{(v/c)}{\sqrt{1 - (v^2/c^2)}}. \tag{30.15}$$

Substitution of these values for $a$ and $b$ back into Eq. (30.11) yields at last the final result we have been painfully groping for:

$$x' = \frac{x - vt}{\sqrt{1 - (v^2/c^2)}},$$
$$t' = \frac{t - (v/c^2)x}{\sqrt{1 - (v^2/c^2)}}. \tag{30.16}$$

Equations (30.16), together with the equations

$$y' = y \quad \text{and} \quad z' = z$$

(which are valid for the case we have discussed, in which all motion is only in the $x$-direction) give the "relativistic transformation equations" by which to translate information obtained in one inertial frame of reference into information valid for another. This, then, is the result of our two postulates, namely, that all inertial frames are equivalent and that the speed of light is equal to $c$ in each.

**Problem 30.10.** (a) By considering a signal which moves in frame $K$ according to the equation $x = ct$, and applying Eqs. (30.16), verify that the speed of light is indeed equal to $c$ in frame $K'$. (b) Compare Eq. (30.16) with the Galilean transformation equations (30.6), which are still found to hold well enough for doing problems in mechanics as long as the relative speeds involved are much smaller than $c$. For example, show that Eq. (30.16) reduces to (30.6) if $v \ll c$.

---

The transformation equations (30.16), which Einstein derived in something like this abstract manner from his two postulates in 1905, had previously been proposed on the basis of different reasoning. As Einstein was aware, the equations correspond to the so-called "Lorentz transformation equations" which H. A. Lorentz had derived earlier in connection with his theory of electrons. The suggestion of an apparent length contraction of a moving object proportional to the factor $\sqrt{1 - (v^2/c^2)}$, as in Eq. (30.9), had been published by FitzGerald in 1889 as a possible explanation of the Michelson-Morley experiment (Section 30.2). However, Lorentz and FitzGerald had used these equations with a different physical interpretation; they had assumed that there really is an absolute frame of reference, fixed in the ether, and that changes in the measured values of space and time intervals occur only in frames of reference that are moving with respect to absolute space.

For Einstein, on the contrary, all frames of reference moving at constant relative velocity with respect to a given inertial frame are equivalent, and none can be singled out as at rest in "absolute space." An observer in *each* frame would observe that lengths appear to be contracted in the other frames moving past him. Moreover, each observer would notice a "time dilation" of clocks carried in the other frames; if each observer carried a watch of identical physical construction (so that they would click off seconds at exactly the same rate if they were initially sitting side by side in the

**Figure 30.6**

$K$
(fixed system)

**Figure 30.7**

same frame of reference), then each observer would find that the intervals between clicks registered on the watches in other frames were longer when compared with his own. The amount of this dilation is given simply by the factor $1/\sqrt{1 - (v^2/c^2)}$. You can prove it by thinking of the case in which both frames momentarily coincide, with watches respectively at $x = 0$ and $x' = 0$, and calculating the ratio of $t'$ to $t$, by means of the second of Eqs. (30.11).

To summarize: *If an inertial frame of reference $K'$ is moving at a constant velocity $v$ with respect to another inertial frame $K$, then an observer in either frame would find from his own measurements that length intervals of objects moving with the other's frame are contracted by a factor $\sqrt{1 - (v^2/c^2)}$, and time intervals between events are dilated (the moving clocks appear to run slow to the fixed observer) by a factor $1/\sqrt{1 - (v^2/c^2)}$.* Note that the effect of apparent contraction and dilation reduces to zero if $v \ll c$; and that it goes to extreme values as $v \to c$ (length of a moving meter stick appears to contract to zero, time interval between one click and the next on the moving clock appears to dilate to $\infty$). Note also that of course the moving observer does not see such effects when looking

only at his own clocks and meter sticks—*they* are at rest with respect to him, and so are quite "normal."

---

**Problem 30.11.** The plot of $v/c$ versus

$$\frac{1}{\sqrt{1 - (v^2/c^2)}}$$

in Fig. 30.6 shows that the relativistic form of the transformation equations is usually not needed if the relative speeds are less than about 20% of the speed of light.

a) Verify that the curve in Fig. 30.6 is correct by actual calculation for some one value of $v/c$ above 0.2.

b) How fast must $K'$ go for $K$ to see a "contraction" by $\frac{1}{2}$ and a "time dilation" by 2?

**Problem 30.12.** Imagine four identically made clocks. Put two at equal distances from the midpoint $O$ in one coordinate system $K$, and the two others at equal distances from the midpoint $O'$ in another, relatively moving coordinate system $K'$. At some moment, all clocks are momentarily lined up in pairs, as indicated in Fig. 30.7. (Assume that all clocks happen to show 12:00 noon as seen from $O$, or are just then set to show 12:00 noon by someone positioned at each clock.) At that instant also, a light flash goes off at $O$ (or at $O'$—they coincide momentarily anyway).

a) Explain why $O$ (in $K$) will find that the two clocks at $A$ and $B$ are synchronized.

b) Explain why $O'$ will find that clocks $A'$ and $B'$ are not synchronized (for example, $A'$ shows 12:04 p.m. on its face when $B'$ shows only 12:02 p.m. if the distances involved are long enough!)

c) Explain why $O'$ is justified in saying that there is something wrong with the setting of his clocks at $A'$ and $B'$, and why he is justified in changing the settings; for example, setting the clocks at $A'$ and $B'$ both to 12:03 p.m.

d) Explain why $O$, seeing this being done, may object to it, but to no avail.

e) Explain why from now on $O$ and $O'$ each have synchronized clocks, *but will never again get the same measurement for the length of any relatively moving "object"* (say for a parade marching in $K$, or simply a stick lying in $K$ and measured both by $O$ and by $O'$, or for a stick lying in $K'$ and measured both by $O$ and $O'$).

f) Explain what has been lost and gained by the procedure of resetting the clocks.

**Problem 13.13.** From Eq. (30.16), show that there exists also a set of inverse transformation equations:

$$x = \frac{x' + vt'}{\sqrt{1 - (v^2/c^2)}},$$

$$t = \frac{t' + vx'/c^2}{\sqrt{1 - (v^2/c^2)}}.$$

[*Hint:* It is arbitrary which coordinate system is considered at rest, but by convention the relative velocity of $K'$ with respect to $K$, shown to go to the

right in Fig. 30.3, is considered positive, and the relative velocity between the systems is reversed and hence is changed to $-v$ if $K'$ is considered stationary and $K$ is considered moving.]

**Problem 30.14.** From Eq. (30.16) or its inverse transformation equation, make the following computations.

a) Show how the time interval $\Delta t$ ($\equiv t_2 - t_1$) measured by the clocks of $O$ is related to the time interval $\Delta t'$ ($\equiv t_2' - t_1'$) that elapses on a clock resting in $K'$ (and hence carried along with $K'$).

b) Show how the time interval $\Delta t'$ ($\equiv t_2' - t_1'$) measured by the clocks of $O'$ is related to the time interval $\Delta t$ ($\equiv t_2 - t_1$) that elapses on a clock resting in $K$.

c) Show that the effect of time dilation is symmetrical, i.e., that $\Delta t' = \Delta t/\sqrt{1 - (v^2/c^2)}$ (for a clock at a fixed $x$ in $K$, and observed from $K'$), and also $\Delta t = \Delta t'/\sqrt{1 - (v^2/c^2)}$ (for a clock at a fixed $x'$ in $K'$, and observed from $K$).

**Problem 30.15.** Whereas the classical velocity addition theorem was given by $u = u' + v$ (or $u' = u - v$), the relativistic addition theorem is

$$u = \frac{u' + v}{1 + \dfrac{u'v}{c^2}} \quad \text{or} \quad u' = \frac{u - v}{1 - \dfrac{uv}{c^2}}.$$

Starting from Eq. (30.16), show that this is true. (It is assumed that $u$ is the speed of an object in the $x$ direction, which is parallel to the $x'$ direction.)

[*Hint:* The speed $u$ of an object is by definition

$$u = \lim_{\Delta t \to 0} \frac{\Delta x}{\Delta t},$$

and from the relativistic transformation equations, we had

$$\Delta x = \frac{\Delta x' + v\,\Delta t'}{\sqrt{1 - (v^2/c^2)}},$$

and

$$\Delta t = \frac{\Delta t' + \dfrac{v \, \Delta x'}{c^2}}{\sqrt{1 - (v^2/c^2)}}.$$

*Notes:* (1) In studying relativity, it is even more advisable than in other subjects to consult other books to get a large variety of examples, etc. This problem is solved in several of the books cited in the bibliography at the end of the chapter. However, terminology sometimes differs from book to book, and it is advisable to look for this carefully. In most books, you will find for example that $1/\sqrt{1 - (v^2/c^2)}$ is symbolized by $\gamma$, and occasionally the relative speed of the two coordinate systems is symbolized not by $v$ but by $V$ or $u$. (2) As you can quickly show, the relativistic addition theorem turns into the classical one, as it should, when $u' \ll c$ (or when $u \ll c$).

**Problem 30.16.** Using the relativistic addition theorem for velocities, show that when $K'$ is moving away from $K$ at a speed $v = c/2$, a particle in $K'$ shot out at speed $u'$ equal to $\frac{3}{4}$ the speed of light will appear from $K$ to be moving at a speed $\frac{10}{11}c$.

Similarly, show in the above case that when $K'$ sends out a beam of light which he finds to be moving at speed $c'$, $K$ will find this same beam to be moving at a speed $c = c'$. Why is this result *not* surprising?

**Problem 30.17.** (Worked example). An airplane is found to be 50 m long when measured on the ground before it starts to fly. While it is traveling at a constant speed of 300 m/sec over the field, its crew also finds it to be 50 m long. (Why?)

a) What is the plane's length in flight as measured by persons on the ground? How much time has elapsed on the ground while observers there see the pilot's clock advance one second?

b) The hangar is also 50 m long as measured on the ground. What is the hangar's length as measured by the crew of the moving plane? How much time elapses on board while the crew observe the control tower clock advance 1 sec?

*Solution:* (a) Consider a coordinate system fixed on the ground ($K$) and one moving with the plane ($K'$). To measure the length $\Delta x$ (where $\Delta x \equiv x_2 - x_1$) of the plane from the ground as it flies by, one has to locate its two ends ($x_1$ and $x_2$) at the same time (at $t_1 = t_2$). This might be done by locating simultaneously the front and back end of its shadow.

Note that the plane has length $\Delta x'$ of 50 m (where $\Delta x' \equiv x_2' - x_1'$) as measured in the coordinate system in which the plane is at rest (namely in $K'$). We also know that

$$x_2' = \frac{(x_2 - vt_2)}{\sqrt{1 - (v^2/c^2)}},$$

$$x_1' = \frac{(x_1 - vt_1)}{\sqrt{1 - (v^2/c^2)}}.$$

Therefore

$$x_2' - x_1' = \frac{1}{\sqrt{1 - (v^2/c^2)}} (x_2 - x_1),$$

since

$$t_1 = t_2.$$

Hence

$\Delta x$ (length of shadow on ground)

$$= \sqrt{1 - (v^2/c^2)} \cdot \Delta x'$$

$$= \sqrt{1 - \left(\frac{3 \times 10^2}{3 \times 10^8}\right)^2} \cdot 50 \text{ m}$$

$$= \sqrt{1 - 10^{-12}} \cdot 50 \text{ m} \approx (1 - \tfrac{1}{2} \cdot 10^{-12}) \cdot 50 \text{ m}$$

or slightly less than 50 m. (It appears "contracted" when seen from $K$.)

For the next part of the problem, we use the Lorentz transformation equations for $x$ and for $t$:

$$x = \frac{1}{\sqrt{1 - (v^2/c^2)}}\, (x' + vt'),$$

$$t = \frac{1}{\sqrt{1 - (v^2/c^2)}}\left(t' + \frac{v}{c^2}\, x'\right).$$

We observe a time interval $(t'_2 - t'_1)$ on a clock in the moving system (1 sec) by "locating" the ends of the interval ($t'_1$ and $t'_2$) at the same point ($x'_1 = x'_2$) in $K'$, and we ask what time interval $(t_2 - t_1)$ in $K$ has elapsed:

$$t_2 = \frac{1}{\sqrt{1 - (v^2/c^2)}}\left(t'_2 + \frac{v}{c^2}\, x'_2\right),$$

$$t_1 = \frac{1}{\sqrt{1 - (v^2/c^2)}}\left(t'_1 + \frac{v}{c^2}\, x'_1\right),$$

$$t_2 - t_1 = \frac{1}{\sqrt{1 - (v^2/c^2)}}\, (t'_2 - t'_1),$$

or, since

$$x'_1 = x'_2,$$

$$\Delta t_{\text{on ground}} = \frac{1}{\sqrt{1 - v^2/c^2}} \cdot \Delta t'_{\text{on plane}},$$

$$\Delta t_{\text{on ground}} = \frac{1}{\sqrt{1 - 10^{-12}}}\, (1\ \text{sec})$$

$$\approx (1 + \tfrac{1}{2} \cdot 10^{-12}) \cdot 1\ \text{sec}.$$

This is longer than $\Delta t'$ on plane ("time dilation").

b) If, from the ground, the plane is seen to move with speed $+v$, then, from the plane, the ground is seen to move with speed $-v$. By substituting $-v$ for $v$ in the relevant equations above, we get

$$\Delta x'_{\text{hangar measured from plane}}$$

$$= \sqrt{1 - \frac{(-v)^2}{c^2}} \cdot \Delta x_{\text{hangar measured on ground}}$$

$$= \sqrt{1 - 10^{-12}} \cdot 50\ \text{m}.$$

$$\Delta t'_{\text{measured on plane}}$$

$$= \frac{1}{\sqrt{1 - \frac{(-v)^2}{c^2}}} \cdot \Delta t_{\text{measured at tower}}$$

$$= \frac{1}{\sqrt{1 - 10^{-12}}} \cdot (1\ \text{sec}).$$

In general, when you are in any inertial system and observe a ruler and a clock which are in motion relative to you, the ruler appears shorter and the clock slower (its time intervals are longer) regardless of the direction of the velocity and regardless of who you think is "really" at rest.

**Problem 30.18.** Imagine that a space probe could be made to go constantly at the speed of $0.8c$, and that it is sent to a star located 5 light years from us. By our clocks, it would therefore take 6.25 years to reach the star. Show that the time elapsed and the distance covered as measured from within the space probe are respectively 3.75 years and 3 light years.

## 30.7  CONSEQUENCES AND EXAMPLES

A well-known predicted consequence of the time-dilation effect is the *twin paradox* (sometimes called the *clock paradox*): If one twin goes off in a spaceship traveling at a speed close to the speed of light, then turns around and returns to earth, he should find that he is younger than the other twin who stayed behind! Seen from the frame of reference of the earth, biological processes of aging, as well as the operation of clocks (including the interval between heartbeats) run more slowly in the spaceship than on earth. Nevertheless the predicted outcome is difficult to believe intuitively; after all, why is the situation not symmetrical in *this* case? Why couldn't the spaceship be taken as the frame of reference, so that the earthbound twin would be younger from the viewpoint of the traveler? One has the feeling that when the traveler returns, all time dilation effects on both sides should have been canceled out, since each cannot objectively be younger than the other when they meet again. But that argument assumes incorrectly that the spaceship is an inertial frame of reference all the time; in fact, unlike the earth, it has to undergo large accelerations in order to start, turn around, and stop; hence the time-dilation effects are *not* the same for both twins. As discussed in detail in some of the readings cited at the end of the chapter, the "traveling twin" will in fact have aged less than the "stay-at-home twin"—though the actual differences for any realizable speeds of travel are completely insignificant for living beings. It is, however, quite expected that such very small differences can be found when one compares the time shown on two initially identical, synchronized clocks, after one of them has been sent through some long journey at high speed.

## 30.8 THE EQUIVALENCE OF MASS AND ENERGY

It had been suggested before 1905 that the mass or inertia of an electron must increase with increasing speed, so that the acceleration, by a given force, of an already fast-moving electron, would be less than that of the same electron when starting from rest. This mass increase had been found experimentally by W. Kaufmann (1902) and others, by deflecting in electric and magnetic fields the high-speed electrons (β-rays) emitted by radioactive nuclei.

Einstein deduced from his special theory of relativity in 1905 the conclusion that the "tangible," ponderable mass of any particle, whether charged like an electron or not, increases with speed according to the equation

$$m = \frac{m_0}{\sqrt{1 - (v^2/c^2)}}, \qquad (30.17)$$

where $m_0$ is the mass of the particle at rest with respect to the observer, called *rest mass*, and $m$

is the mass of the particle measured while it moves at speed $v$ relative to the observer, called *observed* or *relativistic mass*. (This mass determination may be made by supplying a known centripetal force and measuring the radius of curvature of the path, i.e., from $F = mv^2/r$.) Furthermore, $c$ is the velocity of light, a constant, about $3 \times 10^8$ m/sec.

Now it is evident from Eq. (30.17) that for ordinary speed, where $v$ is much less than $c$, $m = m_0$ within the experimental error of measurement. But precisely in the motion of atomic and subatomic particles this restriction no longer holds—some β-rays are ejected from disintegrating nuclei with speeds greater than 90% that of light.

At first glance this increase of mass with speed is deeply disturbing. Surely a moving particle has no more molecules, no more "tangible," ponderable matter than before! If mass increases, mass is not simply a measure of the quantity of matter after all. And then the law of conservation of mass would appear to be no longer strictly true. Such are the speculations that we must consider next.

---

**Problem 30.19.** Evaluate the ratio $m/m_0$ for an object that has successively a speed $v$ of 0.1, 0.3, 0.5, 0.7, 0.9, and 0.97 times the velocity of light $c$. (The experimental check, incidentally, has been found to be very good.) What should be the mass of any object if $v = c$?

**Problem 30.20.** How much would your mass increase if you were flying in a jet craft at the speed of sound (about 330 m/sec)?

---

The ratio $(v/c)$ is usually replaced by the symbol $\beta$. Thus Eq. (30.17) is written $m = m_0(1 - \beta^2)^{-1/2}$. However, the binomial theorem from algebra teaches that $(1 - \beta^2)^{-1/2} = 1 + \beta^2/2 + 3\beta^4/16 +$ other terms of still higher powers. Now if $v$ is considerably smaller than $c$, $\beta^2$ is very small, and

$\beta^4$, together with all higher terms, is negligible. Therefore we can write

$$m \doteq m_0 \left(1 + \frac{\beta^2}{2}\right) = m_0 \left(1 + \frac{v^2}{2c^2}\right)$$

$$= m_0 + \frac{\frac{1}{2}m_0 v^2}{c^2}.$$

This last result opens our eyes to an amazing physical interpretation of the increase in mass with speed, for it points out directly that the increase in mass is

$$\Delta m \equiv m - m_0 = \frac{\frac{1}{2}m_0 v^2}{c^2},$$

where we may at least tentatively identify the term $\frac{1}{2}m_0 v^2$ with kinetic energy. Thus

$$\Delta m = KE/c^2,$$

and in trying to understand the change in mass with speed, we were led to the concept that kinetic energy added to a ponderable object also adds inertia in an amount $KE/c^2$. Whether we say that energy *has* mass or *is* mass or *appears as* mass is only playing with words and adds nothing to the last equation.

Although we have arrived at this result by an approximation, the equation $KE/c^2 = \Delta m$ is true in general. Only we must realize that in such a purely mechanical situation *the kinetic energy of a fast-moving particle might be redefined as*

$$KE = \Delta m c^2 = mc^2 - m_0 c^2$$

and is not separately evaluated by $\frac{1}{2}m_0 v^2$; for the mass increases from $m_0$ with speed, and so more kinetic energy is required for equal changes in speed at the latter stages of reaching $v$ than at the earlier stages.

---

**Problem 30.21.** Einstein's original approach was to postulate that no material body can be made to reach a velocity equal to or larger than the speed of light, with respect to the observer. Prove that the foregoing is in harmony with this postulate. [*Hint:* What kinetic energy would be needed if such speeds were desired?]

---

The idea that it is the increase of energy which changes the mass can now be extended further to include energies other than ordinary kinetic. Consider first a spring with a mass $m_0$ when limply extended. When we compress it and so give it elastic potential energy PE, its mass increases to $(m_0 + PE/c^2)$. A lump of metal likewise increases its observed mass when it is heated; here $\Delta m =$ heat energy supplied$/c^2$. In brief, the principle of mass equivalence of energy was soon extended to cover the increase (or decrease) of any type of energy; *for every unit* (1 joule) *of energy supplied* to a material object, whether it be kinetic energy given to a bullet, gravitational potential energy acquired by the change in relative distance of two stars, or any other form of energy, the *mass of the system thereby increases by*

$$\frac{1 \text{ joule}}{(3 \times 10^8 \text{ m/sec})^2} = 1.1 \times 10^{-24} \text{ kg}.$$

To repeat, this does not mean there are now in the system more molecules than before; what has changed is the observable inertia of the energy-enriched material. Evidently the rate of exchange, i.e., the factor $c^2$ in

$$\Delta m = \text{energy}/c^2,$$

indicates that such changes in mass are beyond direct experience in ordinary mechanical experiments, and therefore there was no need to consider this academic point in the section on the conservation of mass. But in nuclear phenomena or in cyclotrons and other accelerators, where the masses of the particles involved are relatively small to start with and the energies are relatively large, the changes in mass may become very noticeable. If you prefer, *mass* does not measure the amount of tangible or ponderable matter in a system, it measures the amount of matter *together with* the mechanical energies, electrical energies, etc. Therefore we rewrite the law of conservation of mass to read as follows. In a closed system,

$$\sum \left( m_0 + \frac{\text{energy}}{c^2} \right) = \text{constant.} \quad (30.18)$$

A striking illustration of the law in this form is the following example of the complete inter-conversion between crude, coarse matter and what might seem to be disembodied, radiant energy. At times, very intense gamma-rays (namely, x-rays of short wavelength, and so akin to visible light) may be transformed into material particles, more precisely into pairs of charged particles, one an ordinary negative electron, the other a positively charged electron (a *positron*). Here we may say that before the conversion the mass in the system was = (energy of $\gamma$-ray/$c^2$), namely, the mass equivalent of the radiant energy; and after the conversion, the mass in the system is = (rest mass of both particles + total KE of particles/$c^2$); the mass so calculated is the same before and after.

Another typical instance of the use and range of these ideas may be found in comparing the mass of a nucleus with the mass of its component particles (nucleons) after separation. For example, a helium nucleus (or $\alpha$-particle) has a rest mass of about 4.0028 amu; it is composed of four nucleons, namely, two protons and two neutrons. When the helium nucleus is broken up in a nuclear reaction, the rest mass of each proton measures about 1.0076 amu and that of each neutron about 1.0090 amu. But $(2 \times 1.0076 + 2 \times 1.0090)$ amu is larger than 4.0028 amu by about 0.03 amu (or, since 1 amu = $1.66 \times 10^{-27}$ kg, by $5 \times 10^{-29}$ kg), and it would seem that the sum of the parts "weighs" more than the parent nucleus. And so they do indeed, because energy had to be supplied to break up the helium nucleus, specifically

$5 \times 10^{-29}$ kg $\times (3 \times 10^8$ m/sec$)^2$

$$= 4.5 \times 10^{-12} \text{ joule.}$$

The complete breakup of helium nuclei is therefore an energy-absorbing (*endothermic*) reaction, and the energy that must be supplied, called the *binding energy* of the nucleus, is stored, as it were, as potential energy among the fragments. Indeed, it can be retrieved in the reverse reaction of building up helium nuclei from protons and neutrons.

All that has been said so far demonstrates the *mass equivalence of energy*. Now nature generally exhibits a symmetry in her operations; that is, one may reasonably suspect that a particle of matter may lose some of its mass by giving up a corresponding amount of energy. This can, of course, be made to occur most directly by slowing down a high-speed particle. It will get "lighter" by an amount given as (loss of KE/$c^2$) although, of course, whatever agency served to slow the particle must then absorb that much energy and consequently must get that much "heavier" in turn.

But there are more striking ways to realize this exchange between mass and energy: The electron and positron mentioned above may again recombine, and then assume once more the character of $\gamma$-radiation; ordinary ponderable matter now seems to disappear from view and "pure," "free" radiant energy is observed instead. The balance here is as follows: The relativistic mass $m$ of the two particles (total rest mass plus mass equivalent of KE) goes into $mc^2$ energy units of $\gamma$-radiation. Again it is largely a matter of taste whether to say here that matter has been destroyed while energy has been created, or that matter *is* energy, or has assumed the *form* of energy, etc. The physical facts behind such phrases are the same, as stated, for example, by Einstein in 1905 as follows:

"If a body gives off the energy $\Delta E$ in the form of radiation, its mass diminishes by $\Delta E/c^2$ .... The mass of a body is a measure of its energy content; if the energy changes by $\Delta E$, the mass [of that body] changes in the same sense by $\Delta E/(3 \times 10^{10})^2$, the energy being measured in ergs and the mass in grams."

Not always is the change as drastic as in the example of electron-positron "annihilation." In some subatomic phenomena only a small fraction of the mass of the system is transferred into an equivalent quantity of radiant or kinetic energy, as in the release of nuclear energy. But now the factor $c^2$ is favorable—by the relation $\Delta E = (\Delta m)c^2$, any small $\Delta m$ results in a large $\Delta E$.

---

**Problem 30.22.** According to the latest figures, the combined installations for producing electric power in the United States have a capacity of about $10^{11}$ joules of energy per second. If instead of hydroelectric, steam, and internal combustion engines, one central nuclear power plant were to provide this energy by conversion into electrical energy of the nuclear "fuel," how much matter would be converted per day? How much material would be handled per year, if only 0.08% of the material supplied is converted by this process?

---

It stands to reason that the law of conservation of energy must be reformulated if energy-mass conversions are to be included. One simple way of doing this is to consider every object in the system as a potential source for complete annihilation, and therefore to assign to every rest mass $m_0$ a *rest energy* $m_0c^2$, a potential energy which in the course of events may be partly or completely converted into other forms of energy. Then we can safely say:

In a closed system, the total amount of energy, namely, rest energy ($\Sigma m_0 c^2$) plus all *other* forms of energy ($\Sigma E$), is constant, or

$$\Sigma(m_0c^2 + E) = \text{constant}. \qquad (30.19)$$

Now this is really not a new law; Eq. (30.18), the extended law of conservation of mass, read $\Sigma(m_0 + E/c^2) = \text{constant}$, and if Eq. (30.19) is divided through by $c^2$ on both sides we obtain the very same expression. Either equation tells

essentially the same story. As Einstein himself pointed out:

"Pre-relativity physics contains two conservation laws of fundamental importance, namely, the law of conservation of energy and the law of conservation of mass; these two appear there as completely independent of each other. Through relativity theory they melt together into *one* principle."

In this sense one may say that the concepts of mass and of energy also "melt together," as though they were two aspects of one physical reality.

---

**Problem 30.23.** If the laws of conservation of mass and of energy are jointly contained either in Eq. (30.18) or in Eq. (30.19), how shall we reformulate the law of conservation of linear momentum? Surely Eq. (30.17) shows that the old form, which assumed that mass is independent of velocity, is no longer strictly tenable.

### 30.9  NUCLEAR TRANSMUTATIONS

In 1919 the great British physicist Ernest Rutherford observed a then very strange occurrence. During the bombardment of nitrogen gas with alpha particles from a radium preparation, there were produced during the collision some high-speed particles, far longer in range than the stream of incoming alpha particles. After detailed investigation, Rutherford proposed that the following was happening: The alpha particles, known to come in with a kinetic energy $KE_\alpha = 12.3 \times 10^{-13}$ joule, collided (inelastically) with the nuclei of relatively stationary nitrogen atoms. At that instant there would occur a *transmutation* of the nitrogen nucleus into a rare but well-known type (or isotope) of oxygen, observed to recoil with a kinetic energy $KE_{Ox} \doteq 1 \times 10^{-13}$ joule; this transmutation would be accompanied by the simultaneous ejection from the changing nucleus of a positively charged particle, a *proton* with $KE_p \doteq 9.5 \times 10^{-13}$ joule.

Recognizing that alpha particles and protons are respectively the same structures as the nuclei of ordinary helium (He) and hydrogen (H), we can write this nuclear reaction in the usual symbolism (where the superscripts 14, 4, 17, and 1 refer to the *mass numbers*, i.e., the approximate masses in amu, and where the more accurate masses, as later separately and carefully determined for those particles *at rest*, are entered as a last line. These rest masses refer to the inertia which the particles exhibit if they do not also have kinetic energy.)*

$$
\begin{array}{ccc}
 & N^{14} & + & He^4 \\
\text{KE (joules):} & & & (12.3 \times 10^{-13}) \\
\text{Rest mass (amu):} & (14.0075) & & (4.0039)
\end{array}
$$

$$
\left.
\begin{array}{ccc}
\rightarrow & O^{17} & + & H^1 \\
 & (1 \times 10^{-13}) & & (9.5 \times 10^{-13}) \\
 & (17.0045) & & (1.0081)
\end{array}
\right\} . \quad (30.20)
$$

Rutherford's suggestion was epochal. It was the fulfillment of the ancient alchemic dream in unexpected form—the first induced transmutation of matter, although, of course, in quantities too small to be directly recovered. The proposal initiated a whole new way of thinking in nuclear physics. Evidently, the total rest mass of $O^{17}$ and $H^1$ nuclei is larger than the rest mass of $N^{14}$ and $He^4$, precisely by 0.0012 amu; and on the other hand, the kinetic energy of the incident particle is larger (by about $1.8 \times 10^{-13}$ joule) than the sum of kinetic energies for the recoiling particles $O^{17}$ and $H^1$. We expect that the term $\Sigma(m_0 + KE/c^2)$ will be constant before and after transmutation, and this is true. (Check this.) The observed decrease in kinetic energy is compensated for by the increase of rest mass.

*Hydrogen fusion.* The previous historic reaction, in analogy to some chemical processes, may be called *endothermic*. Other nuclear processes can be *exothermic*, or energy-developing. When the nuclei of hydrogen atoms are so agitated (as by the heat in stars or in "atomic" explosions) that they can collide with one another despite their mutual electric repulsion, then they may presumably join or fuse, and thereby two hydrogen nuclei (of the "heavy" variety, made of a proton and a neutron each) may become one nucleus of helium; the latter has, however, less rest mass than that of both building blocks together, and the difference is evidenced as an increase in the kinetic energy available to the end products of the fusion process.

**Problem 30.24.** As first explained by H. A. Bethe (1939), the sun obtains its energy by fusion processes that turn four hydrogen nuclei into one helium nucleus in cycles involving the emission of energy in different radiations. If each of the hydrogen atoms has a rest mass of 1.0081 and each helium atom a rest mass of 4.0039, calculate the energy available in each fusion cycle. [*Note:* The process is so slow that only 1% of the sun's hydrogen is consumed in $10^9$ years; still, many millions of tons of ponderable mass are lost each minute by radiation of energy into space.]

*Uranium fission.* Another kind of exothermic nuclear transmutation has had serious consequences in twentieth-century history. An isotope of uranium, $U^{235}$, can absorb a neutron and break up into isotopes of barium and krypton, plus three neutrons:

$$U^{235} + n^1 \rightarrow Ba^{140} + Kr^{93} + 3n^1. \quad (30.21)$$

(This is only one of several possible reactions.) The emission of several neutrons in this *fission* reaction makes it possible for one such nuclear process to trigger off several like it in the immediate neighborhood, and as each exothermic reaction

---

*    Recall that 1 atomic mass unit = $1.66 \times 10^{-27}$ kg. The numerical values given here are recent and were not available to Rutherford with this accuracy. As is customary, the calculation is made on the basis of the mass of the *neutral atom*, i.e., the nucleus plus the electrons around the nucleus. Why is this permissible?

generates more of the same, an explosive chain reaction is created.

---

**Problem 30.25.** Following are the data for Eq. (30.21). The rest masses of $U^{235}$ and the neutron are 235.116 and 1.0090 amu, respectively. The energy of the incident neutron is negligible. After fission, the fragments have a total of about $3.2 \times 10^{-11}$ joules of kinetic energy. (a) What is the total rest mass of the fragments? (b) What is the total inertia of the fragments as it would be measured right after fission, before they have handed on their kinetic energy by collision with other atoms?

**Problem 30.26.** The nuclear reaction in which Chadwick discovered the neutron is a transmutation that may be written as follows:

$$_4\text{Be}^9 + {}_2\text{He}^4 \rightarrow {}_6\text{C}^{12} + {}_0\text{n}^1.$$

The rest masses of neutral isotopes are $_4\text{Be}^9 =$ 9.01219 amu, $_2\text{He}^4 = 4.00260$, $_6\text{C}^{12} = 12.00000$; mass of neutron $(_0\text{n}^1) = 1.00866$ amu; also the energy of the bombarding $_2\text{He}^4$ nuclei, obtained from polonium, is known to be about $5.3 \times 10^6$ electron volts. Calculate the energy and speed of the emerging neutrons. (Recall that 1 amu $\doteq$ $1.66 \times 10^{-27}$ kg, 1 eV $\doteq 1.60 \times 10^{-19}$ joule. Clearly express any assumptions you make in your calculations.)

## 30.10  REMARKS ON THE GENERAL THEORY OF RELATIVITY*

The special theory of relativity can deal only with physical phenomena in which gravitational forces are not involved. By 1916 Einstein had worked out the basic form of a more comprehensive theory, known as the general theory of relativity. Although some of the details of this theory are still under

---

*    See also Albert Einstein, *Relativity, The Special and General Theory*, Part II, from which quotations in this section are taken.

vigorous discussion, and the details depend on considerable mathematical development, its basic principles and predictions are of great interest and importance, and so require at least a brief summary here.

In explaining his general theory in 1916, Einstein characteristically pointed out first a fact which had been well known, but never properly understood before: *The inertial mass of every object has the same value as its gravitational mass.* This fact was first illustrated by Galileo's discovery that all objects have the same gravitational acceleration at a given location, together with its interpretation by Newton's laws. Thus, by Newton's second law of motion, the net force acting on the object is proportional to its *inertial* mass,

$$\boldsymbol{F}_{\text{net}} = m_{\text{inertial}}\boldsymbol{a}.$$

But if the gravitational pull is the cause of the acceleration, $\boldsymbol{a}$ takes on the value equal to $\boldsymbol{g}$ and the gravitational force is proportional to the gravitational mass of the object:

$$\boldsymbol{F}_G = m_{\text{gravitational}} \cdot \boldsymbol{g},$$

or, to put it into entirely equivalent terms,

$$\boldsymbol{F}_G = m_{\text{gravitational}} \boldsymbol{f}_G,$$

where $\boldsymbol{f}_G$ is the gravitational field strength (see Section 24.8). When we equate these two expressions for the force, we have

$$\boldsymbol{a} = \frac{m_{\text{gravitational}}}{m_{\text{inertial}}} \cdot \boldsymbol{f}_G.$$

The fact that in free fall $\boldsymbol{a}$ is by experiment numerically the same for all bodies at a given location (where the field strength $\boldsymbol{f}_G$ must have a fixed value independent of the nature of the object it acts on) implies that the ratio $m_{\text{gravitational}}/m_{\text{inertial}}$ is the same for all bodies. We can then adopt a system of units in which this ratio is one (as is usually done without explicit discussion). Hence the value of the gravitational mass of a body is equal to that of its inertial mass.

Einstein now argues that this equality must mean that "the *same* quality of a body manifests itself according to circumstances as 'inertia' or as 'weight.'" The significance of this assertion may be illustrated by considering a large box (of negligible mass) in empty space far away from a planet or other gravitating objects. Inside the box is an observer, who is floating freely in space, since there is no gravitational field present. Now suppose someone attaches a hook and rope to the outside of the top* of the box and begins to pull it with a constant force. The box undergoes a uniform acceleration, which is transmitted to the observer inside the box by the floor when it scoops him up or hits him:

"He must therefore take up this pressure by means of his legs if he does not wish to be laid out full length on the floor. He is then standing in the box in exactly the same way as anyone stands in a room of a house on our earth. If he releases a body which he previously had in his hand, the acceleration of the box will no longer be transmitted to this body, and for this reason the body will approach the floor of the box with an accelerated relative motion. The observer will further convince himself that *the acceleration of the body toward the floor of the box is always of the same magnitude, whatever kind of body he may happen to use for the experiment.*

"...the man in the box will thus come to the conclusion that he and the box are in a gravitational field which is constant with regard to time. Of course he will be puzzled for a moment as to why the box does not fall in this gravitational field. Just then, however, he discovers the hook in the middle of the top of the box and the rope which is attached to it, and he consequently comes to the conclusion that the box is suspended at rest in the gravitational field.

---

* Of course up to this point the word "top" is inappropriate, since there is no physical way to determine up or down directions in the box. But as we shall see, the side to which the hook is attached becomes the top.

"Ought we to smile at the man and say that he errs in his conclusion? I do not believe we ought to if we wish to remain consistent; we must rather admit that his mode of grasping the situation violates neither reason nor known mechanical laws."

Thus, even though to an *outside* observer the box may appear to be in accelerated motion in gravitation-free space, the observer *inside* the box may legitimately regard it as being at rest, but subject to a gravitational field. This discussion illustrates that physics can be formulated in a way that frames of reference in uniformly accelerated motion relative to each other may also be considered equivalent (*principle of equivalence*).

Einstein's general theory has been accepted by most physicists, at least as a working hypothesis, primarily because of the successful confirmation of three predictions deduced from it.

a) *Gravitational action on light.* If the box discussed above has a little hole in one side, and a beam of light from a distant source comes in while the box is at rest, the beam will go straight across to the other side as long as the box is at rest. But when the box is accelerated greatly enough, the beam will of course appear to the man inside to curve downward as it passes across the box. To put it differently, the observer in an accelerated frame of reference will notice that light does not travel in straight lines. But his frame of reference, by the principle of equivalence, can be equally well regarded as being fixed, though acted on by a gravitational field. In that case we may conclude that any "real" gravitational field should also have an effect on the gravitation of light. Einstein thus predicted that the light from stars, passing near the sun, would be deflected toward the sun, so that the apparent positions of the stars themselves (as seen from the earth, say) would be shifted away from the sun. Of course this effect could only be observed during a total eclipse of the sun, and its magnitude is so small that it is not

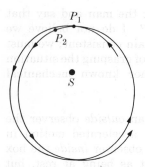

**Fig. 30.8** Advance of the perihelion for the orbit of the planet Mercury from $P_1$ to $P_2$ (greatly exaggerated).

noticeable except by careful observation. As mentioned in Section 30.1, it was the successful test of this prediction by the British eclipse expedition of 1919 that provided the first convincing evidence for Einstein's general theory of relativity.

b) *Advance of the perihelion of Mercury.* Another success of the general theory, which Einstein developed in detail in 1915, was the calculation of a small correction to the theoretical Newtonian orbits of planets in the solar system. The observed fact that the orbit of the planet Mercury slowly rotates around the sun, so that the position of closest approach to the sun (perihelion) gradually shifts, had been a long-standing puzzle in celestial mechanics.

Einstein's explanation depends on the idea that energy is associated with mass, and that even energy in "empty" space, for example the energy represented in the sun's gravitational field, may be affected by and in turn act on other objects that exert gravitational forces. As a planet gets close to the sun, it experiences a gravitational force greater than would be expected from Newton's law of universal gravitation in its classical form; one way of putting it is that the effective mass of the sun, plus its gravitational field, increases. This effect is significant only for planets that come in very close to the sun, such as Mercury, and that have eccentric orbits, so that there is a consider-

able variation in their distance from the sun over which to get differences of the effect.

According to Einstein's theory, when a planet approaches the sun (neighborhood of the perihelion, $P_1$), it feels an extra force due to the additional effective mass of the sun; the effect is to deflect the planet into a changed orbit so that the next perihelion is at $P_2$ (see Fig. 30.8). The result is that the planet still moves very nearly in an ellipse, but the ellipse itself rotates slowly around the sun. The shift from $P_1$ to $P_2$ between one pass and the next is called the "advance of the perihelion" of the orbit. The numerical value of the rate of advance, as calculated from the general theory of relativity, is almost exactly in agreement with the observed value.

c) *Gravitational red shift of light emitted by a massive star.* A further prediction of the general theory of relativity is that time intervals appear to be dilated (clocks are seen to run more slowly) when we look into regions of space in which there are strong gravitational fields. The rate at which a light wave is emitted from an atom also serves as a clock. Hence the frequency of electromagnetic radiation produced in such regions by a particular atomic process (say, in the emission of a Balmer line) is less, as received and measured by an observer in another part of space. The result is an observed shift of spectral-line frequencies toward the red (low-frequency) end of the spectrum. This gravitational "red shift" of light that is coming to us from the sun or from more massive stars was observed in the 1920s by astronomers, who were motivated by Einstein's prediction. Incidentally, this effect is quite different from another "red shift," attributed to a Doppler effect that is caused by the relative motion of a star away from the earth (see below) rather than by its light going through a strong gravitational field.

The gravitational red shift is also involved in another relativistic phenomenon that has recently aroused interest among astrophysicists: the "gravitational collapse" of extremely massive stars. It is believed that a star several times as massive as the sun would be unstable, because the gravitational attractive forces tending to make it contract would overcome the expansive tendency provided by the random motions of its molecules, the pressure of the escaping radiation, and repulsive forces between nuclei. As the star collapsed, its intense gravitational field would impose a greater and greater red shift on all emitted radiation, until finally the frequency tends to zero and no radiation could escape at all; hence the star could no longer be seen. On the other hand, all radiation (and matter) approaching the vicinity of the star would be sucked into it by the strong gravitational field and could not escape. Such a star would thus appear to us as a "black hole" in the sky, blocking our vision of anything directly on the other side of it.

The possible fate of stars contracting to a small volume has recently attracted much scientific speculation, in part because of the discovery of "pulsars" in the 1960s. These are thought to be extremely dense, rapidly rotating stars composed mainly of neutrons and emitting radiation a bit like a rotating light beacon. The behavior of matter and radiation in the intense gravitational field of a collapsing neutron star may well provide a crucial test of the general theory of relativity. Other astronomical applications of the theory have been discussed since the 1920s in connection with the problem of the overall structure and evolution of the universe. Einstein in 1915 proposed a set of equations relating the distribution of matter throughout the entire universe to the properties of local space-time frames of reference, that is, to the so-called "curvature" of space, which may be thought to be responsible for the bending of light rays mentioned above. He suggested that on the cosmic scale this curvature might have the effect of making the universe finite though unbounded. A beam of light traveling in space (but always affected by the gravitation of all the matter in the universe) might eventually return to its starting point, somewhat as if it had been going around the surface of an immense sphere. Thus the universe could be considered to have a definite radius $R$, though all locations in it would be equivalent and none could be called the "center" or the "boundary." This theory was soon developed further by William de Sitter in Holland, A. A. Friedman in the USSR, and Abbé Georges Lemaitre in Belgium. It was Friedman who, in 1922, first showed that Einstein's equations allowed the possibility of an expanding universe (radius $R$ increasing with time).

The idea of an expanding universe emerged independently from astronomical observations of change in the frequencies of spectral lines of radiation received from distant galaxies; as mentioned above, these could be interpreted as a Doppler effect due to the motion of the galaxies. (This is analogous to the apparent decrease in pitch of a moving automobile horn as it recedes from the hearer.) In most cases the frequency was found to be "red-shifted," i.e., lower than that of the corresponding line from terrestrial sources, corresponding to motion *away* from the earth. In 1929, the American astronomer Edwin Hubble inferred from such observations that the galaxies are receding from the earth at speeds proportional to their distance from us. This does not mean that the earth is the center of the universe; the observations are compatible with a uniform expansion of the universe which would be interpreted in the same way by an observer anywhere in the universe.*

---

*    It is customary to explain this reasoning by suggesting a two-dimensional analogy. Consider the surface of a rubber balloon, with dots painted on its surface to represent the galaxies. As the balloon is inflated, an observer at *each* dot sees the rest moving away just as if he were at the center of the expansion. The same analogy also can be used to illustrate how the entire universe (represented by the surface of the balloon) could be finite yet unbounded. One should be warned that the analogy unfortunately also suggests misleading ideas, e.g., that the "curvature" of three-dimensional space is somehow related to a four-dimensional space in which it is embedded, just as the two-dimensional surface of the balloon is curved because it is the surface of a sphere in three-dimensional space.

If the universe is expanding, it must have been smaller in the past. Indeed, Hubble's results imply that it had a very small size about $10^{10}$ years ago, when this initial state may have given way in a "big bang"—an explosive beginning of the expansion.

It is not yet clear whether we can expect the expansion to continue indefinitely. If so, we would have to predict a modern version of the eventual "heat death" of the universe (Section 18.4), since a finite amount of matter and energy would be dissipated into an unlimited volume. On the other hand, it is possible that the gravitational forces between the galaxies, though extremely weak at large distances, would nevertheless act eventually to slow down and stop the expansion and change it into a contraction (just as a ball thrown up from the surface of the earth will eventually fall back down if its velocity is less than the "escape velocity"). In that case the universe would eventually collapse into a "black hole." This catastrophe might be followed by another big bang, leading to a newly expanding universe, and so forth; one would then have a cyclic "oscillating" universe.

These are only a few of the cosmological theories suggested by Einstein's general theory of relativity and its more recent modifications. For that matter, the general theory itself may not be general enough to encompass the full complexities of the real world, and by the time this book is published, still further important changes in our concepts and theories may be coming to light.

## RECOMMENDED FOR FURTHER READING

A. Einstein, "Autobiographical notes," in *Albert Einstein, Philosopher-Scientist* (edited by P. A. Schilpp), New York, Library of Living Philosophers, 1949

A. Einstein, *Relativity: The Special and General Theory*, translated by R. W. Lawson from the German book published in 1917; New York: Crown Publishers; paperback reprint of the 1931 edition

P. Frank, *Einstein: His Life and Times*, New York: Knopf, 1947

*Project Physics Reader,* Vol. 5, pages 25–104. Extracts from writings of Infeld, Gamow, Rogers, Taylor and Wheeler, Einstein and Infeld, and Klein.

J. A. Wheeler, "Our universe: the known and the unknown," *American Scientist*, Vol. 56, pages 1–20 (1968)

L. P. Williams, *Relativity Theory, its Origins and Impact on Modern Thought*, New York: Wiley, 1968. An anthology including artists' responses to relativity as well as extracts from the writings of Einstein and other physicists.

## SOURCES, INTERPRETATIONS, AND REFERENCE WORKS

A. B. Arons, *Development of Concepts of Physics*, Reading, Mass.: Addison-Wesley, 1965, Chapter 36

Albert Einstein *et al.*, *The Principle of Relativity*, translated from German by W. Perrett and G. B. Jeffery, New York: Dover paperback reprint of the 1923 edition. A collection of original research papers.

J. C. Hafele and R. E. Keating, "Around-the-world atomic clocks," *Science*, Vol. 177, pages 166–170 (1972). Four cesium beam atomic clocks were flown on commercial jet flights around the world twice, once eastward and once westward. Relativity theory predicts that the clocks compared with reference clocks at the U.S. Naval Observatory, should have lost $40 \pm 23$ nanoseconds during the eastward trip, and gained $275 \pm 21$ nanoseconds during the westward trip; it was found that they lost $59 \pm 10$ nanoseconds and gained $273 \pm 7$ nanoseconds, respectively, thus confirming the theory of the "twin paradox."

G. Holton *et al.*, *Special Relativity Theory, Selected Reprints*, New York: American Institute of Physics. Includes the article "On the origins of the special theory of relativity" and the annotated bibliography prepared by G. Holton, and reprints of 15 other articles on the clock (twin) paradox and other aspects of relativity.

G. Holton, "Influences on Einstein's early work in relativity theory," *American Scholar*, Vol. 37, pages 59–79 (Winter 1967–68). "Einstein, Michelson, and the 'crucial' experiment," *Isis*, Vol. 60, pages 133–197 (1969). "On trying to understand scientific genius," *American Scholar*, Vol. 41, pages 95–110 (1971–72). All three are included in G. Holton, *Thematic Origins of Scientific Thought: Kepler to Einstein*, Cambridge, Mass.: Harvard University Press, 1973

C. Kittel, W. D. Knight, and M. A. Ruderman, *Berkeley Physics Course*, Vol. 1, *Mechanics*, New York: McGraw-Hill, 1965, Chapters 11–14

A. Koslow, *The Changeless Order*. Extracts from writings of Newton, Leibniz, Kant, Mach, Einstein and others, on space, time, and motion.

M. Munitz, *Space, Time, and Creation*, New York: Collier Books, 1961, paperback reprint of the 1957 edition

J. D. North, *The Measure of the Universe, a History of Modern Cosmology*, Oxford: Clarendon Press, 1965

R. Ruffini and J. A. Wheeler, "Introducing the black hole," *Physics Today*, January 1971, pages 30–41 (and cover illustration)

C. M. Will, "Einstein on the firing line," *Physics Today*, Vol. 25, No. 10, pages 23–29 (October 1972). A survey of tests of general relativity and similar theories.

# APPENDIXES

APPENDIXES

# Appendix I

# DEFINED VALUES, FUNDAMENTAL CONSTANTS, AND ASTRONOMICAL DATA

## DEFINED VALUES

| Name of quantity | Symbol | Value |
|---|---|---|
| Meter | m | 1,650,763.73 wavelengths in vacuum of the radiation corresponding to transition between the $2p_{10}$ and $5d_5$ level of the $Kr^{86}$ atom |
| Second | sec | 9,192,631,770 periods of the radiation corresponding to the transition between the two hyperfine levels of the ground state of the cesium-133 atom |
| Kilogram | kg | Mass of the standard kilogram at International Bureau of Weights and Measures, Sèvres, France |
| Degree Kelvin | °K or kelvin | 1/273.16 of absolute temperature of triple point of water |
| Atomic mass unit | amu | $\frac{1}{12}$ of the mass of an atom of $C^{12}$ isotope |

| Name of quantity | Symbol | Value |
|---|---|---|
| Gravitational acceleration at surface of earth | $g$ | 9.80665 m/sec |
| Normal atmosphere | atm | 101325 newtons/m² |
| Mole | | Amount of substance containing same number of molecules as 12 g of pure $C^{12}$ |
| Liter | $l$ | 1,000.028 cm³ |
| Ampere | amp | Force between two long parallel wires 1 m apart, each carrying a current of one amp, is $2 \times 10^{-7}$ newtons |
| Electrostatic force constant | $k_e$ | $8.9874 \times 10^9$ newton-m²/coul² |

## ASTRONOMICAL DATA

| Name of quantity | Value |
| --- | --- |
| Mass of earth | $5.98 \times 10^{24}$ kg |
| Mean radius of earth | $6.38 \times 10^{6}$ m |
| Mean distance from earth to sun | $1.496 \times 10^{11}$ m |
| Eccentricity of earth's orbit | 0.0167 |
| Mean distance from earth to moon | $3.84 \times 10^{8}$ m |
| Sun's diameter | $1.39 \times 10^{9}$ m |
| Sun's mass | $1.98 \times 10^{30}$ kg |

## MATHEMATICAL CONSTANTS

$\pi = 3.14159$;    $1/\pi = 0.31831$;    $\pi^2 = 9.86960$;
$\sqrt{\pi} = 1.77245$

$e = 2.71828$;    $1/e = 0.36788$

$\sqrt{2} = 1.41421$

$\sqrt{3} = 1.73205$

## GREEK ALPHABET

| A | $\alpha$ | Alpha | N | $\nu$ | Nu |
| --- | --- | --- | --- | --- | --- |
| B | $\beta$ | Beta | $\Xi$ | $\xi$ | Xi |
| $\Gamma$ | $\gamma$ | Gamma | O | $o$ | Omicron |
| $\Delta$ | $\delta$ | Delta | $\Pi$ | $\pi$ | Pi |
| E | $\epsilon$ | Epsilon | P | $\rho$ | Rho |
| Z | $\zeta$ | Zeta | $\Sigma$ | $\sigma$ | Sigma |
| H | $\eta$ | Eta | T | $\tau$ | Tau |
| $\Theta$ | $\theta, \vartheta$ | Theta | $\Upsilon$ | $\upsilon$ | Upsilon |
| I | $\iota$ | Iota | $\Phi$ | $\phi, \varphi$ | Phi |
| K | $\kappa$ | Kappa | X | $\chi$ | Chi |
| $\Lambda$ | $\lambda$ | Lambda | $\Psi$ | $\psi$ | Psi |
| M | $\mu$ | Mu | $\Omega$ | $\omega$ | Omega |

## FUNDAMENTAL CONSTANTS*

| Name of quantity | Symbol | Value |
|---|---|---|
| Velocity of light | $c$ | $2.9979 \times 10^8$ m/sec |
| Charge on electron | $e$ | $(-) 1.602 \times 10^{-19}$ coul |
| Rest mass of electron | $m_e$ | $9.110 \times 10^{-31}$ kg |
| Electron charge to mass ratio | $e/m_e$ | $1.7588 \times 10^{11}$ coul/kg |
| Planck's constant | $h$ | $6.626 \times 10^{-34}$ joule-sec |
| Boltzmann constant | $k$ | $1.3806 \times 10^{-23}$ joule/degree K |
| Avogadro's number | $N$ | $6.022 \times 10^{23}$ molecules/mole |
| Universal gas constant | $R$ | $8.314$ joules/mole-degree K |

| Name of quantity | Symbol | Value |
|---|---|---|
| Mass of neutral hydrogen ($H^1$) | | 1.007825 amu |
| Mass of proton (p) | | 1.00728 amu |
| Mass of neutron (n) | | 1.00867 amu |
| Mass of electron (e) | | $5.486 \times 10^{-4}$ amu |
| Ratio of proton to electron mass | | 1836.11 |
| 1 amu | | $1.6605 \times 10^{-27}$ kg |
| Rydberg constant for infinitely heavy nucleus | $R_\infty$ | $1.09737 \times 10^7$/m |
| Rydberg constant for hydrogen | $R_H$ | $1.09678 \times 10^7$/m |
| Universal gravitational constant | $G$ | $6.670 \times 10^{-11}$ newton-m$^2$/kg$^2$ |
| Faraday constant | $F$ | 96,487 coul/mole |

* Source: E. R. Cohen and J. W. M. DuMond, "Our knowledge of the fundamental constants of physics and chemistry in 1965," *Reviews of Modern Physics*, Vol. 37, pages 537–594 (1965), with slight revisions in some values as recommended by B. N. Taylor, W. H. Parker, and D. N. Langenberg, "Determination of $e/h$, using macroscopic quantum phase coherence in superconductors: implications for quantum electrodynamics and the fundamental physical constants," *Reviews of Modern Physics*, Vol. 41, pages 375–496 (1969). In each case, the figures are known to higher accuracy than implied here, but are given in the form most useful for the purposes of this book. For further data, consult a collection such as the *Handbook of Chemistry and Physics* (Chemical Rubber Publishing Co., Cleveland; issued yearly).

# Appendix II

# TABLE OF CONVERSION FACTORS

## METRIC SYSTEM PREFIXES

| tera | $10^{12}$ | pico | $10^{-12}$ |
|------|-----------|------|------------|
| giga | $10^9$ | nano | $10^{-9}$ |
| mega | $10^6$ | micro | $10^{-6}$ |
| kilo | $10^3$ | milli | $10^{-3}$ |
| hecto | $10^2$ | centi | $10^{-2}$ |
| deka | $10$ | deci | $10^{-1}$ |

## LENGTH

1 meter = 39.37 inches    1 inch = 0.0254 meter
1 meter = 3.281 feet    1 foot = 0.3048 meter
1 kilometer =    1 mile =
     0.6214 mile      1.609 kilometers
1 angstrom (Å) = $10^{-10}$ meter
1 astronomical unit (AU) = $14.96 \times 10^7$ kilometers
       = $9.29 \times 10^7$ miles

1 light year = $9.46 \times 10^{15}$ meters
1 parsec (distance at which 1 AU subtends 1 in. of arc) = 206,265 AU = $3.084 \times 10^{16}$ meters

## AREA

1 $m^2$ = $1.55 \times 10^3$ $in^2$    1 $in^2$ = 6.452 $cm^2$ =
       $6.452 \times 10^{-4}$ $m^2$

1 $m^2$ = 10.76 $ft^2$    1 $ft^2$ = 144 $in^2$ =
       $9.29 \times 10^{-2}$ $m^2$

## VOLUME

1 $m^3$ = 35.31 $ft^3$    1 $ft^3$ = 0.02832 $m^3$
1 liter $\cong$ $10^{-3}$ $m^3$

## VELOCITY

1 m/sec = 3.281 ft/sec    1 ft/sec = 0.3048 m/sec
1 m/sec = 3.6 km/hr    1 mi/hr = 1.467 ft/sec
1 m/sec = 2.237 mi/hr      = 0.4470 m/sec

## ACCELERATION

1 $m/sec^2$ = 3.281 $ft/sec^2$    1 $ft/sec^2$ =
       0.3048 $m/sec^2$

## FORCE

1 newton =    1 pound-force (1bf) =
   $10^5$ dynes = 0.2247 lbf      4.448 newtons

## MASS

1 gram =    1 lb = 453.6 g
     $2.205 \times 10^{-3}$ lb

## PRESSURE

1 atm = $1.013 \times 10^5$ newtons/$m^2$
1 atm = 14.7 lb/$in^2$
1 bar = $10^5$ newtons/$m^2$
1 atm = 760 torr
1 torr = 0.001316 atm = 133.3 newtons/$m^2$

## ENERGY

1 joule = $10^7$ ergs =    1 cal = 4.184 joules
     0.239 cal

1 joule = 0.2778 ×    1 kilowatt-hour =
   $10^{-6}$ kilowatt-hour      $3.6 \times 10^6$ joules

1 joule = $0.625 \times 10^{19}$    1 eV = $1.60 \times 10^{-19}$
     electron volts      joule

# Appendix III

# ALPHABETIC LIST OF THE ELEMENTS

| Element | Symbol | Atomic Number $Z$ | Element | Symbol | Atomic Number $Z$ |
|---|---|---|---|---|---|
| Actinium | Ac | 89 | Curium | Cm | 96 |
| Aluminum | Al | 13 | Dysprosium | Dy | 66 |
| Americium | Am | 95 | Einsteinium | Es | 99 |
| Antimony | Sb | 51 | Erbium | Er | 68 |
| Argon | A | 18 | Europium | Eu | 63 |
| Arsenic | As | 33 | Fermium | Fm | 100 |
| Astatine | At | 85 | Fluorine | F | 9 |
| Barium | Ba | 56 | Francium | Fr | 87 |
| Berkelium | Bk | 97 | Gadolinium | Gd | 64 |
| Beryllium | Be | 4 | Gallium | Ga | 31 |
| Bismuth | Bi | 83 | Germanium | Ge | 32 |
| Boron | B | 5 | Gold | Au | 79 |
| Bromine | Br | 35 | Hafnium | Hf | 72 |
| Cadmium | Cd | 48 | Helium | He | 2 |
| Calcium | Ca | 20 | Holmium | Ho | 67 |
| Californium | Cf | 98 | Hydrogen | H | 1 |
| Carbon | C | 6 | Indium | In | 49 |
| Cerium | Ce | 58 | Iodine | I | 53 |
| Cesium | Cs | 55 | Iridium | Ir | 77 |
| Chlorine | Cl | 17 | Iron | Fe | 26 |
| Chromium | Cr | 24 | Krypton | Kr | 36 |
| Cobalt | Co | 27 | Lanthanum | La | 57 |
| Copper | Cu | 29 | Lawrencium | Lw | 103 |

**Alphabetic list of the elements—*Continued***

| Element | Symbol | Atomic Number Z | Element | Symbol | Atomic Number Z |
|---------|--------|-----------------|---------|--------|-----------------|
| Lead | Pb | 82 | Rhodium | Rh | 45 |
| Lithium | Li | 3 | Rubidium | Rb | 37 |
| Lutetium | Lu | 71 | Ruthenium | Ru | 44 |
| Magnesium | Mg | 12 | Samarium | Sm | 62 |
| Manganese | Mn | 25 | Scandium | Sc | 21 |
| Mendelevium | Md | 101 | Selenium | Se | 34 |
| Mercury | Hg | 80 | Silicon | Si | 14 |
| Molybdenum | Mo | 42 | Silver | Ag | 47 |
| Neodymium | Nd | 60 | Sodium | Na | 11 |
| Neon | Ne | 10 | Strontium | Sr | 38 |
| Neptunium | Np | 93 | Sulphur | S | 16 |
| Nickel | Ni | 28 | Tantalum | Ta | 73 |
| Niobium | Nb | 41 | Technetium | Tc | 43 |
| Nitrogen | N | 7 | Tellurium | Te | 52 |
| Nobelium | No | 102 | Terbium | Tb | 65 |
| Osmium | Os | 76 | Thallium | Tl | 81 |
| Oxygen | O | 8 | Thorium | Th | 90 |
| Palladium | Pd | 46 | Thulium | Tm | 69 |
| Phosphorus | P | 15 | Tin | Sn | 50 |
| Platinum | Pt | 78 | Titanium | Ti | 22 |
| Plutonium | Pu | 94 | Tungsten | | |
| Polonium | Po | 84 | (Wolfram) | W | 74 |
| Potassium | K | 19 | Uranium | U | 92 |
| Praseodymium | Pr | 59 | Vanadium | V | 23 |
| Promethium | Pm | 61 | Xenon | Xe | 54 |
| Protactinium | Pa | 91 | Ytterbium | Yb | 70 |
| Radium | Ra | 88 | Yttrium | Y | 39 |
| Radon | Rn | 86 | Zinc | Zn | 30 |
| Rhenium | Re | 75 | Zirconium | Zr | 40 |

# Appendix IV

# PERIODIC TABLE OF ELEMENTS

Atomic weights are the most recent (1961) adopted by the International Union of Pure and Applied Chemistry. The numbers are all relative to the atomic weight of the principal isotope of carbon, $C^{12}$, defined as exactly 12.00000. For artificially produced elements, the approximate atomic weight of the most stable isotope is given in brackets. The full names of the elements are given in Appendix III.

| Group → | | I | II | III | IV | V | VI | VII | VIII | | | 0 |
|---|---|---|---|---|---|---|---|---|---|---|---|---|
| Period | Series | | | | | | | | | | | |
| 1 | 1 | 1 H 1.00797 | | | | | | | | | | 2 He 4.0026 |
| 2 | 2 | 3 Li 6.939 | 4 Be 9.0122 | 5 B 10.811 | 6 C 12.01115 | 7 N 14.0067 | 8 O 15.9994 | 9 F 18.9984 | | | | 10 Ne 20.183 |
| 3 | 3 | 11 Na 22.9898 | 12 Mg 24.312 | 13 Al 26.9815 | 14 Si 28.086 | 15 P 30.9738 | 16 S 32.064 | 17 Cl 35.453 | | | | 18 Ar 39.948 |
| 4 | 4 | 19 K 39.102 | 20 Ca 40.08 | 21 Sc 44.956 | 22 Ti 47.90 | 23 V 50.942 | 24 Cr 51.996 | 25 Mn 54.9380 | 26 Fe 55.847 | 27 Co 58.9332 | 28 Ni 58.71 | |
| | 5 | 29 Cu 63.54 | 30 Zn 65.37 | 31 Ga 69.72 | 32 Ge 72.59 | 33 As 74.9216 | 34 Se 78.96 | 35 Br 79.909 | | | | 36 Kr 83.80 |
| 5 | 6 | 37 Rb 85.47 | 38 Sr 87.62 | 39 Y 88.905 | 40 Zr 91.22 | 41 Nb 92.906 | 42 Mo 95.94 | 43 Tc [99] | 44 Ru 101.07 | 45 Rh 102.905 | 46 Pd 106.4 | |
| | 7 | 47 Ag 107.870 | 48 Cd 112.40 | 49 In 114.82 | 50 Sn 118.69 | 51 Sb 121.75 | 52 Te 127.60 | 53 I 126.9044 | | | | 54 Xe 131.30 |
| | 8 | 55 Cs 132.905 | 56 Ba 137.34 | 57–71 Lanthanide series* | 72 Hf 178.49 | 73 Ta 180.948 | 74 W 183.85 | 75 Re 186.2 | 76 Os 190.2 | 77 Ir 192.2 | 78 Pt 195.09 | |
| | 9 | 79 Au 196.967 | 80 Hg 200.59 | 81 Tl 204.37 | 82 Pb 207.19 | 83 Bi 208.980 | 84 Po [210] | 85 At [210] | | | | 86 Rn [222] |
| 7 | 10 | 87 Fr [223] | 88 Ra [226] | 89–103 Actinide series† | 104 Rf‡ | 105 Ha‡ | | | | | | |

| *Lanthanide series: | 57 La 138.91 | 58 Ce 140.12 | 59 Pr 140.907 | 60 Nd 144.24 | 61 Pm [147] | 62 Sm 150.35 | 63 Eu 151.96 | 64 Gd 157.25 | 65 Tb 158.924 | 66 Dy 162.50 | 67 Ho 164.930 | 68 Er 167.26 | 69 Tm 168.934 | 70 Yb 173.04 | 71 Lu 174.97 |
|---|---|---|---|---|---|---|---|---|---|---|---|---|---|---|---|
| †Actinide series: | 89 Ac [227] | 90 Th 232.038 | 91 Pa [231] | 92 U 238.03 | 93 Np [237] | 94 Pu [242] | 95 Am [243] | 96 Cm [245] | 97 Bk [249] | 98 Cf [249] | 99 Es [253] | 100 Fm [255] | 101 Md [256] | 102 No [255] | 103 Lr [257] |

‡Proposed symbols for Rutherfordium (Rf) and Hahnium (Ha). See A. Ghiorso, *Science*, Vol. 171, p. 127 (1971).

# Appendix V

## SUMMARY OF SOME TRIGONOMETRIC RELATIONS

1) *Right triangles:* (a) The sum of the angles in any plane triangle being 180°, and angle $\gamma$ in Fig. A.1 being 90°, it follows that $\alpha + \beta = 90°$. (b) By the Pythagorean theorem, $c^2 = a^2 + b^2$.

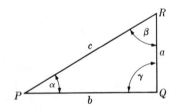

**Figure A.1**

2) *Definitions:* We can *define* six functions of any one angle in a right triangle, namely sine (sin), cosine (cos), tangent (tan), cotangent (cot), secant (sec), and cosecant (cosec). Specifically,

$$\sin \alpha = \frac{\text{opposite side}}{\text{hypotenuse}} = \frac{a}{c},$$

$$\cos \alpha = \frac{\text{adjacent side}}{\text{hypotenuse}} = \frac{b}{c},$$

$$\tan \alpha = \frac{a}{b}, \quad \cot \alpha = \frac{b}{a},$$

$$\sec \alpha = \frac{c}{b}, \quad \operatorname{cosec} \alpha = \frac{c}{a}.$$

(For the first three functions, these ratios are given in Appendix VI.) It follows directly from these definitions:

a) $\quad \tan \alpha = \dfrac{\sin \alpha}{\cos \alpha}, \quad \cot \alpha = \dfrac{\cos \alpha}{\sin \alpha} = \dfrac{1}{\tan \alpha},$

$$\sec \alpha = \frac{1}{\cos \alpha}, \quad \operatorname{cosec} \alpha = \frac{1}{\sin \alpha}.$$

Therefore only the sine and cosine functions of an angle need be discussed in detail.

b) $\quad \sin^2 \alpha = (\sin \alpha)^2 = \dfrac{a^2}{c^2},$

$$\cos^2 \alpha = (\cos \alpha)^2 = \frac{b^2}{c^2}.$$

Thus $\sin^2 \alpha + \cos^2 \alpha = \dfrac{a^2 + b^2}{c^2} = 1$ (by the Pythagorean theorem).

c) Because $\sin \beta = \dfrac{b}{c}$, $\cos \beta = \dfrac{a}{c}$ and so on, therefore

$$\sin \alpha = \cos \beta, \cos \alpha = \sin \beta,$$

$$\sin \alpha = \cos (90 - \alpha), \cos \alpha = \sin (90 - \alpha).$$

**Example 1.** In a right triangle ($\gamma = 90°$), $\alpha = 35°$, length $b = 4$ cm; what is length $a$?

$$\tan \alpha = \frac{a}{b},$$

therefore $a = b \tan \alpha = 4 \times 0.700 = 2.8$ cm.

**Example 2.** In a right triangle, $a = 8.3$ cm, $c = 21$ cm; what are $\alpha$ and $\beta$?

$$\sin \alpha = \frac{a}{c} = \frac{8.3}{21} = 0.395.$$

As seen in the table, this lies between $\sin 23° = 0.391$ and $\sin 24° = 0.407$. But

$$\frac{0.395 - 0.391}{0.407 - 0.391} = \frac{0.004}{0.016} = \frac{1}{4}.$$

Thus angle $\alpha$ is (roughly) $\frac{1}{4}$ of 1° higher than 23°, or 23° 15′. And $\beta = 90° - \alpha = 66° 45′$.

3) One can define the same trigonometric functions for any angle, even for an angle not in a triangle but enclosed between an inclined line and the horizontal (Fig. A.2). If $r$ is the length of the line and $(x,y)$ the $x$ and $y$ coordinates of its end point $P$, the other end being at the origin of the coordinate system, then

Figure A.3

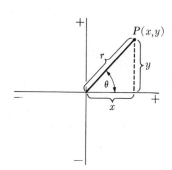

Figure A.2

$$\sin \theta = \frac{y}{r}, \qquad \cos \theta = \frac{x}{r}, \qquad \tan \theta = \frac{y}{x}.$$

By convention, $\theta$ is measured counterclockwise from the right (positive) section of the abscissa.

4) If $\theta$ is larger than 90°, its trigonometric functions are defined as before, for example:

$$\sin \theta = \frac{y}{r}.$$

a) $\theta = 90°$ to 180° (Fig. A.3). Note that $x$ has a negative value, being on the left side of the $xy$ coordinate system, or in the second quadrant. Thus

$$\cos \theta = \frac{x}{r} = -\cos \phi = -\cos (180° - \theta),$$

$$\sin \theta = \frac{y}{r} = \sin \phi = \sin (180° - \theta).$$

b) $\theta = 180°$ to 270° (Fig. A.4); $\phi = \theta - 180°$. Both $x$ and $y$ have negative values:

$$\sin \theta = -\sin \phi = -\sin (\theta - 180°),$$

$$\cos \theta = -\cos \phi = -\cos (\theta - 180°).$$

Figure A.4

c) $\theta = 270°$ to 360° (Fig. A.5); $\phi = 360° - \theta$. Now $y$ has a negative value, $x$ a positive one:

Figure A.5

$$\sin \theta = -\sin \phi = -\sin (360° - \theta),$$

$$\cos \theta = \cos \phi = \cos (360° - \theta).$$

These relations show that trigonometric tables computed for the range 0° to 90° serve for all angles.

**Example 3.** Find $\sin \theta$ and $\tan \theta$ if $\theta = 295°$. Note that $\theta$ is between 270° and 360°;

$$\sin \theta = -\sin (360° - \theta) = -\sin 65° = -0.906.$$

And

$$\tan \theta = \frac{\sin \theta}{\cos \theta} = \frac{-0.906}{0.423} = -2.15.$$

5) *General triangles.* Even when a triangle does not have a right angle, as in Fig. A.6, there still exist several simple relations between the sides and the angles.

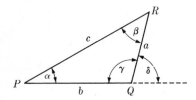

**Figure A.6**

a) Law of Cosines: The square of any one side is equal to the sum of the squares of the other two sides minus twice the product of those two into the cosine of their included angle. For example,

$$c^2 = a^2 + b^2 - 2ab \cos \gamma.$$

Thus one may find one side in terms of two others and one angle, or one angle in terms of three sides. Because the complementary angle $\delta = 180 - \gamma$, and $\cos \delta = \cos (180 - \gamma) = -\cos \gamma$, we may write

$$c^2 = a^2 + b^2 + 2ab \cos \delta.$$

b) Law of Sines: In any triangle, the ratio of any side and the sine of the opposite angle is

constant, or

$$\frac{a}{\sin \alpha} = \frac{b}{\sin \beta} = \frac{c}{\sin \gamma}.$$

Thus two sides and one angle determine the remaining angles and side, and so forth.

6) *Other useful trigonometric relations*

$$\sin (\alpha + \beta) = \sin \alpha \cos \beta + \cos \alpha \sin \beta,$$

$$\cos (\alpha + \beta) = \cos \alpha \cos \beta - \sin \alpha \sin \beta,$$

$$2 \sin \alpha \cos \alpha = \sin 2\alpha.$$

$$\cos 2\alpha = \cos^2 \alpha - \sin^2 \alpha.$$

# Appendix VI

## NATURAL TRIGONOMETRIC FUNCTIONS

**Table 1**    Natural Trigonometric Functions

| Angle | | | | | Angle | | | | |
|---|---|---|---|---|---|---|---|---|---|
| De-gree | Ra-dian | Sine | Co-sine | Tan-gent | De-gree | Ra-dian | Sine | Co-sine | Tan-gent |
| 0° | 0.000 | 0.000 | 1.000 | 0.000 | | | | | |
| 1° | 0.017 | 0.017 | 1.000 | 0.017 | 46° | 0.803 | 0.719 | 0.695 | 1.036 |
| 2° | 0.035 | 0.035 | 0.999 | 0.035 | 47° | 0.820 | 0.731 | 0.682 | 1.072 |
| 3° | 0.052 | 0.052 | 0.999 | 0.052 | 48° | 0.838 | 0.743 | 0.669 | 1.111 |
| 4° | 0.070 | 0.070 | 0.998 | 0.070 | 49° | 0.855 | 0.755 | 0.656 | 1.150 |
| 5° | 0.087 | 0.087 | 0.996 | 0.087 | 50° | 0.873 | 0.766 | 0.643 | 1.192 |
| 6° | 0.105 | 0.105 | 0.995 | 0.105 | 51° | 0.890 | 0.777 | 0.629 | 1.235 |
| 7° | 0.122 | 0.122 | 0.993 | 0.123 | 52° | 0.908 | 0.788 | 0.616 | 1.280 |
| 8° | 0.140 | 0.139 | 0.990 | 0.141 | 53° | 0.925 | 0.799 | 0.602 | 1.327 |
| 9° | 0.157 | 0.156 | 0.988 | 0.158 | 54° | 0.942 | 0.809 | 0.588 | 1.376 |
| 10° | 0.175 | 0.174 | 0.985 | 0.176 | 55° | 0.960 | 0.819 | 0.574 | 1.428 |
| 11° | 0.192 | 0.191 | 0.982 | 0.194 | 56° | 0.977 | 0.829 | 0.559 | 1.483 |
| 12° | 0.209 | 0.208 | 0.978 | 0.213 | 57° | 0.995 | 0.839 | 0.545 | 1.540 |
| 13° | 0.227 | 0.225 | 0.974 | 0.231 | 58° | 1.012 | 0.848 | 0.530 | 1.600 |
| 14° | 0.244 | 0.242 | 0.970 | 0.249 | 59° | 1.030 | 0.857 | 0.515 | 1.664 |
| 15° | 0.262 | 0.259 | 0.966 | 0.268 | 60° | 1.047 | 0.866 | 0.500 | 1.732 |
| 16° | 0.279 | 0.276 | 0.961 | 0.287 | 61° | 1.065 | 0.875 | 0.485 | 1.804 |
| 17° | 0.297 | 0.292 | 0.956 | 0.306 | 62° | 1.082 | 0.883 | 0.469 | 1.881 |
| 18° | 0.314 | 0.309 | 0.951 | 0.325 | 63° | 1.100 | 0.891 | 0.454 | 1.963 |
| 19° | 0.332 | 0.326 | 0.946 | 0.344 | 64° | 1.117 | 0.899 | 0.438 | 2.050 |
| 20° | 0.349 | 0.342 | 0.940 | 0.364 | 65° | 1.134 | 0.906 | 0.423 | 2.145 |
| 21° | 0.367 | 0.358 | 0.934 | 0.384 | 66° | 1.152 | 0.914 | 0.407 | 2.246 |
| 22° | 0.384 | 0.375 | 0.927 | 0.404 | 67° | 1.169 | 0.921 | 0.391 | 2.356 |
| 23° | 0.401 | 0.391 | 0.921 | 0.424 | 68° | 1.187 | 0.927 | 0.375 | 2.475 |
| 24° | 0.419 | 0.407 | 0.914 | 0.445 | 69° | 1.204 | 0.934 | 0.358 | 2.605 |
| 25° | 0.436 | 0.423 | 0.906 | 0.466 | 70° | 1.222 | 0.940 | 0.342 | 2.748 |
| 26° | 0.454 | 0.438 | 0.899 | 0.488 | 71° | 1.239 | 0.946 | 0.326 | 2.904 |
| 27° | 0.471 | 0.454 | 0.891 | 0.510 | 72° | 1.257 | 0.951 | 0.309 | 3.078 |
| 28° | 0.489 | 0.469 | 0.883 | 0.532 | 73° | 1.274 | 0.956 | 0.292 | 3.271 |
| 29° | 0.506 | 0.485 | 0.875 | 0.554 | 74° | 1.292 | 0.961 | 0.276 | 3.487 |
| 30° | 0.524 | 0.500 | 0.866 | 0.577 | 75° | 1.309 | 0.966 | 0.259 | 3.732 |
| 31° | 0.541 | 0.515 | 0.857 | 0.601 | 76° | 1.326 | 0.970 | 0.242 | 4.011 |
| 32° | 0.559 | 0.530 | 0.848 | 0.625 | 77° | 1.344 | 0.974 | 0.225 | 4.332 |
| 33° | 0.576 | 0.545 | 0.839 | 0.649 | 78° | 1.361 | 0.978 | 0.208 | 4.705 |
| 34° | 0.593 | 0.559 | 0.829 | 0.675 | 79° | 1.379 | 0.982 | 0.191 | 5.145 |
| 35° | 0.611 | 0.574 | 0.819 | 0.700 | 80° | 1.396 | 0.985 | 0.174 | 5.671 |
| 36° | 0.628 | 0.588 | 0.809 | 0.727 | 81° | 1.414 | 0.988 | 0.156 | 6.314 |
| 37° | 0.646 | 0.602 | 0.799 | 0.754 | 82° | 1.431 | 0.990 | 0.139 | 7.115 |
| 38° | 0.663 | 0.616 | 0.788 | 0.781 | 83° | 1.449 | 0.993 | 0.122 | 8.144 |
| 39° | 0.681 | 0.629 | 0.777 | 0.810 | 84° | 1.466 | 0.995 | 0.105 | 9.514 |
| 40° | 0.698 | 0.643 | 0.766 | 0.839 | 85° | 1.484 | 0.996 | 0.087 | 11.43 |
| 41° | 0.716 | 0.656 | 0.755 | 0.869 | 86° | 1.501 | 0.998 | 0.070 | 14.30 |
| 42° | 0.733 | 0.669 | 0.743 | 0.900 | 87° | 1.518 | 0.999 | 0.052 | 19.08 |
| 43° | 0.750 | 0.682 | 0.731 | 0.933 | 88° | 1.536 | 0.999 | 0.035 | 28.64 |
| 44° | 0.768 | 0.695 | 0.719 | 0.966 | 89° | 1.553 | 1.000 | 0.017 | 57.29 |
| 45° | 0.785 | 0.707 | 0.707 | 1.000 | 90° | 1.571 | 1.000 | 0.000 | |

# Appendix VII

# COMMON LOGARITHMS

| N | 0 | 1 | 2 | 3 | 4 | 5 | 6 | 7 | 8 | 9 |
|---|---|---|---|---|---|---|---|---|---|---|
| 10 | 0000 | 0043 | 0086 | 0128 | 0170 | 0212 | 0253 | 0294 | 0334 | 0374 |
| 11 | 0414 | 0453 | 0492 | 0531 | 0569 | 0607 | 0645 | 0682 | 0719 | 0755 |
| 12 | 0792 | 0828 | 0864 | 0899 | 0934 | 0969 | 1004 | 1038 | 1072 | 1106 |
| 13 | 1139 | 1173 | 1206 | 1239 | 1271 | 1303 | 1335 | 1367 | 1399 | 1430 |
| 14 | 1461 | 1492 | 1523 | 1553 | 1584 | 1614 | 1644 | 1673 | 1703 | 1732 |
| 15 | 1761 | 1790 | 1818 | 1847 | 1875 | 1903 | 1931 | 1959 | 1987 | 2014 |
| 16 | 2041 | 2068 | 2095 | 2122 | 2148 | 2175 | 2201 | 2227 | 2253 | 2279 |
| 17 | 2304 | 2330 | 2355 | 2380 | 2405 | 2430 | 2455 | 2480 | 2504 | 2529 |
| 18 | 2553 | 2577 | 2601 | 2625 | 2648 | 2672 | 2695 | 2718 | 2742 | 2765 |
| 19 | 2788 | 2810 | 2833 | 2856 | 2878 | 2900 | 2923 | 2945 | 2967 | 2989 |
| 20 | 3010 | 3032 | 3054 | 3075 | 3096 | 3118 | 3139 | 3160 | 3181 | 3201 |
| 21 | 3222 | 3243 | 3263 | 3284 | 3304 | 3324 | 3345 | 3365 | 3385 | 3404 |
| 22 | 3424 | 3444 | 3464 | 3483 | 3502 | 3522 | 3541 | 3560 | 3579 | 3598 |
| 23 | 3617 | 3636 | 3655 | 3674 | 3692 | 3711 | 3729 | 3747 | 3766 | 3784 |
| 24 | 3802 | 3820 | 3838 | 3856 | 3874 | 3892 | 3909 | 3927 | 3945 | 3962 |
| 25 | 3979 | 3997 | 4014 | 4031 | 4048 | 4065 | 4082 | 4099 | 4116 | 4133 |
| 26 | 4150 | 4166 | 4183 | 4200 | 4216 | 4232 | 4249 | 4265 | 4281 | 4298 |
| 27 | 4314 | 4330 | 4346 | 4362 | 4378 | 4393 | 4409 | 4425 | 4440 | 4456 |
| 28 | 4472 | 4487 | 4502 | 4518 | 4533 | 4548 | 4564 | 4579 | 4594 | 4609 |
| 29 | 4624 | 4639 | 4654 | 4669 | 4683 | 4698 | 4713 | 4728 | 4742 | 4757 |
| 30 | 4771 | 4786 | 4800 | 4814 | 4829 | 4843 | 4857 | 4871 | 4886 | 4900 |
| 31 | 4914 | 4928 | 4942 | 4955 | 4969 | 4983 | 4997 | 5011 | 5024 | 5038 |
| 32 | 5051 | 5065 | 5079 | 5092 | 5105 | 5119 | 5132 | 5145 | 5159 | 5172 |
| 33 | 5185 | 5198 | 5211 | 5224 | 5237 | 5250 | 5263 | 5276 | 5289 | 5302 |
| 34 | 5315 | 5328 | 5340 | 5353 | 5366 | 5378 | 5391 | 5403 | 5416 | 5428 |
| 35 | 5441 | 5453 | 5465 | 5478 | 5490 | 5502 | 5514 | 5527 | 5539 | 5551 |
| 36 | 5563 | 5575 | 5587 | 5599 | 5611 | 5623 | 5635 | 5647 | 5658 | 5670 |
| 37 | 5682 | 5694 | 5705 | 5717 | 5729 | 5740 | 5752 | 5763 | 5775 | 5786 |
| 38 | 5798 | 5809 | 5821 | 5832 | 5843 | 5855 | 5866 | 5877 | 5888 | 5899 |
| 39 | 5911 | 5922 | 5933 | 5944 | 5955 | 5966 | 5977 | 5988 | 5999 | 6010 |
| 40 | 6021 | 6031 | 6042 | 6053 | 6064 | 6075 | 6085 | 6096 | 6107 | 6117 |
| 41 | 6128 | 6138 | 6149 | 6160 | 6170 | 6180 | 6191 | 6201 | 6212 | 6222 |
| 42 | 6232 | 6243 | 6253 | 6263 | 6274 | 6284 | 6294 | 6304 | 6314 | 6325 |
| 43 | 6335 | 6345 | 6355 | 6365 | 6375 | 6385 | 6395 | 6405 | 6415 | 6425 |
| 44 | 6435 | 6444 | 6454 | 6464 | 6474 | 6484 | 6493 | 6503 | 6513 | 6522 |
| 45 | 6532 | 6542 | 6551 | 6561 | 6571 | 6580 | 6590 | 6599 | 6609 | 6618 |
| 46 | 6628 | 6637 | 6646 | 6656 | 6665 | 6675 | 6684 | 6693 | 6702 | 6712 |
| 47 | 6721 | 6730 | 6739 | 6749 | 6758 | 6767 | 6776 | 6785 | 6794 | 6803 |
| 48 | 6812 | 6821 | 6830 | 6839 | 6848 | 6857 | 6866 | 6875 | 6884 | 6893 |
| 49 | 6902 | 6911 | 6920 | 6928 | 6937 | 6946 | 6955 | 6964 | 6972 | 6981 |
| 50 | 6990 | 6998 | 7007 | 7016 | 7024 | 7033 | 7042 | 7050 | 7059 | 7067 |
| 51 | 7076 | 7084 | 7093 | 7101 | 7110 | 7118 | 7126 | 7135 | 7143 | 7152 |
| 52 | 7160 | 7168 | 7177 | 7185 | 7193 | 7202 | 7210 | 7218 | 7226 | 7235 |
| 53 | 7243 | 7251 | 7259 | 7267 | 7275 | 7284 | 7292 | 7300 | 7308 | 7316 |
| 54 | 7324 | 7332 | 7340 | 7348 | 7356 | 7364 | 7372 | 7380 | 7388 | 7396 |

| N | 0 | 1 | 2 | 3 | 4 | 5 | 6 | 7 | 8 | 9 |
|---|---|---|---|---|---|---|---|---|---|---|
| 55 | 7404 | 7412 | 7419 | 7427 | 7435 | 7443 | 7451 | 7459 | 7466 | 7474 |
| 56 | 7482 | 7490 | 7497 | 7505 | 7513 | 7520 | 7528 | 7536 | 7543 | 7551 |
| 57 | 7559 | 7566 | 7574 | 7582 | 7589 | 7597 | 7604 | 7612 | 7619 | 7627 |
| 58 | 7634 | 7642 | 7649 | 7657 | 7664 | 7672 | 7679 | 7686 | 7694 | 7701 |
| 59 | 7709 | 7716 | 7723 | 7731 | 7738 | 7745 | 7752 | 7760 | 7767 | 7774 |
| 60 | 7782 | 7789 | 7796 | 7803 | 7810 | 7818 | 7825 | 7832 | 7839 | 7846 |
| 61 | 7853 | 7860 | 7868 | 7875 | 7882 | 7889 | 7896 | 7903 | 7910 | 7917 |
| 62 | 7924 | 7931 | 7938 | 7945 | 7952 | 7959 | 7966 | 7973 | 7980 | 7987 |
| 63 | 7993 | 8000 | 8007 | 8014 | 8021 | 8028 | 8035 | 8041 | 8048 | 8055 |
| 64 | 8062 | 8069 | 8075 | 8082 | 8089 | 8096 | 8102 | 8109 | 8116 | 8122 |
| 65 | 8129 | 8136 | 8142 | 8149 | 8156 | 8162 | 8169 | 8176 | 8182 | 8189 |
| 66 | 8195 | 8202 | 8209 | 8215 | 8222 | 8228 | 8235 | 8241 | 8248 | 8254 |
| 67 | 8261 | 8267 | 8274 | 8280 | 8287 | 8293 | 8299 | 8306 | 8312 | 8319 |
| 68 | 8325 | 8331 | 8338 | 8344 | 8351 | 8357 | 8363 | 8370 | 8376 | 8382 |
| 69 | 8388 | 8395 | 8401 | 8407 | 8414 | 8420 | 8426 | 8432 | 8439 | 8445 |
| 70 | 8451 | 8457 | 8463 | 8470 | 8476 | 8482 | 8488 | 8494 | 8500 | 8506 |
| 71 | 8513 | 8519 | 8525 | 8531 | 8537 | 8543 | 8549 | 8555 | 8561 | 8567 |
| 72 | 8573 | 8579 | 8585 | 8591 | 8597 | 8603 | 8609 | 8615 | 8621 | 8627 |
| 73 | 8633 | 8639 | 8645 | 8651 | 8657 | 8663 | 8669 | 8675 | 8681 | 8686 |
| 74 | 8692 | 8698 | 8704 | 8710 | 8716 | 8722 | 8727 | 8733 | 8739 | 8745 |
| 75 | 8751 | 8756 | 8762 | 8768 | 8774 | 8779 | 8785 | 8791 | 8797 | 8802 |
| 76 | 8808 | 8814 | 8820 | 8825 | 8831 | 8837 | 8842 | 8848 | 8854 | 8859 |
| 77 | 8865 | 8871 | 8876 | 8882 | 8887 | 8893 | 8899 | 8904 | 8910 | 8915 |
| 78 | 8921 | 8927 | 8932 | 8938 | 8943 | 8949 | 8954 | 8960 | 8965 | 8971 |
| 79 | 8976 | 8982 | 8987 | 8993 | 8998 | 9004 | 9009 | 9015 | 9020 | 9025 |
| 80 | 9031 | 9036 | 9042 | 9047 | 9053 | 9058 | 9063 | 9069 | 9074 | 9079 |
| 81 | 9085 | 9090 | 9096 | 9101 | 9106 | 9112 | 9117 | 9122 | 9128 | 9133 |
| 82 | 9138 | 9143 | 9149 | 9154 | 9159 | 9165 | 9170 | 9175 | 9180 | 9186 |
| 83 | 9191 | 9196 | 9201 | 9206 | 9212 | 9217 | 9222 | 9227 | 9232 | 9238 |
| 84 | 9243 | 9248 | 9253 | 9258 | 9263 | 9269 | 9274 | 9279 | 9284 | 9289 |
| 85 | 9294 | 9299 | 9304 | 9309 | 9315 | 9320 | 9325 | 9330 | 9335 | 9340 |
| 86 | 9345 | 9350 | 9355 | 9360 | 9365 | 9370 | 9375 | 9380 | 9385 | 9390 |
| 87 | 9395 | 9400 | 9405 | 9410 | 9415 | 9420 | 9425 | 9430 | 9435 | 9440 |
| 88 | 9445 | 9450 | 9455 | 9460 | 9465 | 9469 | 9474 | 9479 | 9484 | 9489 |
| 89 | 9494 | 9499 | 9504 | 9509 | 9513 | 9518 | 9523 | 9528 | 9533 | 9538 |
| 90 | 9542 | 9547 | 9552 | 9557 | 9652 | 9566 | 9571 | 9576 | 9581 | 9586 |
| 91 | 9590 | 9595 | 9600 | 9605 | 9609 | 9614 | 9619 | 9624 | 9628 | 9633 |
| 92 | 9638 | 9643 | 9647 | 9652 | 9657 | 9661 | 9666 | 9671 | 9675 | 9680 |
| 93 | 9685 | 9689 | 9694 | 9699 | 9703 | 9708 | 9713 | 9717 | 9722 | 9727 |
| 94 | 9731 | 9736 | 9741 | 9745 | 9750 | 9754 | 9759 | 9763 | 9768 | 9773 |
| 95 | 9777 | 9782 | 9786 | 9791 | 9795 | 9800 | 9805 | 9809 | 9814 | 9818 |
| 96 | 9823 | 9827 | 9832 | 9836 | 9841 | 9845 | 9850 | 9854 | 9859 | 9863 |
| 97 | 9868 | 9872 | 9877 | 9881 | 9886 | 9890 | 9894 | 9899 | 9903 | 9908 |
| 98 | 9912 | 9917 | 9921 | 9926 | 9930 | 9934 | 9939 | 9943 | 9948 | 9952 |
| 99 | 9956 | 9961 | 9965 | 9969 | 9974 | 9978 | 9983 | 9987 | 9991 | 9996 |

# Appendix VIII

# SYSTEMS OF UNITS

The system of units we have been using refers to the meter, the kilogram, and the second, as the three fundamental standards of measurement. Therefore it is called the MKS system. But the laws of nature do not, of course, depend on the magnitude of the arbitrarily chosen standards. It is evident that it would make no fundamental difference if one used (as is sometimes done in the text of this book) the centimeter instead of the meter, or the gram instead of the kilogram; those choices correspond to the *cgs* (centimeter-gram-second) system found in a number of books. The whole subject of units is of interest only to a few specialists, though it sometimes seems to loom large to introductory students. Do not feel that the following comments, intended for the curious, are of deep significance, or have to be memorized!

We might for example have referred throughout the book to the units used for daily transactions in the U.S. and Great Britain, i.e., the *foot* instead of the meter, the *pound avoirdupois* (lb) instead of the kilogram. The latter corresponds to the mass of 27.692 in$^3$ of water at 4°C and ordinary pressure— a quantity of matter having an inertia about 0.4536 times that of an object of 1-kg mass. As in the case of the kilogram, the pound is expressed by the equivalent inertia of a cylinder of precious metal, this one being kept by the Bureau of Standards for

the United Kingdom in London.* The relation 1 lb = 0.4536 kg or 1 kg ≅ 2.2 lb is a choice that had no deep profound reason, and is historically rather accidental. At any rate, together with the equivalence between 1 ft and 0.3048 m, it comprises the relation between the MKS system of units and what is called the British Absolute System, or foot-pound-second (fps) system. In a more general sense, the one is called a metric system, the other a British system.

Certainly, spring balances might be calibrated by accelerating a 1-lb standard mass, so that $F_{net} = m$ (in lb) $\times a$ (in ft/sec$^2$). The unit of force so obtained, and used in connection with the fps system, is the lb-ft/sec$^2$, called the *poundal* (abbreviated pdl). It is quite analogous to the name *newton* given to 1 kg-m/sec$^2$, the unit of force used in the MKS system, or to the name *dyne* given to 1 g-cm/sec$^2$, the unit of force used in the cgs system. As you can confirm by direct calculation, 1 pdl =

---

* In the U.S., the relation between the pound and the gram (like that between the foot and the meter) is fixed not by a comparison between two metal pieces at the Bureau of Standards at Washington, but by act of Congress. However, there is no conflict here, for the legal decision was designed to agree with the results of direct comparison of standard masses within the limits of ordinary measurement.

0.1383 newtons $= 1.383 \times 10^4$ dynes, or 1 newton $=$ 7.23 pdl (1 newton $= 10^5$ dynes).

The introduction of electrical quantities further complicates the situation. For example, the MKS and cgs systems are rather similar as long as one considers only combinations of mass, length, and time, since the basic units are related to each other by powers of 10. (See the conversion factors given in Appendix II.) But it used to be customary to associate the cgs system with so-called *electrostatic units* (esu) of charge. In this system, Coulomb's law was written in the form

$$F = \frac{q_1 q_2}{r^2},$$

with the constant $k_e$ in Eq. (24.1) being set equal to 1. The unit of electric charge (1 esu, also called 1 statcoulomb) was defined as the quantity of a point charge which, 1 cm away from an equally large point charge, exerts on it 1 dyne of force. One could also define electric current in *electromagnetic units* (emu), using the *abampere* as the amount of current which, when flowing in two parallel straight wires at a distance of 1 cm, gives rise to a force of 2 dynes on each cm length of one wire. (Compare the definition of the ampere given in Section 25.3.) The *abcoulomb* was then defined as the amount of charge that passes through a wire when a current of 1 abampere is flowing for 1 second. It can then be shown that 1 abcoulomb $= c$ statcoulomb, where $c$ is a conversion factor numerically equal to the speed of light expressed in cm/sec; furthermore one can convert to the MKS system by using the relation: 1 coulomb $= \frac{1}{10}$ abcoulomb.

Although the cgs/esu/emu system does have some advantages, the majority of physicists have decided that the MKS system (or MKSA, when the ampere is added as the fourth fundamental unit) is preferable, and this system will probably be used in nearly all modern textbooks.

The choice of *primary standards* is still under discussion in the scientific community, and several of the definitions given in Appendix I are based on fairly recent but tentative decisions voted at international conferences. The general tendency has been to abandon the use of macroscopic physical objects kept in a particular place (such as the "standard meter") in favor of atomic properties that can easily be reproduced with great accuracy anywhere in the universe. Thus one no longer has to rely on a comparison with the standard objects kept in France to determine the magnitude of the meter or the second; one now measures the wavelength or frequency of a specified electronic transition of a particular atom ($Kr^{86}$ or $Cs^{133}$). Similarly, a Martian scientist could adopt the Kelvin absolute temperature scale without having to know the melting and boiling points of water at the pressure corresponding to the earth's atmosphere; he would only have to locate the triple point of water in terms of his own common units of temperature and pressure. (See the last paragraph of Section 19.3.) And he could use our scale of atomic masses without knowing the precise isotopic composition of terrestrial oxygen, as long as he had available some pure $C^{12}$. Only one of the fundamental units, the kilogram, has so far not been given a definition in terms of universal atomic properties.

# Appendix IX

# VECTOR ALGEBRA

We noted in Section 10.1 that motion with constant speed around a circle implies that the velocity vector is continually changing in direction though not in magnitude. A seemingly paradoxical situation arises at this point and demands a brief mathematical digression: In order to change the direction of a vector quantity at constant magnitude, one must, in general, add to it another vector with far from negligible magnitude.

Consider as an example the case of a disabled ship, adrift in wind and tide at some velocity $v_1$, perhaps 2 m/sec due North. Now a rescue boat throws a line and pulls to give the ship a new velocity $v_2$, again 2 m/sec, but now due West. What is the velocity $v_3$ which the rescue vessel superposes on $v_1$ to modify it to $v_2$? Figure A.7 gives the solution in familiar graphic form: $v_2$ is to be the resultant of $v_1$ and $v_3$, but you notice that the case is a little different from the one in Chapter 8, Fig. 8.8, where we were given $v_3$ and $v_1$ (or $v_a$ and $v_b$) and asked to find $v_2$ (or $v$). Now the unknown is not the resultant, but one side of the vector parallelogram. Accordingly, we enter first $v_1$, then $v_2$—both are known in direction and magnitude—and then the connecting line from the head of the arrow representing $v_1$ to the head of $v_2$. Thus the direction and magnitude of the missing side are found, and by measurement from a scale drawing we should find that $v_3$ in this example is about 2.8 knots to SW.

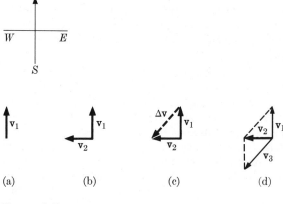

(a)  (b)  (c)  (d)

**Figure A.7**

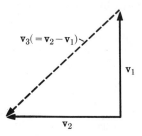

**Fig. A.8** Vector subtraction: $v_3 = v_2 - v_1$.

550

Comparison of parts (c) and (d) of Fig. A.7 indicates that the last stage (d) is really superfluous; the connecting line in (c) adequately represented $v_3$ even before the parallelogram was completed (Fig. A.8). From now on we may find the difference of any two vector quantities $A$ and $B$ by this abbreviated graphical procedure.

Furthermore, we shall occasionally follow the convention of using boldface italic letters to refer to vector quantities ($F$, $v$, $a$). The addition of two vector quantities $P$ and $Q$ may then be written $P + Q = R$. On reading such equations, we immediately realize that, in general, $R$ is not obtained simply by adding the magnitudes of $P$ and $Q$, but that one must find $R$, in direction as well as magnitude, by means of a graphical or a trigonometric solution (see Fig. A.9). In the same manner, the equation $B - A = C$ refers to a vector subtraction such as is graphically and trigonometrically* presented in Fig. A.10.

*Example of vector addition.* Two men are pulling on ropes attached to a heavy object; the first pulls with force $F_1 = 90$ newtons at 40° East of North, the second with force $F_2 = 45$ newtons in a direction 30° West of North. Find the resultant of these two forces, i.e., the vector sum, the single equivalent force $R$ which might just as well be applied to this object instead of $F_1$ and $F_2$.†

*Solution:* First we make a sketch in terms of the chosen arrow symbolism of the relative magnitude and placement of vectors $F_1$ and $F_2$ [Fig. A.11(a)]. Then we complete the parallelogram; the diagonal corresponds to $R$, the vectorial sum of $F_1$ and $F_2$. The right half of the parallelogram is a triangle

---

* See Appendix V for a summary of the elements of trigonometry.

† Our previous study of adding velocities to obtain the true velocity of a projectile and of adding forces to find the net force has illustrated the general truth that several superposed vector quantities provide just the same effect as a *single* vector quantity equal to the vectorial sum of the superposed quantities.

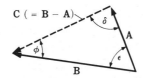

**Fig. A.9** Vector addition graphically and trigonometrically.

**Fig. A.10** Vector subtraction, graphically and trigonometrically.

**Fig. A.11** Vector addition.

**Fig. A.12**  Vector addition.

composed of $R$, $F_1$, and a side corresponding to $F_2$; the angle opposite $R$, indicated as $\gamma$, is directly given by $\gamma = 180° - (\theta_1 + \theta_2)$, in this case $180° - 70°$, or $110°$.* Now we find the magnitude of $R$ by

$$R = \sqrt{F_1^2 + F_2^2 - 2F_1F_2 \cos \gamma}$$

$$= \sqrt{8100 + 2025 - 2 \times 45 \times 90 \times (-0.34)}$$

$$= 114 \text{ newtons.}$$

The direction of $R$ may be fixed in several equivalent ways, but here it would seem most consistent to determine $\alpha$, then $(\theta_2 - \alpha)$, the angle of $R$ with respect to the North direction:

$$\sin \alpha = F_2 \frac{\sin \gamma}{R} = 45 \times \frac{0.94}{114} = 0.37; \qquad \alpha \doteq 22°$$

and

$$(\theta_2 - \alpha) = 18°.$$

The answer to the original question, then, is that the resultant force is 114 newtons along a direction 18° East of North. And incidentally, the same procedure is applicable for calculating the sum of *any* two vectors, no matter at what angle to each other, no matter what their physical nature. Even the specific cases in which a right angle exists between the two vectors are, in substance, the same

---

* Recall that for angles between 90° and 180° the following relations apply: $\sin x = \sin (180° - x)$; $\cos x = -\cos (180° - x)$. Therefore $\sin 110° = \sin 70° = 0.94$; $\cos 110° = -\cos 70° = -0.34$.

problem simplified by the convenient facts that there the factor corresponding to $2F_1F_2 \cos \gamma$ under the square root is zero (since $\cos 90° = 0$), and that $\sin \alpha = F_2 \sin \gamma/R = F_2/R$ (since $\sin 90° = 1$).

Just as examination of Fig. A.7 leads us to a simplification of the parallelogram method of vector subtraction, namely to Fig. A.8, so will a moment's thought prove that one may, if one so desires, similarly abbreviate the process of vector addition. Both in Fig. 5.5 and Fig. A.11(b), only the right half of the parallelogram is really needed; the left half is a congruent triangle, and therefore superfluous for our construction. Therefore, to add two vector quantities, we draw their arrowlike representation and join them head to tail [Fig. A.12(a), (b)], the resultant $R$ then being represented by the arrow running from the tail of the first drawn to the head of the last drawn arrow. When there are more than two vectors to be added, we may do it either by adding two of them, then adding a third one to their resultant, then a fourth to *that* resultant, and so forth; or more conveniently, though less accurately, we may do it graphically by adding the corresponding arrows head to tail, as in Fig. A.12(c). This is generally referred to as the *polygon method* of vector addition; it will soon prove to be an important tool in our argument.

---

**Problem A.1.** Note, describe, and justify carefully the difference between abbreviated vector subtraction (Fig. A.8) and vector addition [Fig. A.12(a)].

**Problem A.2.** (a) Prove, by graphical construction, that the polygon method of vector addition must

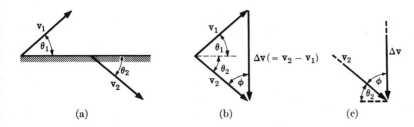

**Fig. A.13**  $v = v_2 - v_1$.

(a)          (b)          (c)

give the same result as adding several vectors one at a time. (b) Prove by graphical construction that in Fig. A.12(c) the same resultant is obtained regardless of the *order* in which the individual arrows are joined together.

**Problem A.3.** (a) $F_1 = 5$ newtons due West, $F_2 = 8$ newtons at 30° S of E. Find the resultant force, in direction and magnitude, by the parallelogram method (graphically), by trigonometric calculation, and by the abbreviated graphical method (polygon method, which here becomes simply a triangle construction). (b) To the two forces in part (a), a third is added: $F_3 = 4$ newtons 10° W of N. Find the resultant of all three forces by the graphical polygon method (in two different sequences of joining the arrows), and then by adding $F_3$ to the resultant of $F_1$ and $F_2$ in a trigonometric calculation.

---

*Example of vector subtraction.* A bullet is shot in a direction 40° above the horizontal plane and with an initial velocity $v_1 = 3.5 \times 10^4$ cm/sec. After 46 sec the bullet returns to the same plane with a velocity $v_2$ having the same magnitude as $v_1$, namely, $3.5 \times 10^4$ cm/sec, but directed 40° below the horizontal. What was the change of velocity $\Delta v$ during that time interval (magnitude and direction)?

*Solution:* Again, first make a sketch of the physical situation [Fig. A.13(a)], and then a sketch of the scheme previously discussed for the subtraction of vector quantities [Fig. A.13(b)]. Here the angle

opposite the side representing $\Delta v$ is $(\theta_1 + \theta_2)$, or 80°; therefore

$$\Delta v = \sqrt{v_1^2 + v_2^2 - 2v_1v_2 \cos(\theta_1 + \theta_2)}$$

$$= \sqrt{\begin{array}{l}(12.25 \times 10^8 + 12.25 \times 10^8 \\ - 2 \times 3.5 \times 10^4 \times 3.5 \times 10^4 \times 0.174)\end{array}}$$

$$= 4.5 \times 10^4 \text{ cm/sec.}$$

The direction of $\Delta v$ is "obviously" vertically downward, but it is important that this be confirmed rigorously. In Fig. A.13(c), angle $(\theta_2 + \phi)$ gives the inclination of $\Delta v$ with the vertical; $\theta_2$ is known, $\phi$ can be easily computed from

$$\sin \phi = v_1 \frac{\sin(\theta_1 + \theta_2)}{\Delta v},$$

and the sum, $\theta_2 + \phi$, is thereby confirmed to be 90°; that is, $\Delta v$ is perpendicular to the horizontal plane.

This problem, which incidentally was designed to illustrate again the result of a change in direction of a vector even at constant magnitude, can be carried one important step further. Knowing now that during 46 sec the change in velocity was $4.5 \times 10^4$ cm/sec, we at once can find the average acceleration. By definition, $a = \Delta v/t$, therefore $a = (4.5 \times 10^4/46)$ cm/sec² = 980 cm/sec², in fact the expected acceleration of gravity. Furthermore, acceleration being a vector, it is necessary to specify its direction, which is that of $\Delta v$, namely, vertically downward.

**Problem A.4.** A sailboat cruises on a river which flows in a direction 15° North of East. If there were no wind, the river current would impart to the boat a speed of 5 mi/hr in that same direction. On the other hand, without current, the wind alone would move the boat at 9 mi/hr due Southeast. Find (a) the resultant of these two vectors, and (b) the difference between these two vectors. The result of (a) represents the total velocity of the boat referred to some fixed point on shore. What is the physical significance of the other result?

**Problem A.5.** Two forces, $F_1 = 10$ newtons, $F_2 = 6$ newtons, act on the same object at a 30° angle with each other. What is the resultant net force? What third force must be applied to establish equilibrium? (The latter force is called the *equilibrant*.)

# BIBLIOGRAPHY

*Note:* the following anthologies are frequently cited by short titles.

*Nature and Nature's Laws* = *Nature and Nature's Laws, Documents of the Scientific Revolution*, edited by Marie Boas Hall, New York: Harper & Row, 1970, paperback

*Origins and Growth of Physical Science* = *The Origins and Growth of Physical Science*, edited by D. L. Hurd and J. J. Kipling, Baltimore: Penguin Books, 1964, paperback

*Project Physics Reader* = *Readers*, volumes 1–6, component parts of *The Project Physics Course*, New York: Holt, Rinehart & Winston, 1970–71

*World of the Atom* = *The World of the Atom*, edited by H. A. Boorse and L. Motz, New York: Basic Books, 1966

The Bibliography is limited to works available in English; no attempt has been made to list all editions or translations, but in most cases a recent American publication has been cited.

Abell, G., *Exploration of the Universe*, New York: Holt, Rinehart, and Winston, second edition, 1969.

Abro, A. d', *The Rise of the New Physics*, New York: Dover Publications, 1951 (reprint of the 1939 edition)

Accademia del Cimento, "The preface to the reader," pages 219–222 in *Nature and Nature's Laws* (excerpted from *Essayes of Natural Experiments*, 1684, translation of *Saggi di Naturali Esperienze*, 1667)

Adams, F. D., *The Birth and Development of the Geological Sciences*, New York: Dover Publications, 1954 (reprint of the 1938 edition)

Adams, Henry, *The Degradation of the Democratic Dogma*, New York: Capricorn Books (Putnam's), 1958, paperback (reprint of the 1919 edition)

Aiton, E. J., "Kepler's second law of planetary motion," *Isis*, Vol. 60, pages 75–90 (1969)

Alexander, H. G., editor, *The Leibniz-Clarke correspondence, with extracts from Newton's Principia and Opticks*, New York: Philosophical Library, 1956

Ampère, A. M., "Actions between currents," pages 218–229 in *Origins and Growth of Physical Science*, Vol. 2 (translation of part of an article in *Annales de Chimie et de Physique*, 1820)

Anderson, D. L., *The Discovery of the Electron*, Princeton, N.J.: D. Van Nostrand, 1964; Momentum Book paperback

Andrade, E. N. da C., *Rutherford and the Nature of the Atom*, Garden City, N.Y.: Doubleday Anchor Books, 1964; paperback

Andrade, E. N. da C., *Sir Isaac Newton, his Life and Work*, Garden City, N.Y.: Doubleday Anchor Books, 1958; paperback

Aristotle, "On cosmology," pages 28–32 in *Origins and Growth of Physical Science*, Vol. 1 (excerpted from *On the Heavens*)

Aristotle, *On the Heavens*, Cambridge, Mass.: Harvard University Press, 1939 (translated from *De Caelo*)

Aristotle, "Natural and Forced Motions," pages 26–41 in *The Changeless Order*, edited by A. Koslow, New York: Braziller, 1967 (translated excerpts from *Physics*)

Aristotle, *Physics*, translated by H. G. Apostle, Bloomington, Indiana: Indiana University Press, 1969; paperback

Armitage, A., *Sun Stand Thou Still*, New York: Henry Schuman, 1947; reprinted as *The World of Copernicus*, Mentor Books paperback. Revised edition, *Copernicus, The Founder of Modern Astronomy*, New York: A. S. Barnes & Co., 1957; Perpetua edition, 1962

Arnheim, Rudolf, *Entropy and Art, an essay on order and disorder*, Berkeley: University of California Press, 1971

Arons, A. B., *Development of Concepts of Physics*, Reading, Mass.: Addison-Wesley, 1965

Avogadro, A., "Essay on a manner of determining the relative masses of the elementary molecules of bodies and the proportions in which they enter into these compounds," pages 65–80 in *Origins and Growth of Physical Science*, Vol. 2 (translated from *Journal de Physique*, 1811)

Bacon, Francis, "The Errors of Learning" and "The Classification of Learning," pages 99–109 in *Nature and Nature's Laws* (excerpted from *The Advancement of Learning*, 1605)

Bacon, Francis, "The nature of forms, and the form of heat," pages 291–293 in *Nature and Nature's Laws* (excerpted from Novum Organum, 1620)

Bacon, Francis, *The New Organon, and Related Writings*, edited by F. H. Anderson, Indianapolis: Bobbs-Merrill, 1960, paperback (translated from *Novum Organum*, 1620)

Baker, R. H., *Astronomy*, Princeton, N.J.: D. Van Nostrand, eighth edition, 1964

Ballentine, L. A., "The statistical interpretation of quantum mechanics," *Reviews of Modern Physics*, Vol. 42, pages 358–381 (1970)

Balmer, J. J., "Note on the spectral lines of hydrogen," pages 365–368 in *World of the Atom* (translated from *Annalen der Physik*, 1885)

Barnett, M. K., "Sadi Carnot and the Second Law of Thermodynamics," *Osiris*, Vol. 13, pages 327–357 (1958)

Becquerel, H., "The Radiation from Uranium," pages 364–367 in *The Origins and Growth of Physical Science*, volume 2 (translated from *Comptes Rendus hebdomadaires des Séances de l'Academie des Sciences*, Paris, 1896)

Behrens, C. E., "Atomic theory from 1904 to 1913," *American Journal of Physics*, Vol. 11, pages 60–66 (1943)

Bell, A. E., *Christian Huygens and the Development of Science in the 17th Century*, London: Arnold, 1947

Bell, E. T., *Men of Mathematics*, New York: Simon & Schuster, 1965 (reprint of the 1937 edition); paperback

Bernal, J. D., *Science in History*, Cambridge, Mass.: MIT Press, fourth edition 1970

Bernatowicz, A. J., "Dalton's rule of simplicity," *Journal of Chemical Education*, Vol. 47, pages 577–579 (1970)

Bernoulli, D., "On the properties and motions of elastic fluids, especially air," pages 57–65 in *Kinetic Theory*, Vol. 1, edited by S. G. Brush, New York: Pergamon Press, 1965 (translated from *Hydrodynamica*, 1738)

Berry, A., *A Short History of Astronomy, from earliest times through the nineteenth century*, New York: Dover Publications, 1961 (reprint of the 1898 edition)

Beveridge, W. I. B., *The Art of Scientific Investigation*, New York: W. W. Norton, 1950; paperback

Blackwell, R. J., "Descartes' laws of motion," *Isis*, Vol. 57, pages 220–234 (1966)

Blake, P. M., Ducasse, C. J., and Madden, E. H., *Theories of Scientific Method*, Seattle, Wash.: University of Washington Press, 1960; paperback

Blin-Stoyle, R. J., *et al.*, *Turning Points in Physics*, Amsterdam: North-Holland, 1959; Harper Torchbook paperback reprint

Boas, M., "Bacon and Gilbert," pages 458–459 in *Roots of Scientific Thought*, edited by P. Wiener and A. Noland, New York: Basic Books, 1957

Boas, M., "The establishment of the mechanical philosophy," *Osiris*, Vol. 10, pages 412–541 (1952)

Boas, M., *The Scientific Renaissance 1450–1630*, New York: Harper & Row, 1962; paperback

Bohm, D., *Causality and Chance in Modern Physics*, London: Routledge and Kegan Paul, 1957; Harper Torchbook reprint

Bohr, Niels, "Can quantum-mechanical description of physical reality be considered complete?" pages 130–142 in *Physical Reality*, edited by S. Toulmin, New York: Harper & Row, 1970 (reprinted from *Physical Review*, 1935)

Bohr, Niels, "Discussion with Einstein on epistemological problems in atomic physics," pages 199–241 in *Albert Einstein, Philosopher-Scientist*, edited by P. A. Schilpp, New York: Library of Living Philosophers, 1949

Bohr, Niels, "On the constitution of atoms and molecules," pages 132–159 in D. ter Haar, *The Old Quantum Theory*, New York: Pergamon Press, 1967 (reprinted from *Philosophical Magazine*, 1913); also pages 751–765 in *World of the Atom*

Boltzmann, Ludwig, *Lectures on Gas Theory*, Berkeley: University of California Press, 1964 (translated from *Vorlesungen über Gastheorie*, 1896–98)

Boltzmann, Ludwig, "On the relation of a general mechanical theorem to the second law of thermodynamics," pages 188–193 in *Kinetic Theory*, Vol. 2, edited by S. G. Brush, New York: Pergamon Press, 1966 (translated from *Sitzungsberichte der kaiserlichen Akademie der Wissenschaften*, Vienna, 1877)

Boorse, H. A., and Motz, L., editors, *The World of the Atom*, New York: Basic Books, 1966

Bork, Alfred M., "Randomness and the Twentieth Century," pages 167–187 in *Project Physics Reader*, Vol. 3 (reprinted from *The Antioch Review*, 1967)

Born, Max, *Atomic Physics*, New York: Hafner, 7th edition (1962) (translated from *Moderne Physik*; first German edition, 1933)

Born, Max, "Max Karl Ernst Ludwig Planck," pages 462–484 in *The World of the Atom* (excerpted from *Obituary Notices of Fellows of the Royal Society*, 1948)

Born, Max, "Quantum mechanics of collision processes," pages 206–225 in *Wave Mechanics*, edited by G. Ludwig, New York: Pergamon Press, 1968 (translated from *Zeitschrift für Physik*, 1926)

Born, Max, and Jordan, Pascual, "Quantum mechanics," pages 183–205 in *Wave Mechanics*, edited by G. Ludwig, New York: Pergamon Press, 1968 (translated from *Zeitschrift für Physik*, 1925)

Born, Max, "Wave corpuscles," pages 1086–1093 in *World of the Atom* (excerpted from *Atomic Physics*, 1962)

Boyer, C. B., "Aristotle's Physics," *Scientific American*, pages 48–51 (May 1950)

Boyle, Robert, "The air pump," pages 202–215 in *Nature and Nature's Laws* (excerpted from *New Experiments Physico-Mechanical*, 1660)

Boyle, Robert, "Of the excellency and grounds of the corpuscular or mechanical philosophy," pages 311–332 in *Nature and Nature's Laws* (excerpted from *The Excellency and Grounds of the Mechanical Hypothesis*, 1674)

Boyle, Robert, *The Sceptical Chymist*, New York: Dutton, 1964 (reprint of the 1661 edition)

Boyle, Robert, "The Spring of the Air," pages 43–51 in *Kinetic Theory*, Vol. 1, edited by S. G. Brush, New York: Pergamon Press, 1965 (excerpt from Boyle's *New Experiments Physico-Mechanical*, 1660)

Bridgman, P. W., *The Logic of Modern Physics*, New York: Macmillan, 1927, reprinted 1960, paperback

Bridgman, P. W., "On scientific method," pages 18–19 in *Project Physics Reader*, Vol. 1 (excerpted from *Reflections of a Physicist*, 1955)

Bridgman, P. W., *Reflections of a Physicist*, New York: Philosophical Library, 1950

Brody, B. and Capaldi, N., editors, *Science: Men, Methods, Goals*, New York: Benjamin, 1968; paperback

Broglie, Louis de, *The Current Interpretation of Wave Mechanics, a critical study*, New York: American Elsevier, 1964

Broglie, Louis de, "The undulatory aspects of the electron," pages 1048–1059 in *World of the Atom* (excerpted from his Nobel Prize Address, 1929)

Bromberg, J., "Maxwell's displacement current and his theory of light," *Archive for History of Exact Sciences*, Vol. 4, pages 218–234 (1967)

Bronowski, Jacob, "The Arrow of Time," pages 127–132 in *Project Physics Reader*, Vol. 3 (excerpted from *Insight*, 1964)

Brougham, Henry, review of the Bakerian Lecture on the theory of Light and Colours by Thomas Young, *Edinburgh Review*, Vol. 1, pages 450–456 (1803)

Brown, Robert, "A brief account of microscopical observations made in the months of June, July, and August 1827 on the particles contained in the pollen of plants; and on the general existence of active molecules in organic and inorganic bodies," pages 208–212 in *World of the Atom* (excerpted from *Philosophical Magazine*, 1828)

Brown, S. C., *Count Rumford: Physicist Extraordinary*, Garden City, N.Y.: Doubleday Anchor Books, 1962; paperback

Brown, S. C., *Benjamin Thompson—Count Rumford: Count Rumford on the Nature of Heat*, New York: Pergamon Press, 1967

Brown, T. M., "Resource Letter Energy Concepts from Gali... *Journal of Physics*, Vol. 33, pa...

Bruno, G., "On the infinite univers... 174–183 in *Theories of the Universe*, Munitz, New York: Free Press, 19... (excerpted from the translation in *Giordan...* D. W. Singer, 1950)

Brush, S. G., "The development of the kinetic theor... gases," *Annals of Science*, Vol. 13, pages 188–19..., 273–282 (1957), Vol. 14, pages 185–196, 243–255 (1958); *American Journal of Physics*, Vol. 29, pages 593–605 (1961), Vol. 30, pages 269–281 (1962)

Brush, S. G., "James Clerk Maxwell and the Kinetic Theory of Gases: A Review Based on Recent Historical Studies," *American Journal of Physics*, Vol. 39, pages 631–640 (1971)

Brush, S. G., *Kinetic Theory*, New York: Pergamon Press, 3 Vols., 1965–72

Brush, S. G., editor, *Resources for the History of Physics*, Hanover, N.H.: University Press of New England, 1972; paperback

Brush, S. G., "Thermodynamics and History," *The Graduate Journal*, Vol. 7, pages 477–565 (1967)

Brush, S. G., "The wave theory of heat: A forgotten stage in the transition from the caloric theory to thermodynamics," *British Journal for the History of Science*, Vol. 5, pages 145–167 (1970)

Buchdahl, G., *The Image of Newton and Locke in the Age of Reason*, New York and London: Sheed and Ward, 1961

Burke, J. G., "Descartes on the refraction and the velocity of light," *American Journal of Physics*, Vol. 34, pages 390–400 (1966)

Burstyn, H. L., "Galileo's attempt to prove that the earth moves," *Isis*, Vol. 53, pages 161–185 (1962)

Burtt, E. A., *The Metaphysical Foundations of Modern Physical Science*, New York: Humanities Press, 1952; Garden City, N.Y., Doubleday & Co., 1954 (paperback reprint of the second edition, 1932)

Butterfield, H., *The Origins of Modern Science*, New York: Free Press, revised edition 1965, paperback

Calder, N., *Violent Universe*, New York: Viking, 1970

Campbell, N. R., *Foundations of Science*, New York: Dover Publications, 1957 (reprint of *Physics, The Elements*, 1920)

Čapek, M., *Philosophical Impact of Contemporary Physics*, Princeton, N.J.: Van Nostrand, 1961

Cardwell, D. S. L., *From Watt to Clausius: The Rise of Thermodynamics in the Early Industrial Age*, Ithaca, N.Y.: Cornell University Press, 1971

Cardwell, D. S. L., editor, *John Dalton and the Progress of Science*, New York: Barnes & Noble, 1968

Carnap, R., *Philosophical Foundations of Physics*, New York: Basic Books, 1966

Carnot, Sadi, *Reflections on the Motive Power of Fire, etc.*, edited by E. Mendoza, New York: Dover Publications, 1960; paperback (translated from *Réflexions sur la Puissance Motrice du Feu*, 1824)

Cassebaum, H. and Kauffman, G. B., "The periodic system of the chemical elements: The search for its discoverer," *Isis*, Vol. 62, pages 314–327 (1971)

Chalmers, T. W., *Historic Researches*, London: Morgan Brothers, 1949

Clagett, M., *Greek Science in Antiquity*, New York: Abelard Schuman, 1955; Collier Books; paperback

Clagett, M., *The Science of Mechanics in the Middle Ages*, Madison, Wisconsin: University of Wisconsin Press, 1959

Clark, G. N., *Science and Social Welfare in the Age of Newton*, New York: Oxford University Press (second edition 1949, reprinted 1969)

Clausius, Rudolf, "The nature of the motion which we call heat," pages 111–134 in *Kinetic Theory*, Vol. 1, edited by S. G. Brush, New York: Pergamon Press, 1965 (translated from *Annalen der Physik*, 1857)

Cline, B., *Men Who Made a New Physics: Physicists and the Quantum Theory*, New York: Crowell, 1965; Signet Science Library paperback (originally published as *The Questioners*)

Cohen, E. R. and DuMond, J. W. M., "Our knowledge of the fundamental constants of physics and chemistry in 1965," *Reviews of Modern Physics*, Vol. 37, pages 537–594 (1965)

Cohen, I. B., *The Birth of a New Physics*, Garden City, N.Y.: Doubleday, Anchor Books, 1960; paperback

Cohen, I. B., "The first explanation of interference," *American Journal of Physics*, Vol. 8, pages 99–106 (1940)

Cohen, I. B., *Franklin and Newton*, Philadelphia: American Philosophical Society, 1956

Cohen, I. B., *An Introduction to Newton's Principia*, Cambridge, Mass.: Harvard University Press, 1971

Cohen, I. B., "Kepler's celestial music," pages 49–61 in *Project Physics Reader*, Vol. 2 (excerpted from *The Birth of a New Physics*, 1960)

Cohen, I. B., "Newton, Hooke, and 'Boyle's Law' (Discovered by Power and Towneley)," *Nature*, Vol. 204, page 618 (1964)

Cohen, I. B., "Roemer and the first determination of the velocity of light," *Isis*, Vol. 31, pages 327–379 (1940)

Cohen, M. R. and Drabkin, I.E., *A Source Book in Greek Science*, New York: McGraw-Hill, 1948; reprinted by Harvard University Press

Collingwood, R. G., *The Idea of Nature*, New York: Oxford University Press, 1945; paperback

Colodny, R. G., editor, *Beyond the Edge of Certainty*, Englewood Cliffs, N.J.: Prentice-Hall, 1965

Conant, J. B., *The Overthrow of the Phlogiston Theory*, pages 65–115 in *Harvard Case Histories in Experimental Science*, edited by J. B. Conant, Cambridge, Mass.: Harvard University Press, 1948

Conant, J. B., "Robert Boyle's experiments in pneumatics," pages 1–63 in *Harvard Case Histories in Experimental Science*, edited by J. B. Conant, Cambridge, Mass.: Harvard University Press, 1948. Also available as a separate pamphlet.

Conant, J. B., *Science and Common Sense*, New Haven: Yale University Press, 1951, reprinted 1961, paperback

Conant, J. B., *On Understanding Science*, New Haven: Yale University Press, 1947

Copernicus, Nicolaus, "Commentariolus," pages 55–90 in *Three Copernican Treatises*, translated with notes by E. Rosen, New York: Dover Publications, second edition, 1959

Copernicus, Nicolaus, *On the Revolutions of the Heavenly Spheres*, pages 501–838 in *Great Books of the Western World*, edited by R. M. Hutchins, Vol. 16, Chicago: Encyclopaedia Britannica, 1952 (translated from *De Revolutionibus Orbium Caelestium*, 1543)

Copernicus, Nicolaus, "On the revolutions of the heavenly spheres," pages 149–173 in *Theories of the Universe*, edited by M. K. Munitz, New York: Free Press, 1957, paperback (excerpted from a translation of the preface and Book I of *De Revolutionibus*, 1543); see also pages 99–111 in *Origins and Growth of Physical Science*, Vol. 1 (the excerpts given there are erroneously attributed to *Commentariolus*)

Cornford, F. M., *Principium Sapientiae*, London: Cambridge University Press, 1952

Coulomb, C. A., "Law of Electric Force," "Fundamental Law of Electricity," "Second Memoir on Electricity and Magnetism," and "Law of Magnetic Force," pages 195–209 in *Origins and Growth of Physical Science*, Vol. 2 (excerpts from *Memoires de l'Académie Royale des Sciences*, Paris, for 1785)

Cowling, T. G., *Molecules in Motion*, London: Hutchinson, 1950; Harper Torchbook paperback reprint

Crombie, A. C., "Descartes," *Scientific American*, pages 160–173 (October 1959)

Crombie, A. C., *Medieval and Early Modern Science*, Cambridge, Mass.: Harvard University Press, 1959, paperback reprint by Doubleday Anchor Books

Cropper, W. H., *The Quantum Physicists and an Introduction to their Physics*, New York: Oxford University Press, 1970

Curie, Pierre and Marie, "Polonium and Radium are Discovered," pages 370–382 in *Origins and Growth of the Physical Sciences*, Vol. 2

Dalton, John, *A New System of Chemical Philosophy*, London: Dawson 1950 (reprint of the 1808–1827 edition); New York: Philosophical Library, 1964 (first volume only); excerpts from Vol. 1 in pages 24–26, *Origins and Growth of Physical Science*, Vol. 2; also pages 333–338 in *Nature and Nature's Laws*

Dampier, W. C., *A History of Science*, New York: Cambridge University Press, 4th edition, 1948; paperback

Darwin, Charles Robert, *The Origin of Species by Means of Natural Selection of the Preservation of Favoured Races in the Struggle for Life*, Cambridge, Mass.: Harvard University Press, 1964 (reprint of the 1859 edition); several reprints of later editions are available

Descartes, René, "The author's letter to the translator, which may serve by way of preface," "The principles of material things," "The visible world," and "The earth," pages 110–121, 294–311 in *Nature and Nature's Laws* (translated from *Principes de la Philosophie*, 1647)

Descartes, René, "The Cartesian Laws of Motion," pages 264–270 in *Nature and Nature's Laws* (translated from *Principes de la Philosophie*, 1647)

Descartes, René, *Discourse on Method, Optics, Geometry and Meteorology*, Indianapolis: Bobbs-Merrill Co., 1965 (translated from French edition of 1637)

Descartes, René, *Principles of Philosophy*, in *The Philosophical Works of Descartes*, Cambridge, England: Cambridge University Press, 1911; paperback reprint (partial translation of *Principia Philosophiae*, 1644)

DeWitt, B. S., "Quantum mechanics and reality," *Physics Today*, September 1970, pages 30–35

Dibner, B., *Oersted and the Discovery of Electromagnetism*, Norwalk, Conn.: Burndy Library, 1961

Dibner, B., *Wilhelm Conrad Röntgen and the Discovery of X-Rays*, New York: Watts, 1968

Dicks, D. R., *Early Greek Astronomy to Aristotle*, Ithaca, N.Y.: Cornell University Press, 1970

Dijksterhuis, E. J., *The Mechanization of the World Picture*, New York: Oxford University Press, 1961; paperback (translated from the Dutch edition of 1950)

Dirac, P. A. M., "The evolution of the physicist's picture of nature," pages 131–140 in *Project Physics Reader*, Vol. 5 (reprinted from *Scientific American*, 1963)

Donne, John, "An Anatomy of the World, The First Anniversary," pages 186–197 in *Complete Poetry and Selected Prose*, New York: Modern Library, 1952

Drake, S., *Galileo Studies, Personality, Tradition, and Revolution*, Ann Arbor, Michigan: University of Michigan Press, 1970

Drake, S., "Galileo and the law of inertia," *American Journal of Physics*, Vol. 32, pages 601–608 (1964)

Drake, S., and Drabkin, I. E., *Mechanics in Sixteenth Century Italy: Selections from Tartaglia, Benedetti, Guido Ubaldo and Galileo*, Madison, Wisconsin: University of Wisconsin Press, 1969

Drake, S., "Uniform acceleration, space, and time," *British Journal for the History of Science*, Vol. 5, pages 21–43 (1970)

Dreyer, J. L. E., *A History of Astronomy*, New York: Dover, 1953 (reprint of *A History of the Planetary Systems from Thales to Kepler*, 1905); see also chapter on "Medieval Cosmology" reprinted in *Theories of the Universe*, edited by M. K. Munitz, New York: Free Press, 1957

Dugas, R., *A History of Mechanics*, New York: Central Book Company, 1955

Dugas, R., *Mechanics in the Seventeenth Century*, New York: Central Book Co., 1958 (translation of *Le Mécanique au 17ᵉ Siècle*, 1954)

Duhem, Pierre, *The Aim and Structure of Physical Theory*, translated from the second French edition of 1914 by P. P. Wiener, Princeton, N.J.: Princeton University Press, 1954; Atheneum Books paperback

Duhem, Pierre, *To Save the Phenomena, an Essay on the Idea of Physical Theory from Plato to Galileo*, Chicago: University of Chicago Press, 1969 (translated from the French edition of 1908)

Eddington, A. S., *The Nature of the Physical World*, Cambridge University Press, 1928

Ehrenfest, Paul and Tatiana, *The Conceptual Foundations of the Statistical Approach in Mechanics*, Ithaca, New York: Cornell University Press, 1959 (translated from *Encyklopadie der mathematischen Wissenschafter*, 1912)

Einstein, Albert, "Autobiographical notes," pages 1–97 in *Albert Einstein, Philosopher-Scientist* (edited by P. A. Schilpp), New York: Library of Living Philosophers, 1949 (German original and English translation on facing pages)

Einstein, Albert, "Can quantum-mechanical description of physical reality be considered complete?" pages 123–130 in *Physical Reality*, edited by S. Toulmin, New York: Harper & Row, 1970 (reprinted from *Physical Review*, 1935)

Einstein, Albert and Infeld, L., *The Evolution of Physics*, New York: Simon and Schuster, 1938; paperback reprint 1961; excerpts in *Project Physics Reader*, Vols. 4 and 5

Einstein, Albert, "Maxwell's influence on the development of the conception of physical reality," pages 66–73 in *James Clerk Maxwell: A Commemoration Volume, 1831–1931*, London: Cambridge University Press, 1931

Einstein, Albert, "On a heuristic point of view about the creation and conversion of light," pages 91–107 in D. ter Haar, *The Old Quantum Theory*, New York: Pergamon Press, 1967; also pages 544–557 in *World of the Atom* (translated from *Annalen der Physik*, 1905)

Einstein, Albert, *et al.*, *The Principle of Relativity*, New York: Dover Publications (n.d.), reprint of the 1923 edition (translation of papers published in German, 1895–1919)

Einstein, Albert, "Relativity and the problem of space," pages 166–180 in *The Changeless Order*, edited by A. Koslow, New York: Braziller, 1967 (excerpted from *Relativity*, 1920; original German edition, 1917)

Einstein, Albert, "Remarks to the Essays appearing in this collective volume," pages 665–683 in *Albert*

*Einstein, Philosopher-Scientist*, edited by P. A. Schilpp, New York: Library of Living Philosophers, 1949

Einstein, Albert, "The simultaneity of events," pages 157–167 in *The Changeless Order*, edited by A. Koslow, New York: Braziller, 1967 (translated excerpt from *Annalen der Physik*, 1905)

Eliade, M., *Cosmos and History: the Myth of the Eternal Return*, translated from French by W. R. Trask, New York: Harper Torchbook paperback reprint, 1959

Elkana, Y., "Helmholtz' 'Kraft': An illustration of concepts in flux," *Historical Studies in the Physical Sciences*, Vol. 2, pages 263–298 (1970)

Epstein, P. S., "Centennial of the undulatory theory of light," *Science*, Vol. 63, pages 387–393 (1926)

Faraday, M., "Induced currents," pages 231–240 in *The Origins and Growth of Physical Science*, Vol. 2, edited by D. L. Hurd and J. J. Kipling, Baltimore: Penguin Books, 1964 (excerpted from *Philosophical Transactions of the Royal Society of London*, 1832)

Faraday, M., "On electro-chemical decomposition," pages 276–285 in *The Origins and Growth of Physical Science*, Vol. 2, edited by D. L. Hurd and J. J. Kipling, Baltimore: Penguin Books, 1964 (excerpted from *Philosophical Transactions of the Royal Society of London*, 1834)

Farrington, B., *Science in Antiquity*, New York: Oxford University Press, second edition, 1969; paperback

Feigl, H., "Operationism and scientific method" and "Some remarks on the meaning of scientific explanation," pages 498–514 in *Readings in Philosophical Analysis*, edited by H. Feigl and W. Sellars, New York: Appleton-Century-Crofts, 1949 (reprinted from *Psychological Review*, 1945 and 1948)

Feynman, R. P., *The Character of Physical Law*, Cambridge, Mass.: MIT Press, 1967, paperback

Feynman, R. P., Leighton, R. B., and Sands, M., "Newton's Laws of Dynamics," pages 77–90 in *Project Physics Reader*, Vol. 1 (excerpted from *The Feynman Lectures on Physics*, 1963)

Feynman, R. P., Leighton, R. B., and Sands, M., "Scientific Imagination," *Project Physics Reader*, Vol. 4, pages 285–288 (excerpted from *The Feynman Lectures on Physics*, 1964)

Feynman, R. P., Leighton, R. B., and Sands, M., "Universal Gravitation," pages 87–90 in *Project Physics Reader*, Vol. 2 (excerpted from *The Feynman Lectures on Physics*, 1963)

Feynman, R. P., "The values of science," pages 1–6 in *Project Physics Reader*, Vol. 1 (reprinted from *Frontiers in Science*, edited by Edward Hutchings, Jr., 1958)

Ford, K. W., "Entropy and the Second Law of Thermodynamics," pages 59–86 in *Project Physics Reader*, Vol. 3 (excerpted from *Basic Physics*, 1968)

Forman, P., "Weimar culture, causality, and quantum theory, 1918–1927. Adaptation by German physicists and mathematicians to a hostile intellectual environment," *Historical Studies in the Physical Sciences*, Vol. 3, pages 1–115 (1971)

Frank, P., *Einstein: His Life and Times*, New York: Knopf, 1947

Frank, P., *Modern Science and its Philosophy*, Cambridge, Mass.: Harvard University Press, 1949; reprinted by Collier Books, 1961; paperback

Franklin, Benjamin, *Experiments and Observations on Electricity* (edited by I. B. Cohen), Cambridge, Mass.: Harvard University Press, 1941

Franklin, B., "The kite experiment," pages 192–193 in *The Origins and Growth of Physical Science*, Vol. 2, edited by D. L. Hurd and J. J. Kipling, Baltimore: Penguin Books, 1964 (from a letter dated 1752, first published in his *Experiments and Observations on Electricity*)

Freund, Ida, *The Study of Chemical Composition, An Account of its Method and Historical Development*, New York: Dover Publications, 1968 (reprint of the 1904 edition)

Galilei, Galileo, *Abjuration*, printed as Appendix E in Sedgwick and Tyler, *A Short History of Science*, New York: Macmillan, second edition, 1939

Galilei, Galileo, *Dialogues Concerning Two New Sciences*, New York: Dover Publications, 1952 (reprint of the Crew-De Salvio translation, 1914, of *Discorsi e dimostrazioni matematichè intorno à due nuove scienze*, 1638)

Galilei, Galileo, *Dialogue Concerning the Two Chief World Systems—Ptolemaic and Copernican*, Berkeley: University of California Press, second edition 1967 (translated by S. Drake from *Dialogo . . . sopra i due massimi sistemi del mondo, Tolemaico, e Copernicano*, 1632); another translation of the same work is *Dialogue on the Great World Systems*, Chicago: University of Chicago Press, 1953 (Salusbury's translation revised by G. De Santillana)

Galilei, Galileo, *Discoveries and Opinions of Galileo*, translated with notes by S. Drake, Garden City, N.Y.: Doubleday Anchor Books, 1957, paperback

Galilei, Galileo, *On Motion and On Mechanics*, Madison, Wisconsin: University of Wisconsin Press, 1960 (translated from *De Motu*, ca. 1590, and *Le Meccaniche*, ca. 1600)

Galilei, Galileo, "Naturally accelerated motion," pages 163–177 in *Origins and Growth of Physical Science*, Vol. 1 (extract from translation of *Two New Sciences*, 1638)

Galilei, Galileo, *The Starry Messenger*, pages 21–58 in *Discoveries and Opinions of Galileo* (translated from *Sidereus Nuncius*, 1610)

Gal-Or, B., "The crisis about the origin of irreversibility and time anisotropy," *Science*, Vol. 176, page 11–17 (1972)

Gamow, George, "The Law of Disorder," pages 87–124 in *Project Physics Reader*, Vol. 3 (excerpted from *One, Two, Three . . . Infinity*, 1947)

Gamow, G., "Mr. Tompkins and simultaneity," pages 43–48 in *Project Physics Reader*, volume 5 (excerpted from *Mr. Tompkins in Paperback*, 1965)

Gamow, G., *Thirty Years that Shook Physics*, Garden City, N.Y.: Doubleday, 1966

Garber, E. W., "Clausius' and Maxwell's kinetic theory of gases," *Historical Studies in the Physical Sciences*, Vol. 2, pages 299–319 (1970)

Gardner, M., "Can time go backward?" pages 193–200 in *Project Physics Reader*, Vol. 6 (reprinted from *Scientific American*, 1967)

Gay-Lussac, J., "Memoir on the Combination of Gaseous Substances with Each Other," pages 48–62 in *The Origins and Growth of Physical Science*, Vol. 2, edited by D. L. Hurd and J. J. Kipling, Baltimore: Penguin Books, 1964; also pages 160–170 in *World of the Atom* (translated from *Memoires de la Société d'Arcueil*, 1809)

Gerhart, J. B. and Nussbaum, R. H., "Motion in Words," pages 28–30 in *Project Physics Reader*, Vol. 1 (excerpted from *Motion*, 1966)

Geymonat, L., *Galileo Galilei, a Biography and Inquiry into his Philosophy of Science*, translated from Italian, New York: McGraw-Hill, 1965; paperback

Gillispie, C. C., editor, *Dictionary of Scientific Biography*, New York: Scribner, 1970

Gillispie, C. C., *The Edge of Objectivity*, Princeton, N.J.: Princeton University Press, 1960; paperback

Gillmor, C. S., *Coulomb and the Evolution of Physics and Engineering in Eighteenth-Century France*, Princeton, N.J.: Princeton University Press, 1971

Gingerich, O., "The computer versus Kepler," *American Scientist*, Vol. 52, pages 218–226 (1964)

Gingerich, O., *et al.*, *Frontiers in Astronomy*, selections from *Scientific American* with introduction by O. Gingerich, San Francisco: Freeman, 1970

Grant, E., "Medieval and seventeenth-century conceptions of an infinite void space beyond the cosmos," *Isis*, Vol. 60, pages 39–60 (1969)

Greaves, R. L., "Puritanism and science: the anatomy of a controversy," *Journal of the History of Ideas*, Vol. 30, pages 345–368 (1969)

Greenaway, R., *John Dalton and the Atom*, Ithaca: Cornell University Press, 1967

Guerlac, H., "Some Daltonian Doubts," *Isis*, Vol. 52, pages 544–554 (1961)

Haar, D. ter, *The Old Quantum Theory*, New York: Pergamon Press, 1967

Haber, F. C., *The Age of the World from Moses to Darwin*, Baltimore: Johns Hopkins University Press, 1959

Hall, A. R., *From Galileo to Newton 1630–1720*, New York: Harper, 1963

Hall, A. R., "Mechanics and the Royal Society, 1668–70," *British Journal for the History of Science*, Vol. 3, pages 24–38 (1966)

Hall, A. R., "Merton revisited, or science and society in the seventeenth century," *History of Science*, Vol. 2, pages 1–16 (1963)

Hall, A. R., *The Scientific Revolution 1500–1800*, Boston: Beacon Press, 1954; paperback

Hall, Marie Boas, editor, *Nature and Nature's Laws*, New York: Harper, 1970; paperback

Hall, Marie Boas, "Robert Boyle," *Scientific American*, Vol. 217, No. 2, pages 97–102 (August 1967)

Hall, Marie Boas, *Robert Boyle on Natural Philosophy: An Essay with Selections from his Writings*, Bloomington Ind.: Indiana University Press, 1965

Hall, Marie Boas, see also Boas, M.

Halley, Edmund, "A discussion of elliptical orbits of comets," pages 210–212 in *Origins and Growth of Physical Science* (excerpted from *The Elements of Astronomy*, 1715)

Hamilton, W. R., "On a general method of expressing the paths of light, and of the planets, by the coefficients of a characteristic function," pages 1030–1040 in *World of the Atom* (excerpted from *Dublin University Review*, 1833)

*Handbook of Chemistry and Physics*, Cleveland: Chemical Rubber Publishing Company, issued yearly

Hankins, T. L., "The reception of Newton's second law of motion in the eighteenth century," *Archives Internationales d'Histoire des Sciences*, Vol. 20, pages 43–65 (1967)

Hanson, N. R., "Contra-equivalence: a defense of the originality of Copernicus," *Isis*, Vol. 55, pages 308–325 (1964)

Hanson, N. R., "The Copernican disturbance and the Keplerian revolution," *Journal of the History of Ideas*, Vol. 22, pages 169–184 (1961)

Hanson, N. R., *Patterns of Discovery*, New York: Cambridge University Press, 1958; paperback

Hartner, W., "Galileo's contribution to astronomy," *Vistas in Astronomy*, Vol. 11, pages 31–43 (1968)

Harvard Project Physics, *The Project Physics Course*, Reader, 6 volumes. New York: Holt, Rinehart and Winston, 1970–71

Harvard University, Committee on the Objectives of a General Education in a Free Society, *General Education in a Free Society*, Cambridge, Mass.: Harvard University Press, 1945

Heilbron, J. L., "Quantum historiography and the archive for history of quantum physics," *History of Science*, Vol. 7, pages 90–111 (1968)

Heilbron, J. L., and Kuhn, T. S., "The genesis of the Bohr atom," *Historical Studies in the Physical Sciences*, Vol. 1, pages 211–290 (1969)

Heisenberg, W., "The Copenhagen interpretation of the

quantum theory," pages 1113–1122 in *World of the Atom* (excerpted from *Physics and Philosophy*, 1958)

Heisenberg, W., "Critique of the physical concepts of the corpuscular theory of matter," pages 1107–1113 in *World of the Atom* (excerpted from *The Physical Principles of the Quantum Theory*, 1930)

Heisenberg, Werner, "The development of the interpretation of the quantum theory," pages 12–29 in *Niels Bohr and the Development of Physics*, edited by W. Pauli *et al.*, London: Pergamon Press, and New York: McGraw-Hill, 1955

Heisenberg, Werner, "The interpretation of kinematic and mechanical relationships according to the quantum theory," pages 168–182 in *Wave Mechanics*, edited by G. Ludwig, New York: Pergamon Press, 1968 (translated from *Zeitschrift für Physik*, 1925)

Heisenberg, Werner, *The Physical Principles of the Quantum Theory*, New York: Dover Publications (n.d.) (translated from *Die physikalischen Prinzipien der Quantentheorie*, 1930)

Heisenberg, Werner, *Physics and Philosophy*, New York: Harper and Row, paperback reprint of the 1958 edition

Helmholtz, Hermann von, "The conservation of force," pages 89–110 in *Kinetic Theory*, Vol. 1, edited by S. G. Brush, New York: Pergamon Press, 1965 (excerpt from translation of *Über die Erhaltung der Kraft*, 1847)

Helmholtz, Hermann von, "On the conservation of force" (translated from a lecture given in 1862), pages 186–222 in his *Popular Scientific Lectures*, New York: Dover Publications, 1962

Helmholtz, Hermann von, "On the interaction of natural forces" (translated from a lecture given in 1854), pages 59–92 in his *Popular Scientific Lectures*, New York, Dover Publications, 1962

Helmholtz, Hermann von, *Selected Writings of Hermann von Helmholtz*, edited by R. Kahl, Middletown, Conn.: Wesleyan University Press, 1971

Herapath, John, "A mathematical inquiry into the causes, laws, and principal phenomena of heat, gases, gravitation, etc.," pages 198–205 in *World of the Atom* (excerpted from *Annals of Philosophy*, 1821)

Herivel, J., *The Background to Newton's Principia*, New York: Oxford University Press, 1965

Herivel, J., "Newton's discovery of the law of centrifugal force," *Isis*, Vol. 51, pages 546–553 (1960)

Hermann, A., *The Genesis of Quantum Theory*, Cambridge, Mass.: MIT Press, 1971 (translated from *Fruhgeschichte der Quantentheorie, 1899–1913*, 1969)

Hertz, H., "Electric radiation," pages 255–266 in *Origins and Growth of Physical Science*, Vol. 2 (translated from *Sitzungsberichte der Preussischen Akademie der Wissenschaften*, Berlin, 1888)

Hesse, Mary, *Forces and Fields*, London: Nelson, 1961; New York: Philosophical Library, 1962

Hiebert, E., *Historical Roots of the Principle of Conserva-*

*tion of Energy*, Madison, Wisconsin: State Historical Society of Wisconsin, 1962

Hiebert, E., "The uses and abuses of thermodynamics in religion," *Daedalus*, Vol. 95, pages, 1046–1080 (1966)

Hindmarsh, W. R., *Atomic Spectra*, New York: Pergamon Press, 1967

Hirosige, T., and Nisio, S., "The genesis of the Bohr atom model and Planck's theory of radiation," *Japanese Studies in the History of Science*, No. 9, pages 35–47 (1970)

Holton, G., "Einstein, Michelson, and the 'crucial' experiment," *Isis*, Vol. 60, pages 133–197 (1969)

Holton, G., "Influences on Einstein's early work in relativity theory," *American Scholar*, Vol. 37, pages 59–79 (Winter 1967–68)

Holton, G., "Johannes Kepler's universe: its physics and metaphysics," *American Journal of Physics*, Vol. 24, pages 340–351 (1956), reprinted in *Toward Modern Science*, edited by R. M. Palter, New York: The Noonday Press, 1961, volume 2, pages 192–216

Holton, G., "Kepler," pages 62–64 in *Project Physics Reader*, Vol. 2 (reprinted from *Scientific American*, 1960)

Holton, G., "On the origins of the special theory of relativity," *American Journal of Physics*, Vol. 28, pages 627–636 (1960)

Holton, G., "On trying to understand scientific genius," *American Scholar*, Vol. 41, pages 95–110 (1971–72)

Holton, G., "Presupposition in the construction of theories," pages 77–108 in *Science as a Cultural Force* (edited by H. Woolf), Baltimore: Johns Hopkins Press. 1964

Holton, G., "The relevance of physics," *Physics Today*, Vol. 23, No. 11, pages 40–47 (November 1970)

Holton, G., "Resource Letter SRT-1 on Special Relativity Theory," *American Journal of Physics*, Vol. 30, pages 462–469 (1962)

Holton, G., "The roots of complementarity," *Daedalus*, Vol. 99, pages 1015–1055 (1970)

Holton, G., *et al.*, *Special Relativity Theory, Selected Reprints*, New York: American Institute of Physics, 1963

Holton, G., "The thematic imagination in science," pages 88–108 in *Science and Culture*, edited by G. Holton, Boston: Beacon Press, 1967, paperback

Holton, G. *Thematic Origins of Scientific Thought: Kepler to Einstein*, Cambridge, Mass.: Harvard University Press, 1973

Home, R. W., "The third law in Newton's mechanics," *British Journal for the History of Science*, Vol. 4, pages 39–51 (1968)

Hoyle, Fred, "Close reasoning" and "Opening scenes," *Project Physics Reader*, Vol. 1, pages 7–17 and Vol. 2, pages 1–19 (both excerpted from *The Black Cloud*, 1957)

Humphreys, W. C., "Galileo, falling bodies, and inclined planes. An attempt at reconstructing Galileo's discovery of the law of squares," *British Journal for the History of Science*, Vol. 3, pages 225–244 (1967)

Humphries, Rolfe, *The Way Things Are*, Bloomington: Indiana University Press, 1968

Hurd, D. L., and Kipling, J. J., *The Origins and Growth of Physical Science*, Baltimore: Penguin Books, 1964 (based on *Moments of Discovery*, edited by G. Schwartz and P. W. Bishop, 1958)

Hutton, James, *Theory of the Earth*, Weinheim: H. R. Engelmann; Codicote, Herts.: Wheldon & Wesley, 1959 (reprint of the 1795 edition)

Huygens, Christian, "Preface" and "Chapter I, On Rays Propagated in Straight Lines," pages 165–181 in *Origins and Growth of Physical Science*, Vol. 2 (excerpted from translation *Traité de la lumière*, 1690)

Huygens, Christian, *Treatise on Light*, New York: Dover Publications, 1962 (translated from *Traité de la lumière*, 1678, pub. 1790)

Ihde, A. J., *The Development of Modern Chemistry*, New York: Harper & Row, 1964

Infeld, L., "Einstein," pages 25–42 in *Project Physics Reader*, Vol. 5 (excerpted from *Quest*, 1941)

Infeld, L., "Dirac and Born," pages 141–150 in *Project Physics Reader*, Vol. 5 (excerpted from *Quest*, 1941)

Jaeger, W., *Aristotle*, Oxford: Clarendon Press, second edition, 1948

Jaffe, B., *Moseley and the Numbering of the Elements*, Garden City, N.Y.: Doubleday, 1971

James, W. S., "The discovery of the gas laws," *Science Progress*, Vol. 23, pages 261–279 (1928), and Vol. 24, pages 57–71 (1929)

Jammer, M., *Concepts of Force*, Cambridge, Mass.: Harvard University Press, 1957; Harper Torchbook paperback reprint

Jammer, M., *Concepts of Mass*, Cambridge, Mass.: Harvard University Press, 1961; Harper Torchbook paperback reprint

Jammer, M., *Concepts of Space*, Cambridge, Mass.: Harvard University Press, 1954; Harper Torchbook paperback reprint

Jammer, M., *The Conceptual Development of Quantum Mechanics*, New York: McGraw-Hill, 1966

Johnson, D., "Mysterious craters of the Carolina coast," *American Scientist*, January 1944

Johnson, F. R., "Gresham College: Precursor of the Royal Society," pages 328–353 in *Roots of Scientific Thought*, edited by P. Wiener and A. Noland, New York: Basic Books, 1957

Jones, T. B., *The Figure of the Earth*, Lawrence, Kansas: Coronado Press, 1967

Joule, James Prescott, "Free expansion of Gases," pages 172–173 in *Source Book in Physics*, edited by W. F. Magie, Cambridge, Mass.: Harvard University Press, 1963 (excerpted from *Philosophical Magazine*, 1845)

Joule, James Prescott, "A letter to the Editors of the *Philosophical Magazine*," and "On the mechanical equivalent of heat," pages 243–255 in *World of the Atom* (excerpted from *Philosophical Magazine*, 1845, and *Philosophical Transactions of the Royal Society of London*, 1850)

Joule, James Prescott, "On matter, living force, and heat," pages 78–88 in *Kinetic Theory*, Vol. 1, edited by S. G. Brush, New York: Pergamon Press, 1965 (reprinted from a lecture given in 1847)

Joule, James Prescott, "Some remarks on heat, and the constitution of elastic fluids," *Memoirs of the Literary and Philosophical Society of Manchester*, Vol. 9, pages 107–114 (1848, pub. 1851) (reprinted in *Philosophical Magazine*, 1857)

Kant, I., "On the reality of space" and "Time and space as conditions of knowledge," pages 126–138 in *The Changeless Order*, edited by A. Koslow, New York: Braziller, 1967 (excerpted from *Kant's Inaugural and Early Writings on Space*, translated by J. Handyside, 1929)

Kelvin, William Thomson, "Electrical Units of Measurement," pages 80–143 in *Popular Lectures and Address*, Vol. 1, second edition, London: Macmillan, 1891 (reprint of a lecture given in 1883)

Kelvin, William Thomson, "The kinetic theory of the dissipation of energy," pages 176–187 in *Kinetic Theory*, Vol. 2, edited by S. G. Brush, New York: Pergamon Press, 1966 (reprinted from *Proceedings of the Royal Society of Edinburgh*, 1874)

Kemble, E. C., *Physical Science: Its Structure and Development*, Cambridge, Mass., MIT Press, 1966

Kepler, J., "The discovery of the laws of planetary motion," excerpt from a translation of *Harmonies of the World*, pages 122–134 in *Origins and Growth of Physical Science*, Vol. 1

Kepler, Johannes, *Epitome of Copernican Astronomy*, Books IV and V, pages 843–1004 in *Great Books of the Western World*, edited by R. M. Hutchins, Vol. 16, Chicago: Encyclopedia Britannica, 1952 (translated from the Latin edition of 1618–21)

Kepler, Johannes, *The Harmonies of the World*, Book V, pages 1005–1085 in *Great Books of the Western World*, edited by R. M. Hutchins, Vol. 16, Chicago: Encyclopedia Britannica, 1952 (translated from the Latin edition of 1619)

Kepler, J., *Kepler's Conversation with Galileo's Sidereal Messenger*, New York: Johnson Reprint, 1965 (translated from *Dissertatio cum Nuncio Sidereo*, 1610)

Kepler, Johannes, *Kepler's Somnium, The Dream, or Posthumous Work on Lunar Astronomy*, Madison, Wis.: University of Wisconsin Press, 1967 (translated from the Latin edition, 1634)

Keynes, J. M., "Newton, the Man," pages 310–323 in *Essays in Biography*, New York: Norton, new edition, 1963 (reprinted from *Newton Tercentenary Celebrations,* 1947)

King, W. J., "The quantification of the concepts of electric charge and electric current," *The Natural Philosopher*, Vol. 2, pages 107–127 (1963)

Kittel, C., Knight, W. D., and Ruderman, M. A., *Berkeley Physics Course*, Vol. 1, Mechanics, New York: McGraw-Hill, 1965

Kirkpatrick, P., "Bad physics in athletic measurements," pages 95–100 in *Project Physics Reader*, Vol. 1 (reprinted from *American Journal of Physics*, 1944)

Klein, M. J., "Einstein and some civilized discontents," pages 99–104 in *Project Physics Reader*, Vol. 5 (reprinted from *Physics Today*, 1965)

Klein, M. J., "Einstein and the wave-particle duality," *The Natural Philosopher*, Vol. 3, pages 1–49 (1964)

Klein, M. J., "Einstein's first paper on quanta," *The Natural Philosopher*, Vol. 2, pages 57–86 (1963)

Klein, M. J., "Max Planck and the beginnings of the quantum theory," *Archive for History of Exact Sciences*, Vol. 1, pages 459–479 (1962)

Klein, M. J., "Maxwell, his Demon, and the second law of thermodynamics," *American Scientist*, Vol. 58, pages 84–97 (1970)

Klein, M. J., "Thermodynamics and quanta in Planck's work," *Physics Today*, Vol. 19, No. 11, pages 23–32 (November 1966)

Kline, J. V., "The case of the counting balls," *American Journal of Physics*, Vol. 28, pages 102–103 (1960)

Knight, D. M., *Atoms and Elements, a Study of Theories of Matter in England in the Nineteenth Century*, London: Hutchinson, 1967

Knight, D. M., editor, *Classical Scientific Papers—Chemistry* (second series), New York: American Elsevier, 1971

Koenig, F. O., "On the history of science and of the second law of thermodynamics," pages 57–111 in *Men and Moments in the History of Science*, edited by H. M. Evans, Seattle: University of Washington Press, 1959

Kompaneyets, A., *Basic Concepts in Quantum Mechanics*, New York: Reinhold, 1966

König, Franz, "Religion and Natural Sciences," *New York Times*, July 7, 1968, page E11 (excerpted from a speech at a meeting of Nobel Prize winners)

Koslow, A., *The Changeless Order, the Physics of Space, Time and Motion*, New York: Braziller, 1967

Koyré, A., *Documentary History of the Problem of Fall from Kepler to Newton*, Philadelphia: American Philosophical Society, 1955

Koyré, A., *From the Closed World to the Infinite Universe*, Baltimore: Johns Hopkins Press, 1952; Harper Torchbook paperback

Koyré, A., "Galileo and Plato," pages 147–175 in *Roots of Scientific Thought*, edited by P. Wiener and A. Noland, New York: Basic Books, 1957

Koyré, A., *Metaphysics and Measurement*, Cambridge: Harvard University Press, 1968

Koyré, A., *Newtonian Studies*, Cambridge, Mass.: Harvard University Press, 1965

Koyré, A., "The significance of the Newtonian synthesis" and other essays in his *Newtonian Studies*, Cambridge, Mass.: Harvard University Press, 1965; Phoenix Books paperback

Kubrin, D., "Newton and the cyclical cosmos: Providence and the mechanical philosophy," *Journal of the History of Ideas*, Vol. 28, pages 325–346 (1967)

Kuhn, T. S., *The Copernican Revolution*, Cambridge, Mass.: Harvard University Press, 1957; paperback

Kuhn, T. S., "Energy conservation as an example of simultaneous discovery," pages 321–356 in *Critical Problems in the History of Science* (edited by M. Clagett), Madison, Wisconsin: University of Wisconsin Press, 1959; reprinted in *The Sociology of Science* (edited by B. Barber and W. Hirsch), New York: The Free Press of Glencoe, 1962

Kuhn, T. S., *The Structure of Scientific Revolutions*, Chicago: University of Chicago Press, second edition, 1970

Laplace, P. S., "The nebular hypothesis," pages 214–226 in *Origins and Growth of Physical Science*, Vol. 1 (excerpt from Harte's translation of *Exposition du Monde*, 5th edition, 1824; first published 1796)

Laplace, P. S., *Celestial Mechanics*, Bronx, N.Y.: Chelsea Publishing Co., 1966 (reprint of the Bowditch translation, 1829–1839, of the first four volumes of *Traité de mécanique Celeste*, 1798–1825)

Laplace, P. S., *A Philosophical Essay on Probabilities*, New York: Dover Publications, 1951 (reprint of the Truscott-Emory translation of *Essai philosophique sur les probabilités*, 1814)

Laudans, L., "Theories of scientific method from Plato to Mach," *History of Science*, Vol. 7, pages 1–63 (1969)

Lavoisier, Antoine-Laurent, *Elements of Chemistry*, New York: Dover Publications 1965 (reprint of the English translation, 1790, of *Traité Élémentaire de Chimie*, 1789)

Lawrence, E. O., "The evolution of the cyclotron," pages 51–64 in *Project Physics Reader*, Vol. 6 (reprinted from *The Development of High-Energy Accelerators*, edited by M. S. Livingston, 1966)

Leibniz, G. W. von, "Absolute and relational theories of space and time," and "Quantity of Motion," pages 87–103, 226–229 in *The Changeless Order*, edited by A. Koslow, New York: Braziller, 1967 (from *The Leibniz-Clarke Correspondence*, edited by H. G. Alexander, 1956, and *Acta Eruditorum*, 1686, resp.)

Lodge, O., *Pioneers of Science*, New York: Dover Publications, 1960 (reprint of the 1893 edition)

Lucretius Carus, Titus, *The Way Things Are*, Bloomington: Indiana University Press, 1908 (translated from *De Rerum Natura*)

Ludwig, G., *Wave Mechanics*, New York: Pergamon Press, 1968

MacDonald, D. K. C., *Faraday, Maxwell, and Kelvin*, Garden City, N.Y.: Anchor Books, 1964; paperback

Mach, Ernst, "The definition of mass," pages 148–154 in *The Changeless Order*, edited by A. Koslow, New York: Braziller, 1967 (excerpted from *History and Root of the Principle of the Conservation of Energy*, 1911)

Mach, Ernst, "On the economical nature of physical inquiry," pages 186–213 in his *Popular Scientific Lectures*, LaSalle, Ill.: Open Court Publishing Co., fifth edition, 1943 (translated from a German lecture, 1882)

Mach, Ernst, "The guiding principles of my scientific theory of knowledge and its reception by my contemporaries," pages 30–43 in *Physical Reality*, edited by S. Toulmin, New York: Harper & Row, 1970 (translated from *Scientia*, 1910)

Mach, Ernst, *The History and Root of the Principle of Conservation of Energy*, translated from German; Chicago: Open Court Publishing Co., 1911

Mach, Ernst, "On the law of inertia," pages 141–147 in *The Changeless Order*, edited by A. Koslow, New York: Braziller, 1967 (excerpted from *The Science of Mechanics*; original German edition, 1883)

Mach, Ernst, "On the principle of the conservation of energy," in his *Popular Scientific Lectures*, pages 137–185 (reprinted from *The Monist*, 1894)

Mach, Ernst, *The Science of Mechanics*, LaSalle, Ill.: Open Court Publishing Co., 6th edition 1960; paperback (translated from the 9th German edition of *Die Mechanik*; first edition 1883)

Magie, W. F. (editor), *A Source Book in Physics*, Cambridge, Mass.: Harvard University Press, 1963 (reprint of the 1935 edition)

Manuel, F. E., *A Portrait of Isaac Newton*, Cambridge, Mass.: Harvard University Press, 1968

Mason, S. F., *A History of the Sciences*, New York: Collier Books (Macmillan), 1962 (revised edition of *Main Currents of Scientific Thought*, 1956)

Maxwell, James Clerk, "A dynamical theory of the electromagnetic field," pages 242–253 in *Origins and Growth of Physical Science*, Vol. 2 (excerpted from *Philosophical Transactions of the Royal Society of London*, 1865)

Maxwell, James Clerk, "Illustrations of the Dynamical Theory of Gases," pages 148–171 in *Kinetic Theory*, Vol. 1, edited by S. G. Brush, New York: Pergamon Press, 1965 (excerpted from *Philosophical Magazine*, 1860)

Maxwell, James Clerk, "On the induction of electric currents," pages 229–232 in *Project Physics Reader*, Vol. 4 (excerpt from his *Treatise on Electricity and Magnetism*, 1873)

Maxwell, James Clerk, "Molecules," pages 361–378 in *The Scientific Papers of James Clerk Maxwell*, New York: Dover Publications, 1965 (reprinted from *Nature*, 1873)

Mayer, Julius Robert, "The forces of inorganic nature," pages 71–77 in *Kinetic Theory*, Vol. 1, edited by S. G. Brush, New York: Pergamon Press, 1965 (translated from *Annalen der Chemie und Pharmacie*, 1842)

McCormmach, R., "The atomic theory of John William Nicholson," *Archive for History of Exact Sciences*, Vol. 3, pages 160–184 (1966)

McCormmach, R., "John Michell and Henry Cavendish: weighing the stars," *British Journal for History of Science*, Vol. 4, pages 126–155 (1968)

McGucken, W., *Nineteenth-Century Spectroscopy*, Baltimore: The Johns Hopkins Press, 1969

McGuire, J. E. and Rattansi, P. M., "Newton and the 'Pipes of Pan,'" *Notes and Records of the Royal Society of London*, Vol. 21, pages 108–143 (1966)

McMullin, E. (editor), *Galileo, Man of Science*, New York: Basic Books, 1968

Mead, Margaret, *Coming of Age in Samoa*, New York: Morrow, 1928

Mendeléeff, Dmitri Ivanovitch, *The Principles of Chemistry*, New York: Kraus Reprint Co., 1969 (reprint of the 1905 edition)

Mendeléeff, Dmitri Ivanovitch, "The relations between the properties of elements and their atomic weights," pages 302–312 in *The World of the Atom*; also pages 83–100 in *Origins and Growth of Physical Science*, Vol. 2 (translated from *Journal of the Russian Chemical Society*, 1869)

Mendoza, E., editor, *Reflections on the Motive Power of Fire, by Sadi Carnot, and Other Papers on the Second Law of Thermodynamics, by E. Clapeyron and R. Clausius*, New York: Dover Publications, 1960; paperback

Merton, R. K., "Behavior patterns of scientists," in *American Scientist*, Vol. 57, pages 1–23 (1969)

Merton, R. K., *On the Shoulders of Giants*, New York: Free Press, 1965

Merton, R. K., "Singletons and multiples in scientific discovery: A chapter in the sociology of science," *Proceedings of the American Philosophical Society*, Vol. 105, pages 470–486 (1961)

Meyer, Herbert W., *A history of electricity and magnetism*, Cambridge, Mass.: MIT Press, 1971

Michelson, A. A., "Velocity of light," pages 51–68 in *Project Physics Reader*, Vol. 4 (excerpted from *Studies in Optics*, 1927)

Milton, John, *Paradise Lost*, New York: Macmillan, 1966, paperback (first published 1667)

Moody, E. A., "Galileo and Avempace: The Dynamics of the Leaning Tower Experiment," pages 176–206 in *Roots of Scientific Thought*, edited by P. P. Wiener and

A. Noland, New York: Basic Books, 1957 (reprinted from *Journal of the History of Ideas*, 1951)

Moore, R., *Niels Bohr: The Man, His Science, and the World They Changed*, New York: Knopf, 1966

More, L. T., *Isaac Newton, a Biography*, New York: Dover Publications, 1962 (reprint of the 1934 edition)

More, L. T., *The Life and Works of the Honourable Robert Boyle*, London and New York: Oxford University Press, 1944

Morrison, P. and E., "Heinrich Hertz," *Scientific American*, December 1957, pages 98–106

Munitz, M., *Space, Time, and Creation*, New York: Collier Books, 1961, paperback reprint of the 1957 edition

Munitz, M. K. (editor), *Theories of the Universe*, New York: Free Press (Macmillan), 1957; paperback

Murdoch, J. E., "Thomas Bradwardine," pages 390–397 in *Dictionary of Scientific Biography*, edited by C. C. Gillispie, New York: Charles Scribner's Sons, Vol. II, 1970

Nakayama, S., "On the alleged independent discovery of Kepler's law by Asada Goryu," *Japanese Studies in History of Science*, No. 7, pages 55–59 (1968)

Nash, L. K., "The atomic-molecular theory," pages 215–321 in *Harvard Case Histories in Experimental Science*, edited by J. B. Conant, Cambridge, Mass.: Harvard University Press, 1948; also available as a separate pamphlet

Nash, L. K., "The origins of Dalton's chemical atomic theory," *Isis*, Vol. 47. pages 101–116 (1956)

Neugebauer, O., *The Exact Sciences in Antiquity*, Providence, R.I.: Brown University Press, second edition, 1957; New York: Dover Publications, 1969, paperback

Neugebauer, O., "On the planetary theory of Copernicus," *Vistas in Astronomy*, Vol. 10, pages 89–103 (1968)

Newman, J. R., "James Clerk Maxwell," *Scientific American*, June 1955, reprinted in *Project Physics Reader*, Vol. 3, pages 133–154 and Vol. 4, pages 195–227

Newton, Isaac, "Conclusion," pages 323–333 in *Nature and Nature's Laws* (excerpted from *Unpublished Scientific Papers of Isaac Newton*, edited by A. R. & M. B. Hall, 1962)

Newton, Isaac, "Dispersion of light," pages 156–163 in *Origins and Growth of Physical Science*, Vol. 2 (excerpted from *Philosophical Transactions of the Royal Society of London*, 1672)

Newton, Isaac, Four Letters to Richard Bentley, pages 46–58 in *Newton's Philosophy of Nature*, edited by H. S. Thayer, New York: Hafner Publishing Company, 1953 (letters written 1692–93)

Newton, Isaac, *Mathematical Principles of Natural Philosophy*, Berkeley, Calif.: University of California Press, 1934 (F. Cajori's revision of A. Motte's translation, 1729, of the third edition of *Philosophiae Naturalis Principia Mathematica*; the first edition was in 1687);

excerpts from introductory sections in *Origins and Growth of Physical Science*, Vol. 1, pages 180–208 and in *Nature and Nature's Laws*, pages 271–283

Newton, Isaac, *New Theory about Light and Colors*, München: W. Fritsch, 1965 (reprinted from *Philosophical Transactions of the Royal Society of London*, 1672); also pages 47–59 in *Isaac Newton's Papers & Letters on Natural Philosophy*, edited by I. B. Cohen, Cambridge, Mass.: Harvard University Press, 1958

Newton, Isaac, *Papers and Letters on Natural Philosophy and Related Documents*, edited by I. B. Cohen *et al.*, Cambridge, Mass.: Harvard University Press, 1958

Newton, Isaac, *Newton's Philosophy of Nature, selections from his writings*, edited by H. S. Thayer, New York: Hafner, 1953; paperback

Newton, Isaac, *Newton's Unpublished Scientific Papers*, edited by A. R. and M. B. Hall, London: Cambridge University Press, 1962

Newton, Isaac, *Opticks*, London, 4th edition 1730, reprinted by Dover, New York, 1952

Newton, Isaac, "The Repulsion Theory," pages 52–56 in *Kinetic Theory*, Vol. 1, edited by S. G. Brush, New York: Pergamon Press, 1965 (excerpts from the Motte-Cajori translation of Newton's *Principia*, 3rd ed. 1726)

Nicolson, M., "Kepler, the *Somnium*, and John Donne," pages 306–327 in *Roots of Scientific Thought*, edited by P. Wiener and A. Noland, New York: Basic Books, 1957

North, J. D., *The Measure of the Universe, a History of Modern Cosmology*, Oxford: Clarendon Press, 1965

Northrop, F. S. C., *Logic of the Sciences and the Humanities*, New York: Macmillan, 1947; Meridian Books, paperback

Odom, H., "The estrangement of celestial mechanics and religion," *Journal of the History of Ideas*, Vol. 27, pages, 533–548 (1966)

Oldenberg, O., *Introduction to Atomic Physics*, New York: McGraw-Hill, second edition, 1954

Palmer, W. G., *A History of the Concept of Valency to 1930*, New York: Cambridge University Press, 1965

Palter, R. M. (editor), *Toward Modern Science*, New York: The Noonday Press, 1961

Partington, J. R., *A History of Chemistry*, Vol. IV, London: Macmillan, 1964

Patterson, E. C., *John Dalton and the Atomic Theory*, Garden City, N.Y.: Anchor Books, 1970; paperback

Pauli, W., "On the connection between the completion of electron groups in an atom with the complex structure of spectra," pages 184–203 in D. ter Haar, *The Old Quantum Theory*, New York: Pergamon Press, 1967 (translated from *Zeitschrift für Physik*, 1925)

Pecquet, Jean, "Demonstrations of the vacuum," pages 187–202 in *Nature and Nature's Laws* (excerpted from *New Anatomical Experiments*, 1653)

Perl, M. R., "Newton's justification of the laws of motion," *Journal of the History of Ideas*, Vol. 27, pages 585–592 (1966)

Planck, Max, "On Mach's theory of physical knowledge," pages 44–52 in *Physical Reality*, edited by S. Toulmin, New York: Harper & Row, 1970 (translated from *Physikalische Zeitschrift*, 1910)

Planck, Max, "The origin and development of the quantum theory," pages 491–501 in *World of the Atom*, (excerpted from *Nobel Prize Lectures in Physics, 1901–1921*)

Planck, Max, *Planck's Original Papers in Quantum Physics (1900)*, edited by H. Kangro, London: Taylor & Francis, 1972 (text and annotations in German and English)

Planck, Max, *The Philosophy of Physics*, New York: W. W. Norton, 1936; paperback reprint, 1963 (collection of translated articles)

Planck, Max, *Scientific Autobiography and Other Papers*, New York: Philosophical Library, 1949

Planck, Max, *A Survey of Physical Theory*, New York: Dover Publications, 1960 (collection of translated articles)

Planck, Max, "The unity of the physical world-picture," pages 3–27 in *Physical Reality*, edited by S. Toulmin, New York: Harper & Row, 1970 (translated from a lecture given in 1908)

Plato, *Phaedo*, Indianapolis, Ind.: Bobbs-Merrill, 1960, paperback

Plato, *Timaeus*, Baltimore: Penguin Books, 1965, 1969, paperback; excerpts, pages 67–88 in *Theories of the Universe*, edited by M. K. Munitz, New York: Free Press, 1957, paperback

Playfair, John, *Illustrations of the Huttonian Theory of the Earth*, New York: Dover Publications, 1964 (reprint of the 1802 edition)

Poincaré, Henri, *Mathematics and Science; Last Essays*, New York: Dover Publications, 1963 (translated from *Dernières Pensées*, 1913)

Poincaré, Henri, "Mechanism and experience," pages 203–207 in *Kinetic Theory*, Vol. 2, edited by S. G. Brush, New York: Pergamon, 1966 (translated from *Revue de Metaphysique*, 1893)

Poincaré, Henri, *Science and Hypothesis*, New York: Dover Publications 1952 (translated from *La Science et l'Hypothèse*, 1902)

Poincaré, Henri, *Science and Method*, New York: Dover Publications, 1952 (translated from *Science et Méthode*, 1908)

Poincaré, Henri, *The Value of Science*, New York: Dover Publications 1958 (translated from *La Valeur de la Science*, 1905)

Polya, George, "How to solve it," page 20 in *Project Physics Reader*, Vol. 1 (excerpted from *How to Solve It*, 1957)

Pomper, P., "Lomonosov and the discovery of the law of conservation of matter in chemical transformations," *Ambix*, Vol. 10, pages 119–127 (1962)

Popper, K. R., *Conjectures and Refutations*, New York: Basic Books, 1962

Popper, K. R., *The Logic of Scientific Discovery*, London: Hutchinson, 1959; New York: Science Editions paperback, 1961 (translated from *Logik der Forschung*, 1934)

Posin, D. Q., *Mendeléev: The Story of a Great Scientist*, New York: McGraw-Hill, 1948

Power, H., "The preface to the ingenious reader," and "The conclusion. To the Generous Virtuosi, and Lovers of Experimental Philosophy," pages 122–130 in *Nature and Nature's Laws* (excerpted from *Experimental Philosophy*, 1664)

Price, D. J., "Contra-Copernicus: a critical re-estimation of the mathematical planetary theory of Ptolemy, Copernicus, and Kepler," pages 197–218, in *Critical Problems in the History of Science*, edited by M. Clagett, Madison, Wisconsin: University of Wisconsin Press, 1962

Priestley, Joseph, *Experiments and Observations on Different Kinds of Air*, New York: Kraus Reprint, 1970 (reprint of the 1790 edition)

Priestley, Joseph, *The History and Present State of Electricity, with Original Experiments*. New York: Johnson Reprint Corporation, 1966

*Project Physics Reader*, see Harvard Project Physics

Prout, William, "Article I" and "Gaseous Bodies and the weight of their atoms," pages 184–191 in *World of the Atom* (excerpted from *Annals of Philosophy*, 1815, 1816)

Przibram, K. (editor), *Letters on Wave Mechanics: Schrödinger, Planck, Einstein, Lorentz*, New York: Philosophical Library, 1967

Ptolemy, *The Almagest*, pages 5–478 in *Great Books of the Western World*, edited by R. M. Hutchins, Vol. 16, Chicago: Encyclopedia Britannica, 1952

Ptolemy, "On the order of the Theorems," "That the Heaven rotates as a sphere," and "The absolute immobility of the earth," pages 64–71 in *Origins and Growth of Physical Science*, Vol. 1 (excerpted from *The Almagest*)

Raman, V. V., and Forman, P., "Why was it Schrödinger who developed de Broglie's ideas?" *Historical Studies in the Physical Sciences*, Vol. 1, pages 291–314 (1969)

Randall, J. H., Jr., *The Career of Philosophy*, New York: Columbia University Press, 1962; paperback

Randall, J. H., Jr., *The Making of the Modern Mind*, Boston, Mass.: Houghton Mifflin, 1940

Ravetz, J., "The origins of the Copernican revolution," *Scientific American*, October 1966, pages 88–98

Rogers, E. M., "Mathematics and relativity," pages 49–82 in *Project Physics Reader*, Vol. 5 (excerpted from *Physics for the Inquiring Mind*, 1960)

Rogers, E. M., "The Great Molecular Theory of Gases,"

pages 46–58 in *Project Physics Reader*, Vol. 3 (excerpted from *Physics for the Inquiring Mind*, 1960)

Roller, D., and Roller, D. H. D., "The development of the concept of electric charge: electricity from the Greeks to Coulomb," pages 541–639 in *Harvard Case Histories in Experimental Science*, edited by J. B. Conant, Cambridge, Mass.: Harvard University Press, 1948; also available as a separate pamphlet

Roller, D., "The early developments of the concepts of temperature and heat: the rise and decline of the caloric theory," pages 117–214 in *Harvard Case Histories in Experimental Science*, edited by J. B. Conant, Cambridge, Mass.: Harvard University Press, 1948

Romer, A., *The Discovery of Radioactivity and Transmutation*, New York: Dover Publications, 1964

Romer, A., *Radiochemistry and the Discovery of Isotopes*, New York: Dover Publications, 1970

Ronchi, Vasco, *The Nature of Light*, Cambridge, Mass.: Harvard University Press, 1970

Röntgen, W., "On a new kind of rays," pages 324–338 in *Origins and Growth of Physical Science*, Vol. 2 (translated from *Sitzungsberichte der Wurzburger Physikalische-medizinischen Gesellschaft*, 1895–96)

Rosen, E., *Three Copernican Treatises*, New York: Dover Publications, 1959

Rozental, S., editor, *Niels Bohr: His Life and Work as Seen by his Friends and Colleagues*, New York: Wiley, 1967; paperback

Ruffini, R., and Wheeler, J. A., "Introducing the black hole," *Physics Today*, January 1971, pages 30–41 (and cover illustration)

Russell, Bertrand, *The ABC of Atoms*, New York: E. P. Dutton, 1923

Russell, Bertrand, *Human Knowledge, its Scope and Limits*, New York: Simon and Schuster, 1948

Russell, Bertrand, *Scientific Outlook*, New York: W. W. Norton, 1931

Rutherford, Ernest, and Soddy, Frederick, "The Cause and Nature of Radioactivity," pages 384–401 in *The Origins and Growth of Physical Science*, Vol. 2 (excerpted from *Philosophical Magazine*, 1902)

Rutherford, Ernest, "The development of the theory of atomic structure," pages 61–74 in *Background to Modern Science*, edited by J. Needham and W. Pagel, New York: Macmillan, 1938

Rutherford, Ernest, "The scattering of $\alpha$ and $\beta$ particles by matter and the structure of the atom," pages 108–131 in *The Old Quantum Theory*, edited by D. ter Haar, New York: Pergamon Press, 1967 (reprinted from *Philosophical Magazine*, 1911)

Sabra, A. I., *Theories of light from Descartes to Newton*, London: Oldbourne Book Company, and New York: American Elsevier, 1967

Sachs, Robert G., "Time reversal," *Science*, Vol. 176, pages 587–597 (1972)

Santillana, G. de., *The Crime of Galileo*, Chicago: University of Chicago: University of Chicago Press, 1955; Phoenix Books paperback

Santillana, G. de., *The Origins of Modern Scientific Thought, from Anaximander to Proclus, 600 B.C. to 500 A.D.*, Chicago: University of Chicago Press, 1961 Mentor Books paperback

Sarton, George, *Ancient Science and Modern Civilization*, Lincoln, Neb.: University of Nebraska Press, 1954; paperback

Sarton, George, *A History of Science*, Cambridge, Mass.: Harvard University Press, 1952, 1959

Sarton, George, *The Study of the History of Science*, New York: Dover Publications, 1957 (reprint of the 1936 edition)

Schaffner, K. F., *Nineteenth-Century Aether Theories*, New York: Pergamon Press, 1972

Schilpp, Paul Arthur, editor, *Albert Einstein, Philosopher-Scientist*, New York: Harper, 1949

Schlegel, R., *Time and the Physical World*, New York: Dover Publications, 1968

Schlegel, R., "Statistical explanation in physics: the Copenhagen interpretation," *Synthese*, Vol. 21, pages 65–82 (1970)

Schneer, C. J., *The evolution of physical science*, New York: Grove Press, 1960 (originally published as *The Search for Order*)

Schneer, C. J., *Mind and Matter: Man's changing concepts of the material world*, New York: Grove Press, 1969

Schonland, Basil, *The atomists 1805–1933*, New York: Oxford University Press, 1968

Schrödinger, Erwin, "The fundamental idea of wave mechanics," pages 161–172 in *Project Physics Reader*, Vol. 5 (reprinted from *Nobel Prize Lectures in Physics, 1922–1941*, 1965)

Schrödinger, Erwin, "Quantization as an eigenvalue problem," pages 94–126, 151–167 in *Wave Mechanics*, edited by G. Ludwig, New York: Pergamon Press, 1968 (translated from *Annalen der Physik*, 1926)

Schrödinger, Erwin, "On the relationship of the Heisenberg-Born-Jordan quantum mechanics to mine," pages 127–150 in *Wave Mechanics*, edited by G. Ludwig, New York: Pergamon Press, 1968 (translated from *Annalen der Physik*, 1926)

Schrödinger, Erwin, *Science, Theory, and Man*, New York: Dover Publications, 1957 (reprint of *Science and the Human Temperament*, 1935)

Schrödinger, Erwin, "Wave mechanics," pages 1067–1076 in *World of the Atom* (excerpted from *Four Lectures on Wave Mechanics*, 1928)

Scott, J. F., *Scientific Works of René Descartes*, London: Taylor & Francis, 1952

Scott, W. T., "Resource letter FC-1 on the evolution of the electromagnetic field concept," *American Journal of Physics*, Vol. 31, pages 819–826 (1963)

Shankland, R. S., "Michelson-Morley experiment," *American Journal of Physics*, Vol. 32, pages 16–35 (1964)

Shapley, H., Wright, H., and Rapport, S., *Readings in the Physical Sciences*, New York: Appleton-Century-Crofts, 1948

Shea, W. R. J., "Galileo's claim to fame: The proof that the earth moves from the evidence of the tides," *British Journal for the History of Science*, Vol. 5, pages 111–127 (1970)

Shurcliff, W. A., and Ballard, S., "Popular applications of polarized light," pages 69–87 in *Project Physics Reader*, Vol. 4 (excerpted from *Polarized Light*, 1964)

Simpson, T. K., "Some Observations on Maxwell's Treatise on Electricity and Magnetism," *Studies in History and Philosophy of Science*, Vol. 1, pages 249–263 (1970)

Singer, Charles, *A Short History of Science*, Oxford: Clarendon Press, 1941

Small, R., *An Account of the Astronomical Discoveries of Kepler*, Madison, Wisconsin: University of Wisconsin Press, 1963 (reprint of the 1804 edition)

Smedslund, J., "The acquisition of conservation of substance and weight in children: I. Introduction," *Scandinavian Journal of Psychology*, Vol. 2, pages 11–20, 85–87 (1961), reprinted in *Readings in Child Behavior and Development* (C. B. Stendler, editor), New York: Harcourt, Brace and World, second edition, 1964

Smyth, A. L., *John Dalton 1766–1844: A Bibliography of Works by and about him*, Manchester, England: Manchester University Press, 1966

Smyth, H. D., *Atomic Energy for Military Purposes*, Washington, D. C.: U.S. Government Printing Office, 1945; Princeton: Princeton University Press, 1946

Snow, C. P., "Failure and Success," pages 1–9 in *Project Physics Reader*, Vol. 5 (excerpted from *The Search*, 1934)

Snow, C. P., "Rutherford," pages 1–8 in *Project Physics Reader*, Vol. 6 (excerpted from *Variety of Men*, 1967)

Snow, C. P., *The Two Cultures and the Scientific Revolution*, New York: Cambridge University Press, 1959; reprinted with supplement as *The Two Cultures and A Second Look*, 1964; paperback

Solmsen, F., *Aristotle's System of the Physical World*, Ithaca, N.Y.: Cornell University Press, 1960

Sparberg, E., "Misinterpretation of theories of light," *American Journal of Physics*, Vol. 34, pages 377–389 (1966)

Spronsen, J. W. von, *The Periodic System of Chemical Elements*, New York: American Elsevier, 1969

Stauffer, R. C., "Speculation and experiment in the background of Oersted's discovery of electromagnetism," *Isis*, Vol. 48, pages 33–50 (1957)

Stewart, A. B., "The discovery of stellar aberration," *Scientific American*, March 1964, pages 100–108

Stimson, D., *Scientists and Amateurs*, New York: Henry Schuman, 1948

Strong, E. W., "Newton's 'Mathematical Way,'" pages 412–432 in *Roots of Scientific Thought*, edited by P. Wiener and A. Noland, New York: Basic Books, 1957

Süsskind, C., "Observations of electromagnetic-wave radiation before Hertz," *Isis*, Vol. 55, pages 32–42 (1964)

Swenson, L. S., Jr., *The Ethereal Aether, A History of the Michelson-Morley-Miller Aether-Drift Experiments, 1880–1930*, Austin, Texas: University of Texas Press, 1972

Tangherlini, F. R., "On Snell's law and the gravitational deflection of light," *American Journal of Physics*, Vol. 36, pages 1001–1004 (1968)

Taylor, B. N., Parker, W. H., and Langenberg, D. N., "Determination of $e/h$, using macroscopic quantum phase coherence in superconductors: implications for quantum electrodynamics and the fundamental physical constants," *Reviews of Modern Physics*, Vol. 41, pages 375–496 (1969)

Taylor, E. F., and Wheeler, J. A., "Parable of the surveyors," pages 83–88 in *Project Physics Reader*, Vol. 5 (excerpted from *Spacetime Physics*, 1966)

Taylor, F. S., *A Short History of Science and Scientific Thought*, New York: Norton, 1949 (reprint of *Science Past and Present*, 1945)

Thackray, A. W., "Documents relating to the origins of Dalton's chemical atomic theory," *Memoirs of the Manchester Literary and Philosophical Society*, Vol. 108, No. 2, pages 1–2 (1966)

Thackray, A. W., "The emergence of Dalton's chemical atomic theory," *British Journal for the History of Science*, Vol. 3, pages 1–23 (1966)

Thackray, A. W., "The origin of Dalton's chemical atomic theory: Daltonian doubts resolved," *Isis*, Vol. 57, pages 35–55 (1966)

Theobald, D. W., *The Concept of Energy*, New York: Barnes and Noble, 1966

Thomson, G. P., *The Inspiration of Science*, New York: Oxford University Press, 1961; Anchor Books paperback reprint

Thomson, William; see Kelvin

Thorndike, L., "Newness and novelty in seventeenth-century science," pages 443–475 in *Roots of Scientific Thought*, edited by P. Wiener and A. Noland, New York: Basic Books, 1957

Torricelli, E., "The principle of the barometer," pages 184–187 in *Nature and Nature's Laws*, edited by M. B. Hall, New York: Harper & Row, 1970, paperback (translated from his *Letter to Ricci*, 1644)

Toulmin, S., and Goodfield, J., *The Architecture of Matter*, New York: Harper, 1962; paperback

Toulmin, S., and Goodfield, J., *The Discovery of Time*, New York: Harper, 1965; paperback

Toulmin, S., "The evolutionary development of natural science," *American Scientist*, Vol. 55, pages 456–471 (1967)

Toulmin, S., and Goodfield, J., *The Fabric of the Heavens*, New York: Harper, 1961

Toulmin, S. (editor), *Physical Reality*, New York: Harper and Row, 1970; paperback

Toulmin, S., *Human Understanding*, Princeton, N. J.: Princeton University Press, 1972

Tricker, R. A. R., *The Contributions of Faraday and Maxwell to Electrical Science*, New York: Pergamon Press, 1966

Tricker, R. A. R., *Early Electrodynamics, the First Law of Circulation*, New York: Pergamon Press, 1965

Truesdell, C., "Early kinetic theories of gases," pages 272–304 in his *Essays in the History of Mechanics*, New York: Springer-Verlag, 1968

Truesdell, C., "Reactions of late Baroque mechanics to success, conjecture, error, and failure in Newton's Principia," pages 138–183 in his *Essays in the History of Mechanics*, New York: Springer-Verlag, 1968

Tyndall, John, *Heat Considered as a Mode of Motion*, London: Longmans, Green and Co., 1863, seventh edition 1887

van der Waerden, B. L., *Sources of Quantum Mechanics*, Amsterdam: North-Holland, 1967; Dover paperback reprint

Velikovsky, Immanuel, *Worlds in Collision*, New York: Macmillan, 1950

Volta, A., "The Voltaic Pile," pages 211–216 in *The Origins and Growth of Physical Sciences*, Vol. 2 (excerpted from *Philosophical Transactions of the Royal Society of London*, 1800)

Walker, W. C., "The detection and estimation of electric charges in the eighteenth century," *Annals of Science*, Vol. 1, pages 66–100 (1936)

Wallace, W., "The enigma of Domingo de Soto: Uniformiter difformus and falling bodies in late medieval physics," *Isis*, Vol. 59 pages 384–401

Wallis, John, "The origins of the Royal Society," pages 223–224 in *Nature and Nature's Laws* (excerpted from *A Defence of the Royal Society*, 1678)

Wartofsky, M. W., "All fall down: the development of the concept of motion from Aristotle to Galileo," pages 419–473 in his *Conceptual Foundations of Scientific Thought*, New York: Macmillan, 1968

Waterston, J. J., "On the physics of media that are composed of free and perfectly elastic molecules in a state of motion," pages 217–235 in *World of the Atom* (excerpted from *Philosophical Transactions of the Royal Society of London*, 1892)

Watson, F. G., *Between the Planets*, Cambridge, Mass.: Harvard University Press, revised edition 1956

Webster, C., "The discovery of Boyle's law and the concept of the elasticity of air in the seventeenth century," *Archive for History of Exact Sciences*, Vol. 2, pages 441–502 (1965)

Weiner, C., "1932—Moving into the new physics," *Physics Today*, Vol. 25, No. 5, pages 40–49 (May 1972)

Wertheimer, M., *Productive Thinking*, New York: Harper, 1945

Westfall, R. S., *Force in Newton's Physics*, New York: American Elsevier, 1971

Wheeler, J. A., "Our universe: the known and the unknown," *American Scientist*, Vol. 56, pages 1–20 (1968)

Whipple, F. L., *Earth, Moon and Planets*, Cambridge, Mass.: Harvard University Press, revised edition 1963

Whitehead, Alfred North, *The Aims of Education*, New York: Macmillan, 1929, 1959; Free Press paperback, 1967

Whitehead, A. N., *Modes of Thought*, New York: Macmillan, 1938

Whitehead, A. N., *Science and the Modern World*, New York: Macmillan, 1925; Mentor paperback reprint

Whiteside, D. T., "Before the Principia: the maturing of Newton's thoughts on dynamical astronomy, 1664–1684," *Journal of the History of Astronomy*, Vol. 1, pages 5–19 (1970)

Whittaker, E. T., *From Euclid to Eddington, A Study of Conceptions of the External World*, New York: Dover Publications, 1958 (reprint of the 1947 edition)

Whittaker, E. T., *A History of the Theories of Aether and Electricity*, London: Nelson, revised edition, 1951

Wiener, P., and Noland, A. (editors), *Roots of Scientific Thought*, New York: Basic Books, 1957

Wightman, W. P. D., *The Growth of Scientific Ideas*, New Haven: Yale University Press, 1951; paperback

Will, C. M., "Einstein on the firing line," *Physics Today*, Vol. 25, No. 10, pp. 23–29 (October 1972)

Williams, L. P., *Michael Faraday*, New York: Basic Books, 1965; Simon and Schuster paperback

Williams, L. P., *Relativity Theory, its Origins and Impact on Modern Thought*, New York: Wiley, 1968

Wilson, C. A., "From Kepler's laws, so-called, to universal gravitation. Empirical factors," *Archive for History of Exact Sciences*, Vol. 6, pages 89–170 (1970)

Wolf, A., *A History of Science, Technology, and Philosophy in the XVIth and XVIIth Centuries*, and same for XVIIIth Century, New York: Macmillan, 1935 and 1939; Harper Torchbook reprint, paperback

Wright, S., *Classical Scientific Papers—Physics*, New York: American Elsevier, 1965

Ziman, J., *Public Knowledge, an Essay Concerning the Social Dimension of Science*, New York: Cambridge University Press, 1968, paperback

# INDEX